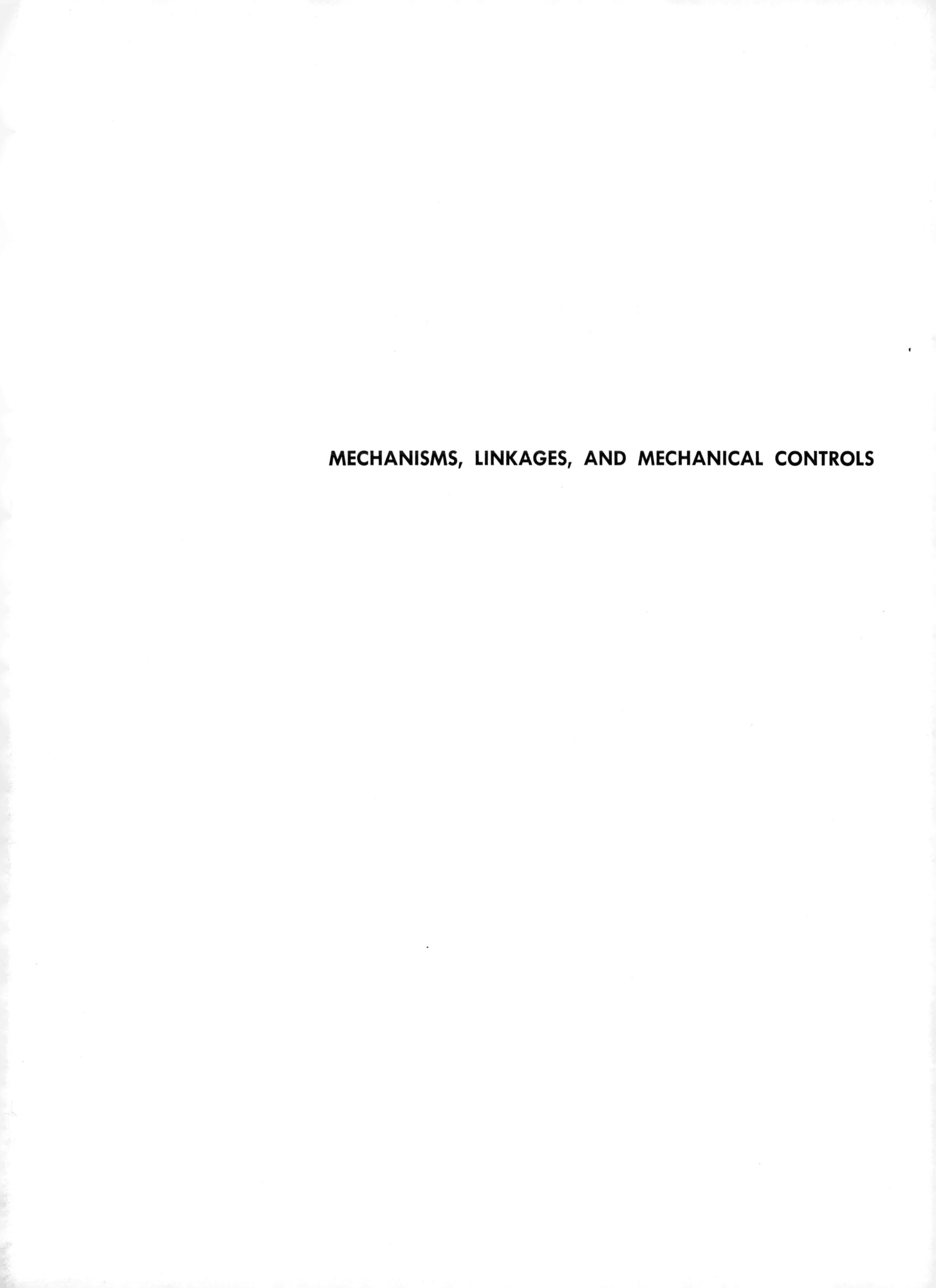

MECHANISMS, LINKAGES, AND MECHANICAL CONTROLS

MECHANISMS, LINKAGES, AND MECHANICAL CONTROLS

Edited by

NICHOLAS P. CHIRONIS

Associate Editor, PRODUCT ENGINEERING

McGRAW-HILL BOOK COMPANY

New York
San Francisco
Toronto
London
Sydney

MECHANISMS, LINKAGES, AND MECHANICAL CONTROLS

 Library of Congress Catalog Card Number: 65-23552

10775

4567891011 HDBP 7543210698

PREFACE

Here is a compilation of classical and modern mechanisms and devices for providing a wide variety of motions and functions. This wealth of material has been drawn largely from Product Engineering magazine and has been compiled, edited, and thoroughly indexed to permit quick access to an item by its application or by its principle.

The book is designed for students of mechanical movements as well as machinery and instrument designers. It covers not only devices for performing such classical functions as providing intermittent motion (and the book has the widest collection of geneva mechanisms for this purpose), but others for more sophisticated requirements such as multiplying, differentiating, torque-limiting, governing, and tension-controlling.

Also included are such mechanical components as special-function cams, unusually shaped gears, bellow devices, spring arrangements, escapements, star wheels, genevas, friction devices, differentials, variable-speed drives, and many more.

Several thousand illustrations show how to arrange these components and devices to provide the necessary functions. The accompanying descriptions are clear but concise. There are practically no theoretical analyses; the book for the most part confines itself to the presentation of practical design ideas.

Nicholas P. Chironis

CONTENTS

Chapter 1 VARIABLE-SPEED DRIVES

Your guide to variable-speed

Mechanical drives

D. Z. DVORAK, analytical engineer, Pratt & Whitney Co, West Hartford, Conn

Most of the mechanical types have a *limited* variable-speed range in that they cannot produce a zero or near-zero output speed. Those that include zero speed in their range are said to have an *infinite* variable-speed range. These terms are only trade jargon because all do produce an infinite variety of speeds.

Generally, the outputs are irreversible—if reversibility is required, the direction of input rotation must be changed. Because drive motors are directly coupled to the input shafts of the drive, split arrangements which permit the mechanical separation of the motor from the adjustable output speed unit are not possible.

Mechanical drives have been grouped in nine major classes, based on their operating principle:

1) Cone drives
2) Disk drives
3) Ring drives
4) Spherical drives
5) Multiple-disk drives
6) Impulse drives
7) Controlled-differential drives
8) Belt drives
9) Chain drives

Many of the drives are available as *commercial* units; they can be obtained in specific sizes and horsepowers from one or more manufacturers. Some drives are not available commercially and must be custom designed.

Table I includes the horsepower *capabilities* of the drive. The horsepower *requirements* of the application are determined by the basic equation

$$\text{hp} = \frac{T\,n}{63{,}000}$$

where T = torque, in.-lb; n = speed, rpm.

The drive must also be matched to the torque requirements of the application. The three basic types of horsepower-torque characteristics of variable-speed drives are illustrated in Fig 1:

Constant horsepower application—The torque decreases almost hyperbolically with increase in speed. This requirement is frequently found with machine tools, particularly on spindle drives. The critical condition here is at minimum speed when the torque and the stress of the mechanical parts is at a maximum.

Constant torque application—The horsepower requirements increase proportionally to speed, as is the case with many conveyors, reciprocating compressors, printing presses, and machine-tool feed drives, or where the load is almost a pure friction load. The selection of the drive should be based on the power requirement at the maximum speed.

Variable horsepower and speed—This requirement is common with propellers and centrifugal pumps. The critical condition is at maximum speed. The power delivered at low speed is usually more than is needed.

Only stepless drives are described here. Most of the drives are of medium capacity; however, some are limited to 5 hp or less. All are over 1 hp except the impulse drives.

All drives, with the exception of the PIV drive, are based on the friction principle. Therefore, a certain amount of slip, which increases with torque, can be expected. Slippage serves as a safety device to prevent damage from overload. But high slippage is undesirable because it decreases speed regulation, efficiency, and life of the system. All drives except the belt-driven units operate in an oil bath or mist.

1.

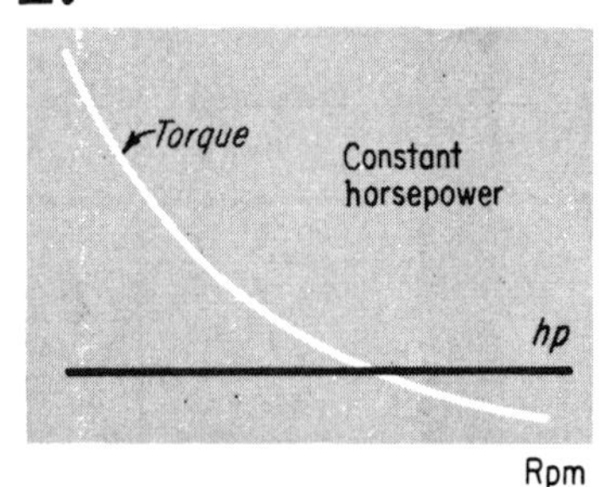

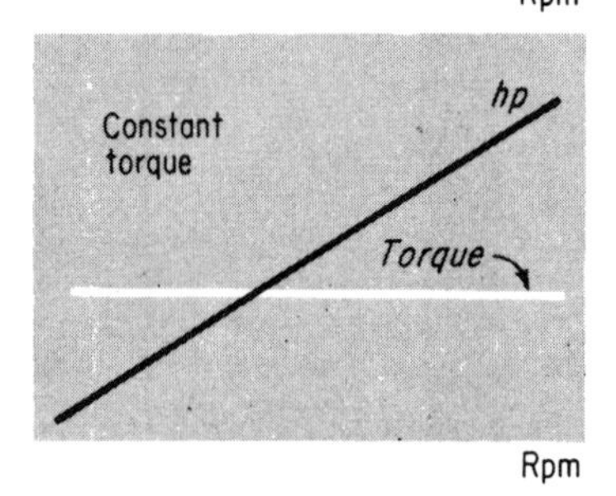

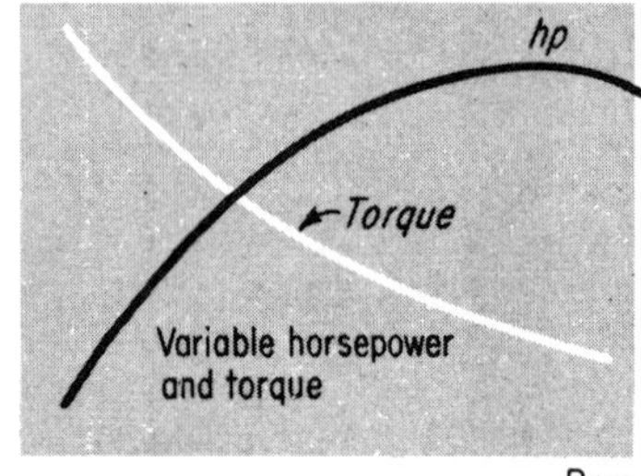

Table I

Type of Drive	Manufacturer
Cone drives	Graham Transmissions Inc. Menomonee Falls, Wis.
	Simpo Kogyo Co Ltd Karahashi, Minami-ku Kyoto, Japan
Disk drives	Sentinel (Shrewsbury) Ltd Shrewsbury, England
	Block and Vaupel Wuppertal, Germany
Ring drives	Master Electric Div of Reliance Electric Dayton 1, Ohio
	H Stroeter Dusseldorf, Germany
	Excelermatic Inc Rochester, N.Y.
Spherical drives	New Departure Div of General Motors Bristol, Conn.
	Perbury Engineering Ltd. England
	Cleveland Worm & Gear Div of Eaton Mfg Co Cleveland
	Excelecon Corp
	Friedr. Cavallo Berlin-Neukoelln Germany
Multiple-disk drives	Ligurtecnica Genoa, Italy
	Reeves Pulley Div of Reliance Electric Co Columbus, Ind
Impulse drives	Morse Chain Co Ithaca, NY
	Zero Max Co Minneapolis 8, Minn
Differential drives	Link-Belt Co
	Stratos Div of Fairchild Engine & Airplane Corp Babylon, NY
	Lombard Governor Corp Ashland, Mass

Commercial variable-speed drives

Trade name	Fig No.	Basic elements	Maximum horsepower capacity	Maximum speed variation	Horsepower-torque characteristics	Peak efficiency	Comments
Graham	4	Tapered rollers Fixed ring Planetary gears carrier	5 hp	Non-reversing type: ⅓ of input rpm to zero Reversing type: 1/5 of rpm in either direction	See curve in Fig 4 High torque at high speed also available	85% at full load High	For low speed and zero-speed applications
Ring cone RC	5	Tapered rollers Preloaded ring	10 hp	4:1	constant hp	High	Similar to ring drive. Max output speed for 10 hp units =2400 rpm
Ring cone SC	6	Planetary cones Track ring	20 hp	From 4:1 to 24:1	Combination constant torque and constant hp	85%	Employs a planetary cone system in place of planetary gears, with movable track ring (in place of ring gear)
F.U.	8	Disks beveled roller	20 hp	6:1 (10:1)	constant hp	90%	Also made in France by La Filiere Unicum, Paris
—	9	Planetary friction disks	—	—	—	—	Friction wheels act as planets. Orbits are adjustable
Speed ranger	10	Steel ring Variable-ptch sheaves	3 hp	8:1 (16:1)	constant torque	90%	Available with output speeds from 2 to 4100 rpm. Made in Germany by Hans Heynay, Munich
—	11	Specal-shape rings Inverted sheave wheels	—	10:1	constant torque	—	Similar in principle to the Speed Ranger (above), except that special shape rings are employed
—	12	Rings Beveled disks	5 hp	12:1	—	96%	Novel principle
Transitorq	14	Spherical disks Tilting rollers	20 hp	6:1 (10:1)	constant hp or constant torque	—	Not in production
—	15	Double spherical disks Tilting rollers	—	—	—	—	Reportedly still in development stage
Cleveland	16	Spherical beveled disks Axle-mounted balls	15 hp	9:1	constant hp constant torque over low range also available	To 90%	High torque at low speed. Made in Germany by Eisenwerk Wulfel, in Switzerland by: Aciera SA, Le Locle
Excelecon	17	Input concave disks Beveled rollers	15 hp	9:1	Constant hp	90%	High torque at low speed
Cavallo	18	Axially-free ball Cone disks	—	—	—	—	Very simple and compact
—	19	Multiple-disks Balls	33 hp	5:1	—	To 95%	Multiple-disks permit high horsepower
Beier	20	Input tapered disks Output rimmed disks	60 hp	4:1	Combination constant torque and constant hp, see Fig 33	85%	Units up to several hundreds hp built in England by Beier Infinitely Variable Gear Co, London SE1
Morse	22	Gear-linkage system One way clutch	1.5 hp	4½:1 to 120:1 (180 rpm max)	constant torque 175 ft-lb	over 95%	Slight pulsating motion of output is of distinct advantage to feeders and mixer
Zero-Max	23	Linkage system One way clutch	¾ hp	0 to ¼ input rpm (2000 rpm max)	constant torque	—	
—		Differential gears Speed variator	25 hp	A great variety of torque-rpm characteristics can be achieved depending on arrangement.		Varies	For very accurate control applications. Expensive
—		Differential gears Speed variator	75 hp				
—		Differential gears Speed variator	15 hp				

Table includes only drives in the 1 to 100 hp range. Maximum hp capacity column gives maximum hp of standard units only. Units are frequently available in smaller sizes, sometime with fraction hp output.

Cone drives

The simpler types in this group have a cone or tapered roller in combination with a wheel or belt (Fig 2). They have evolved from the stepped-pulley system. Even the more sophisticated designs are capable of only a limited (although infinite) speed range and generally must be spring-loaded to reduce slippage.

Adjustable cone drive (Fig 2A). This is perhaps the oldest variable-speed friction system and is usually custom built. Power from the motor-driven cone is transferred to the output shaft by the friction wheel, which is adjustable along the cone side to change the output speed. The speed depends upon the ratio of diameters at point of contact.

Two-cone drive (Fig 2B). The adjustable wheel is the power transfer element, but this drive is difficult to preload because both input and output shafts would have to be spring loaded. The second cone, however, doubles the speed reduction range.

Cone-belt drives (Fig 2C and D). In (C) the belt envelops both cones; in (D) a long-loop endless belt runs between the cones. Stepless speed adjustment is obtained by shifting the belt along the cones. The cross section of the belt must be large enough to transmit the rated force, but the width must be kept to a minimum to avoid a large speed differential over the belt width.

Electrically coupled cones (Fig 3). This patented device (US Patent 3,048,046) is composed of thin laminates of paramagnetic material. The laminates are separated with semidielectric materials which also localize the effect of the inductive field. There is a field generating device within the driving cone. Adjacent to the cone is a positioning motor for the field generating device. The field created in a

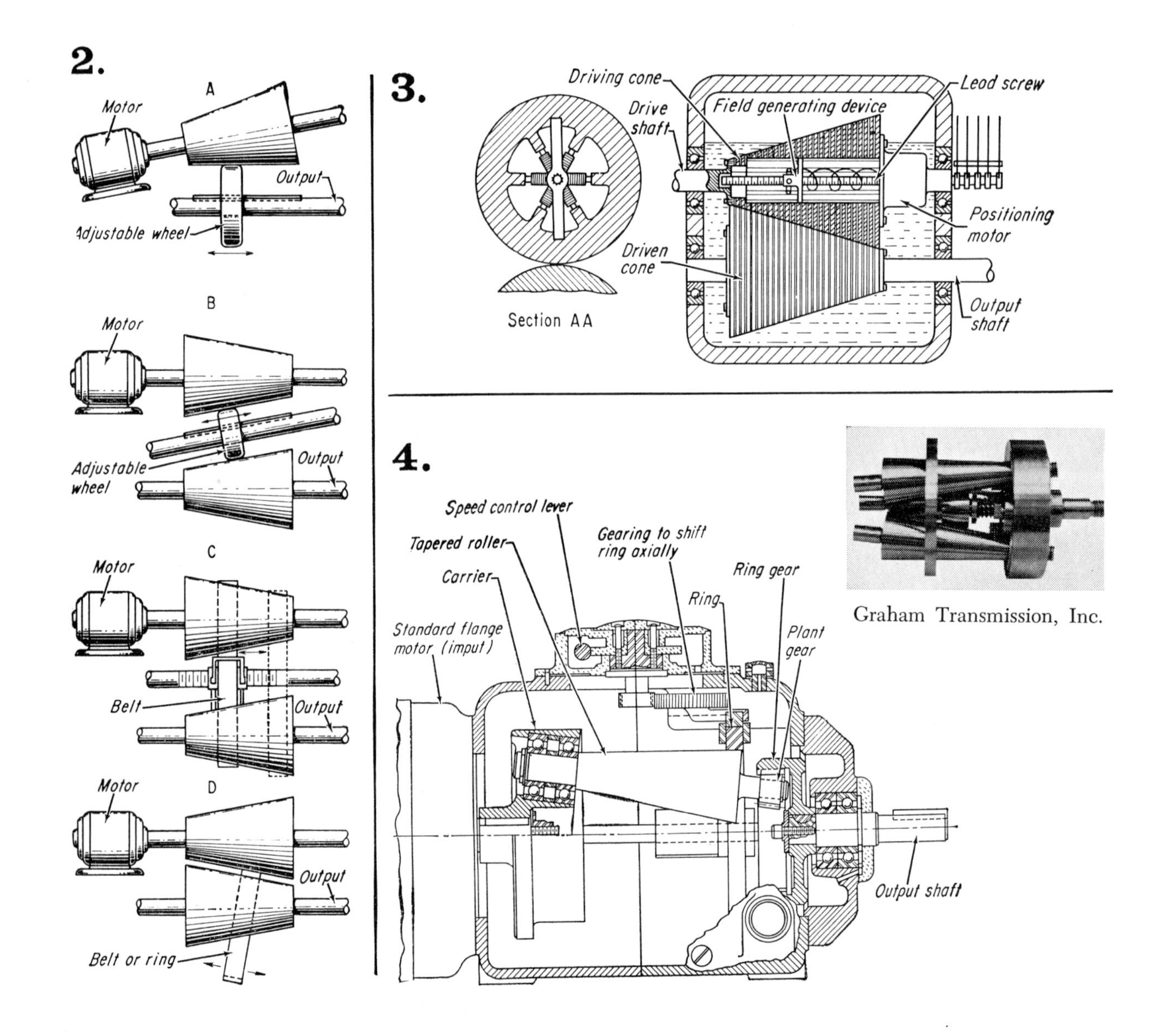

Graham Transmission, Inc.

particular section of the driving cone induces a magnetic effect in the surrounding lamination. This causes the laminate—and its opposing lamination—to couple and rotate with the drive shaft. The ratio of diameters of the cones, at the point selected by positioning the field generating component, determines the speed ratio.

Graham drive (Fig 4). This commercial unit combines a planetary-gear set and three tapered rollers (only one of which is shown). The ring is positioned axially by a cam and gear arrangement. The drive shaft rotates the carrier with the tapered rollers, which are inclined at an angle equal to their taper so that their outer edge is parallel to the centerline of the assembly. Traction pressure between rollers and ring is created by centrifugal force, or spring loading of the rollers. At the end of each roller a pinion meshes with a ring gear. The ring gear is part of the planetary gear system and is coupled to the output shaft.

The speed ratio depends on the ratio of the diameter of the fixed ring to the effective diameter of the roller at the point of contact, and is set by the axial position of the ring. The output speed, even at its maximum, is always reduced to about one-third of input speed because of the differential feature. When the angular speed of the driving motor equals the angular speed of the centers of the tapered rollers around their mutual centerline (which is set by the axial position of the nonrotating friction ring), the output speed is zero. This drive is manufactured up to 3 hp; efficiency is to 85%.

Cone-and-ring drive (Fig 5). Here, two cones are encircled by a preloaded ring. Shifting the ring axially varies the output speed. This principle is similar to that of the cone-and-belt drive (Fig 2C). In this case, however, the contact pressure between ring and cones increases with load to limit slippage.

Planetary-cone drive (Fig 6). This is basically a planetary gear system but with cones in place of gears. The planet cones are rotated by the sun cone which, in turn, is driven by the motor. The planet cones are pressed between an outer nonrotating ring and the planet holder. Axial adjustment of the ring varies the rotational speed of the cones around their mutual axis. This varies the speed of the planet holder and the output shaft. Thus the mechanism resembles that of the Graham drive (Fig 4) pictured on the facing page.

Speed adjustment range of the unit illustrated is from 4:1 to 24:1. The system is built in Japan in capacities of 2 hp and under.

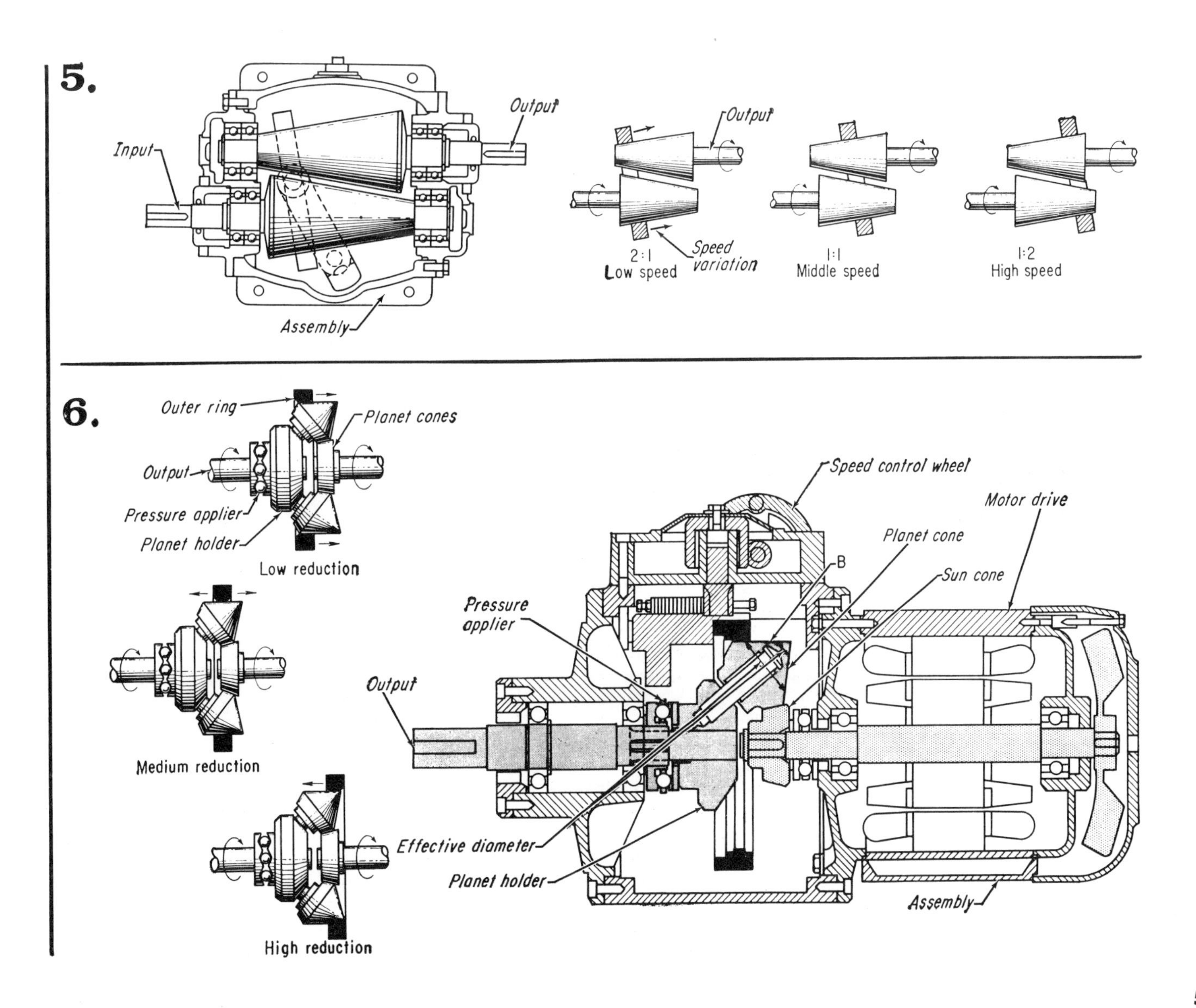

Disk drives

Adjustable disk drives (Fig 7A and B). The output shaft in (A) is perpendicular to the input shaft. If the driving power, the friction force, and the efficiency stay constant, then the output torque decreases in proportion to increasing output speed. The wheel is made of friction material, the disk of steel. Because of relatively high slippage, only small torques can be transmitted. The wheel can move over the center of the disk as this system has infinite speed adjustment.

To increase the speed range, a second disk can be added. This arrangement (Fig 7B) also makes input and output shafts parallel.

Spring-loaded disk drive (Fig 8). To reduce slippage, the contact force between the rolls and disks in this commercial drive is increased by means of the spring assembly in the output shaft. Speed adjustments are made by rotating the leadscrew to shift the cone roller in the vertical direction. The unit illustrated is built to 4-hp capacity. With a double assembly of rollers, units are available to 20 hp. Units operate to 92% efficiency. Standard speed range is 6:1, but units of 10:1 have been built. The power transferring components, which are made from hardened steel, operate in an oil mist, thus minimizing wear.

Planetary disk drive (Fig 9). Four planet disks replace planet gears in this friction drive. Planets are mounted on levers which control radial position and therefore control the orbit. Ring and sun disks are spring loaded.

7.

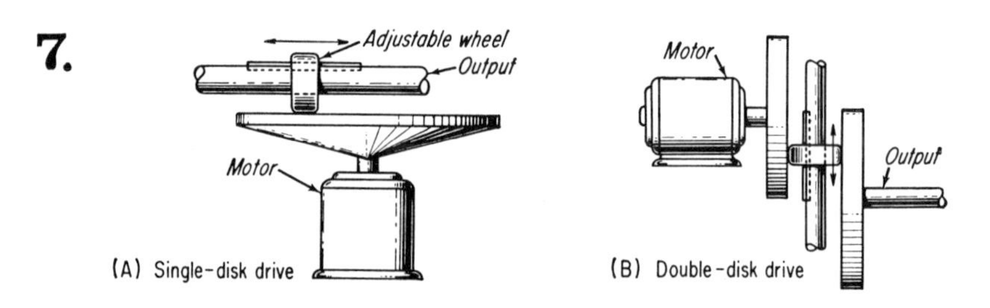

10a.

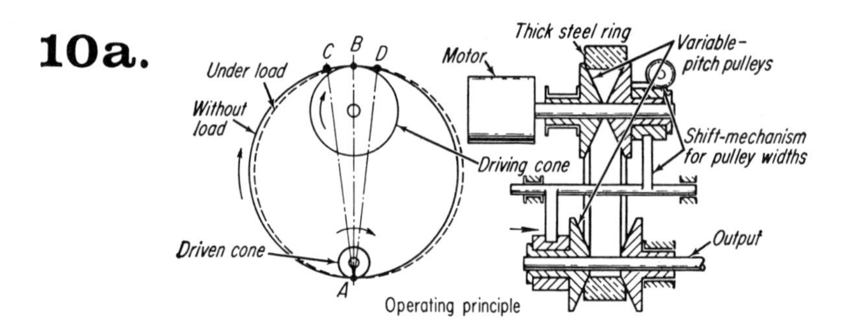

Ring drives

Ring-and-pulley drive (Fig 10). A thick steel ring in this type drive encircles two variable pitch (actually *variable width*) pulleys. A novel gear-and-linkage system simultaneously changes the width of both pulleys (see Fig 10 B). For example, when the top pulley opens, the sides of the bottom pulley close up. This reduces the effective pitch diameter of the top pulley and increases that of the bottom pulley, thus varying the output speed.

Normally, the ring engages the pulleys at point *A* and *B*. However, under load, the driven pulley resists rotation and the contact point moves from *B* to *D* because of the very small elastic deformation of the ring. The original circular shape of the ring is changed to a slightly oval form and the distance between points of contact decreases. This wedges the ring between the pulley cones and increases the contact pressure between ring and pulleys in proportion to the load applied, so that constant horsepower at all speeds is obtained. The unit is available to 3-hp capacity; speed variations to 16:1, with practical range about 8:1.

Some manufacturers employ rings with special-shaped cross section, (Fig 11) by inverting one of the sets of sheaves.

Double-ring drive (Fig 12). Power transmission is through two steel traction rings that engage two sets of disks mounted on separate shafts. Such a drive requires that the outer disks be under a compression load by a spring system (not illlustrated). The rings are hardened and convex-ground to reduce wear. Speed is changed by tilting the ring support cage, forcing the rings to move the desired position.

8.

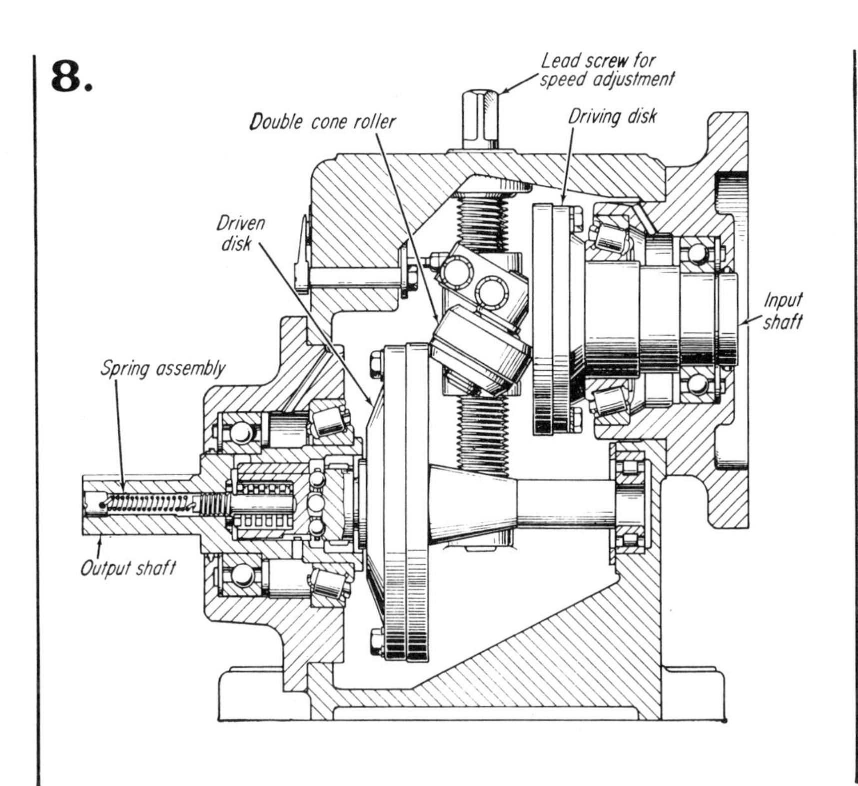

Lead screw for speed adjustment
Double cone roller
Driving disk
Driven disk
Input shaft
Spring assembly
Output shaft

9

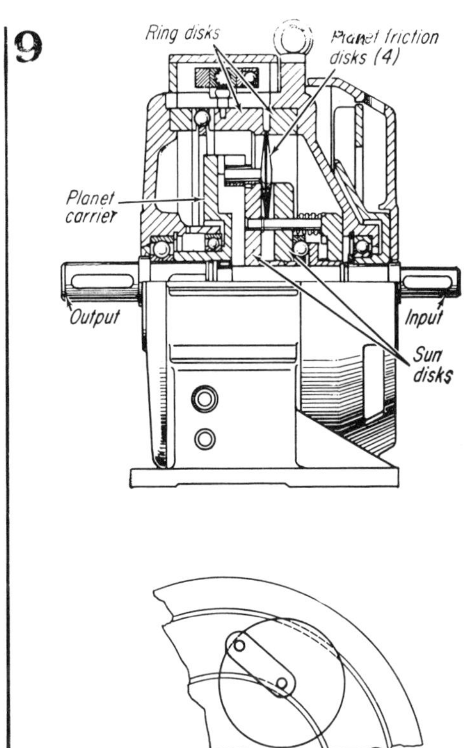

Ring disks
Planet friction disks (4)
Planet carrier
Output
Input
Sun disks

10b.

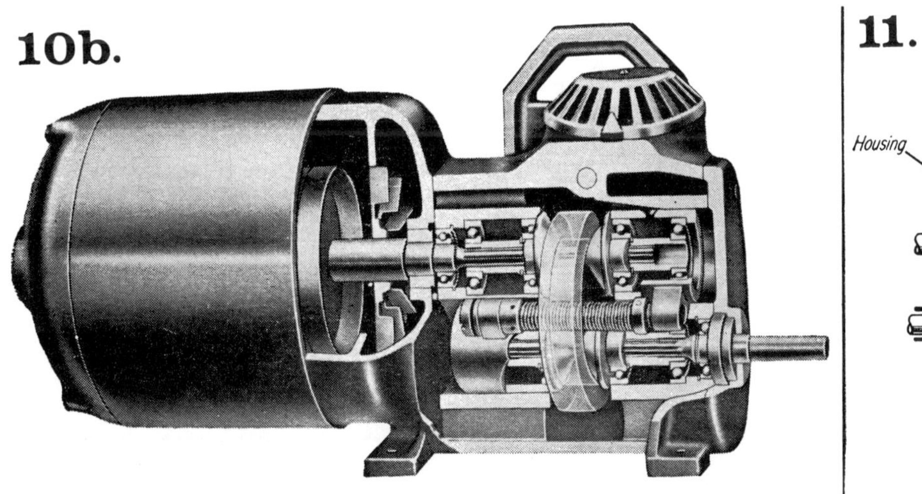

11.

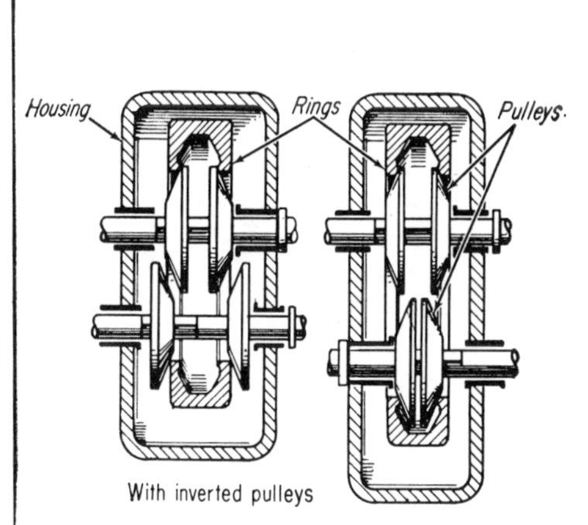

Housing
Rings
Pulleys
With inverted pulleys

12.

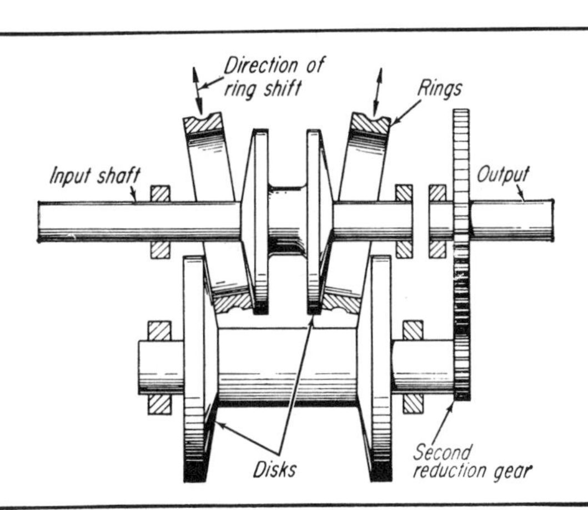

Direction of ring shift
Rings
Input shaft
Output
Disks
Second reduction gear

Spherical drives

Sphere-and-disk drives (Fig 13 and 14). Speed variations in Fig 13 are obtained by changing the angle that the rollers make in contacting spherical disks. As illustrated, the left spherical disk is keyed to the driving shaft, the right contains the output gear. The sheaves are loaded together by a helical spring.

One commercial unit, Fig 14, is a coaxial input and output shaft version of the previous arrangement. The rollers are free to rotate on bearings and can be adjusted to any speed between the limits of 6:1 to 10:1. An automatic device regulates the contact pressure of the rollers, maintaining the pressure exactly in proportion to the imposed torque load.

Double-sphere drive (Fig 15). Higher speed reductions are obtained by grouping a second set of spherical disks and rollers. This also reduces operating stresses and wear. The input shaft runs through the unit and carries two opposing spherical disks. The disks drive the double-sided output disk through two sets of three rollers. To change the ratio, the angle of the rollers is varied. The disks are axially loaded by hydraulic pressure.

Tilting-ball drive (Fig 16). Power is transmitted between disks by steel balls whose rotational axes can be tilted to change the relative lengths of the two contact paths around the balls, and hence the output speed. The ball axes can be tilted uniformly in either direction; the effective rolling radii of balls and disks produce speed variations up to 3:1 increase, or 1:3 decrease, with the total up to 9:1 variation in output speed.

Tilt is controlled by a cam plate through which all ball axes project. To prevent slippage under starting or shock load, torque responsive mechanisms are located on the input and output sides of the drive. The axial pressure created is proportional to the applied torque. A worm drive positions the plate. The drives are manufactured to 15-hp capacity. Efficiency characteristics are shown in the chart.

Sphere-and-roller drive (Fig 17). The roller, with spherical end surfaces, is eccentrically mounted between the coaxial input and output spherical disks. Changes in speed ratio are

13.

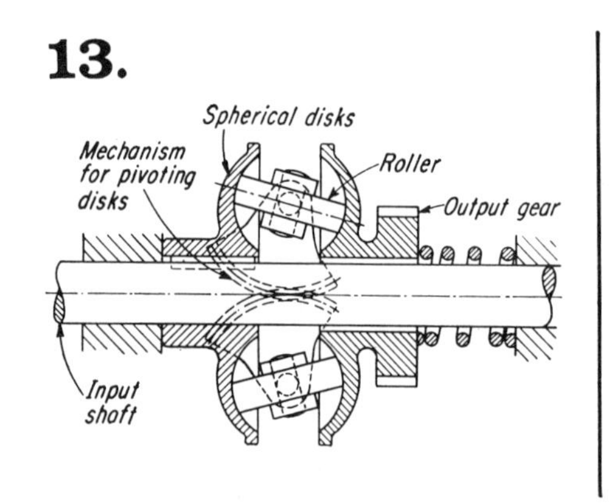

15.

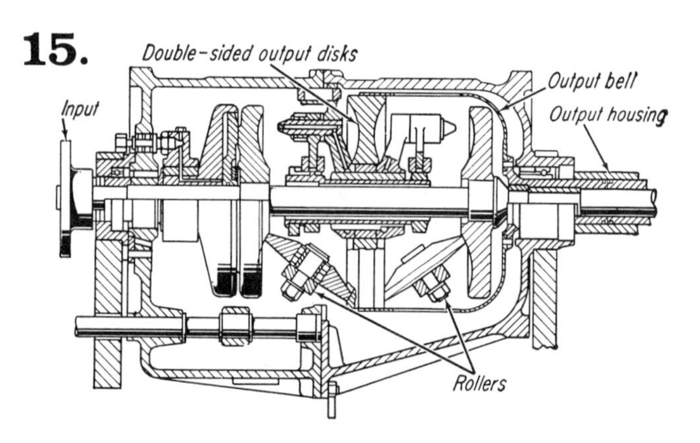

14.

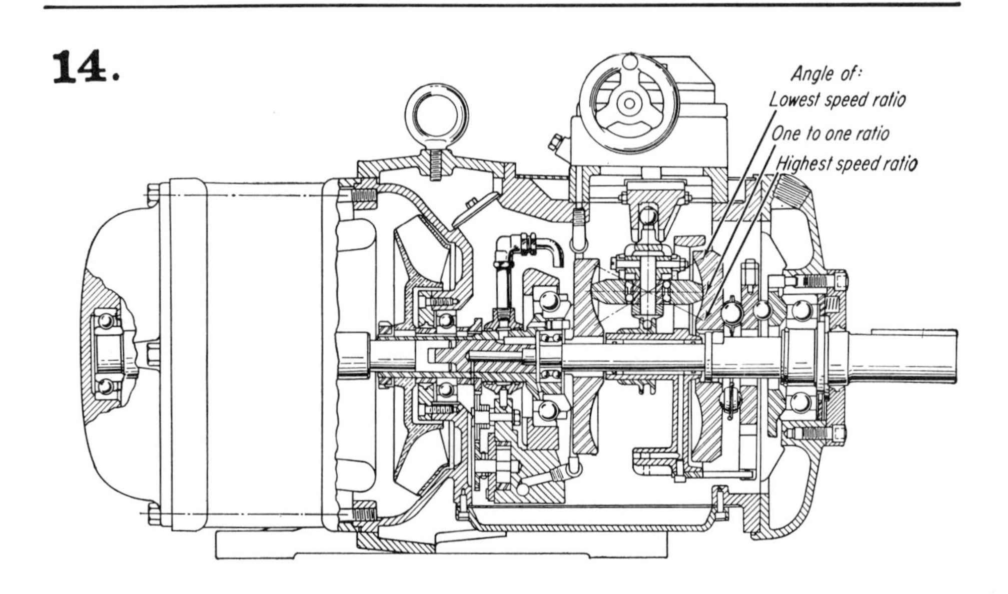

made by changing the angular position of the roller.

The output disk rotates at the same speed as the input when the roller centerline is parallel to the disk centerline, as in (A). When the contact point is nearer the centerline on the output disk and further from the centerline on the input disk, as in (B), the output speed exceeds that of the input. Conversely, when the roller contacts the output disk at a large radius, as in (C), the output speed is reduced.

A loading cam maintains the necessary contact force between the disks and power roller. Speed range is to 9:1; efficiency close to 90%.

Ball-and-cone drive (Fig 18). This is a remarkably simple drive. Input and output shafts are offset. Two opposing cones with 90-deg internal vertex angles are fixed to each shaft. The shafts are preloaded against each other. Speed variation is obtained by positioning the ball which contacts the cones. Note that the ball can shift laterally in relation to the ball plate. The conical cavities, as well as the ball, have hardened surfaces and the drive operates in an oil bath.

17.

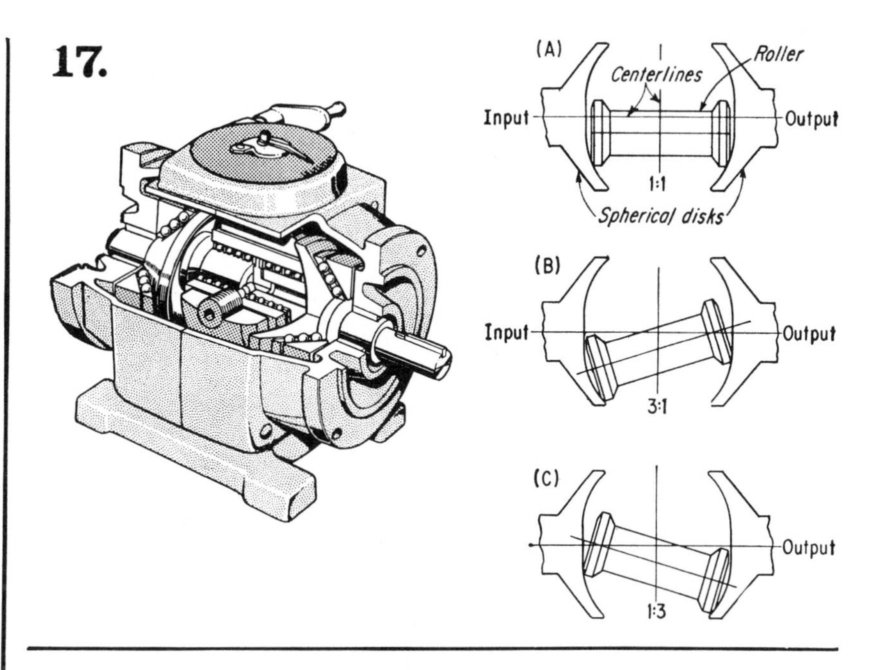

18.

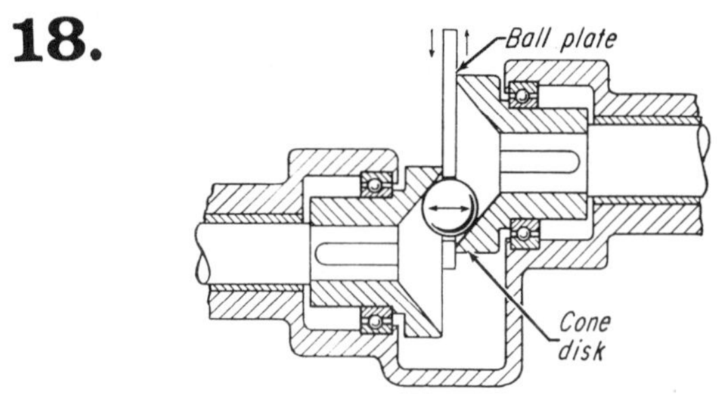

16.

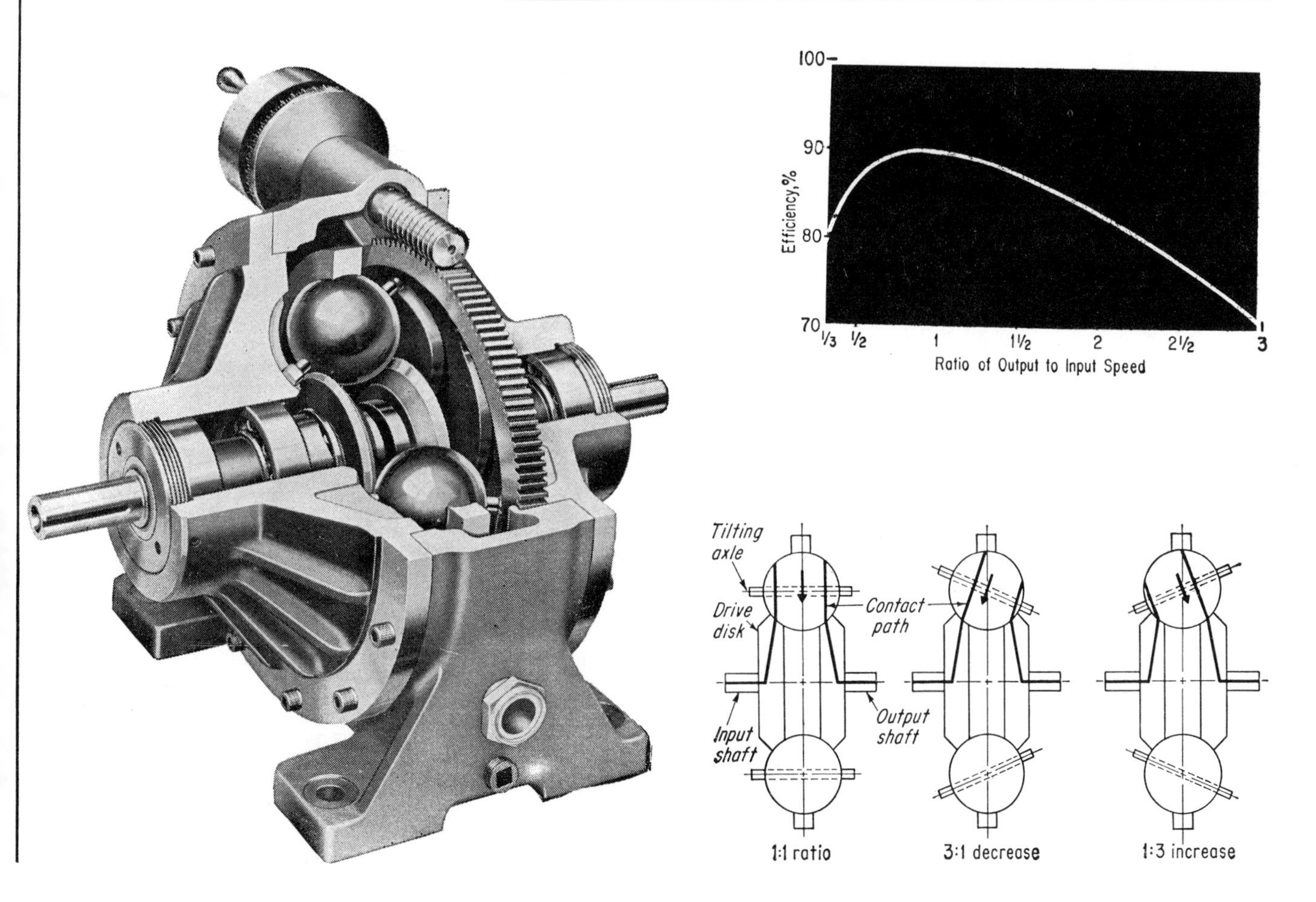

Multiple disk drives

Ball-and-disk drive (Fig 19). Friction disks are mounted on splined shafts to allow axial movement. Steel balls are carried by swing arms, rotate on guide rollers, and are in contact with driving and driven disks. Belleville springs provide the loading force between the balls and the disks. Position of balls controls the ratio of contact radii, and thus the speed.

Only one pair of disks is required to provide the desired speed ratio—purpose of multiple disks is to increase the torque capacity. And, if the load changes, a centrifugal loading device increases or decreases the axial pressure in proportion to the speed. The helical gears permit the output shaft to be coaxial with the input shaft. Speed ratios are from 1:1

19.

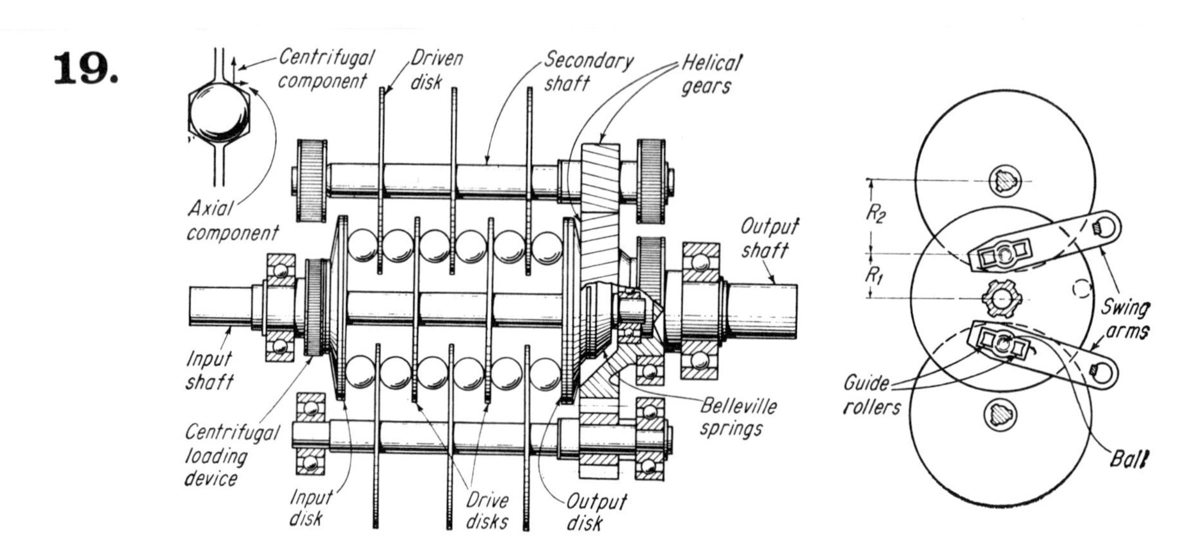

21. **22.**

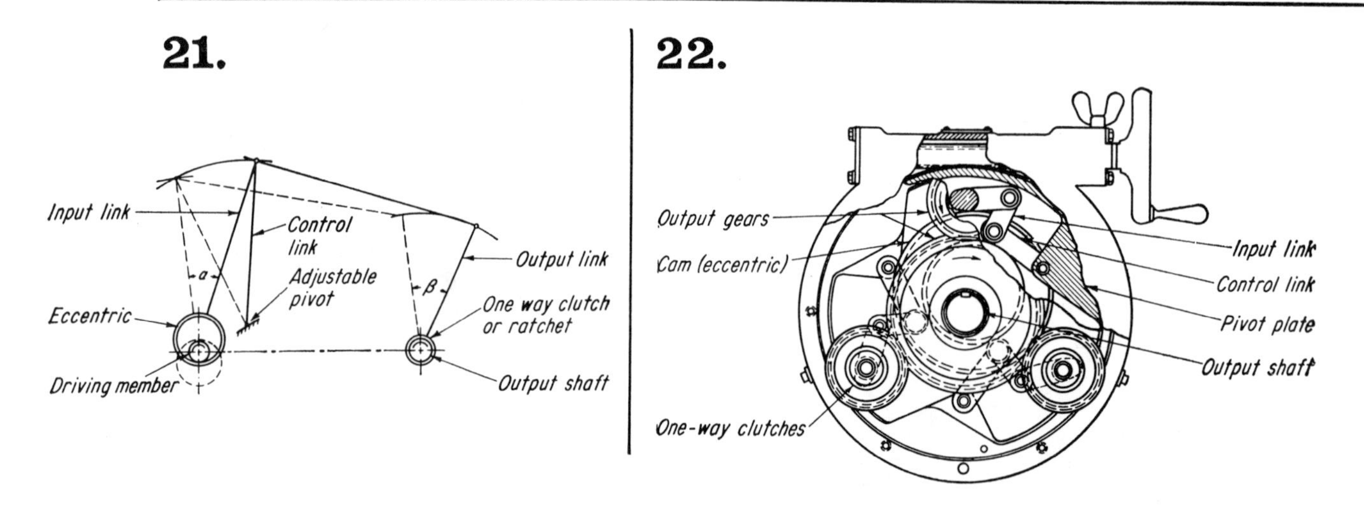

Impulse drives

Variable-stroke drive (Fig 21). This is a combination of a four-bar linkage with a one-way clutch or ratchet. The driving member rotates the eccentric which, through the linkage, causes the output link to rotate a fixed amount. On the return stroke, the output link overrides the output shaft. Thus a pulsating motion is transmitted to the output shaft which in many applications such as feeders and mixers is a distinct advantage. Shifting the adjustable pivot varies the speed ratio. By adding eccentrics, cranks, and clutches in the system, the frequency of pulsations per revolution can be increased to produce a smoother drive.

Morse drive (Fig 22). The oscillating motion of the eccentric on the output shaft imparts motion to the input link, which in turn rotates the output gears. Travel of the input link is regulated by the control link which oscillates around its pivot and carries the roller, which rides in the eccentric cam track. Usually, three linkage systems and gear assemblies overlap the motions, two linkages on return, while the third is driving. Turning the hanto 1:5, efficiency to 92%. Small units to 9 hp; large units to 38 hp.

Oil-coated disks (Fig 20). Power is transmitted without metal-to-metal contact at 85% efficiency. The interleaved disk sets are coated with oil during operation. At the points of contact axial pressure applied by the rim disks compresses the oil film, increasing its viscosity. The cone disks transmit motion to the rim disks by shearing the molecules of the high-viscosity oil film.

Three stacks of cone disks (only one stack shown) surround the central rim stack. Speed changes are produced by moving the cone disks radially toward (output speed increases) or away (decreases) from the rim disks. A spring and cam on the output shaft maintain pressure on the disks at all times.

The drive is available to 60 hp, but much larger units have been built. For small units, air cooling is provided; for big units, water cooling is required.

Under normal conditions the drive has the capacity to transmit its rated power with negligible slip—1% at high speed, 3% at low speeds.

20.

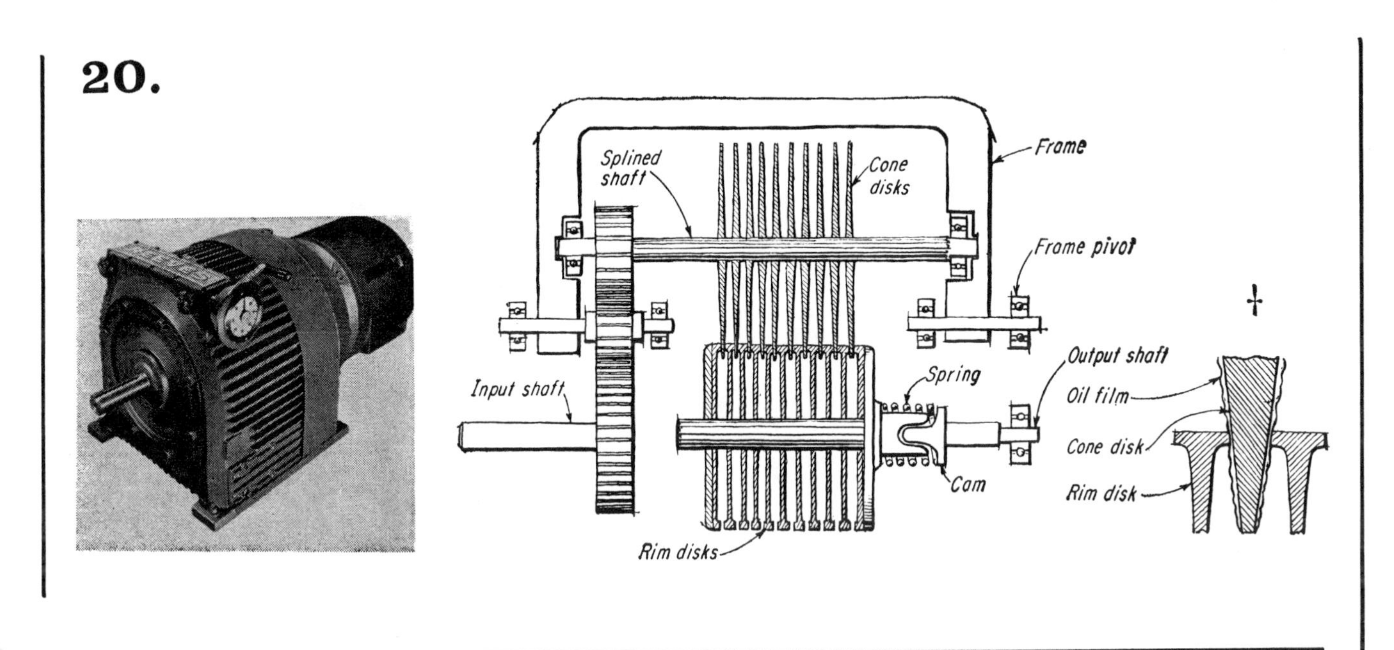

23.

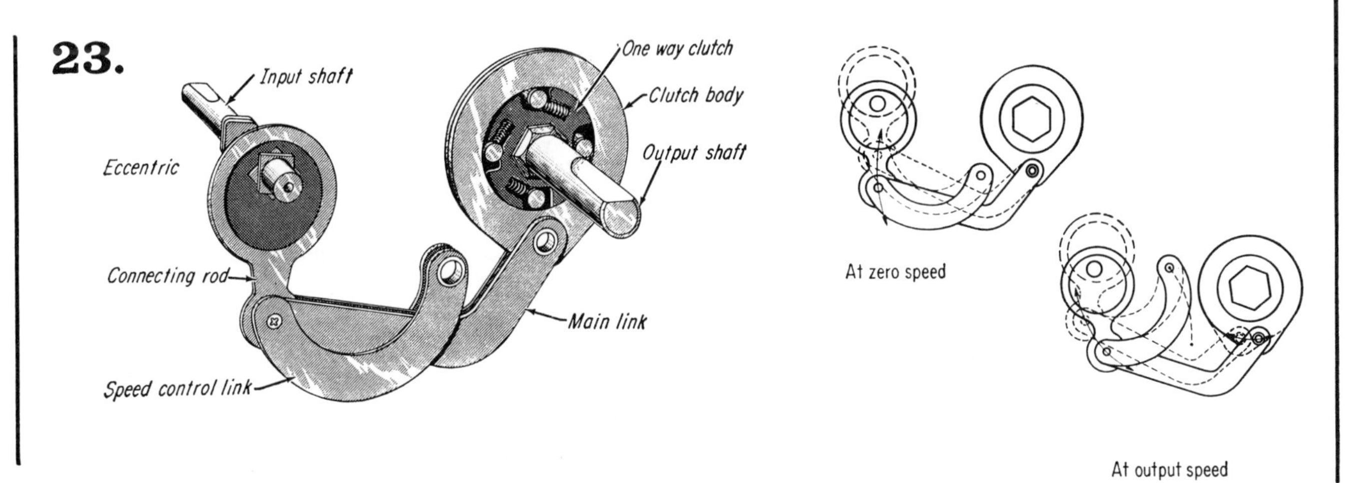

dle repositions the control link and changes the oscillation angles of the input link, intermediate gear, and input gear. This drive is specified as constant-torque device with limited range. Maximum torque output is 175 ft-lb at maximum input speed of 180 rpm. The speed variation is between 4½:1 to 120:1.

Zero-Max drive (Fig 23). This drive, which is also based on the variable-stroke principle, delivers with 1800-rpm input, 7200 or more impulses per min to the output shaft at all speed settings above zero. The pulsating nature of this drive is again damped by the number of parallel working sets of mechanisms between the input and output shaft. The illustration above shows only one of these working sets.

At zero output speed, the eccentric on the input shaft moves the connecting rod up and down through an arc. There is no back and forth motion to the main link. To set the output speed, the pivot is moved (upwards in the illustration), thereby changing the direction of the connecting rod motion and imparting a back and forth motion to the main link. The one-way clutch mounted on the output shaft provides the ratchet action. Reversing the input shaft rotation will not reverse the output. However, the reversing can be accomplished in two ways—by a special reversible clutch, or by a bell-crank mechanism in gearhead models.

This drive belongs to the infinite speed range class, because its output speed passes through zero. Maximum input speed: 2000 rpm. Speed range: from zero to ¼ of input speed. Maximum capacity: to ¾ hp.

Variable-Speed Drives — Additional Variations

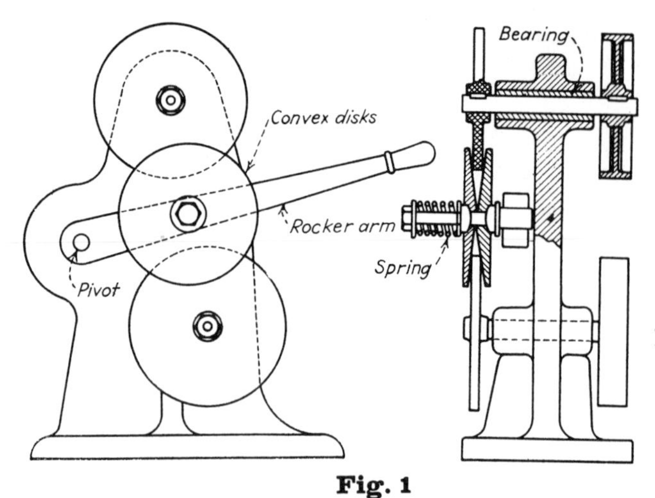

Fig. 1

Fig. 1 —The well-known Sellers' disks consist of a device for transmitting power between fixed parallel shafts. Convex disks mounted freely on a rocker arm and pressing firmly against the flanges of the shaft wheels by a coiled spring form the intermediate sheave. Speed ratio changed by moving rocker lever. No reverse possible, but driven shaft may rotate above or below driver speed. Convex disk must be mounted on self aligning bearings to ensure good contact at all positions.

Fig. 2 —A curved disk device made possible by motorization. Motor is swung on its pivot in such a manner as to change the effective diameters of the contact circles. A compact drive for a small drill press.

Fig. 3 —Another motorized modification of an older device. Principle similar to Fig. **3,** but with only two shafts. Ratio changed by sliding motor in Vee guides.

Fig. 4 —Two cones mounted close together and making actual contact through a squeezed belt. Speed ratio changed by shifting belt longitudinally. Taper on cones must be moderate in order to avoid excessive wear on sides of belt.

Fig. 5 —These speed cones are mounted at any convenient distance apart and connected by a Condor Whipcord belt, whose outside edges consist of an envelope of tough, flexible, wear-resisting rubberized fabric built to withstand the wear caused by the belt edge travelling at a slightly different velocity to the part of the cone in actual contact. Speed ratio changed by sliding belt longitudinally.

Fig. 2

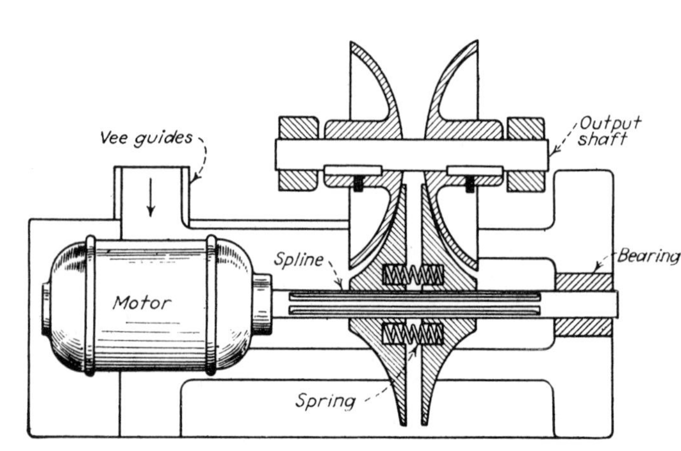

Fig. 3

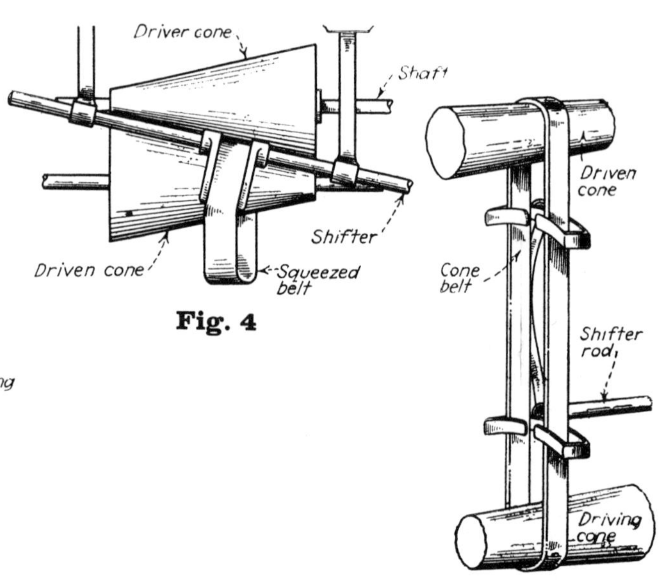

Fig. 4

Fig. 5

Fig. 6 —Another device to avoid belt "creep" and wear in speed cone transmissions. The inner bands are tapered on the inside and present a flat or crowned surface to the belts in all positions. Speed ratio changed by moving inner bands rather than main belts.

Fig. 7 —Another device for avoiding belt wear when using speed cones. Creeping acting of belt not entirely eliminated, and universal joints present problem of cost and maintenance.

Fig. 8 —An extension of the principle used in Fig. 7 whereby a roller is substituted for the belt, giving more compactness.

Fig. 9 —The main component of this drive is a hollow cone driven by a conical roller. Speed ratio changed by sliding driving unit in Vee guides. Note that when roller is brought to the center of the hollow cone, the two run at identical speed with the same characteristics as a cone clutch. This feature makes system very attractive where heavy torque at motor speed is required in combination with lower speeds for light preliminary operations.

Fig. 10—In this transmission, the cones are mounted in line and supported by the same shaft. One cone is keyed to main shaft and the other is mounted on a sleeve. Power transmitted by series of rocking shafts and rollers. Pivoting rocking shafts and allowing them to slide changes speed ratio.

Fig. 11 —This J. F. S. transmission uses curved surfaces on its planetary rollers and races. The cone-shaped inner races revolve with the drive shaft, but are free to slide longitudinally on sliding keys. Strong compression springs keep these races in firm contact with the three planetary rollers.

Fig. 12—Featuring simplicity with only five major parts, this Graham transmission employs three tapered rollers carried by a spider fastened to the drive shaft. Each roller has a pinion meshing with a ring gear connected to the output shaft. The speed of the rollers, and in turn, the speed of the output shaft is varied by moving contact ring longitudinally, thus changing the ratio of the diameters in contact.

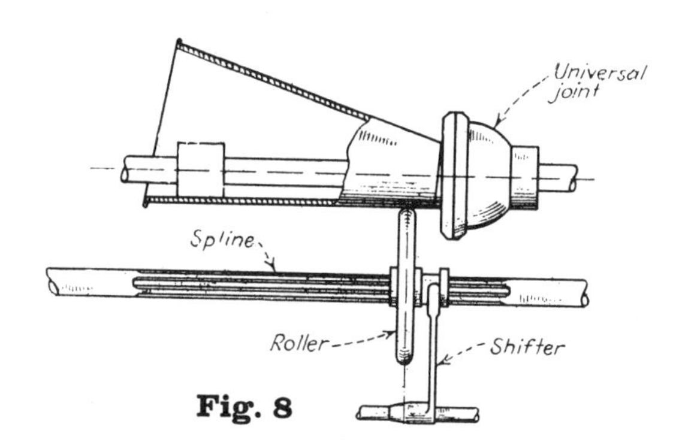

Fig. 8

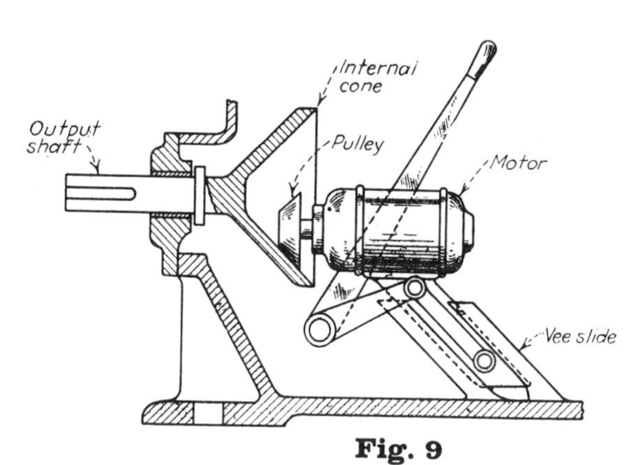

Fig. 9

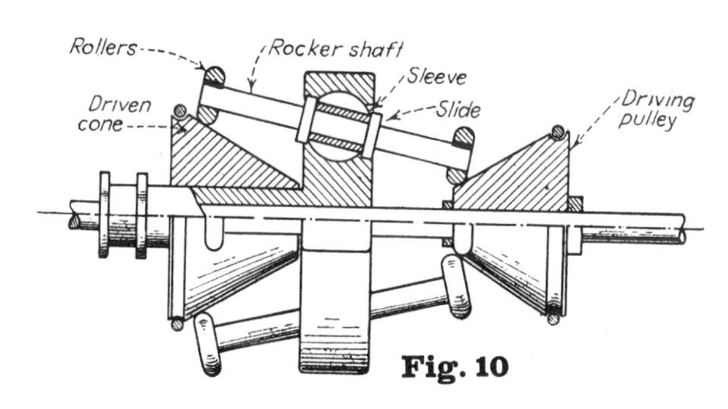

Fig. 10

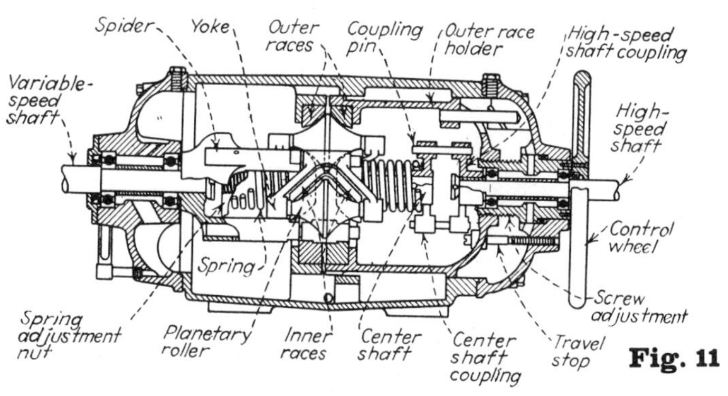

Fig. 11

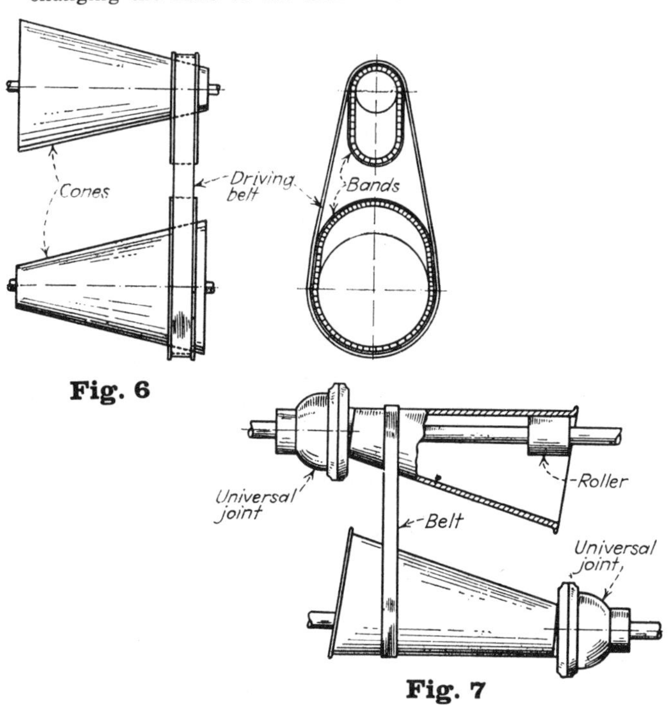

Fig. 6

Fig. 7

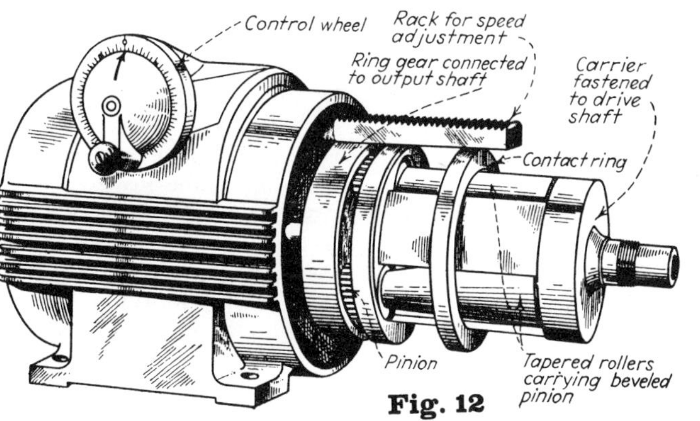

Fig. 12

Adjustable sphere drive

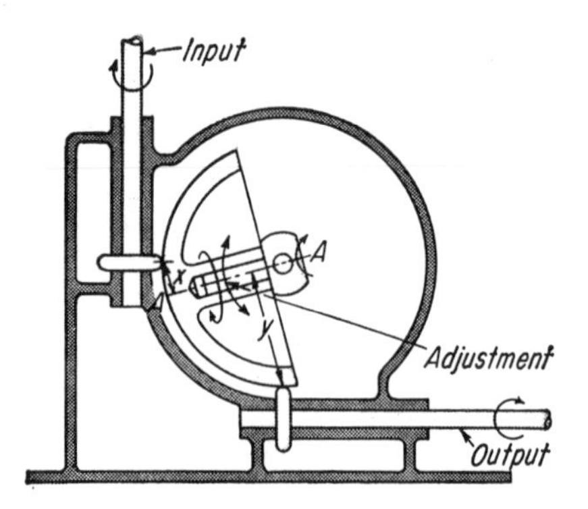

The input rotates the sphere around axis A, which in turn drives the output. The speed decrease is proportional to the ratio of x to y. Changing the direction of the sphere axis changes the contact points on the sphere and hence distances x and y. This results in a change in the output ratio.

Double sphere drive

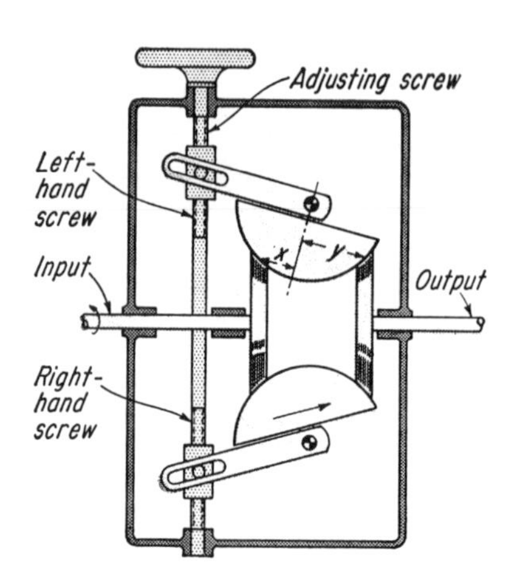

The ratio of distance x to y, which determines the output speed, is changed in this case by turning the adjusting screw to tilt both spheres simultaneously.

VARIABLE SPEED DRIVE for sub-fhp applications is stepless from 0 to 250 rpm. It uses a rocker arm to convert rotary motion to variable linear motion in minimum space. The complete drive incorporates four rocker arms, each connected to a one-way driving clutch. Four eccentrics transmit rotary input motion to the rocker arms. The linear motion of each rocker arm is delivered to its attached clutch in the output section in varying degrees, depending on the position of an adjustable pivot block that rides on the rocker arms. The Speed-Mate Co

The Cone-Trol drive shown right by Cone-Drive Gears, Detroit, is stepless, over a 6:1 speed range, and easily controlled by handwheel or servo

Heart of the drive is a cone roller which is mounted between two circular plates. As the position of the cone roller is changed, the circumference of the rolling paths on the input and output disks is increased and decreased respectively. The infinitely variable output speed is produced by an adjusting screw which changes the roller's path.

Controlled differential drives

By coupling a differential gear assembly to one of the variable speed drives you can increase the horsepower capacity—at the expense of the speed range—or you can increase the speed range—at the expense of the horsepower range. Numerous combinations of the variables are possible.

The type of differential depends on the manufacturer. Some systems have bevel gears, others have planetary gears. Both single and double differentials are employed. Variable-speed drives with differential gears are commercially available up to 30 hp.

Horsepower - increase differential (Fig 1). The differential is coupled so that the output of the motor is fed into one side and the output of the speed variator into the other side. An additional gear pair is employed as illustrated.

Output speed $n_4 = \frac{1}{2}\left(n_1 + \frac{n_2}{R}\right)$

Output torque

$$T_4 = 2T_3 = 2RT_2$$

Output hp

$$\text{hp} = \left(\frac{Rn_1 + n_2}{63{,}025}\right) T_2$$

hp increase

$$\Delta\text{hp} = \left(\frac{Rn_1}{63{,}025}\right) T_2$$

Speed variation

$$n_{4\,\text{max}} - n_{4\,\text{min}}$$

$$= \frac{1}{2R}\,(n_{2\,\text{max}} - n_{2\,\text{min}})$$

Speed range increase differential (Fig 2). This arrangement achieves a wide range of speed with the low limit at zero or in reverse direction.

1

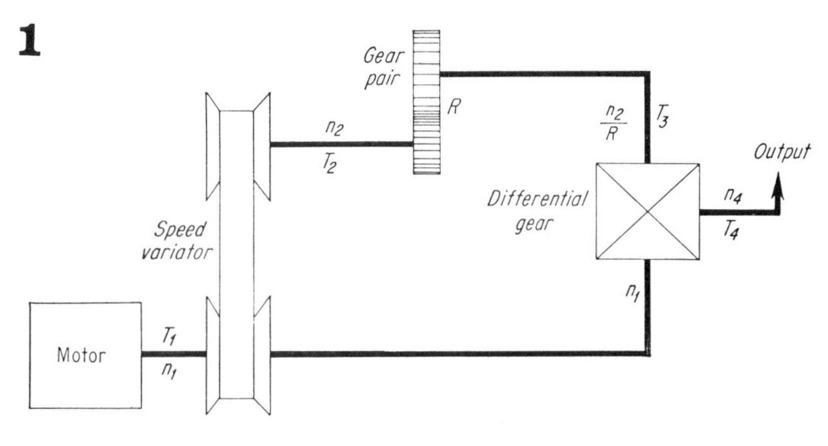

2

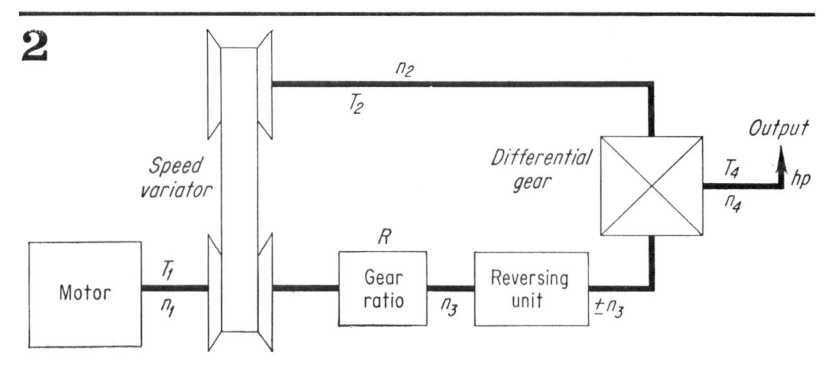

3

Bruning Model 50 Whiteprinter

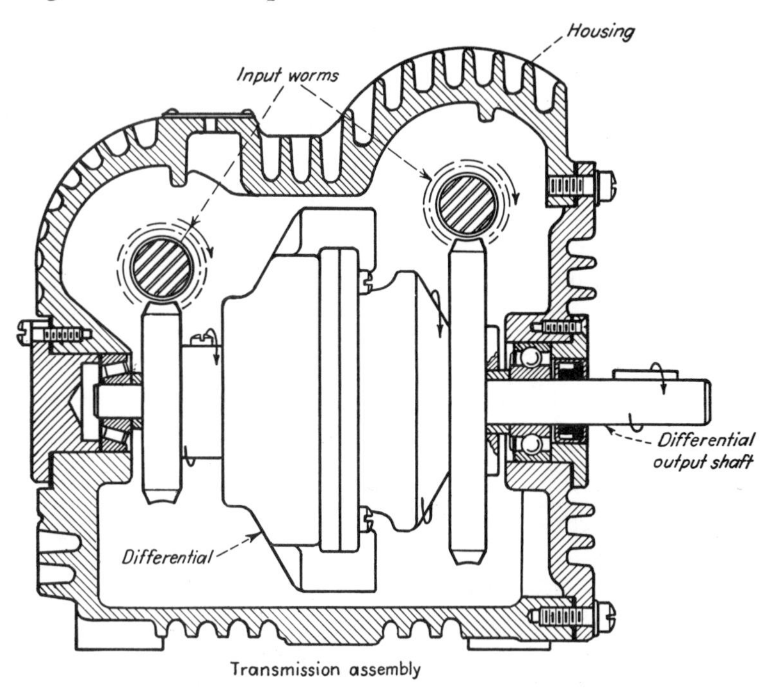

Transmission assembly

VARIABLE SPEED TRANSMISSION consists of two sets of worm gearing feeding into a differential mechanism. Output shaft speed depends on difference in rpm between the two input worms. When worm speeds are equal, output is zero. Each worm shaft carries a cone-shaped pulley. These pulleys are mounted so that their tapers are in opposite directions. Shifting the position of the drive belt on these pulleys has a compound effect on output speed.

Ratchet and Inertia Variable-Speed Drives

Ratchet and inertia type mechanisms for variable-speed driving of heavy or light loads

CYRIL DONALDSON
Rochester Athenæum & Mechanics Institute

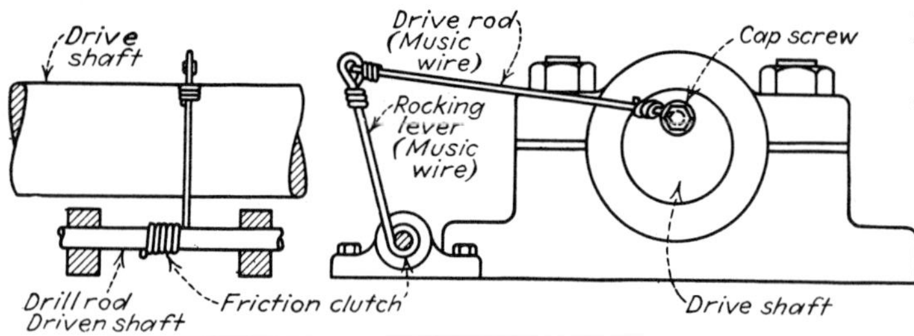

Fig. 1

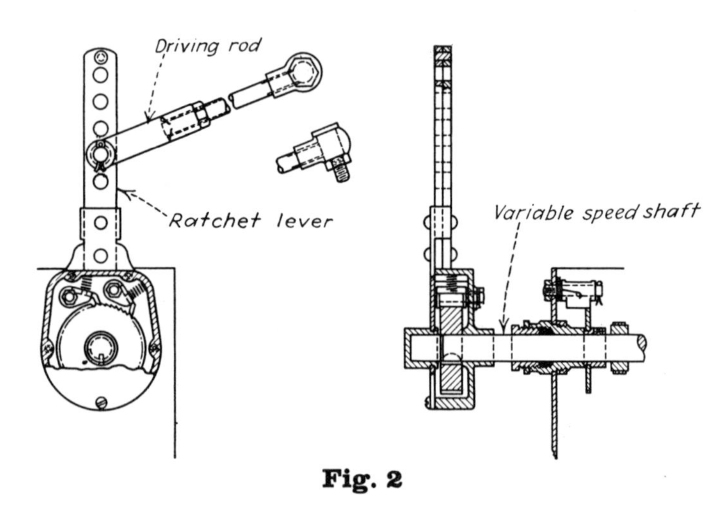

Fig. 2

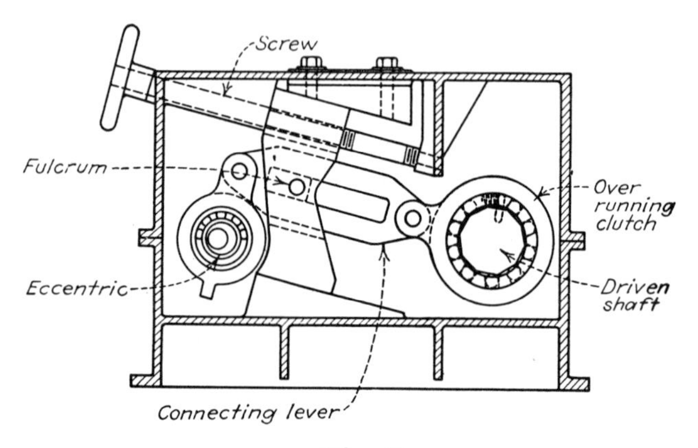

Fig. 3

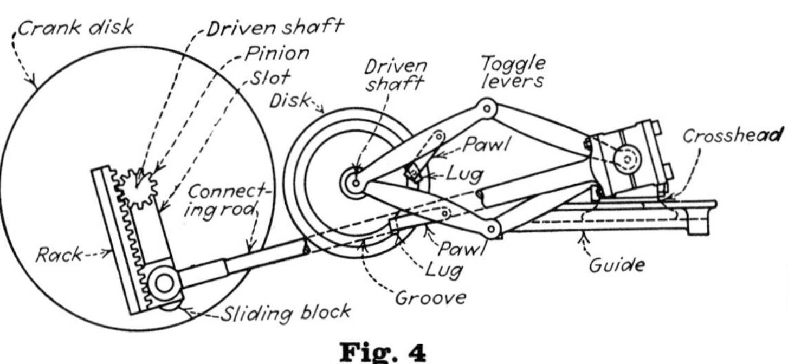

Fig. 4

Fig. 1 —A handy variable speed device suitable only for very light drives in laboratory or experimental work. Drive rod receives motion from drive shaft and rocks lever. Friction clutch made by winding wire around drill-rod in lathe with diameter slightly smaller than diameter of driven shaft. Speed ratio changed when unit is stationary, by varying length of rods, or throw of eccentric.

Fig. 2 —This Torrington lubricator drive illustrates the general principles of ratchet transmission devices. Reciprocating motion from a convenient sliding part, or from an eccentric, rocks the ratchet lever, which gives the variable speed shaft an intermittent uni-directional motion. Speed ratio can be changed only when unit is stationary. Placing fork of driving rod in different hole varies the throw of the ratchet lever.

Fig. 3 —An extension of the principle illustrated above, this Lenney transmission replaces the ratchet with an over-running clutch. Speed of the driven shaft can be varied while the unit is in motion by changing the position of the connecting lever fulcrum.

Fig. 4 —Another transmission employing the principle shown above. Crank disk imparts motion to connecting rod. Crosshead moves toggle levers which in turn give uni-directional motion to clutch wheel by means of friction pawls engaging in groove. Speed ratio changed by varying throw of crank with the aid of the rack and pinion.

Fig. 5 —A variable-speed transmission used on gasoline railroad section cars. The connecting rod from the crank, mounted on the constant speed shaft, rocks the oscillating lever and actuates the over-running clutch, thus giving intermittent but uni-directional motion to the variable speed shaft. The toggle link keeps the oscillating lever within a prescribed path. Speed ratio changed by swinging bell crank towards position shown in dotted lines, around pivot attached to frame. This varies the movement of the over-running clutch. Several units out of phase with each other are necessary for continuous motion of the shaft.

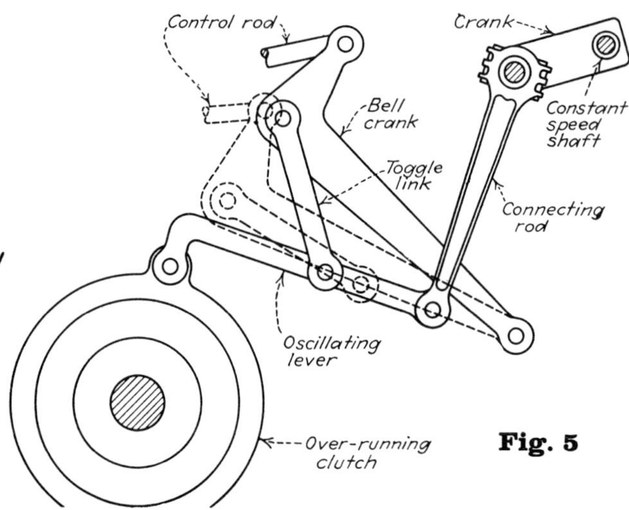

Fig. 5

Fig. 6 —This Thomas transmission is an integral part of an automobile engine in which piston motion is transferred by means of the conventional connecting rod to long arm of a bell-crank lever oscillating about a fixed fulcrum. Attached to the short arm of the bell crank lever is a horizontal connecting rod which in turn rotates the crankshaft. Crankshaft motion is rendered continuous and steady by means of a flywheel, but no power other than that required to operate auxiliaries is taken from this shaft. The main power output is transferred from the bell-crank lever to the over-running clutch by means of a third connecting rod. Speed ratio is changed by sliding the top end of the third connecting rod within the bell-crank lever by means of a cross-head and guide mechanism. Highest ratio is obtained when the crosshead is farthest from the fulcrum and movement of the cross-head toward the fulcrum reduces the ratio until a "neutral" position is reached when the center line of the connecting rod coincides with the fulcrum.

Fig. 7 —Another automobile transmission system built as an integral part of the engine, this Constantinesco torque converter features an inherently automatic change of speed ratio according to the speed and load on the engine. The constant speed shaft rotates a crank which in turn operates two oscillating levers having inertia weights at one end, while the other ends are attached by links to the rocking levers. Incorporated in these rocker levers are over-running clutches. Since at low engine speeds the inertia weights oscillate through a wide angle at low speed, the reaction of the inertia force on the other end of the lever is very slight, and the link imparts no motion to the rocker lever. Speed increase of engine causes inertia weight reaction to increase thereby rocking small end of oscillating lever as the crank rotates. Consequent motion rocks rocking lever by means of link and the variable shaft is driven in one direction.

Fig. 8 —Featuring a differential gear with an adjustable escapement, this transmission by-passes a variable proportion of the drive shaft revolutions. Constant speed shaft rotates freely mounted worm wheel carrying two pinion shafts. The firmly fixed pinions on these shafts in turn rotate the sun gear which meshes with other planetary gears, rotating the small worm gear attached to the variable speed output shaft.

Fig. 9 —In this Morse transmission, an eccentric cam, integral with the constant speed input shaft, rocks three ratchet clutches by means of a series of linkage systems containing three rollers running in a circular groove cut in the cam face. Uni-directional motion is conveyed to the output shaft from the clutches by planetary gearing. Speed ratio is changed by rotating anchor ring containing fulcrum of links, thus varying the stroke of the levers.

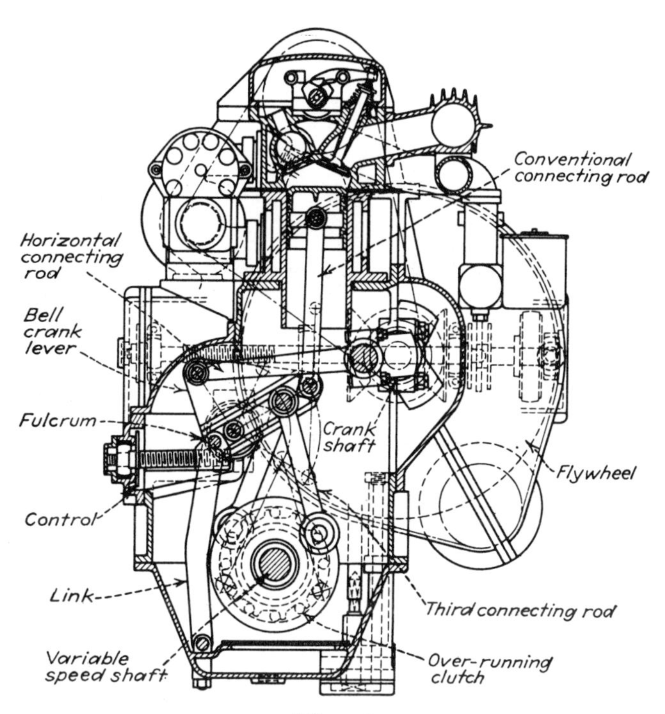

Fig. 6

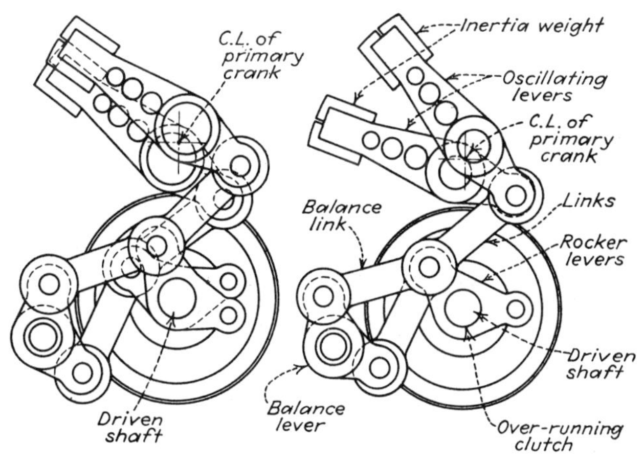

Fig. 7

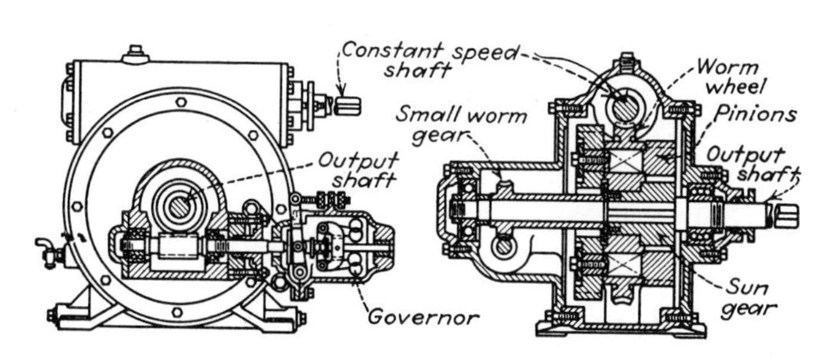

Fig. 8

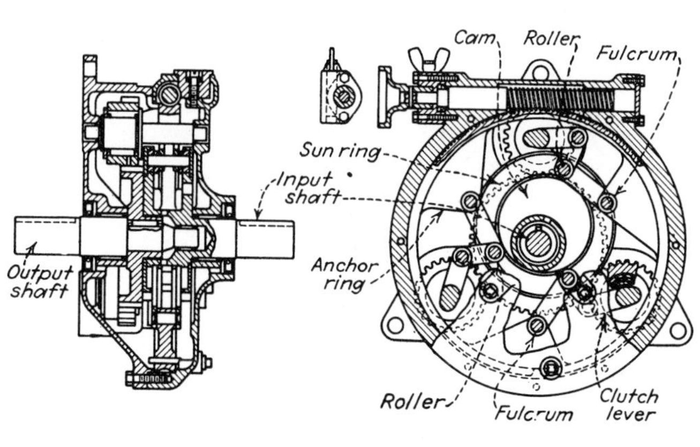

Fig. 9

Your guide to variable-speed

Belt and chain drives

Second of a two-part series on mechanical drives, this article covers basic types, arrangements, and sheave designs

D. Z. DVORAK, analytical engineer, Pratt & Whitney Co, West Hartford, Conn

1.

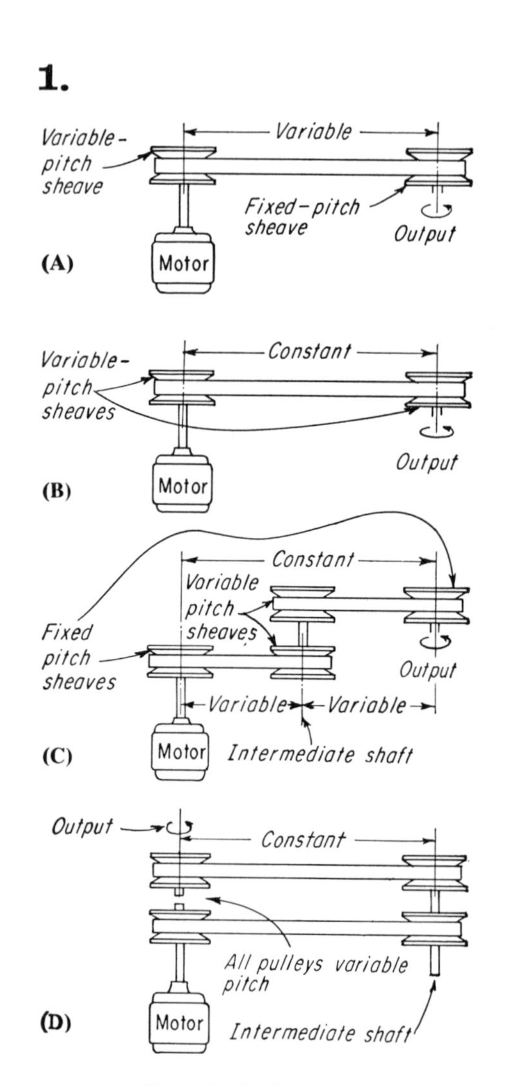

Four basic belt arrangements for varying output speed.

2.

Sliding motor base for varying the center-to-center distance.

5.

Handwheel adjusts sheave on motor shaft, spring-loaded sheave on output shaft then self adjusts to a new pitch diameter.

VARIABLE-SPEED drives provide an infinite number of speed ratios within a specific range. They differ from the stepped-pulley drives in that the stepped drives offer only a discrete number of velocity ratios.

The first article in this series (see the article on p. 2) covered mechanical "all-metal" drives. For the most part, these employ friction or preloaded cones, disks, rings, and spheres, which do involve a certain amount of slippage. Belt drives, on the other hand, have little slippage or frictional losses, and chain has none—it maintains a fixed phase relationship between the input and output shafts.

Belt drives

Belt drives have high efficiency and are relatively low in price. Most use V-belts, reinforced by steel wires, of up to 3-in. width.

Speed adjustment in belt drives is obtained through one of the four basic arrangements shown in Fig 1.

Variable-distance system (Fig 1A). A variable-pitch sheave on the input shaft opposes a solid (fixed-pitch) sheave on the output shaft. To vary the speed, the center distance is varied, usually by means of an adjustable base motor of the tilting or sliding types (Fig 2).

Speed variations up to 4:1 are easily achieved, but torque and horsepower characteristics depend on the location of the variable-diameter sheave (Fig 3).

Fixed-distance system (Fig 1B). Variable-pitch sheaves on both input and output shafts maintain a constant center distance between shafts. The sheaves are controlled by linkage. Either the pitch diameter of one sheave is positively controlled and the disks of the other sheave, being under spring tension, adjust automatically (Fig 5), or the pitch diameters of both sheaves are positively controlled by the linkage system (Fig 4). Pratt & Whitney has applied the system in Fig 5 to the spindle drive of numerically controlled machines.

Speed variations up to 11:1 are obtained, which means that with a 1200-rpm motor, the maximum output speed will be $1200 (11)^{\frac{1}{2}} = 3984$ rpm, and the minimum output speed $= 3984/11 = 362$ rpm.

Double-reduction system (Fig 1C). Solid sheaves are on both the input and output shafts; but both sheaves on the intermediate shaft are of variable-pitch type. Center distance between input and output is constant.

Coaxial shaft system (Fig 1D). The intermediate shaft in this arrangement permits the output shaft to be coaxial with that of the input shaft. To maintain a fixed center distance, all four sheaves must be of the variable-pitch type and controlled by linkage, similar to the system in Fig 6. Speed variation up to 16:1 is available.

Packaged belt units (Fig 7). These combine the motor and variable-pitch transmissions as an integral unit. The belts are usually ribbed, and speed ratios can be dialed by a handle.

text continued, next page

3.

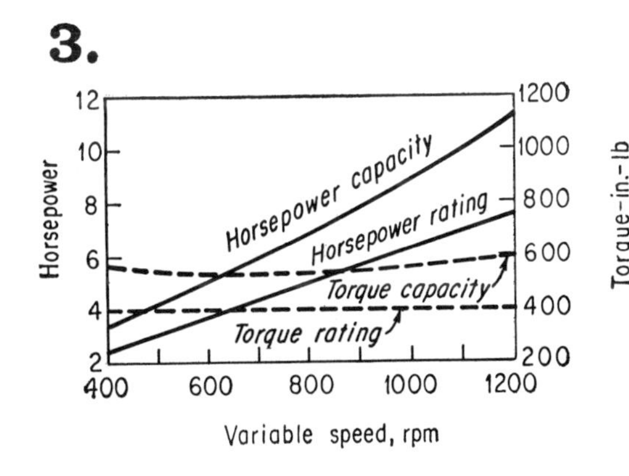

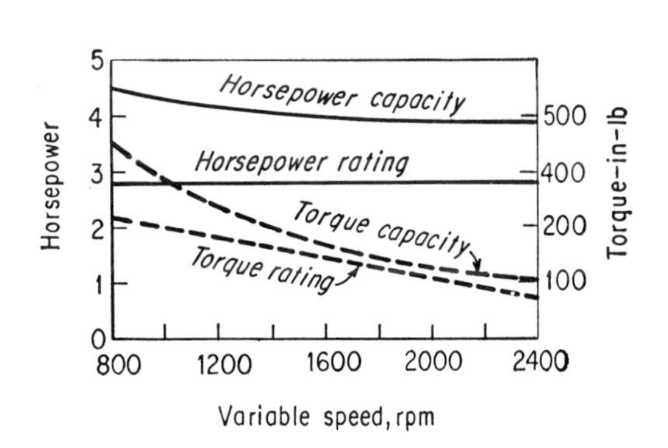

Horsepower vs speed for drives with variable pulley on input shaft (left), and output shaft (right).

4.

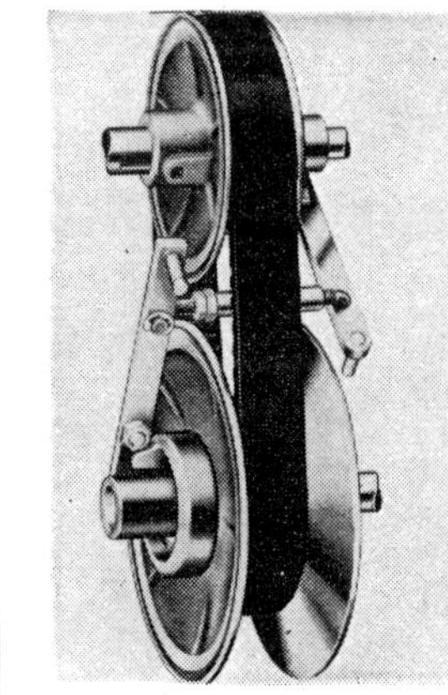

Linkage controlled pulleys.

6.

Tandem arrangement employs dual belt-system to produce high speed-reduction.

7.

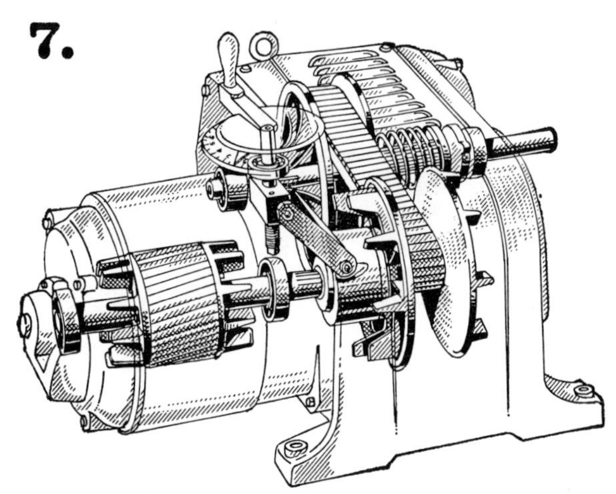

Packaged belt unit.

Sheave designs

The axial shifting of variable-pitch sheaves is controlled by one of four methods:

Linkage actuation. The sheave assemblies in Fig 5 are directly controlled by linkages which, in turn, are manually adjusted.

Spring pressure. The cones of the sheaves in Fig 2 and 4 are axially loaded by spring force. A typical pulley of this type is illustrated in Fig 8. Such pulleys are used in conjunction with directly controlled sheaves, or with variable center-distance arrangements.

Cam-controlled sheave. The cones of this sheave (Fig 9) are mounted on a floating sheave free to rotate on the pulley spindle. Belt force rotates the cones whose surfaces are cammed by the incline plane of the spring. The camming action wedges the cones against the belt, thus providing sufficient pressure to prevent slippage at the higher speeds, as shown in the curve.

Centrifugal-force actuator. In this unique sheave arrangement (Fig 10) the pitch diameter of driving sheave is controlled by the centrifugal force of steel balls. Another variable-pitch pulley mounted on the driven shaft is responsive to the torque. As the drive speed increases, the centrifugal force of balls forces the sides of the driving sheave together. With a change in load, the movable flange of the driven sheave rotates in relation to the fixed flange. The differential rotation of the sheave flanges cams them together and forces the V-belt to the outer edge of the driven sheave, which is a lower transmission ratio. The driving sheave is also shifted as the load rises with decreasing speed. With a stall load, it is moved to the idling position. When the torque responsive sheave is the driving member, any increase in drive speed closes its flanges and opens the flanges of the centrifugal member, thus maintaining a constant output speed.

The drive has been employed in transmissions ranging from 2 to 12 hp.

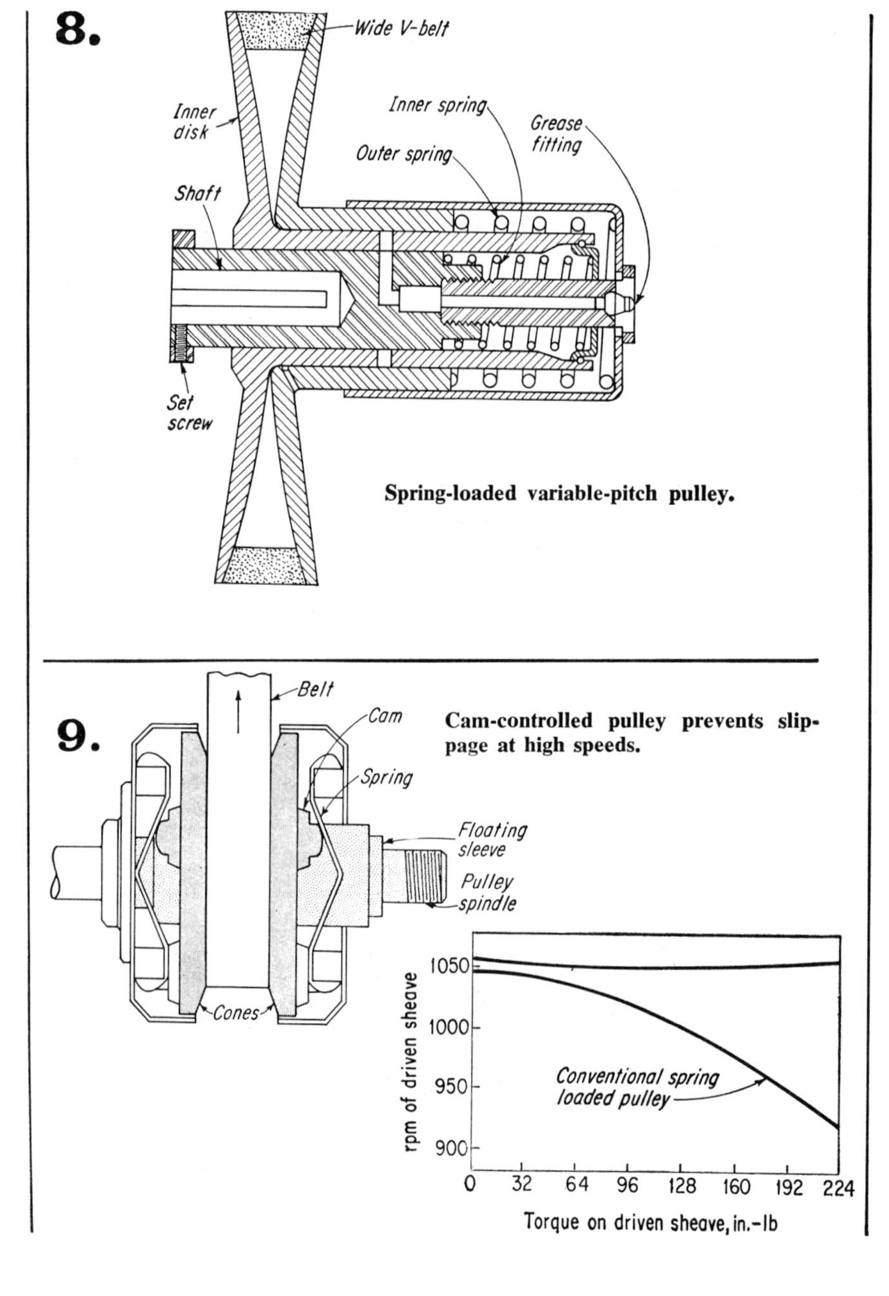

Spring-loaded variable-pitch pulley.

Cam-controlled pulley prevents slippage at high speeds.

10.

Ball-controlled pulley has sides pressured by centrifugal force. Morse Chain Co.

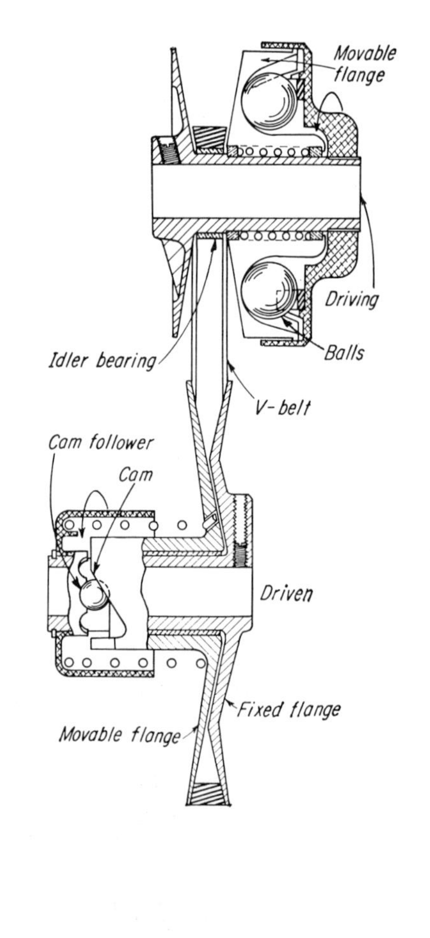

Chain drives

PIV drive (Fig 11). This chain drive (positive, infinitely variable) eliminates any slippage between the self-forming laminated chain teeth and the chain sheaves. The individual laminations are free to slide laterally so as to take up the full width of the sheave. The chain runs in radially grooved faces of conical surface sheaves which are located on the input and output shafts. The faces are not straight cones, but have a slight convex curve to maintain proper chain tension at all positions. The pitch diameters of both sheaves are positively controlled by the linkage. Tooth action is positive throughout operating range. Capacity: to 25 hp, speed variation to 6:1.

Double-roller chain drive (Fig 12). This specially developed chain is built for capacities to 22 hp. The hardened rollers are wedged between the hardened conical sides of the variable-pitch sheaves. Radial rolling friction results in smooth chain engagement.

Single-roller chain drive (Fig 13). The double strand of this chain boosts the capacity to 50 hp. The scissor-lever control system maintains proper proportion of forces at each pair of sheave faces throughout the range.

11.

PIV drive chain grips radially grooved faces of variable-pitch sheave to prevent slippage.

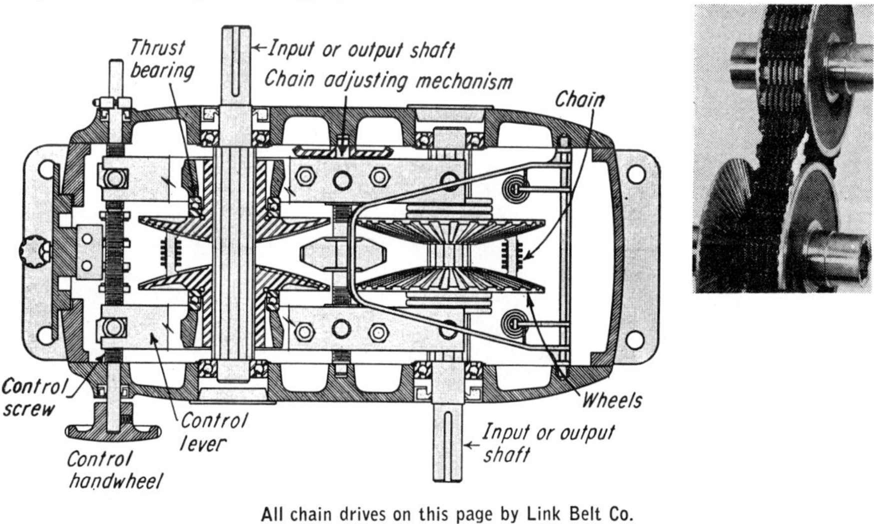

All chain drives on this page by Link Belt Co.

12. RS-E drive combines strength with ease in changing speed.

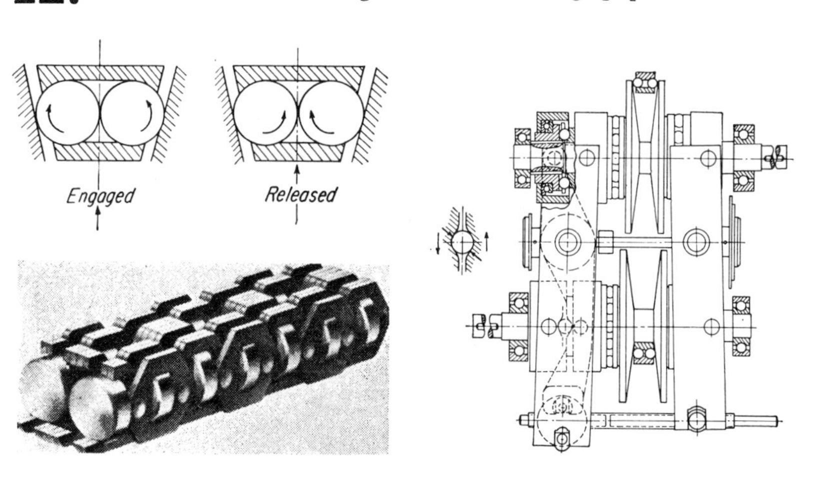

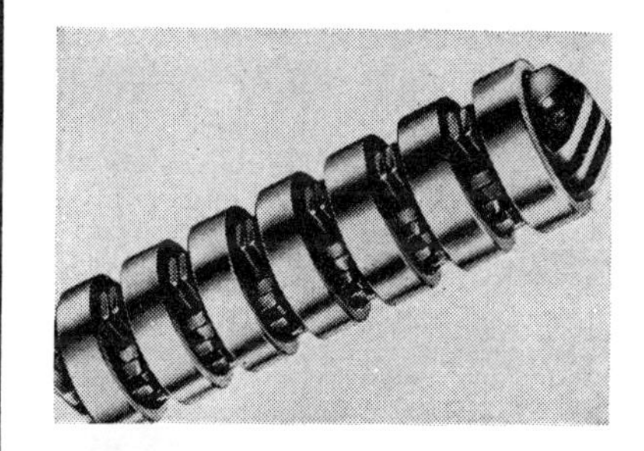

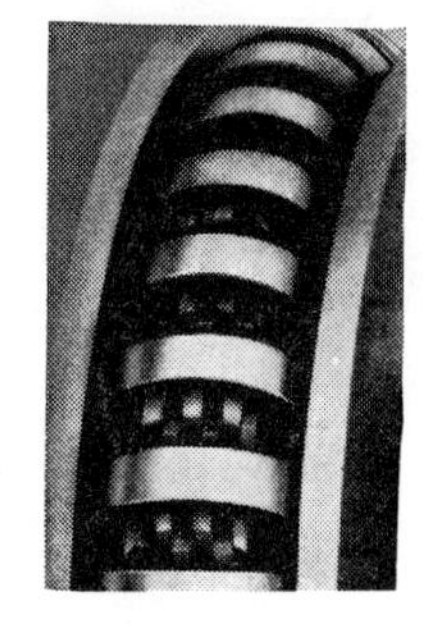

13.

RS drive for high horsepower applications.

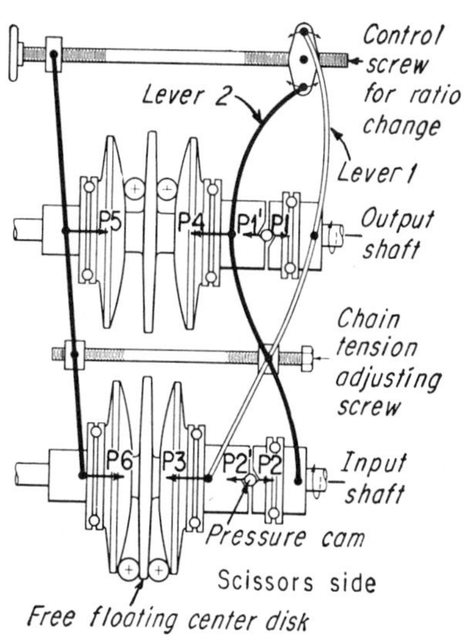

Pushbutton Selects Exact Speed Quickly

Motor-controlled spindle-speed selection permits pushbutton control. Belts for high speeds and gears for high torque are combined in variable-speed power transmission.

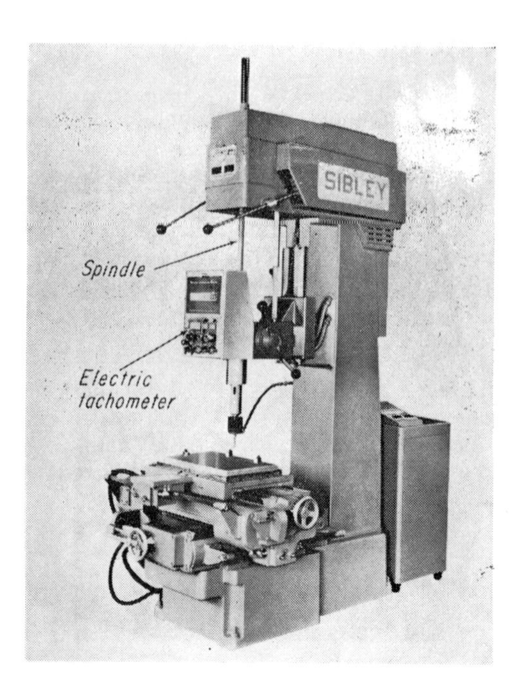

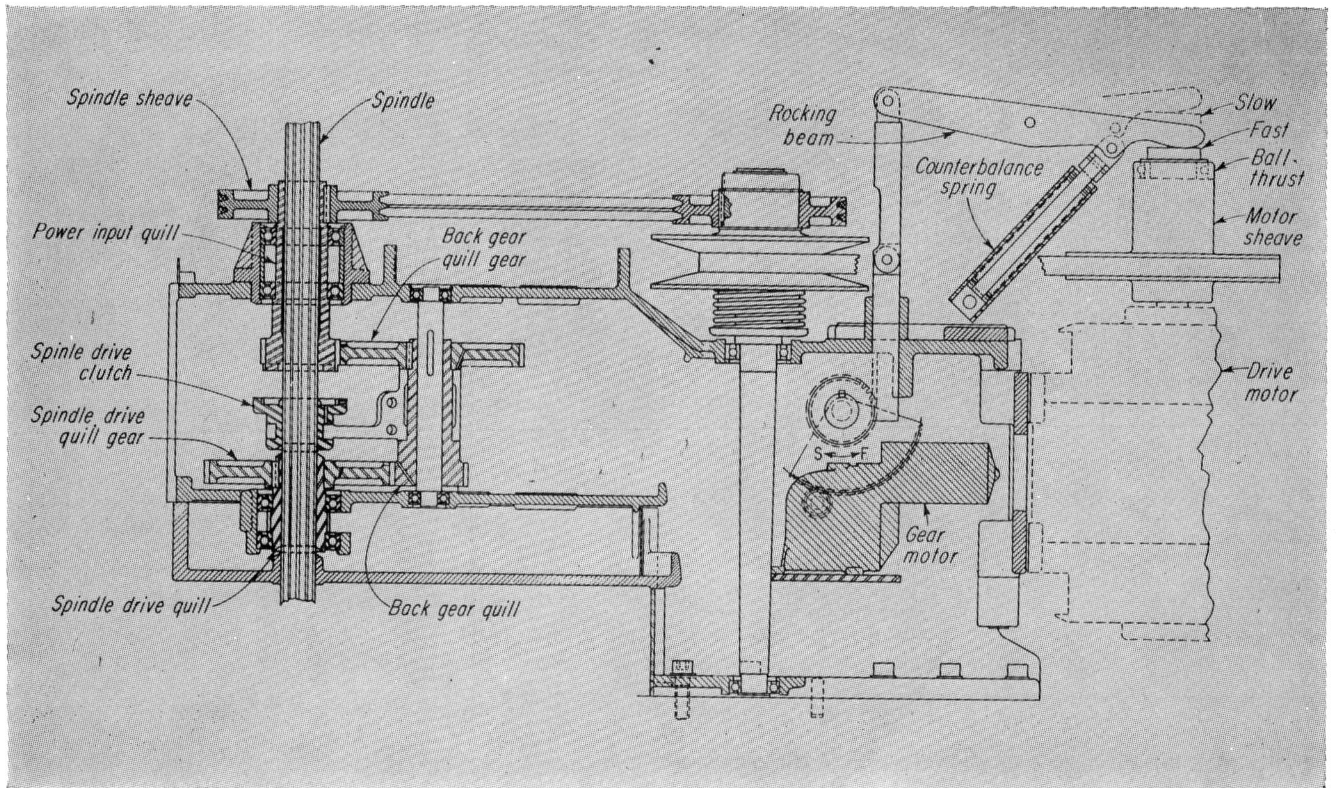

INFINITELY VARIABLE V-belt transmission is used for direct drive between motor and spindle for higher speeds. For lower speeds a back-gear system is interposed between transmission and spindle. Variable-speed transmission and back gears both have a reduction of 6 to 1 in order that the fastest speed in back gear (low range) shall be equal to the slowest speed in belt drive (direct). This eliminates any gap between highest and lowest spindle speeds.

Transmission speed changes are controlled by movement of upper half of the motor sheave. Spring-loaded, variable-diameter sheave on countershaft maintains belt tension as sheave ratio is altered. Rocking beam contacts thrust bearing to control movement of upper flange and effective diameter of motor sheave. A switch-controlled, 9-rpm, reversible gear motor drives the rocking beam through a gear segment and rack. Double worm reduction makes gear motor self-locking when shut off. Electric tachometer permits operator to observe spindle speeds during operation or setting.

With the back gears engaged, spindle is driven by the back-gear quill and quill gears. When the back-gear control lever is shifted, the back gears are lifted out of mesh and the spindle clutch engaged. Teeth on upper edge of the internally splined clutch engage teeth on bottom of the input quill, locking the spindle sheave to the spindle. A mercury governor prevents clutch engagement until motor speed is slowed almost to a stop.

PS . . . The 40-in. swing drilling machine uses punched metal tape for automatic programming of table positions. All operator controls are placed on a Drilling Information Center panel at eye level. Spindle drive and feed mechanism were designed by Prof Carl C. Stevason, Dept of Mechanical Engineering, Univ of Notre Dame, for Sibley Machine and Foundry Corp,

Smoother Drive Without Gears

The transmission in this motor scooter is torque-sensitive; motor speed controls continuously variable drive ratio. Operator merely works throttle and brake.

Scooter is powered by a 2-cycle, single-cylinder engine of 10-cu-in. displacement. The 1.7-gal. fuel tank gives a cruising range up to 170 miles. Fully automatic transmission eliminates manual clutch and shifting, provides smooth acceleration. "Topper" is produced by Harley-Davidson Motor Co.

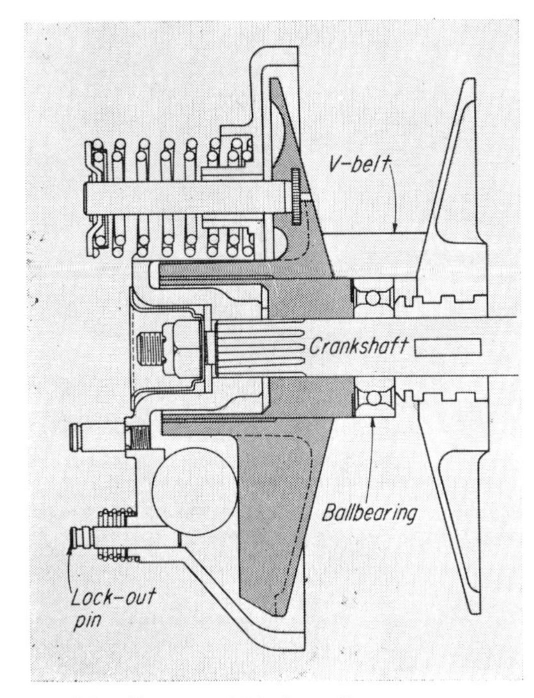

Variable-diameter V-belt pulleys . . .

connect motor and chain drive sprocket to give a wide range of speed reduction. The front pulley incorporates a 3-ball centrifugal clutch which forces the flanges together when the engine speeds up. At idle speed the belt rides on ballbearing between retracted flanges of the pulley. During starting and warmup a lockout prevents the forward clutch from operating.

Upon initial engagement the over-all drive ratio is approximately 18:1. As engine speed increases, belt rides higher up on the forward-pulley flanges until over-all drive ratio becomes approximately 6:1. Resulting variations in belt tension are absorbed by the spring-loaded flanges of the rear pulley. When clutch is in idle position, the V-belt is forced to the outer edge of the rear pulley by spring force. When the clutch engages, the floating half of the front pulley moves inward increasing its effective diameter and pulling the belt down between flanges of the rear pulley.

The transmission is torque-responsive. A sudden engine acceleration increases effective diameter of the rear pulley, lowering the drive ratio. It works this way. Increase in belt tension rotates the floating flange ahead in relation to the driving flange. The belt now slips slightly on its driver. It's at this time that nylon rollers on the floating flange engage cams on the driving flange, pulling the flanges together and increasing effective diameter of the pulley.

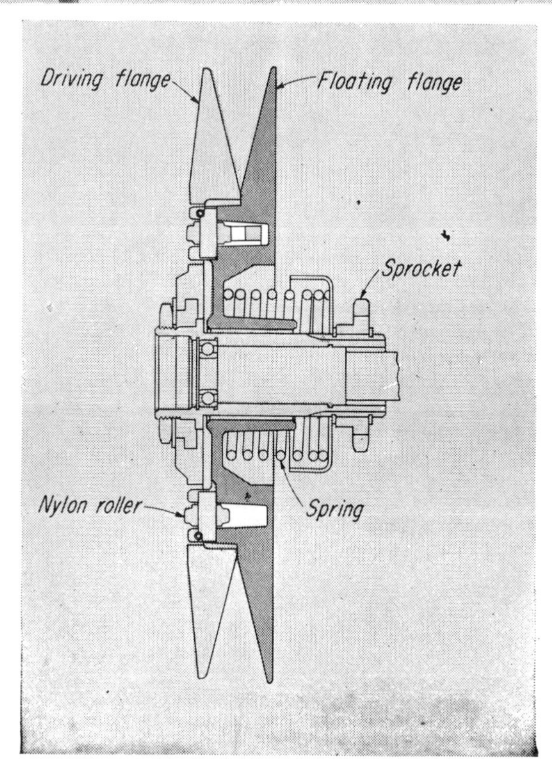

Special-shape Cams Reduce Speed Smoothly

They oscillate two parallel power trains, but blend these motions perfectly to produce constant speed output. Just a twist of a dial sets any ratio from 9:1 up.

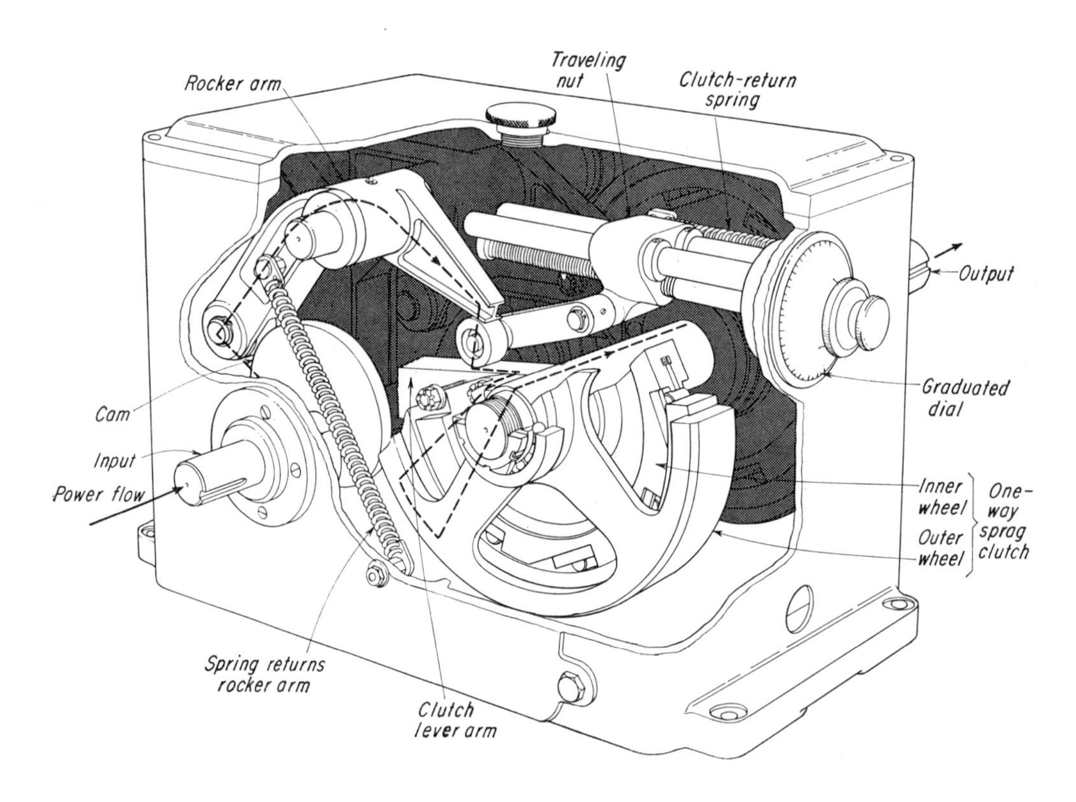

SPEED-REDUCER converts single rotary input into oscillations in two parallel power trains and recombines these oscillations into a single rotary output of the desired speed. Input shaft carries one cam for each power train. Cams are shaped to move rocker arms up and down in proper phase relationship to make the oscillations overlap and produce a smooth output rotation. Each rocker arm transmits movement through a roller to a clutch lever arm. This arm drives the inner wheel of a one-way sprag clutch. Sprags between the inner and outer wheels grip during forward motion of inner wheel and release during its return travel. Outer wheels on each of the two clutches turn the output shaft, maintaining the preset output speed without fluctuation.

Length of output rotation for each stroke of the rocker arm—and hence the output speed—depends on position of the roller between the rocker arm and clutch lever arm. Reduction ratio is infinity (output is stationary) when the roller is held directly under the rocker-arm pivot. Reduction ratio is 9:1 and output speed is maximum when roller is placed under end of the rocker arm as shown in drawing. An infinite variety of intermediate ratios can be set by positioning the roller between these extremes. Dial positions moving nut which locates both rollers simultaneously while unit is stopped or while it is operating.

PS . . . Max output torque at any speed is 2520 lb-in. Highest permissible input speed is 450 rpm; highest possible output speed is 50 rpm. Power limit for the unit is 2 hp, which is reached at 50 rpm and 2520 lb-in. output torque. Units are available with clockwise or counterclockwise drive. With its self-contained oil supply, speed-reducer weighs 385 lb. CONtorq is made by Fairbairn, Lawson, Combe and Barbour Ltd, Leeds, England.

Chapter 2 MECHANISMS FOR INTERMITTENT, DWELL, AND RECIPROCATING MOTION

GENEVA MECHANISMS

Locking-arm geneva

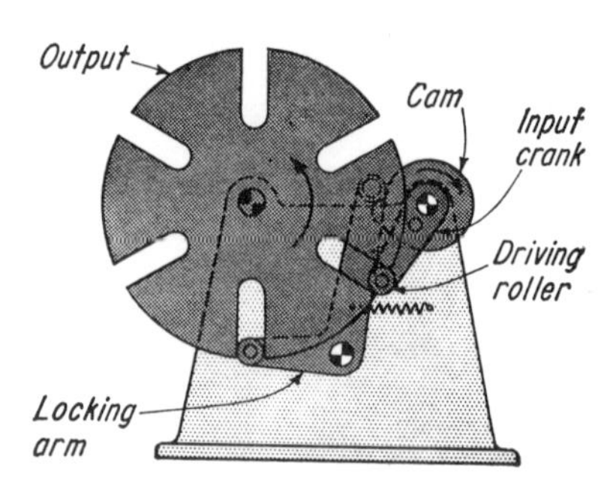

As with most genevas, the driving follower on the rotating input crank enters a slot and rapidly indexes the output. In this version, the roller of the locking-arm (shown leaving the slot) enters slot to prevent the geneva from shifting when not indexing.

Twin geneva drive

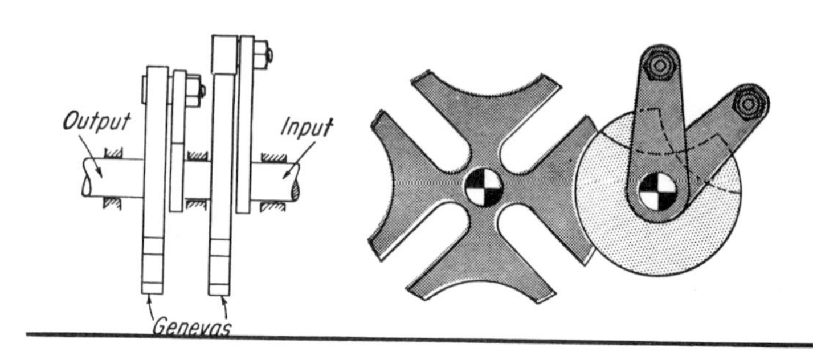

The driven member of the first geneva acts as the driver for the second geneva. This produces a wide variety of output motions including very long dwells between rapid indexes.

Groove cam geneva

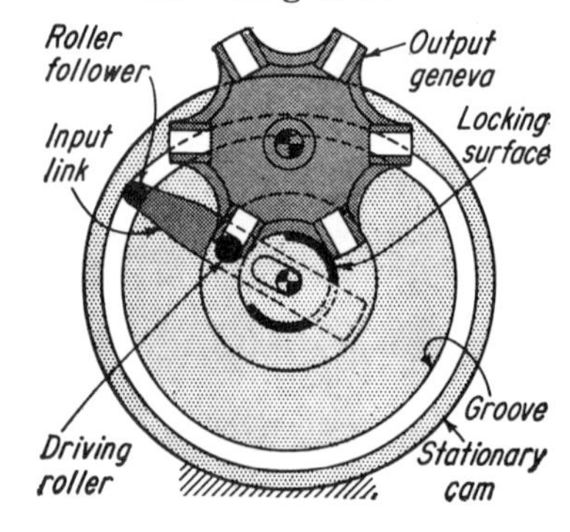

When a geneva is driven by a roller rotating at a constant speed it tends to have very high acceleration and deceleration characteristics. In this modification the input link, which contains the driving roller, can move radially while rotating by means of the groove cam. Thus, as the driving roller enters the geneva slot it moves radially inward, which reduces the geneva acceleration force.

Planetary gear geneva

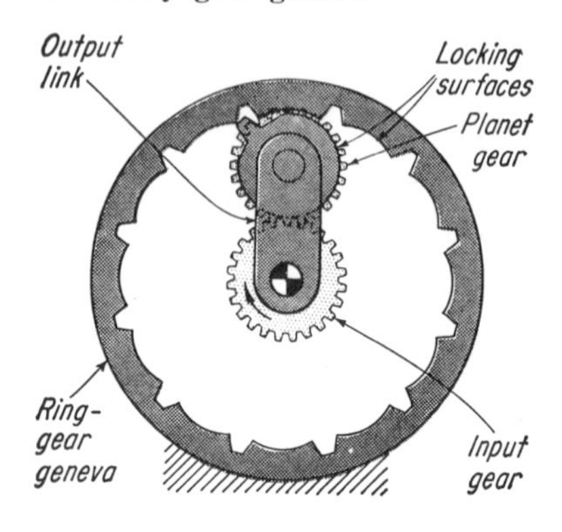

The output link remains stationary while the input gear drives the planet gear with single tooth on the locking disk, which is part of the planet gear, and which meshes with the ring-gear geneva to index the output link one position.

Locking-slide geneva

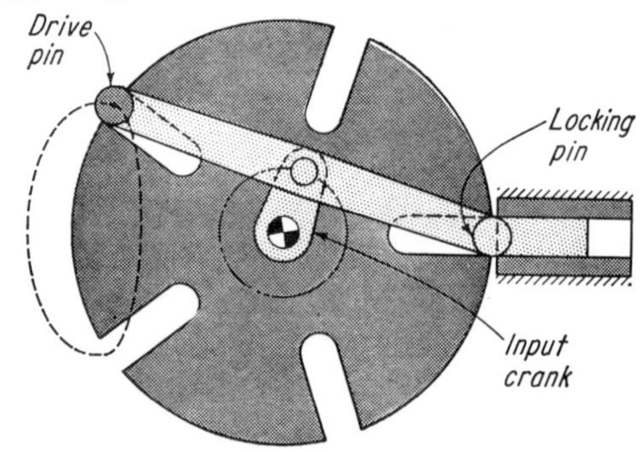

One pin locks and unlocks the geneva; the second rotates the geneva during the unlocked portion. In the position shown the drive pin is about to enter the slot to index the geneva, whereas the locking pin is just clearing the slot.

Symbol indicates a pivot point that is fixed to frame

Four-bar geneva

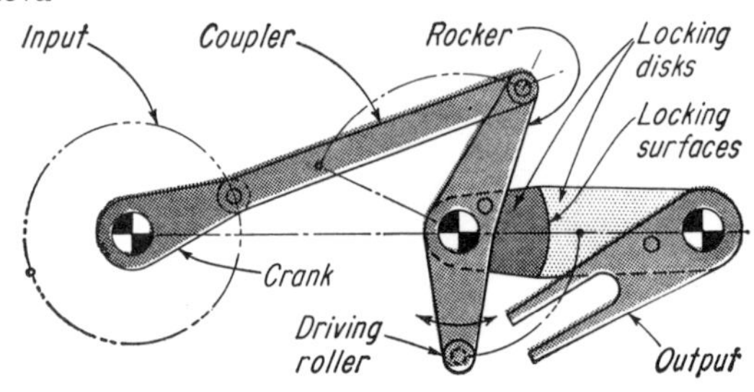

For obtaining a long-dwell motion form an oscillating output. Rotation of the input causes a driving roller to reciprocate in and out of the slot of the output link. The two disk surfaces keep the output in the position shown during the dwell period.

Rapid-transfer geneva

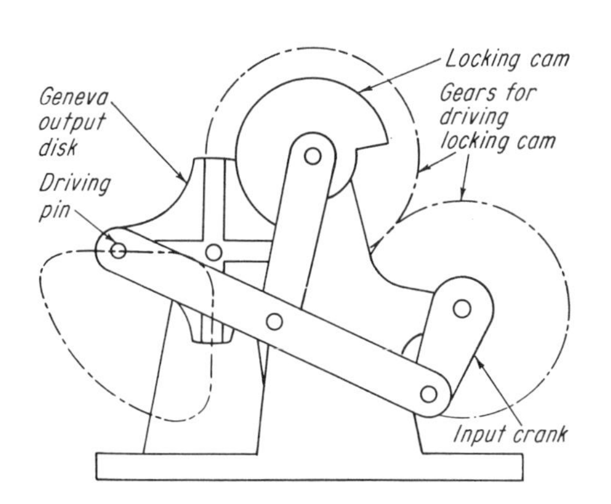

The coupler point at the extension of the connecting link of the 4-bar mechanism describes a curve with two approximately straight lines, 90 deg apart. This provides a good entry situation in that there is no motion in the geneva while the driving pin moves deeply into the slot — then there is an extremely rapid index. A locking cam which prevents the geneva from shifting when it is not indexing, is connected to the input shaft through gears.

Long-dwell geneva

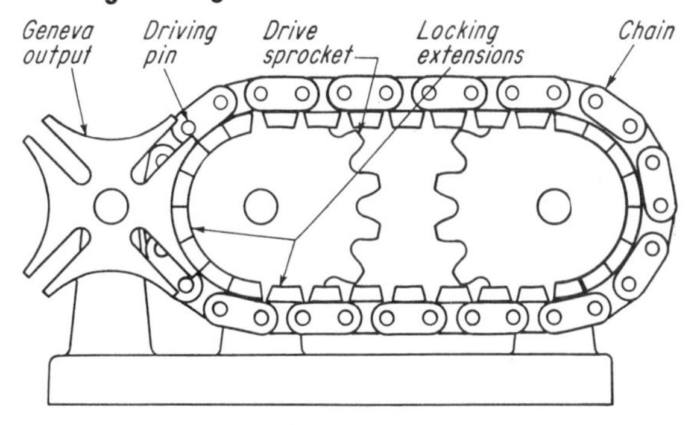

This arrangement employs a chain with an extended pin in combination with a standard geneva. This permits a long dwell between each 90-deg shift in the position of the geneva. The spacing between the sprockets determines the length of dwell. Note that some of the links have special extensions to lock the geneva in place between stations.

Modified motion geneva

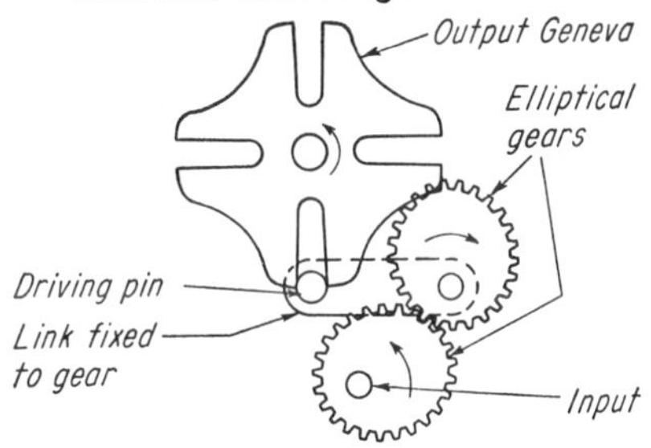

With a normal geneva drive the input link rotates at constant velocity, which restricts flexibility in design. That is, for given dimensions and number of stations the dwell period is determined by the speed of the input shaft. Use of elliptical gears produces a varying crank rotation which permits either extending or reducing the dwell period.

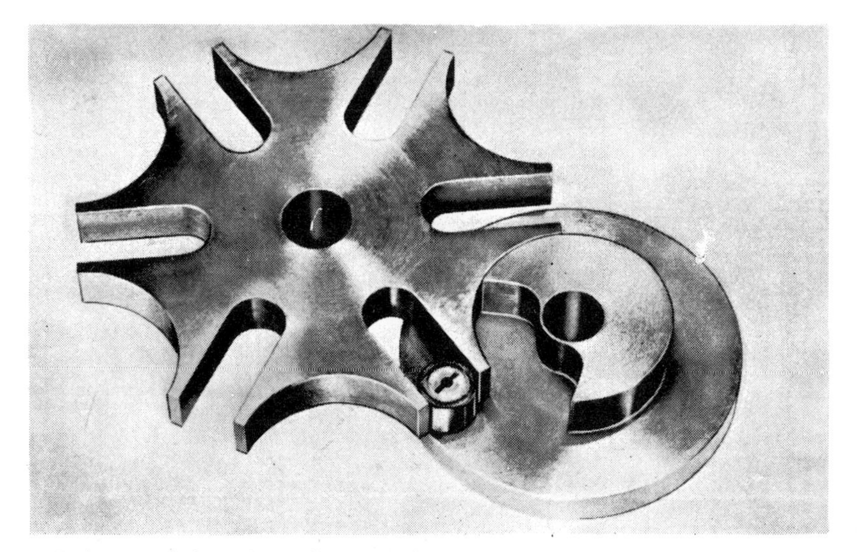

Drive consisting of ductile iron driving wheel having needle bearing-mounted cam follower and alloy steel indexing wheel is said to provide smooth shock-free indexing of assembly tables, transfer machines, feeding devices, etc. Available in four, five, six, or 8-station models with 3 to 6-in. center distances. Capable of producing to 2000 indexes per minute. Geneva Motions Corp.

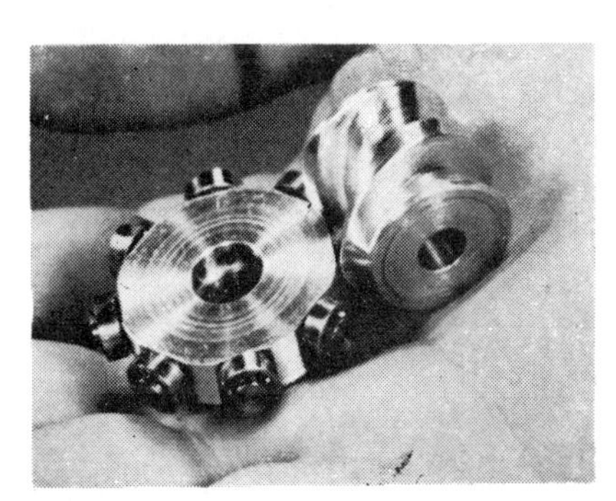

Miniature indexing drive . . . permits speeds as high as 5000 indexes per min. Modified trapezoidal acceleration cam curve is used with dwell time that may be anywhere from 90° to 270° of cam shaft rotation. Output torque can vary from 10 to 50 lb-in., depending on size, which ranges upward from ½ in. center distance. Followers can be equipped with ball, roller bearings, or nonmetallic rollers. **R. Petroff Assoc.**

Modified Geneva Drives

These sketches were selected as practical examples of uncommon but often useful mechanisms. Most of them serve to add a varying velocity component to the conventional Geneva motion. The data

Fig. 1—(Below) In the conventional external Geneva drive, a constant-velocity input produces an output consisting of a varying velocity period plus a dwell. In this modified Geneva, the motion period has a constant-velocity interval which can be varied within limits. When spring-loaded driving roller *a* enters the fixed cam *b*, the output-shaft velocity is zero. As the roller travels along the cam path, the output velocity rises to some constant value, which is less than the maximum output of an unmodified Geneva with the same number of slots; the duration of constant-velocity output is arbitrary within limits. When the roller leaves the cam, the output velocity is zero; then the output shaft dwells until the roller re-enters the cam. The spring produces a variable radial distance of the driving roller from the input shaft which accounts for the described motions. The locus of the roller's path during the constant-velocity output is based on the velocity-ratio desired.

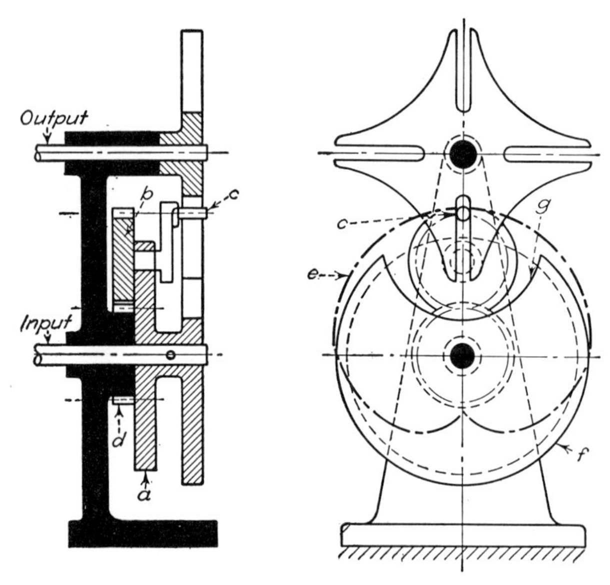

Fig. 2—(Above) This design incorporates a planet gear in the drive mechanism. The motion period of the output shaft is decreased and the maximum angular velocity is increased over that of an unmodified Geneva with the same number of slots. Crank wheel *a* drives the unit composed of plant gear *b* and driving roller *c*. The axis of the driving roller coincides with a point on the pitch circle of the planet gear; since the planet gear rolls around the fixed sun gear *d*, the axis of roller *c* describes a cardioid *e*. To prevent the roller from interfering with the locking disk *f*, the clearance arc *g* must be larger than required for unmodified Genevas.

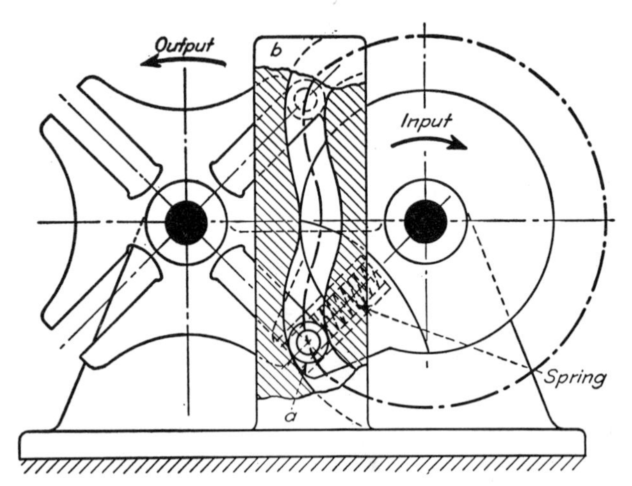

Saxonian Carton Machine Co., Dresden, Germany

Fig. 3—A motion curve similar to that of Fig. 2 can be derived by driving a Geneva wheel by means of a two-crank linkage. Input crank *a* drives crank *b* through link *c*. The variable angular velocity of driving roller *d*, mounted on *b*, depends on the center distance *L*, and on the radii *M* and *N* of the crank arms. This velocity is about equivalent to what would be produced if the input shaft were driven by elliptical gears.

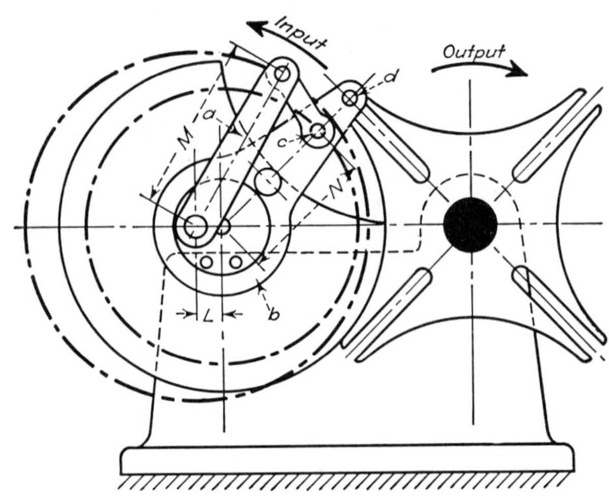

SIGMUND RAPPAPORT
Ford Instrument Company

were based in part on material and figures in AWF and VDMA Getriebeblaetter, published by Ausschuss fuer Getriebe beim Ausschuss fuer wirtschaftiche Fertigung, Leipzig, Germany.

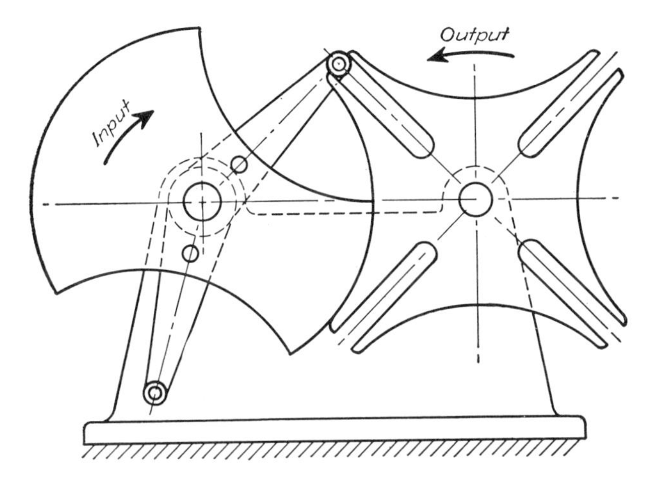

Fig. 4—(Left) The duration of the dwell periods is changed by arranging the driving rollers unsymmetrically around the input shaft. This does not affect the duration of the motion periods. If unequal motion periods are desired as well as unequal dwell periods, then the roller crank-arms must be unequal in length and the star must be suitably modified; such a mechanism is called an "irregular Geneva drive."

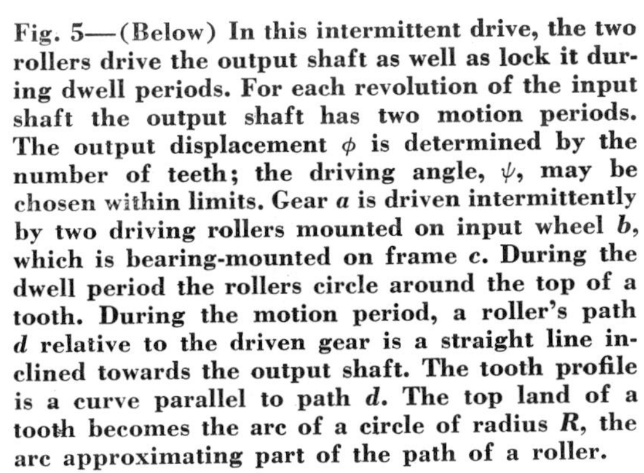
Fig. 5—(Below) In this intermittent drive, the two rollers drive the output shaft as well as lock it during dwell periods. For each revolution of the input shaft the output shaft has two motion periods. The output displacement ϕ is determined by the number of teeth; the driving angle, ψ, may be chosen within limits. Gear *a* is driven intermittently by two driving rollers mounted on input wheel *b*, which is bearing-mounted on frame *c*. During the dwell period the rollers circle around the top of a tooth. During the motion period, a roller's path *d* relative to the driven gear is a straight line inclined towards the output shaft. The tooth profile is a curve parallel to path *d*. The top land of a tooth becomes the arc of a circle of radius *R*, the arc approximating part of the path of a roller.

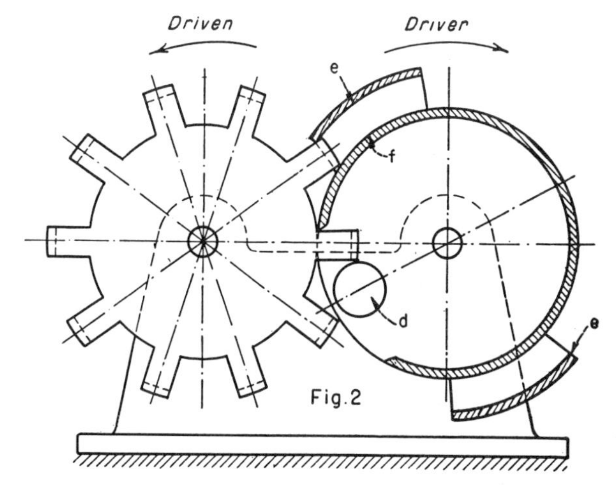

Fig. 6—Intermittent drive with cylindrical lock. Shortly before and after engagement of two teeth with driving pin *d* at the end of the dwell period, the inner cylinder *f* is not adequate to effect positive locking of the driven gear. A concentric auxiliary cylinder *e* is therefore provided with which only two segments are necesary to get positive locking. Their length is determined by the circular pitch of the driven gear.

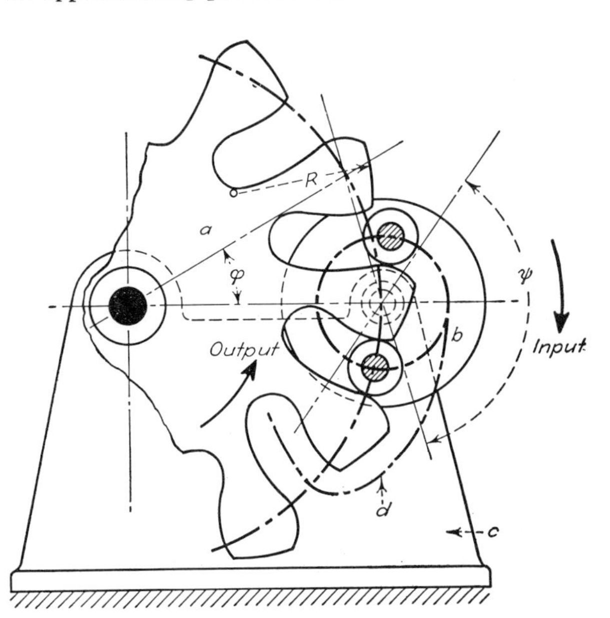

Geneva Operations Synchronized Electrically

Two simple switches make possible the use of this indexing fixture with almost any spindle-type machine tool. The switches synchronize the modified Geneva indexing mechanism with operation of the spindle. To make their fixture complete in itself, engineers of the Ettco Tool Company of Brooklyn incorporated an independent electric motor drive. Use of this separate motor created the problem of timing indexing with spindle traverse. By clever application of the small switches, the designers avoided a mechanical tie-in that would make the fixture expensive and difficult to install.

INDEX FIXTURE is completely self-contained. This makes it possible to use the fixture on many types of machine tools without modifying their design. A 1/20 hp motor drives the eight-station indexing mechanism.

FOR INDEXING, a hardened steel worm drives the bronze gear and cam. This brings the locating pin down and out of the indexing plate. The index roller is located so as to engage a slot in the index shaft as soon as the locating pin leaves the plate. When the index plate has rotated to its next position, the cam lifts the locating pin, fixing the position of the plate. Two switches synchronize the indexing mechanism with the machine tool spindle. One of these mounted on the tool energizes the indexing motor relay as soon as the spindle returns to its rest position. A spring-return, double-throw switch within the index fixture housing deenergizes this relay when it is contacted by the indexing roller. But while the motor coasts to a stop, this switch completes its other throw, then snaps back ready for the next cycle. The second throw starts spindle motor.

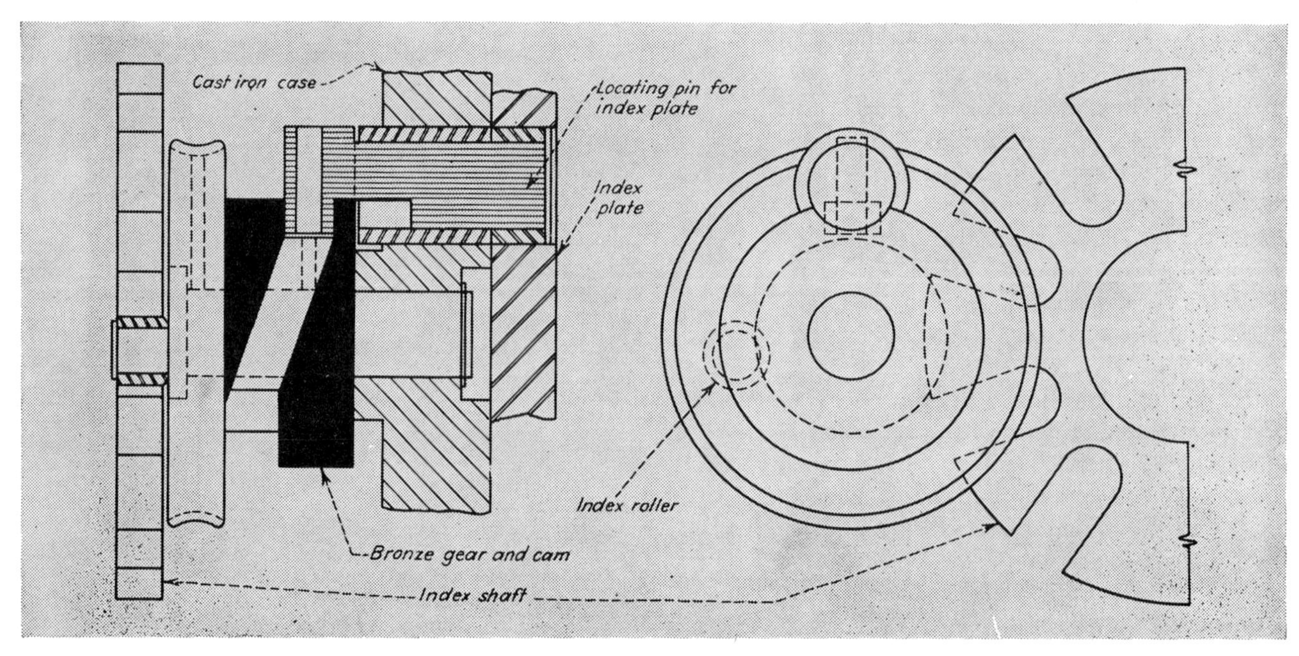

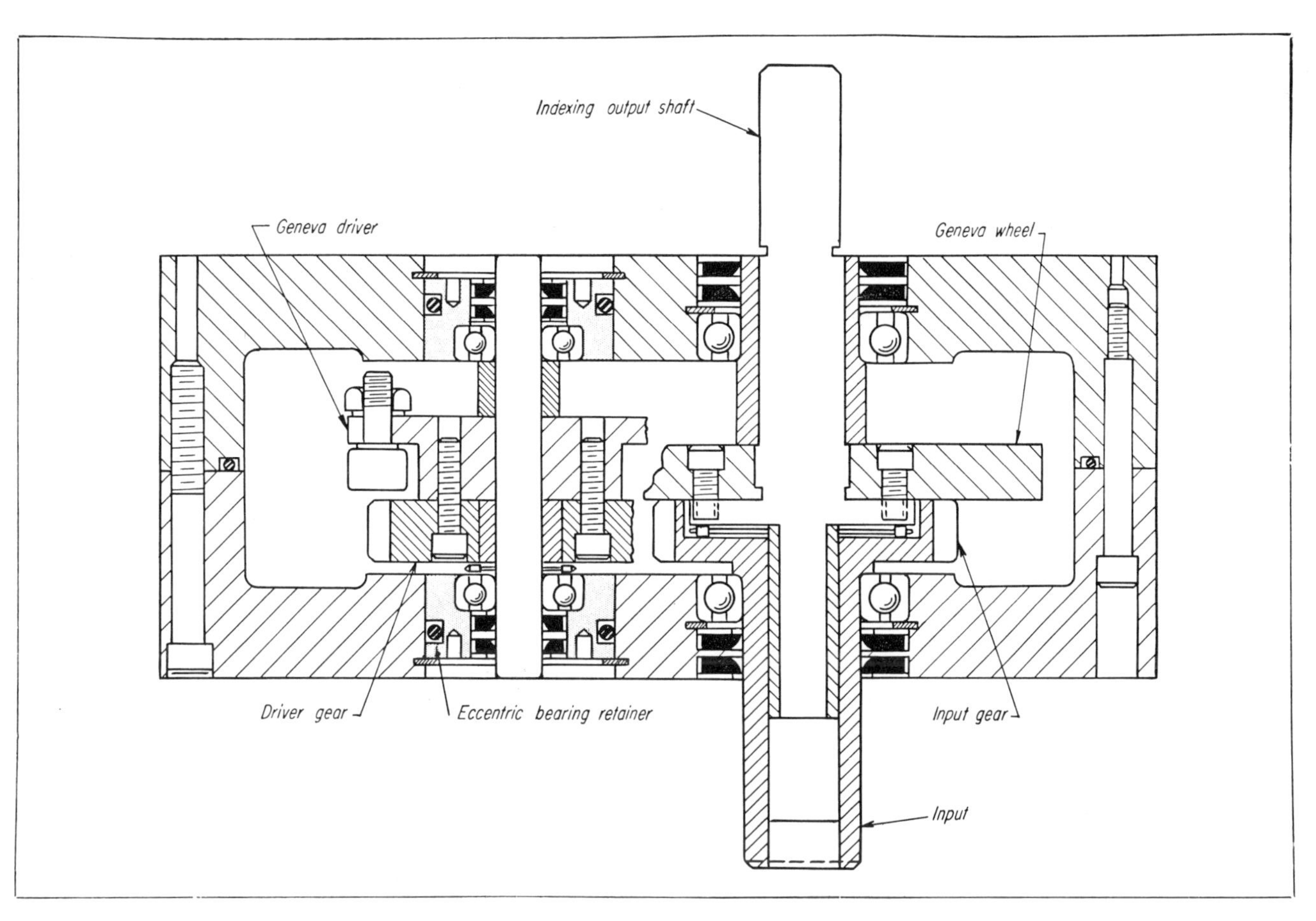

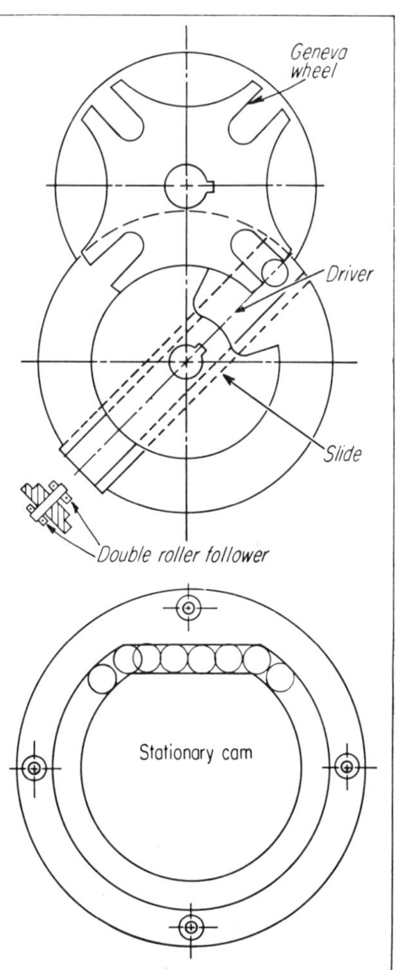

INDEXING

Parts, parts, parts

A small plant in the middle of Brooklyn, Chase-Logeman Corp, had a problem. What can you standardize when you produce a range of special automatic bottle filling machines with low sales volume per unit? One approach was to look for the common denominator shared by all units. It turned out to be the intermittent motion that positioned bottles and vials. Result of the search: a packaged geneva mechanism and a new Mechanical Drives Div to market the drive independently.

The same aluminum housing encloses all models, and can be mounted in any position. Single-roller driver with 2, 3, 4, 5, 6, 7, and 8 station geneva wheels can produce up to 2000 indexes per min. A four-roller driver can produce up to 8000 indexes per min. With 8 possible patterns of acceleration and deceleration, the unit indexes and locks to within ±0.001 in. The enclosed hardened and ground mechanism operates in an oil bath.

Co-axial input-output with unidirectional drive-indexing is achieved by incorporating a pair of gears in the drive train. The input gear is an integral part of the input shaft while the drive gear has the geneva driver mounted on it with cap screws. The driver shaft is held between eccentric bearing retainers which permit center-to-center distance between geneva wheel and driver to be varied a maximum of 0.002 in. The retainer is scribed with a series of radial lines which indicate increments of 0.0002 in. center-to-center distance. A double-pin wrench engages both eccentric retainers to rotate them in unison.

For indexing motions other than those produced by the normal geneva drive a fixed cam guides the driver follower. The double-roller follower is carried on a cross slide in the driver. One side of the roller follower follows the cam while other side drives the geneva wheel to produce the required acceleration-deceleration curve.

INDEXING MECHANISMS

Wheel and slider drive

For each revolution of the input disk the slider moves in to engage the wheel and index it one tooth. A flat spring keeps the wheel locked while stationary.

Double-station star wheel

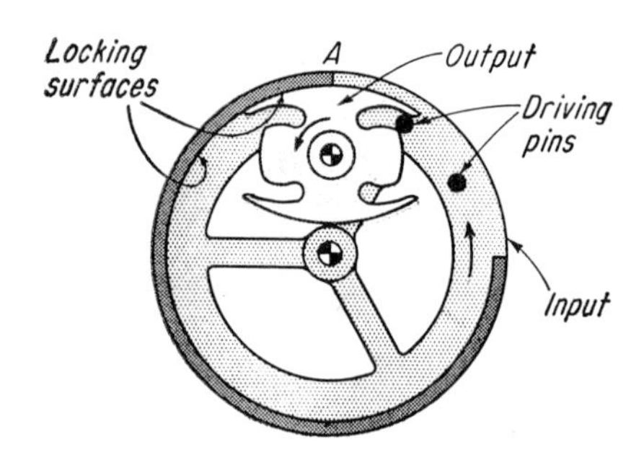

This is a more compact drive than the genevas. The driving wheel has two pins and a locking surface. Locking is just about to end. First the first pin engages the star wheel and begins to rotate it, then the second pin takes over to complete the rotation. The output indexes a full 180 deg during only a 90-deg rotation of the input. It is then locked during the remaining 270 deg of input rotation.

FIVE STAR-WHEEL MECHANISMS

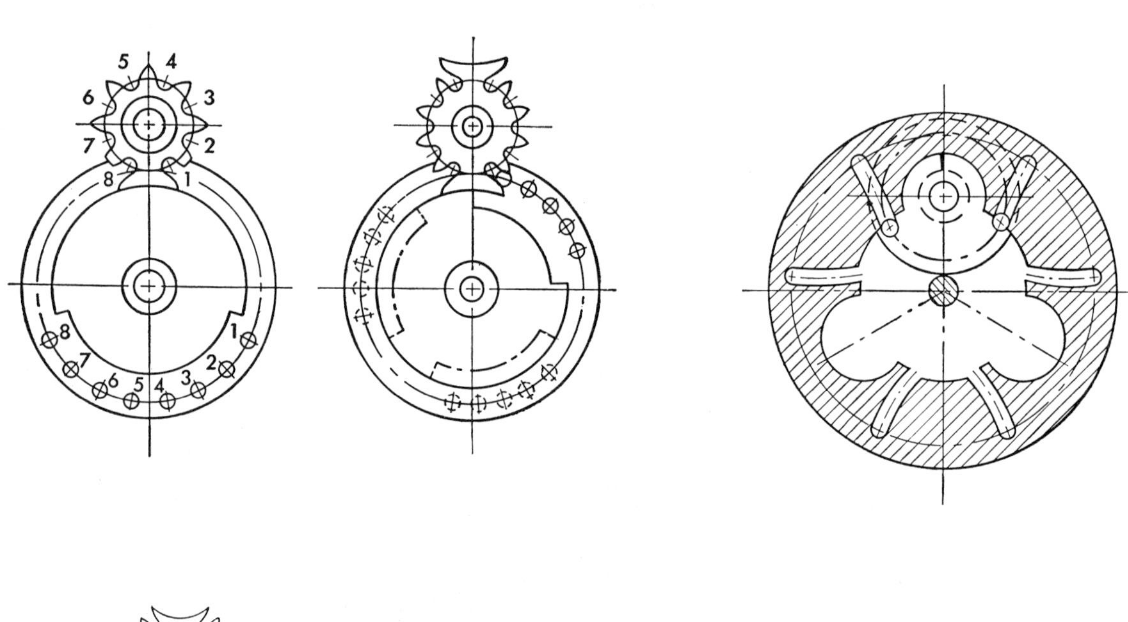

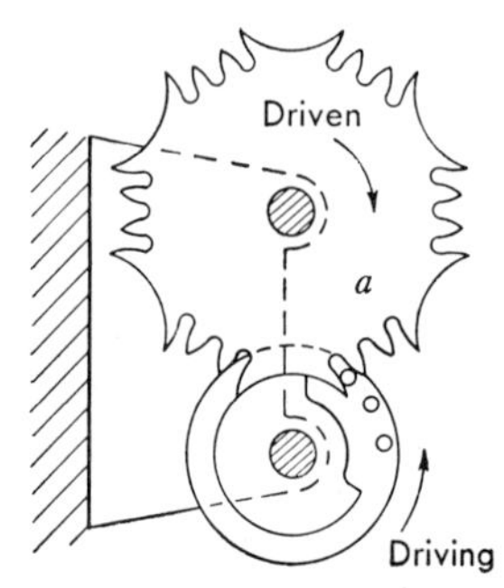

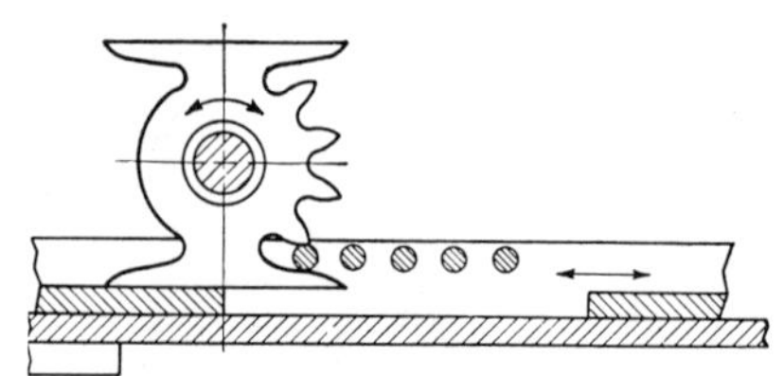

Star wheels are intermittent gears with specially shaped teeth to avoid shock loads. The star wheel in each case is the driven member. These wheels have the advantage that the angle of indexing may vary around the wheel.

Reference: J. S. Beggs, *Mechanism*, McGraw-Hill, N. Y., 1955.

Mechanism for transmitting intermittent motion between two skewed shafts. The shafts need not be at right angles to one another. Angular displacement of the output shaft per revolution of input shaft equals the circular pitch of the output gear wheel divided by its pitch radius. The duration of the motion period depends on the length of the angular join *a* of the locking disks *b*. This drive was used extensively on motion picture projectors.

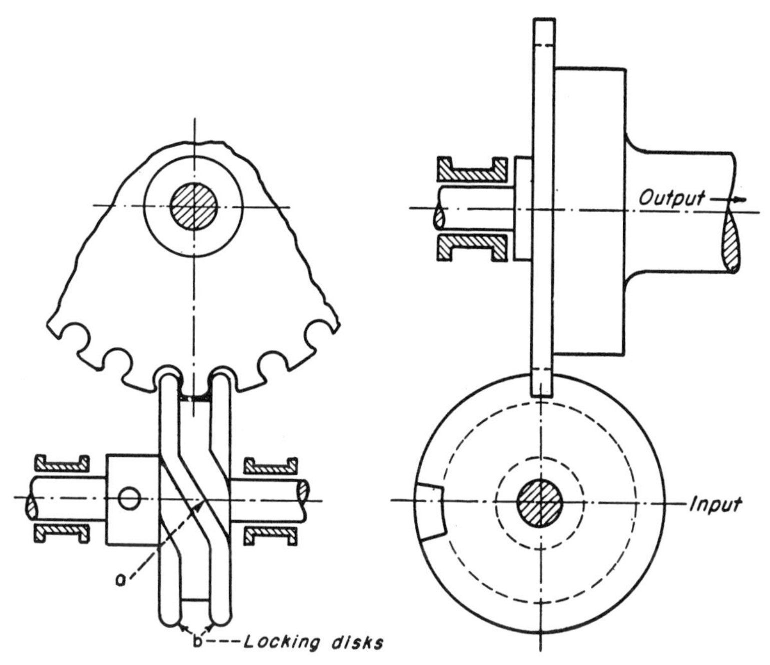

"Multilated tooth" intermittent drive. Driver *b* is a circular disk of width *w* with a cutout *d* on its circumference and carries a pin *c* close to the cutout. The driven gear, *a* of width *2w* has standard spur gear teeth, always an even number, which are alternately of full and of half width (mutilated). During the dwell period two full width teeth are in contact with the circumference of the driving disk, thus locking it; the multilated tooth between them is behind the driver. At the end of the dwell period pin *c* comes in contact with the mutilated tooth and turns the driven gear for one circular pitch. Then, the full width tooth engages the cutout *d* and the driven gear moves one more pitch, whereupon the dwell period starts again and the cycle is repeated. Used only for light loads primarily because of high accelerations encountered.

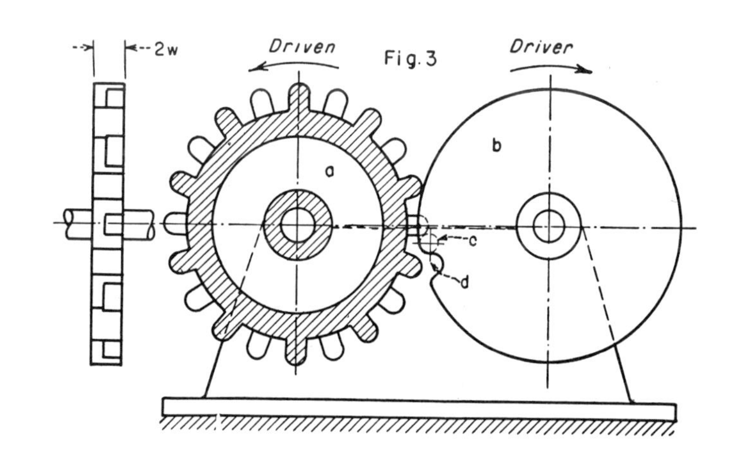

An operating cycle of 180 deg motion and 180 deg dwell is produced by this mechanism. The input shaft drives the rack which is engaged with the output shaft gear during half the cycle. When the rack engages, the lock teeth at the lower end of the coulisse are disengaged and, conversely, when the rack is disengaged, the coulisse teeth are engaged, thereby locking the output shaft positively. The change-over points occur at the dead-center positions so that the motion of the gear is continuously and positively governed. By varying *R* and the diameter of the gear, the number of revolutions made by the output shaft during the operating half of the cycle can be varied to suit requirements.

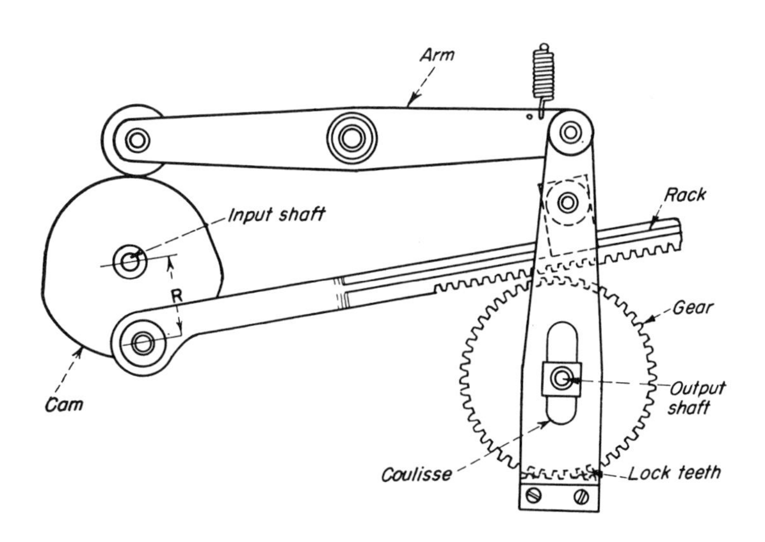

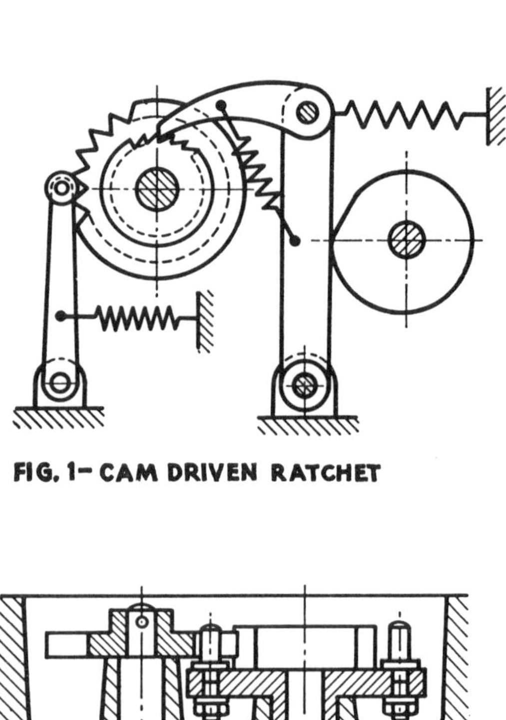

FIG. 1– CAM DRIVEN RATCHET

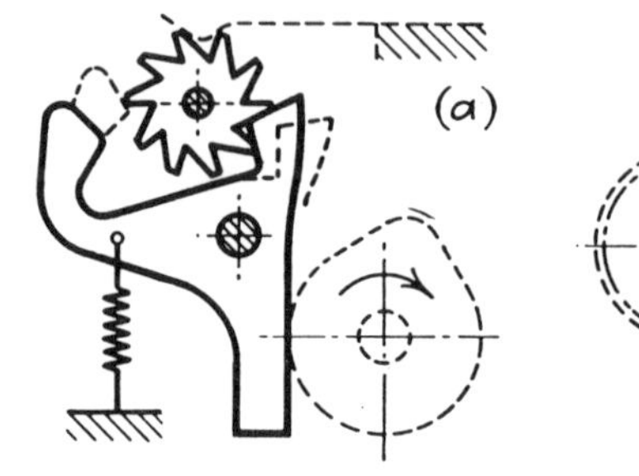

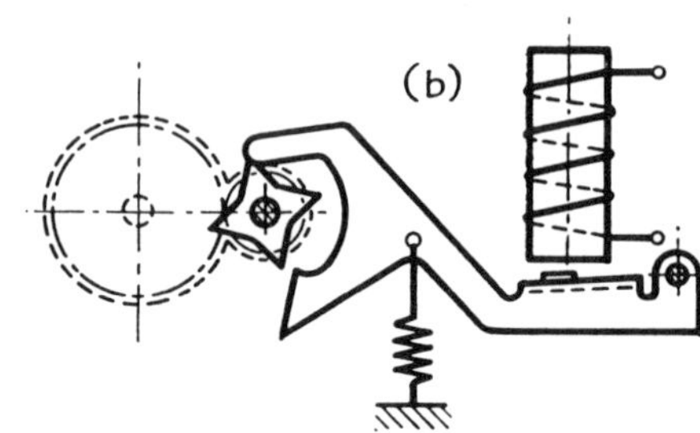

FIG. 3– (a) CAM OPERATED ESCAPEMENT ON A TAXIMETER
(b) SOLENOID OPERATED ESCAPEMENT

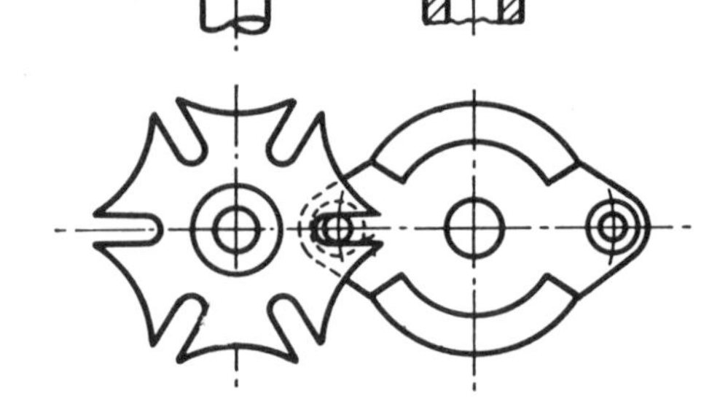

FIG. 2– SIX-SIDED MALTESE CROSS AND DOUBLE DRIVER GIVES 3:1 RATIO

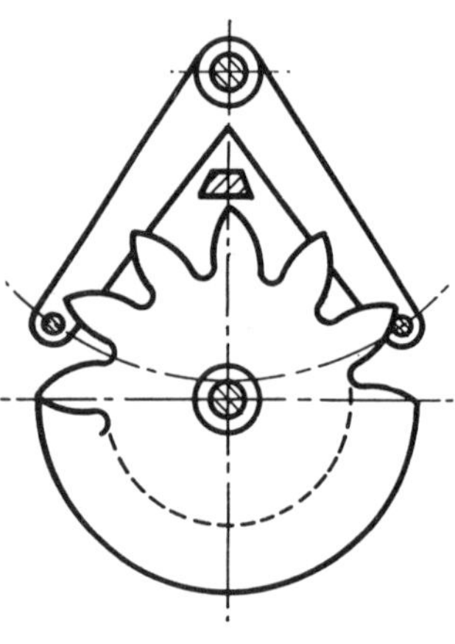

FIG. 4– ESCAPEMENT USED ON AN ELECTRIC METER

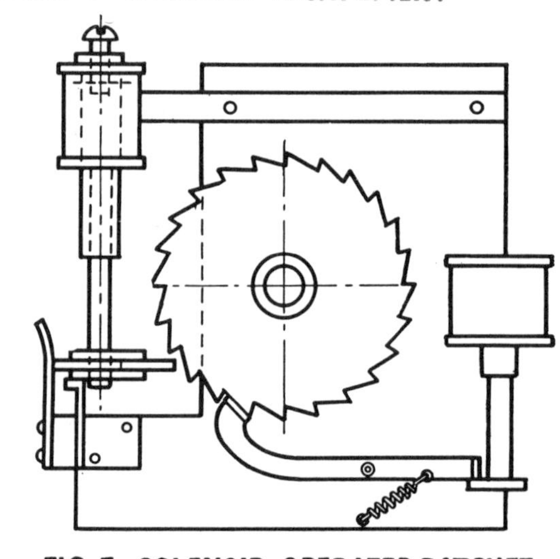

FIG. 5– SOLENOID-OPERATED RATCHET WITH SOLENOID RESETING MECHANISM. A SLIDING WASHER ENGAGES TEETH

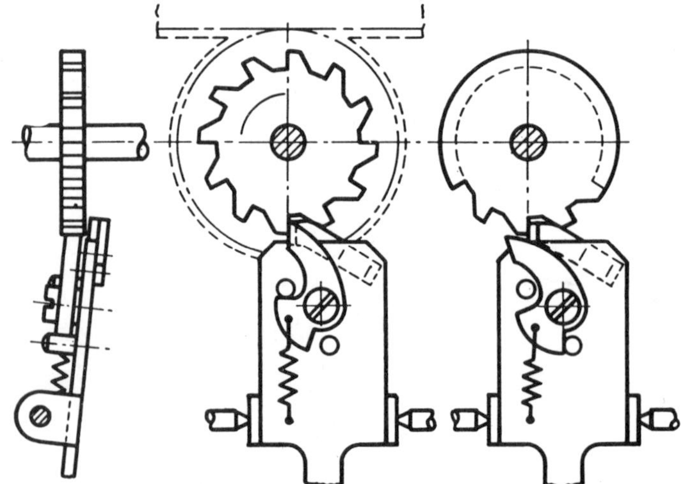

FIG. 6– PLATE OSCILLATING ACROSS PLANE OF RATCHET GEAR ESCAPEMENT CARRIES STATIONARY AND SPRING HELD PAWLS

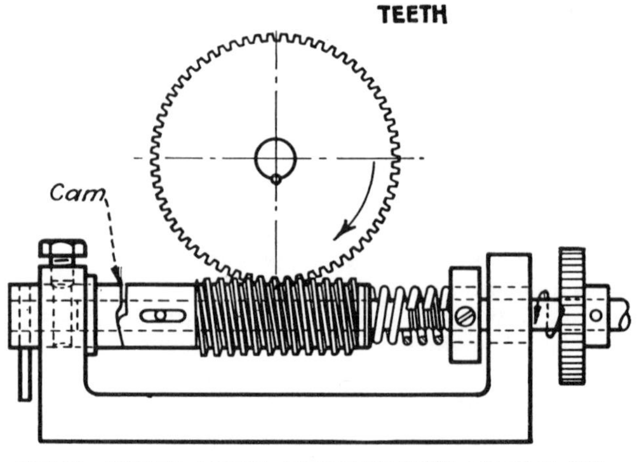

FIG. 7– WORM DRIVE COMPENSATED BY CAM ON WORK SHAFT, PRODUCES INTERMITTENT MOTION OF GEAR

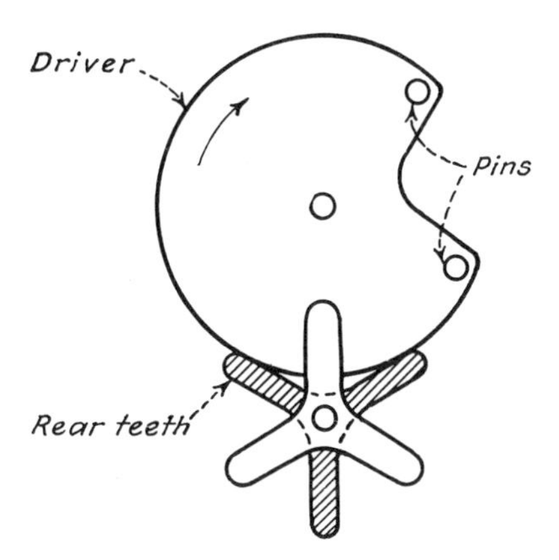

INTERMITTENT COUNTER MECHANISM. One revolution of driver advances driven wheel 120 degrees. Driven wheel rear teeth locked on cam surface during dwell.

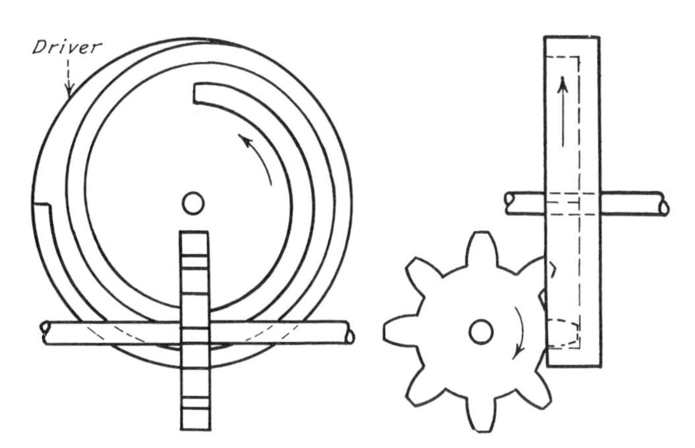

SPIRAL AND WHEEL. One revolution of spiral advances driven wheel 1 tooth. Driven wheel tooth locked in driver groove during dwell.

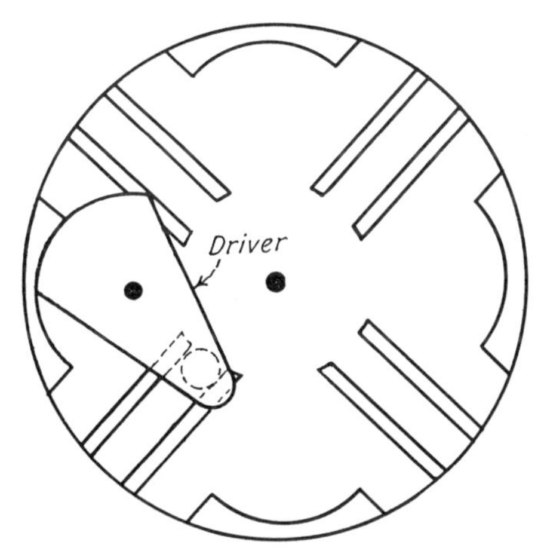

INTERNAL GENEVA MECHANISM. Driver and driven wheel rotate in same direction. Duration of dwell is more than 180 deg of driver rotation.

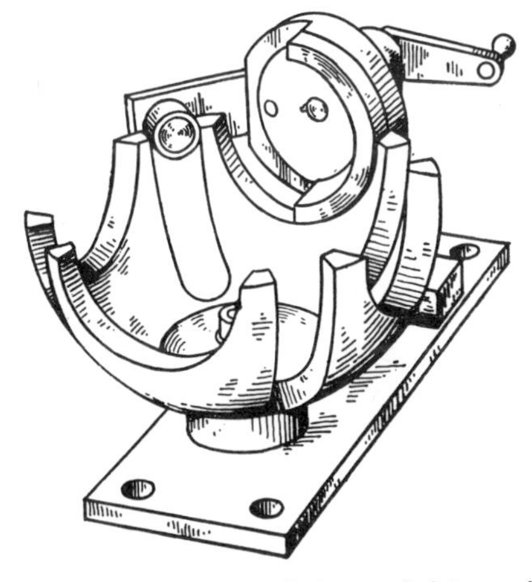

SPHERICAL GENEVA MECHANISM. Driver and driven wheel are on perpendicular shafts. Duration of dwell is exactly 180 deg of driver rotation.

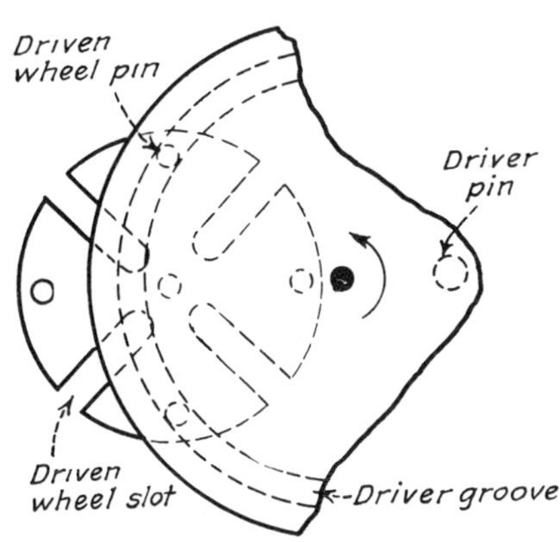

EXTERNAL GENEVA MECHANISM. Driver grooves lock driven wheel pins during dwell. During movement, driver pin mates with driven wheel slot.

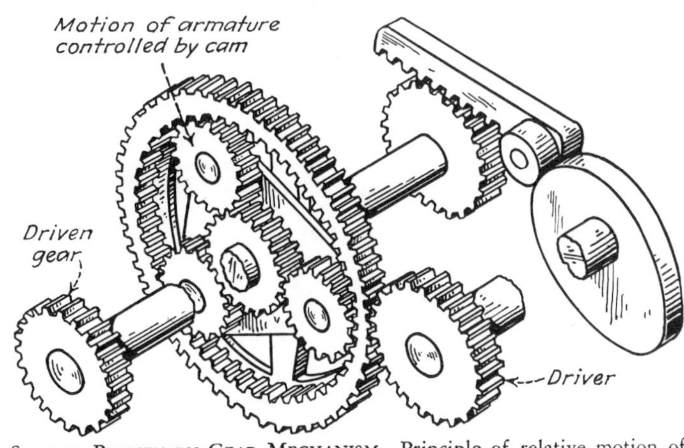

SPECIAL PLANETARY GEAR MECHANISM. Principle of relative motion of mating gears illustrated in method can be applied to spur gears in planetary system. Motion of normally fixed planet centers produces intermittent motion of sum gear.

Indexing sprag-clutch

A one-way clutch, a cam and a solenoid combine to convert a continuously rotating input motion into intermittent output in half or full turns. The clutch is of the sprag type manufactured by Formsprag Co. Signal to the solenoid may be by timer, limit switch, or pushbutton. The input drives the clutch outer race; the output is connected to the load.

The control member includes a solenoid-operated step cam, and a torsion spring which acts in the sprag-engaging direction. The stop, however, which engages the step cam, acts to disengage the sprags. Thus the output is stationary until the solenoid retracts the stop to release the cam. The output then revolves with the cam until the stop (spring returned) strikes the cam step after 180 or 360 deg.

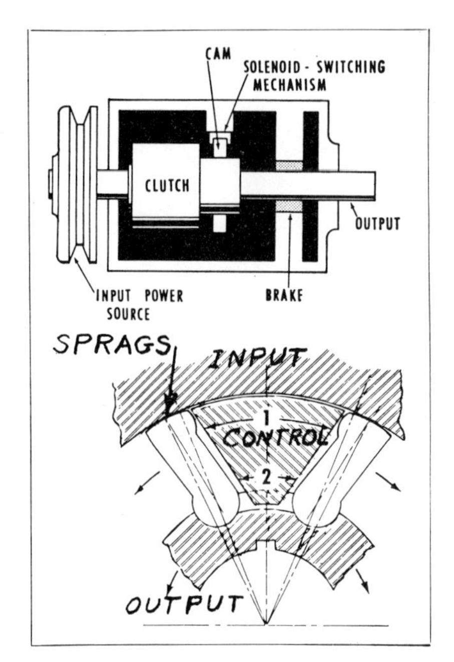

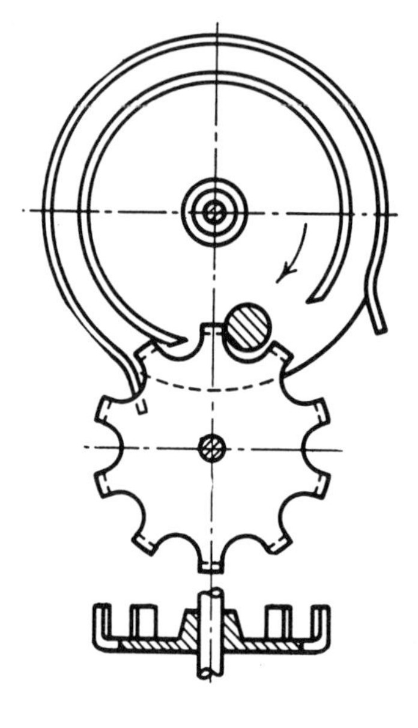

MODIFICATION OF THE GENEVA

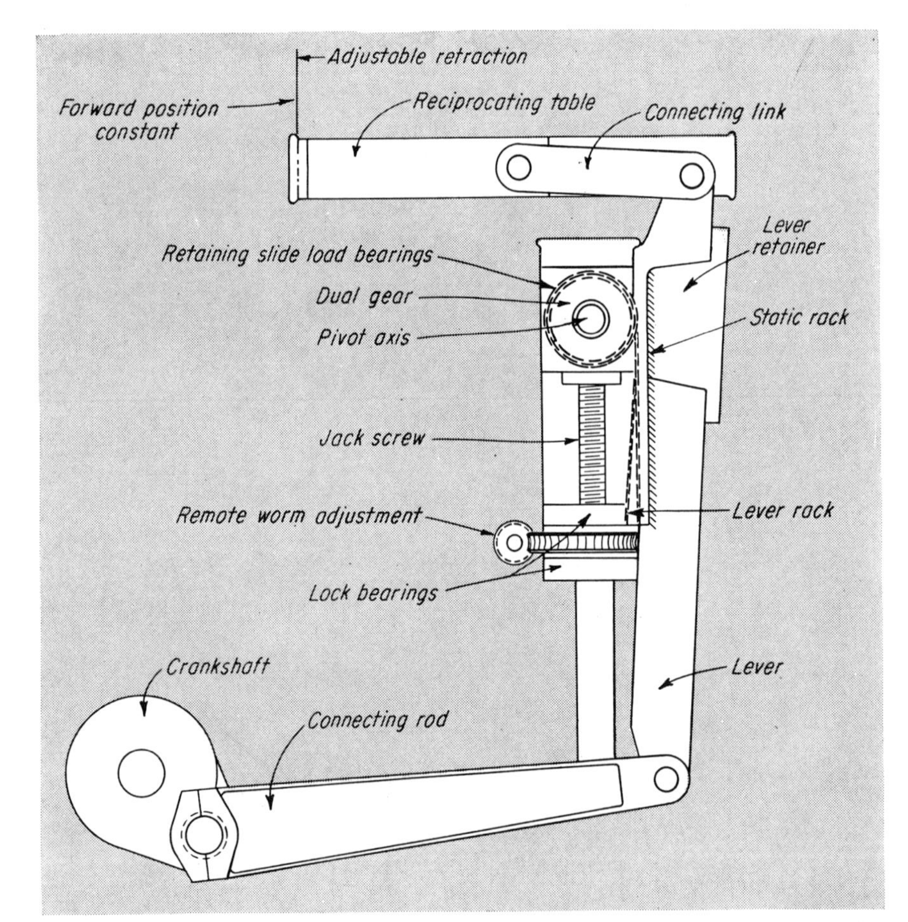

Indexing setup time is cut . . . up to 8 hr in many metalworking applications by method of transferring rotary motion to variable linear travel, which achieves constant pickoff point at one end of linear stroke. Achieves variable pullout from fixed initial pickoff point through adjustable fulcrum pivoting on rack to change angle of lever activated by rotary power source. Fulcrum consists of dual integral gear engaged to two independently positioned racks. Other components include crankshaft (input power); connecting rod, and a perpendicular lever, linked with connecting rods and reciprocating work table. Originally developed for wire fabric welder indexing table, mechanism is suitable for applications where existing methods result in equidistant travel on both sides of pickoff point. National Electric Welding Machines Co.

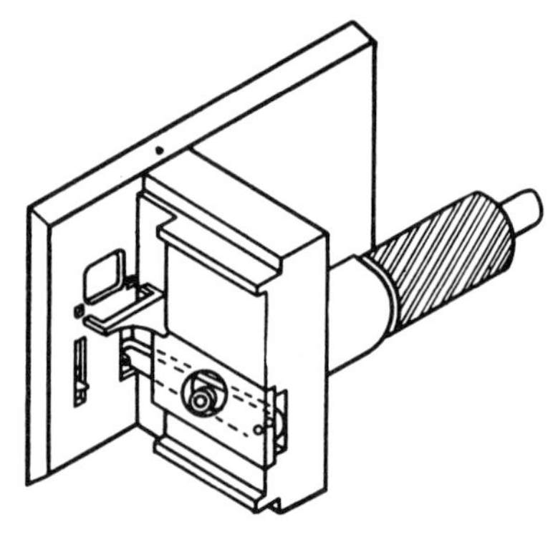

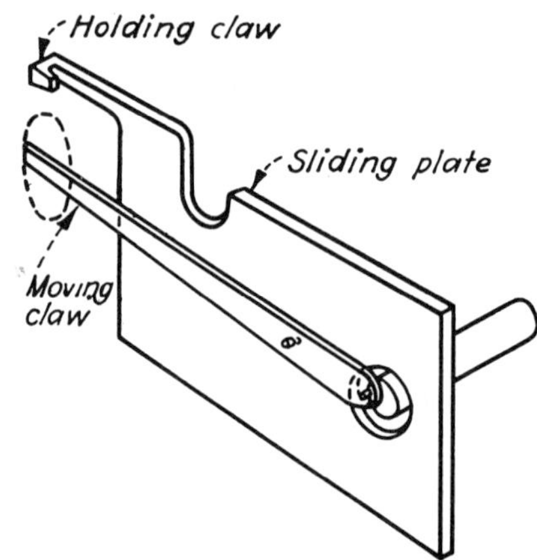

MOTION PICTURE FILM MOVEMENTS FEATURING SIMPLICITY, FEW OPERATING PARTS, QUIETNESS

In this indexing operation, a hydraulically actuated pawl releases escapement that controls two racks also powered hydraulically. Racks position pair of drilling heads on machine that puts holes in tubesheets.

Positioning system also moves worktable. Machine was designed by Walter P. Hill, Inc., Detroit, to produce hole patterns in tubesheets and baffles for heat-transfer equipment. It handles work up to 6 ft dia and 10 in. thick; drills holes in 1/20 the time previously required.

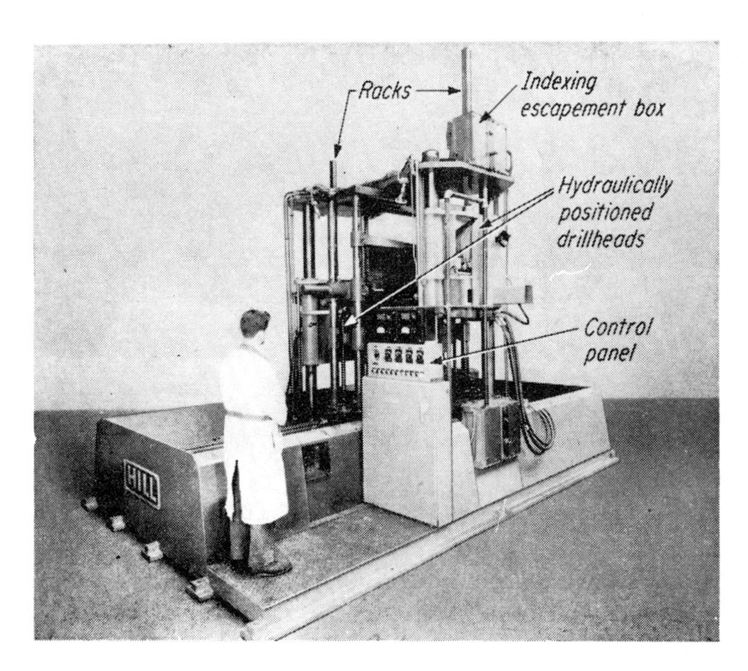

Drills Obey Escapement

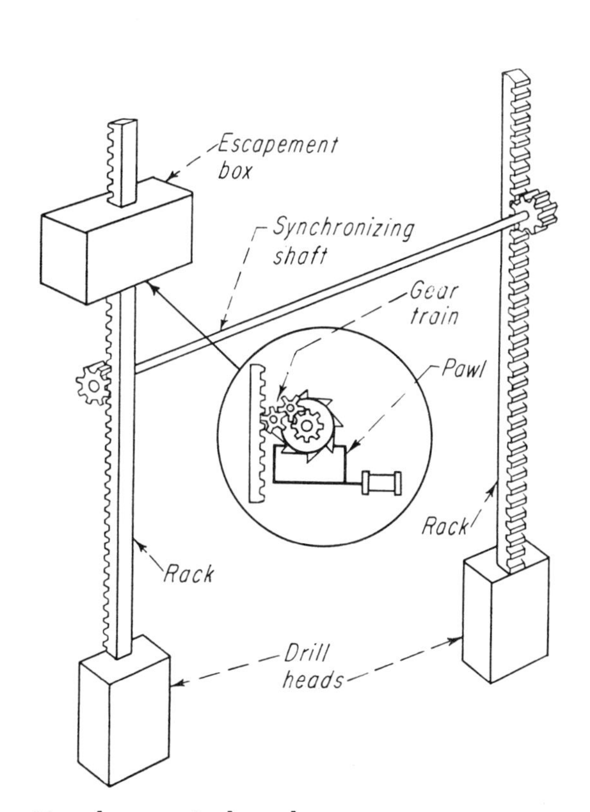

Heads are indexed . . .

to next row when electric controls actuate a hydraulic cylinder that releases the escapement wheel in the head-positioning mechanism. This lets other hydraulic cylinders, one under each drill head, raise and lower the heads in unison. A connecting shaft and pinions keep them synchronized. Hole spacing is varied by changing gears in escapement mechanism.

Hydraulic-powered heads . . .

on each side of machine move drilling spindles up and down 6-in.-dia. columns. The heads produce 18,000-lb thrust and carry 4-in.-dia spindles driven by 15-hp motors; each spindle has a 4-speed drive. Fixture that holds the work is on a table moved longitudinally by hydraulic cylinders. Drilling heads are opposed to increase accuracy by balancing cutting forces. They drill two horizontal rows of holes simultaneously, then index for the next two rows in the preset pattern. Worktable travels 15 ft so fixture can be loaded out of the working area; heads move 36 in. vertically, spindles 12 in. horizontally.

FRICTION DEVICES
For Intermittent Rotary Motion

W. M. HALLIDAY
Southport, England

FIG. 1—WEDGE AND DISK. Consists of shaft *A* supported in bearing block *J*; ring *C* keyed to *A* and containing an annular groove *G*; body *B* which can pivot around the shoulders of *C*; lever *D* which can pivot about *E*; and connecting rod *R* driven by an eccentic (not shown). Lever *D* is rotated counterclockwise about *E* by the connecting rod moving to the left until surface *F* wedges into groove *G*. Continued rotation of *D* causes *A*, *B* and *D* to rotate counterclockwise as a unit about *A*. Reversal of input motion instantly swivels *F* out of *G*, thus unlocking the shaft which remains stationary during return stroke because of friction induced by its load. As *D* continues to rotate clockwise about *E*, node *H*, which is hardened and polished to reduce friction, bears against bottom of *G* to restrain further swiveling. Lever *D* now rotates with *B* around *A* until end of stroke.

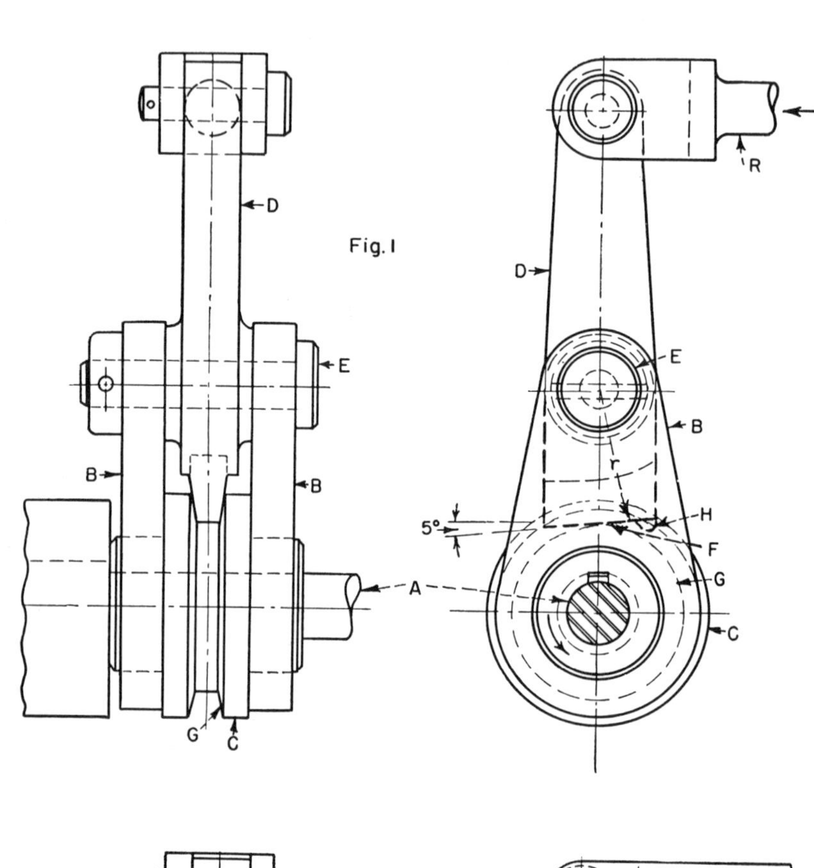

FIG. 2—PIN AND DISK. Lever *D*, which pivots around *E*, contains pin *F* in an elongated hole *K* which permits slight vertical movement of pin but prevents horizontal movement by means of set screw *J*. Body *B* can rotate freely about shaft *A* and has cut-outs *L* and *H* to allow clearances for pin *F* and lever *D*, respectively. Ring *C*, which is keyed to shaft *A*, has an annular groove *G* for clearance of the tip of lever *D*. Counterclockwise motion of lever *D* actuated by the connecting rod jams pin between *C* and the top of cut-out *L*. This occurs about seven degrees from the vertical axis. *A*, *B* and *D* are now locked together and rotate about *A*. Return stroke of *R* pivots *D* around *E* clockwise and unwedges pin until it strikes side of *L*. Continued motion of *R* to the right rotates *B* and *D* clockwise around *A* while the uncoupled shaft remains stationary because of its load.

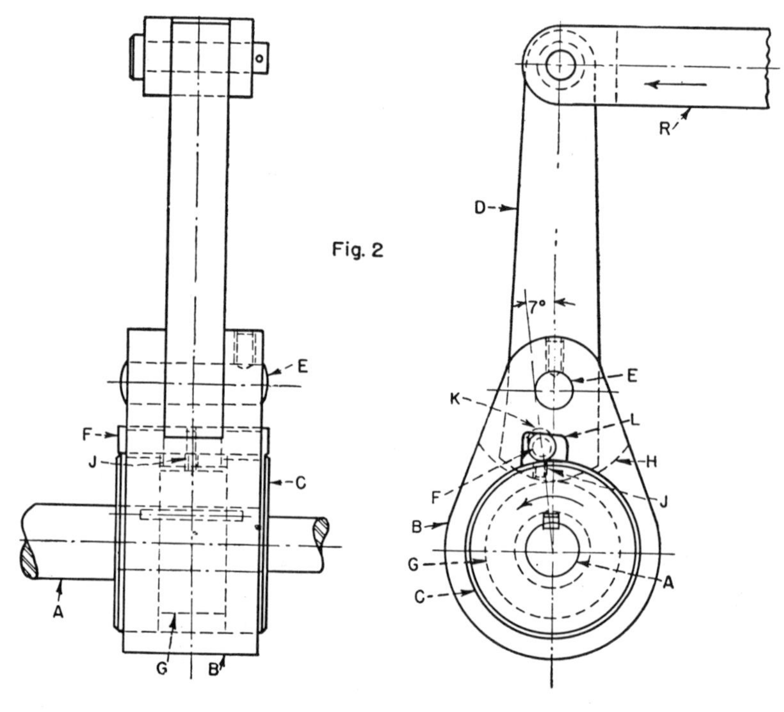

Friction devices are free from the common disadvantages inherent in conventional pawl and ratchet drives such as: (1) noisy operation; (2) backlash needed for engagement of pawl; (3) load concentrated on one tooth of the ratchet; and (4) pawl engagement dependent on an external spring.

The five mechanisms presented here convert the reciprocating motion of a connecting rod into intermittent rotary motion. The connecting rod stroke to the left drives a shaft counterclockwise; shaft is uncoupled and remains stationary during the return stroke of connecting rod to the right.

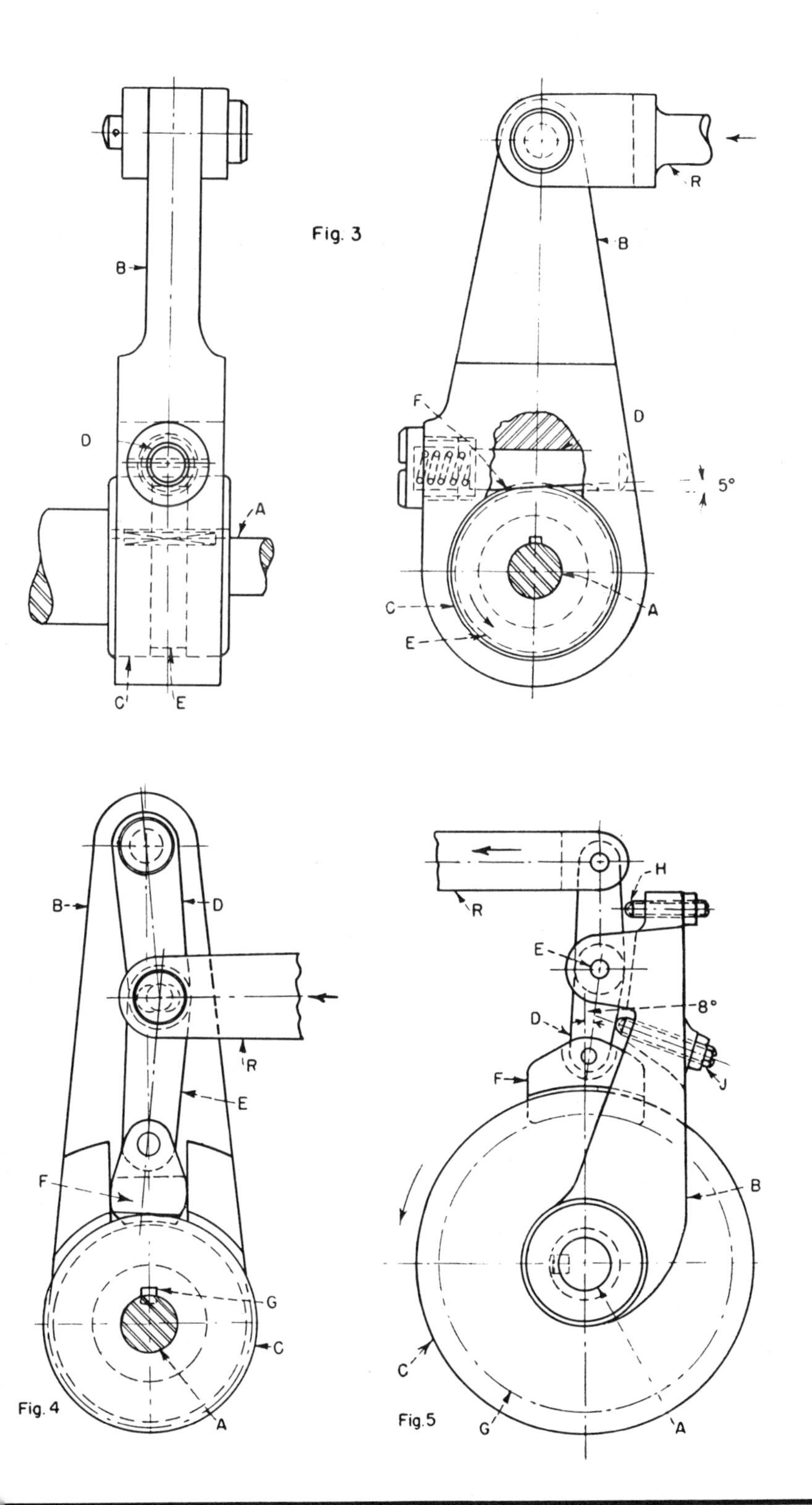

FIG. 3—SLIDING PIN AND DISK. Counterclockwise movement of body *B* about shaft *A* draws pin *D* to the right with respect to body *B*, aided by spring pressure, until the flat bottom *F* of pin is wedged against annular groove *E* of ring *C*. Bottom of pin is inclined about five degrees for optimum wedging action. Ring *C* is keyed to *A* and parts *A, C, D* and *B* now rotate counterclockwise as a unit until end of connecting rod's stroke. Reversal of *B* draws pin out of engagement so that *A* remains stationary while body completes its clockwise rotation.

FIG. 4—TOGGLE LINK AND DISK. Input stroke of connecting rod *R* to the left wedges block *F* in groove *G* by straightening toggle links *D* and *E*. Body *B*, toggle links and ring *C* which is keyed to shaft *A*, rotate counterclockwise together about *A* until end of stroke. Reversal of connecting rod motion lifts block, thus uncoupling shaft, while body *B* continues clockwise rotation until end of stroke.

FIG. 5—ROCKER ARM AND DISK. Lever *D*, activated by the reciprocating bar *R* moving to the left, rotates counterclockwise on pivot *E* thus wedging block *F* into groove *G* of disk *C*. Shaft *A* is keyed to *C* and rotates counterclockwise as a unit with body *B* and lever *D*. Return stroke of *R* to the right pivots *D* clockwise about *E* and withdraws block from groove so that shaft is uncoupled while *D*, striking adjusting screw *H*, travels with *B* about *A* until completion of stroke. Adjusting screw *J* prevents *F* from jamming in groove.

LONG DWELL MECHANISMS

PREBEN W. JENSEN

Oscillating crank and planetary drive

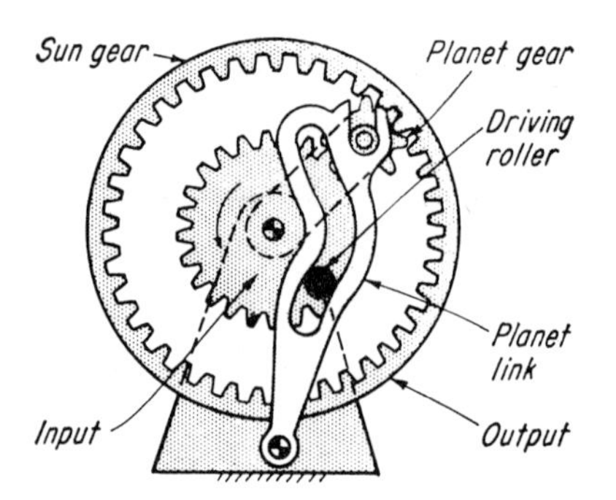

Here the planet is driven with a stop-and-go motion. The driving roller is shown entering the circular-arc slot on the planet link, hence the link and the planet remain stationary while the roller travels this portion of the slot. Result: a rotating output motion with a progressive oscillation.

Chain-slider drive

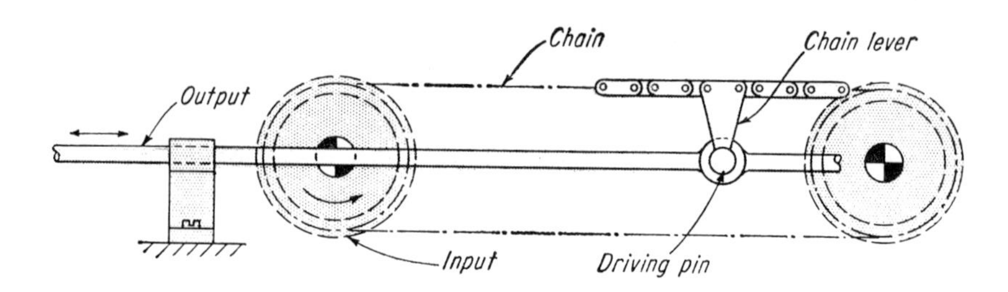

The output reciprocates back and forth with a constant velocity but comes to a long dwell at either end as the chain lever, whose length is equal to the radius of the sprockets, goes around either sprocket.

Chain-oscillating drive

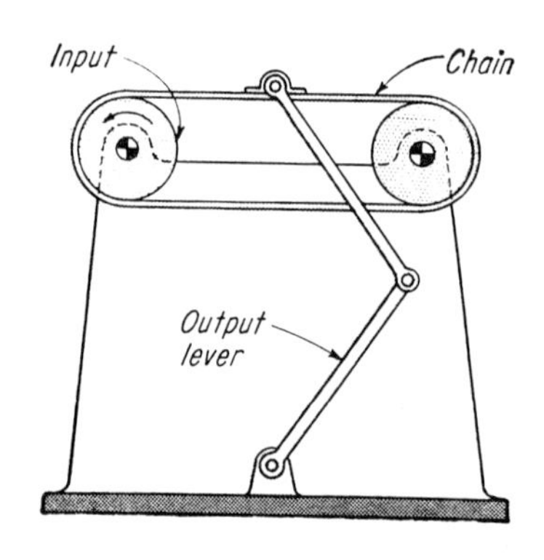

Same principle as before except that the chain link drives a lever which oscillates. The slowdown-dwell occurs as the chain pin goes around the left sprocket.

Chain and slider drive

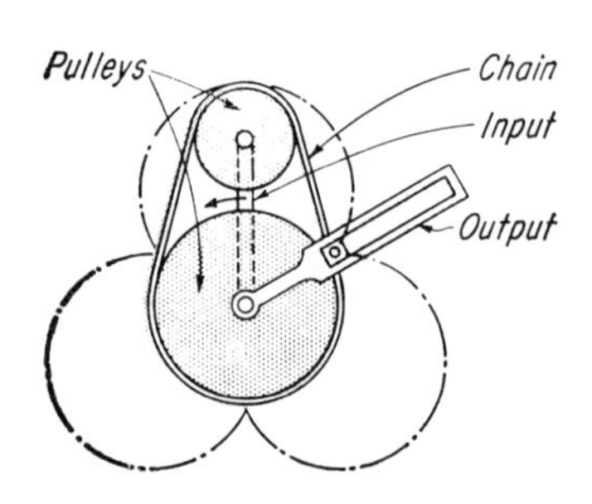

The input crank causes the small pulley to orbit around the stationary larger pulley. A pivot point in the chain slides inside the groove of the output link. In the position shown the output is about to enter a long dwell period of about 120 deg.

Epicyclic dwell mechanism

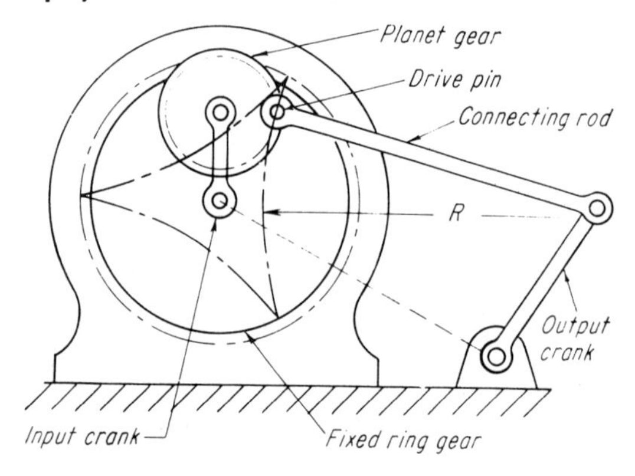

Here the output crank pulsates back and forth with a long dwell at the extreme right position. Input is to the planet gear by means of the rotating crank. The pin on the gear traces the epicyclic three-lobe curve shown. The right portion of the curve is almost a circular arc of radius R. If the connecting rod is made equal to R, the output crank comes virtually to a standstill during a third of the total rotation of the input. The crank then reverses, comes to a stop at left position, reverses, and repeats dwell.

Cam-worm dwell mechanism

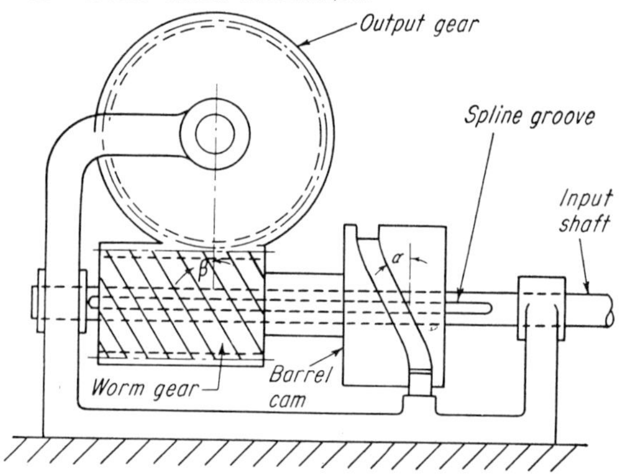

Without the barrel cam, the input shaft would drive the output gear by means of the worm gear at constant speed. The worm and the barrel cam, however, are permitted to slide linearly on the input shaft. Rotation of the input shaft now causes the worm gear to be cammed back and forth, thus adding or subtracting motion to the output. If barrel cam angle α is equal to the worm angle β, the output comes to a stop during the limits of rotation illustrated, then speeds up to make up for lost time.

Cam-helical dwell mechanism

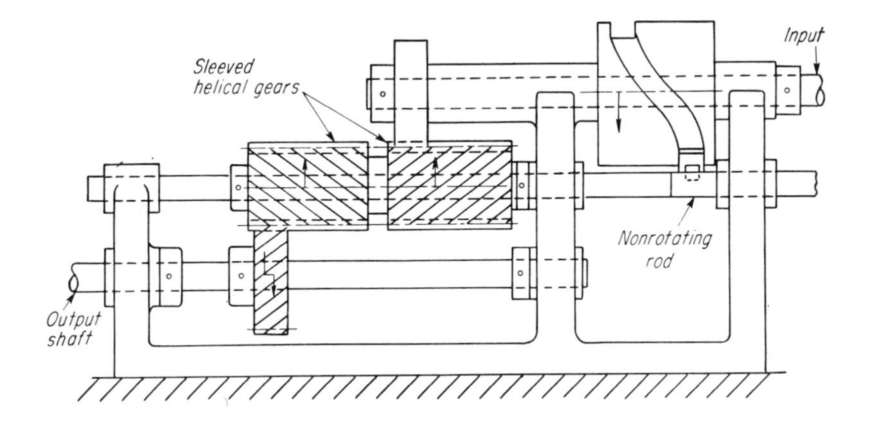

When one helical gear is shifted linearly (but prevented from rotating) it will impart rotary motion to the mating gear because of the helix angle. This principle is used in the mechanism illustrated. Rotation of the input shaft causes the intermediate shaft to shift to the left, which in turn adds or subtracts from the rotation of the output shaft.

6-Bar dwell mechanism

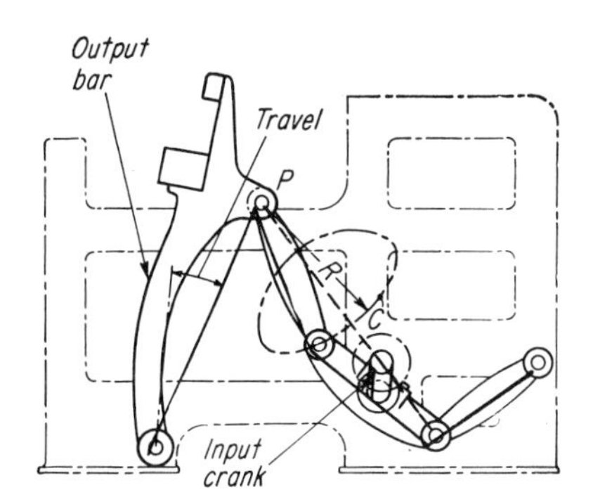

Rotation of the input crank causes the output bar to oscillate with a long dwell at the extreme right position. This occurs because point C describes a curve that is approximately a circular arc (from C to C') with center at P. The output is practically stationary during that portion of curve.

Cam-roller dwell mechanism

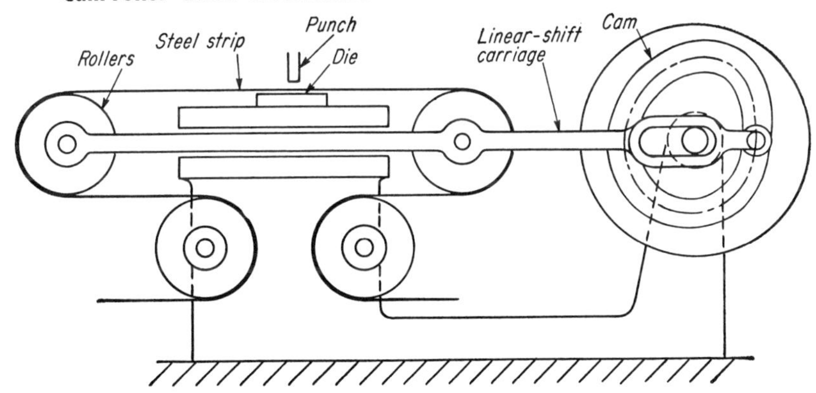

A steel strip is fed at constant linear velocity. But at the die station (illustrated), it is desired to stop the strip so that the punching operation can be performed. The strip passes over movable rollers which, when shifted to the right, cause the strip to move to the right. Since the strip is normally fed to the left, proper design of the cam can nullify the linear feed rates so that the strip stops, and then speeds to catch up to the normal rate.

Three-gear drive

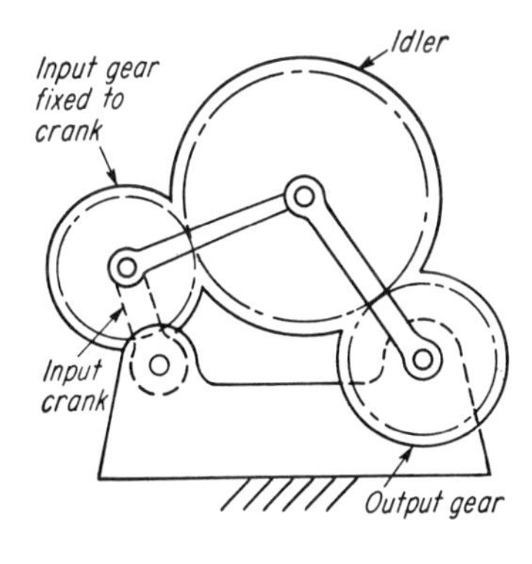

This is actually a four-bar linkage combined with three gears. As the input crank rotates it takes with it the input gear which drives the output gear by means of the idler. Various output motions are possible. Depending on proportions of the gears, the output gear can pulsate, or come to a short dwell—or even reverse briefly.

Double-crank dwell mechanism

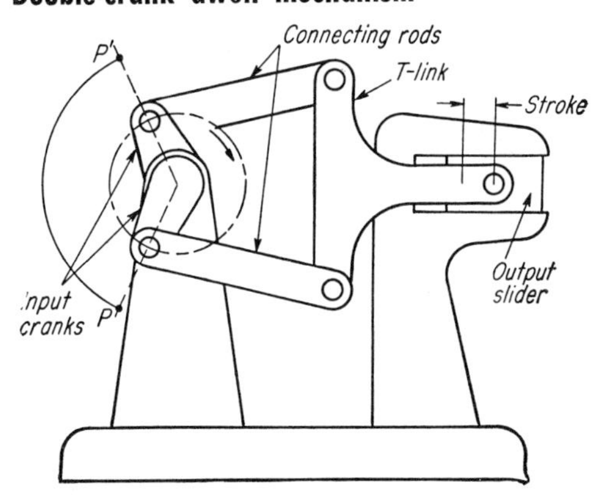

Both cranks are connected to a common shaft which also acts as the input shaft. Thus they always remain a constant distance apart from each other. There are only two frame points—the center of the input shaft and the guide for the output slider. As the output slider reaches the end of its stroke (to the right), it remains practically at standstill while one crank rotates through angle PP'. Used in textile machinery for cutting.

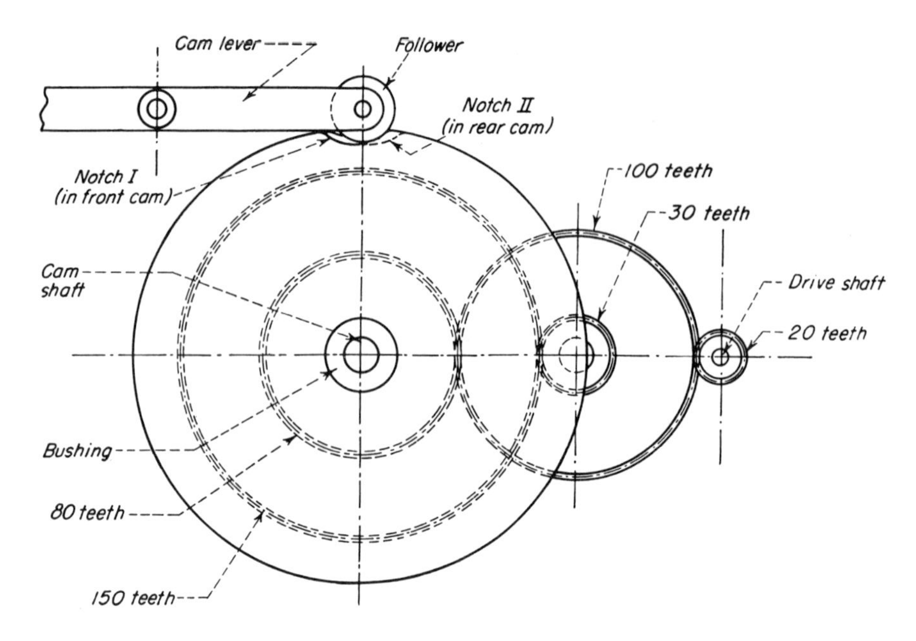

Fast Cam-Follower Motion

Fast cam action every n cycles when n is a relatively large number, can be obtained with this manifold cam and gear mechanism. A single notched cam geared $1/n$ to a shaft turning once a cycle moves relatively slowly under the follower. The double notched-cam arrangement shown is designed to operate the lever once in 100 cycles, imparting to it a rapid movement. One of the two identical cams and the 150-tooth gear are keyed to the bushing which turns freely around the cam shaft. The latter carries the second cam and the 80-tooth gear. The 30- and 100-tooth gears are integral, while the 20-tooth gear is attached to the one-cycle drive shaft. One of the cams turns in the ratio of 20/80 or 1/4; the other in the ratio 20/100 times 30/150 or 1/25. The notches therefore coincide once every 100 cycles (4×25). Lever movement is the equivalent of a cam turning in a ratio of 1/4 in relation to the drive shaft. To obtain fast cam action, n must be broken down into prime factors. For example, if 100 were factored into 5 and 20, the notches would coincide after every 20 cycles.

Intermittent Motion

This mechanism, developed by the author and to his knowledge novel, can be adapted to produce a stop, a variable speed without stop or a variable speed with momentary reverse motion. Uniformly rotating input shaft drives the chain around the sprocket and idler, the arm serving as a link between the chain and the end of the output shaft crank. The sprocket drive must be in the ratio N/n with the cycle of the machine, where n is the number of teeth on the sprocket and N the number of links in the chain. When point P travels around the sprocket from point A to position B, the crank rotates uniformly. Between B and C, P decelerates; between C and A it accelerates;; and at C there is a momentary dwell. By changing the size and position of the idler, or the lengths of the arm and crank, a variety of motions can be obtained. If in the sketch, the length of the crank is shortened, a brief reverse period will occur in the vicinity of C; if the crank is lengthened, the output velocity will vary between a maximum and minimum without reaching zero.

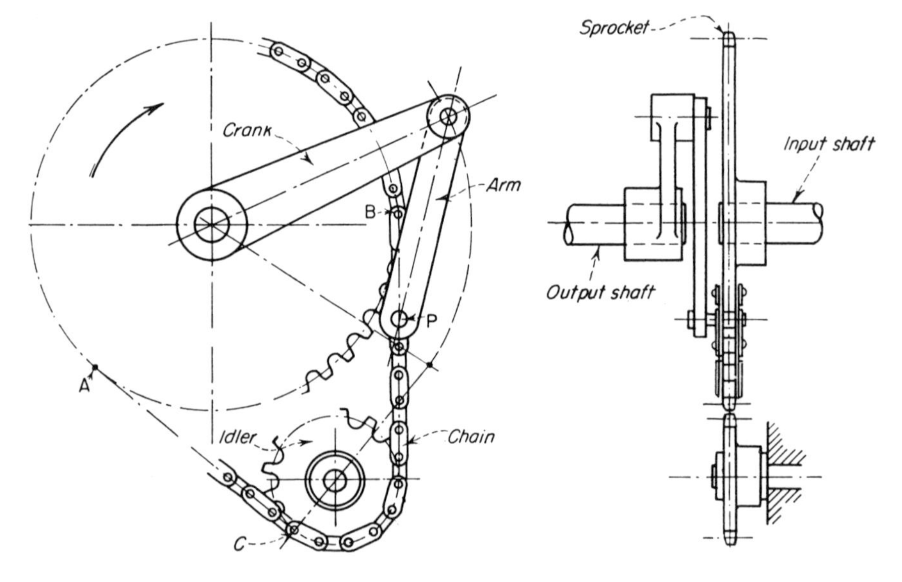

SHORT DWELL MECHANISMS

Gear-slider crank

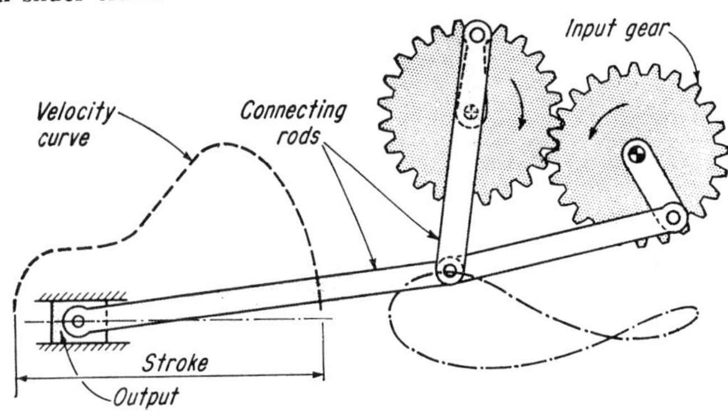

Employed in metal-drawing presses where the piston must move with a low constant velocity. The input drives both gears which in turn drive the connecting rods to produce the velocity curve shown.

Gear oscillating crank

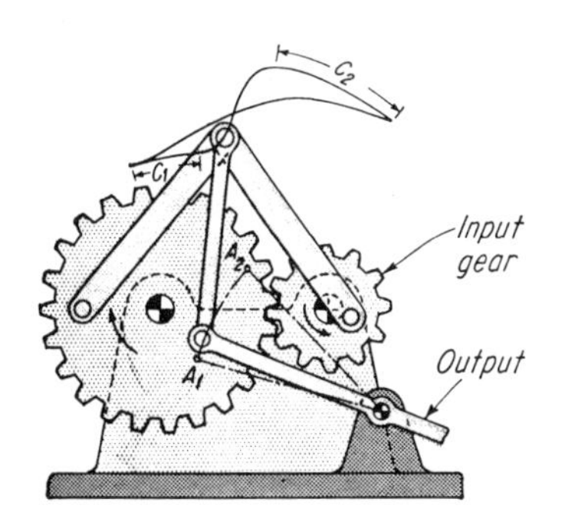

Similar arrangement to the one previously shown, but the curve described by the pin connection has two portions, C_1 and C_2, which are very close to circular arc with centers at A_1 and A_2. Hence the driven link will have a dwell at both extreme positions.

Curve slider drive

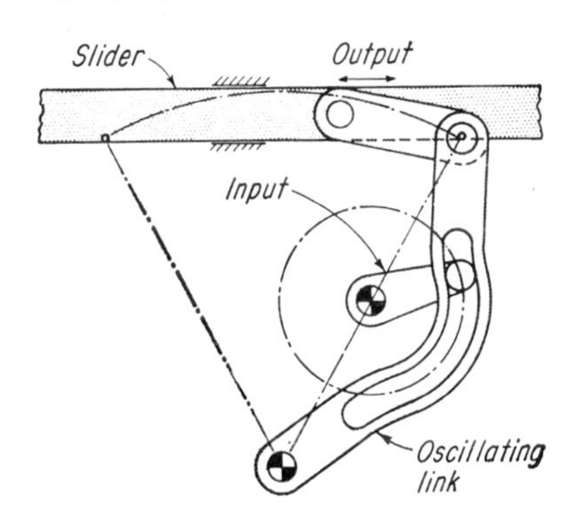

Here, too, the circular arc on the oscillating link permits the link to come to a dwell during the right position of the output slider.

Triple-harmonic drive

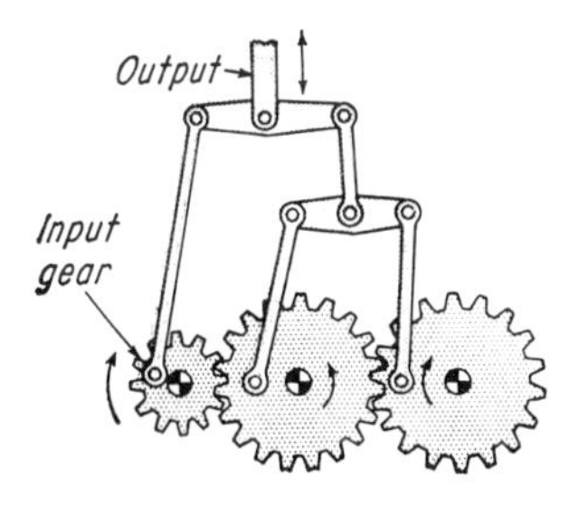

The input drives three gears with connecting rods. A wide variety of reciprocating motions of the output can be obtained by selective proportioning of the linkages, including one to several dwells per cycle.

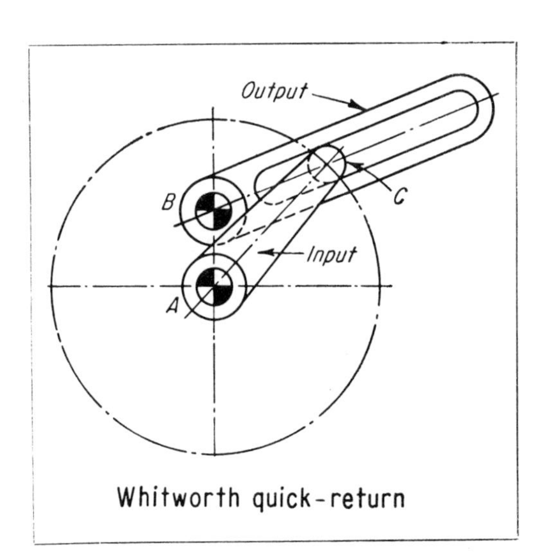

Whitworth quick-return

This is a simple way of imparting a varying motion to the output shaft B. However, the axes, A and B, are not colinear.

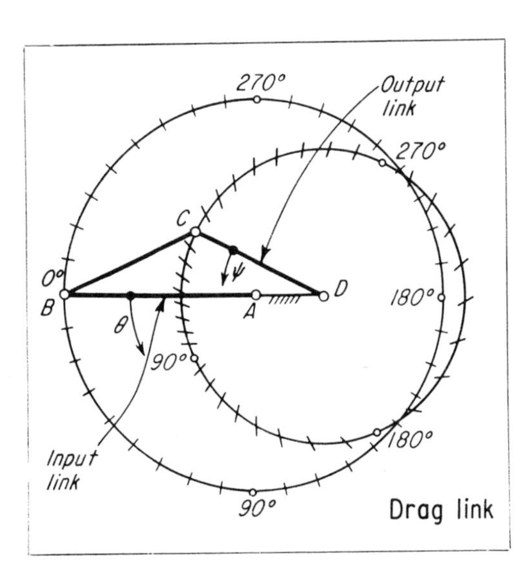

Drag link

This is another simple device for varying the output motion of shaft D. Shaft A rotates with uniform speed.

A look at what's new in

Space mechanisms

There are potentially hundreds of them, but only a few have been discovered so far. Here are the best of one class—the four-bar space mechanisms

NICHOLAS P. CHIRONIS, Associate Editor

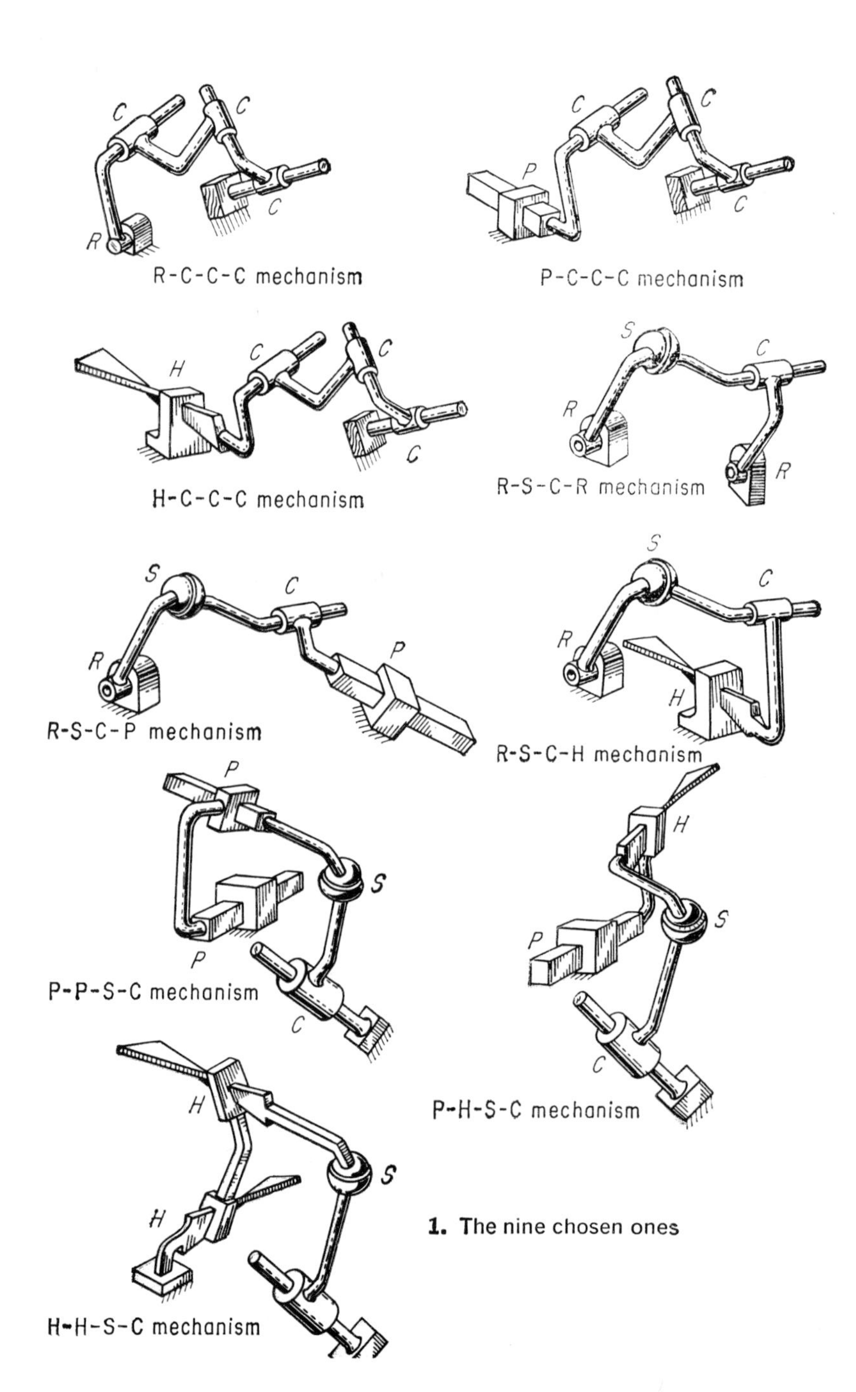

1. The nine chosen ones

A virtually unexplored area of mechanism research is the vast domain of the three-dimensional linkage, frequently called the space mechanism. Only a comparatively few kinds have been investigated or described, and little has been done to classify those that are known. Result: Many engineers do not know much about them and applications of space mechanisms have not been as widespread as might be.

One researcher, however, is making headway in this field. As part of a long-range study of space mechanisms at Oklahoma State University, Prof. Lee Harrisberger set out to discover and identify the kinds of space mechanisms possible within the existing mobility criteria and to study those which have practical characteristics.

Because a space mechanism can exist with a wide variety of connecting joints or "pair" combinations, Harrisberger identifies the mechanisms by the type and sequence of their joints. He established a listing of all the physically realizable kinematic pairs (opposite page) based on the number of degrees of freedom a joint may have. *These pairs are all the known ways of connecting two bodies together for every possible freedom of relative motion between them.*

The practical nine

The next step was to find the combination of pairs and links that would produce practical mechanisms. Using the "Kutzbach criterion" (the only known mobility criterion—it determines the degree of freedom of a mechanism due to the constraints imposed by the pairs), Harrisberger came up with 417 different kinds of space

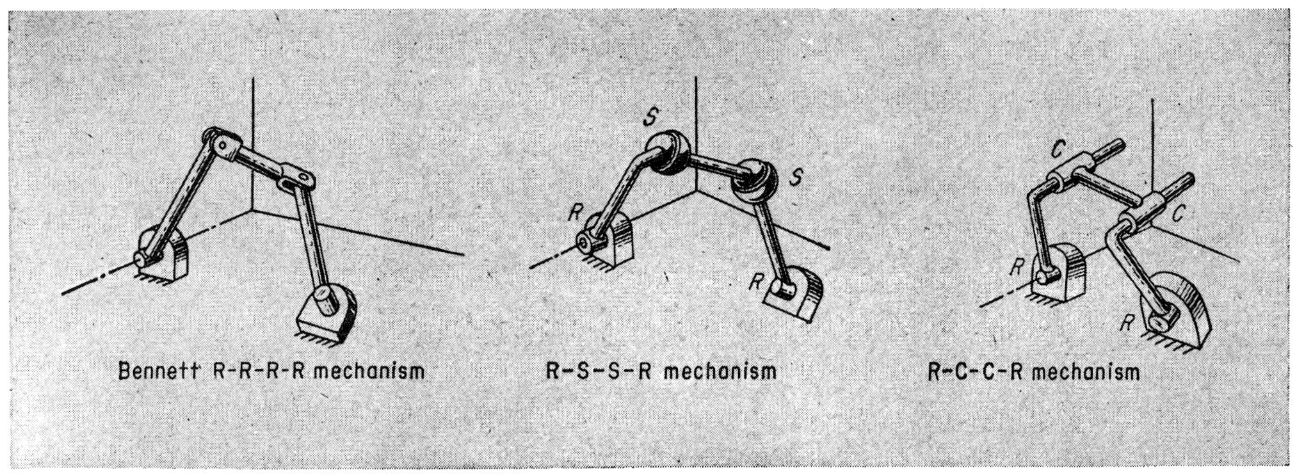

2. The three mavericks

mechanisms. Detailed examination showed many of these to be mechanically complex and of limited adaptability. But the four-link mechanisms had particular appeal because of their mechanical simplicity. At the recent ASME Mechanism Conference held at Purdue University, Harrisberger reported that he found 138 kinds of four-bar space mechanisms, nine of which have particular merit (Fig 1).

These nine, four-link mechanisms have superior physical realizability because they contain only those joints which have area contact and are self-connecting. In the table, these joints are the five closed, lower pair types:

- *R* = Revolute joint, which permits rotation only
- *P* = Prism joint, which permits sliding motion only
- *H* = Helix or screw type of joint
- *C* = Cylinder joint, which permits both rotation and sliding (hence has two degrees of freedom)
- *S* = Sphere joint, which is the common ball joint permitting rotation in any direction (three degrees of freedom)

All these mechanisms can produce rotary or sliding output motion from a rotary input—the most common mechanical requirement for which linkage mechanisms are designed.

The type letters of the kinematic pairs in the table are used to identify the mechanism by ordering the letter symbols consecutively around the closed kinematic chain. The first letter identifies the pair connecting the input link and the fixed link; the last letter identifies the output link, or last link, with the fixed link. Thus a mechanism labeled *R-S-C-R* is a double-crank mechanism with a spherical pair between the input crank and the coupler, and a cylinder pair between the coupler and output crank.

The mavericks

The Kutzbach criterion, Harrisberger notes, is inadequate for the job since it cannot predict the existence of such mechanisms as the Bennett *R-R-R-R* mechanism, the double-ball joint *R-S-S-R* mechanism, and the *R-C-C-R* mechanism (Fig 2). These are "special" mechanisms which require special geometric conditions to have a single degree of freedom. The *R-R-R-R* mechanism requires a particular orientation of the revolute axes and a particular ratio of link lengths to function as a single degree of freedom space mechanism. The *R-S-S-R* configuration, when functioning as a single degree of freedom mechanism, will have a passive degree of freedom of the coupler link. When properly constructed, the configuration *R-C-C-R* also will have a passive degree of freedom of the coupler and will function as a single degree mechanism.

Of these three special four-link mechanisms, Harrisberger feels that the *R-S-S-R* mechanism is an outstanding choice as the most versatile and practical configuration for meeting double-crank motion requirements. As we go to press, however, our editors have dug up another maverick—the Space Crank (Fig 3), described exclusively in PRODUCT ENGINEERING (Mar 2 '59 pp. 50-53). It would seem this also is an *R-R-R-R* mechanism. One wonders how many other mavericks are running around.

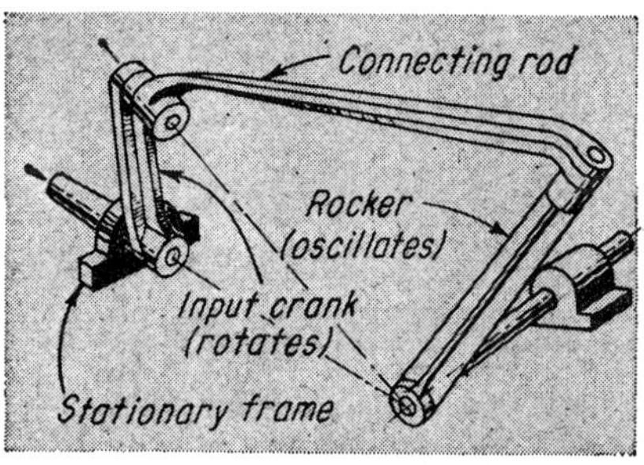

3. Space crank—another maverick?

Classification of kinematic pairs

Degree of freedom	Type number*	Type of joint: Symbol	Type of joint: Name
1	100	R	Revolute
	010	P	Prism
	001	H	Helix
2	200	T	Torus
	110	C	Cylinder
	101	T_H	Torus-helix
	020	..	....
	011	..	....
3	300	S	Sphere
	210	S_S	Sphere-slotted cylinder
	201	S_{SH}	Sphere-slotted helix
	120	P_L	Plane
	021	..	....
	111	..	....
4	310	S_G	Sphere-groove
	301	S_{GH}	Sphere-grooved helix .
	220	C_p	Cylinder-plane
	121	..	....
	211	..	....
5	320	S_p	Sphere-plane
	221	..	....
	311	..	....

*Number of freedoms, given in the order of N_R, N_T, N_H.

Seven popular types of

THREE-DIMENSIONAL DRIVES

Main advantage of three-dimensional drives is their ability to transit motion between nonparallel shafts. They can also generate other types of helpful motion. With this roundup are descriptions of industrial applications.

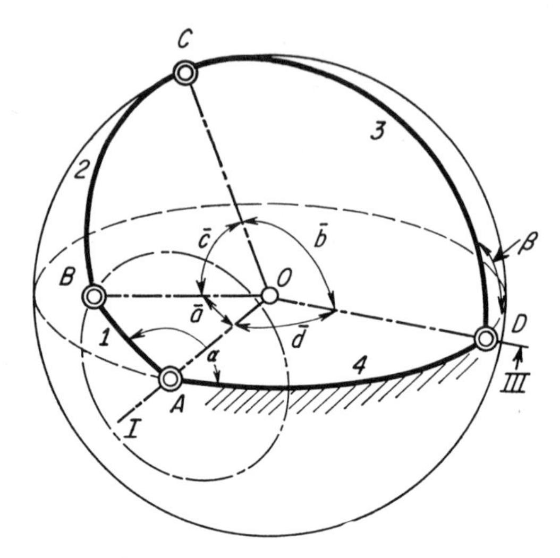

The Spherical Crank

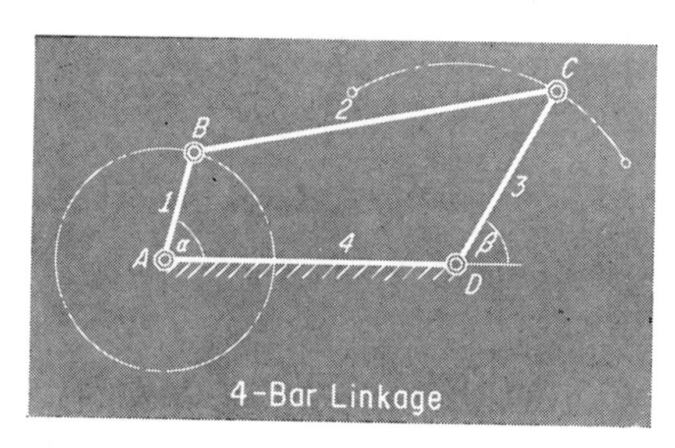

1 spherical crank drive

This type of drive is the basis for most 3-D linkages, much as the common 4-bar linkage is the basis for the two-dimensional field. Both mechanisms operate on similar principles. (In the accompanying sketches, α is the input angle, and β the output angle. This notation has been used throughout the article.)

In the 4-bar linkage, the rotary motion of driving crank *1* is transformed into an oscillating motion of output link 3. If the fixed link is made the shortest of all, then you have a double-crank mechanism, in which both the driving and driven members make full rotations.

In the spherical crank drive, link *1* is the input, link *3* the output. The axes of rotation intersect at point *O*; the lines connecting *AB*, *BC*, *CD* and *DA* can be thought of as part of great circles of a sphere. The length of the link is best represented by angles *a*, *b*, *c* and *d*.

DR W MEYER ZUR CAPELLEN, *professor of kinetics, Aachen Institute of Technology, Germany*

2 spherical-slide oscillator

The two-dimensional slider crank is obtained from a 4-bar linkage by making the oscillating arm infinitely long. By making an analogous change in the spherical crank, you can obtain the spherical slider crank shown at right.

The uniform rotation of input shaft I is transferred into a nonuniform oscillating or rotating motion of output shaft III. These shafts intersect at an angle δ corresponding to the frame link *4* of the spherical crank. Angle γ corresponds to length of link *1*. Axis II is at right angle to axis III.

The output oscillates when γ is smaller than δ; the output rotates when γ is larger than δ.

Relation between input angle α and output angle β is (as designated in skewed Hook's joint, below)

$$\tan \beta = \frac{(\tan \gamma)(\sin \alpha)}{\sin \delta + (\tan \gamma)(\cos \delta)(\cos \alpha)}$$

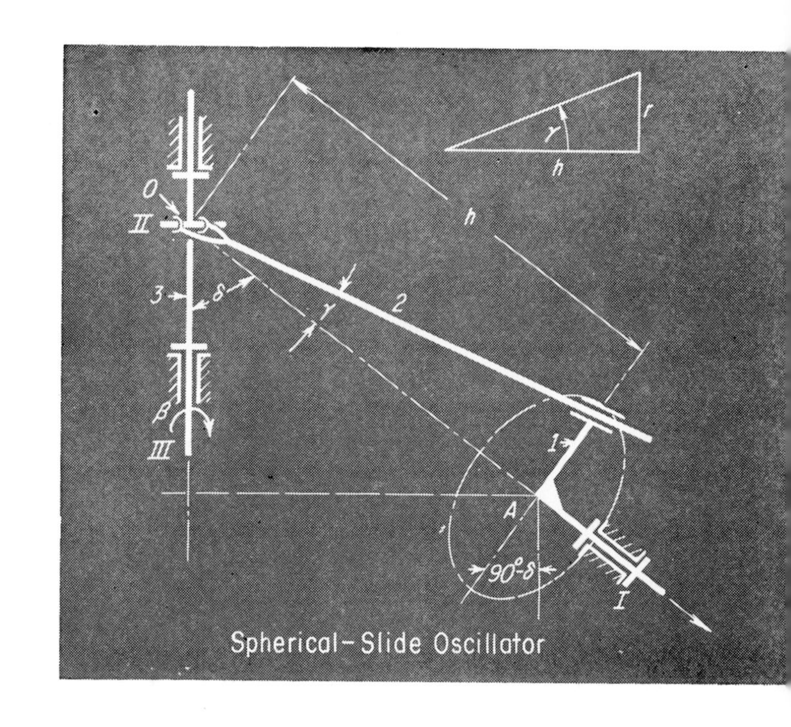
Spherical-Slide Oscillator

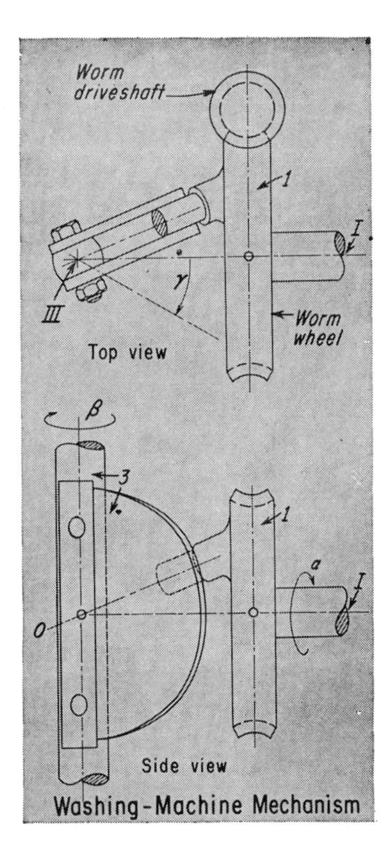

Washing-Machine Mechanism

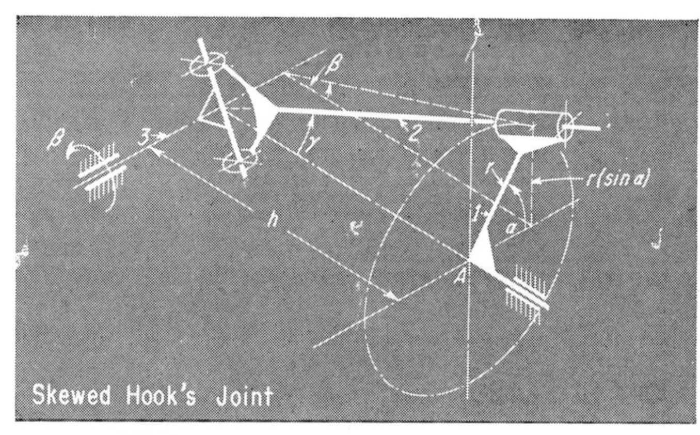
Skewed Hook's Joint

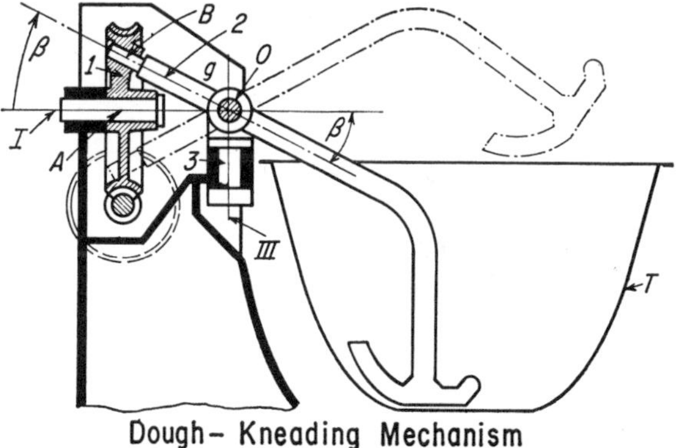
Dough-Kneading Mechanism

3 skewed hook's joint

This variation of the spherical crank is often used where an almost linear relation is desired between input and output angles for a large part of the motion cycle.

The equation defining the output in terms of the input can be obtained from the above equation by making $\delta = 90°$. Thus $\sin \delta = 1$, $\cos \delta = 0$, and

$$\tan \beta = \tan \gamma \sin \alpha$$

The principle of the skewed Hook's joint has been recently applied to the drive of a washing machine (see sketch at left).

Here, the driveshaft drives the worm wheel *1* which has a crank fashioned at an angle γ. The crank rides between two plates and causes the output shaft III to oscillate in accordance with the equation above.

The dough-kneading mechanism at right is also based on the Hook's joint, but utilizes the path of link 2 to give a wobbling motion that kneads dough in the tank.

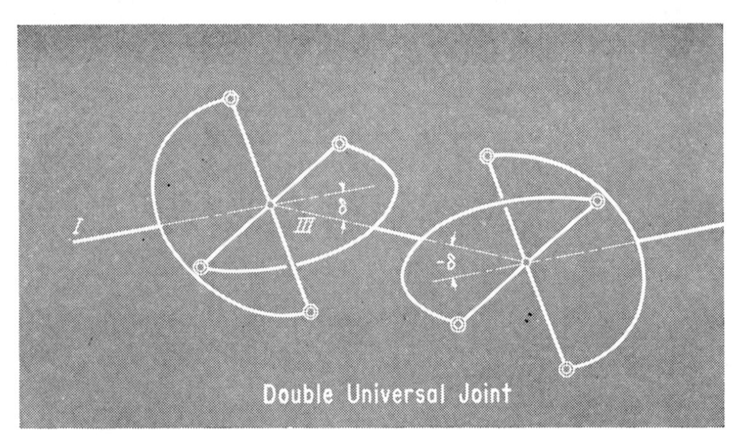

Double Universal Joint

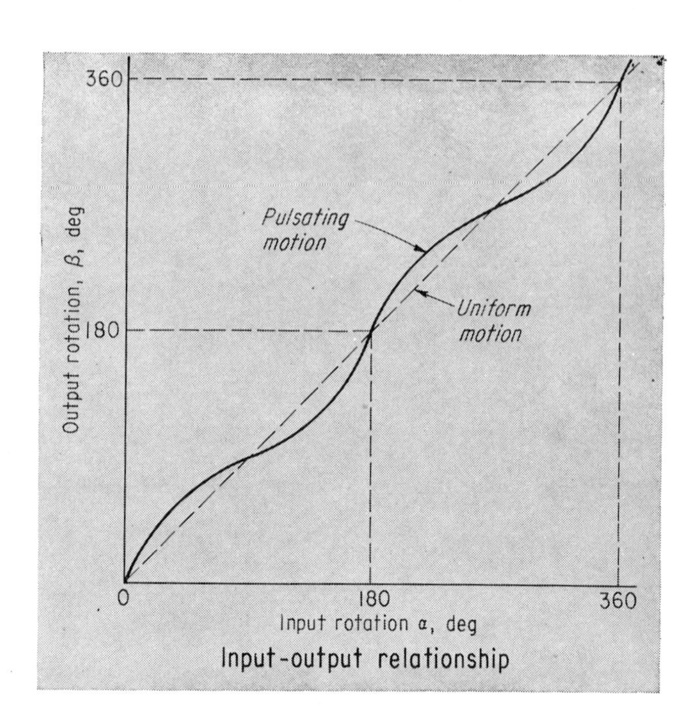

Input-output relationship

4 the universal joint

The universal joint is a variation of the spherical-slide oscillator, but with angle $\gamma = 90°$. This drive provides a totally rotating output and can be operated as a pair, as shown above.

Equation relating input with output for a single universal joint, where δ is angle between connecting link and shaft I:

$$\tan \beta = \tan \alpha \cos \delta$$

Output motion is pulsating (see curve) unless the joints are operated as pairs to provide a uniform motion.

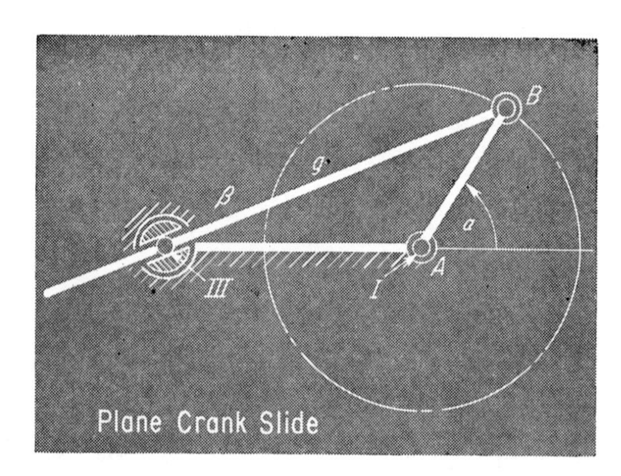

Plane Crank Slide

5 the 3-D crank slide

The three-dimensional crank slide is a variation of a plane crank slide (see sketch), with a ball point through which link g always slides, while a point B on link g describes a circle. A 3-D crank is obtained from this mechanism by making output shaft III not normal to the plane of the circle; another way is to make shafts I and III nonparallel.

A practical variation of the 3-D crank slide is the agitator mechanism (right). As input gear I rotates, link g swivels around (and also lifts) shaft III. Hence, vertical link has both an oscillating rotary motion and a sinusoidal harmonic translation in the direction of its axis of rotation. The link performs what is essentially a screw motion in each cycle.

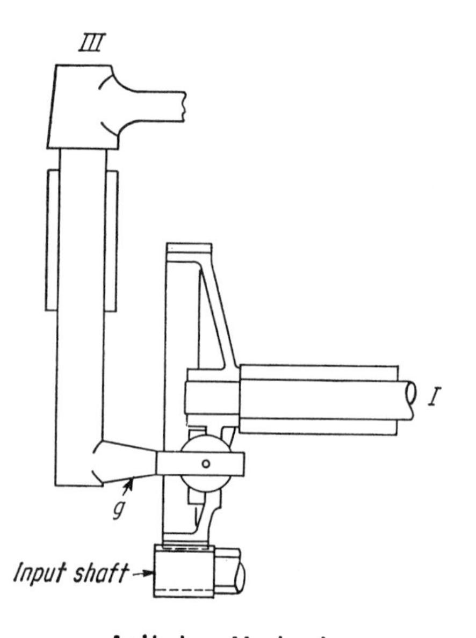

Agitator Mechanism

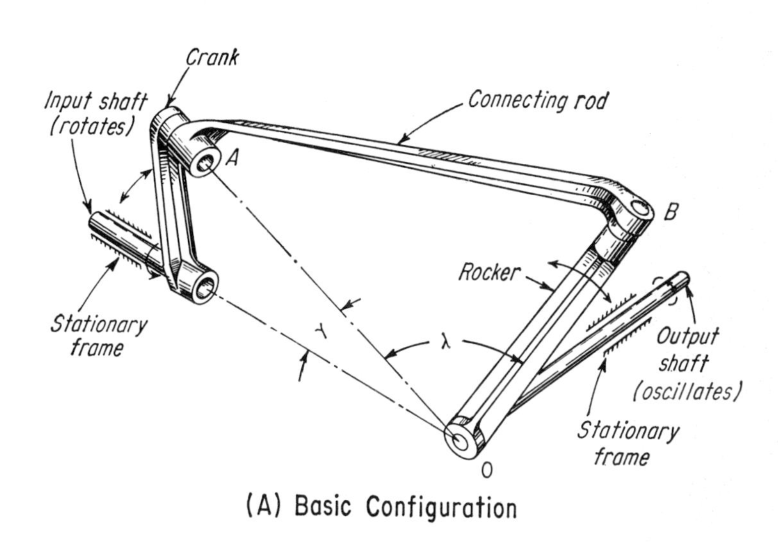

(A) Basic Configuration

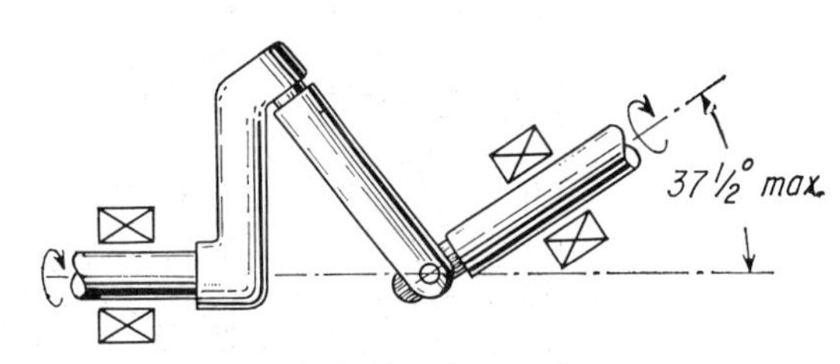

(B) Its Inversion

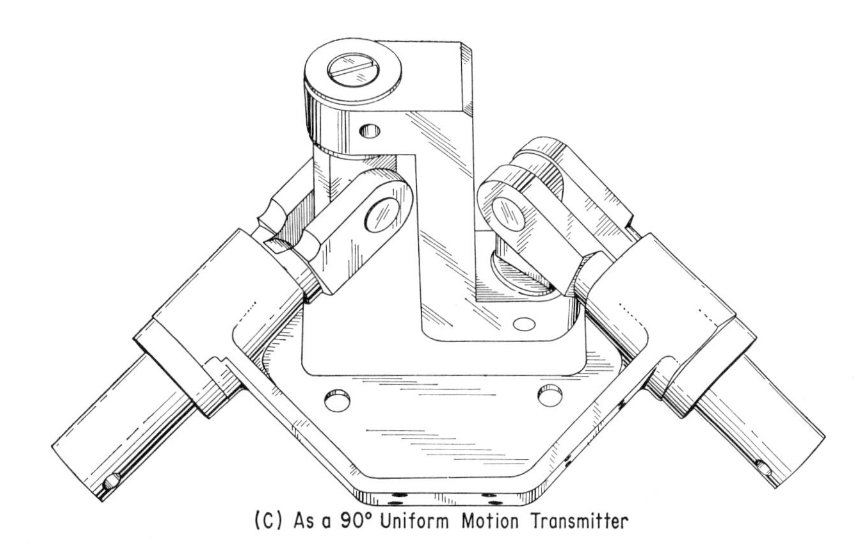
(C) As a 90° Uniform Motion Transmitter

6 the space crank

One of the most recent developments in 3-D linkages is the space crank shown in (A) see also PE—"Introducing the Space Crank—a New 3-D Mechanism," Mar 2 '59. It resembles the spherical crank discussed on page 76, but has different output characteristics. Relationship between input and output displacements is:

$$\cos\beta = (\tan\gamma)(\cos\alpha)(\sin\beta) - \frac{\cos\lambda}{\cos\gamma}$$

Velocity ratio is:

$$\frac{\omega_o}{w_i} = \frac{\tan\gamma \sin\alpha}{1 + \tan\gamma \cos\alpha \cot\beta}$$

where ω_o is the output velocity and ω_i is the constant input velocity.

An inversion of the space crank is shown in (B). It can couple intersecting shafts, and permits either shaft to be driven with full rotations. Motion is transmitted up to 37½° misalignment.

By combining two inversions, (C), a method for transmitting an exact motion pattern around a 90° bend is obtained. This unit can also act as a coupler or, if the center link is replaced by a gear, it can drive two output shafts; in addition, it can be used to transmit uniform motion around two bends.

VARIATIONS OF THE SPACE CRANK

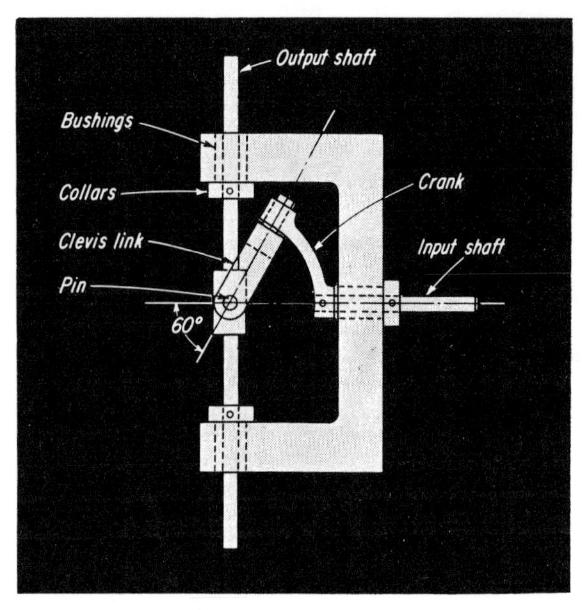

Oscillating motion . . .
powered at right angles. Input shaft making full rotations causes output shaft to oscillate 120°.

Constant-speed-ratio . . .
universal is obtained by using two "inversions" back-to-back. Motion transmitted up to 75° misalignment.

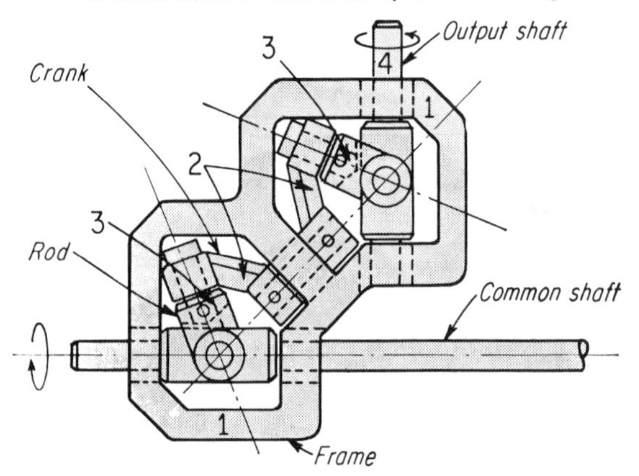

Right-angle . . .
limited-stroke drive transmits exact motion pattern. A multiplicity of fittings can be operated from common shaft.

7 the elliptical slide

The output motion, β, of a spherical slide oscillator, can be duplicated by means of a two-dimensional "elliptical slide." The mechanism has a link g which slides through a pivot point D and is fastened to a point P moving along an elliptical path. The ellipse can be generated by a Cardan drive, which is a planetary gear system with the planet gear half the diameter of the internal gear. The center of the planet, point M, describes a circle; any point on its periphery describes a straight line, and any point in between, such as point P, describes an ellipse.

There are certain relationships between the dimensions of the 3-D spherical slide and the 2-D elliptical slide: $\tan \gamma / \sin \delta = a/d$ and $\tan \gamma / \cot \delta = b/d$, where a is the major half-axis, b the minor half-axis of the ellipse, and d is the length of the fixed link DN. The minor axis lies along this link.

If point D is moved within the ellipse, a completely rotating output is obtained, corresponding to the rotating spherical crank slide.

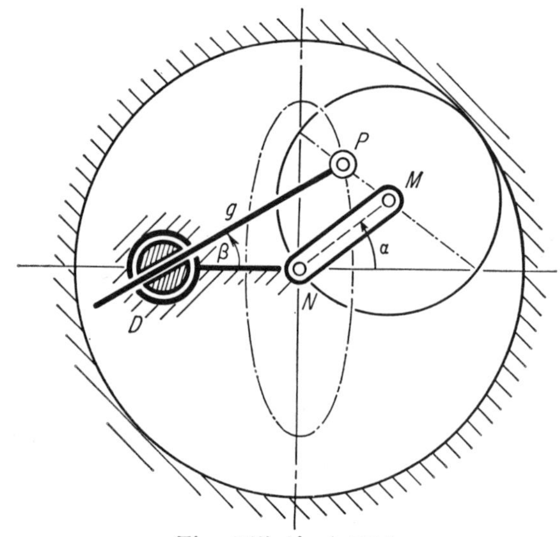

The Elliptical Slide

Spherical linkage, produces an oscillating rotational output from a constant speed input for an input angular displacement.

where ψ=output angular displacement

ρ=tan α

α=constant angle between roller shaft *b* and input axis *c*

$\tan \psi = \rho \sin \phi$

For mechanism to function properly, axes of semicircular link *d*, of roller *a*, and of input shaft *c* must intersect at point *M*.

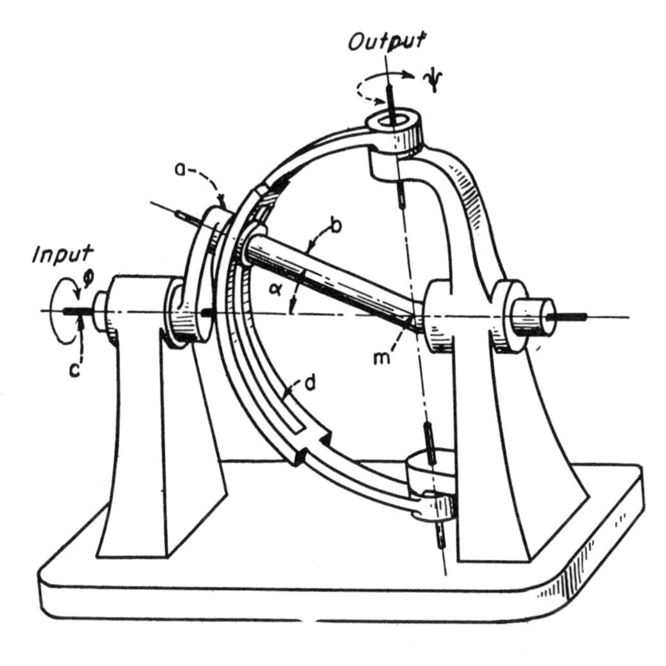

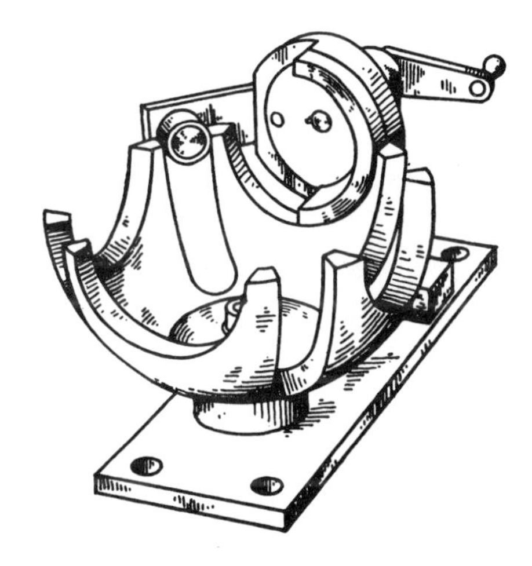

Spherical Geneva Mechanism. Driver and driven wheel are on perpendicular shafts. Duration of dwell is exactly 180 deg of driver rotation.

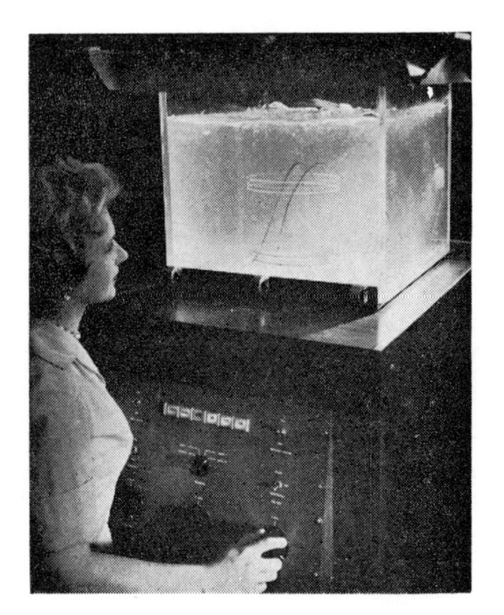

3-D Plotter Freezes Missile Tracks

This three-dimensional plotter coupled to radar and an analog computer shows instantly the trajectory of an enemy missile still thousands of miles away. It was developed initially by Chrysler Corp after noting that a three-dimensional plotter was included in the 10 most wanted devices in a 1958 listing issued by the National Inventors Council.

The plotter consists of a stylus suspended in an optically clear tank containing a transparent gel. Signals from an analog computer to servomotors move the stylus in any direction. Various colored inks are fed under pressure through the stylus as it moves through the gel; the gel closes behind the stylus retaining the ink tracings precisely in suspension.

Production version of the plotter will have a stylus with three ink channels. Traces may be retained as long as desired and then erased with a suction tube; or ink with varying degrees of persistence may also be used.

The plotter is also adaptable to weather mapping, satellite tracking, and machine-tool control.

ROTARY TO RECIPROCATING

PREBEN W. JENSEN

Four-bar slider mechanism

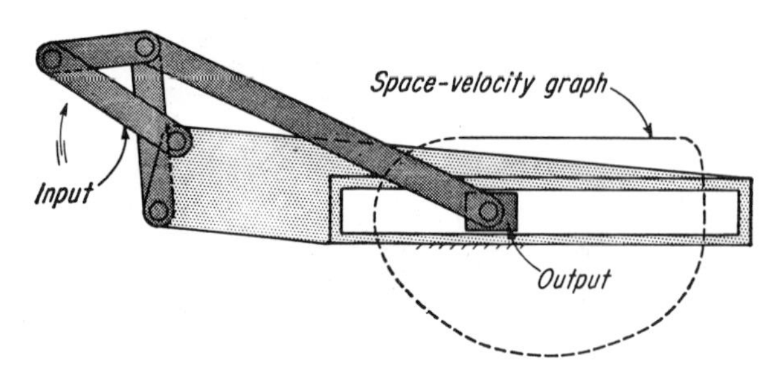

With proper proportions, the rotation of the input link can impart an almost-constant velocity motion to the slider.

Three-gear stroke multiplier

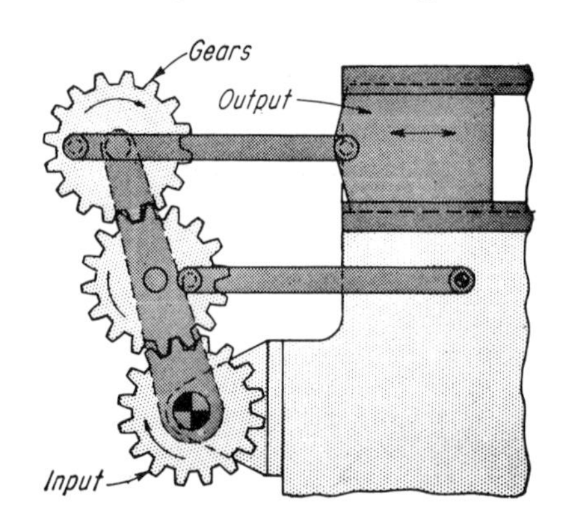

Rotation of the input gear causes the connecting link, attached to the machine frame, to oscillate. This produces a large-stroke reciprocating motion in the output slider.

Oscillating-chain mechanism

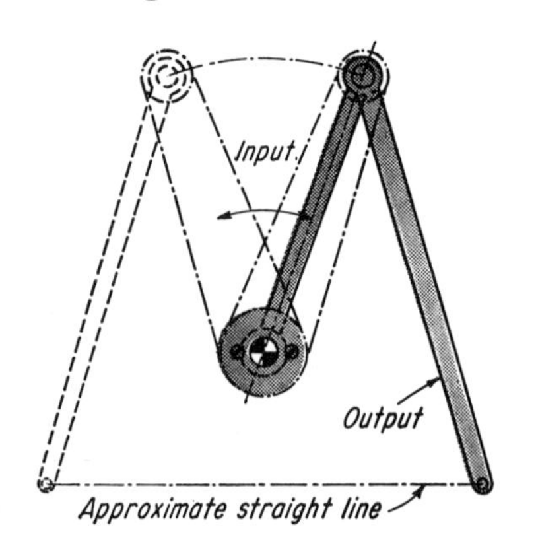

Rotary motion of the input is translated into linear motion of the linkage end. The linkage is fixed to the smaller sprocket, and the larger sprocket fixed to the frame. For instrumentation.

Rack and gear sector

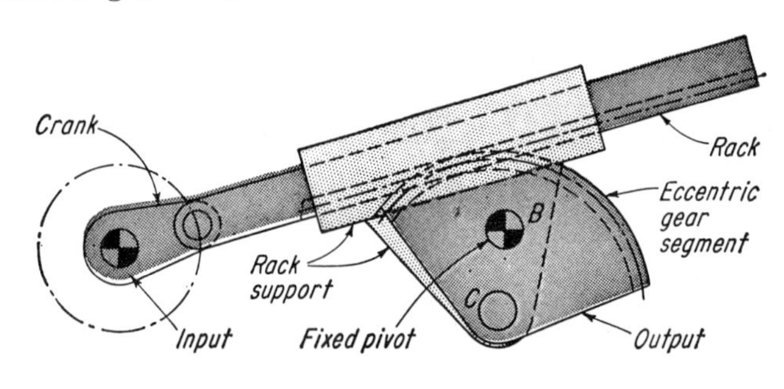

Rotary motion of the input is translated into oscillating motion of the output. The rack support and gear sector are pinned at *C*, but the gear itself oscillates around *B*.

Right angle oscillator

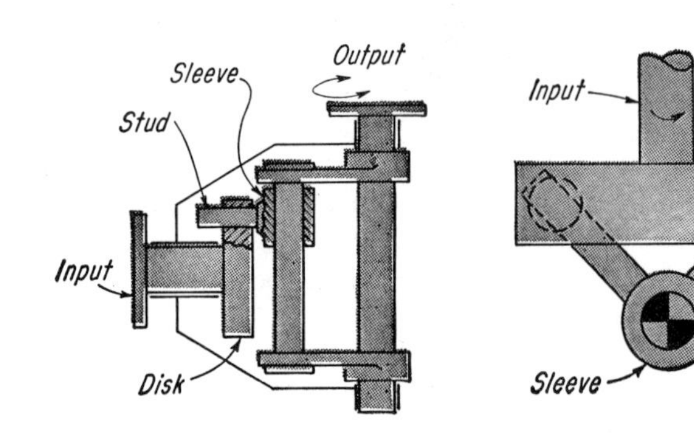

The input shaft and disk (extreme left) drives the stud and sleeve in a circular path. This causes the sleeve to move up and down and imparts an oscillating motion to the output shaft. A second variation is also shown.

Linear reciprocator

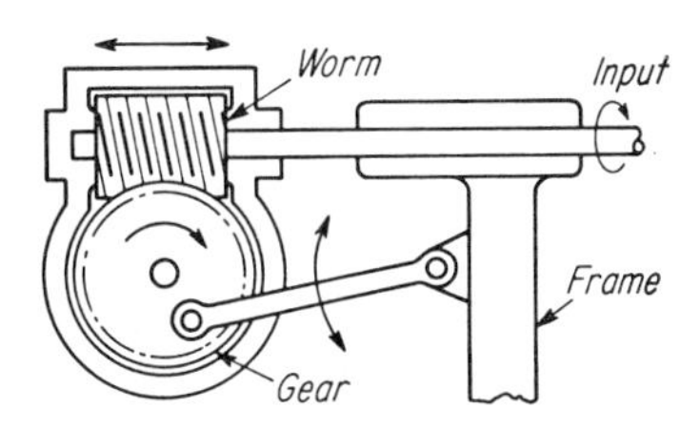

The objective here is to convert a rotary motion into a reciprocating motion that is *in line* with the input shaft. Rotation of the shaft drives the worm gear which is attached to the machine frame by means of a rod. Thus input rotation causes the worm gear to draw itself (and the worm) to the right—thus providing a back and forth motion. Employed in connection with a color-transfer cylinder in printing machines.

Disk and roller drive

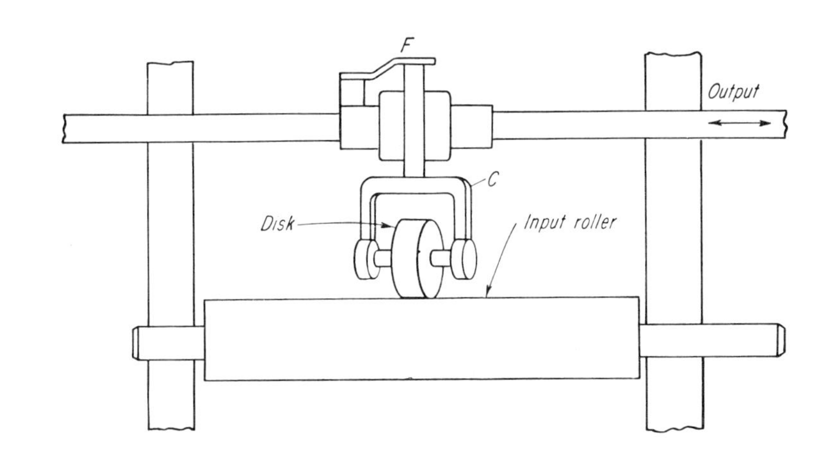

Here a hardened disk, riding at an angle to an input roller, transforms the rotary motion into linear motion *parallel to the axis of the input.* The roller is pressed against the input shaft with the help of a flat spring, *F*. Feed rate is easily varied by changing the angle of the disk. Arrangement can produce an extremely slow feed with a built-in safety factor in case of possible jamming.

Bearings and roller drive

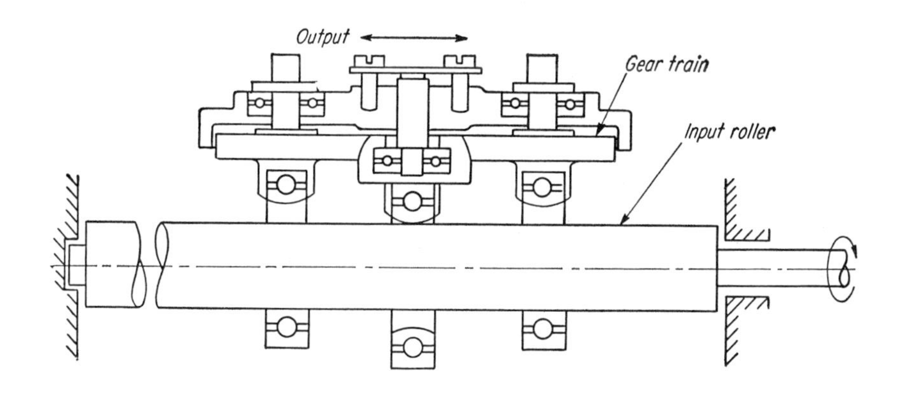

Similar to the previous one, this arrangement avoids large Hertzian stresses between disk and roller by employing three ball bearings in place of the single disk. The inner races of the bearings make contact on one side or the other. Hence a gearing arrangement is required to alternate the angle of the bearings. This arrangement also reduces the bending moment on the shaft.

Reciprocating space crank

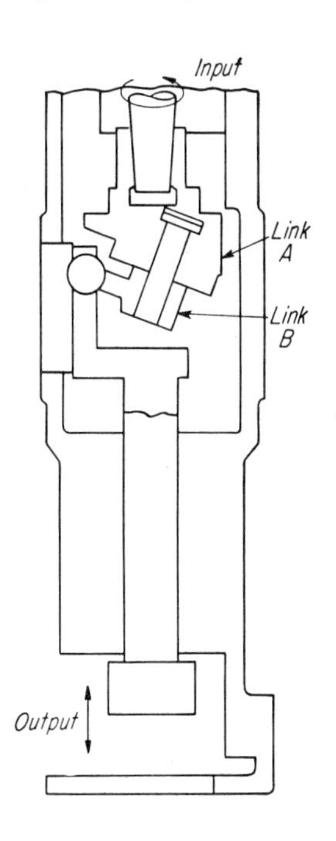

Rotary input causes the bottom surface of link *A* to wobble in reference to the center line. Link *B* is free of link *A* but restrained from rotating by the slot. This causes the output member to reciprocate linearly. Employed for filing machine.

Sinusoidal reciprocator

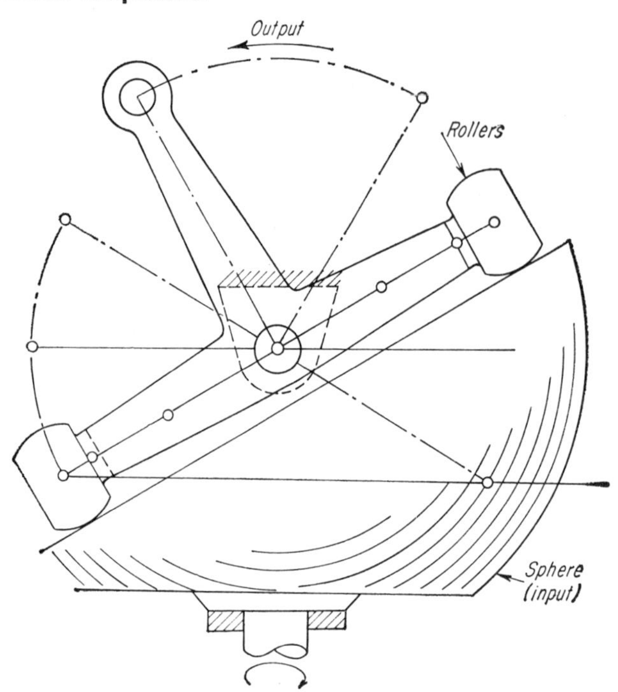

This transforms rotary motion into a reciprocating motion in which the oscillating output member is in the same plane as the input shaft. The output member has two arms with rollers which contact the surface of the truncated sphere. Rotation of the sphere causes the output to oscillate.

Cross-bar reciprocator

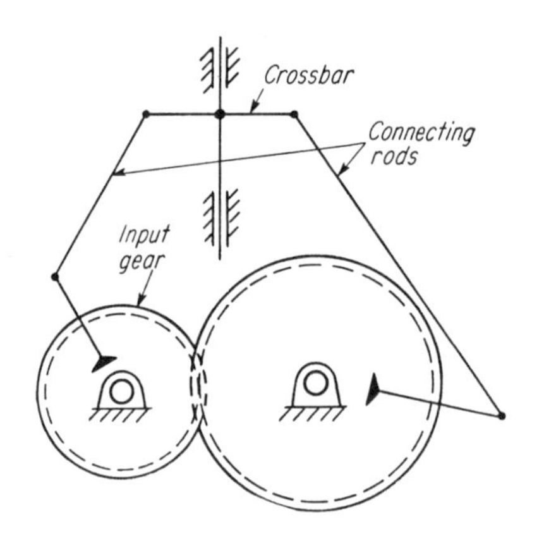

Although complex-looking, this device has been successful in high speed machines for transforming rotary motion into a high-impact linear motion. Both gears contain cranks connected to the cross bar by means of connecting rods.

New drive mechanism that promises greater thermodynamic efficiency . . .

. . . for hot-gas (Stirling cycle) engines has been developed by Philips of Eindhoven (Netherlands). Essentially, it's a multiple four-bar linkage system that makes it possible to separate power and gas-transfer mechanisms and eliminate the need for pressurizing the crankcase. With previous systems, in which power and gas-transfer functions were combined, the crankcase had to be pressurized to limit leakage of gas past the piston and to limit downward forces on the drive mechanism.

In the new system, called a rhombic drive (see diagram) there are twin cranks and control-rod mechanisms, identical in design and offset from the central axis of the engine. The cranks rotate in opposite directions and are coupled by twin gear wheels. Says R J Meijer of Philips: "The symmetry of the system and the coaxial arrangement of power-piston and displacer-piston rods make it an easy matter to avoid putting the crankcase under high pressure . . . The power-piston rod stuffing-box is subject to no lateral thrust, and the friction forces are low." Only the buffer space need be filled with gas under pressure; and this, he says, is much more easily done than pressurizing the whole crankcase.

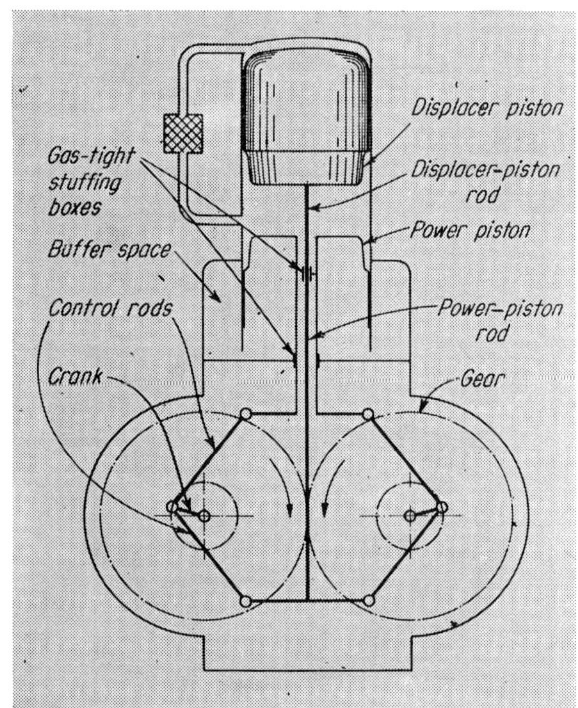

Rhombic drive takes its name from crankcase and control-rod arrangement. Note that power-piston rod is hollow and displacer-piston rod runs through it.

Two-speed reciprocator

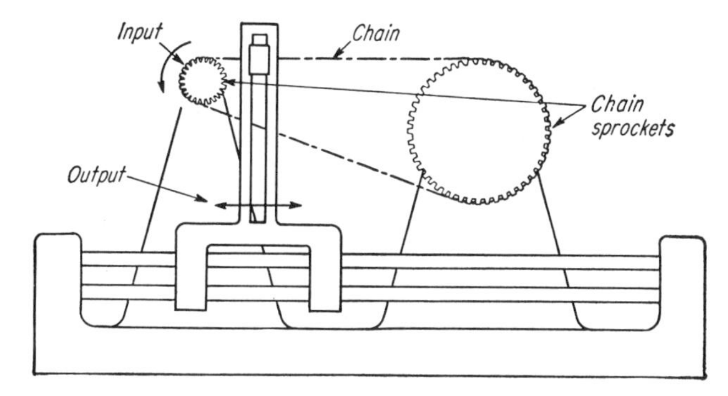

A drive pin, fastened to a chain, rides in a vertical slot to reciprocate a carriage in a horizontal direction. Velocity in both directions is constant and depends on the angle of slope that the chain makes with the vertical slot. Thus, velocity to the left is higher than to the right because the top of the chain is horizontal.

Scotch-Yolk Mechanism

Ram action . . .
is provided by mainshaft eccentric working through a scotch-yoke setup. This consists of a rectangular block, mounted on the eccentric and located between two horizontal ways in the ram. The block transmits vertical component of eccentric action directly to the ram. Ram moves between bronze guides, transmitting thrust to top of the die. Contact surfaces between ram and block run submerged in oil. Ram extends above mainshaft centerline, providing long guides that reduce angular cocking of ram.

Doing away with a pitman reduces over-all press height more than 3 ft—a practical asset in press operation. Because stretch of side members and compression of pitman arm is considerable in heavy-duty presses, some of the flywheel energy goes to creating this deflection rather than doing useful work.

Shortened Press Reduces STRETCH

Thanks to scotch yoke that transmits thrust, this automated forging press also has shorter, stiffer frame. Auxiliary guides eliminate need for guide pins in die sets.

The high-volume 2500-ton press is designed to shape such parts as connecting rods, tractor track links, and wheel hubs. Simple automatic-feed mechanism raises production to a possible 2400 forgings per hour. Produced by Erie Foundry Co, Erie, Penna.

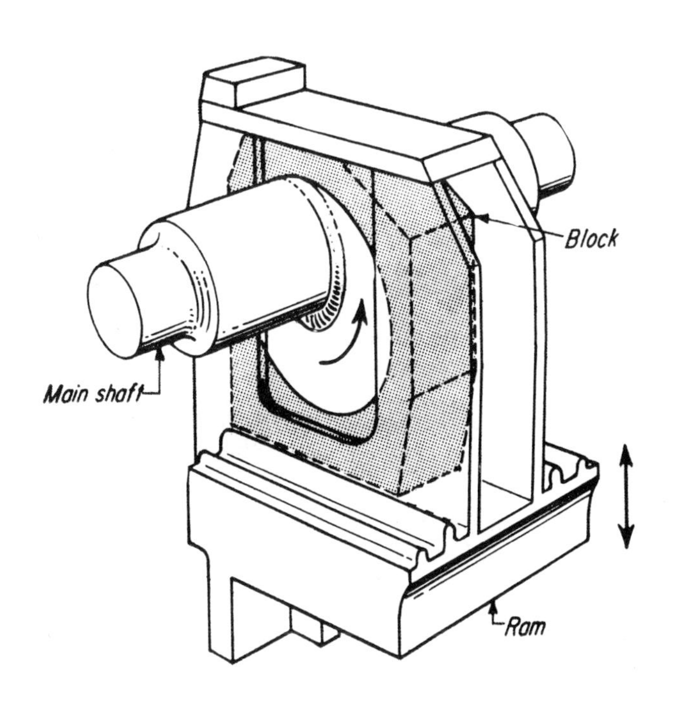

See-saw cam motion smoothly converts

Rotary to linear motion

RUSSELL C BALL Jr, president, Philadelphia Gear Corp, King of Prussia, Pa

HERE is a formidable new competitor for power screws, ball screws, hydraulic cylinders and gear and rack units. It is a sophisticated device with cams, rollers and gears, but tests so far show it offers:

1 . . high thrust loads, to 5,000,000 lb.

2 . . self-locking characteristics under load

3 . . negligible wear on long strokes

4 . . high linear speeds, to 200 in./min

The "Roll-Ramp" actuator is the result of 15 years of investigation by its inventor, Arthur Maroth. A commercial version has been under development and test by Philadelphia Gear Corp for several years and the actuator is now in the production stage.

Principle of operation

The mechanism is enclosed in a stationary housing; an Acme-threaded shaft moves through the housing as the thrusting member. The thread, however, provides only a gripping surface for the internal mechanisms of the actuator.

The basic driving element is an inclined plane—actually two inclined planes mounted in opposition which are wrapped around a disk. This forms a face cam with a succession of long inclines and short declines. Rollers are placed on each incline between two such opposed cams and the assembly is set on a thrust bearing. Rotating the lower cam, while keeping the upper cam from rotating, will cause the upper cam to generate a straight-line, up-and-down motion. This system, however, will yield a very limited stroke (twice the use of one wedge) before the cams must be reset to lift again.

Continuous motion

To generate a continuous lifting motion, two such cam assemblies work in phased relationship. A nut with gear teeth cut in its periphery is set on top of each cam assembly. A screw of a length suitable for the desired travel is threaded through the two nuts. The nuts are free to rotate until they snub down against the upper cam.

In the position shown, rollers in the lower cam assembly are at the bottoms of the inclines. As cam A rotates, cam B lifts the nut and screw. The rollers in the upper cam assembly are almost at the crests of the inclines. As cam C rotates further (driven by the same input as A) its rollers go over the crests and down the cam declines so cam D falls away from the top nut. The top nut is unloaded but it is carried upward by the advance of the screw (still being driven by the bottom rollers). Now, the nut is counter-rotated on the screw by the intermittent motion mechanism.

The counter-rotation returns the nut until it again meets cam D. This allows cam D to re-engage the nut when the rollers of the upper assembly are at the bottoms of the inclines and are about to climb. At this time, the rollers of the lower assembly pass the crests of the inclines and begin to return. The bottom nut is then counter-rotated by the intermittent mechanism to its new starting position in the same manner as the top nut.

The required relationship between cam assemblies is maintained by a common pinion shaft which meshes with gears on the outer rim of both rotating cams. Because the output motion is generated by inclined planes of fixed slope, the rate of lift is proportional to the input speed, and a continuous and smooth output linear motion is obtained.

The threads of the leadscrew support the load only when there is no relative motion between nut and screw. Thus, only negligible wear can occur on the threads. The nuts, when loaded during actuation, act as pushing members with the threads in shear. The screw is merely a positioning device.

Advance of the screw per input revolution is entirely dependent upon the slope and number of inclines of the cams; it is not related to the lead of the screw. Consequently, the mechanical advantage of the actuator is determined by the cam incline and can be made almost as high as desired. The actuator is directly driven by a flange-mounted motor bolted to the housing.

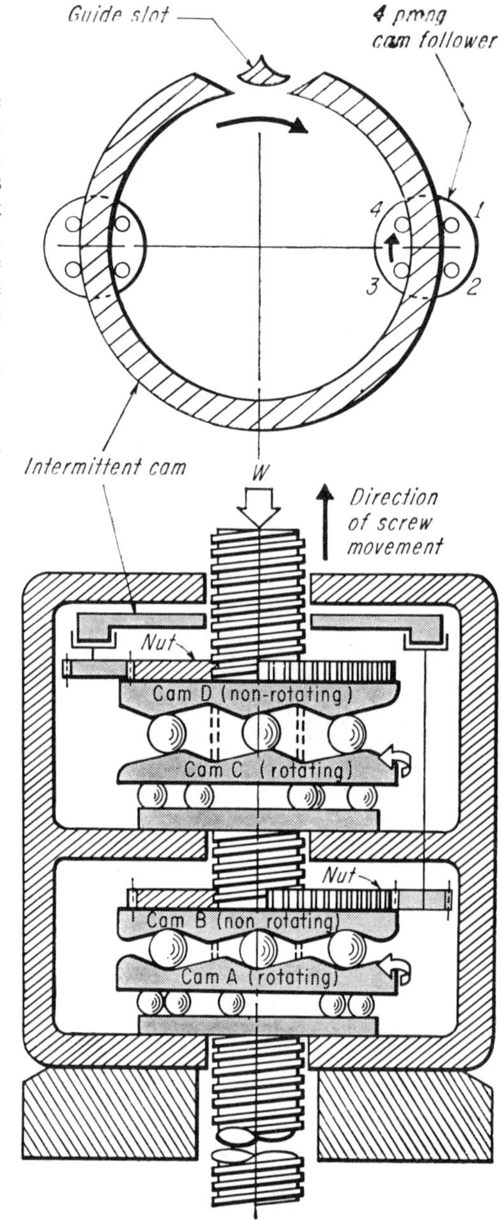

CONVERSION OF ROTARY TO LINEAR MOTION. Two cam assemblies operate in phased relationship. When one pair is driving, the second pair is retracting. The nuts are alternately rewound by the intermittent cam, shown also in the top view, which is driven continuously with cams A and C. As the deflector strikes prong 1 of the 4-prong cam follower at right (which is meshed with one of the nuts), the follower is rotated one quarter turn clockwise. Prong 4 then moves into the position of prong 1. This motion rewinds the nut to the top of cam B. A half revolution later, the operation is repeated for the follower on the left side.

Three typical applications

SLUICE-GATE OPERATION requires lifting and lowering forces of about 75,000 lb. Actuator must have self-locking characteristics to prevent weight of gate from reversing the drive, and accurate positioning control. With new actuator, long stem need not be rigidly restrained to prevent twisting.

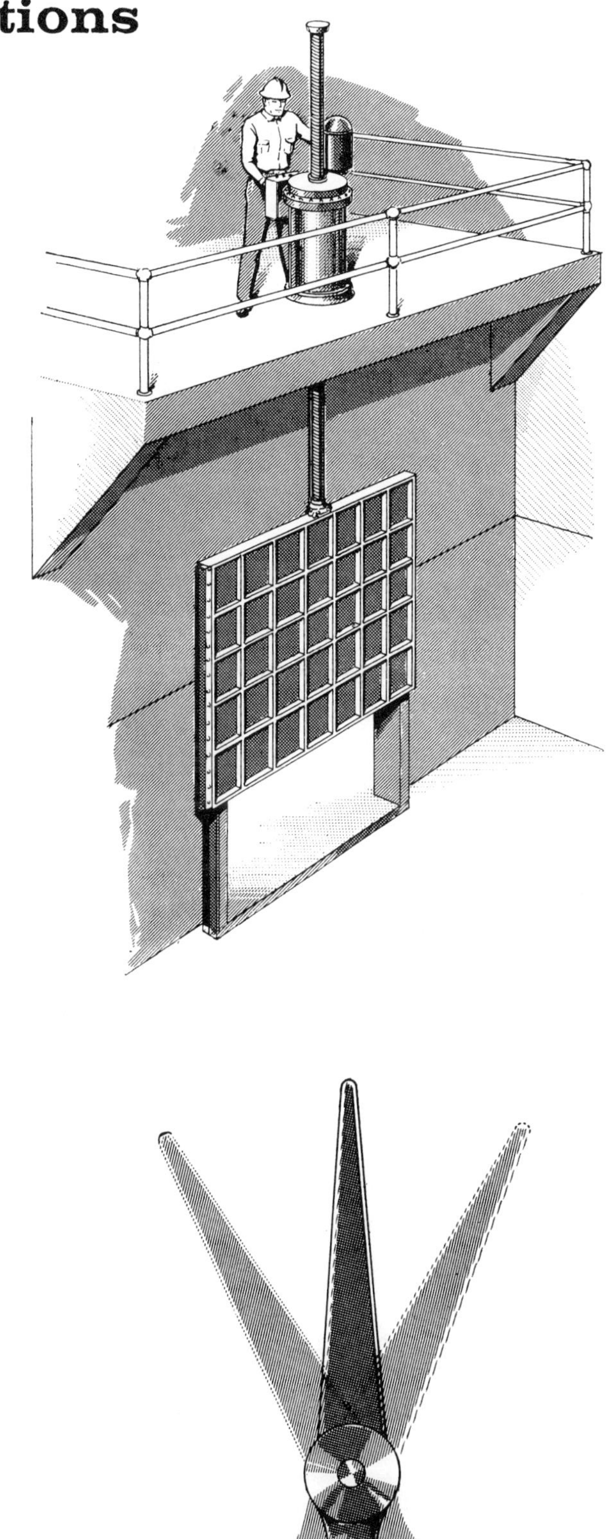

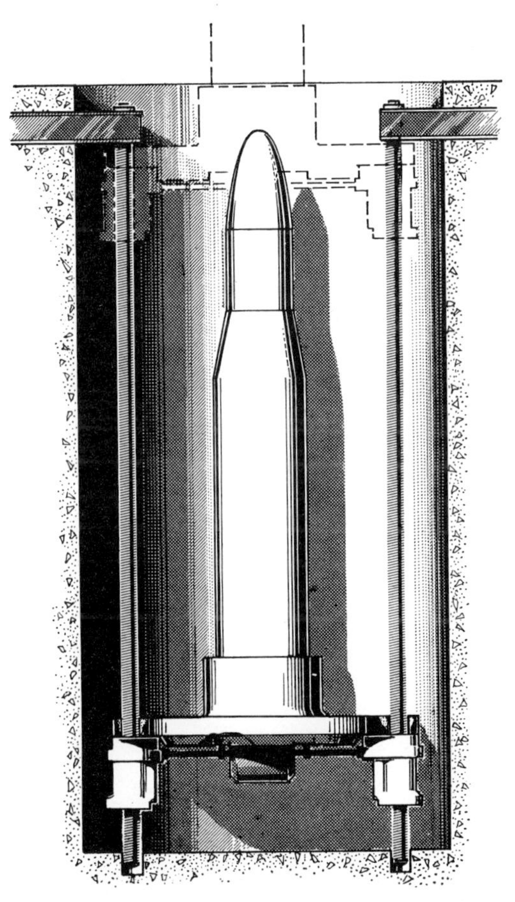

MISSILE LIFT MECHANISM has four actuators powered from a single source to move the platform uniformly at all points. Screw is locked in place while the actuator travels with its prime mover along the screw. The platform can be positioned with precision and very high lifts can be accomplished.

RUDDER-CONTROL UNIT employs actuator to provide continuous back and forth movement, usually of small displacement. Friction forces must be kept to a minimum. Power source can be a reversible hydraulic motor. Because of self-locking characteristics, no backloading from the rudder can move the threaded stem.

Ball-bearing screws

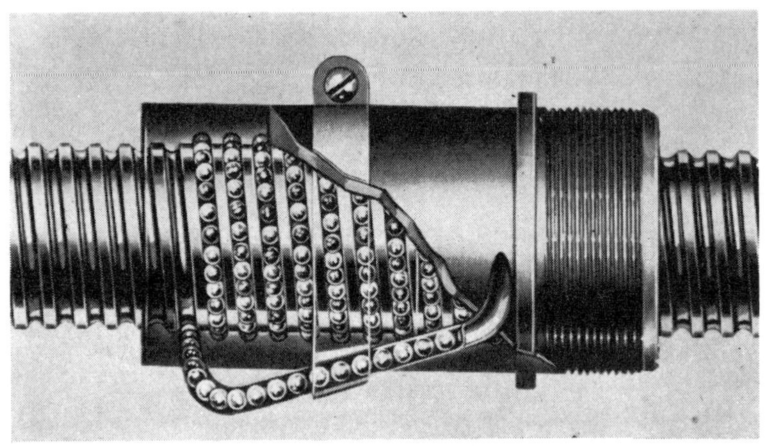

1. CONTINUOUS RECIRCULATING BALLS, rolling in helical grooves, provide ability to carry heavy loads with low friction and high efficiency.

Beaver Precision Products Inc, Detroit

In many applications a single circuit design—in which one tubular ball guide is used—is adequate. In critical applications where constant or very frequent operation is required the multiple circuit—utilizing two or more tubular ball guides—is used.

Saginaw Steering Gear Div, General Motors Corp, Saginaw, Mich.

How a ball screw operates

A typical ball-screw assembly, Fig 1, consists of two components: a screw with circular-form threads, and a nut assembly with an internal helical ball groove to permit the flow of a continuous row of steel balls. Rotation of either the screw or nut causes the rolling balls to move along the helical path. The balls travel at approximately half the speed of the races and exit at the trailing end of the nut. To keep the balls rolling continuously in the system, a return tube deflects and recirculates the balls to the leading end of the nut.

Basic operation of a ball screw is illustrated in Fig 2:

- (A) Rotary motion applied to the screw (manually or by a power source) drives the nut linearly along the screw axis.
- (B) Rotary motion applied to the nut (manually or by a transmission drive such as gears) drives the screw linearly.

Because of the low coefficient of friction, both rotary-to-linear functions are reversible; that is, a linear motion of either the screw or nut can cause the other member to rotate.

Threads on most types are machined or rolled. When machined, they are rough cut, hardened and then ground to obtain highest positioning accuracy, efficiency, load capacity and life. Rolled threads, on the other hand, are cheaper to produce and offer considerable cost savings if quantities are sufficiently high.

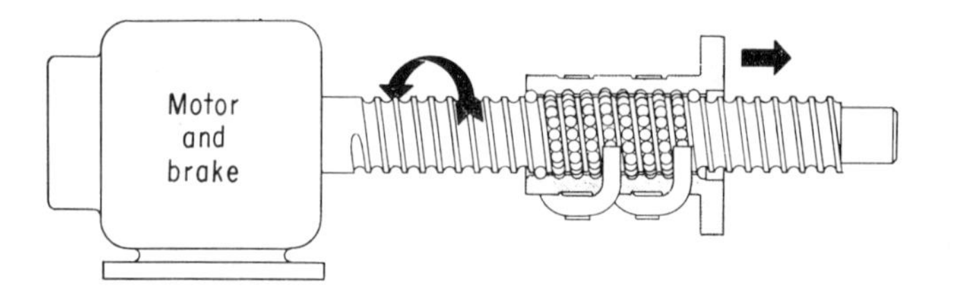

Motor and brake

(B)

2. BASIC OPERATIONS: Rotary motion to (A) screw, or (B) nut, causes other member to travel linearly or provide linear force.

As converters of rotary to linear motion, they are replacing the acme power screw in many applications.

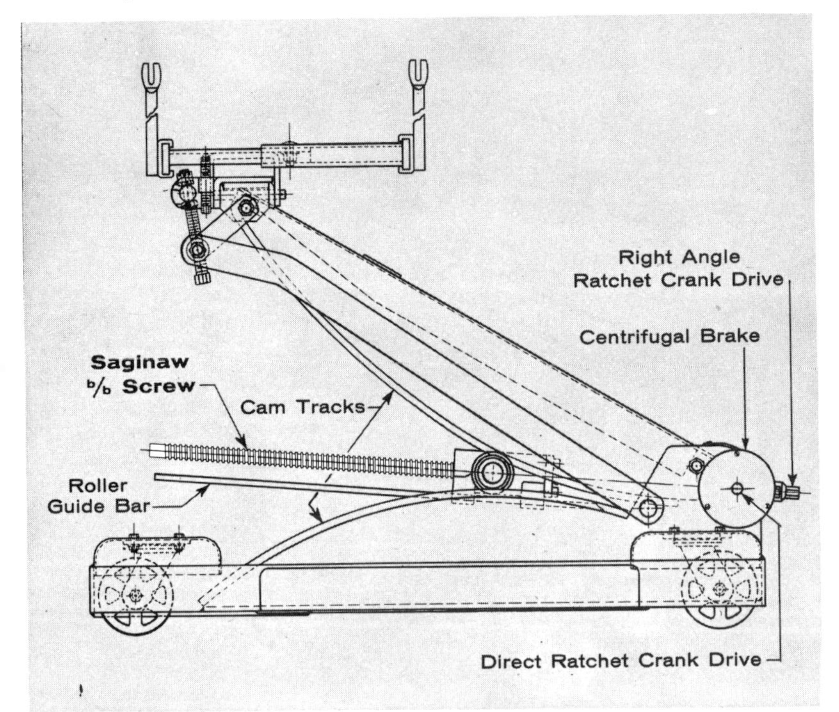

This automotive transmission jack uses a single b/b screw to move rollers between scissor-like cams. Drive speed of the screw is relatively slow and the load is in tension. The brake prevents backdriving under load.

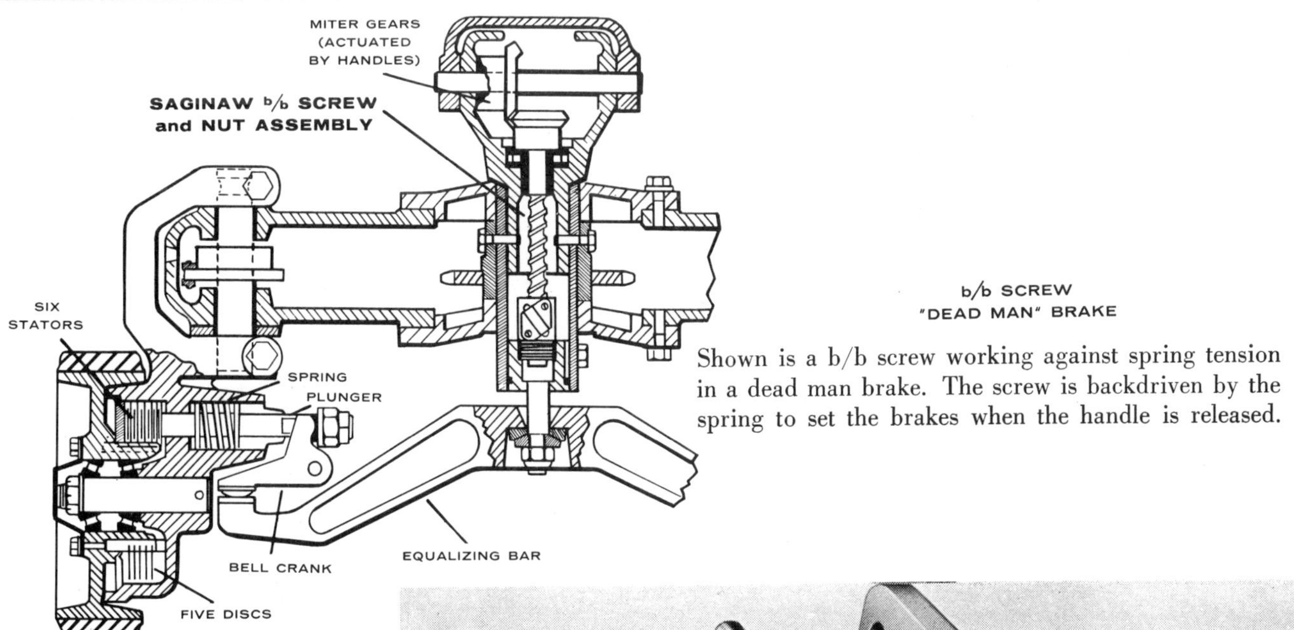

b/b SCREW
"DEAD MAN" BRAKE

Shown is a b/b screw working against spring tension in a dead man brake. The screw is backdriven by the spring to set the brakes when the handle is released.

MECHANICAL linear actuator is a single-package, positive-action unit that gives fast operation and long linear movement, requires low power, and reduces frictional heat while assuring safe operation. A ball bearing type nut acts as a low torque ball screw to reduce friction in converting torque to thrust. Units come with an upright screw, inverted screw, or relating screw in capacities from 2 to 50 tons. Duff-Norton Co, Charlotte, NC.

More Ball-Bearing Screw Applications

Cartridge-operated rotary actuator quickly retracts webbing to forcibly separate a pilot from his seat as the seat is ejected in emergencies. Tendency of pilot and seat to tumble together after ejection prevented opening of chute. Gas pressure from ejection device fires the cartridge in the actuator to force ball-bearing screw to move axially. Linear motion of screw is translated into rotary motion of ball nut. This rapidly rolls up the webbing (stretching it as shown) which snaps the pilot out of his seat. *Talley Industries.*

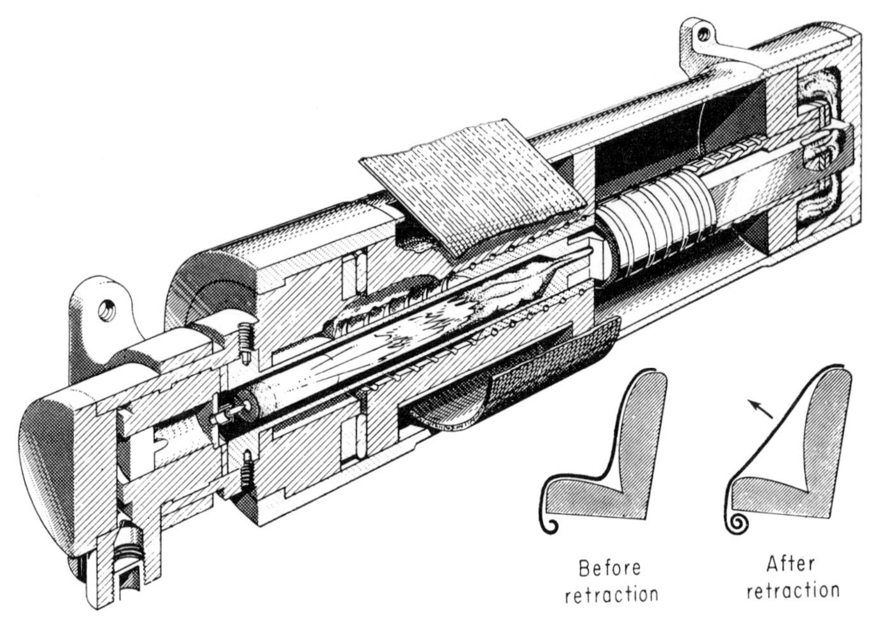

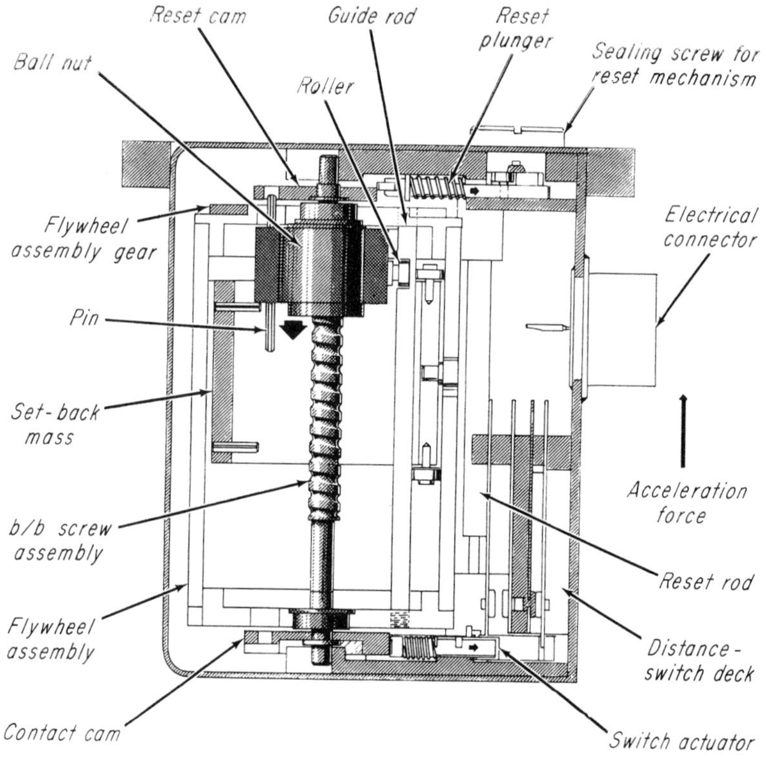

Time-delay switching device integrates time function with missile's linear travel. Purpose is to safely arm the warhead. A strict "minimum G-time" 'system may arm a slow missile too soon for adequate protection of own forces; a fast missile may arrive before warhead is fused. Weight of nut, plus inertia under acceleration will rotate the ball-bearing screw which has a fly wheel on the end. Screw pitch is such that a given number of revolutions of flywheel represents distance traveled. *Globe Industries.*

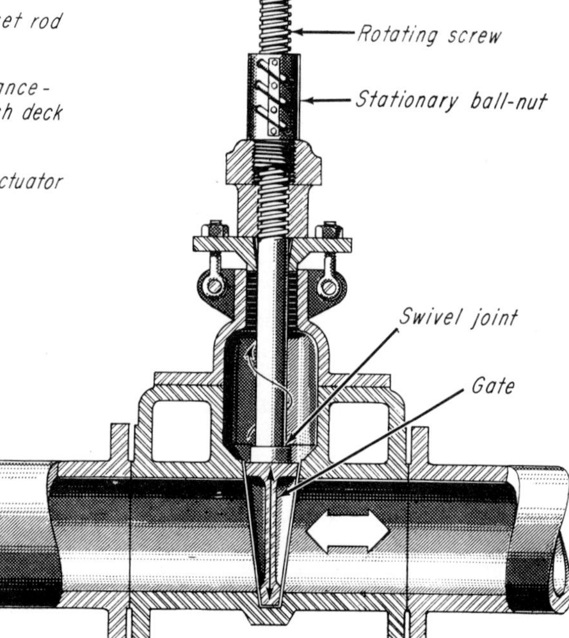

Speedy, easily operated, but more accurate control of flow through valve obtained by rotary motion of screw in stationary ball nut. Screw produces linear movement of gate. The swivel joint eliminates rotary motion between screw and gate.

Chapter 3 STROKE MULTIPLIERS, STRAIGHT-LINERS, AND PARALLEL LINKS

FORCE AND STROKE MULTIPLIERS

PREBEN W. JENSEN

Wide angle oscillator

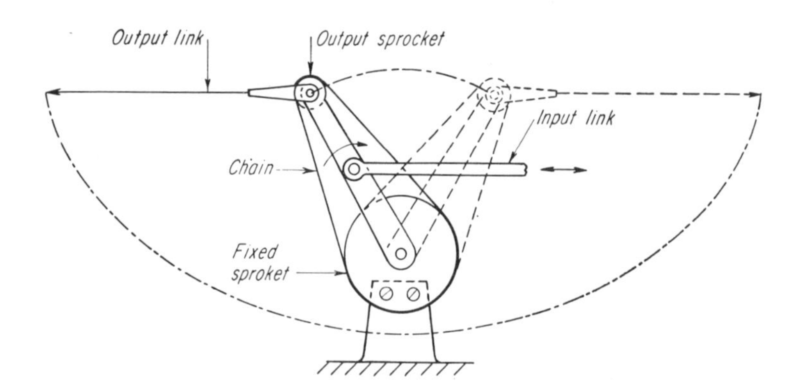

Motion of the input linkage is converted into a wide angle oscillation by means of the two sprockets and chain in the diagram illustrated. An oscillation of 60 deg is converted into 180-deg oscillation.

Gear-sector drive

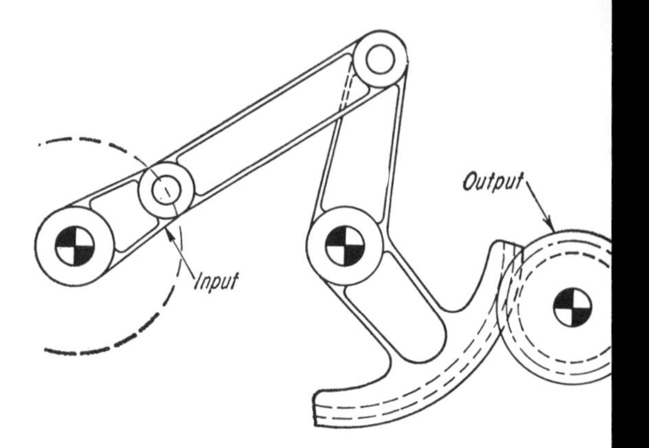

This is actually a four-bar linkage combined with a set of gears. A four-bar linkage usually can get no more than about 120-deg maximum oscillation. The gear segments multiply the oscillation in inverse proportion to the radii of the gears. For the proportions shown, the oscillation is boosted 2½ times.

Angle-doubling drive

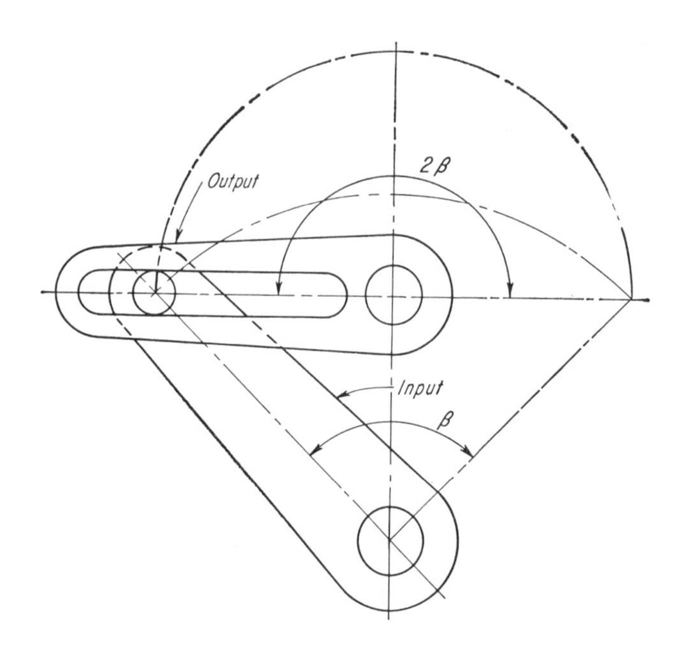

Frequently it is desired to enlarge the oscillating motion β of one machine member into an output oscillation of, say, 2β. If gears are employed, the direction of rotation cannot be the same unless an idler gear is used, in which case the centers of the intput and output shafts cannot be too close. Rotating the input link clockwise causes the output to also follow in a clockwise direction. For a particular set of link proportions, distance between the shafts determines the gain in angle multiplication.

Pulley drive

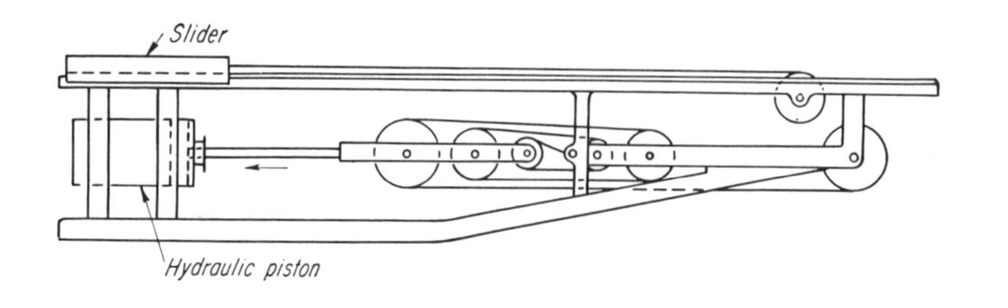

A simple arrangement which multiplies the stroke of a hydraulic piston, causing the slider to propel rapidly to the right for catapulting.

Typewriter drive

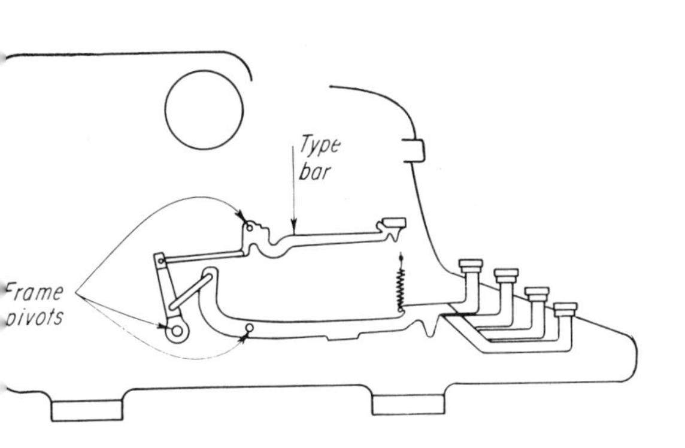

Multiplies the finger force in a typewriter, thus producing a strong hammer action at the roller from a light touch. There are three pivot points attached to the frame. Arrangement of the links is such that the type bar can move in free flight after a key has been struck. The mechanism illustrated is actually two four-bar linkages in series. In certain typewriters, as many as four 4-bar linkages are used in a series.

Double-toggle puncher—

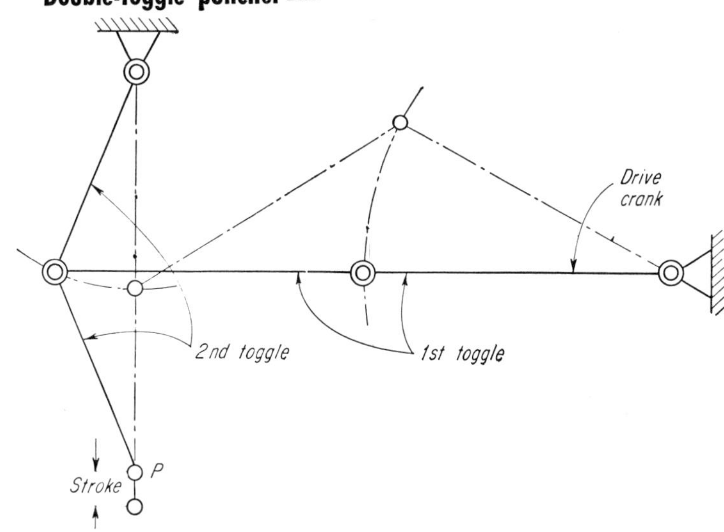

The first toggle keeps point *P* in the raised position even though its weight may be exerting a strong downward force (as in a heavy punch weight). When the drive crank rotates clockwise (driven, say, by a reciprocating mechanism), the second toggle begins to straighten to create a strong punching force.

Gear-rack drive

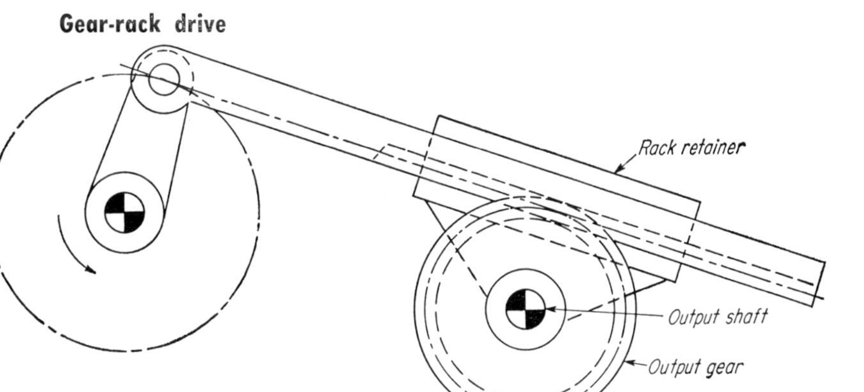

This mechanism is frequently employed to convert the motion of an input crank into a much larger rotation of the output (say, 30 to 360 deg). The crank drives the slider and gear rack, which in turn rotates the output gear.

Chain drive

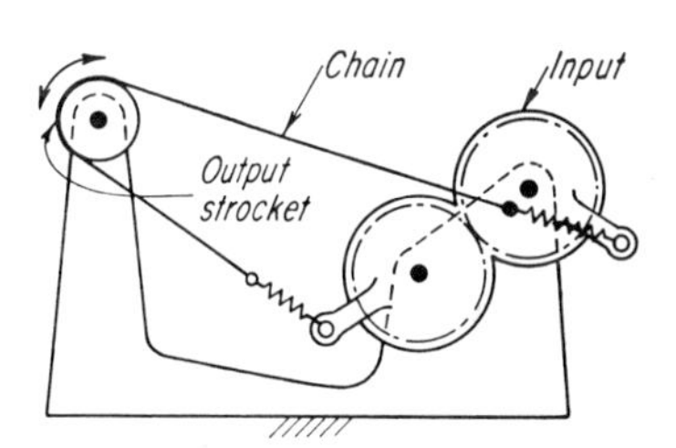

Springs and chains are attached to geared cranks to operate a sprocket output. Depending on the gear ratio, the output will produce a specified oscillation, say two revolutions of output in each direction for each 360 deg of input.

Linkage-train drive

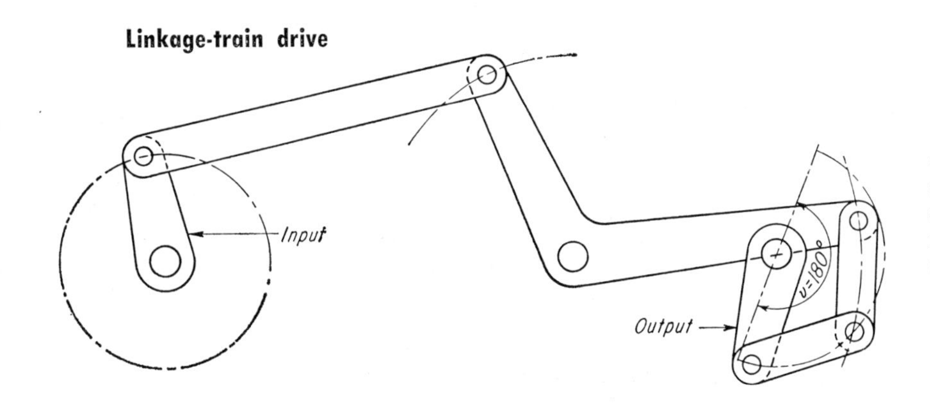

Arranging linkages in series can increase the angle of oscillation. In the case illustrated the oscillating motion of the L-shaped rocker is the input for the second linkage. Final oscillation is 180 deg.

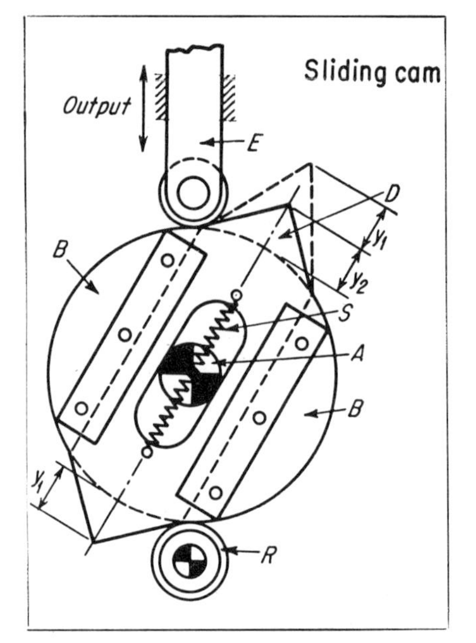

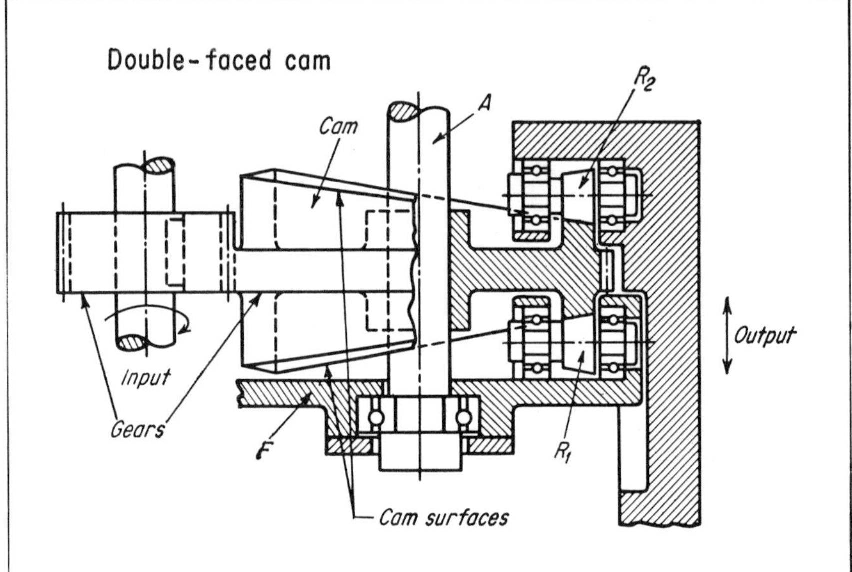

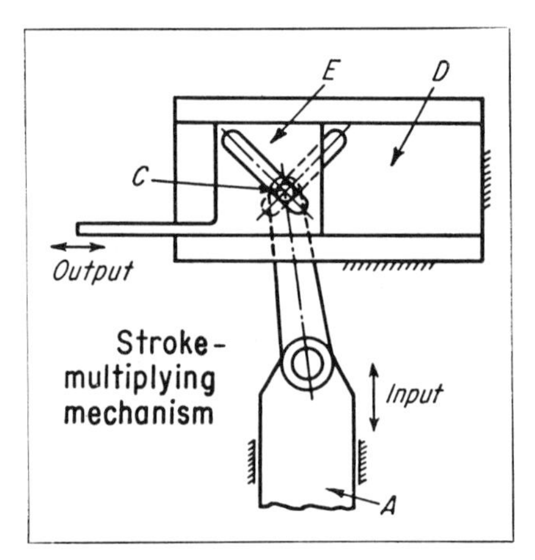

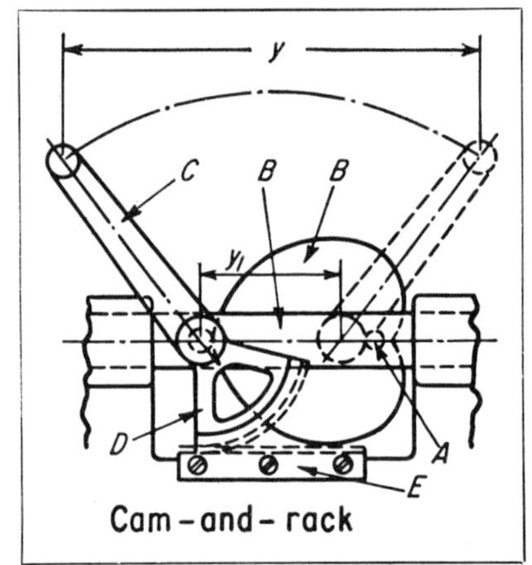

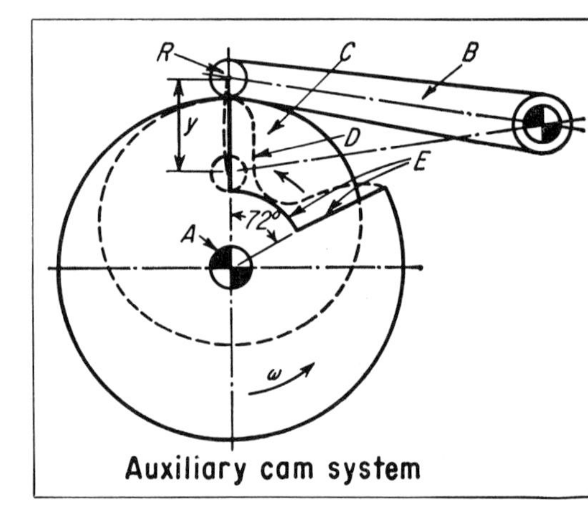

When the pressure angles are too high to satisfy the design requirements, and it is undesirable to enlarge the cam size, then certain devices can be employed to reduce the pressure angles:

Sliding cam — This device is used on a wire-forming machine. Cam *D* has a rather pointed shape because of the special motion required for twisting wires. The machine operates at slow speeds, but the principle employed here is also applicable to high-speed cams.

The original stroke desired is $(y_1 + y_2)$ but this results in a large pressure angle. The stroke therefore is reduced to y_2 on one side of the cam, and a rise of y_1 is added to the other side. Flanges *B* are attached to cam shaft *A*. Cam *D*, a rectangle with the two cam ends (shaded), is shifted upward as it cams off stationary roller *R*, during which the cam follower *E* is being cammed upward by the other end of cam *D*.

Stroke multiplying mechanism — This device is used in power presses. The opposing slots, the first in a fixed member *D* and the second in the movable slide *E*, multiply the motion of the input slide *A* driven by the cam. As *A* moves upward, *E* moves rapidly to the right.

Double-faced cam — This device doubles the stroke, hence reduces the pressure angles to one-half of their original values. Roller R_1 is stationary. When the cam rotates, its bottom surface lifts itself on R_1, while its top surface adds an additional motion to the movable roller R_2. The output is driven linearly by roller R_2 and thus is approximately the sum of the rise of both surfaces.

Cam-and-rack — This device increases the throw of a lever. Cam *B* rotates around *A*. The roller follower travels at distances y_1, during which time gear segment *D* rolls on rack *E*. Thus the output stroke of lever *C* is the sum of transmission and rotation giving the magnified stroke *y*.

Cut-out cam — A rapid rise and fall within 72 deg was desired. This originally called for the cam contour, *D*, but produced severe pressure angles. The condition was improved by providing an additional cam *C* which also rotates around the cam center *A*, but at five times the speed of cam *D* because of a 5:1 gearing arrangement (not shown). The original cam is now completely cut away for the 72 deg (see surfaces *E*). The desired motion, expanded over 360 deg (since 72 x 5 = 360), is now designed into cam *C*. This results in the same pressure angle as would occur if the original cam rise occurred over 360 deg instead of 72 deg.

AUSTRIAN APPROACH to the international typewriter market is the new Austro Standard S-20

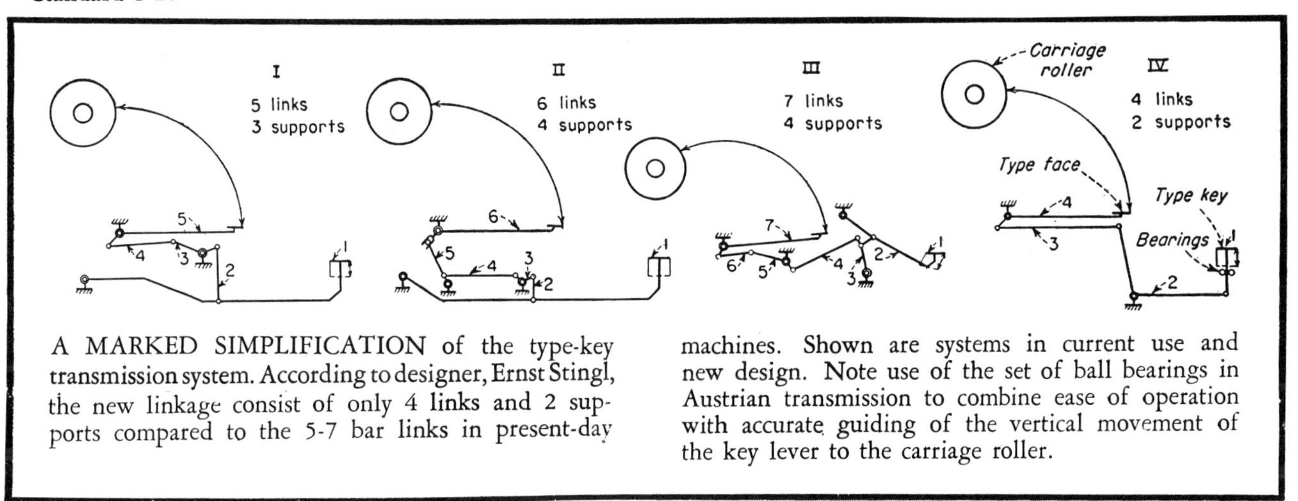

A MARKED SIMPLIFICATION of the type-key transmission system. According to designer, Ernst Stingl, the new linkage consist of only 4 links and 2 supports compared to the 5-7 bar links in present-day machines. Shown are systems in current use and new design. Note use of the set of ball bearings in Austrian transmission to combine ease of operation with accurate guiding of the vertical movement of the key lever to the carriage roller.

Mechanical Linkage Varies Compression In Radial Engine

Through mechanical linkage, the pilot can vary the travel of the pistons during flight. All connecting rods in the new design have identical motions and centrifugal force can be properly balanced. In the conventional radial engine with master connecting rod construction, only one rod has true connecting rod motion and vibrations are necessarily set up.

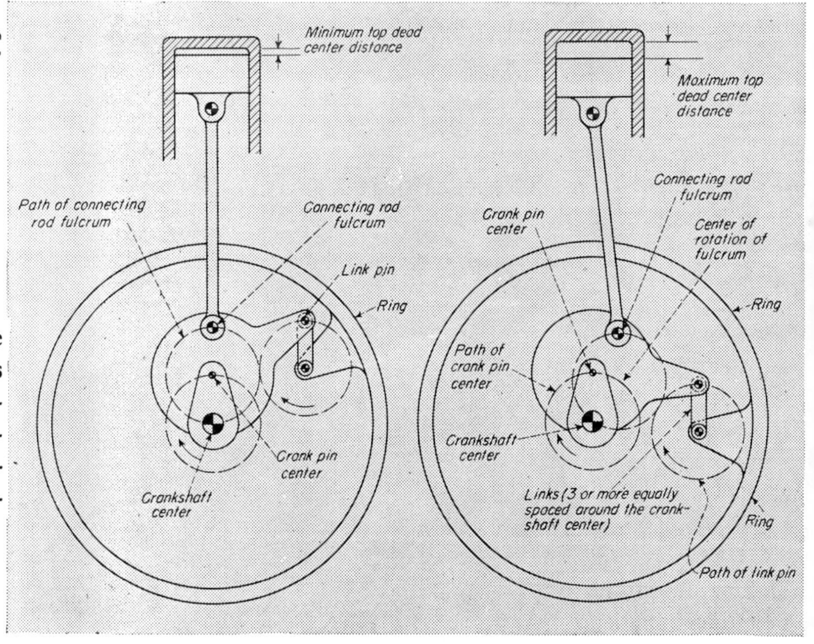

DIAGRAMMATIC VIEW of linkage for one cylinder shows the two extremes to give maximum and minimum compression ratios. Connections for other cylinders are exactly similar. A single link pin connection is shown here; the actual design includes several spaced to maintain balance.

RING IS ROTATED to displace fulcrums angularly around the crankpin center and force the connecting rods to operate in unsymmetrical position to lower compression ratio in cylinders. Linkage shown is for one cylinder only.

Linkages for Multiplying Short Motions

THE ACCOMPANYING SKETCHES show typical mechanisms for multiplying short linear motions, usually converting the linear motion into rotation. Although the particular mechanisms shown are designed to multiply the movements of diaphragms or bellows, the same or similar constructions have possible applications wherever it is required to obtain greatly multiplied motions. These patented transmissions depend on cams, sector gears and pinions, levers and cranks, cord or chain, spiral or screw feed, magnetic attraction, or combinations of these devices.

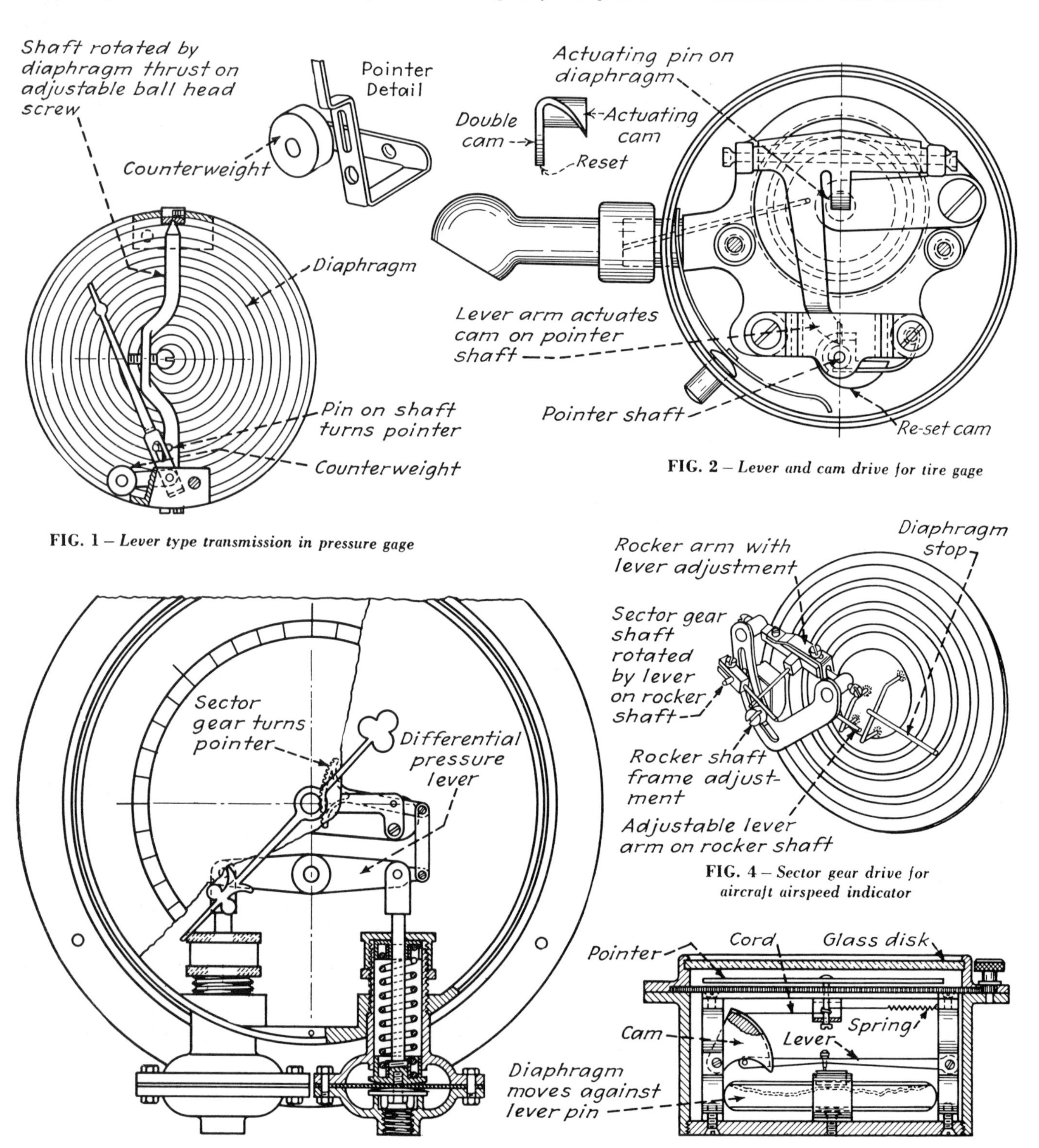

FIG. 1 – *Lever type transmission in pressure gage*

FIG. 2 – *Lever and cam drive for tire gage*

FIG. 3 – *Lever and sector gear in differential pressure gage*

FIG. 4 – *Sector gear drive for aircraft airspeed indicator*

FIG. 5 – *Lever, cam and cord transmission in barometer*

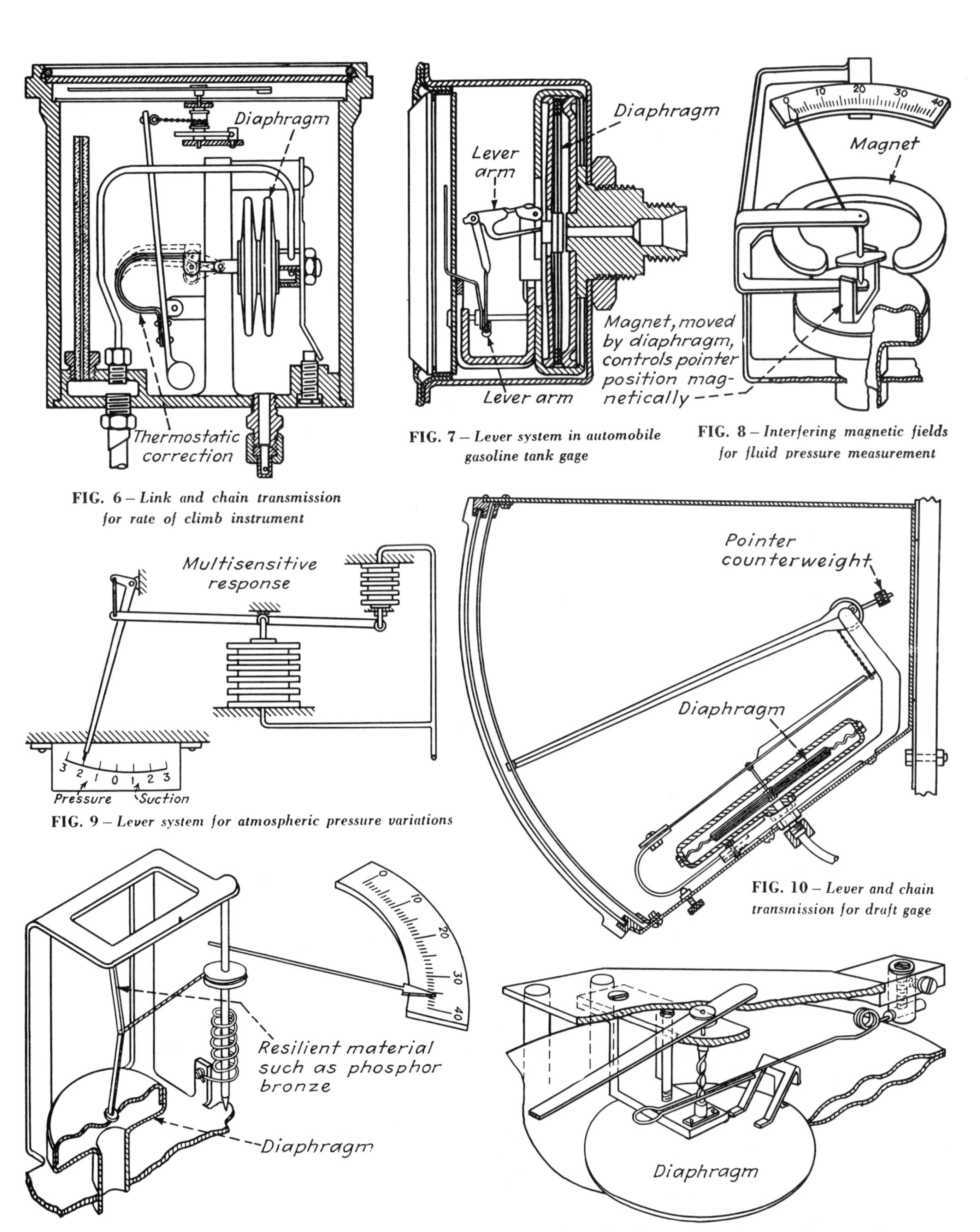

FIG. 6 – *Link and chain transmission for rate of climb instrument*

FIG. 7 – *Lever system in automobile gasoline tank gage*

FIG. 8 – *Interfering magnetic fields for fluid pressure measurement*

FIG. 9 – *Lever system for atmospheric pressure variations*

FIG. 10 – *Lever and chain transmission for draft gage*

FIG. 11 – *Toggle and cord drive for fluid pressure instrument*

FIG. 12 – *Spiral feed transmission for general purpose instrument*

Converting Impulses to Mechanical Movements

TRANSMISSION OF MECHANICAL MOVEMENTS or impulses and conversion of electrical circuit variations to mechanical motion may employ any of a wide variety of mechanical linkages as well as hydraulic, pneumatic, or electrical forces. Patent records show all of the typical devices illustrated on these two pages. While these were designed for transmission of small forces, it is obvious that the parts can be made heavier and modified to withstand heavier loads.

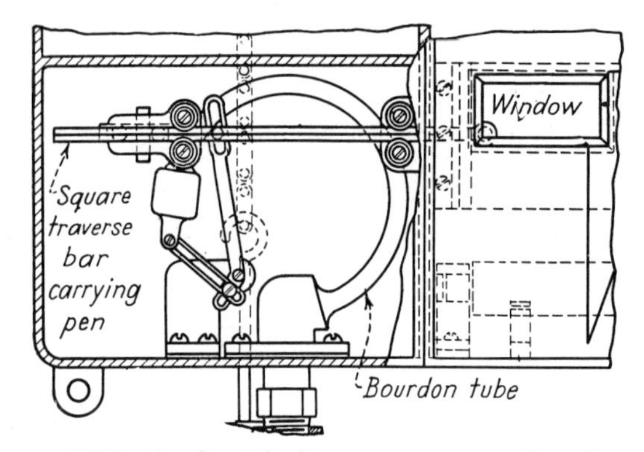

FIG. 1—*Straight line movement produced by linkage from Bourdon tube*

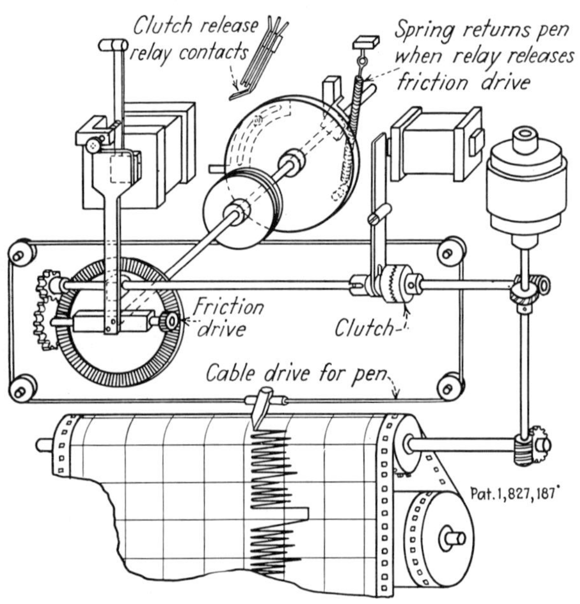

FIG. 2—*Cable drive on recorder of workman efficiency*

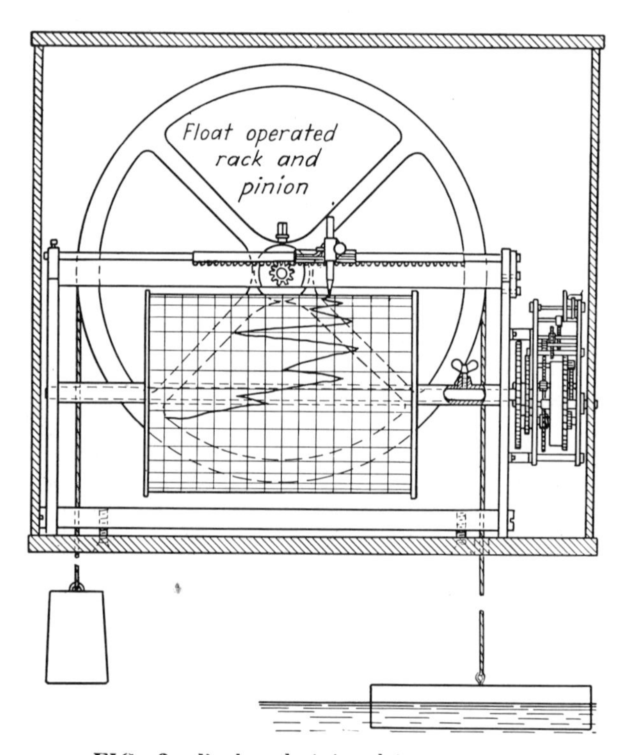

FIG. 3—*Rack and pinion drive to pen*

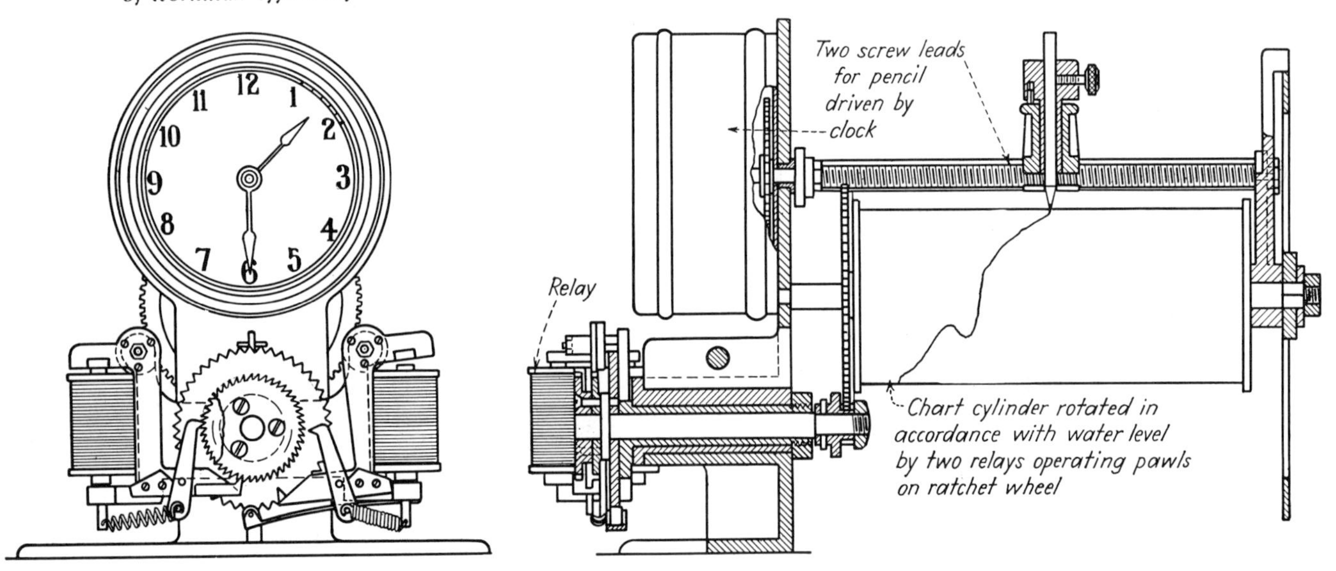

FIG. 4—*Variable reversing chart movement controlled by relays; clock driven pencil*

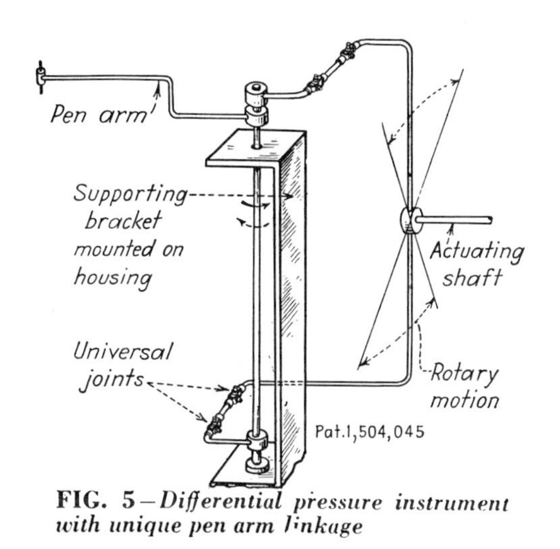

FIG. 5—*Differential pressure instrument with unique pen arm linkage*

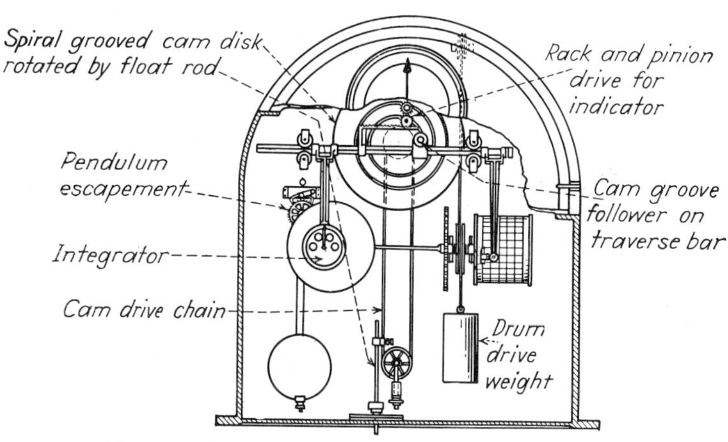

FIG. 6—*Spiral groove cam rotated by float moves pen, integrator arm, and indicator hand*

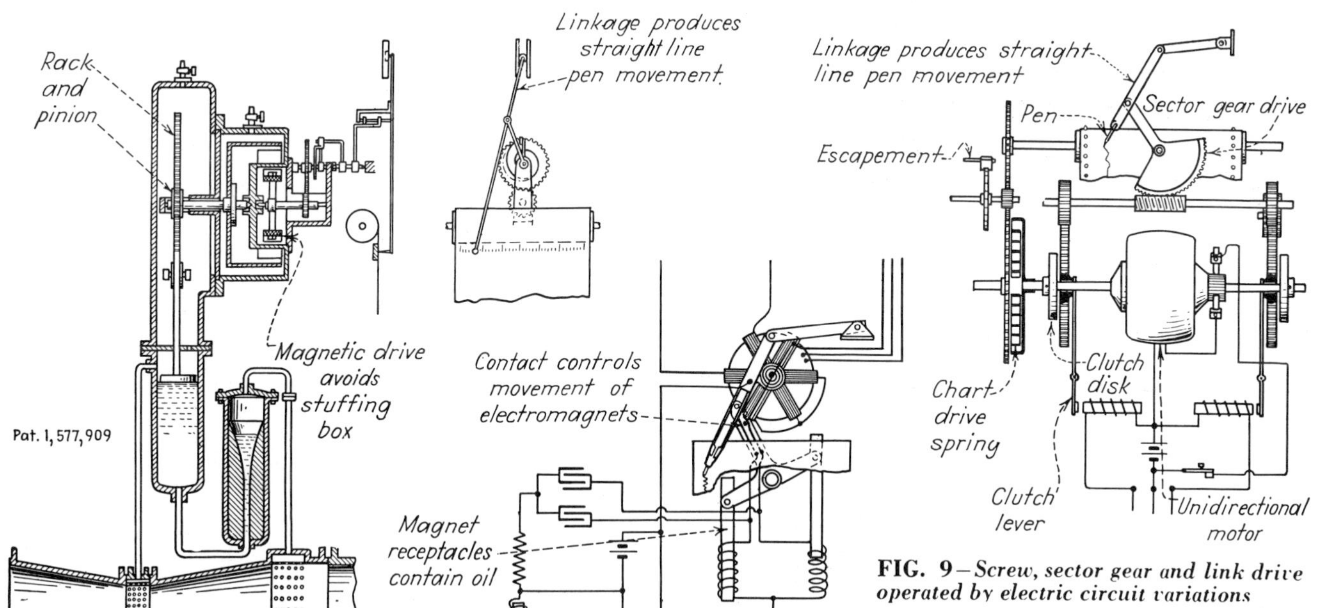

FIG. 7—*Magnetic drive from rack and pinion with straight line movement of pen*

FIG. 8—*Magnetically controlled pen activated by contacts on T-arm and armature*

FIG. 9—*Screw, sector gear and link drive operated by electric circuit variations*

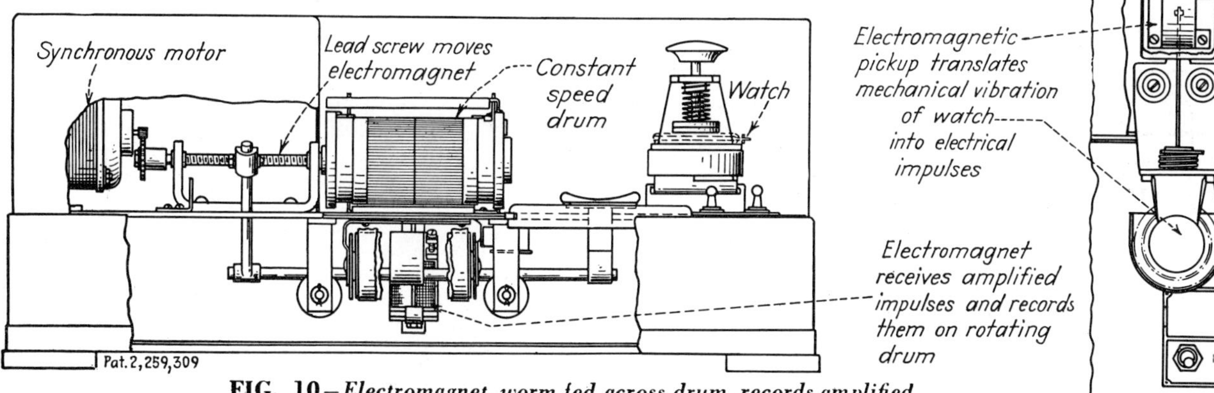

FIG. 10—*Electromagnet, worm fed across drum, records amplified impulses of watch ticks to check time keeping accuracy*

Linkages for Accelerating and Decelerating Linear Strokes

When ordinary rotary cams cannot be conveniently applied, the mechanisms here presented, or adaptations of them, offer a variety of interesting possibilities for obtaining either acceleration or deceleration, or both

JOHN E. HYLER
Peoria, Ill.

FIG. 1—Slide block moves at constant rate of reciprocating travel and carries both a pin for mounting link *B* and a stud shaft on which the pinion is freely mounted. Pinion carries a crankpin for mounting link *D* and engages stationary rack, the pinion may make one complete revolution at each forward stroke of slide block and another in opposite direction on the return—or any portion of one revolution in any specific instance. Many variations can be obtained by making the connection of link *F* adjustable lengthwise along link that operates it, by making crankpin radially adjustable, or by making both adjustable.

FIG. 2—Drive rod, reciprocating at constant rate, rocks link *BC* about pivot on stationary block and, through effect of toggle, causes decelerative motion of driven link. As drive rod advances toward right, toggle is actuated by encountering abutment and the slotted link *BC* slides on its pivot while turning. This lengthens arm *B* and shortens arm *C* of link *BC*, with decelerative effect on driven link. Toggle is spring-returned on the return stroke and effect on driven link is then accelerative.

FIG. 3—Same direction of travel for both the drive rod and the driven link

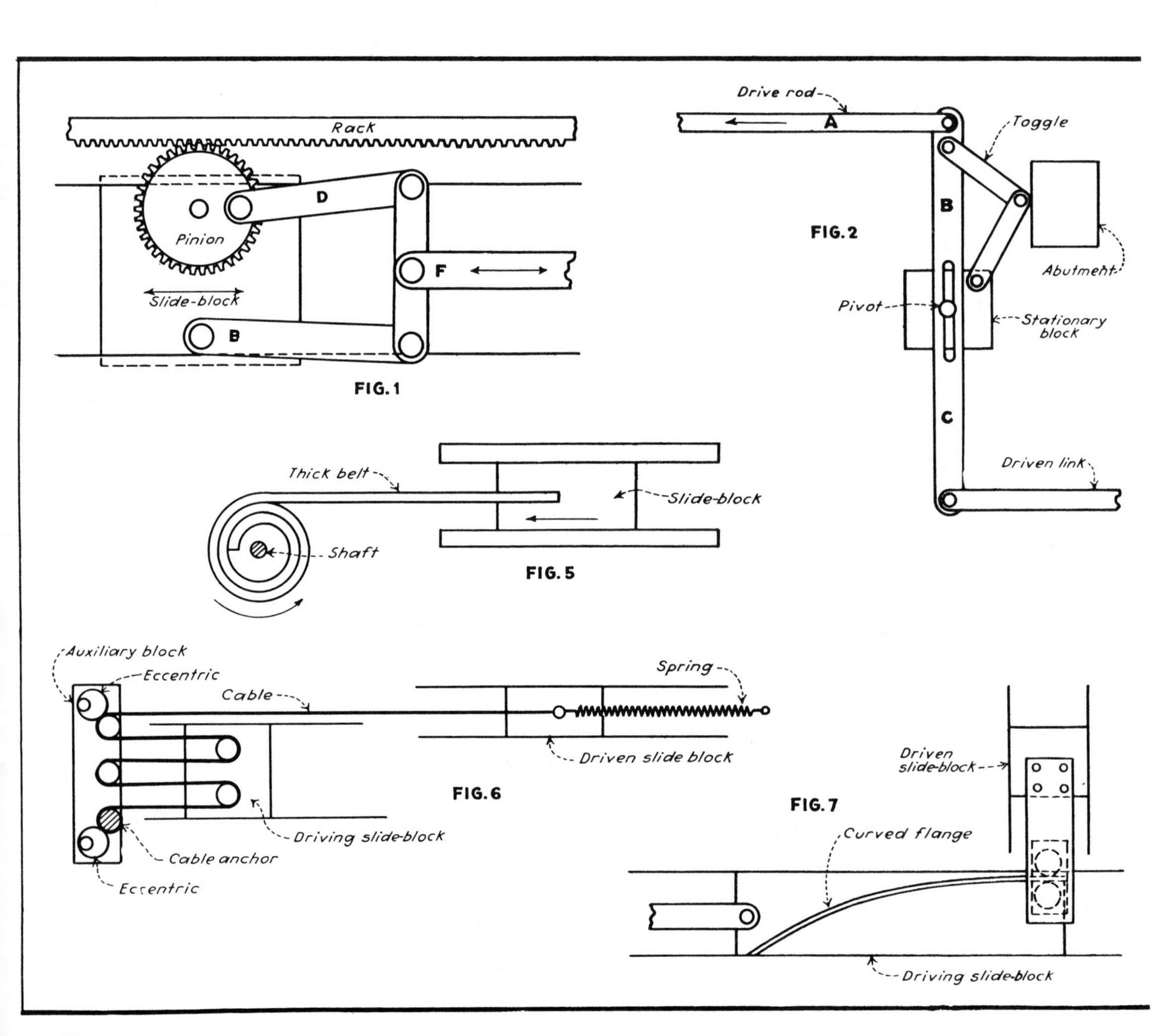

is provided by this variation of the preceding mechanism. Here, acceleration is in direction of arrows and deceleration occurs on return stroke. Accelerative effect becomes less as toggle flattens.

Fig. 4—Bellcrank motion is accelerated as rollers are spread apart by curved member on end of drive rod, thereby in turn accelerating motion of slide block. Driven elements must be spring-returned to close system.

Fig. 5—Constant-speed shaft winds up thick belt, or similar flexible member, and increase in effective radius causes accelerative motion of slide block. Must be spring or weight-returned on reversal.

Fig. 6 — Auxiliary block, carrying sheaves for cable which runs between driving and driven slide blocks, is mounted on two synchronized eccentrics. Motion of driven block is equal to length of cable paid out over sheaves resulting from additive motions of the driving and the auxiliary blocks.

Fig. 7—Curved flange on driving slide block is straddled by rollers pivotally mounted in member connected to driven slide block. Flange can be curved to give desired acceleration or deceleration, and mechanism is self-returned.

Fig. 8—Stepped acceleration of the driven slide block is effected as each of the three reciprocating sheaves progressively engages the cable. When the third acceleration step is reached, the driven slide block moves six times faster than the drive rod.

Fig. 9—Form-turned nut, slotted to travel on rider, is propelled by reversing screw shaft, thus moving concave roller up and down to accelerate or decelerate slide block.

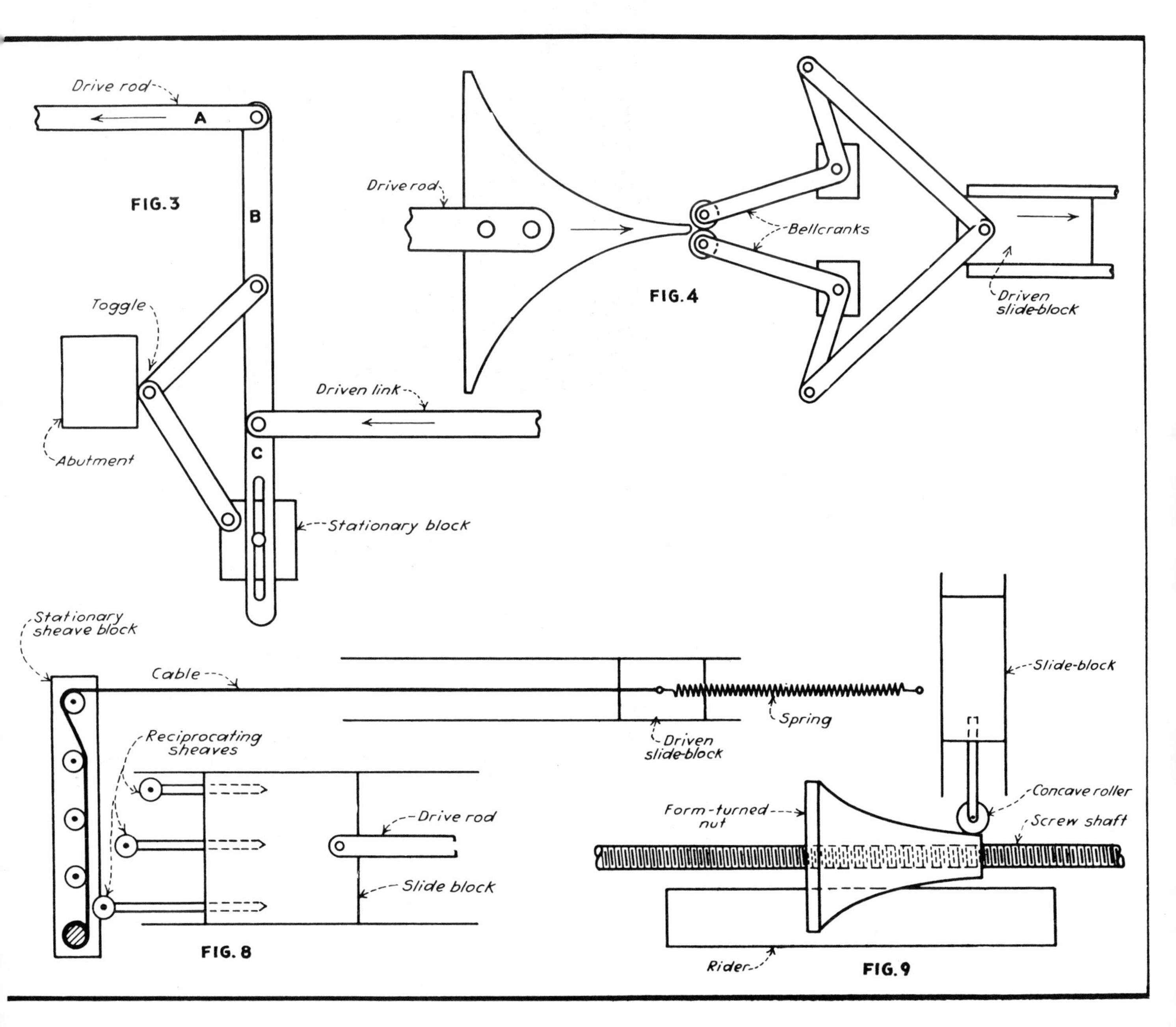

5 LINKAGES for

These devices convert rotary to straight-line motion without the need for guides.

SIGMUND RAPPAPORT, *kinematician*
Ford Instrument Co, Div of Sperry Rand Co

Evans' linkage . . .
has oscillating drive-arm that should have a maximum operating angle of about 40°. For a relatively short guide-way, the reciprocating output stroke is large. Output motion is on a true straight line in true harmonic motion. If an exact straight-line motion is not required, however, a link can replace the slide. The longer this link, the closer does the output motion approach that of a true straight line—if link-length equals output stroke, deviation from straight-line motion is only 0.03% of output stroke.

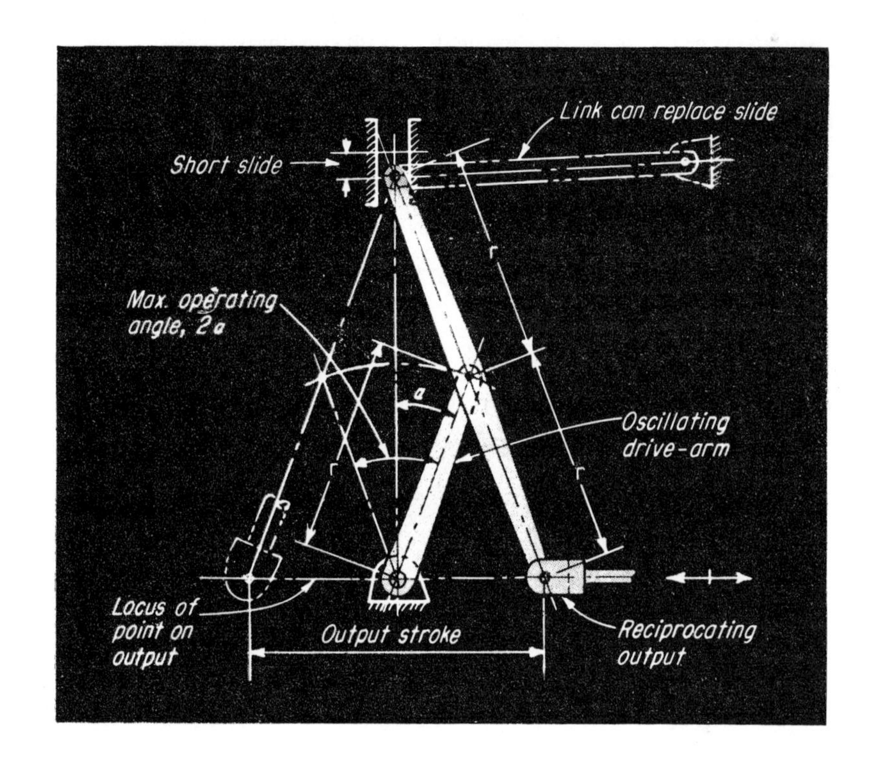

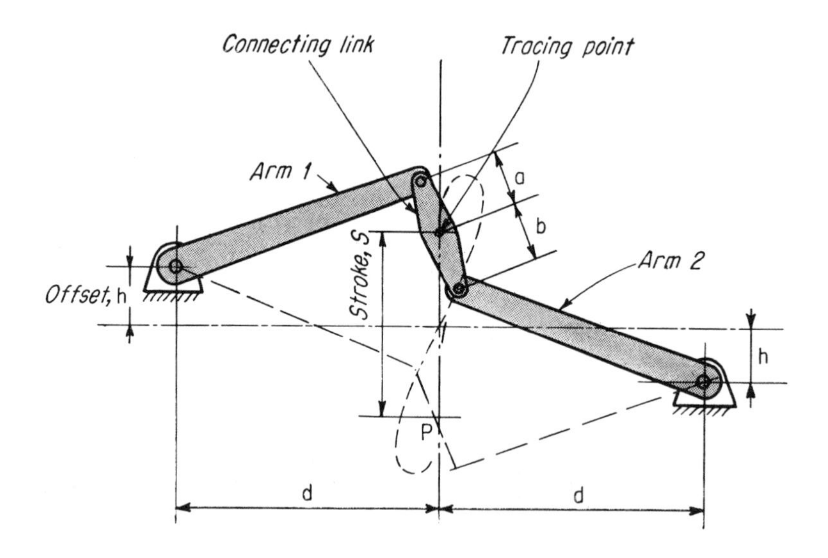

Simplified Watt's linkage . . .
generates an approximate straight-line motion. If the two arms are equally long, the tracing point describes a symmetrical figure 8 with an almost straight line throughout the stroke length. The straightest and longest stroke occurs when the connecting-link length is about ⅔ of the stroke, and arm length is 1.5S. Offset should equal half the connecting-link length. If the arms are unequal, one branch of the figure-8 curve is straighter than the other. It is straightest when a/b equals (arm 2)/(arm 1).

STRAIGHT-LINE MOTION

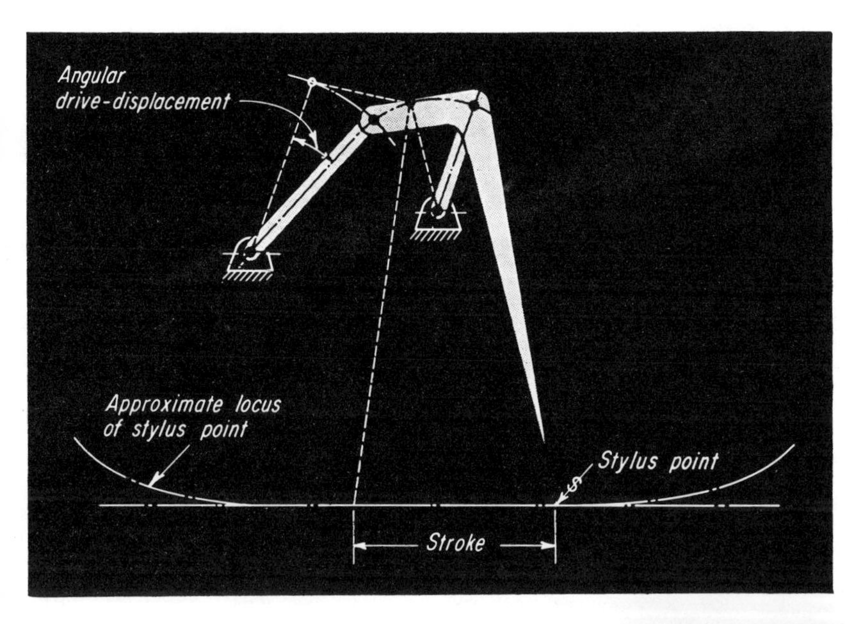

Four-bar linkage . . .
produces approximately straight-line motion. This arrangement provides motion for the stylus on self-registering measuring instruments. A comparatively small drive-displacement results in a long, almost-straight line.

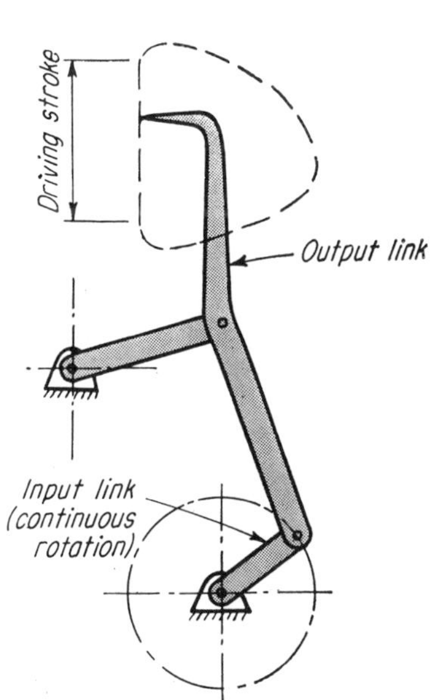

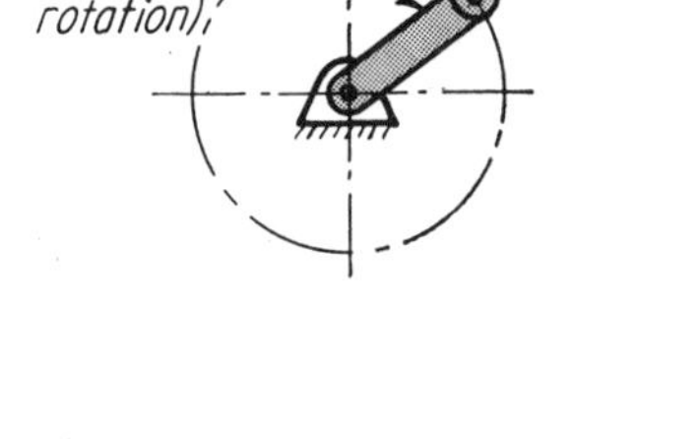

D-drive . . .
results when linkage arms are arranged as shown here. Output-link point describes a path resembling the letter *D*, thus it contains a straight portion as part of its cycle. Motion is ideal for quick engagement and disengagement before and after a straight driving-stroke. Example, the intermittent film-drive in movie-film projectors.

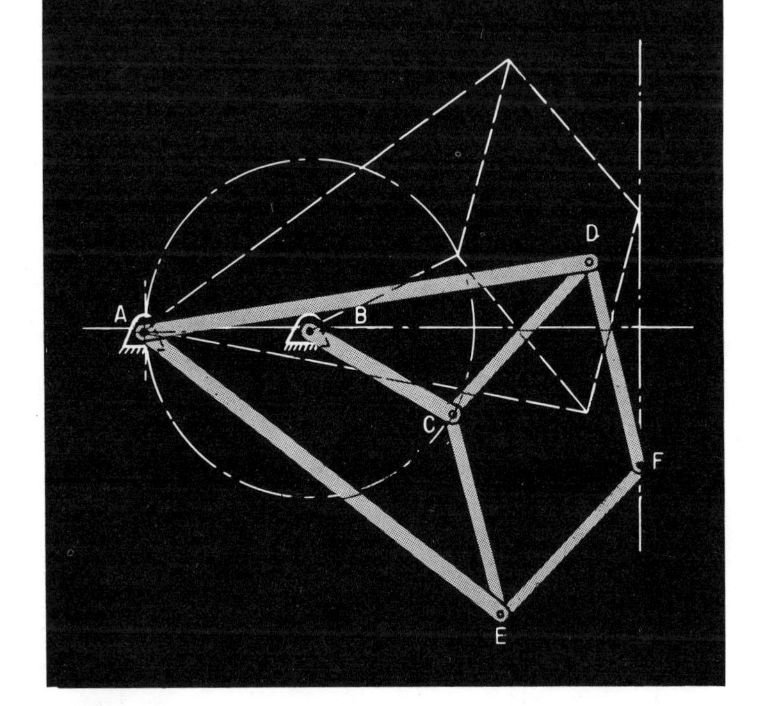

The "Peaucellier cell" . . .
was first solution to the classical problem of generating a straight line with a linkage. Its basis: within the physical limits of the motion, $AC \times AF$ remains constant. Curves described by C and F are, therefore, inverse; if C describes a circle that goes through A, then F will describe a circle of infinite radius—a straight line, perpendicular to *AB*. The only requirements are: $AB = BC$; $AD = AE$; and *CD, DF, FE, EC* are all equal. The linkage can be used to generate circular arcs of large radius by locating A outside the circular path of C.

Linkage Ratios for Straight-Line Mechanisms

ALL straight-line motion mechanisms are based on a few simple geometrical relations, yet offer considerable variety in design. In various modifications of the isosceles and pantograph linkages, resultant motion is usually parallel to motivating force, and lengths of travel remain proportional. Some are designed for short ranges of motion, and paths are only approximately proportional. Where short arcs are involved it is possible to use a bell-crank and link to change direction of motion with little effect upon proportionality of movements at input and output points of linkage.

Straight-line linkages of the Watt, Roberts, Tchebicheff and Peaucellier cells have a more complex relation between lengths of path. The Walschaert valve gear linkage illustrates one means of controlling length of travel and timing of a guided straight-line movement by linkage adjustment rather than the usual cams.

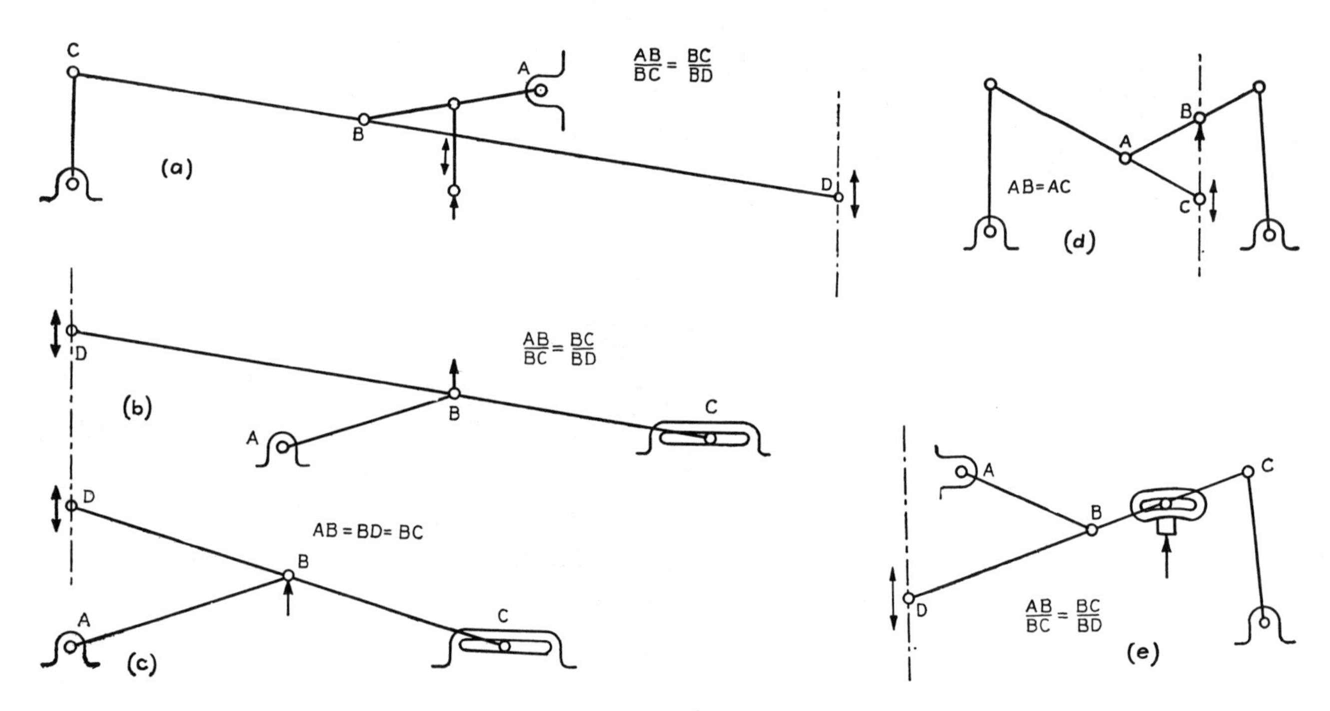

Fig. 1—(a), (b), (c), (d), (e)—Isosceles linkages.

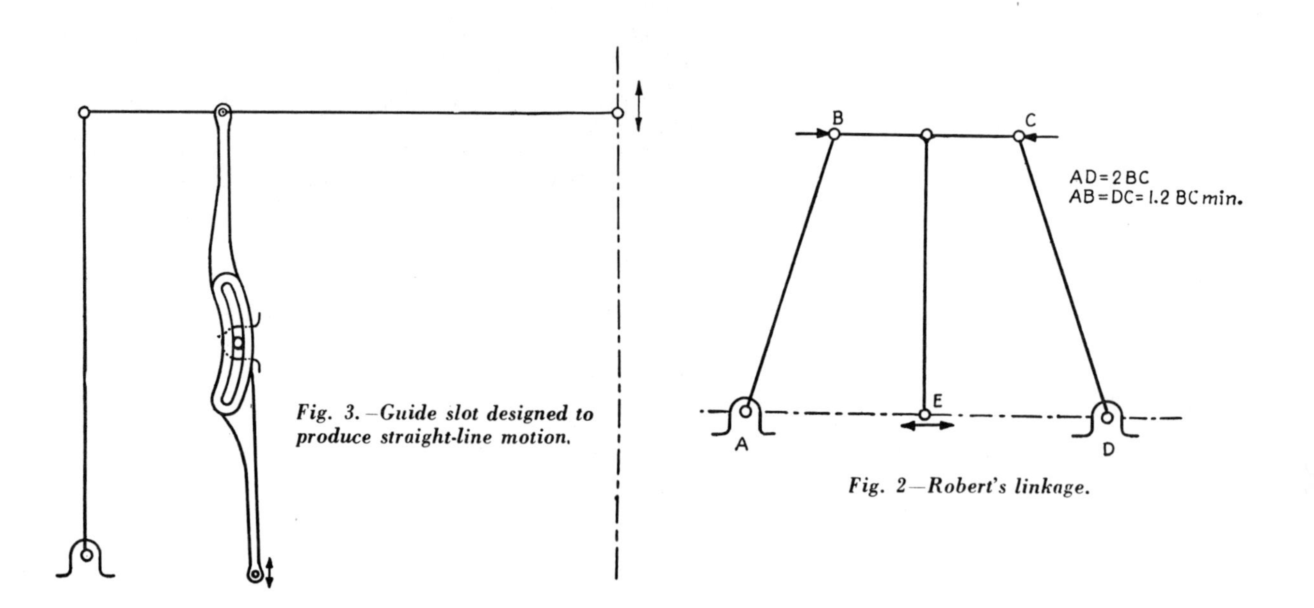

Fig. 3.—Guide slot designed to produce straight-line motion.

Fig. 2—Robert's linkage.

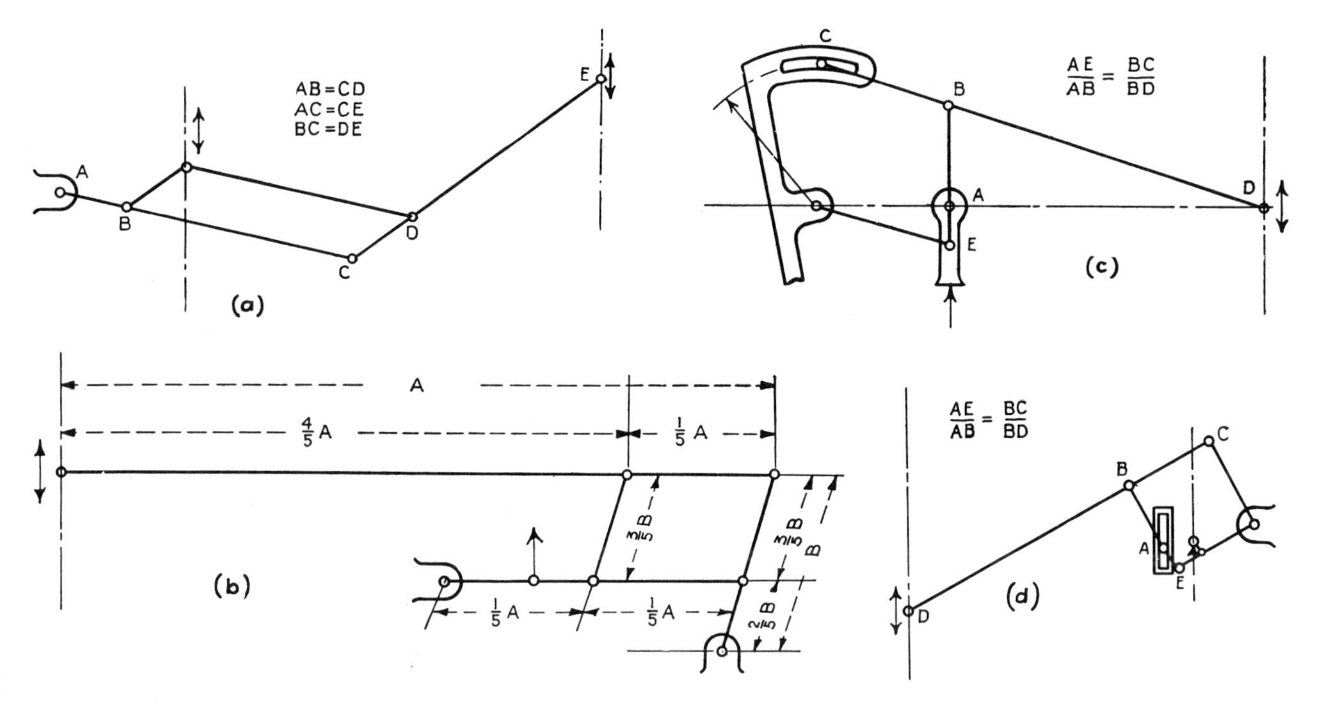

Fig. 4.—(a), (b), (c), (d)—Pantograph linkages.

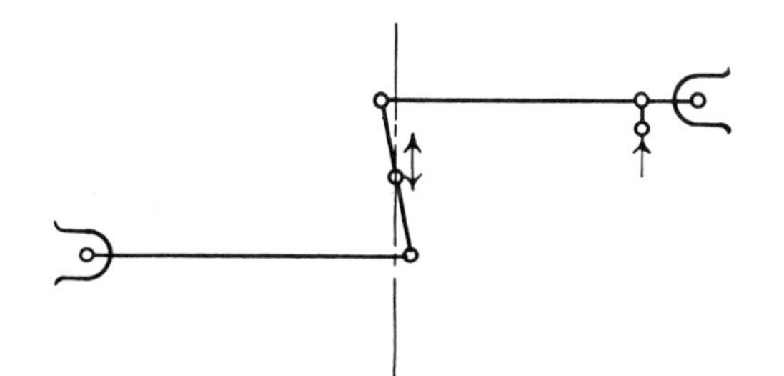

Fig. 5—Watt's linkage.

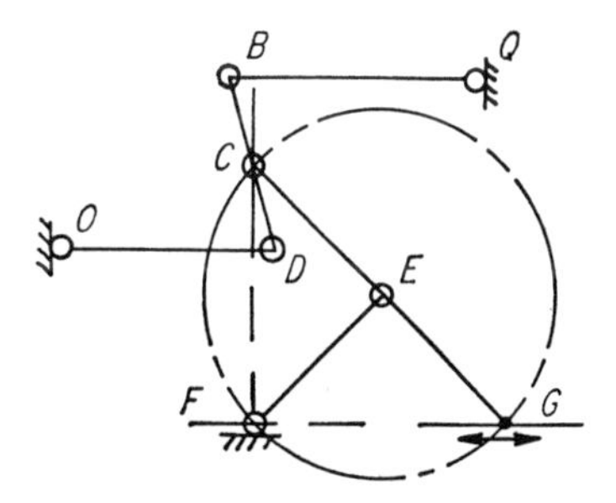

Fig. 6—The Tchebicheff combination of the Watt and Evans mechanisms.
ODCBQ ***the Watt linkage.***
CEFG ***the Evans linkage.***

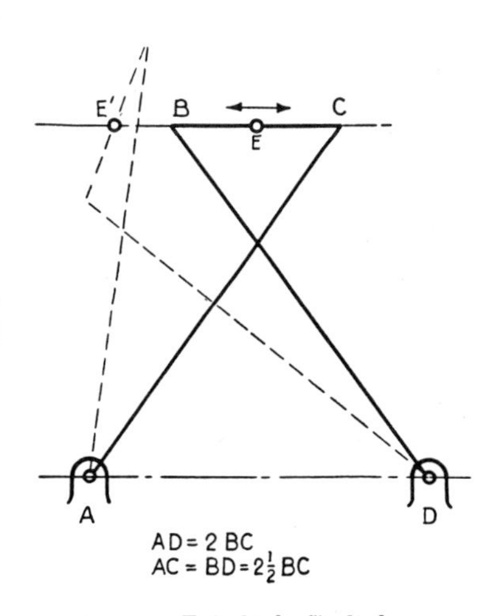

Fig. 7—Tchebicheff's linkage.

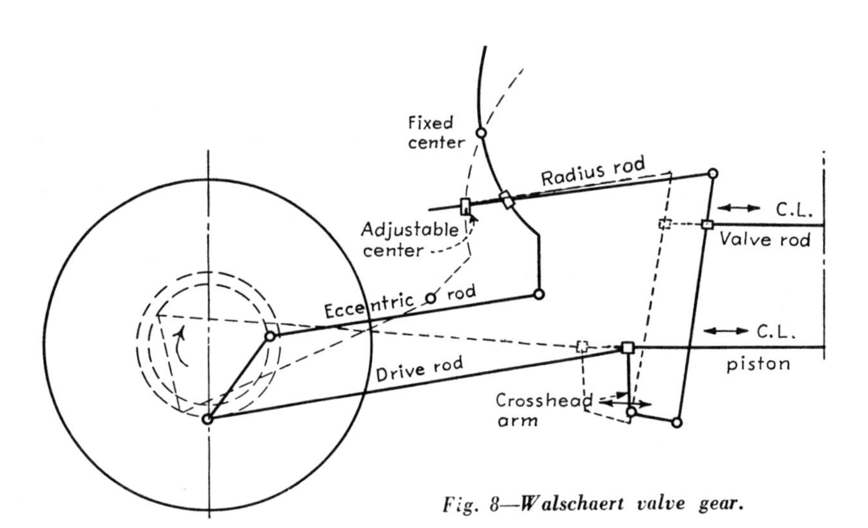

Fig. 8—Walschaert valve gear.

FIG. 1—No linkages or guides are used in this modified hypocyclic drive which is relatively small in relation to the length of its stroke. The sun gear of pitch diameter D is stationary. The drive shaft, which turns the T-shaped arm, is concentric with this gear. The idler and planet gears, the latter having a pitch diameter of $D/2$, rotate freely on pivots in the arm extensions. Pitch diameter of the idler is of no geometrical significance, although this gear does have an important mechanical function. It reverses the rotation of the planet gear, thus producing true hypocyclic motion with ordinary spur gears only. Such an arrangement occupies only about half as much space as does an equivalent mechanism containing an internal gear. Center distance R is the sum of $D/2$, $D/4$ and an arbitrary distance d, determined by a particular application. Points A and B on the driven link, which is fixed to the planet, describe straight-line paths through a stroke of $4R$. All points between A and B trace ellipses, while the line AB envelopes an astroid.

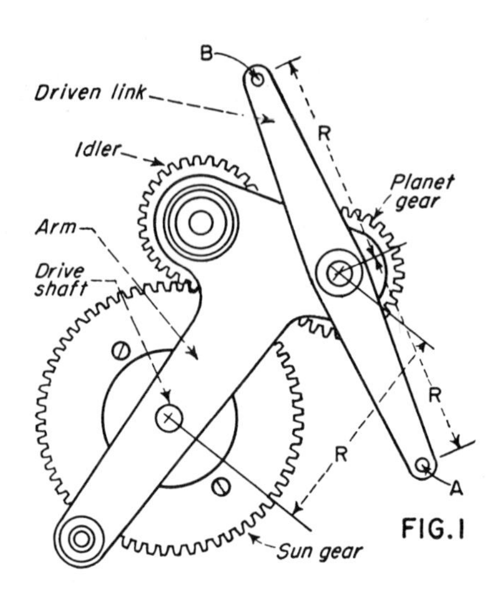

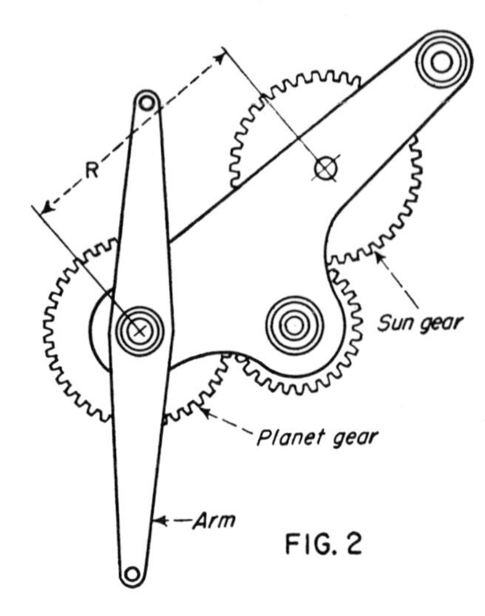

Parallel Motion

FIG. 2—A slight modification of the mechanism in Fig. 1 will produce another type of useful motion. If the planet gear has the same diameter as that of the sum gear, the arm will remain parallel to itself throughout the complete cycle. All points on the arm will thereby describe circles of radius R. Here again, the position and diameter of the idler gear are of no geometrical importance. This mechanism can be used, for example, to cross-perforate a uniformly moving paper web. The value for R is chosen such that $2\pi R$, or the circumference of the circle described by the needle carrier, equals the desired distance between successive lines of perforations. If the center distance R is made adjustable, the spacing of perforated lines can be varied as desired.

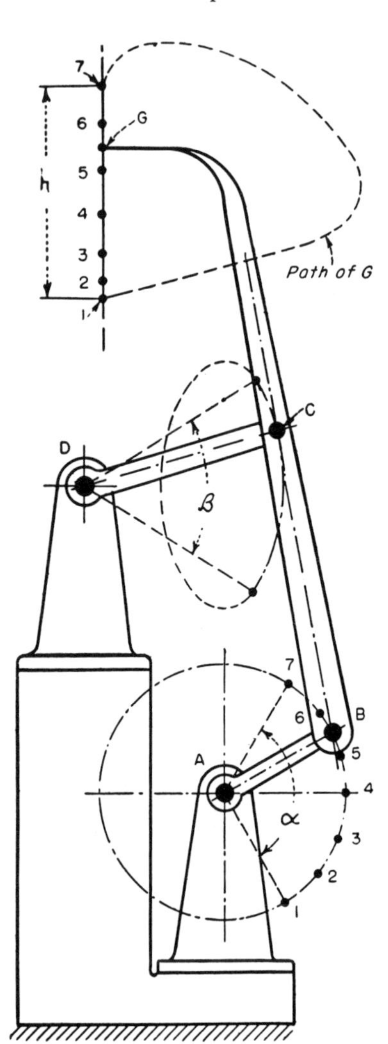

—The task of describing a "D" curve.

Hint for designing: Start with a straight line portion of the path of G, replace the oval arc of C by an arc of the osculating circle, thus determining the length of link DC.

This mechanism is intended to accomplish the following: (1) Film hook, while moving the film strip, must describe very nearly a straight line; (2) Engagement and disengagement of the hook with the perforation of the film must take place in a direction approximately normal to the film; (3) Engagement and disengagement should be shock free. Slight changes in the shape of the guiding slot f enable the designer to vary the shape of the output curve as well as the velocity diagram appreciably.

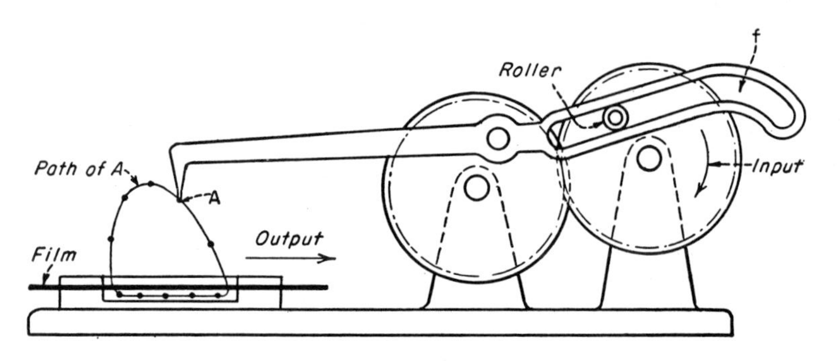

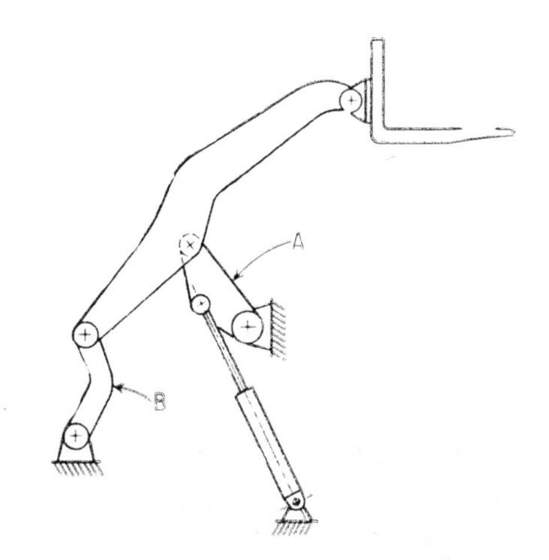

No Mast on this Lift Truck...

. . . Yet the load moves straight up instead of swinging through an arc. It's done with compensating links that shift the pivots as load arm is raised.

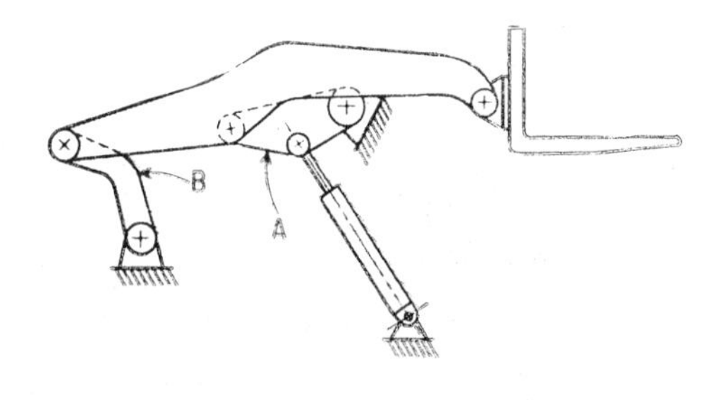

"Elbolift" truck weighs less; stability, maneuverability and visibility are improved. The lift was designed to meet a particular need of the canning industry—handling coils of rolled tinplate stock shipped on pallets. Continental Can Co. wanted a truck that would have sufficient head and elbow room inside boxcars. It presented the problem to Automatic Transportation Co., Chicago, and B. I. Ulinski, director of design and development, devised this lifting mechanism that met requirements. The truck is also used for loading the can-making machines—its fork blades fold to form a ram that fits the coil's eye.

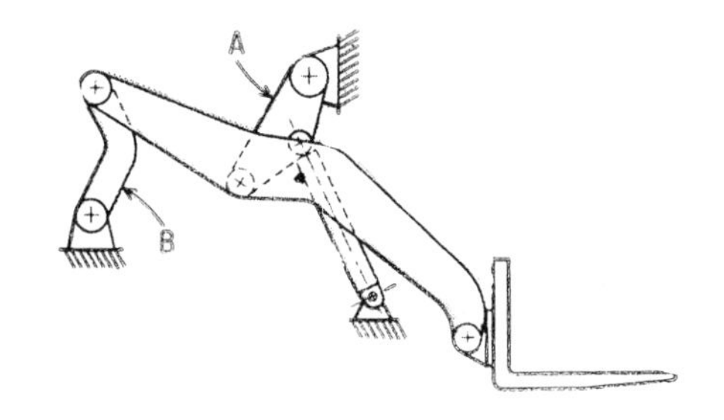

Coils are strapped . . .
to 52-in.-sq pallets that are loaded in boxcars. Truck had to be designed to enter the 8 x 8-ft boxcar door, pick up pallets as near as 1 ft, 7 in. to the doors and maneuver inside the car's 9-ft, 2-in. width. Coils weigh 15,000 lb each and are stacked three-high when unloaded. Stacking required lifting forks to tilt from 10° back to 5° forward and have an 8-in. side-shifting motion, for stack alignment. Maneuverability was obtained with rear-wheel steering.

Straight-line lifting action is result of adding links *A* and *B* to the lifting mechanism. The load arm is raised by link *A* which arcs rearward as the hydraulic cylinder pushes it upward. Relative lengths of links *A* and *B* and the load arm let this action counteract the forward arcing motion of the swinging load arm. Load actually moves to rear as it is lifted.

5 CARDAN-GEAR

SIGMUND RAPPAPORT, *kinematician*
Ford Instrument Co, Div of Sperry Rand Co

These gearing arrangements convert rotation into straight-line motion, without need for slideways.

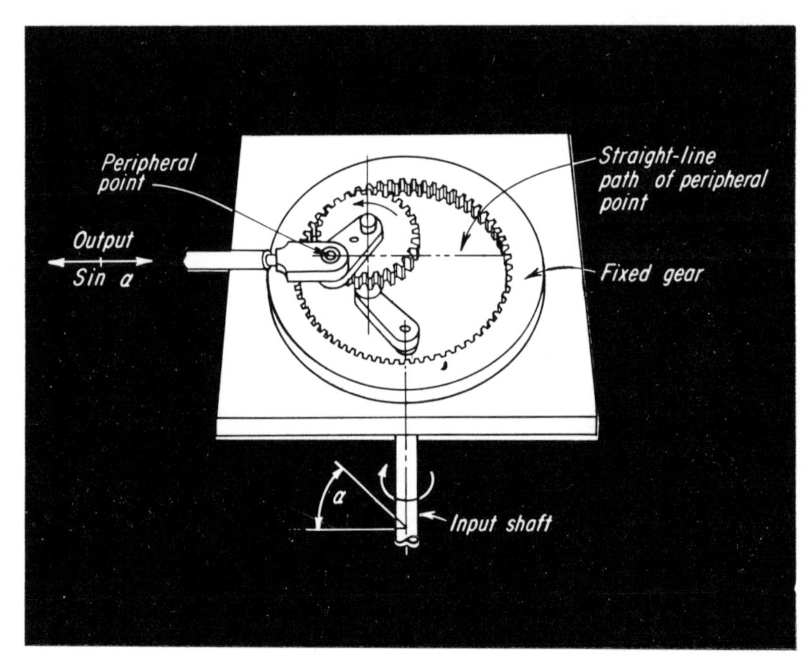

Cardan gearing . . .
works on the principle that any point on the periphery of a circle rolling on the inside of another circle describes, in general, a hypocycloid. This curve degenerates into a true straight line (diameter of the larger circle) if diameters of both circles are in the ratio of 1:2. Rotation of input shaft causes small gear to roll around the inside of the fixed gear. A pin located on pitch circle of the small gear describes a straight line. Its linear displacement is proportional to the theoretically true sine or cosine of the angle through which the input shaft is rotated. Among other applications, Cardan gearing is used in computers, as a component solver (angle resolver).

Cardan gearing and Scotch yoke . . .
in combination provide an adjustable stroke. Angular position of outer gear is adjustable. Adjusted stroke equals the projection of the large dia, along which the drive pin travels, upon the Scotch-yoke centerline. Yoke motion is simple harmonic.

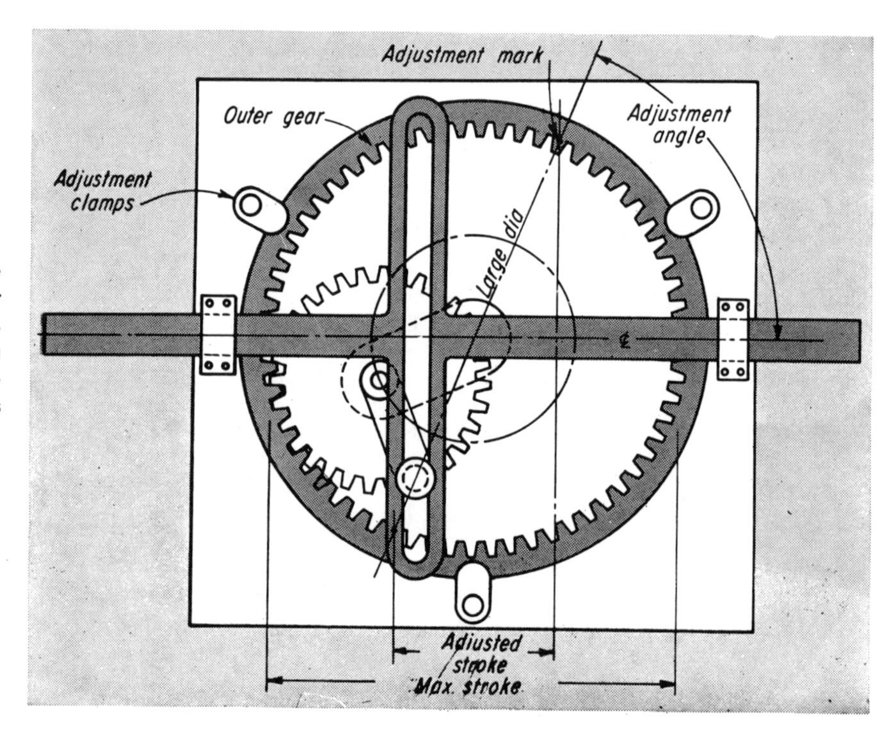

MECHANISMS

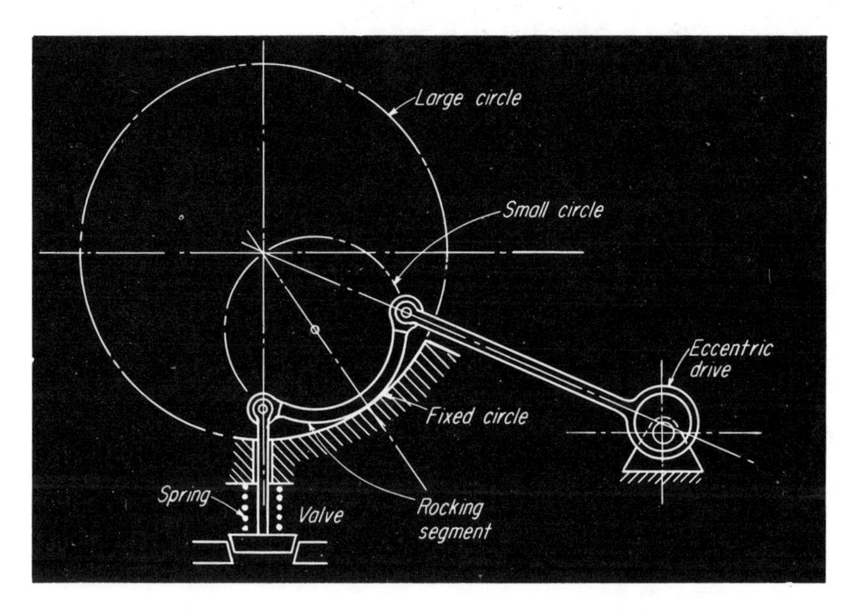

Valve drive . . .
exemplifies how Cardan principle may be applied. A segment of the smaller circle rocks to and fro on a circular segment whose radius is twice as large. Input and output rods are each attached to points on the small circle. Both these points describe straight lines. Guide of the valve rod prevents the rocking member from slipping.

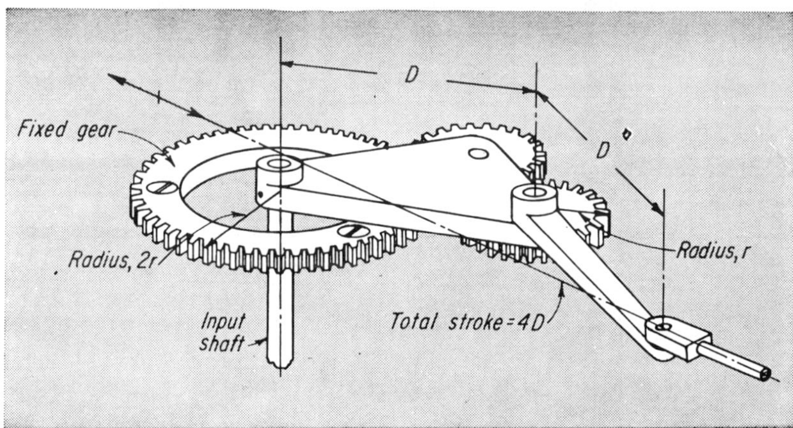

Simplified Cardan principle . . .
does away with need for the relatively expensive internal gear. Here, only spur gears may be used and the basic requirements should be met, i.e. the 1:2 ratio and the proper direction of rotation. Latter requirement is easily achieved by introducing an idler gear, whose size is immaterial. In addition to cheapness, this drive delivers a far larger stroke for the comparative size of its gears.

Rearrangement of gearing . . .
in (4) results in another useful motion. If the fixed sun-gear and planet pinion are in the ratio of 1:1, then an arm fixed to the planet shaft will stay parallel to itself during rotation, while any point on it describes a circle of radius *R*. An example of application: in conjugate pairs for punching holes on moving webs of paper.

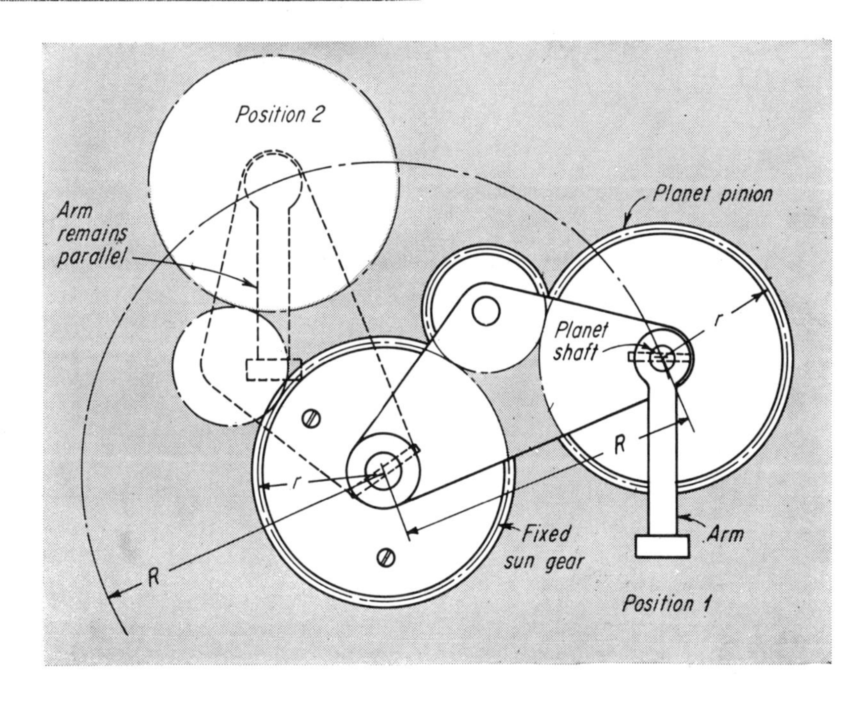

10 ways to change
STRAIGHT-LINE DIRECTION

Arrangements of linkages, slides, friction drives and gears that can be the basis of many ingenious devices.

FEDERICO STRASSER

Linkages

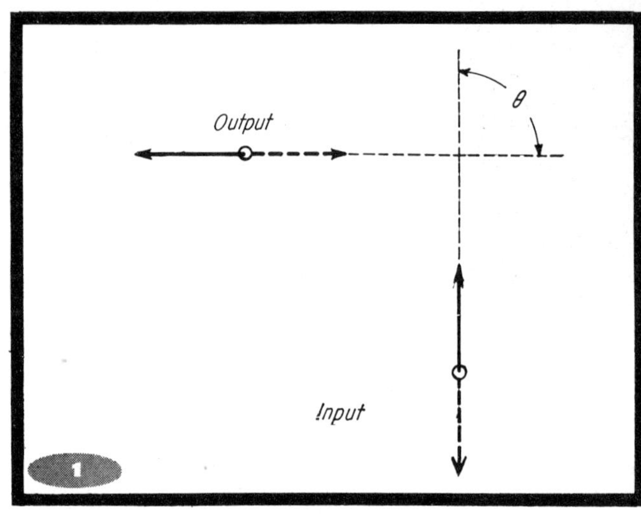

Basic problem (θ is generally close to 90°)

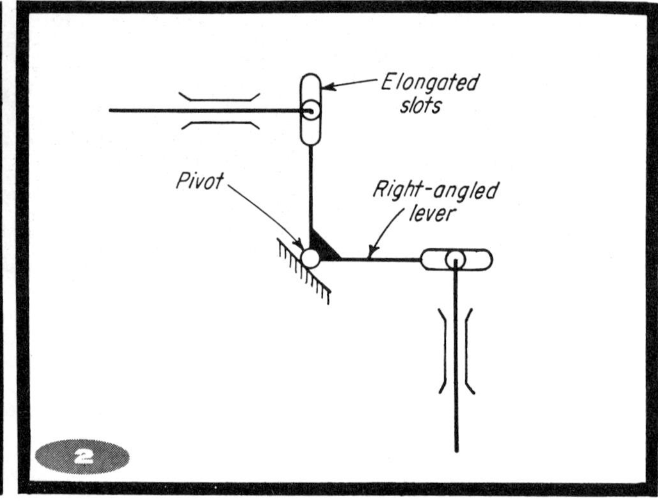

Slotted lever

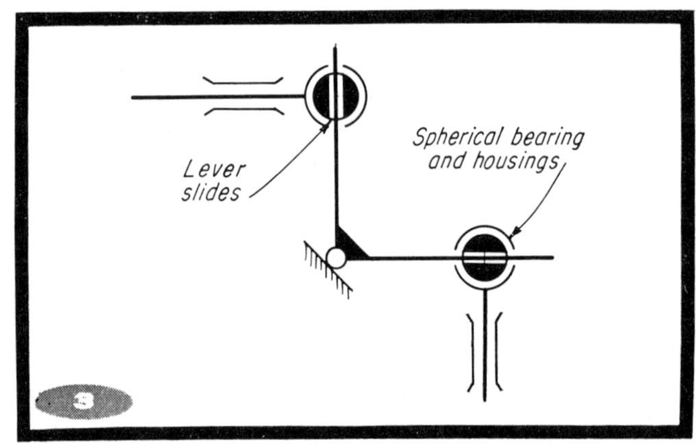

Spherical bearings

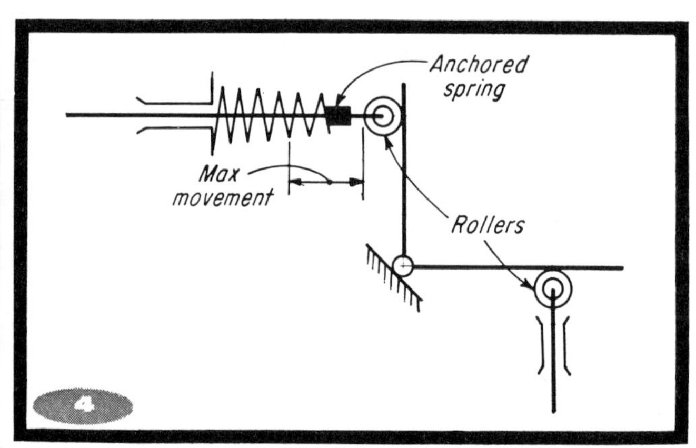

Spring-loaded lever

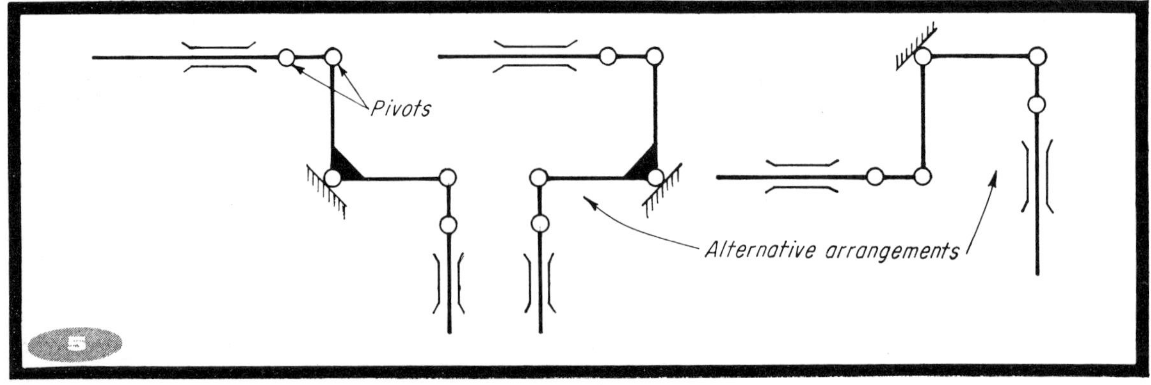

Pivoted levers with alternative arrangements

Guides

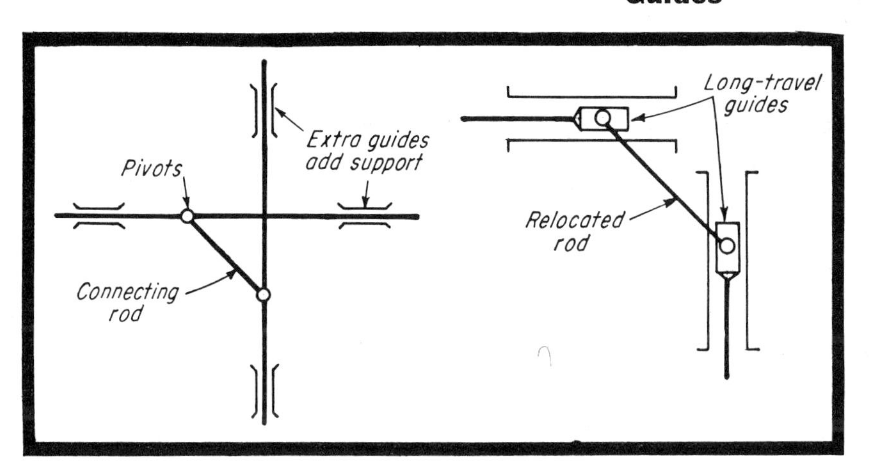

6

Single connecting rod (left) is relocated (right) to get around need for extra guides

Friction Drives

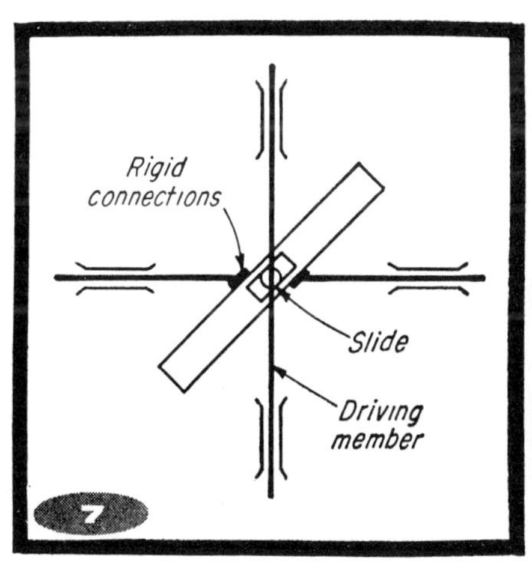

7

Inclined bearing-guide

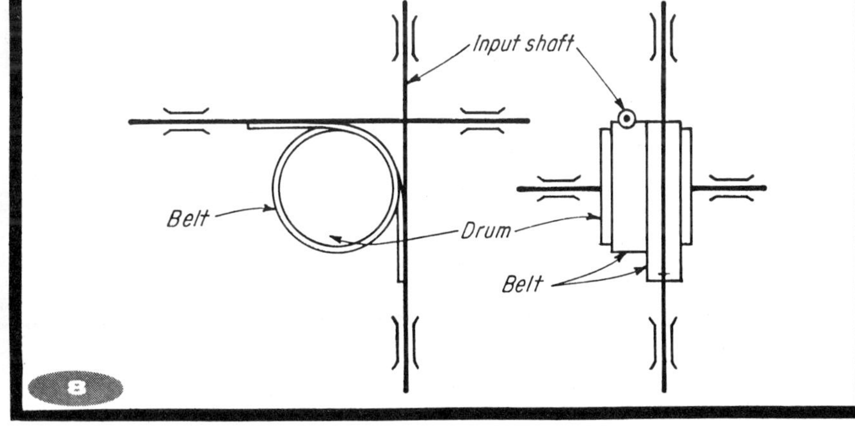

8

Belt, steel band, or rope around drum, fastened to driving and driven members; sprocket-wheels and chain can replace drum and belt

Gears

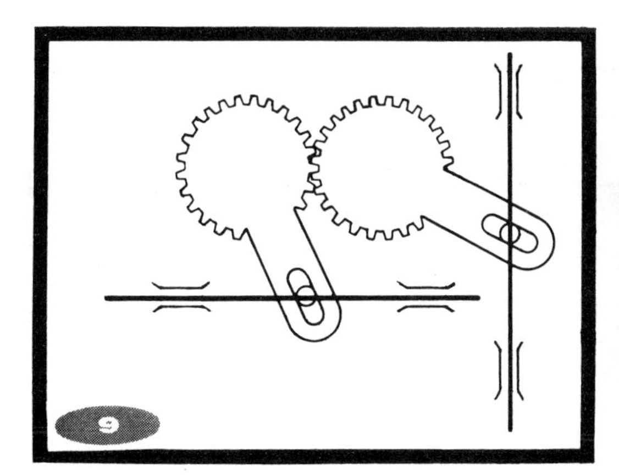

9

Matching gear-segments

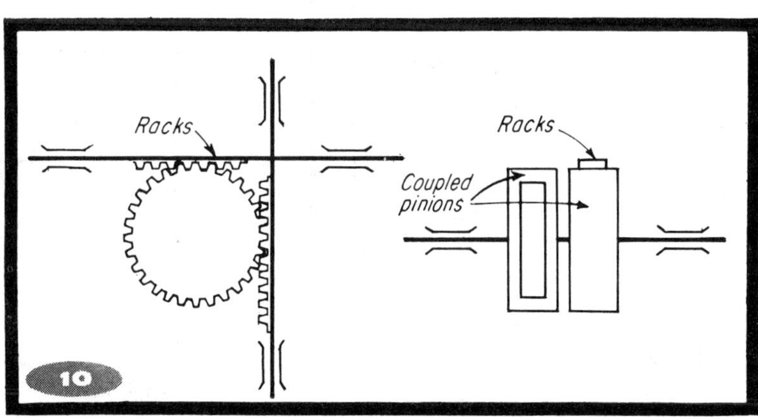

10

Racks and coupled pinions (can be substituted by friction surfaces for low-cost setup)

9 more ways to

CHANGE STRAIGHT-LINE DIRECTION

These devices, using gears, cams, pistons, and solenoids, supplement 10 similar arrangements employing linkages, slides, friction drives, and gears.

FEDERICO STRASSER

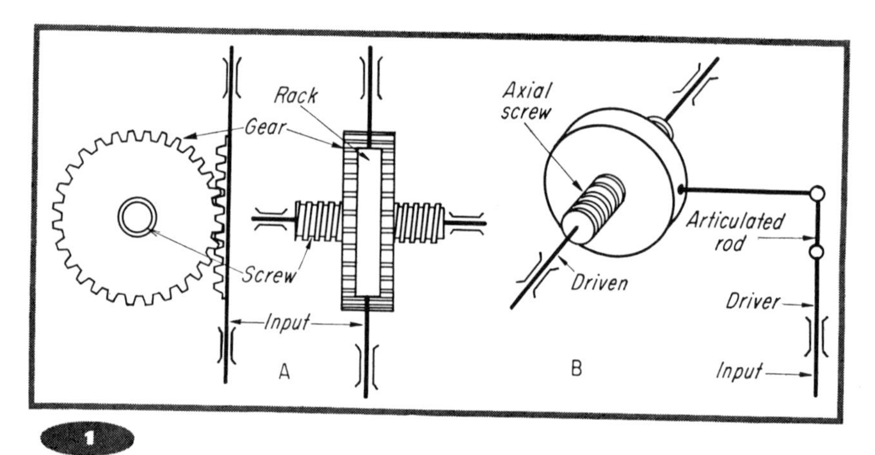

1

Axial screw with rack-actuated gear (A) and articulated driving rod (B) are both irreversible movements, i.e. driver must always drive.

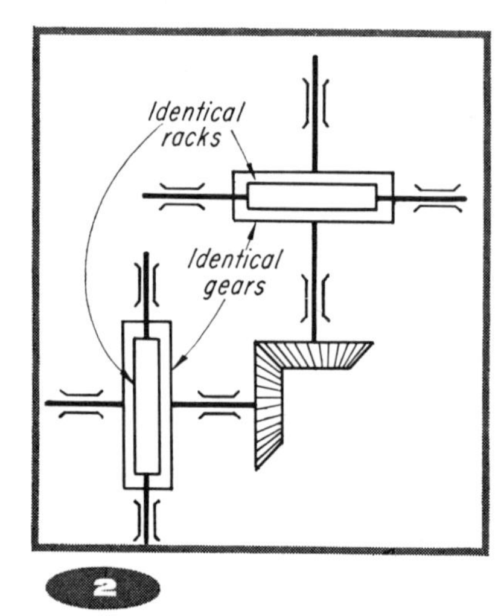

2

Rack-actuated gear with associated bevel gears is reversible.

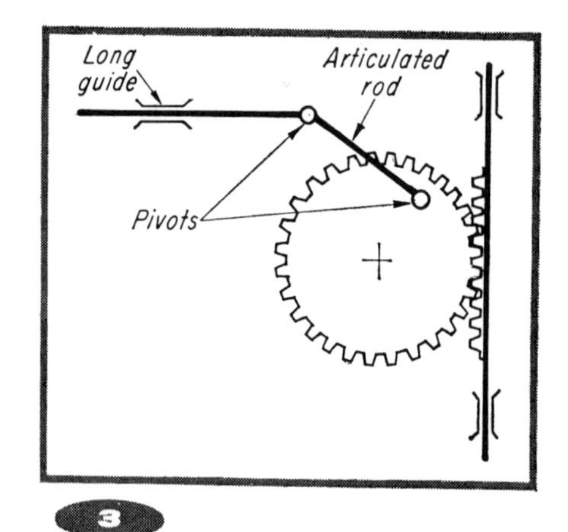

3

Articulated rod on crank-type gear with rack driver. Action is restricted to comparatively short movements.

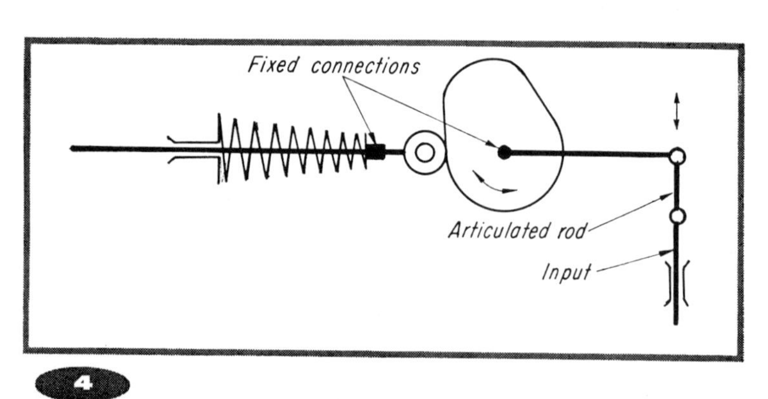

4

Cam and spring-loaded follower allow input/output ratio to be varied according to cam rise. Movement is usually irreversible.

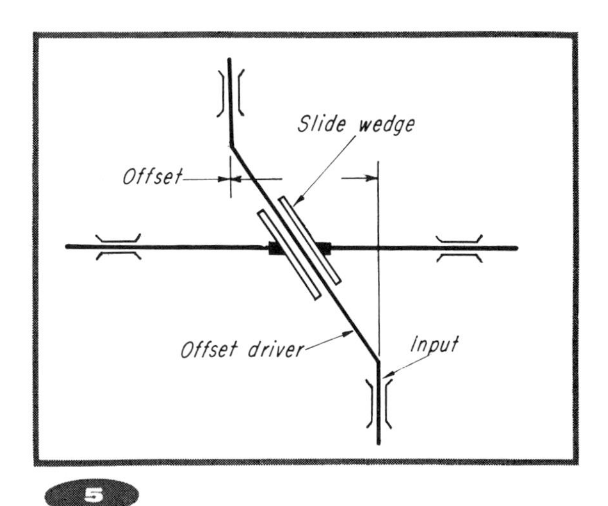

5

Offset driver actuates driven member by wedge action. Lubrication and low coefficient of friction help to allow max offset.

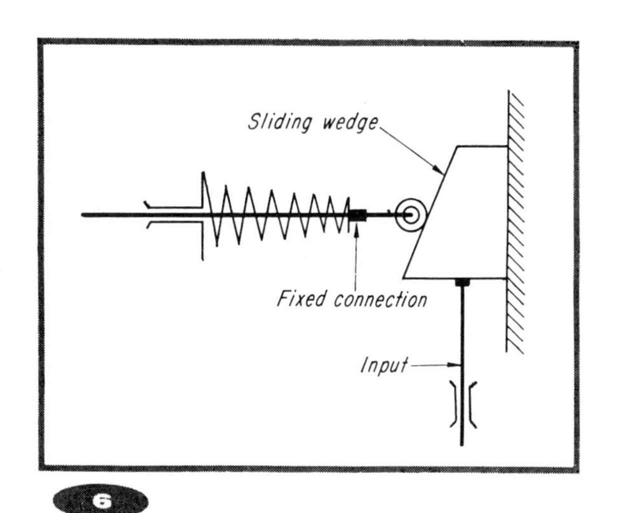

6

Sliding wedge is similar to previous example but requires spring-loaded follower; also, low friction is less essential with roller follower.

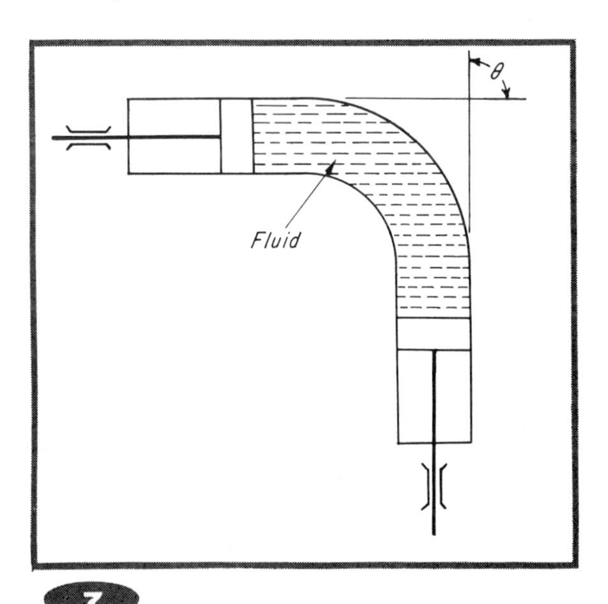

7

Fluid coupling is simple, allows motion to be transmitted through any angle. Leak problems and accurate piston-fitting can make method more expensive than it appears to be. Also, although action is reversible it must always be a compressive one for best results.

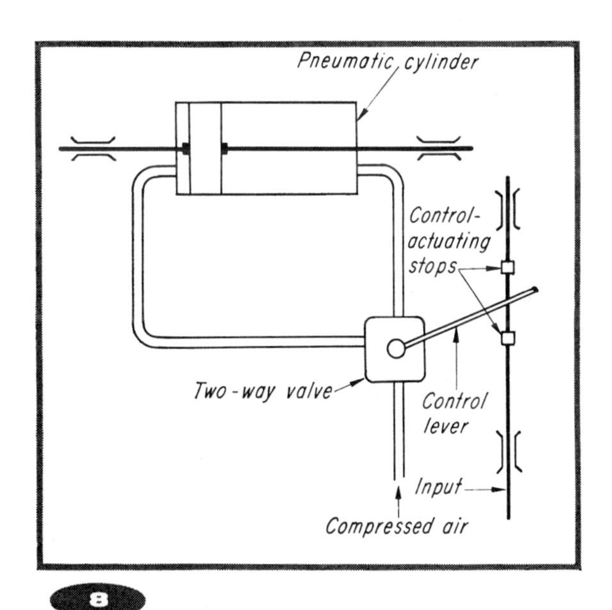

8

Pneumatic system with two-way valve is ideal when only two extreme positions are required. Action is irreversible. Speed of driven member can be adjusted by controlling input of air to cylinder.

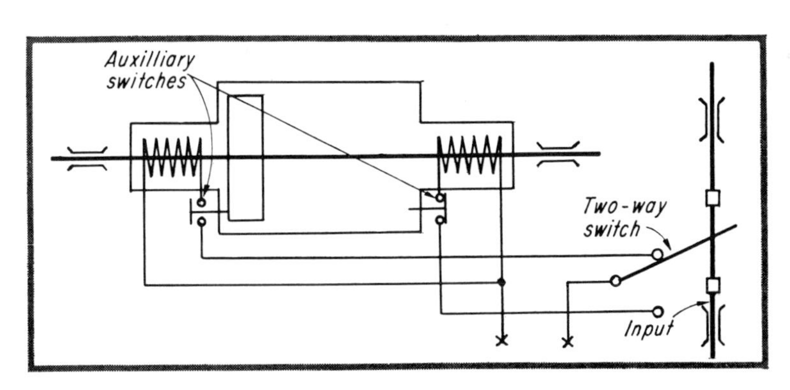

9

Solenoids and two-way switch are here arranged in analogous device to previous example. Contact to energized solenoid is broken at end of stroke. Again, action is irreversible.

PARALLEL LINK MECHANISMS

PREBEN W. JENSEN
Associate Professor, University of Bridgeport

Eight-bar linkage

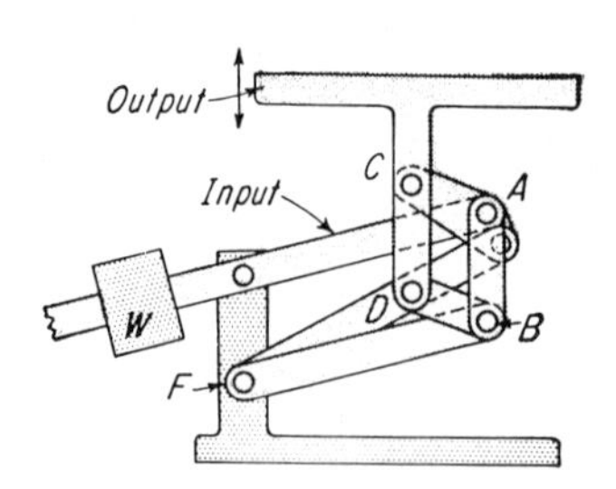

Link AB in this interesting arrangement will always be parallel to EF, and link CD parallel to AB. Hence CD will always be parallel to EF. Also, the linkages are so proportioned that point C moves in approximately a straight line. Final result is that the output plate will be kept horizontal while moving almost straight up or down. The weight permitted this device to function as a disappearing platform in a stage.

Double-handed screw mechanism

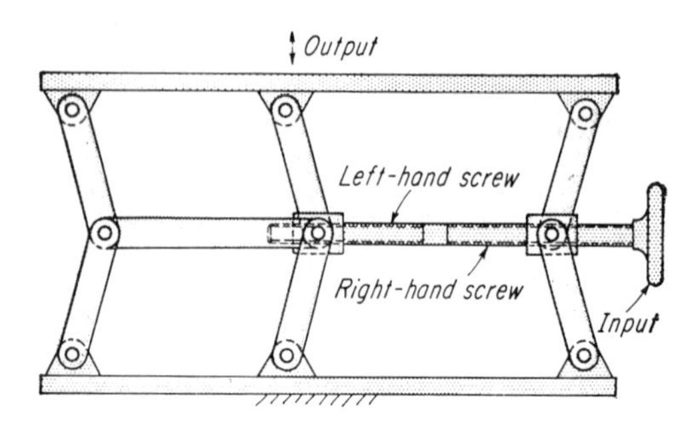

Turning the adjusting screw spreads or contracts the linkage pairs to raise or lower the table. Six parallel links are shown but the mechanism can be built with four, eight, or more links.

Tensioning mechanism

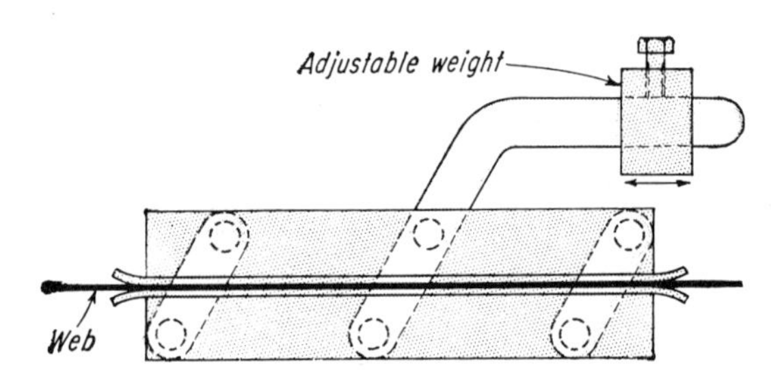

A simple parallel-link device that produces needed tensioning in webs, wires, tapes and strip steels. Adjusting the weight varies the drag.

Triple-pivot mechanism

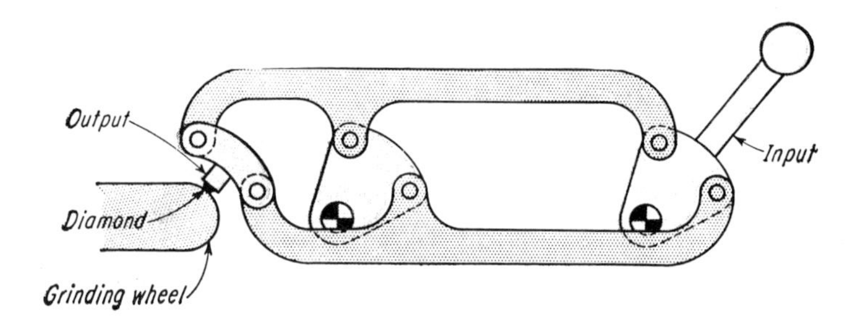

Two triangular plates pivot around fixed points on a machine frame. The output point produces a circular-arc curve. Built for rounding out the ends of the grinding wheels.

STROKE MULTIPLIER

Reciprocating-table drive

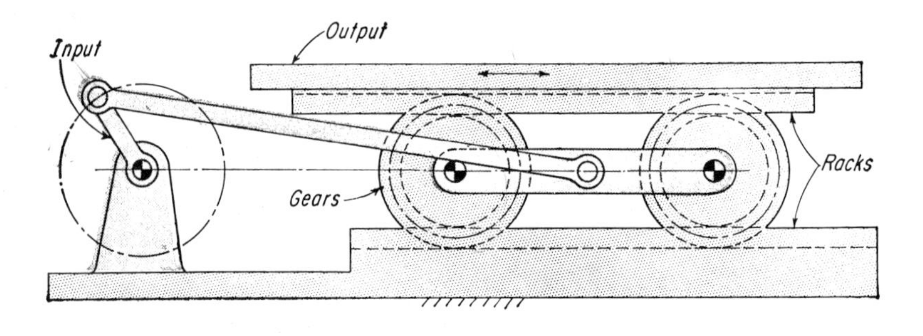

Two gears rolling on stationary bottom rack, drive the movable top rack which is attached to a printing table. When the input crank rotates, the table will move to a distance of four times the crank length.

Parallel link feeder

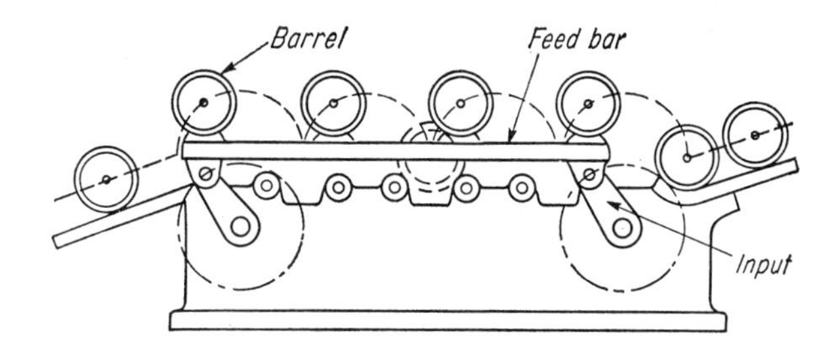

One of the cranks is the input, the other follows to keep the feeding bar horizontal. Employed for moving barrels from station to station.

Parallelogram linkage

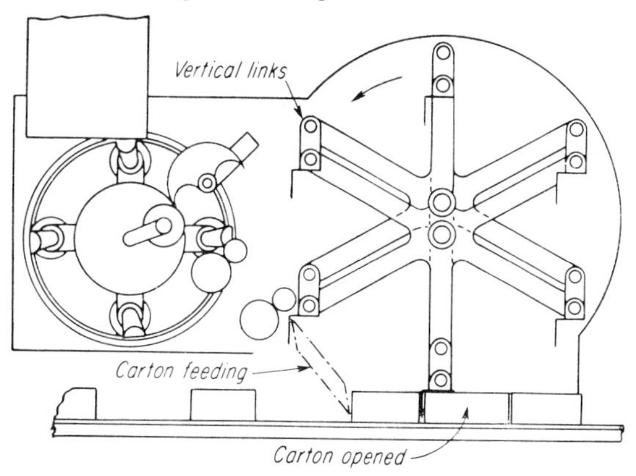

All seven short links are kept in a vertical position while rotating. The center link is the driver. This particular application feeds and opens cartons, but the device is capable of many diverse applications.

Parallel link driller

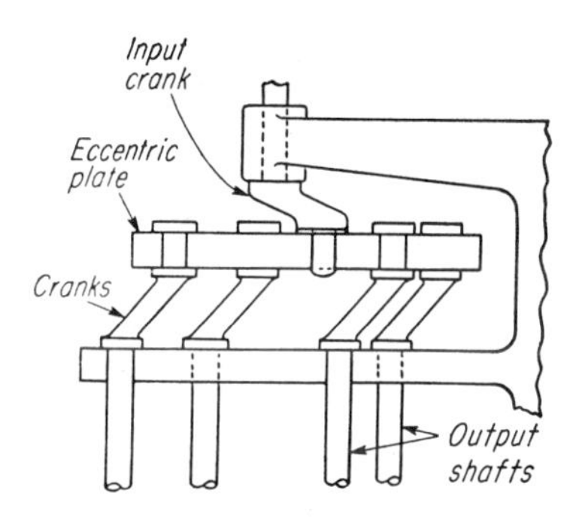

For powering a group of shafts. The input crank drives the eccentric plate. This in turn rotates the output cranks, all of which are of the same length and rotate at the same speed. Gears would require more room between shafts and are usually more expensive.

Parallel plate driver

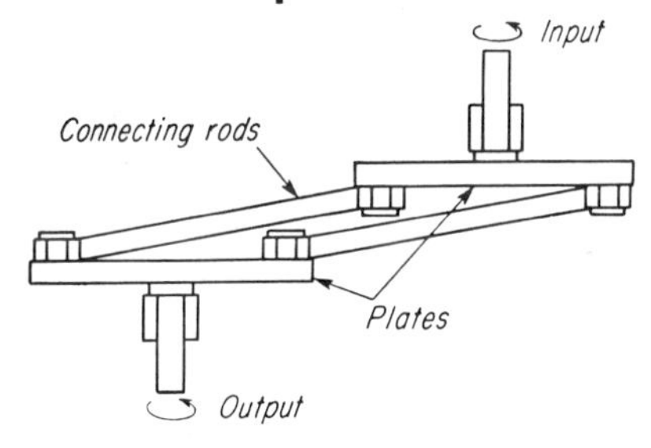

Here again, the input and output rotate with the same angle of relationship. The position of the shafts, however, can vary to suit other requirements without affecting the input-output relationship between shafts.

Curve-scribing mechanism

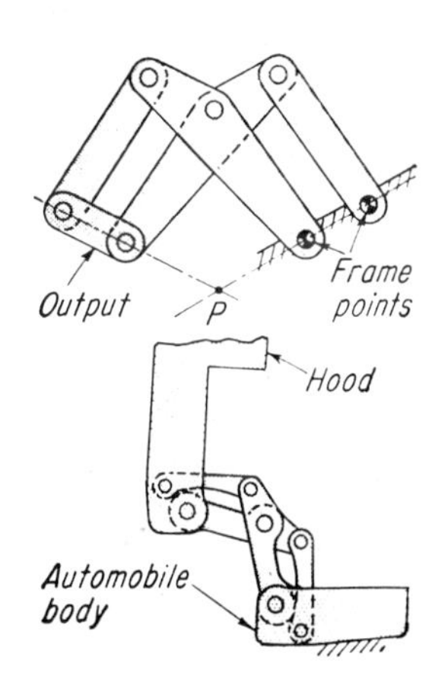

The output link rotates in such a way as to appear revolving around a point moving in space (*P*). This avoids the need for hinges at distant or inaccessible spots. Known also for hinging the hood of automobiles.

Parallel-link coupling

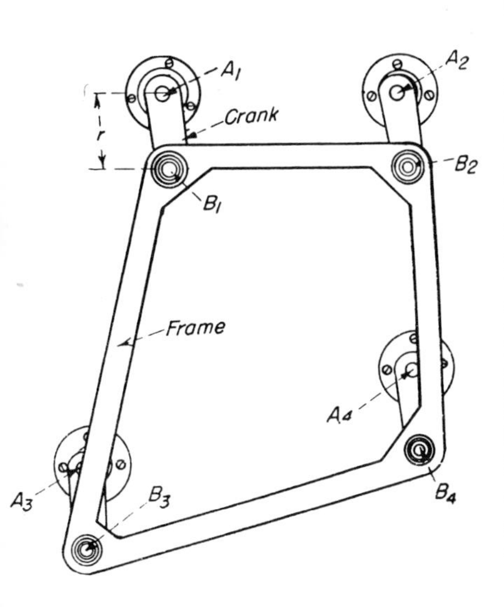

The absence of backlash makes this old but little used mechanism a precision, low-cost replacement for gear or chain drives otherwise used to rotate parallel shafts. Any number of shafts greater than two can be driven from any one of the shafts, provided two conditions are fulfilled: (1) All cranks must have the same length r; and (2) the two polygons formed by the shafts A and frame pivot centers B must be identical. The main disadvantage of this mechanism is its dynamic unbalance, which limits the speed of rotation. To lessen the effect of the vibrations produced, the frame should be made as light as is consistent with strength requirements.

Linkage Keeps Table Flat

Sine table for large vacuum chuck rests on center trunnions and differential links spaced along its sides. The synchronized links adjust table to compound angles, keeping all points in one plane—so that table flatness no longer depends upon its stiffness.

Table and chuck are 24 ft, 2 in. by 10 ft, 2 in., and were made by O. S. Walker Co., Worcester, Mass. The assembly is used with skin mill made by Onsrud Machine Works, Inc., Niles, Ill.

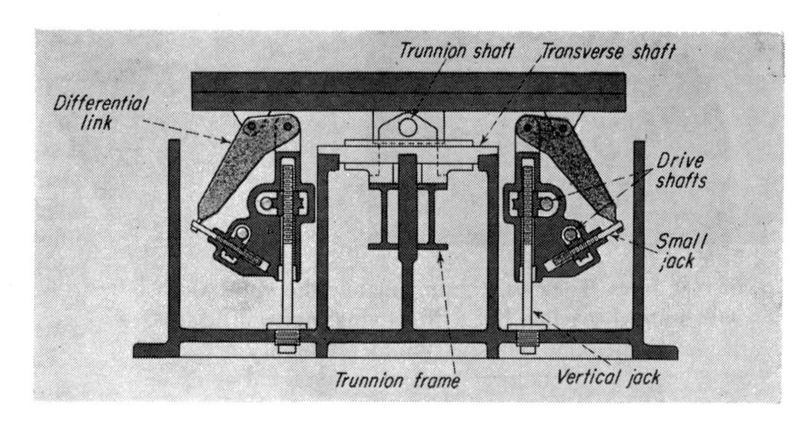

Table is supported . . .

along center on trunnion shafts; along sides by jacks that position differential links. Frame supporting the trunnions pivots on center transverse shaft. Differential links along table edges carry over half the weight of table, chuck and workpiece; are positioned with preloaded micrometer jacks that have a common drive.

The compound jacks form two positioning systems: One controls longitudinal tilt; the other controls transverse tilt. Vertical jackscrews along each side move through equal increments, tilting table about long axis. Those on one side move up while those on opposite side move down. Small jacks are connected to long link arms and control tilt about transverse axis. Jacks at table ends move twice as far as intermediate jacks for a given adjustment. There are no links at jacks in line with transverse shaft.

Table tilts 3° about long axis and ½° about short axis. Position is indicated by revolution counters calibrated in 0.0005-in. increments of vertical displacement.

Limited rotation

Here's a ingenious linkage for rotating an object 180 deg around a fixed point without the need for ring bearings. Such rotational requirements are specified for certain large radar antenna today, but are needed for smaller devices as well.

As the schematic shows rotation is controlled by two parallelogram linkages. The geometrical key to this design is quite simple: The outside points of the parallelogram (B_1, B_2) form a triangle with (and rotate around) a third point (B_0). This triangle must be congruent with the second triangle formed by the inside points of the parallelogram (B_3, B_4) and the rotational axis (A_0).

This design was conceived by Preben W. Jensen, an assistant professor of engineering at the Univ of Bridgeport. He believes the linkage is unique and will apply for a patent on it.

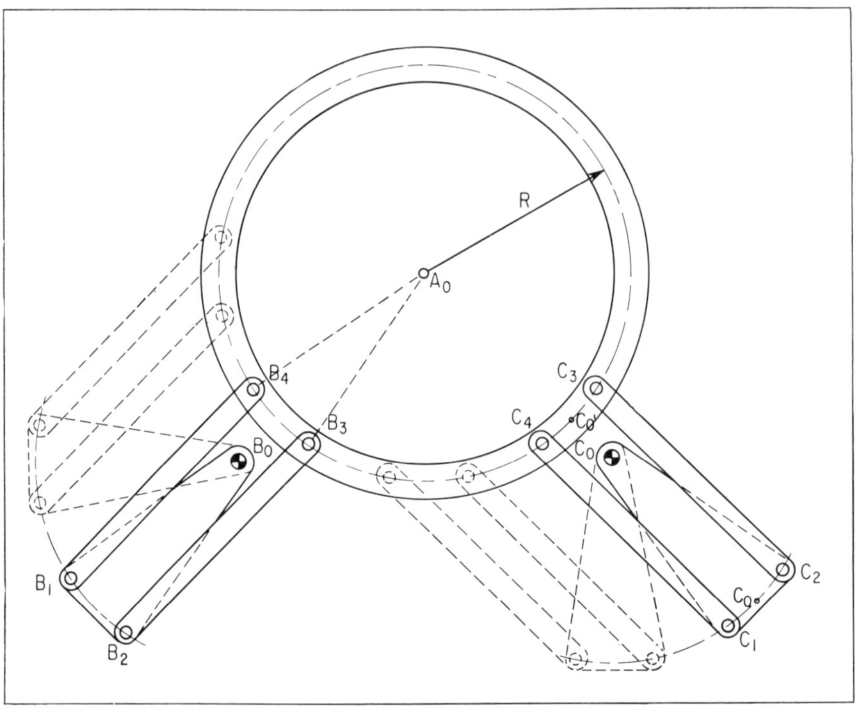

Bearings are out, linkages in

The major advantage of such an arrangement, he says, is that large ring bearings for supporting antennas can be eliminated. One such bearing in a radar operating today is about 13 ft in diameter. Jensen's linkage could quite possibly handle even larger antenna, possibly up to 26 ft dia. Another advantage is that the forces are concentrated in the pivots; so the linkage has relatively small losses from friction when suspended in a vertical position.

The linkage is inspired by the known, but often overlooked, simple parallelogram linkage. In one version of this linkage a pair of parallel arms, connected by crossmembers, is attached to a ring to be rotated. In all there are six pivot points—four at corners of the parallelogram and two on the ring. A change in the shape of the parallelogram thus rotates the ring.

Letter Size Adjusted Through Variable Linkages

Stylus motion of the Varigraph lettering instrument is divided into horizontal and vertical components which are independently reduced, then recombined to move a lettering pen. Equal reductions of both components reproduces proportions of characters on templet in any desired size between 0.150 in. and 0.750 in. Unequal reductions produce compressed or extended characters to fit text into available space or to give special effects. The instrument is manufactured by the Varigraph Company Inc., Lincoln, Neb.

INFINITE VARIETY in sizes and proportions of characters can be obtained from one templet. Flat spring on underside of instrument holds templet in position as device is moved along straight edge to space characters. Other templets are available in a large selection of lettering styles. Parallel links that support stylus plate allow complete freedom of movement within tracing area. Plate, links and majority of other parts of instrument are stamped from hard yellow brass, then nickel plated. When fingers are removed from stylus, small spring lifts point clear of templet.

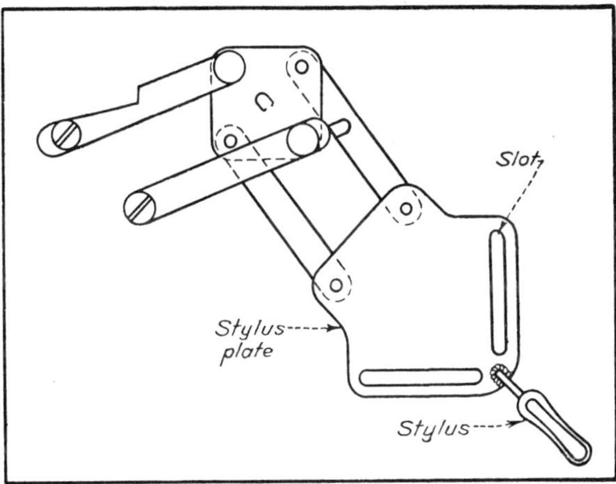

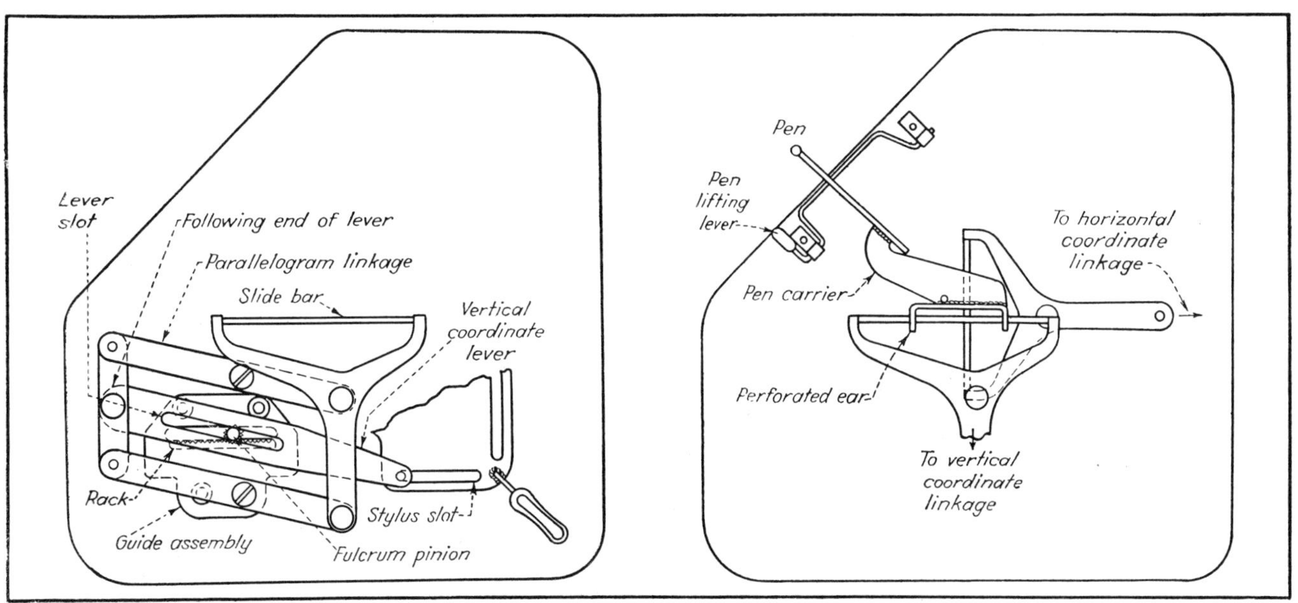

VERTICAL MOVEMENT of stylus plate actuates leading end of vertical coordinate lever through pin in horizontal slot. Motion of following end of lever is reduced to a degree depending on position of moveable fulcrum. Rotation of ratio knob causes pinion on lower end of fulcrum and rack in guide assembly to displace fulcrum along slot in lever. Friction springs maintain desired setting. Parallelogram linkage actuated by following end of coordinate lever moves horizontal slide bar. Horizontal movements of stylus is similarly translated into horizontal movement of vertical slide bar. Perforated ears connect pen carrier to horizontal bar at two points to transmit vertical motion to pen. Notch in lower side of carrier picks up horizontal motion from vertical slide bar. Bent wire lifts pen from paper when desired. Standard lettering pens fit into clutch at end of pen carrier.

Push-pull Linkages

FRANK WILLIAM WOOD JR

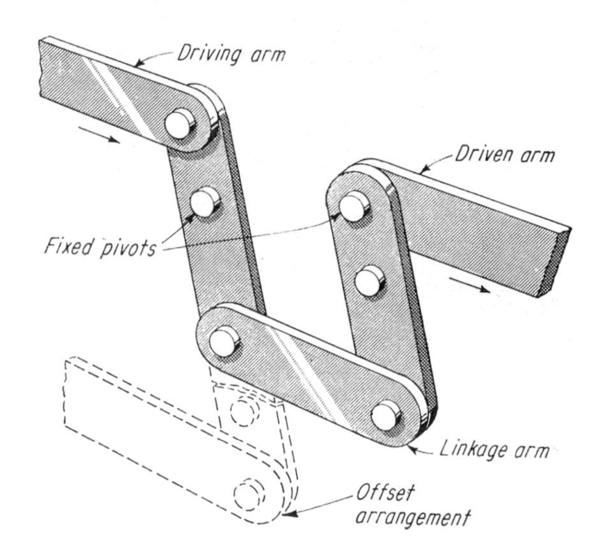

PUSH-PULL LINKAGE for same direction of motion can be obtained by adding linkage arm to previous design. In both cases, if arms are bars it might be best to make them forked rather than merely flatted at their linkage ends.

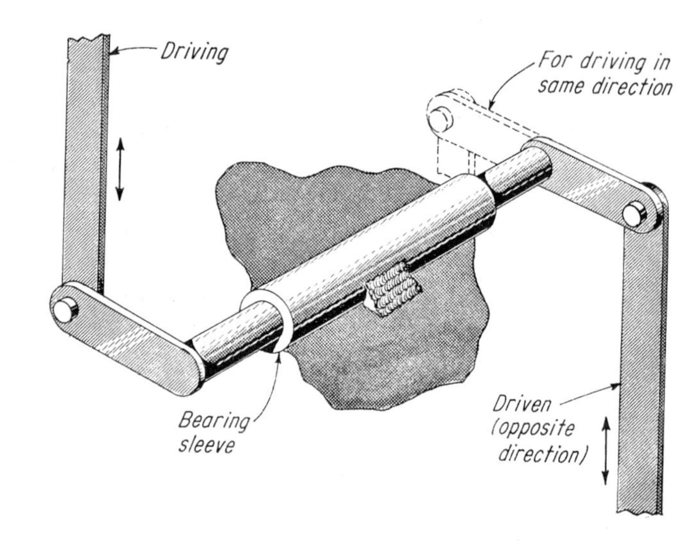

SAME-DIRECTION MOTION is given by this rotary-actuated linkage when end arms are located on the same sides; for opposite-direction motion, locate the arms on opposite sides. Use when a crossover is required between input and output.

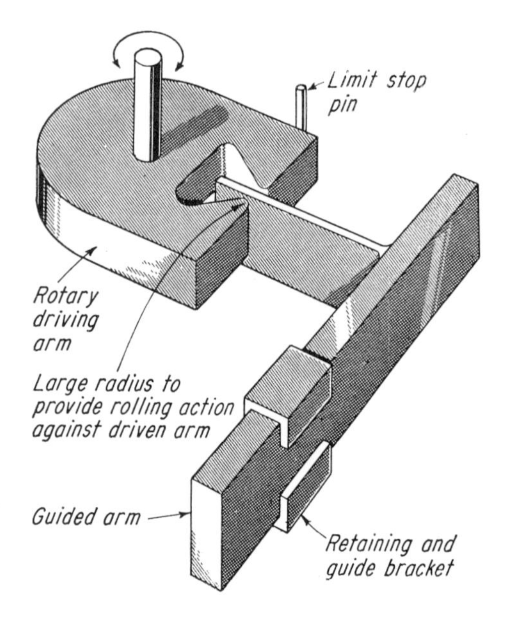

THIS ROTARY-ACTUATED linkage for straight-line 2-direction motion has rotary driving arm with a modified dovetail opening that fits freely around a flat sheet or bar arm. Driven arm reciprocates in slot as rotary driving arm is turned.

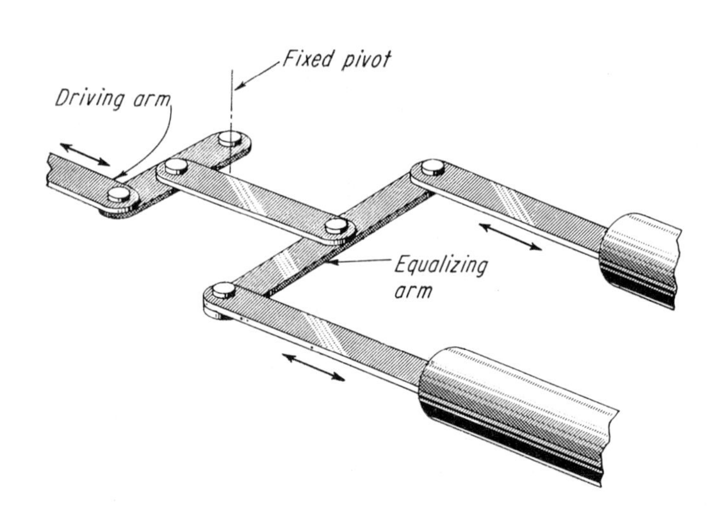

EQUALIZING LINKAGE here has an equalizing arm that balances the input force to two output arms. This arrangement is most suitable for air or hydraulic systems where equal force is to be exerted on the pistons of separate cylinders.

Chapter 4 REVERSING, ADJUSTING, AND SHIFT MECHANISMS

Double-link reverser

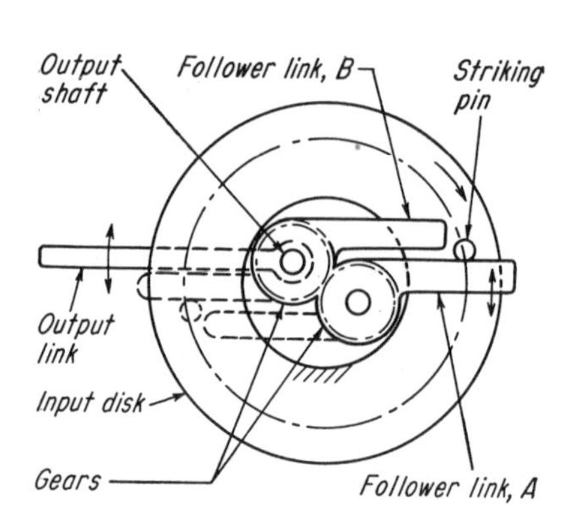

Automatically reverses the output drive every 180-deg rotation of the input. Input disk has press-fit pin which strikes link *A* to drive it clockwise. Link *A* in turn drives link *B* counterclockwise by means of their respective gear segments (or gears pinned to the links). The output shaft and the output link (which may be the working member) are connected to link *B*.

After approximately 180 deg of rotation, the pin slides past link *A* to strike link *B* coming to meet it—and thus reverses the direction of link *B* (and of the output). Then after another 180-deg rotation the pin slips past link *B* to strike link *A* and start the cycle over again.

Toggle-link reverser

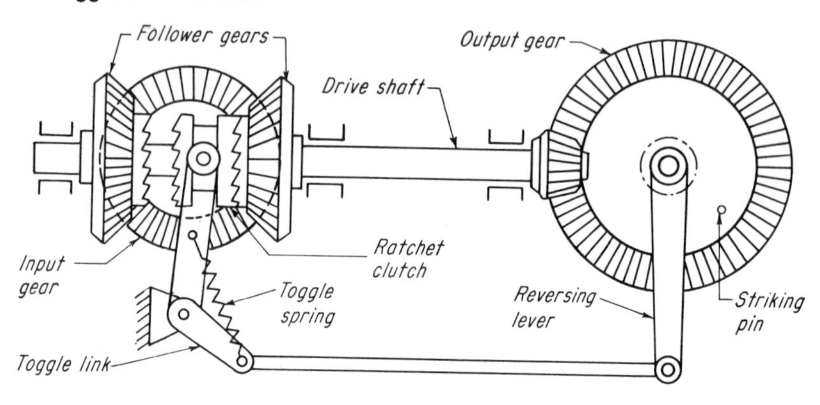

This mechanism also employs a striking pin—but in this case the pin is on the output member. The input bevel gear drives two follower bevels which are free to rotate on their common shaft. The ratchet clutch, however, is spline-connected to the shaft—although free to slide linearly. As shown, it is the right follower gear that is locked to the drive shaft. Hence the output gear rotates clockwise, until the pin strikes the reversing level to shift the toggle to the left. Once past its center, the toggle spring snaps the ratchet to the left to engage the left follower gear. This instantly reverses the output which now rotates counterclockwise until the pin again strikes the reversing level. Thus the mechanism reverses itself every 360-deg rotation of the input.

Modified-Watt's reverser

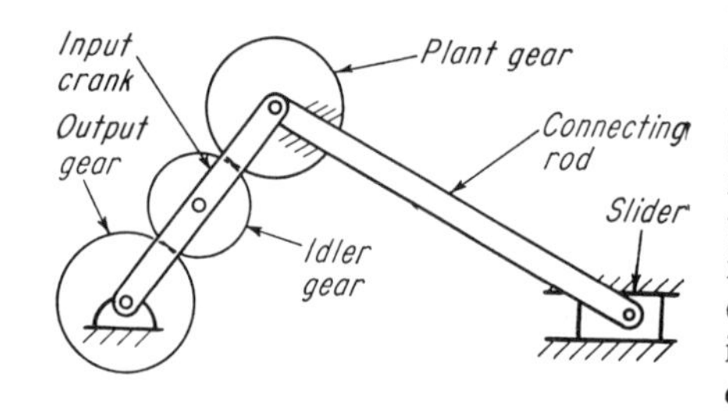

This is actually a modification of the well-known Watt crank mechanism. The input crank causes the planet gear to revolve around the output gear. But because the planet gear is fixed to the connecting rod it causes the output gear to continually reverse itself. If the radii of the two gears are equal then each full rotation of the input link will cause the output gear to oscillate through same angle as rod.

automatically switching from one pivot point to another in midstroke. by Corpet Louvet & Co.

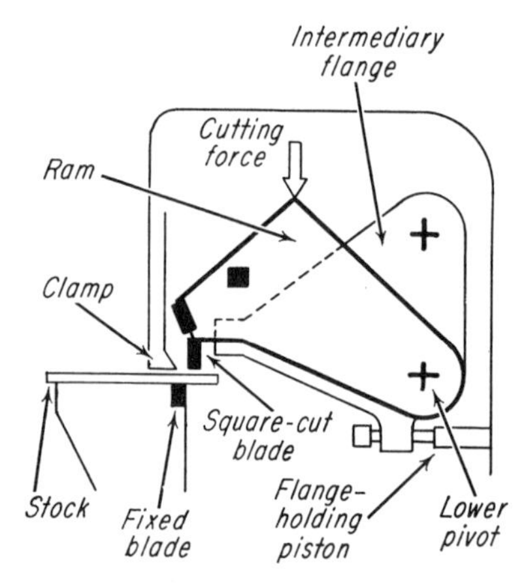

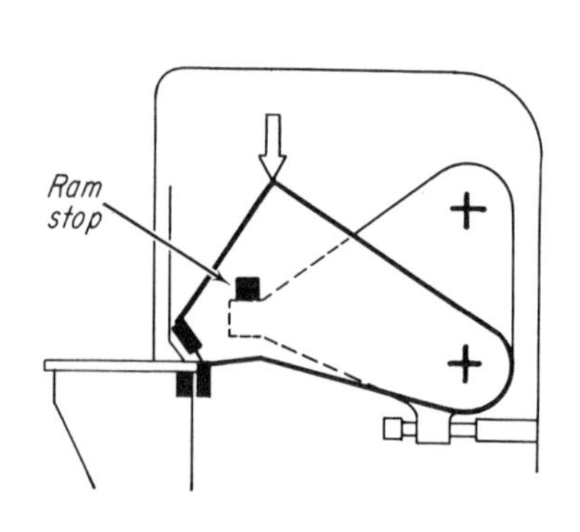

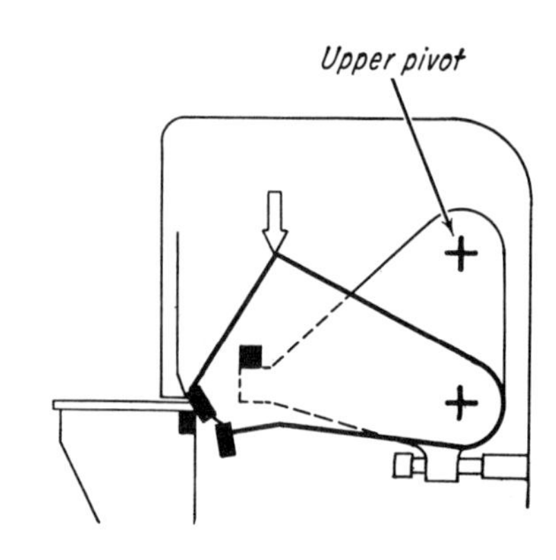

TWO PIVOTS and the intermediary flange govern sequence of cutting action. Flange is connected to the press frame at the upper pivot. The cutting ram is connected to the flange at lower pivot. In first part of cycle, the ram turns around the lower pivot and shears the plate with the square-cut blade; motion of the intermediary flange has been restrained by the flange-holding piston.

After the shearing cut, the ram stop bottoms on the flange. This overcomes the restraining force of the flange-holding piston and the ram turns around upper pivot. This brings the beveling blade into contact with the plate for the bevel cut.

Zooming Is Motorized

Push one button and camera "moves in"; push another and it "backs up." Bell and Howell Co, Chicago.

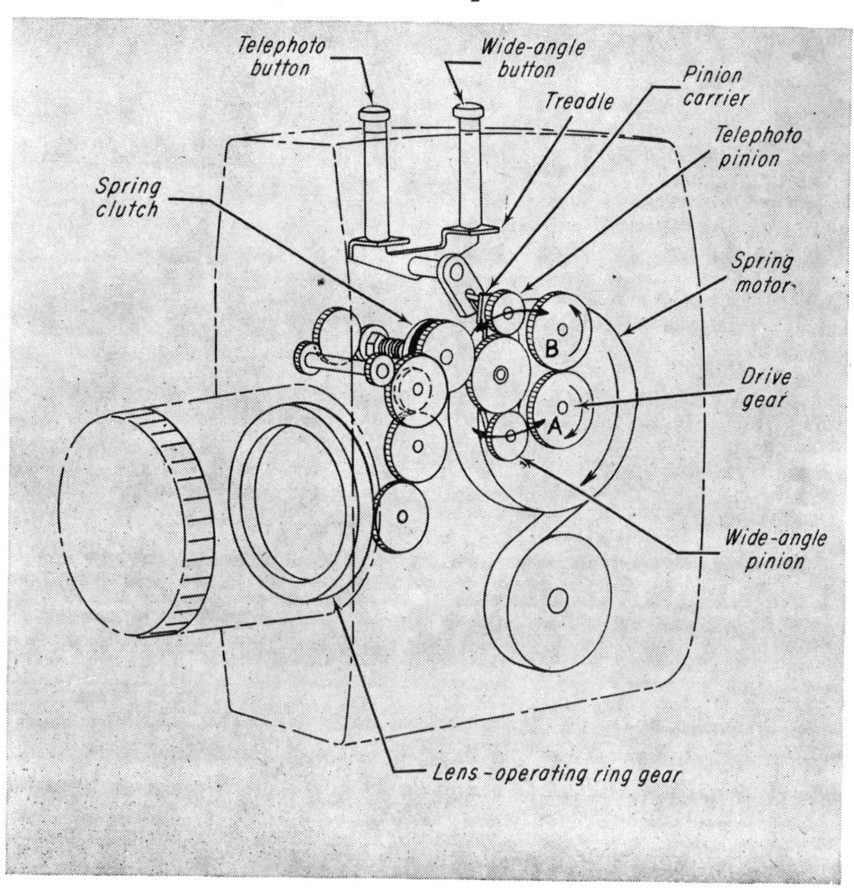

MOVABLE ELEMENTS of lens are controlled by two pushbuttons located at top of camera. The forward button, marked "TELE," adjusts the lens towards the forward or telephoto position; the rear button, marked "WIDE," adjusts it back to the wide-angle position. The buttons actuate a treadle that shifts the drive-pinion carrier. Gear A is driven by the spring motor and meshed with gear B. Both rotate, in opposite directions, whenever the spring motor is operating. Depressing the TELE button pivots the carrier so that the telephoto pinion engages gear B which shifts lens elements forward; pressing the WIDE button engages the wide-angle pinion with gear A, which shifts lens back.

Mounting both pinions on common pinion carrier prevents jams or stoppages. Spring clutch prevents damage when the lens reaches the limits of its travel. When zooming at regular film speed (16 frames per sec) the lens switches from telephoto to wide angle in six seconds. At slow-motion speed (48 frames per sec) the cycle is completed in two sec. The result is that zoom scenes appear the same on the screen whether taken at normal or slow motion.

Dual Clutches Speed Response of Missile Controls

DRIVE MOTOR rotates continuously, so response is fast enough for control applications. Solenoid-operated clutches respond to signals and connect output to the end of motor that will give the desired direction of movement. An idler in one of the gear trains reverses direction of drive to one clutch. Solenoid armatures are mechanically interconnected, so only one clutch can be engaged at a time. When both clutches are deenergized, spring-loaded ball detent centers solenoids, and locks the non-reversible worm-and-gear segment.

. . . Electromechanical actuator is designed for the Bumblebee missile project, a joint effort of the Applied Physics Lab of Johns Hopkins Univ and Convair. The unit has a response time of 8 millisec from start of command signal to full torque.

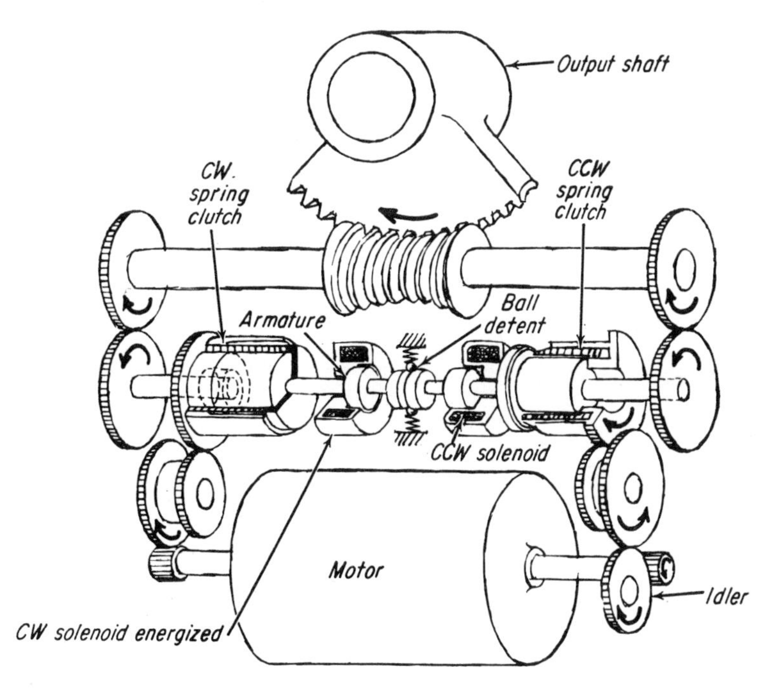

Fast-Reversal Pulley Drive

projector is produced by Baskon Corp

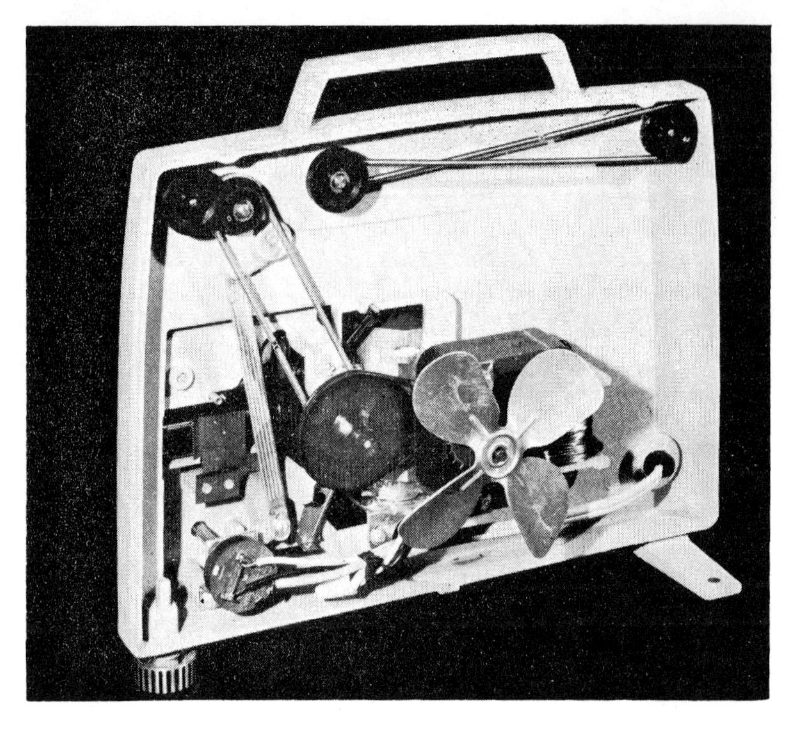

Reel drive . . .

for both forward movement and rewind is shifted by the rotary switch; it also controls lamp and drive motor. A short lever on the switch shaft is linked to an over-center mechanism on which the drive wheel is mounted. During the shift from forward to rewind, the drive pulley crosses its pivot point so that spring tension of the drive belt maintains pressure on the driven wheel. Drive from shutter pulley is 1:1 by spring belt to drive pulley and through a reduction when the forward pulley is engaged. When rewind is engaged, the reduction is eliminated and film rewinds at several times forward speed.

Rewind pulley
Drive pulley
Forward pulley
Spring belt
Shutter pulley
Motor
Forward position
Rotary switch

Rewind pulley
Drive pulley
Forward pulley
Spring belt
Shutter pulley
Motor
Rotary switch
Rewind position

Two-speed operation provided by new cam clutch

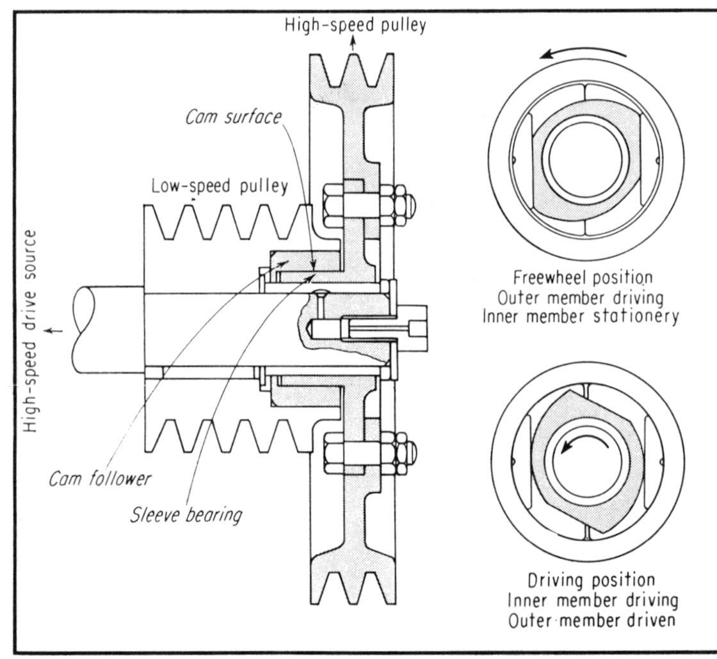

The clutch consists of two rotary members (see diagrams), so arranged that the outer (follower) member acts on its pulley only when the inner member is driving. When the outer member is driving, the inner member idles. First application is in a dry-cleaning machine where the clutch is used as an intermediate between an ordinary and a high-speed motor to provide two output speeds that are used alternately.

Input reversal shifts speeds

Reversing single-speed input to sprag clutches changes output but not direction.

Two speeds required for wash and spin-dry cycles are provided by this transmission, designed primarily for washing machine drives. Constant mesh design uses sprag clutches to engage output, speeding shifts and reducing wear. Single-speed reversible motor provides two output speeds in one direction. Two-speed reversible motor gives four speeds in one direction. Two-speed transmission was designed by Machine Development Dept, Engineering Div, Corning Glass Works, Corning, NY.

INPUT TORQUE is applied in a counter-clockwise direction, and power is transferred through the sprockets and chain to the one-way CCW sprag clutch. The clutch drives the output shaft in a CCW direction at input speed. During this cycle the CW sprag clutch is overrunning and the large gear is freewheeling.

Reversing the motor and torque input to a CW direction engages the one-way CW clutch and drives the gear. The pinion engaged with the gear drives the output counterclockwise as before, but output speed is modified by the gear ratio. During this cycle the CCW clutch is overrunning and the output sprocket freewheels.

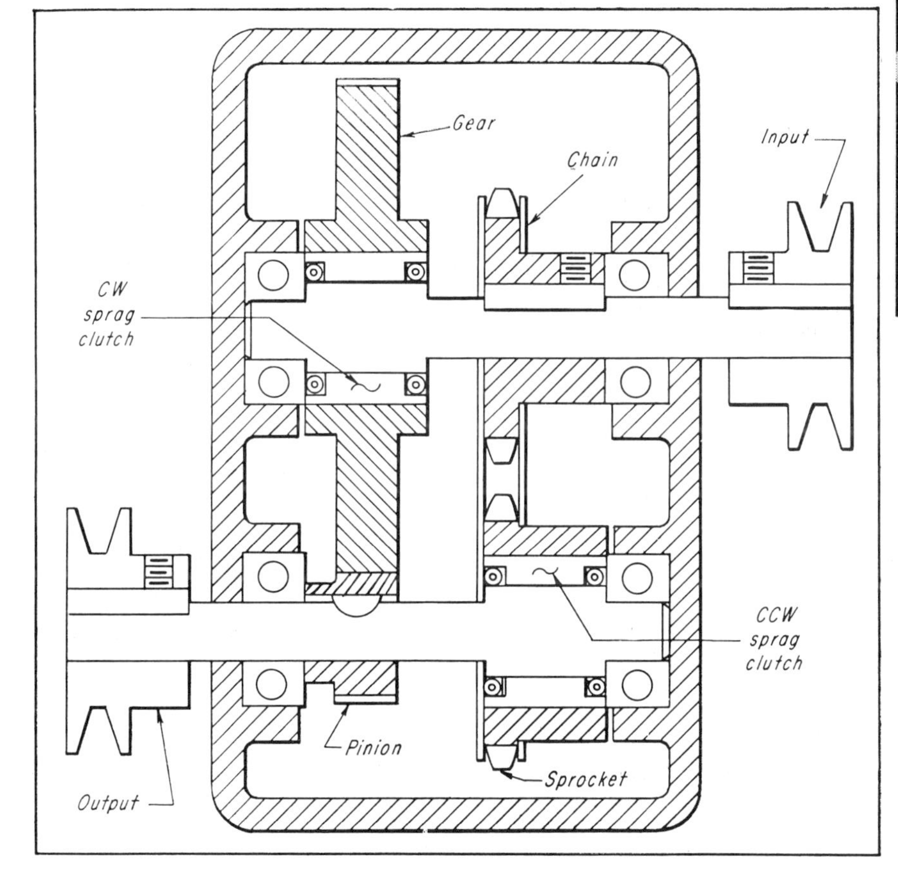

One way only

On a pure chance basis, induction motors will start rotating in the wrong direction 50% of the time. Since time mustn't go backwards timing motors have built-in "no-back" mechanisms. Many different mechanisms are used, most of which produce a hum during start-up or while the motor is running. Unfortunately, the impact and wear produced by these mechanisms are the limiting factors in timer life. To increase motor life and decrease noise, Haydon Div of General Time Corp, Torrington, Conn, developed a simple no-back mechanism which should outlast the motor.

Kicker. A plastic pawl, spring-loaded against a rubber 0-ring is the answer. If the motor starts in the wrong direction, the pawl is cammed against the rubber ring, compressing it slightly. This prevents the motor from starting in the wrong direction. As the compressed rubber springs back it kicks the rotor in the desired direction, and the pawl pivots away from the rotor, maintaining only light contact with the ring. Since the pawl is in continuous contact with the rotor, starts are instantaneous, and the oscillation associated with the starting of induction motors is eliminated.

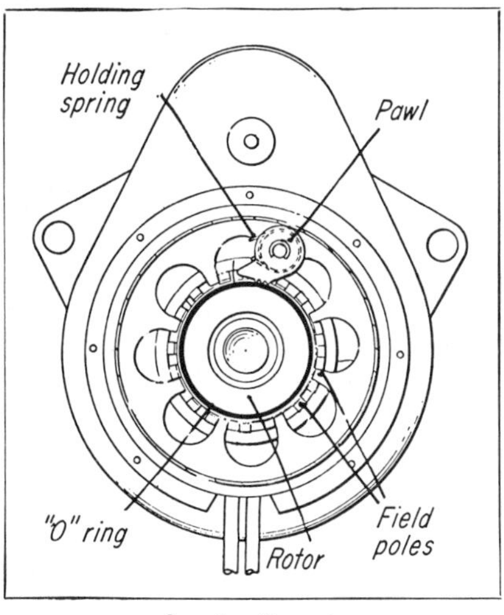

One direction only

In any no-back mechanism the rotor must be allowed enough freedom to oscillate slightly when it is energized. This permits the direction of rotation to be reversed when it is blocked. The rubber 0-ring provides freedom as well as quietness.

Gears and friction disks make a Fast-reversing drive

A slight shift of linkage causes friction disks to reverse the output shaft even under full load.

J BOEHM, A L HERRMANN, J F BLANCHE, George C Marshall Space Flight Center, Huntsville, Ala

HERE is a mechanical device capable of reversing the direction of a high-speed rotating shaft within a fraction of a second. It employs friction disks in combination with gears which eliminate the need for a clutch system to absorb the shock caused by an abrupt change in direction. Reversal can be accomplished while the shaft is under full load.

Key characteristics of the new device are:

- Capable of handling high power at high speeds. It's been tested at 2.5 hp and 10,000 rpm—but the principle can be employed for either higher or lower power requirements.
- Designed for fast start and reversal times. Because of its low inertia it can reverse a 300 ft-lb torque in 0.009 sec.
- Slippage in the friction disks is greatly reduced. A linkage arrangement automatically increases the friction force between disks when load torque is increased.
- Linkage movement is kept to a minimum. A shift in linkage of only a few thousands of an inch will cause reversal in direction. This helps reduce reversal time.

The gear and friction disk arrangement permits easy adjustment, low-cost manufacturing and long-time storage ability. The device was specifically designed for control and guidance systems of missiles. The controls must respond promptly to error signals from computers—even when high torques are required for operation of the control surfaces. This new device should also find wide application in industrial servos.

Basic configuration

It is best to insert the reversing transmission at the high-speed level of the driver. This keeps transmitted torques to a minimum, and the elements accomplishing the reversal can be extremely lightweight which improves the response time.

In the basic arrangement of the device (below), the input shaft continuously drives two hardened, wear-resistant steel disks in opposite directions. Direction of rotation of the driven shaft is reversed by shifting the driven disk to the right or left.

In the actual layout a motor drives twin pairs of opposite-turning disks, (bottom of opposite page). The disks (numbers 1 through 4) are grooved wheels set just enough apart so that when the output wheel 5 contacts the counterclockwise-turning pair (1 and 3) there is a few thousandths of an inch clearance between wheel *5* and the clockwise-turning pair (*2* and 4). Solenoids 17 and 18 turn shaft 15 through a small arc, and the eccentrically mounted pivot controls the linear position of the linkage and disk 5. When solenoid 17 is energized, disk 5 is forced against wheels 2 and 4. When solenoid 18 is energized, disk 5 is shifted over against wheels 1 and 3. When neither solenoid is energized the output disk assumes a neutral position engaging neither pair of wheels and there is no power transmission.

The output torque produces a lateral force on gear 12 and shaft 13. This makes a driven disk bear more heavily

1 •• Basic arrangement of gears and disks

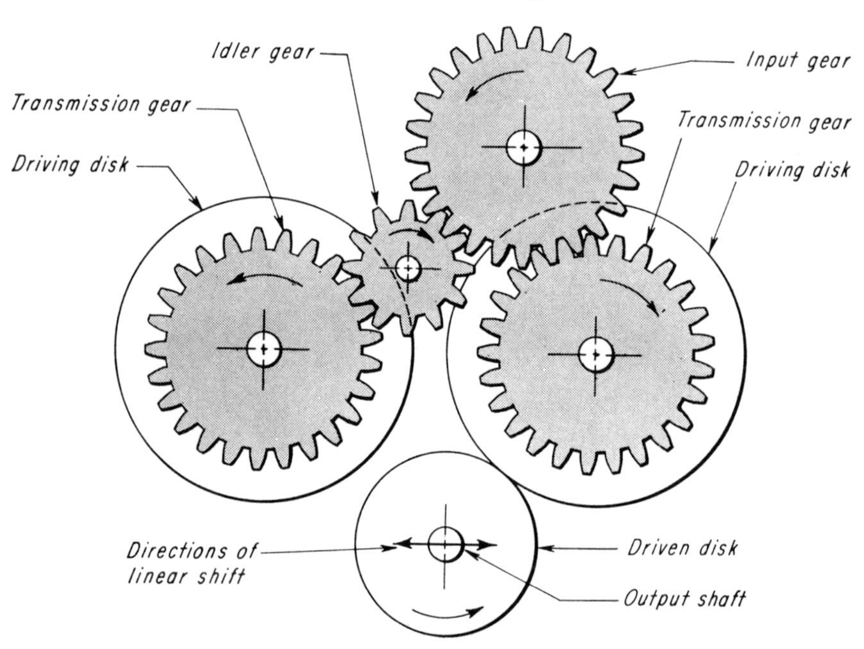

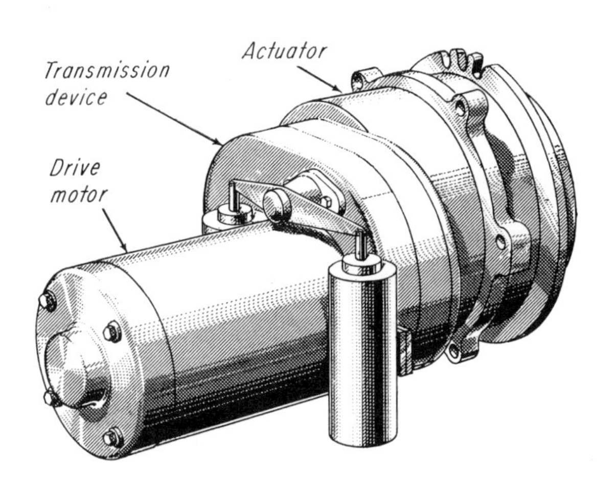

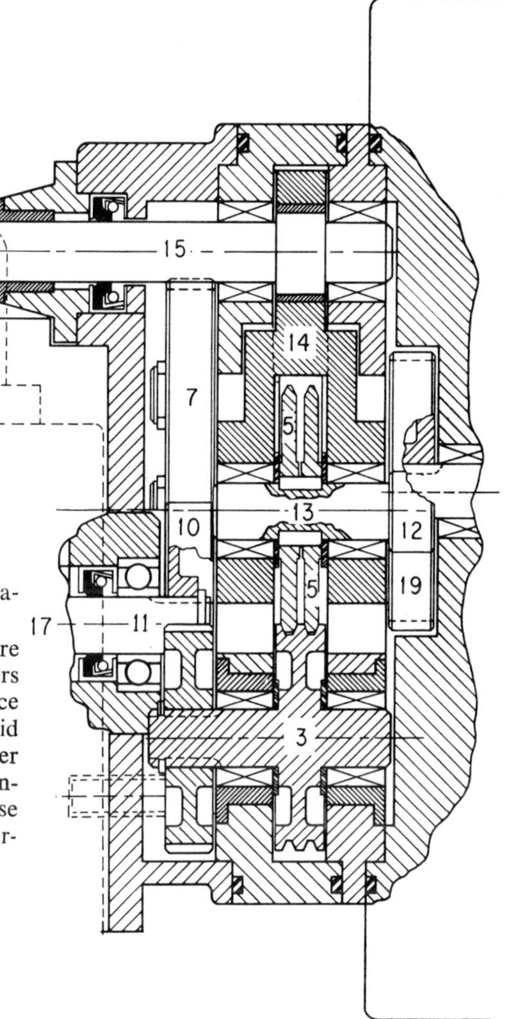

3 · · Mechanical layout of transmission device

against its driving wheels (2 and 4 in this position) as the output torque increases. This effect increases as the diameter of gear 12 is reduced. A value of $\theta = 60°$ produces an optimum force holding the driving and driven disks in contact. A smaller angle provides a higher holding force but at the expense of greater distance between the two pairs of driving disks, greater solenoid travel, and therefore an increase in the reversal time.

Solenoid characteristics

The arrangement of the friction disks also reduces the amount of force that must be applied by the solenoid for fast start, stop and reversing characteristics. However for optimum solenoid performance other modifications must be made.

Standard commercial solenoids were modified as follows: The plungers were drilled out and slotted to reduce inertia and eddy currents. A solenoid with a long coil was selected to further reduce self-induction. Use of a transistor and capacitor allows an increase of the starting current without overheating the solenoid.

2 · · Double-pair arrangement

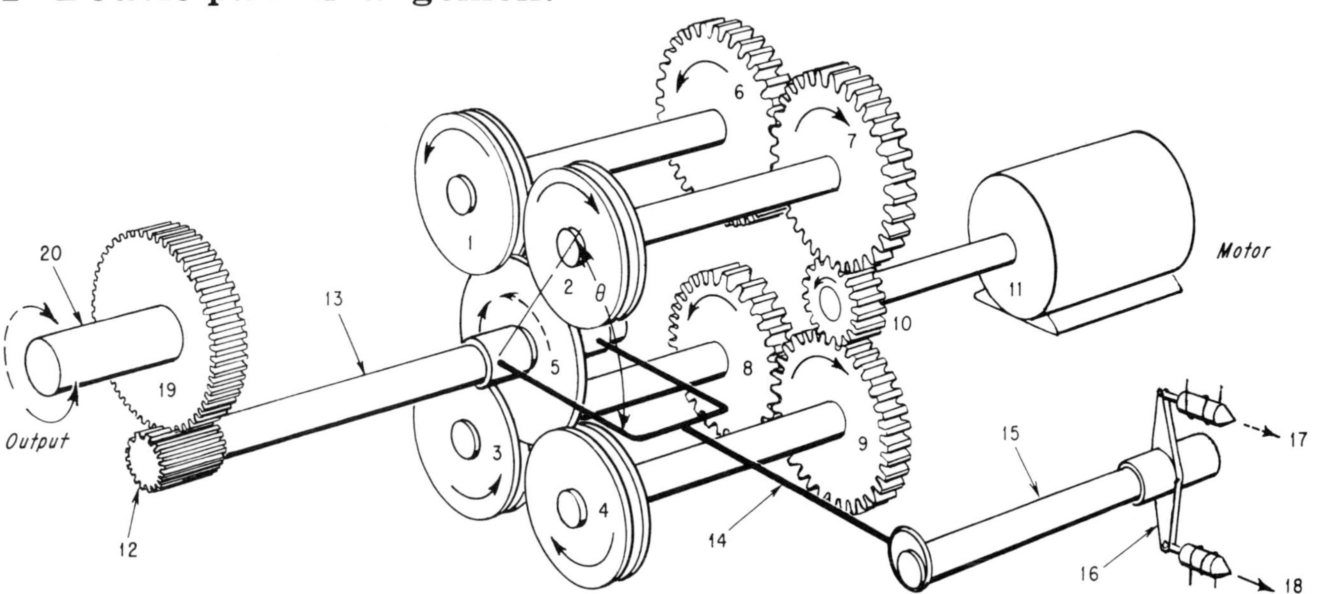

Reversing Gear Mechanisms—I

ADAM FREDERICKS

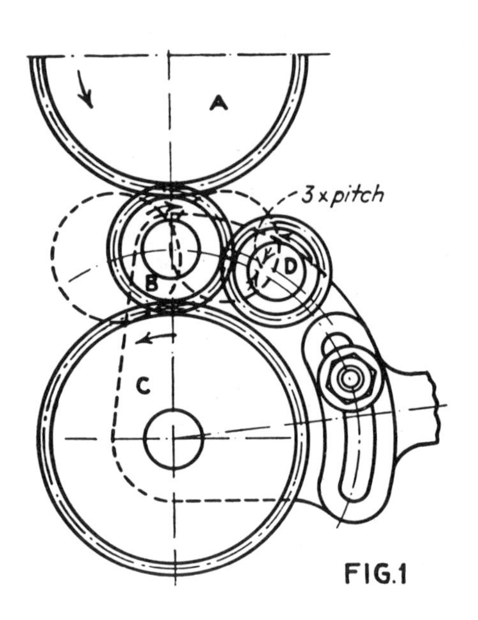

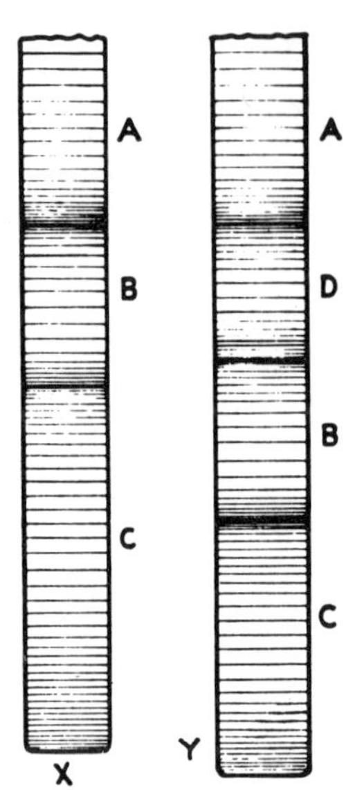

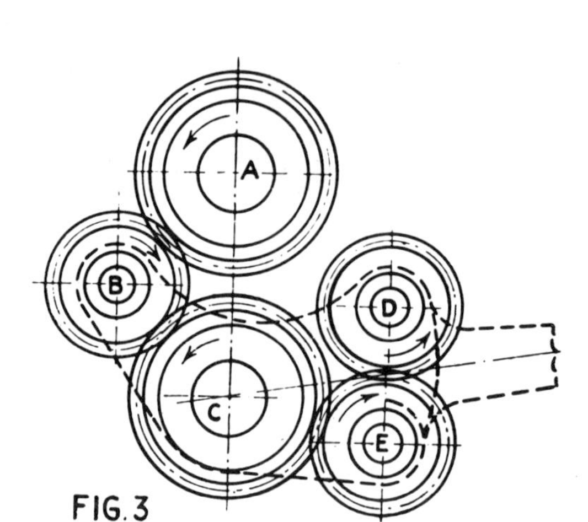

IN MANY kinds of machinery, reversal of direction is accomplished by means of gears. There are several forms of gearing from which to select, namely, the tumbler type of gear, bevel or spur gears with tooth or friction clutches, and planetary gearing.

When using a tumbler gear train for reversing direction of rotation, the clamping force on the swinging arm must be sufficiently great to withstand wedging of the gears when they run in one direction and the disengaging force when the gears run in the opposite direction. When an angular drive is required, bevel gears have an additional advantage of permitting a change of direction, by using a gear on each side of the center line of the central gear with one rotating and two stationary tooth clutches between them. Although being compact, bevel gearing is expensive because it requires an angular drive, making motor placement difficult. When tooth clutches are used for reversing, the power must be shut off when engagement takes place to avoid shock and noise.

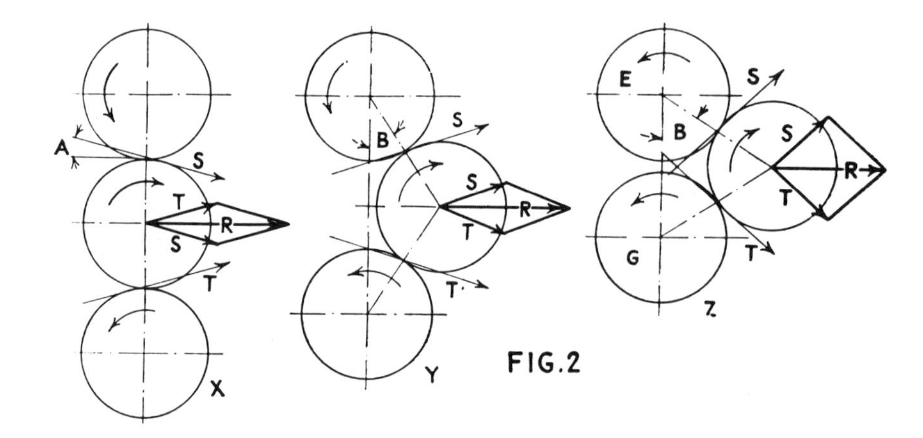

Fig. 1—Illustrates a tumbler reversing gear. Driven by gear *A*, forward drive is through pinion *B* to gear *C* shown at *X*. Pinion *D* meshing constantly with *B*, is made 3 teeth less than *B* and placed 3 tooth pitches above center line of *B*. Pinion *D* will clear *C* by 1 gear pitch. When rolled upward, *D* engages *A* while *B* disengages *A*. The drive will then be *A-D-B-C* as at *Y*. Dotted circles show the neutral position.

Fig. 2—At *X*, when three meshing gears are on the same center line the load on the center gear bearing is approximately double. The bearing load is 2 *cos (tooth pressure angle)* times tangential load. This is proved by a parallelogram of forces at *X*, the resultant *R* being the load on the center bearing. If one gear is placed on a different center line, as shown at *Y*, the resultant bearing load *R* becomes less. At *Z* is the best possible method where the outside diameters of gears *E* and *G* just clear, 1/32 in. being sufficient. In both instances of *Y* and *Z* the resultant *R* can be calculated by the formula, *R* equals 2 *cos (angle B-angle A)* × *tangential load.*

Fig. 3—A modification of Fig. 1. Although one more gear is used it is a better design as it incorporates the principle illustrated at *Z* in Fig. 2. This type of tumbler is locked in position by a spring-backed plunger in a knurled handle. Forward motion is through gears *A-B-C* and reverse is through gears *A-D-E-C*. Gear *A* is the driver. It should be so designed that the least number of gears are running for the direction used most, to save wear and eliminate additional backlash.

Fig. 4—A simple form of reverse gear is shown, operating as back gears of a lathe. In Figs. 1 and 3 the pinions are mounted on studs, overhanging the arm. At *Z* there is a bearing each side of the gear resulting in a sturdier construction. Only three gears are needed. At *X* the forward motion is illustrated by the arrow and the reverse at *Y*.

The throw *E* of the eccentric is regulated by the gear pitch. A 90 deg. movement of the handle at *Y* moves pinion *B* into engagement with driving gear *C*, while gear *D* drops from engagement. A 45 deg. movement places gears in neutral. Pinion *B* at all times engages gear *D*. The assembly may be locked by engaging a lock pin with the eccentric.

Fig. 5—A reverse drive is shown which may also be used as a hurry-up motion in one direction. Power comes through a sliding, keyed clutch *A* in the direction shown. When shifted to the right into the free running pinion *B*, double bevel gear *C* rotates toward the right. Shifting the clutch to the left drives gear *C* in the opposite direction through change gears *D* and pinion *E*. A one to one ratio is shown, but this can be altered by change gears *D* either increasing or decreasing the reverse driving speed. Only clutch *A* and gear *C* are keyed, other members running free.

Fig. 6—Although more compact than that in Fig. 5, it has not the flexibility in speed ranges. The drive may be transmitted through either shaft *A* or *B*. When shaft *B* is the driver and clutch *C* engages free pinion *D*, gear *E* revolves at a slow forward speed, and engaging clutch *C* with free gear *F*, gear *G* rotates at a faster reverse speed. Gear *E* is cupped out sufficiently to accommodate gear *F*.

Fig. 7—A reverse bevel gear with power coming through gear *A* which runs in direction of arrow. Pinions are mounted loosely on clutch shaft, the center or sliding clutch being keyed. When clutch is engaged on either side, the loose pinion clutch drives the sliding clutch and in turn the clutch shaft. The pinions are bronze bushed and held in place by collars or shoulders in the main casting. The surface speed of the bearing in the free pinion is double that of the pinion being driven because it is rotating in the opposite direction.

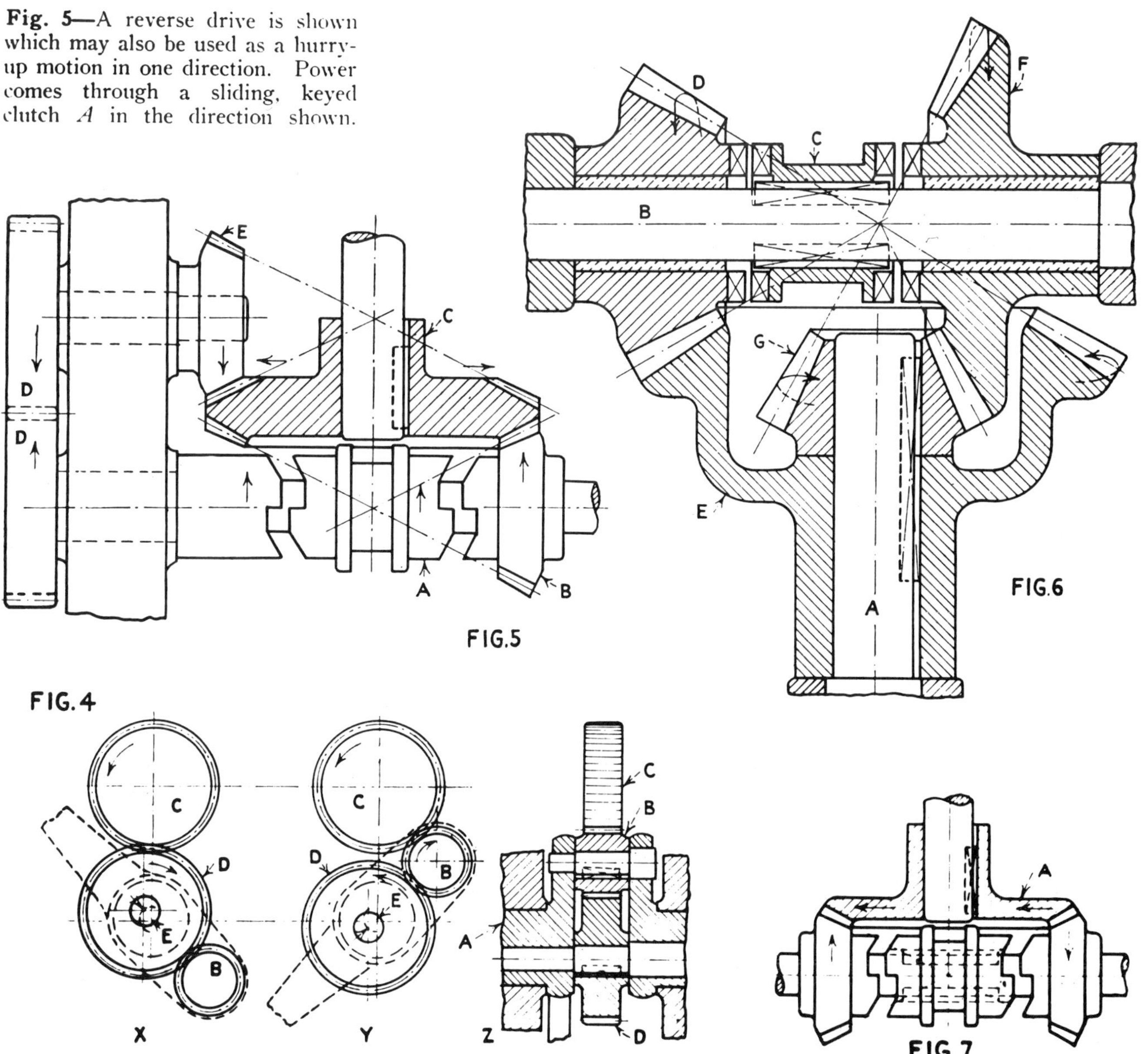

Reversing Gear Mechanisms—II

ADAM FREDERICKS

In these examples, various types of clutches are used with spur gear drives. With the use of friction clutches, it is not necessary to stop the gears when shifting the clutch as the load is picked up gradually, thus shock and noise are avoided. Several examples show sliding gears.

Fig. 8—A spur gear drive requiring two friction clutches and five gears, shaft Z being the driving member. At X is the forward drive and at Y the reverse drive. While the forward and reverse speeds are not equal in this drive, a close approximation is obtained. Both sets of gears having an equal center distance, the sum of the teeth in C and D must be at least 5 less than those of A plus B, to get proper tooth clearance. To keep the ratio as close as possible, these 5 teeth are divided between driver and driven gears. Assume gears of 34 and 54 teeth for the forward drive, making a total of 88 which gives a ratio of 0.629 to 1. With a difference of 5 teeth, gears of 32 and 51 teeth will give a ratio of 0.627 to 1 in the reverse train. The idler gear E changes the direction of rotation.

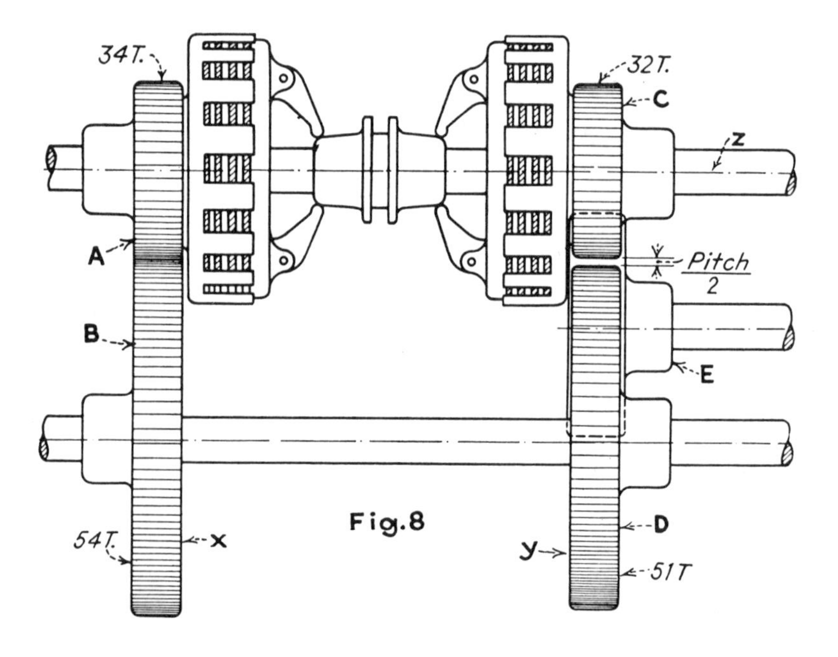

Fig. 9—The reversing mechanism is removed two or three shafts from the driving clutch. The driving cluster gears A and C move longitudinally on a splined shaft. Forward motion is through gears A and B (see end view). The cluster gears are moved so that gear C engages intermediate E which is always is mesh with gear D for reverse drive. To maintain a close ratio as stated in Fig. 8, there is a difference of 5 teeth in the reverse train, giving a clearance of (pitch/2) between gears C and D. The longitudinal space between gears B and D is sufficient to allow complete disengagement before the cluster enters the other pair of gears. The neutral position will give $\frac{1}{16}$ in. clearance each side of the cluster gear. The reverse gear E is shown wider than necessary, but is actually the same width as the other gears.

Fig. 10—This is a modification of Fig. 9 in which one gear is eliminated. This method is more compact, using only three gear widths plus clearance where in Fig. 9 four gear widths plus clearance are needed. By keeping the space between bearings as short as possible, shaft deflection is reduced. If gear A has 34 teeth and gear B has 54 teeth, then gear C must have 5 teeth less or 49 teeth. The ratio then is 0.629 to 0.694 or an increase in speed of approximately 10 per cent when A is the driver. Gear D is the reversing gear.

Fig. 11—A compact reversing drive embodying a friction clutch of expanding ring and plunger type, in which only four gears are used. The wide pinion A fully engages gear D but engages pinion B only partially. The other half of pinion B

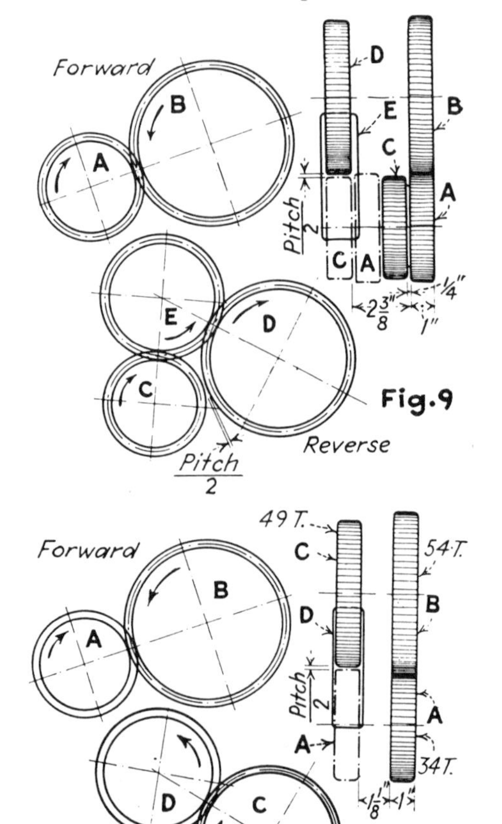

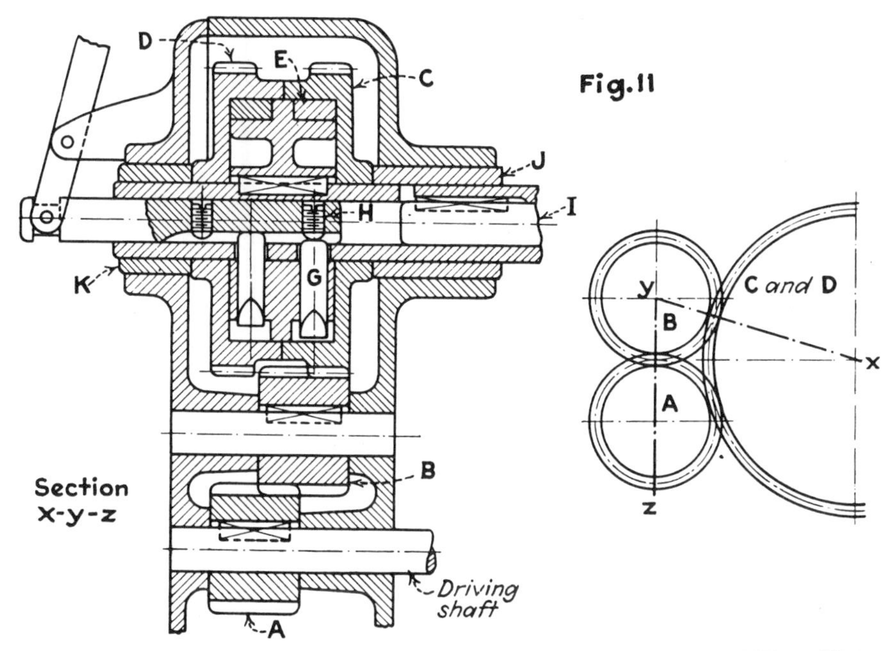

meshes with gear C. The reverse drive is shown in the cross-section view, pinion A driving pinion B. By expanding the friction ring E with plunger G which is actuated by the spherical pointed screw H, shaft I rotates in the same direction as the driving pinion A. All gears are constantly in mesh, gears C and D being mounted loosely which always rotate in the opposite directions, the bearing surface speed being doubled. The assembly is held in place by bushings J and K. The forward motion is through pinion A and gear D by shifting the clutch into engagement with gear D.

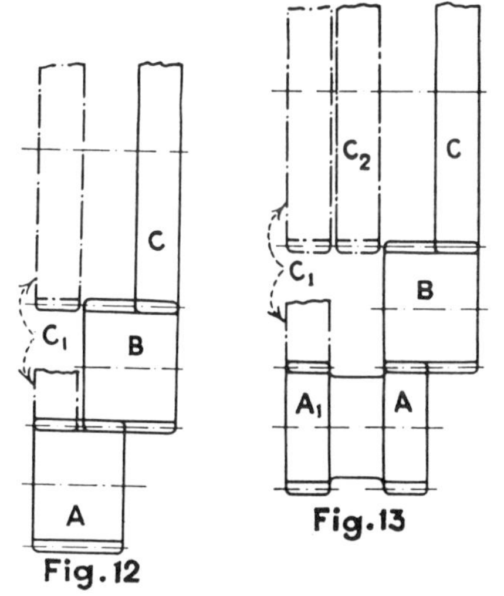

Fig. 12—A modification of Fig. 11. Pinions A and B are the same as in Fig. 11, but the use of gear C, as a splined sliding gear, eliminates gear D (Fig. 11). Reverse drive is as shown A-B-C. Forward drive is through gears A-C_1. One objection occurs in that no neutral is provided. The gear mechanism must be stopped before gear C is shifted.

Fig. 13—To overcome the tendency of breakage of the method in Fig. 12, one more gear width is added. Integral gears A and A_1 have the same number of teeth. Neutral position is shown at C_2, forward is A_1 to C_1 and reverse is through gears A-B-C.

Fig. 14—All gears are inclosed in a built-up planetary-gear driving-pulley with inclosed clutch, and is used for light duty only. Gear A is mounted in the machine casting and does not revolve. Three pairs of planetary gears rotate around gear A driving the free gear B. Clutch pin C is shown in a neutral position in a circular groove in the pulley flange. When pulled to the left, the pin engages the pulley and rotates the shaft D. For reverse, the pin is pushed inward to engage gear B. By varying the ratio of the two gears in the three sets of planets, the required speed is transmitted to gear B. Direction of reverse motion is shown by the arrows.

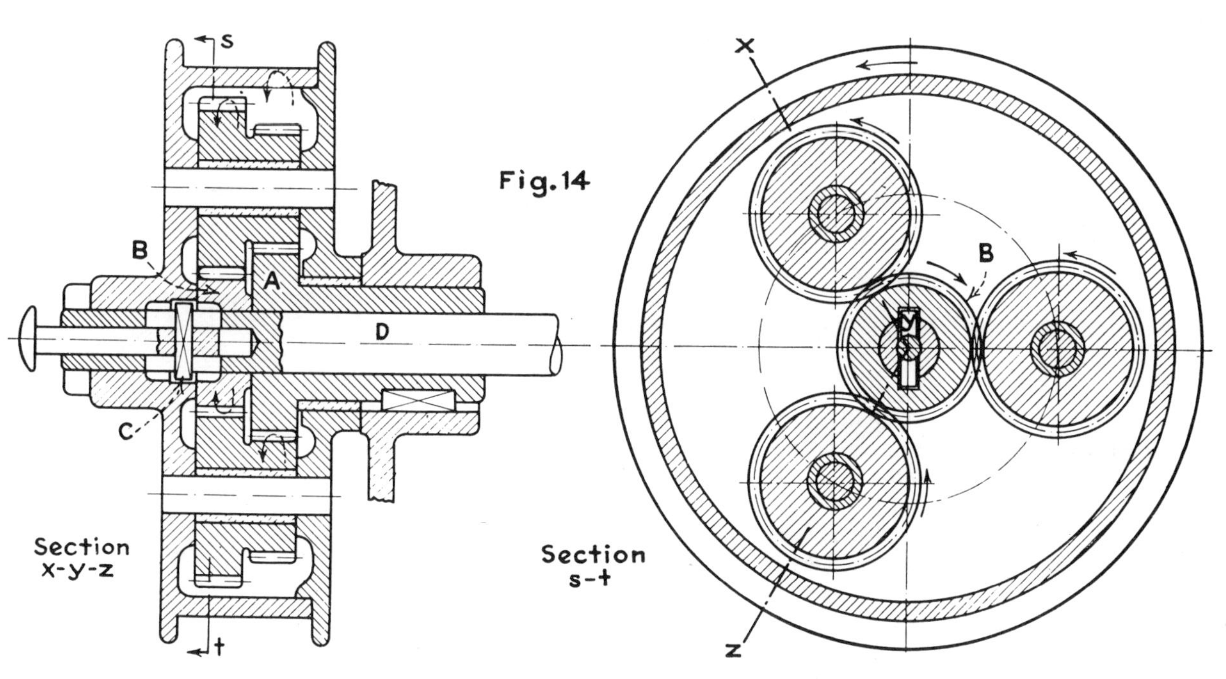

How to Prevent Reverse Rotation

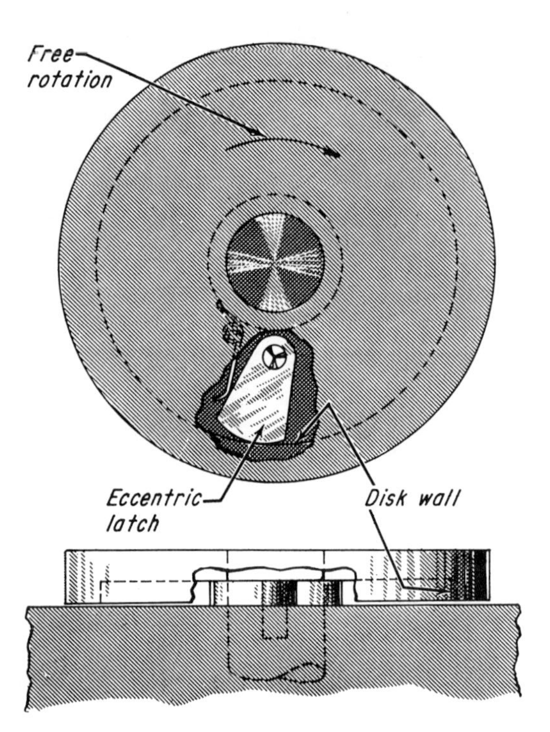

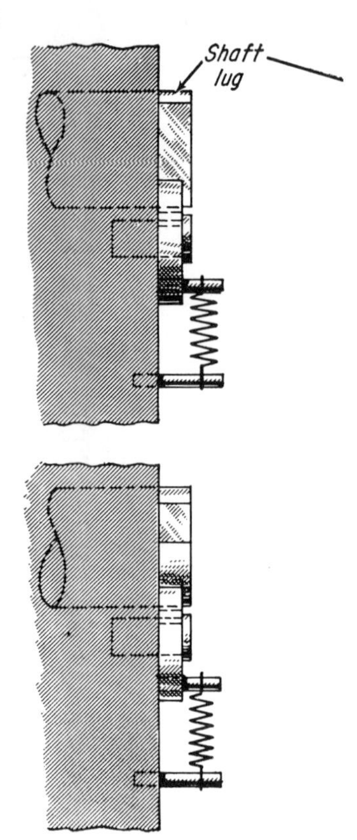

ECCENTRIC LATCH allows shaft to rotate in one direction; attempted reversal immediately causes latch to wedge against disk wall.

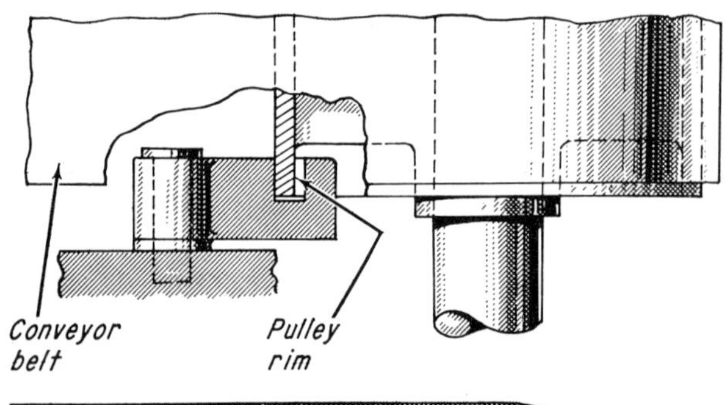

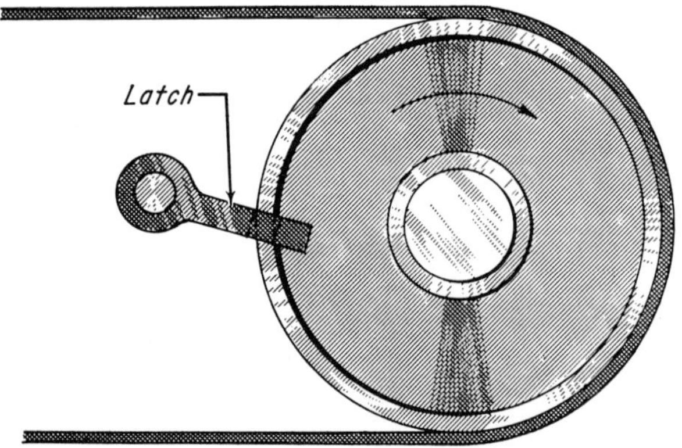

LATCH ON RIM of pulley is free only when rotation is in direction shown. This arrangement is ideal for conveyor-belt pulleys.

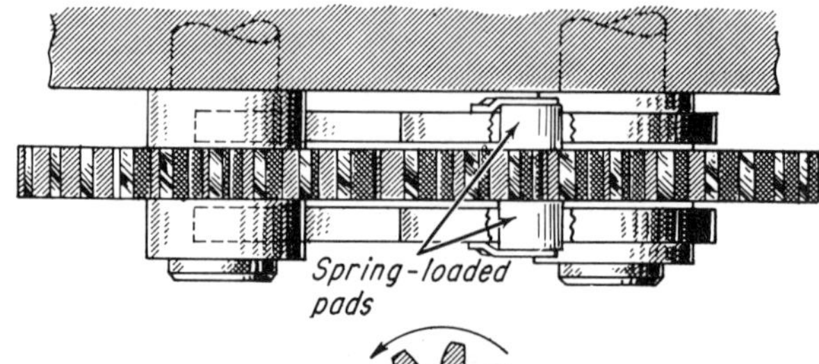

SPRING-LOADED FRICTION PADS contact the right gear. Idler meshes and locks gear set when rotation is reversed.

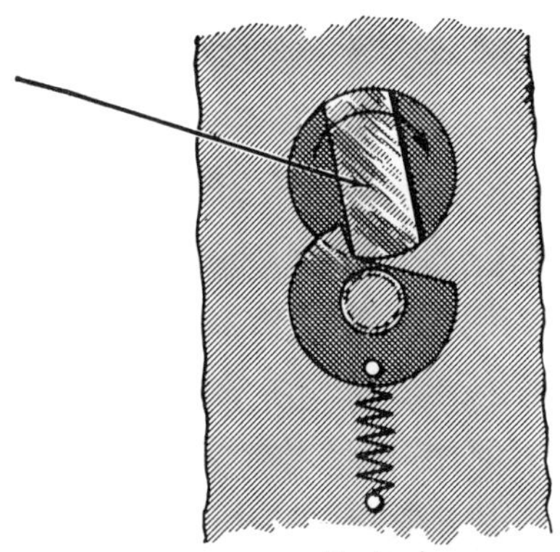

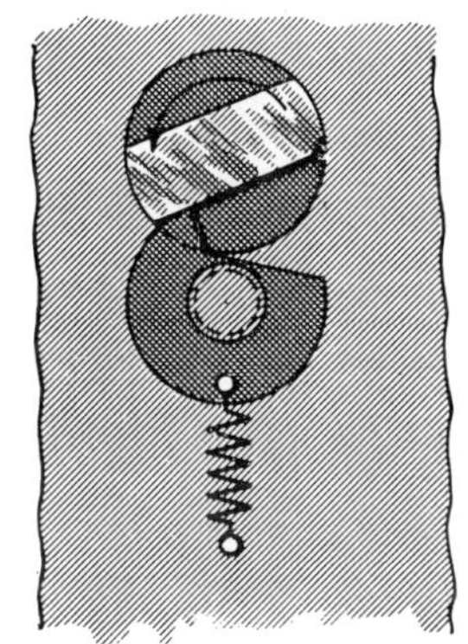

LUG ON SHAFT pushes the notched disk free during normal rotation. Disk periphery stops lug to prevent reverse rotation.

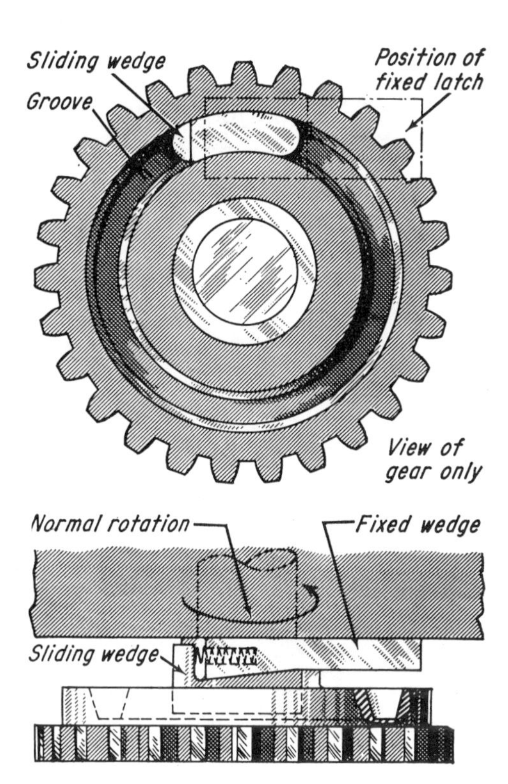

FIXED WEDGE AND SLIDING WEDGE tend to disengage when the gear is turning clockwise. The wedges jam in reverse direction.

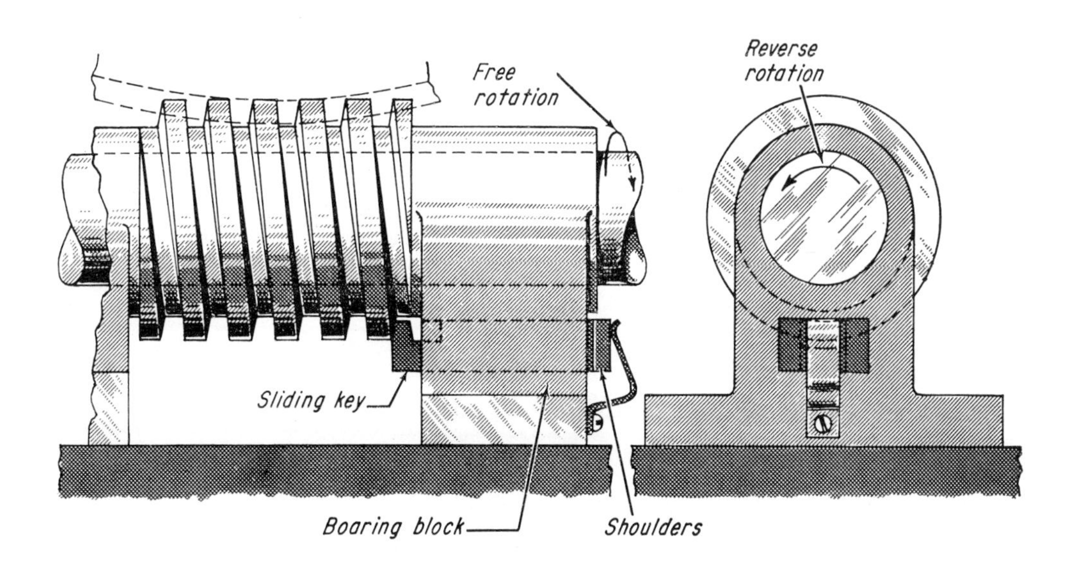

SLIDING KEY has tooth which engages the worm threads. In reverse rotation key is pulled in until its shoulders contact block.

ADJUSTABLE STROKE MECHANISMS

PREBEN W. JENSEN

Adjustable slider drive

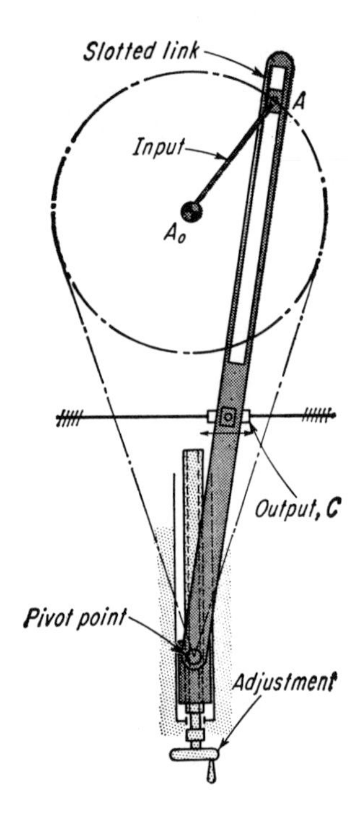

Shifting the pivot point by means of the adjusting screw changes the stroke of the output rod.

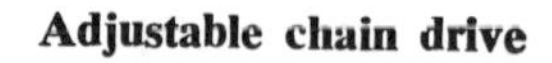

Adjustable chain drive

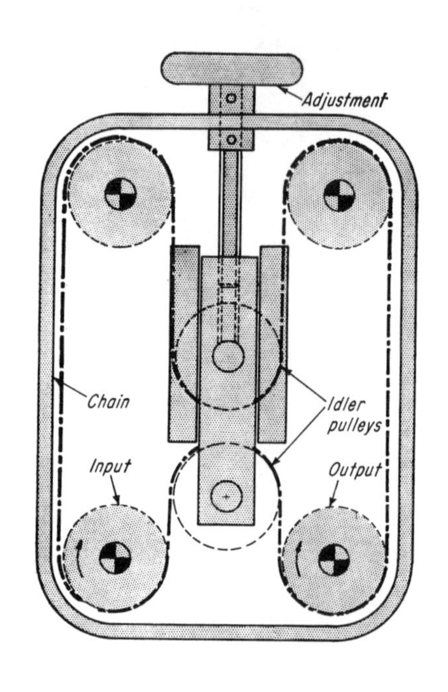

Synchronization between input and output shafts is varied by shifting the two idler pulleys by means of the adjusting screw.

Adjustable clutch drive

As the input crank makes a full rotation, the one-way clutch housing oscillates to produce an output rotation consisting of a series of pulses in one direction. Moving the adjusting block to right or left changes the length of the strokes.

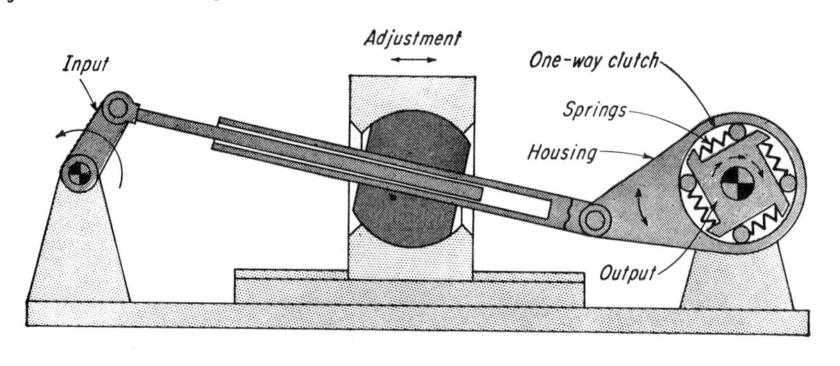

Adjustable-pivot drive

The driving pin rotates around the input center, but because the pivot is stationary with respect to the frame, the end of the slotted link produces a non-circular coupler curve and a fast advance and slow return in the output link. The stroke is varied by rotating the pivot to another position.

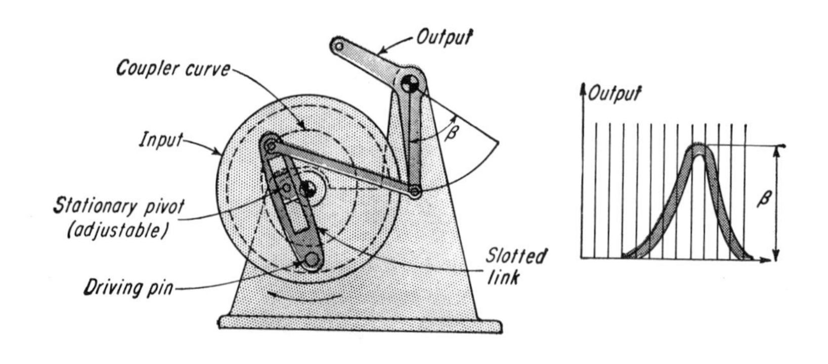

ADJUSTABLE-OUTPUT MECHANISMS

Linkage motion adjuster

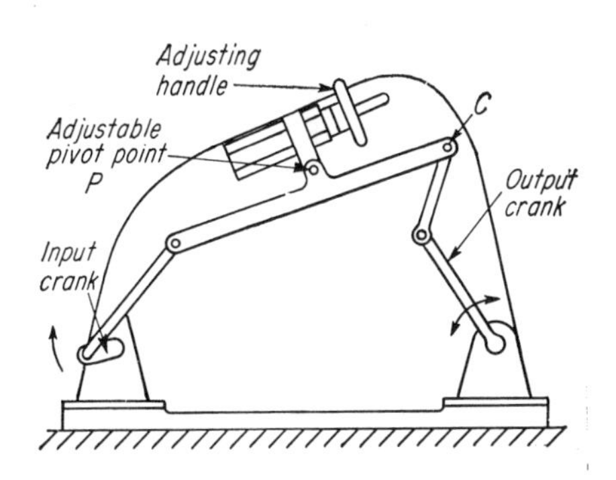

Here the motion and timing of the output link can be varied *during operation* by shifting the pivot point of the intermediate link of the 6-bar linkage illustrated. Rotation of the input crank causes point C to oscillate around the pivot point P. This in turn imparts an oscillating motion to the output crank. Shifting of point P is accomplished by a screw device.

Cam motion adjuster

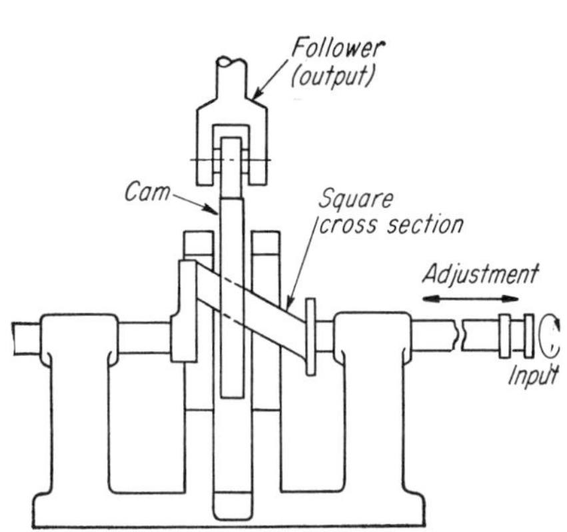

The output motion of the cam follower is varied by linearly shifting the input shaft to the right or left during operation. The cam has a square hole which fits over the square cross section of the crank shaft. Rotation of the input shaft causes eccentric motion in the cam. Shifting the input shaft to the right, for instance, causes the cam to move radially outward, thus increasing the stroke of the follower.

Double-cam mechanism

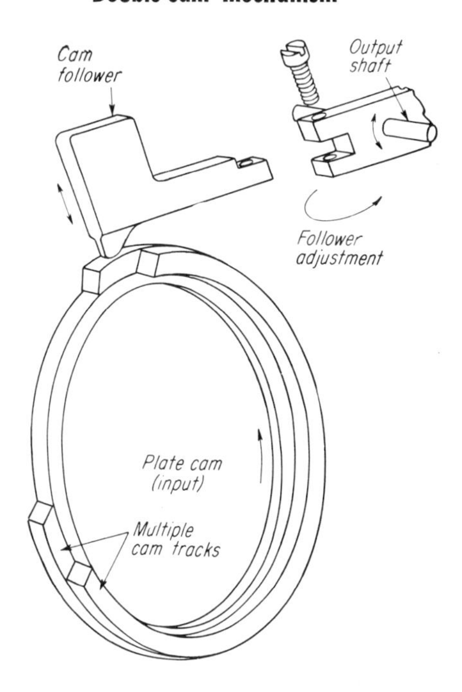

This is a simple but effective technique for changing the timing of a cam. The follower can be adjusted in the horizontal plane but is restricted in the vertical plane. The plate cam contains two or more cam tracks.

Valve stroke adjuster

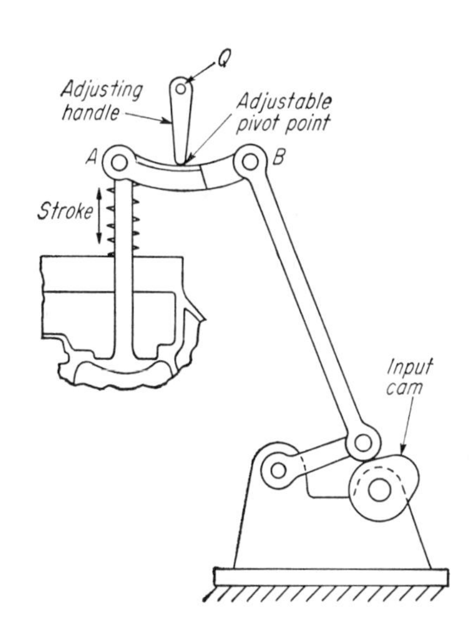

This mechanism adjusts the stroke of valves of combustion-engines. One link has a curved surface and pivots around an adjustable pivot point. Rotating the adjusting link changes the proportion of strokes of points A and B and hence of the valve. The center of curvature of the curve link is at point Q.

3-D mechanism

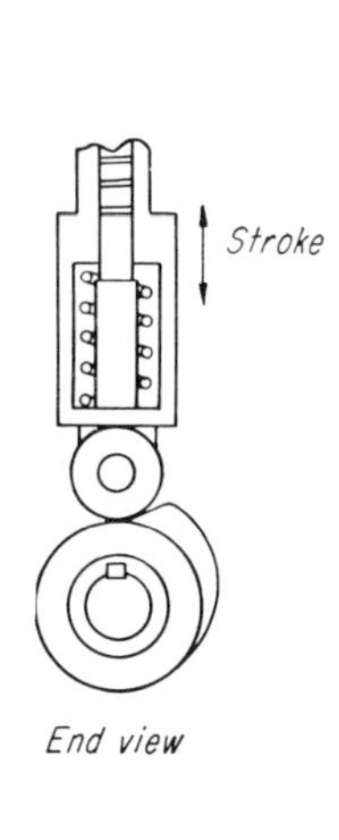

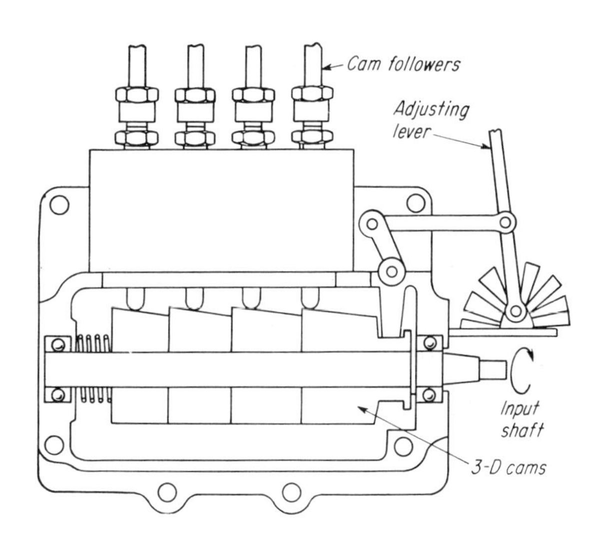

Output motions of four followers are varied during the rotation by shifting the quadruple 3-D cam to the right or left. Linear shift is made by means of adjustment lever which can be released in any of the 6 positions.

Piston-stroke adjuster

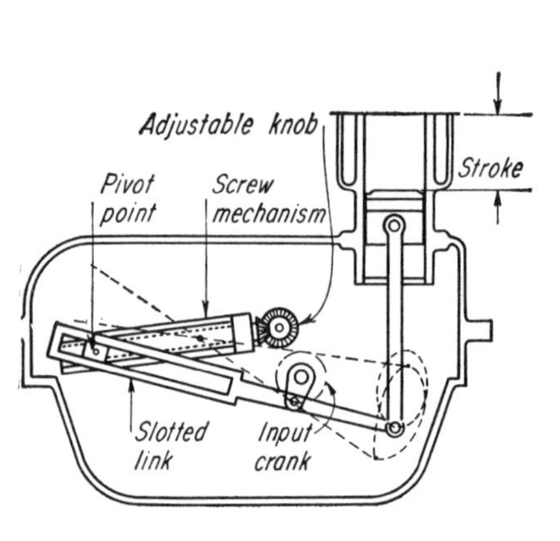

The input crank oscillates the slotted link to drive the piston up and down. The position of the pivot point can be adjusted by means of the screw mechanism even while the piston is under full load.

Shaft synchronizer

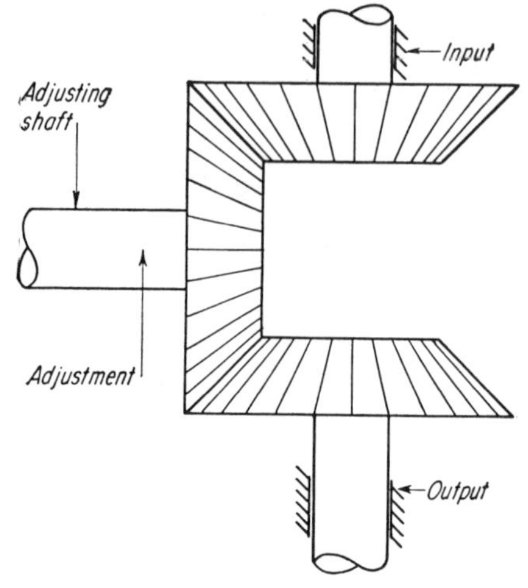

Actual position of the adjusting shaft is normally kept constant. The input then drives the output by means of the bevel gears. Rotating the adjusting shaft in a plane at right angle to the input-output line changes the relative radial position of the input and output shafts, used for introducing a torque into the system while running, synchronizing the input and output shafts, or changing the timing of a cam on the output shaft.

Eccentric pivot point

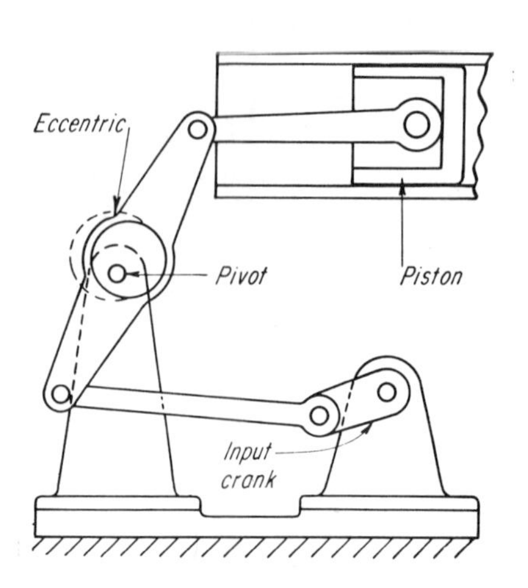

Rotation of the input crank reciprocates the piston. The stroke depends on position of the pivot point which is easily adjusted, even during rotation, by rotating the eccentric shaft.

Mechanical positioner varies machine element relation

Claimed to be highly accurate, the Phase Variator is a 1:1 ratio mechanical unit for varying the relative position of one machine or machine element to another. Installed between the element to be controlled and its drive, the endless chain device assures positive, simple adjustment of timing, registration, sequencing, indexing, stroking, cam positioning and synchronizing.

It is infinitely adjustable within a 360-degree cycle and is easily installed in any position. Adjustments can be made while unit is running or stopped. Size is 3¼ x 8¼ x 12 in.

Candy Mfg Co.

Gear shift mechanism

Variable-speed strip recorders make the tester's life simpler all round: First off, you can choose a chart speed that will produce easily interpreted data. Then you save on expensive paper since you get a longer recording period. These various advantages are to be had with a new strip chart recorder made by F. L. Mosley Co, Pasadena, Calif — model 7100A — which provides 12 chart speeds at the twist of a knob.

Drive mechanism. With 22 nylon gears in continuous mesh, a series of cam-controlled idler spur gears are used to engage selected speed reduction gear sets. The synchronous motor is connected either to the optional 10 to 1 speed reducer or through transfer gears to the motor drive gear.

Any of the 12 output gears may be coupled to the output spline and chart drive by the manually selected idlers. Mounted on spring-loaded arms, the idlers are controlled by a 12-position detented cam. By rotating the cam the appropriate idler is pivoted into simultaneous mesh with the output of the selected speed reducer and the output drive spline.

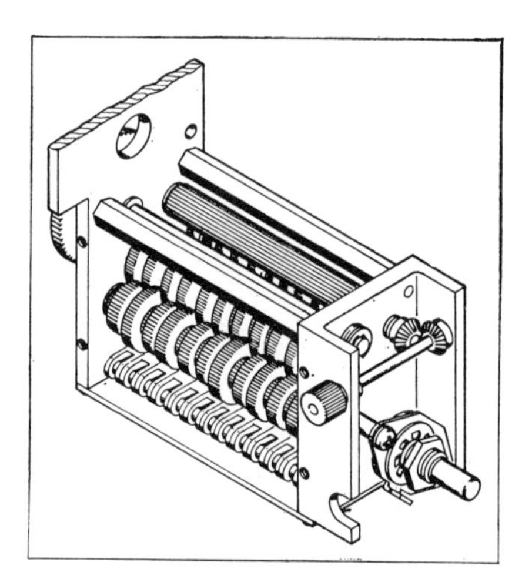

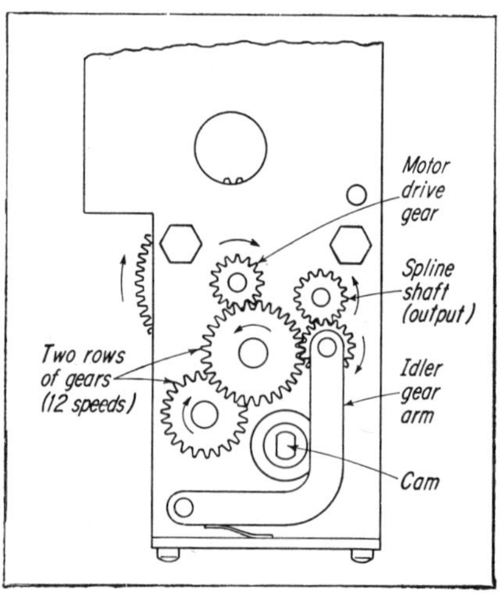

Shifter Designs for Sliding Components

FRED ROGERS

Gears, clutches, valve mechanisms, index plungers, lock bolts are but a few of the many components that require shifting. To a large degree these components depend upon shifting shoes, rollers, rings or yokes for motion. Classification of shifting methods are: 1) where an oscillating movement of the fork causes lateral movement of the part to be shifted; and 2) where the combination of the yoke and shoe is moved parallel to the shifted member.

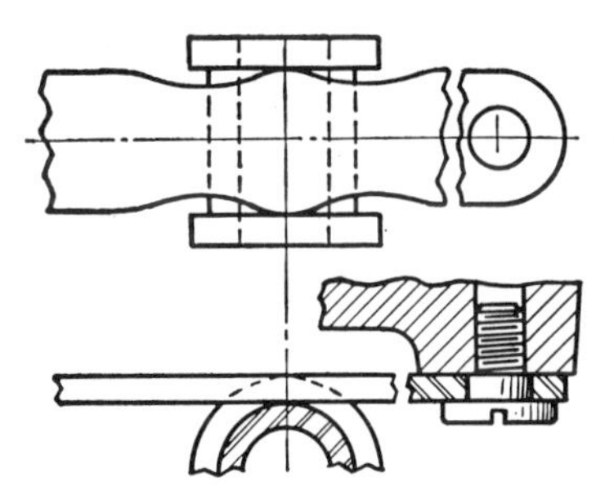

Fig. 1—Radiused rectangular steel bar contacts the spool flanges on one side only. Bar pivots on a shouldered screw. Handle end of bar rests in a housing wall slot. A detent may be added to hold bar in engaged or disengaged position while taking thrust of spool.

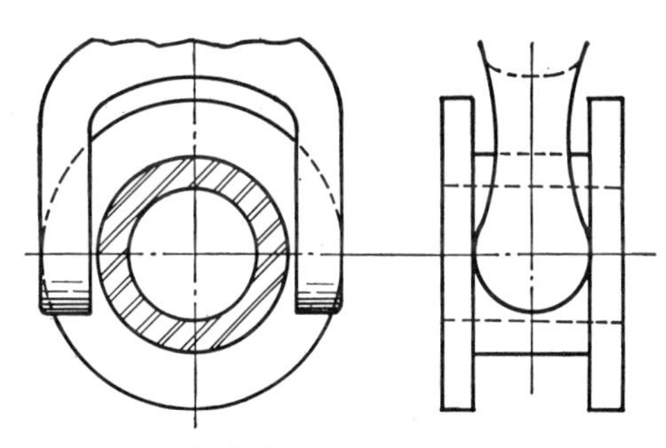

Fig. 2—Cast forked lever makes contact on two sides of the spool. Method is for light work and where longer spool movements than can be obtained in Fig. 1 are required. The longer the movement of the spool, the more circular should be the contacting lever surfaces.

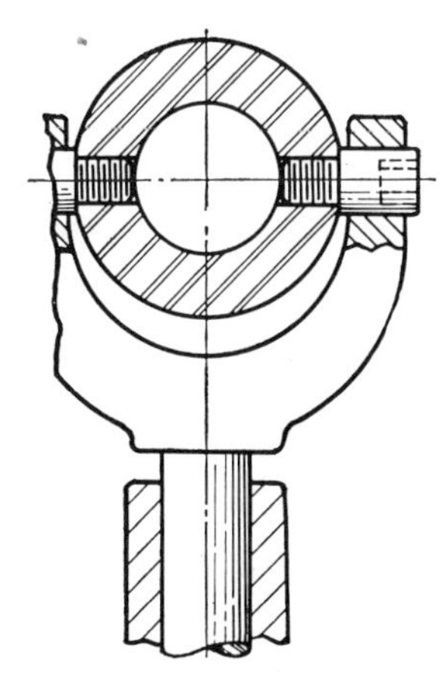

Fig. 3—Shifting may be done by two diametrically placed pins. The pins may be staked in yoke, or they may be screwed into spool. Latter method necessitates a slide fit of the yoke in a slot in the shifting lever since the center of the sleeve moves in a straight line.

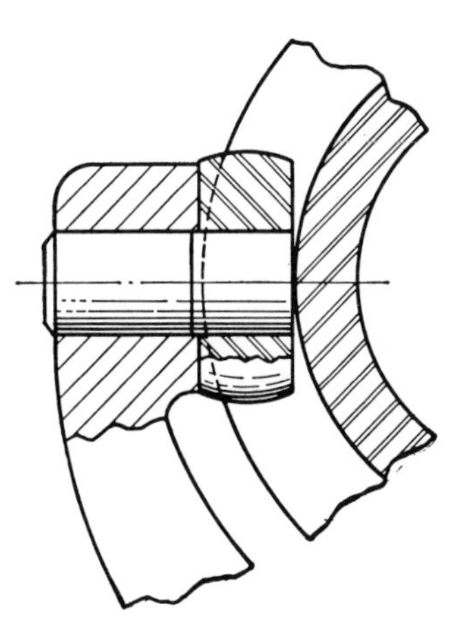

Fig. 4—Crowned rollers are used for heavy work and friction-free shifting effort. Flat face rollers are not recommended since flange surface speed increases from the center outward causing slippage and wear. To prevent binding, roll diameter is slightly less than groove width.

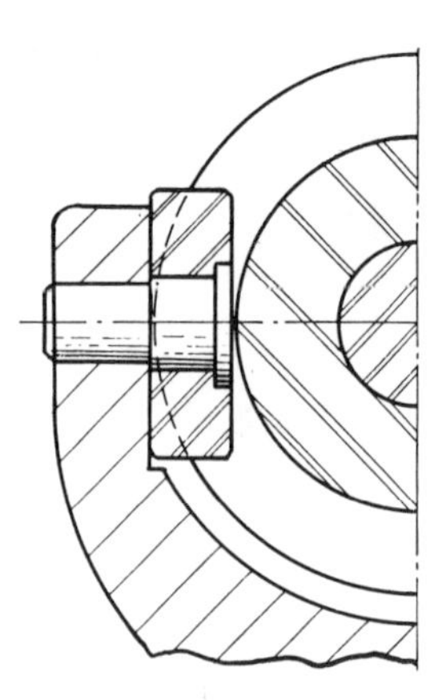

Fig. 5—Square and rectangular shaped block shoes are used for shifting gears and clutches. One- and two-diameter pins are used. One-diameter pin is staked to yoke ear. Shouldered pin controls clearance and counter-bored block prevents loss of shoe in assembling.

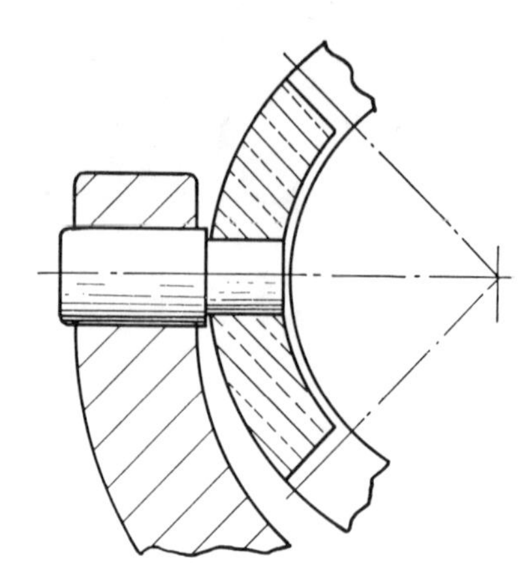

Fig. 6—Curved shoes are used with large diameter shifting spools requiring short lateral movement only. Spool must be set so that the sliding distance of the shoes perpendicular to their lateral displacement is the same above and below the horizontal center line.

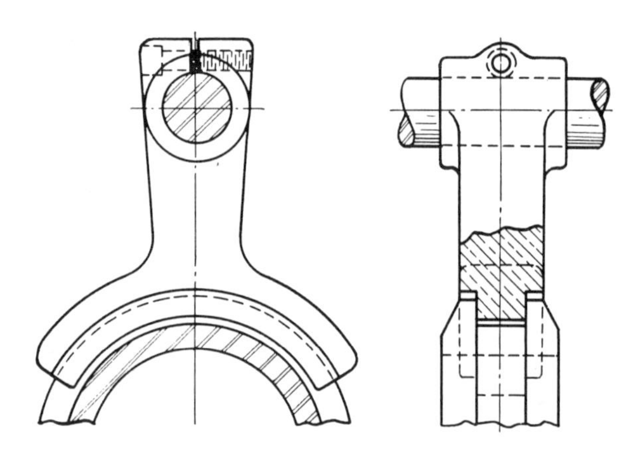

Fig. 7—One piece yoke and shoe. Unit moves parallel to the shifted element rather than pivoting about a center. Cross sections of shank between the hub and shoe of fork should be proportioned to resist excessive lateral deflection. Shoe covers a 90 deg arc on the spool.

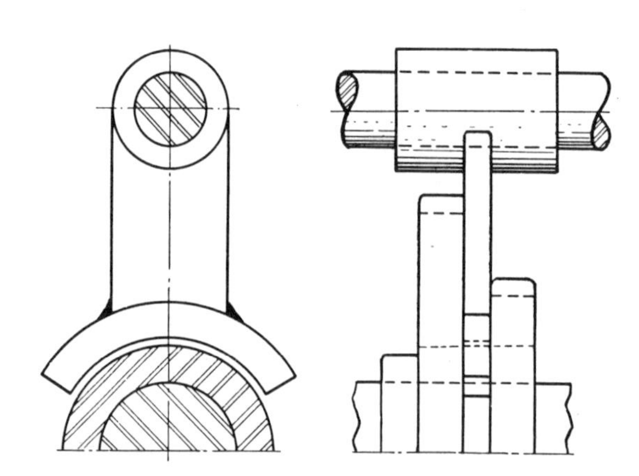

Fig. 8—Welded yoke and shoe for shifting a gear cluster. Hub and shoe are welded to a shank. Clearance between sides of gear teeth and shank is required to avoid rubbing action. Material is soft and operating in an oil bath will give satisfactory service for light work.

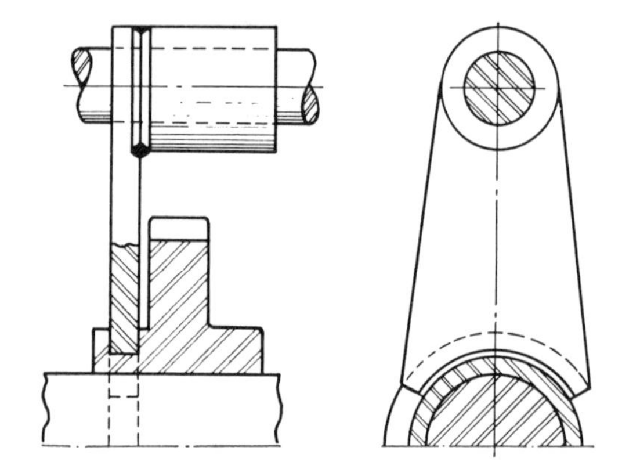

Fig. 9—Shank used as shoe to engage spool cut in hub of gear to be shifted. Shoe end contact with spool is 90 deg. Shank and shoe material is soft. Operating in an oil bath method is satisfactory for light work. Shank is shown welded to hub but other methods may be used.

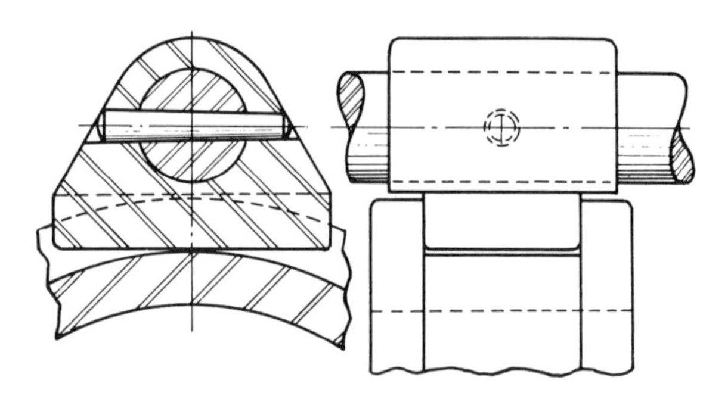

Fig. 10—Compact arrangement that may be suitable where clearance space between shifter rod and spool is small. Provision for preventing shifter rod from twisting may be desirable to avoid shoe binding on bottom of spool groove or shoe may be curved to fit groove.

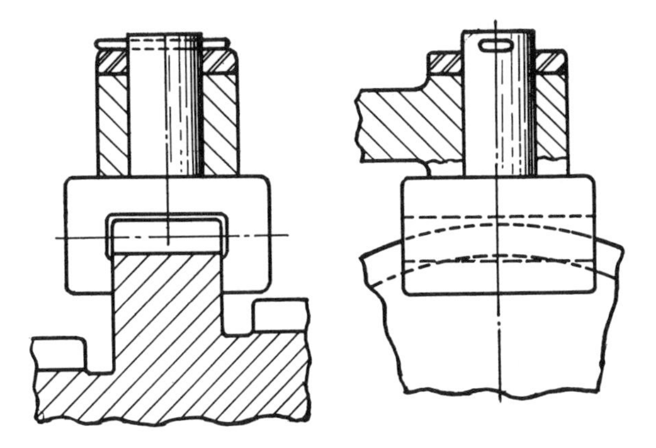

Fig. 11—Straddle type shifting shoes for gear clusters. Either the shoe should be undercut or sides of gear teeth relieved to avoid contact. A straddle shoe may be used over a cluster of gears or over a flange depending upon available space. As shown shoe pivots on pin in shank.

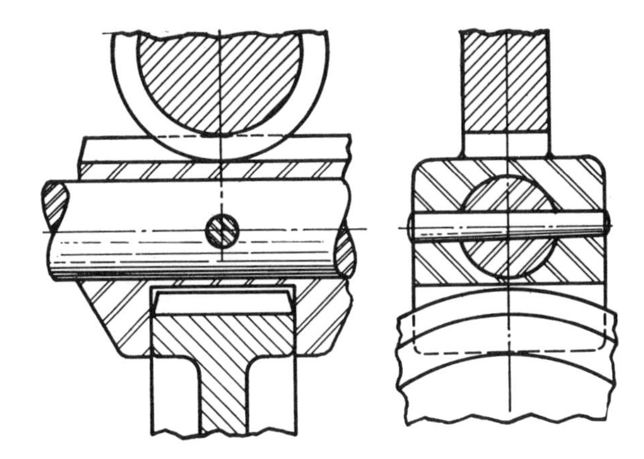

Fig. 12—Compact straddle type shifting design. Rack teeth cut integrally mesh with a pinion. Pinion is shaft connected to a crank rod riding a quadrant with positioning stop pins outside the machine. Gear to be shifted has angularly relieved teeth. Large motions are possible.

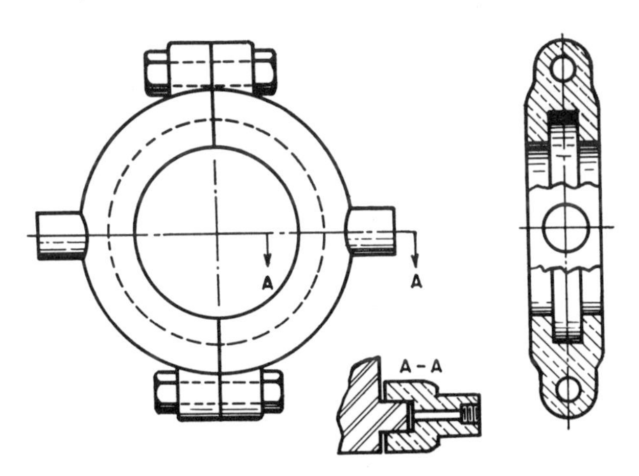

Fig. 13—Thrust or shifting collars offer 360 deg bearing area. Forked shifting lever engages trunnion mounts. Split collar permits assembly over spool flange. Collar fits sides of the single flange of the shifting member. Drilled hole in trunnion is for oil or grease fitting.

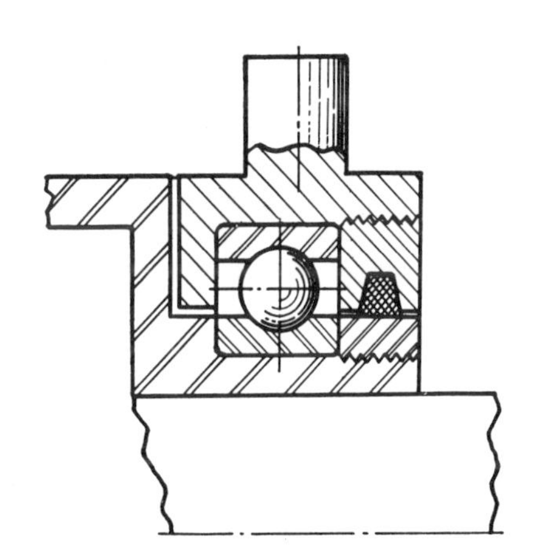

Fig. 15—To minimize sliding friction between shoe and spool, ball bearing may be used in shifting collar. Precautions must be taken to ensure a dustproof and well lubricated assembly. Deep groove or angular contact bearings are used to absorb thrust in one direction.

Fig. 14—Solid shifting collar. Hub of element being shifted is grooved to receive a split bronze ring. Ring is placed in hub groove and the shifting collar is fastened with socket head cap screws. Collar bore is 1/64 in. oversize for freedom.

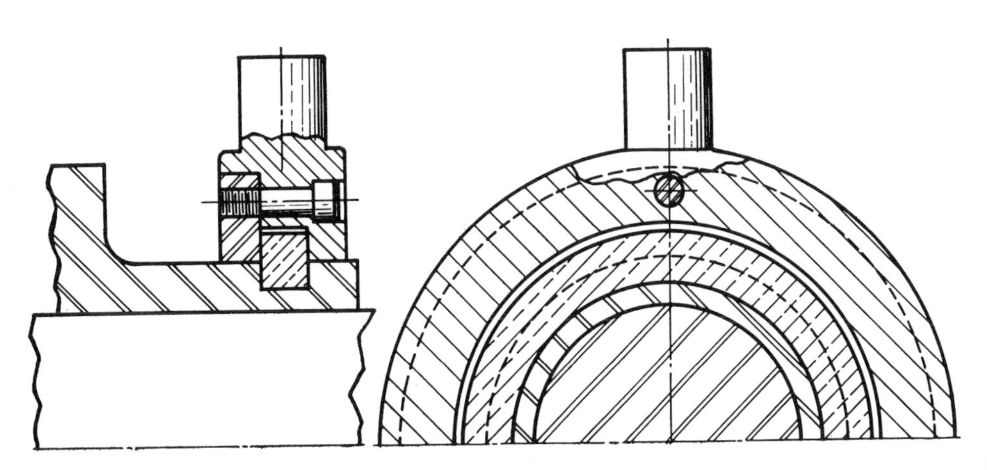

SHIFTING MECHANISMS FOR GEARS AND CLUTCHES

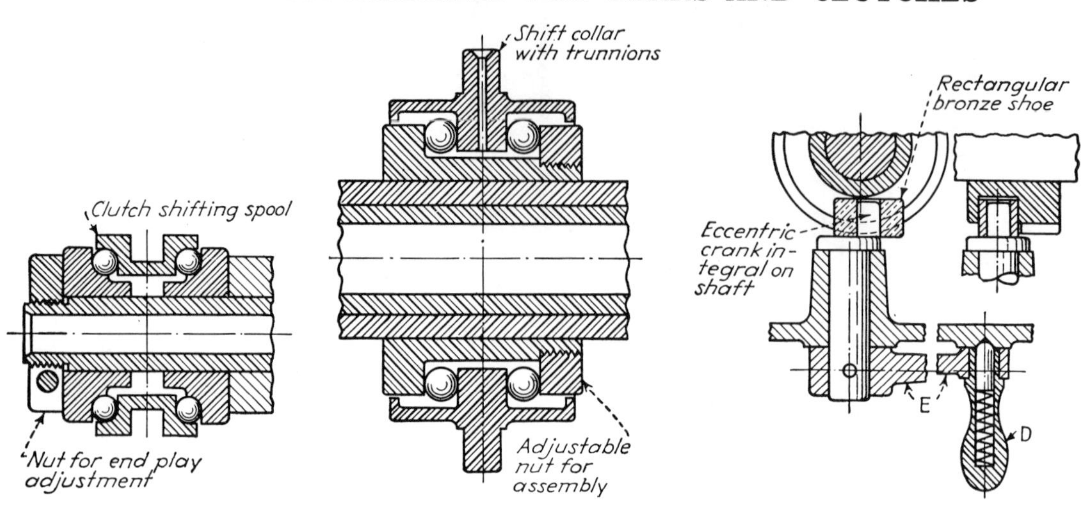

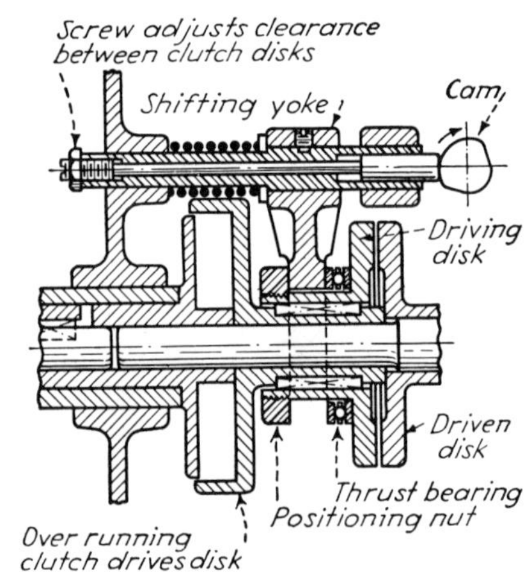

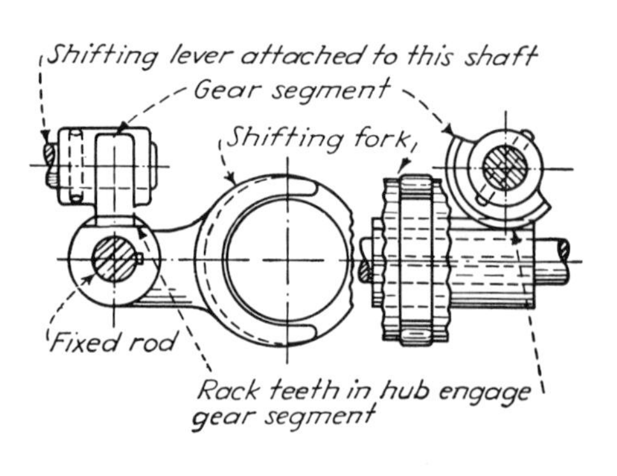

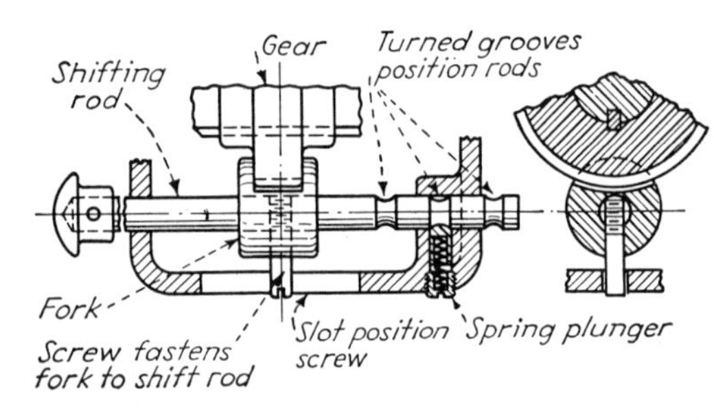

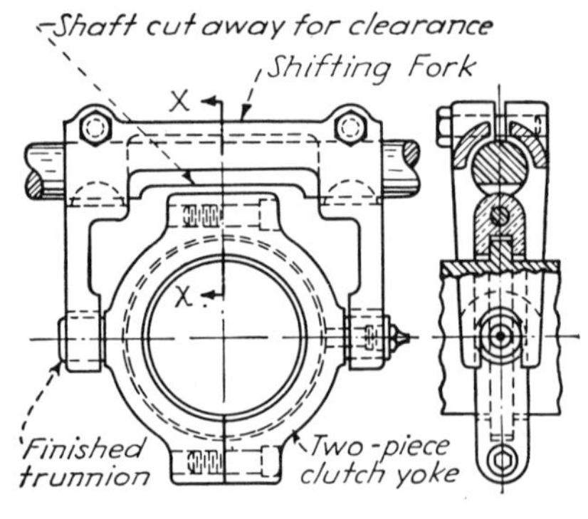

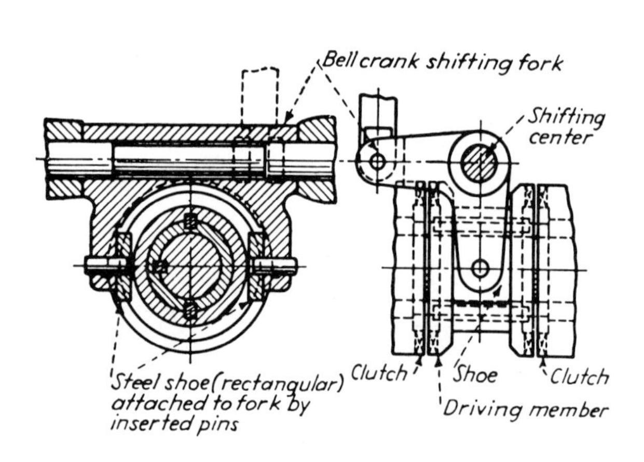

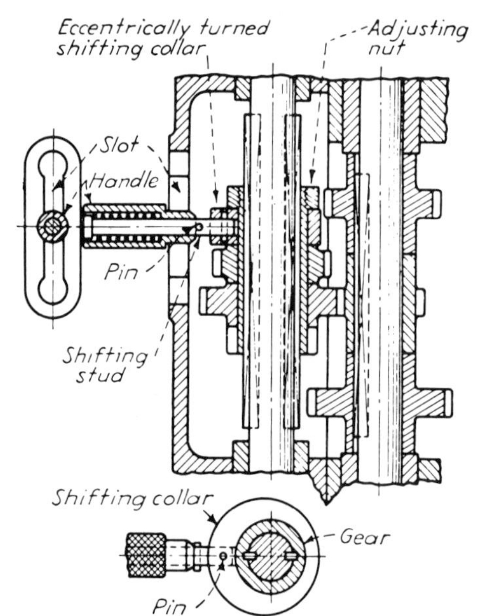

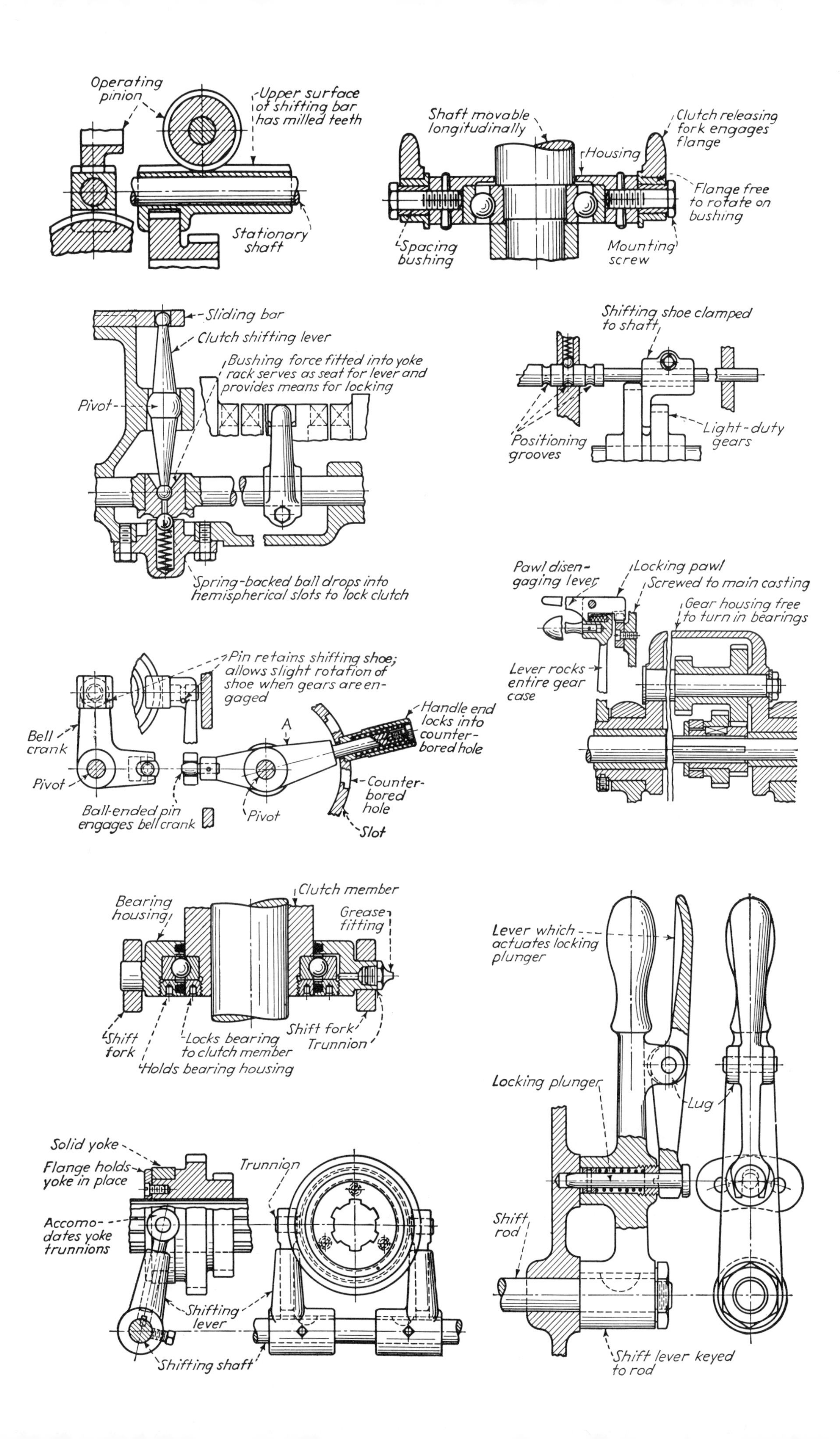
Operating pinion
Upper surface of shifting bar has milled teeth
Stationary shaft
Shaft movable longitudinally
Housing
Clutch releasing fork engages flange
Flange free to rotate on bushing
Spacing bushing
Mounting screw
Sliding bar
Clutch shifting lever
Bushing force fitted into yoke rack serves as seat for lever and provides means for locking
Pivot
Spring-backed ball drops into hemispherical slots to lock clutch
Shifting shoe clamped to shaft
Light-duty gears
Positioning grooves
Pawl disengaging lever
Locking pawl
Screwed to main casting
Gear housing free to turn in bearings
Lever rocks entire gear case
Pin retains shifting shoe; allows slight rotation of shoe when gears are engaged
Bell crank
Pivot
A
Handle end locks into counter-bored hole
Counter-bored hole
Ball-ended pin engages bell crank
Pivot
Slot
Bearing housing
Clutch member
Grease fitting
Shift fork
Locks bearing to clutch member
Shift fork
Trunnion
Holds bearing housing
Lever which actuates locking plunger
Locking plunger
Lug
Shift rod
Shift lever keyed to rod
Solid yoke
Flange holds yoke in place
Accomodates yoke trunnions
Trunnion
Shifting lever
Shifting shaft

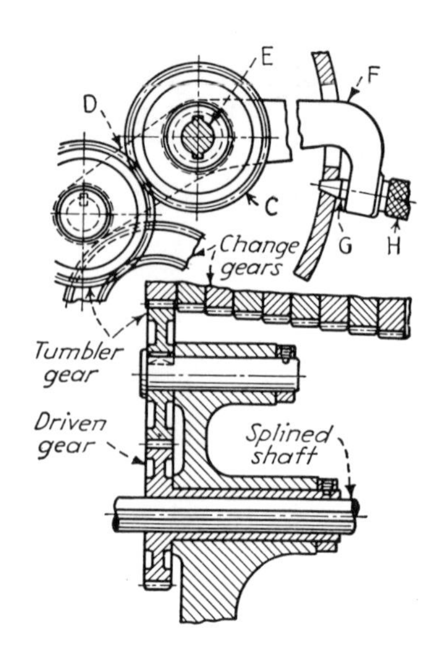
E
F
D
C
Change gears
G
H
Tumbler gear
Driven gear
Splined shaft

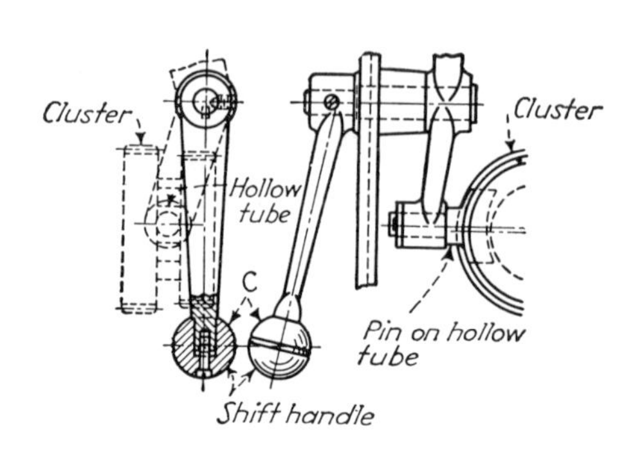
Cluster
Cluster
Hollow tube
C
Pin on hollow tube
Shift handle

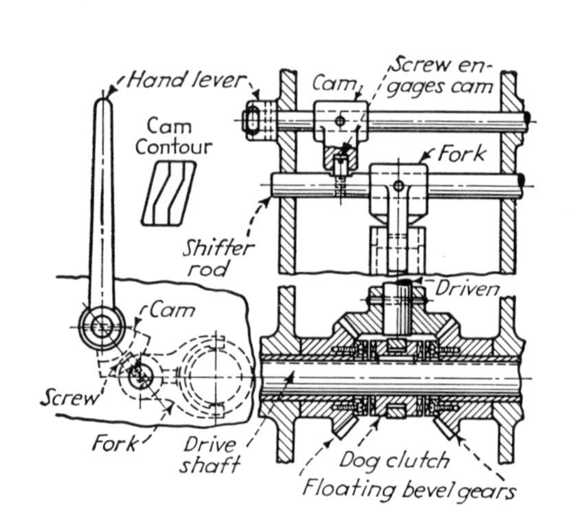
Hand lever
Cam
Screw engages cam
Cam Contour
Fork
Shifter rod
Driven
Cam
Screw
Fork
Drive shaft
Dog clutch
Floating bevel gears

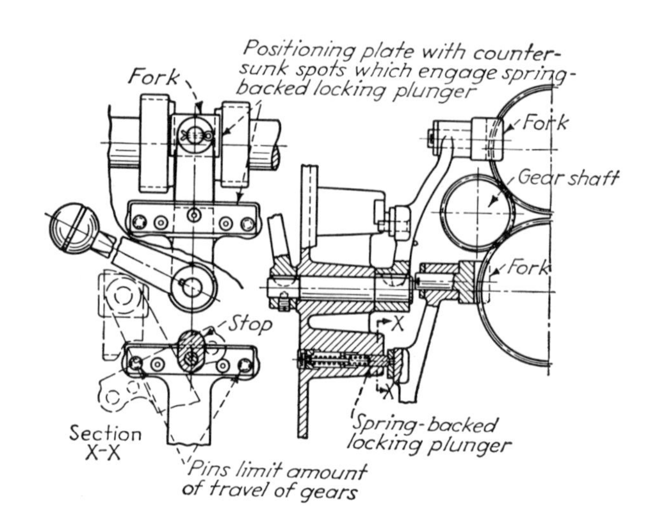
Positioning plate with countersunk spots which engage spring-backed locking plunger
Fork
Fork
Gear shaft
Fork
Stop
X
X
Section X-X
Spring-backed locking plunger
Pins limit amount of travel of gears

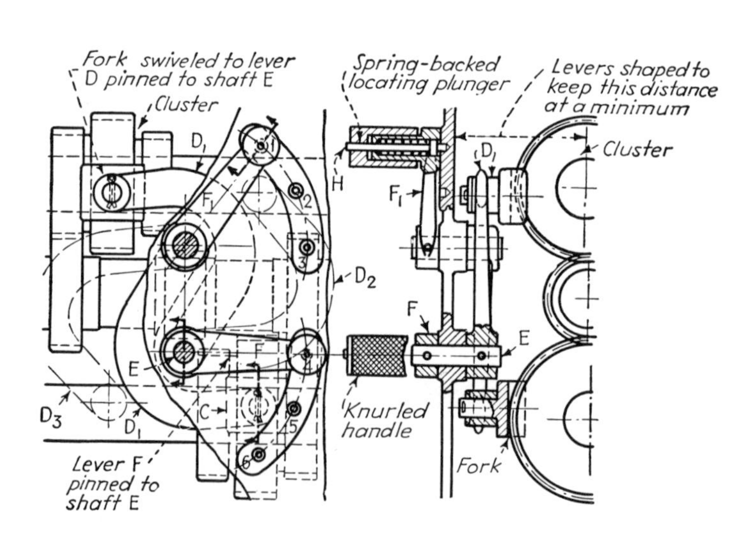
Fork swiveled to lever D pinned to shaft E
Cluster
Spring-backed locating plunger
Levers shaped to keep this distance at a minimum
Cluster
H
F
E
Knurled handle
Fork
Lever F pinned to shaft E

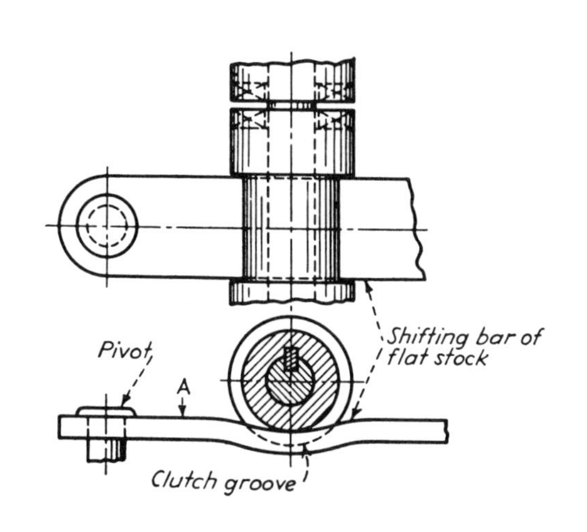
Pivot
A
Shifting bar of flat stock
Clutch groove

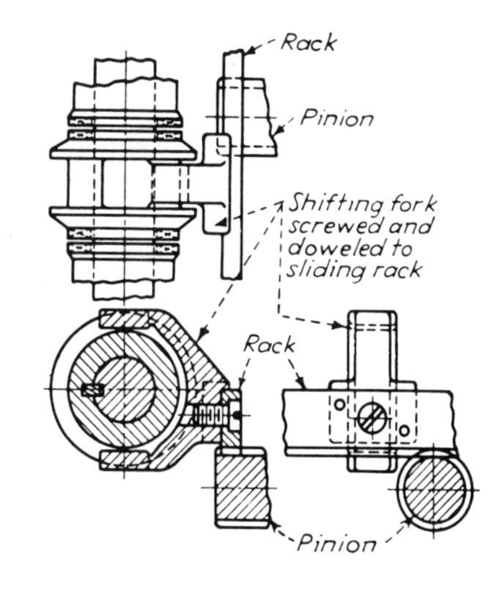
Rack
Pinion
Shifting fork screwed and doweled to sliding rack
Rack
Pinion

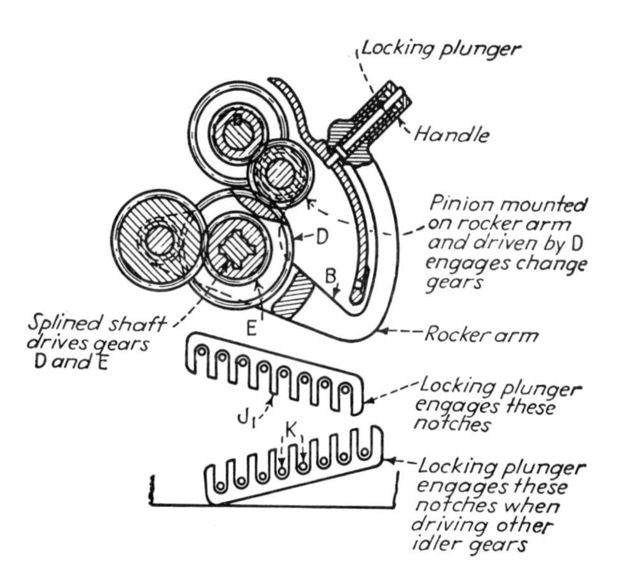
Locking plunger
Handle
Pinion mounted on rocker arm and driven by D engages change gears
D
B
E
Splined shaft drives gears D and E
Rocker arm
Locking plunger engages these notches
J1
K1
Locking plunger engages these notches when driving other idler gears

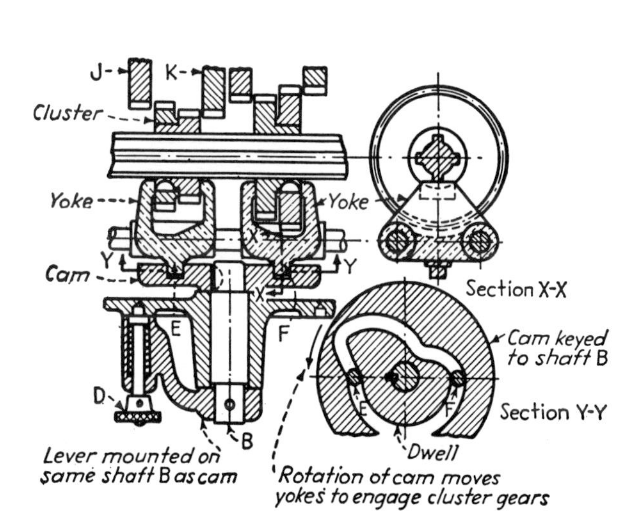
J
K
Cluster
Yoke
Yoke
Y
Y
Cam
X
Section X-X
E
F
D
B
Cam keyed to shaft B
Section Y-Y
Dwell
Lever mounted on same shaft B as cam
Rotation of cam moves yokes to engage cluster gears

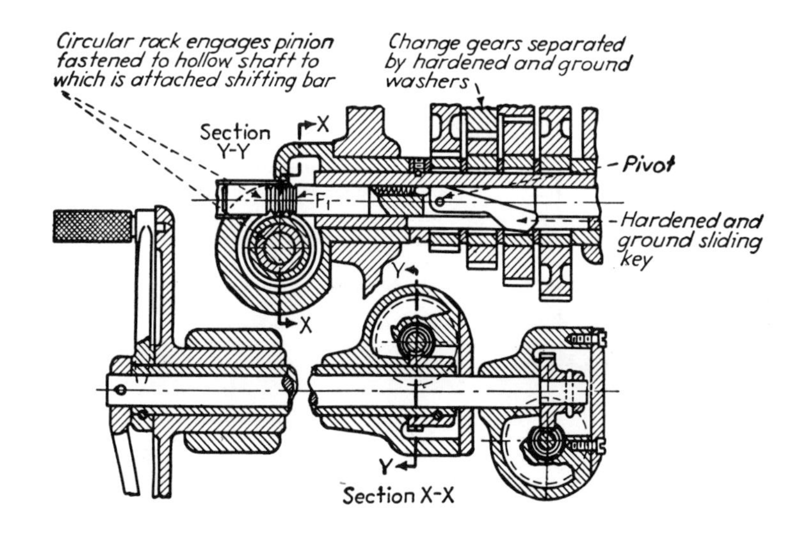
Circular rack engages pinion fastened to hollow shaft to which is attached shifting bar
Change gears separated by hardened and ground washers
Section Y-Y
X
Pivot
F1
Hardened and ground sliding key
Y
X
Y
Section X-X

Gear-Shift Arrangements

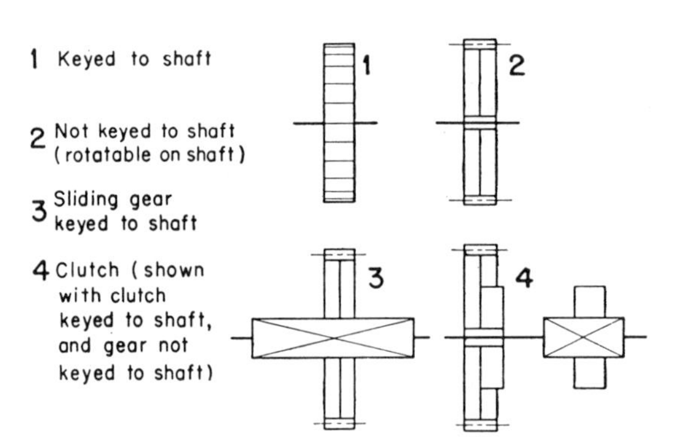

Fig. 1. Schematic symbols used in the illustrations to represent gears and clutches.

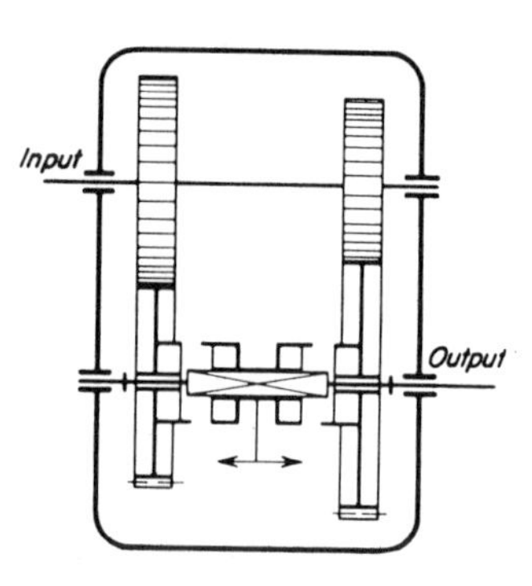

Fig. 2. Double-clutch drive. Two pairs of gears permanently in mesh. Pair I or II transmits motion to output shaft depending on position of coupling; other pair idles. Coupling shown in neutral position with both gear pairs idle. Herring-bone gears recommended for quiet running.

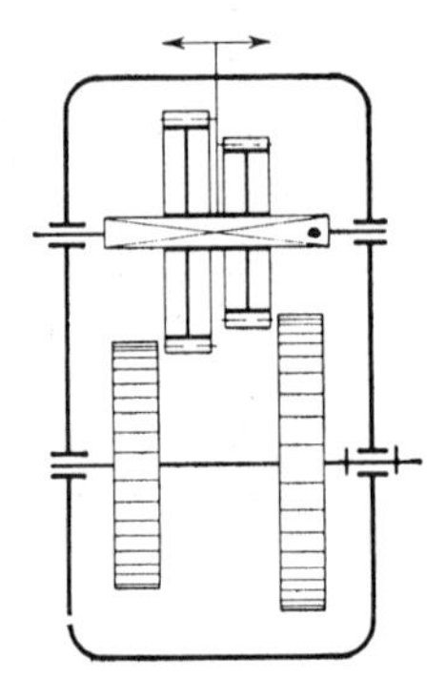

Fig. 3. Sliding-change drive. Gears meshed by lateral sliding. Up to three gears can be mounted on sliding sleeve. Only one pair in mesh in any operating position. Drive simpler, cheaper and more extensively used than drive of Fig. 2. Chamfering side of teeth facilitates engagement.

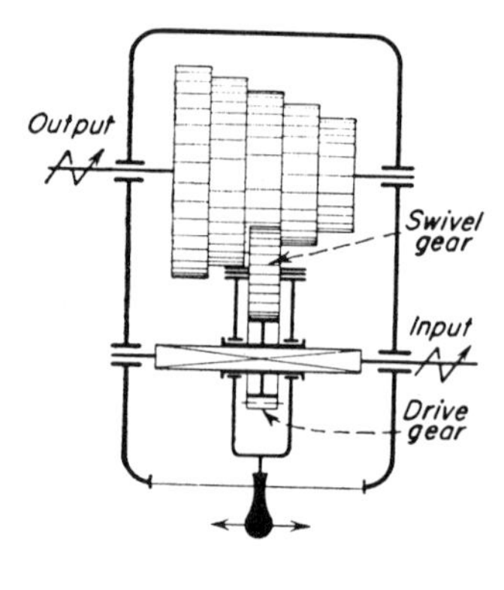

Fig. 4. Swivel-gear drive. Output gears are fastened to shaft. Handle is pushed down, then shifted laterally to obtain transmission through any output gear. Not suitable for transmission of large torques because swivel gear tends to vibrate. Over-all ratio should not exceed 1:3.

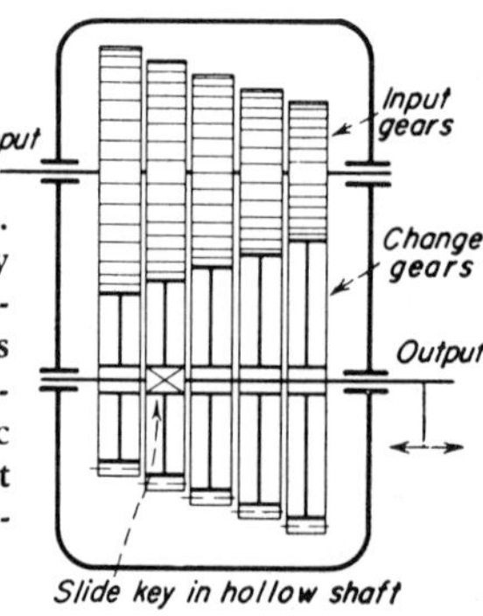

Fig. 5. Slide-key drive. Spring-loaded slide key rides inside hollow output shaft. Slide key snaps out of shaft when in position to lock a specific change gear to output shaft. No central position is shown.

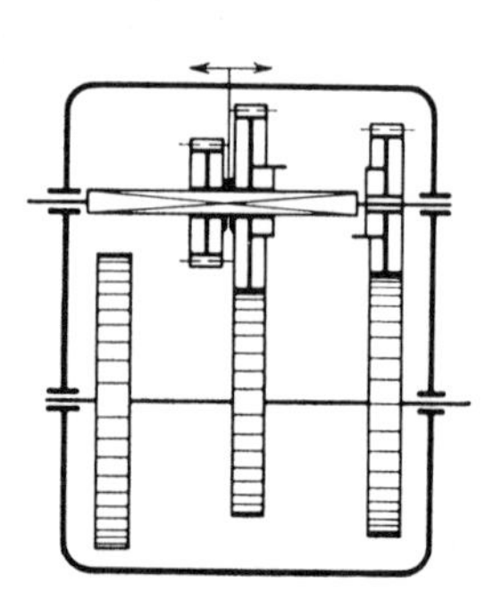

Fig. 6. Combination coupling and slide gears. Three ratios: direct mesh for ratios I and II; third ratio transmitted through gears II and III which couple together.

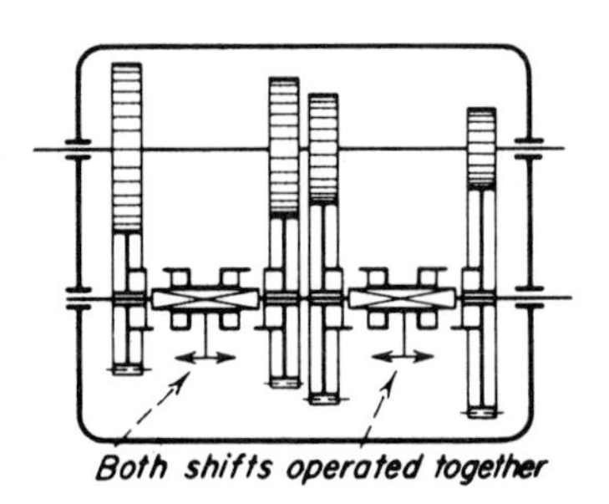

Fig. 7. Double-shift drive. One shift must always be in a neutral position which may require both levers to be shifted when making a change. However only two shafts are used to achieve four ratios.

SIGMUND RAPPAPORT

Project Supervisor, Ford Instrument Company
Adjunct Professor of Kinematics
Polytechnic Institute of Brooklyn

13 ways of arranging gears and clutches to obtain changes in speed ratios.

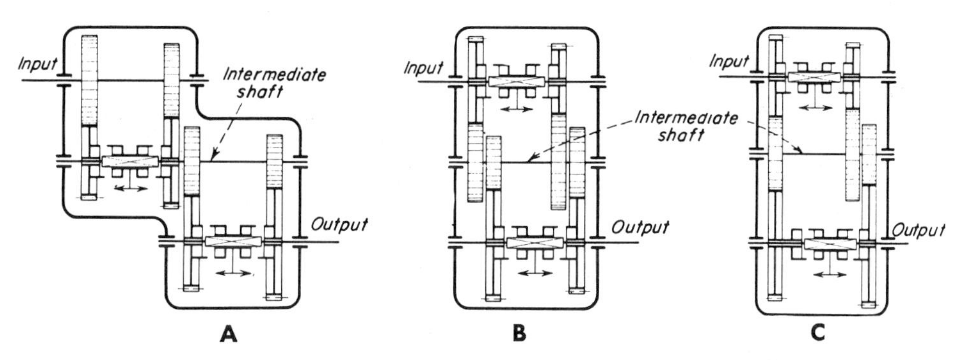

Fig. 8. (A) Triple shaft drive gives four ratios. Output of first drive serves as input for second. Presence of intermediate shaft obviates necessity for always insuring that one shift is in neutral position. Wrong shift lever position can not cause damage. (B) Space-saving modification. Coupling is on shaft *A* instead of intermediate shaft. (C) Still more space saved if one gear replaces a pair on intermediate shaft. Ratios can be calculated to allow this.

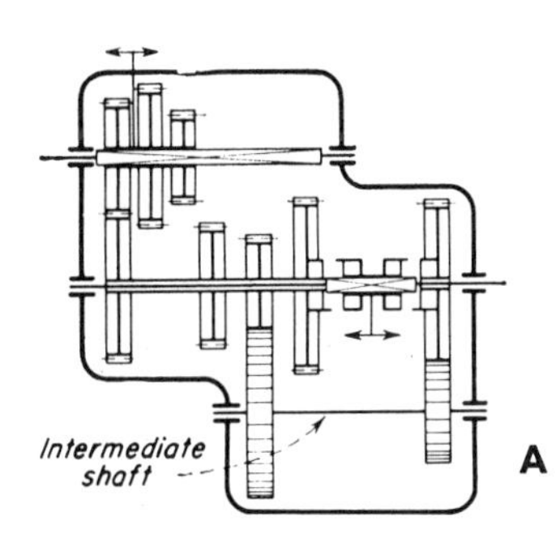

Fig. 9. Six ratios available with two couplings and (A) ten gears, (B) eight gears. Up to six gears in permanent mesh. It is not necessary to insure that one shift is in neutral.

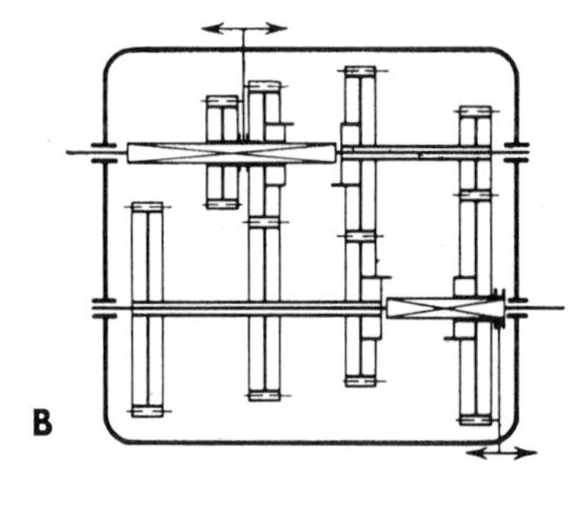

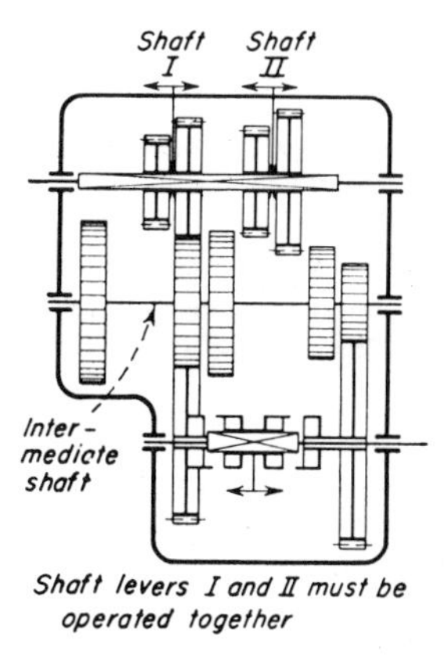

Fig. 10. Eight-ratio drive uses two slide gears and a coupling. This arrangement reduces number of parts and meshes. Position of shifts I and II are interdependent. One shift must be in neutral if other is in mesh.

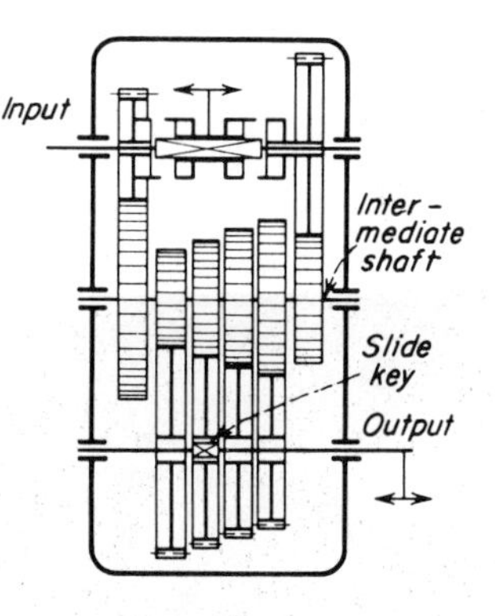

Fig. 11. Eight ratios; coupled gear drive and slide-key drive in series. Comparatively low strength of slide key limits drive to small torque.

Data based on material and sketches in AWF und VDMA Getrieblaetter, published by Ausschuss fuer Getriebe beim Ausschuss fuer Wirtschaftiche Fertigung, Leipzig, Germany.

14 ADJUSTING

Here is a selection of some basic arrangements that provide and hold mechanical adjustment.

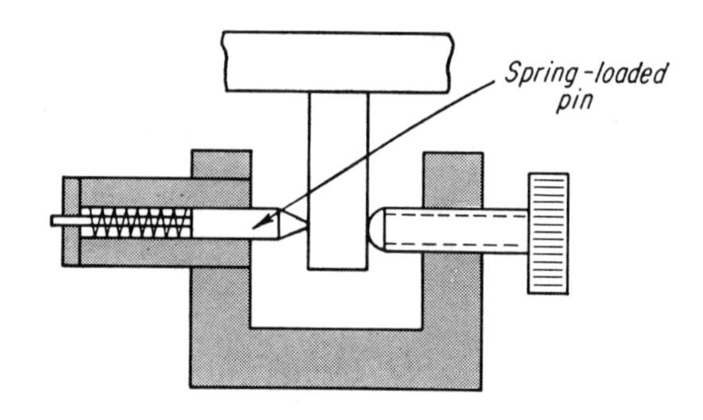

Spring-loaded pin . . .
supplies counterforce against which adjustment force must always act. A levelling foot would work against gravity—but for most other setups a spring is needed to give counterforce.

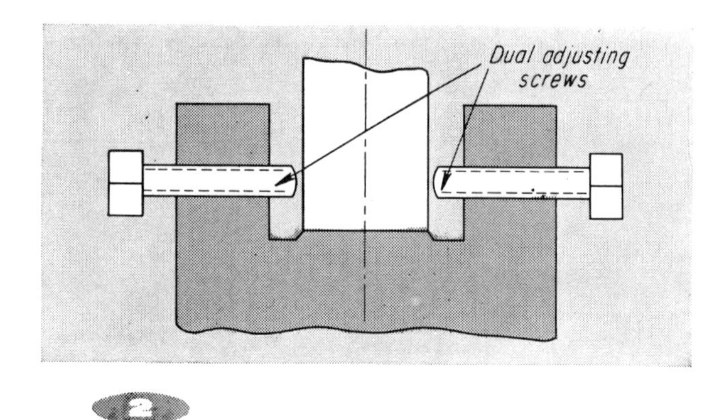

Dual screws . . .
provide inelastic counterforce. Backing-off one screw and tightening the other allows extremely small adjustments to be made. Also, once adjusted, the position remains solid against any forces tending to move device out of adjustment.

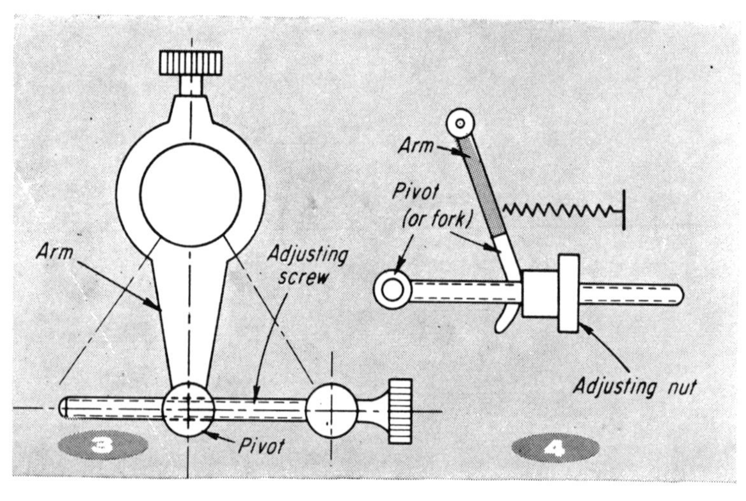

Swivel motion . . .
is necessary in (3) between adjusting screw and arm because of circular locus of female thread in the actuated member. Similar action (4) requires either the screw to be pivoted or the arm to be forked.

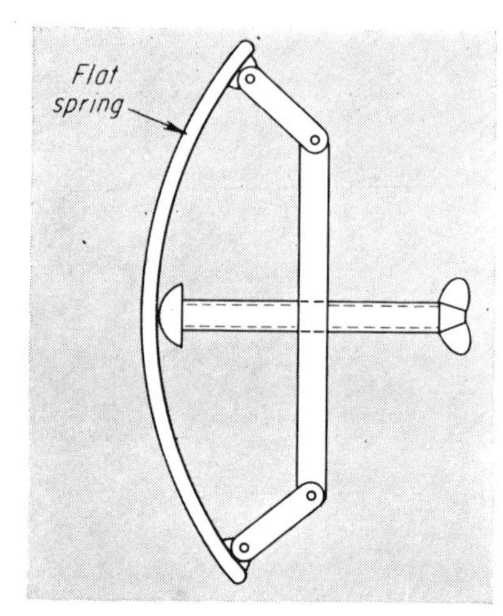

Arc-drafting guide . . .
is example of adjusting device where one of its own components, the flat spring, both supplies the counterforce and performs the mechanism's main function—guiding the pencil.

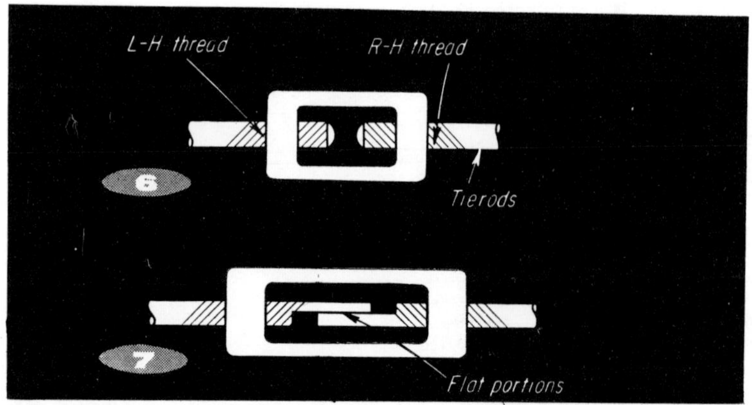

Tierods . . .
with opposite-hand threads at ends require (6) only a similarly threaded nut to provide simple, axial adjustment. Flats on rod ends (7) make it unnecessary to restrain both rods against rotation when adjusting screw is turned—restraining one rod is enough.

DEVICES

FEDERICO STRASSER

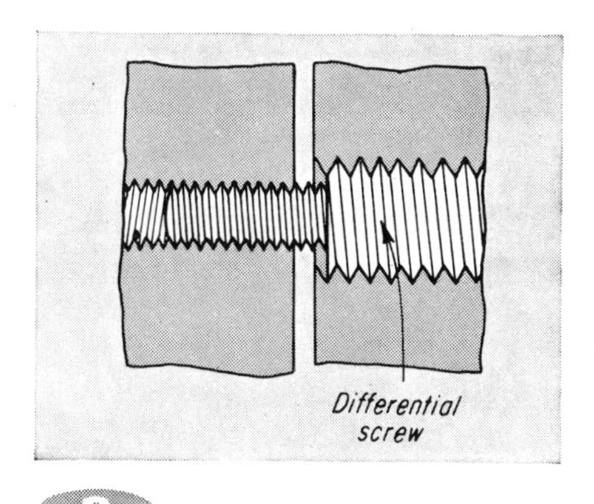

8

Differential screw . . .
has same-hand thread but with different pitches. Relative distance between two components can be adjusted with high precision by differential screws.

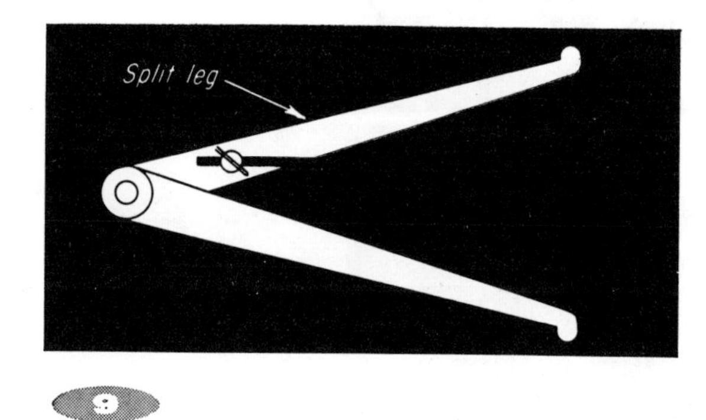

9

Split-leg caliper . . .
is example of simple but highly efficient adjusting-device design. Tapered screw forces split leg apart, thus enlarging opening betwen the two legs.

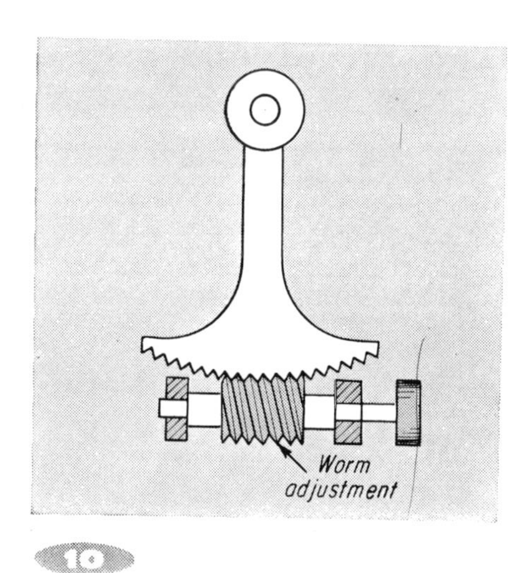

10

Worm adjustment . . .
is shown here in device for varying position of an arm. Measuring instruments, and other designs requiring fine adjustments, usually need this type of adjusting device.

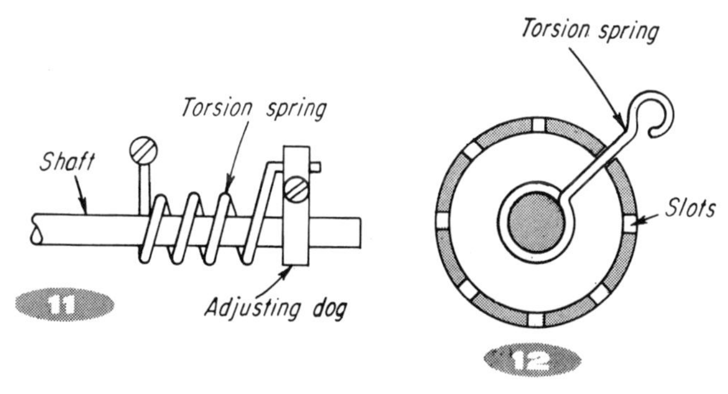

Shaft torque . . .
is adjusted (11) by rotating the spring-holding collar relative to shaft, and locking collar at position of desired torque. Adjusting slots (12) accommodate torsion-spring arm after spring is wound to desired torque.

Rack and toothed stop . . .
(13) are frequently used to adjust heavy louvers, boiler doors and similar equipment. Adjustment is not continuous, depends on the rack pitch. Large counter-adjustment forces may necessitate weighted rack to prevent tooth disengagement. Indexing holes (14) provide similar adjustment to rack. Pin makes position less liable to be accidentally moved.

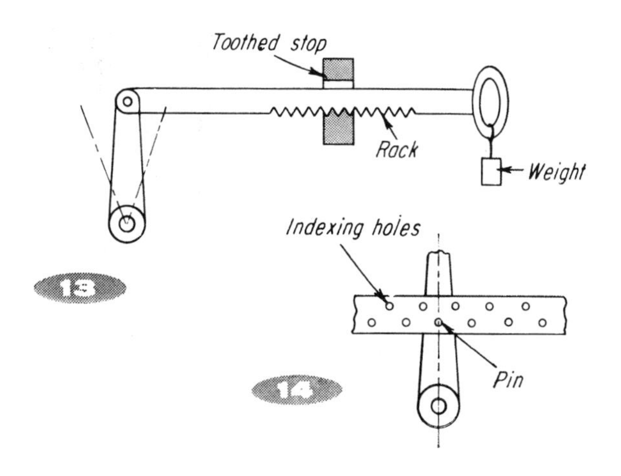

12 EXPANDING and

Parallel bars, telescoping slides and other mechanisms that can spark answers to many design problems.

FEDERICO STRASSER

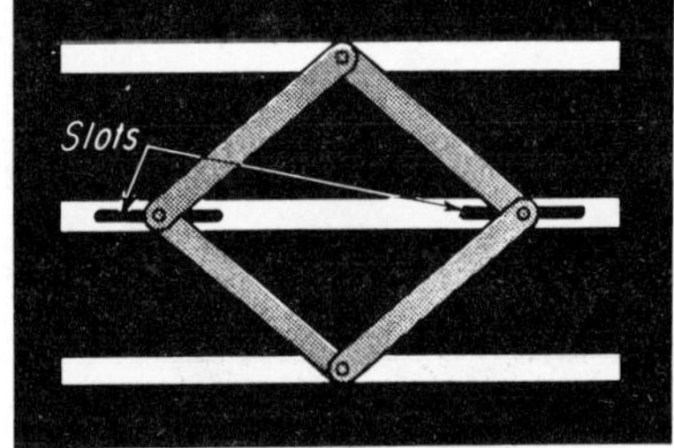

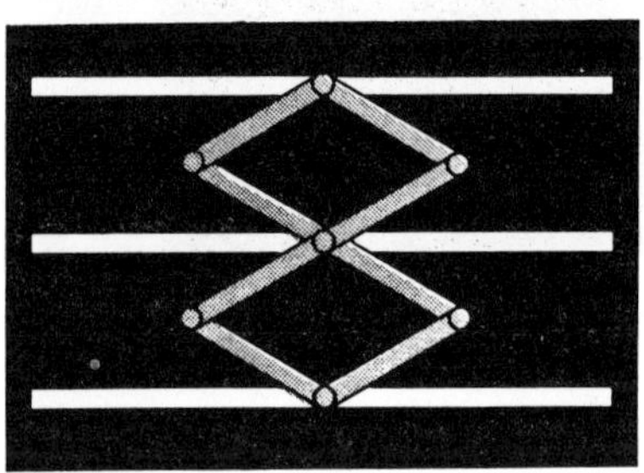

Expanding grilles . . .
are often put to work as a safety feature. Single parallelogram (1) requires slotted bars; double parallelogram (2) requires none—but middle grille-bar must be held parallel by some other method.

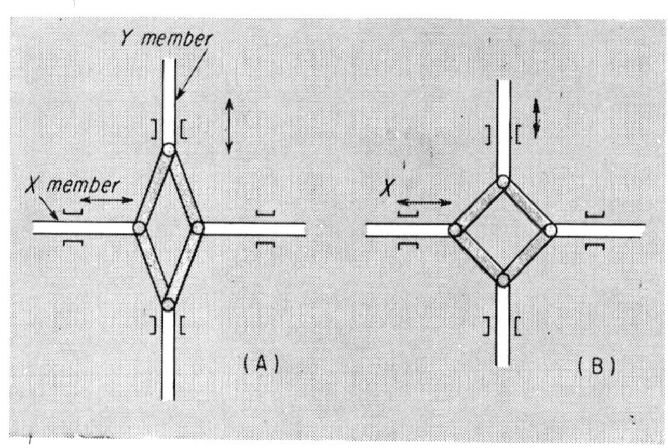

Variable motion . . .
can be produced with this arrangement. In (A) position, the Y member is moving faster than X member. In (B), speeds of both members are instantaneously equal. If the motion is continued in same direction, speed of X will become the greater.

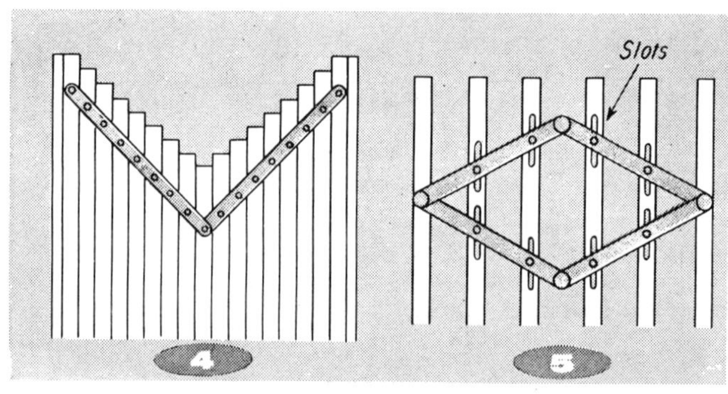

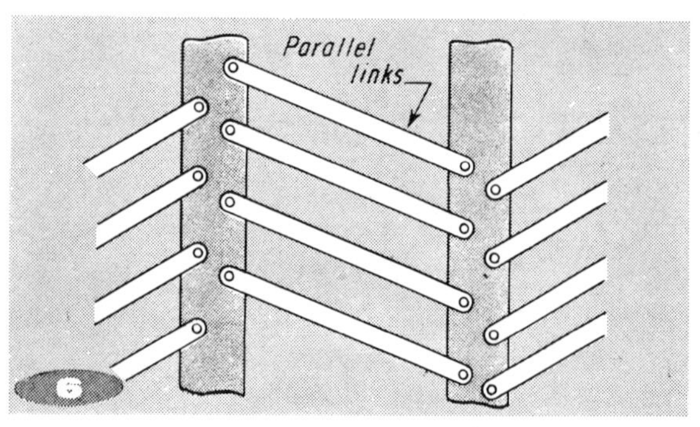

Multi-bar . . .
shutters, gates, etc. (4) can take various forms. Slots (5) allow for vertical adjustment. Space between bars may be made adjustable (6) by connecting the vertical bars with parallel links.

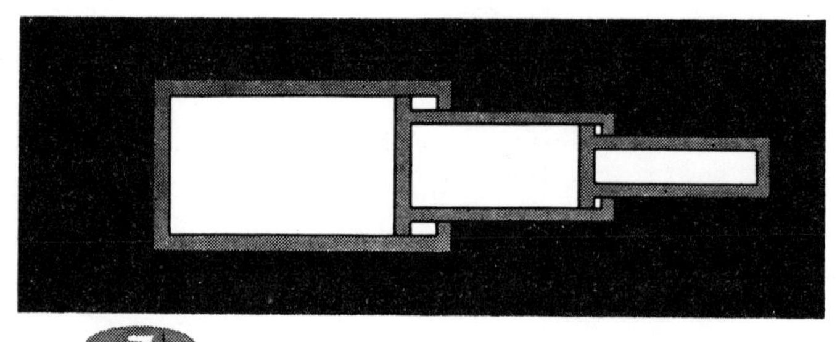

7

Telescoping devices . . .
are basis for many expanding and contracting mechanisms. In arrangement shown, nested tubes can be sealed and filled with a highly temperature-responsive medium such as a volatile liquid.

CONTRACTING DEVICES

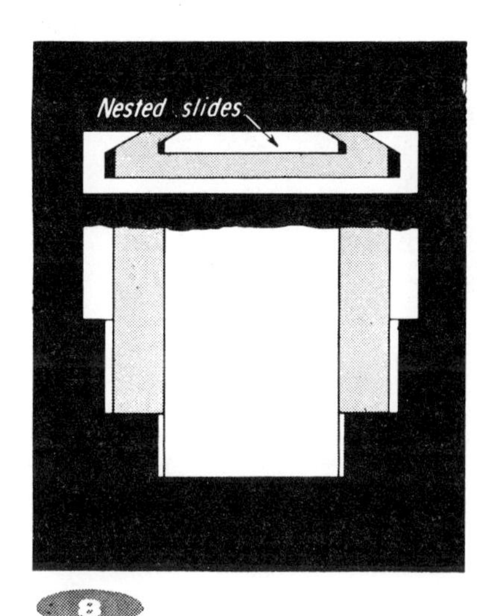

8

Nested slides . . .
can provide extension for machine-tool table or other designs where accurate construction is necessary. In this design adjustments to obtain smooth sliding must be made first, then the table surface must be levelled.

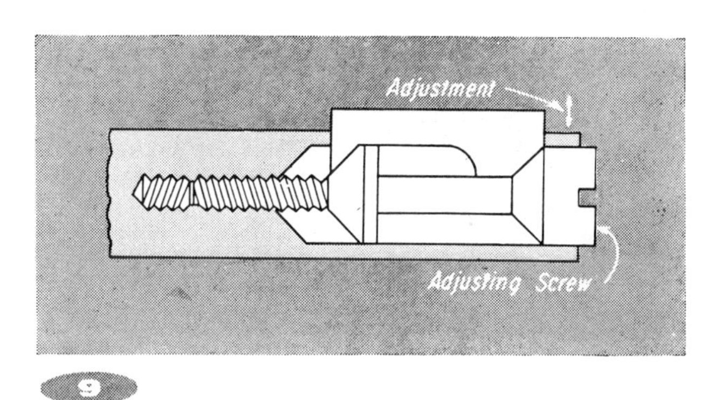

9

Expanding mandrels . . .
of the circular type are well-known. Shown here is a less common mandrel-type adjustment. Parallel member, adjusted by two tapered surfaces on screw, can exert powerful force if taper is small.

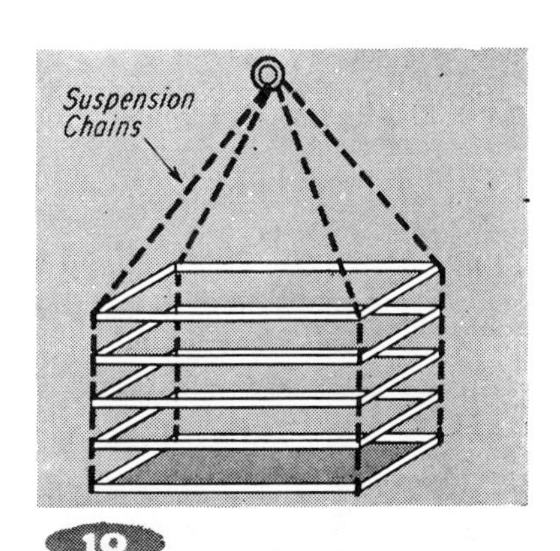

10

Expanding basket . . .
is opened when suspension chains are lifted. Baskets take up little space when not in use. Typical use for these baskets is for conveyor systems. As tote baskets they also allow easy removal of contents because they collapse clear of the load.

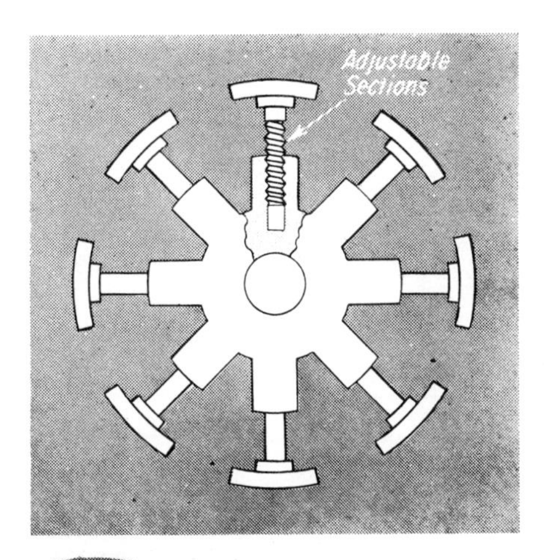

11

Expanding wheel . . .
has various applications besides acting as pulley or other conventional wheel. Examples: electrical contact on wheel surfaces allow many repetitive electrical functions to be performed while wheel turns; dynamic and static balancing is simplified when expanding wheel is attached to non-expanding main wheel. As a pulley, expanding wheel may have a steel band fastened to only one section and passing twice around the circumference, thereby allowing adjustment.

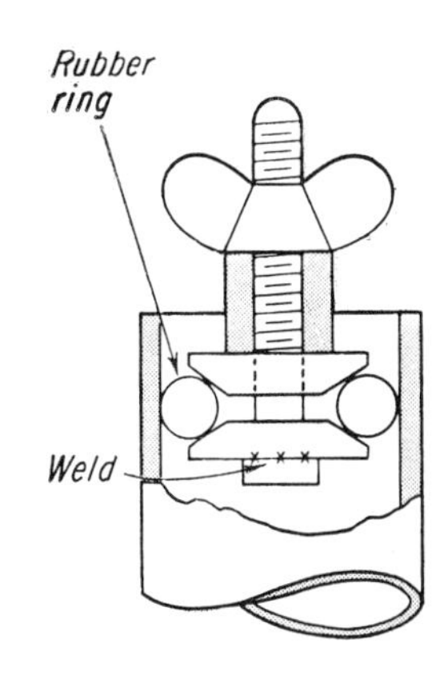

12

Pipe stopper . . .
relies on bulging of rubber ring for its action—soft rubber will allow greater adjustment than hard rubber, and also conform more easily to rough pipe-surfaces. Hard rubber, however, withstands higher pressures. Weld screw head to washer for leaktight joint.

Arms that extend

IRWIN N. SCHUSTER, Design Consultant, Secane, Pa.

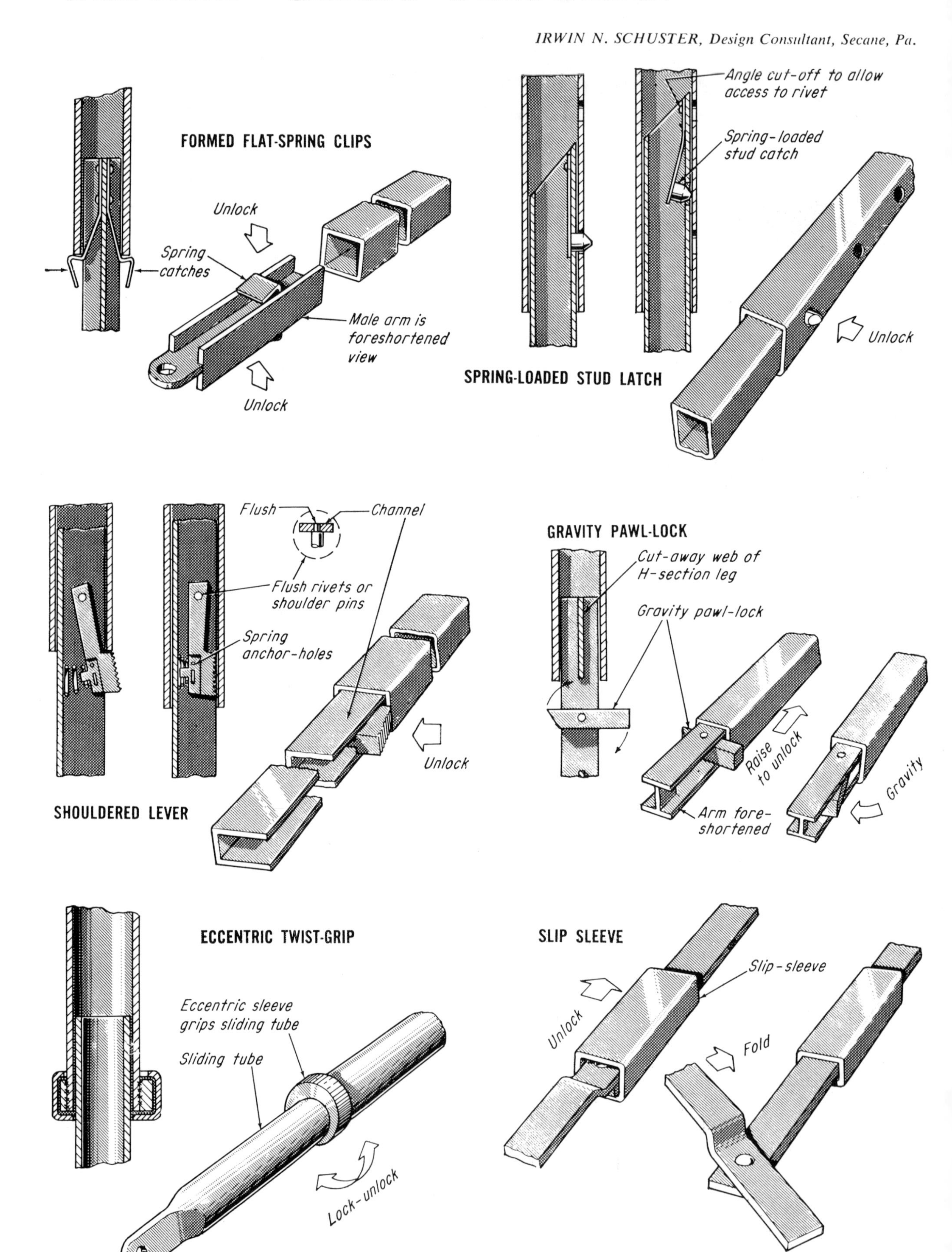

Chapter 5 COMPUTING MECHANISMS AND COUNTERS

Computing mechanisms—I

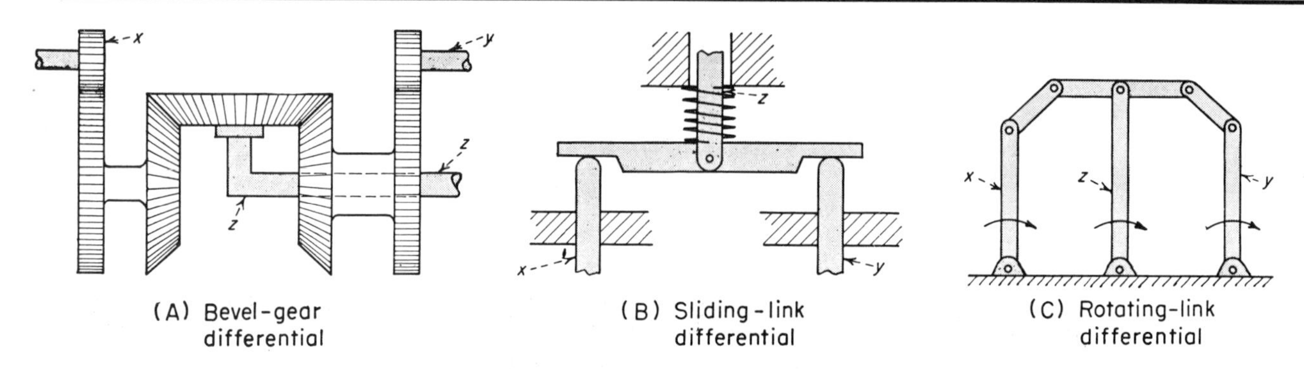

Fig. 1—ADDITION AND SUBTRACTION. Usually based on the differential principle; variations depend on whether inputs: (A) rotate shafts, (B) translate links, (C) angularly displaced links. Mechanisms solve equation: $z=c\ (x \pm y)$, where c is scale factor, x and y are inputs, and z is the output. Motion of x and y in same direction results in addition; opposite direction—subtraction.

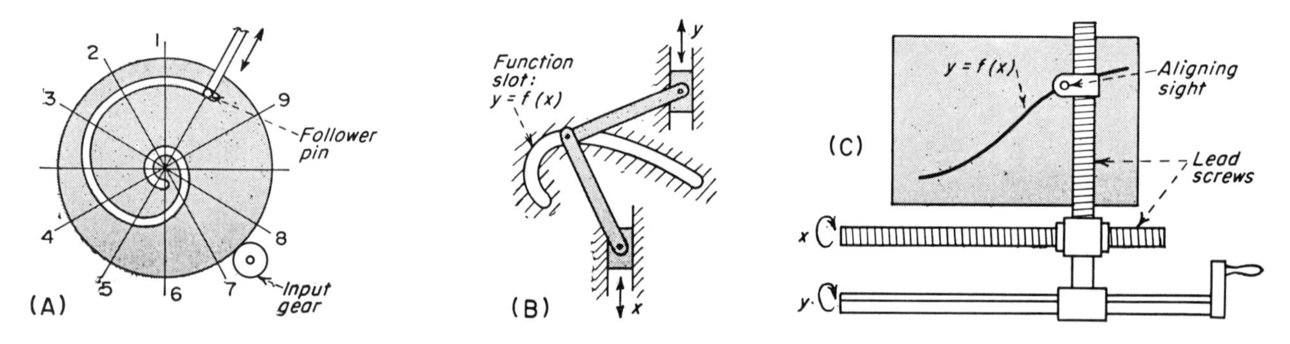

Fig. 2—FUNCTION GENERATORS mechanize specific equations. (A) Reciprocal cam converts a number into its reciprocal. This simplifies division by permitting simple multiplication between a numerator and its denominator. Cam rotated to position corresponding to denominator. Distance between center of cam to center of follower pin corresponds to reciprocal. (B) Function-slot cam. Ideal for complex functions involving one variable. (C) Input table. Function is plotted on large sheet attached to table. Lead screw for x is turned at constant speed by an analyzer. Operator or photoelectric follower turns y output to keep sight on curve.

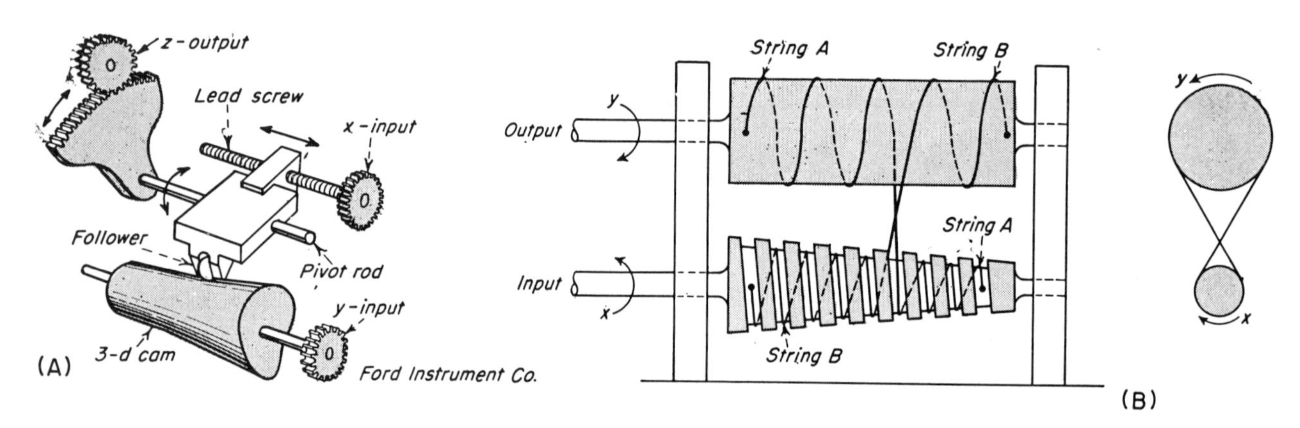

Fig. 3—(A) THREE-DIMENSIONAL CAM generates functions with two variables: $z=f\ (x,\ y)$. Cam rotated by y-input; x-inputs shifts follower along pivot rod. Contour of cam causes follower to rotate giving angular displacement to z-output gear. (B) Conical cam for squaring positive or negative inputs: $y=c\ (\pm x)^2$. Radius of cone at any point is proportional to length of string to right of point; therefore, cylinder rotation is proportional to square of cone rotation. Output is fed through a gear differential to convert to positive number.

Analog computing mechanisms are capable of virtually instantaneous response to minute variations in input. Basic units, similar to the types shown, are combined to form the final computer. These mechanisms add, subtract, resolve vectors, or solve special or trigonometric functions.

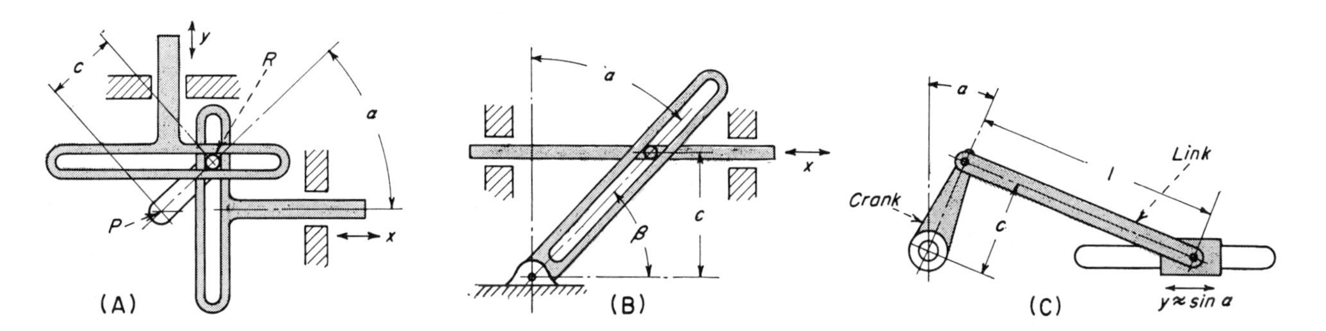

Fig. 4—TRIGONOMETRIC FUNCTIONS. (A) Scotch-yoke mechanism for sine and cosine functions. Crank rotates about fixed point *P* generating angle α and giving motion to arms: $y = c \sin \alpha$; $x = c \cos \alpha$. (B) Tangent-cotangent mechanism generates: $x = c \tan \alpha$ or $x = c \cot \beta$. (C) Eccentric and follower is easily manufactured but sine and cosine functions are approximate. Maximum error is: $e \max = l - \sqrt{l^2 - c^2}$; error is zero at 90 and 270 deg. *l* is the length of the link and *c* is the length of the crank.

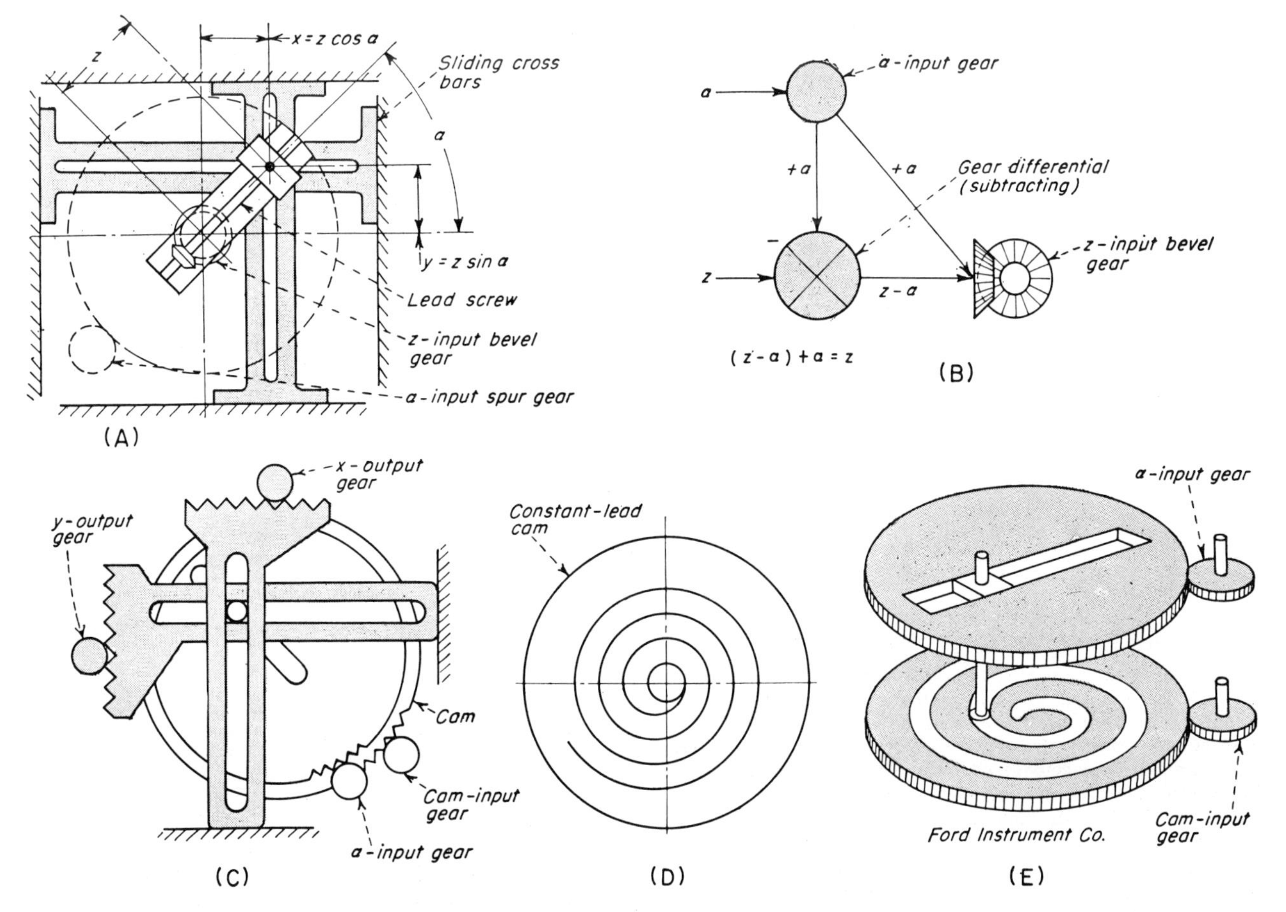

Fig. 5—COMPONENT RESOLVERS for obtaining x and y components of vectors that are continuously changing in both angle and magnitude. Equations are: $x = z \cos \alpha$, $y = z \sin \alpha$ where z is magnitude of vector, and α is vector angle. Mechanisms can also combine components to obtain resultant. Input in (A) are through bevel gears and lead screws for z input, and through spur gears for α-input. Compensating gear differential (B) prevents α-input from affecting z-input. This problem solved in (C) by using constant-lead cam (D) and (E).

Computing mechanisms—II

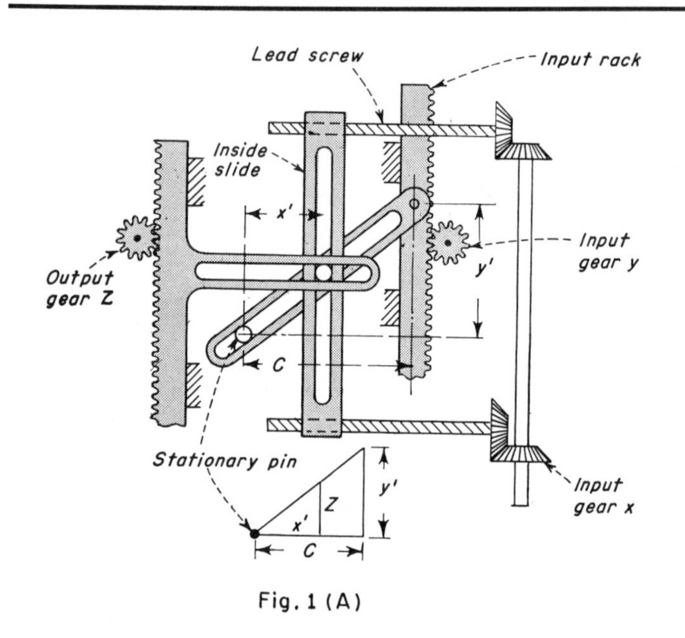

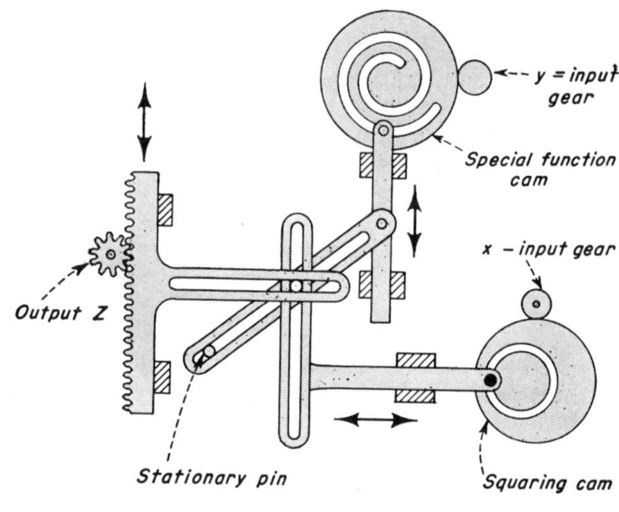

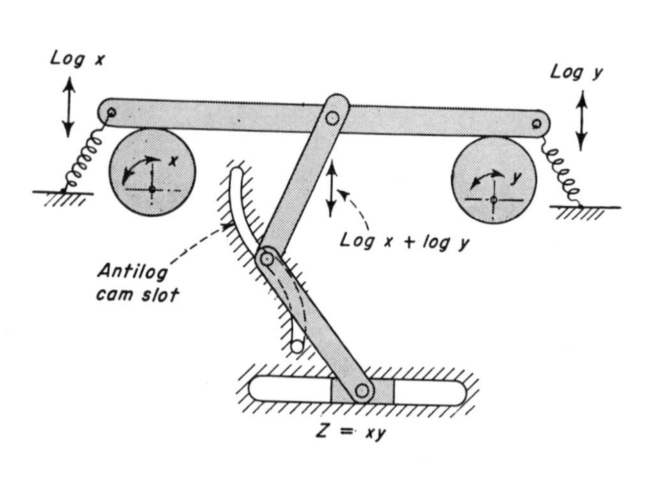

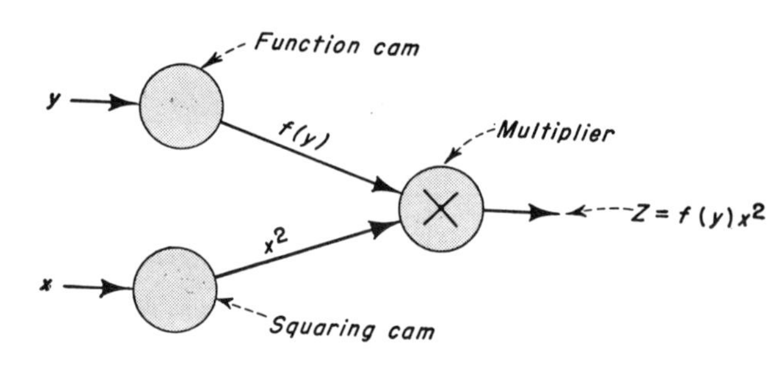

Fig. 1—MULTIPLICATION OF TWO TABLES x and y usually solved by either: (A) Similar triangle method, or (B) logarithmic method. In (A), lengths x' and y' are proportional to rotation of input gears x and y. Distance c is constant. By similar triangles: $z/x = y/c$ or $z = xy/c$, where z is vertical displacement of output rack. Mechanism can be modified to accept negative variables. In (B), input variables are fed through logarithmic cams giving linear displacements of log x and log y. Functions are then added by a differential link giving $z = \log x + \log y = \log xy$ (neglecting scale factors). Result is fed through antilog cam; motion of follower represents $z = xy$.

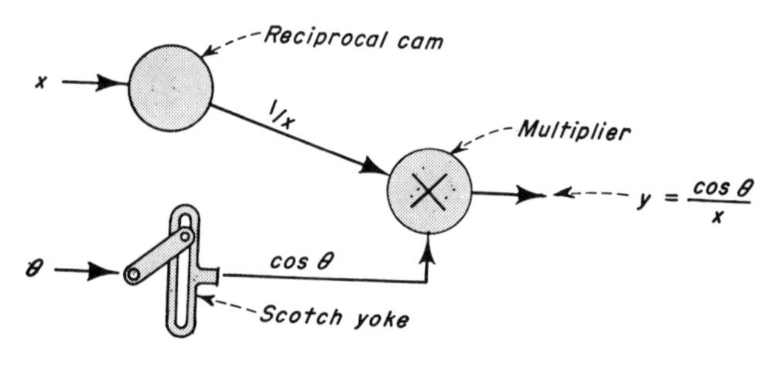

Fig. 2—MULTIPLICATION OF COMPLEX FUNCTIONS can be accomplished by substituting cams in place of input slides and racks of mechanism in Fig. 1. Principle of similar triangles still applies. Mechanism in (A) solves the equation: $z = f(y)\, x^2$. Schematic is shown in (B). Division of two variables can be done by feeding one of the variables through a reciprocal cam and then multiplying it by the other. Schematic in (C) shows solution of $y = \cos \theta / x$.

Several typical computing mechanisms for performnig the mathematical operations of multiplication, division, differentiation, and integration of variable functions.

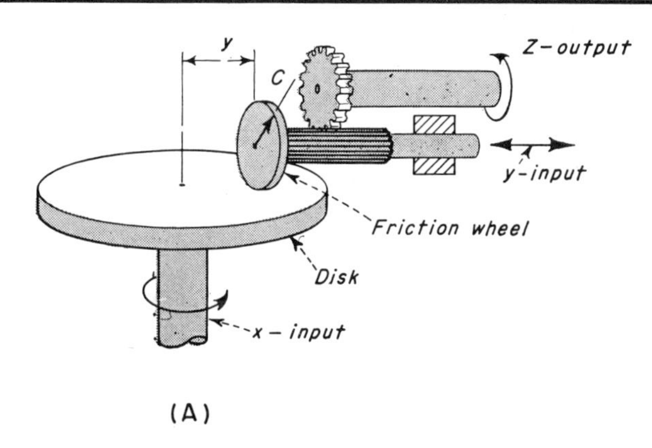

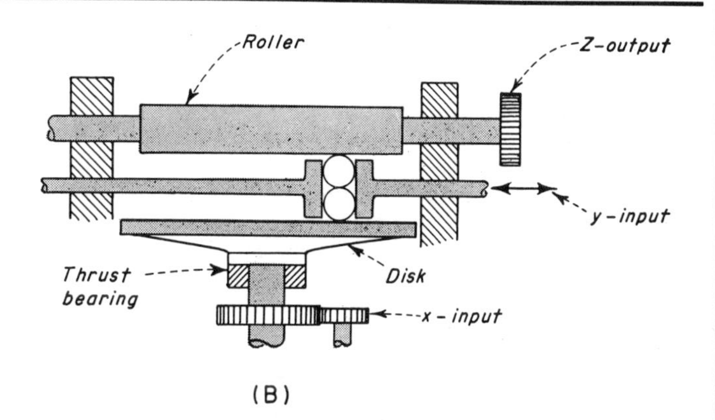

Fig. 3—INTEGRATORS are basically variable speed drives. Disk in (A) is rotated by x-input which, in turn, rotates the friction wheel. Output is through gear driven by spline on shaft of friction wheel. Input y varies the distance of friction wheel from center of disk. For a wheel with radius c, rotation of disk through infinitesmal turn dx causes corresponding turn dz equal to: dz=(1/c) y dx. For definite x revolutions, total z revolutions will be equal to the integral of (1/c) y dx, where y varies as called for by the problem. Ball integrator in (B), gives pure rolling in all directions.

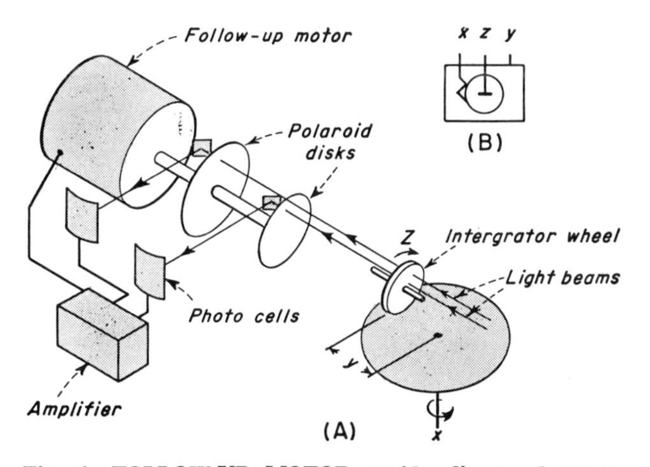

Fig. 4—FOLLOW-UP MOTOR avoids slippage between wheel and disk of integrator in Fig. 3 (A). No torque is taken from wheel except to overcome friction in bearings. Web of integrator wheel is made of polaroid. Light beams generate current to amplifier which controls follow-up motor. Symbol at upper right corner is schematic representation of integrator. For more information see, "Mechanism," by Joseph S. Beggs, McGraw-Hill Book Co., N. Y., 1955.

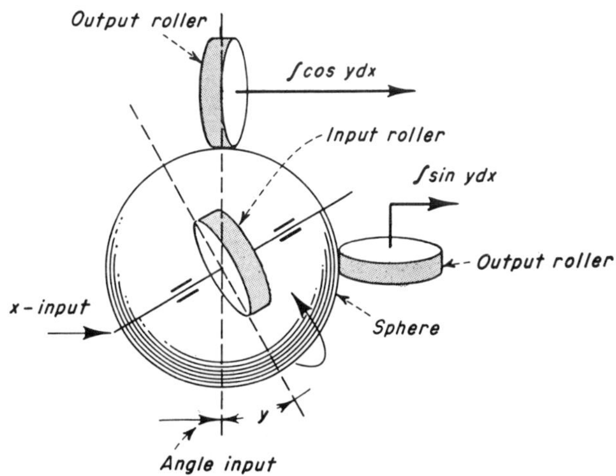

Fig. 5—COMPONENT INTEGRATOR uses three disks to obtain x and y components of a differential equation. Input roller x spins sphere; y input changes angle of roller. Output rollers give integrals of components paralleling x and y axes. Ford Instrument Company.

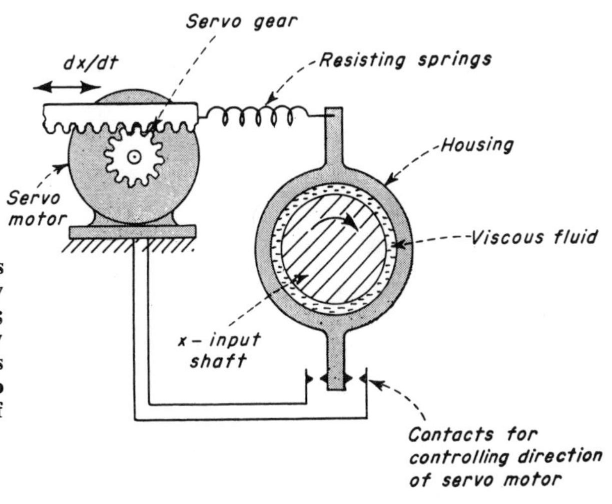

Fig. 6—DIFFERENTIATOR uses principle that viscous drag force in thin layer of fluid is proportional to velocity of rotating x-input shaft. Drag force counteracted by spring; spring length regulated by servo motor controlled by contacts. Change in shaft velocity causes change in viscous torque. Shift in housing closes contacts causing motor to adjust spring length and balance system. Total rotation of servo gear is proportional to dx/dt.

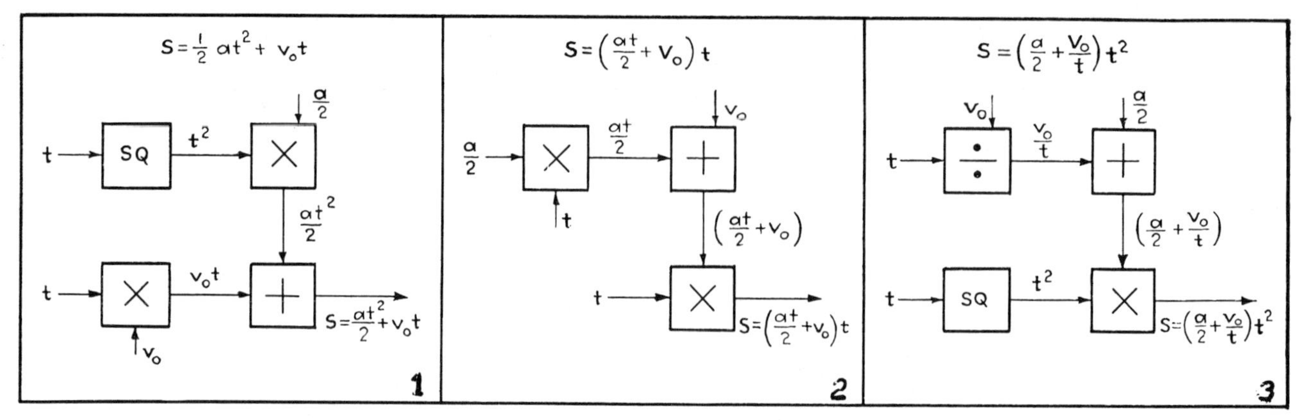

Figs. 1–3 — Block diagram analysis: the preliminary step in mechanizing an equation. Symbols in the blocks indicate the mathematical operation to be performed. (1) Arrangement for the displacement-time equation as usually written. (2) A rearrangement that eliminates the squaring device. (3) Another arrangement of the equation, resulting in more complexity.

Mechanical Computing Mechanisms

ROBERT R. REID and DU RAY E. STROMBACK
Research Engineers, The Franklin Institute Laboratories for Research and Development

This installment begins the analyses of individual mechanisms for function generation, including trigonometric functions. Also, four mechanisms for adding and subtracting two variables are described.

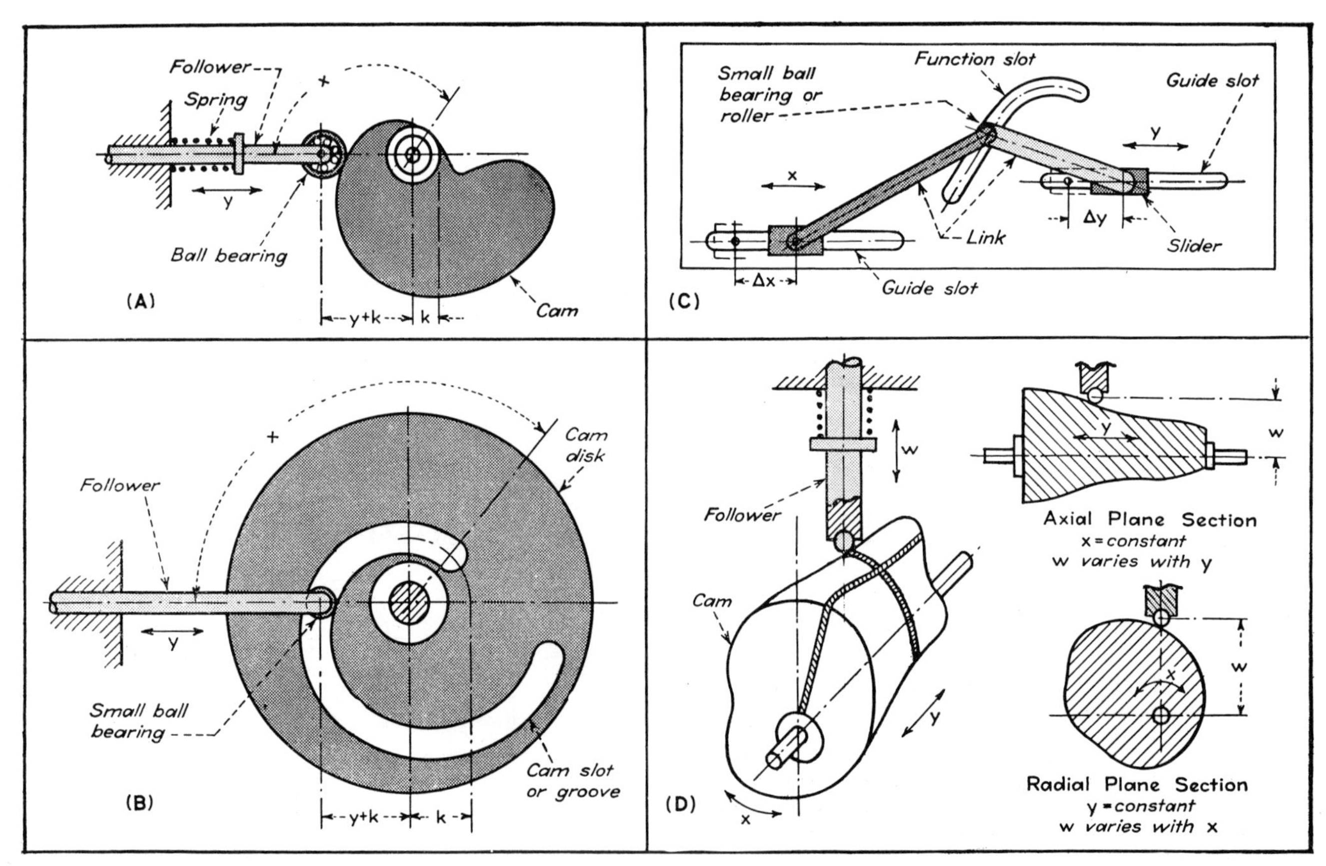

Fig. 4—Some cam mechanisms for function generation: (A) Rotary plate cam; (B) Rotary groove cam, which avoids spring loading the mechanism; (C) Function slot cam, which is frequently useful; (D) Three-dimensional cam, for two variables.

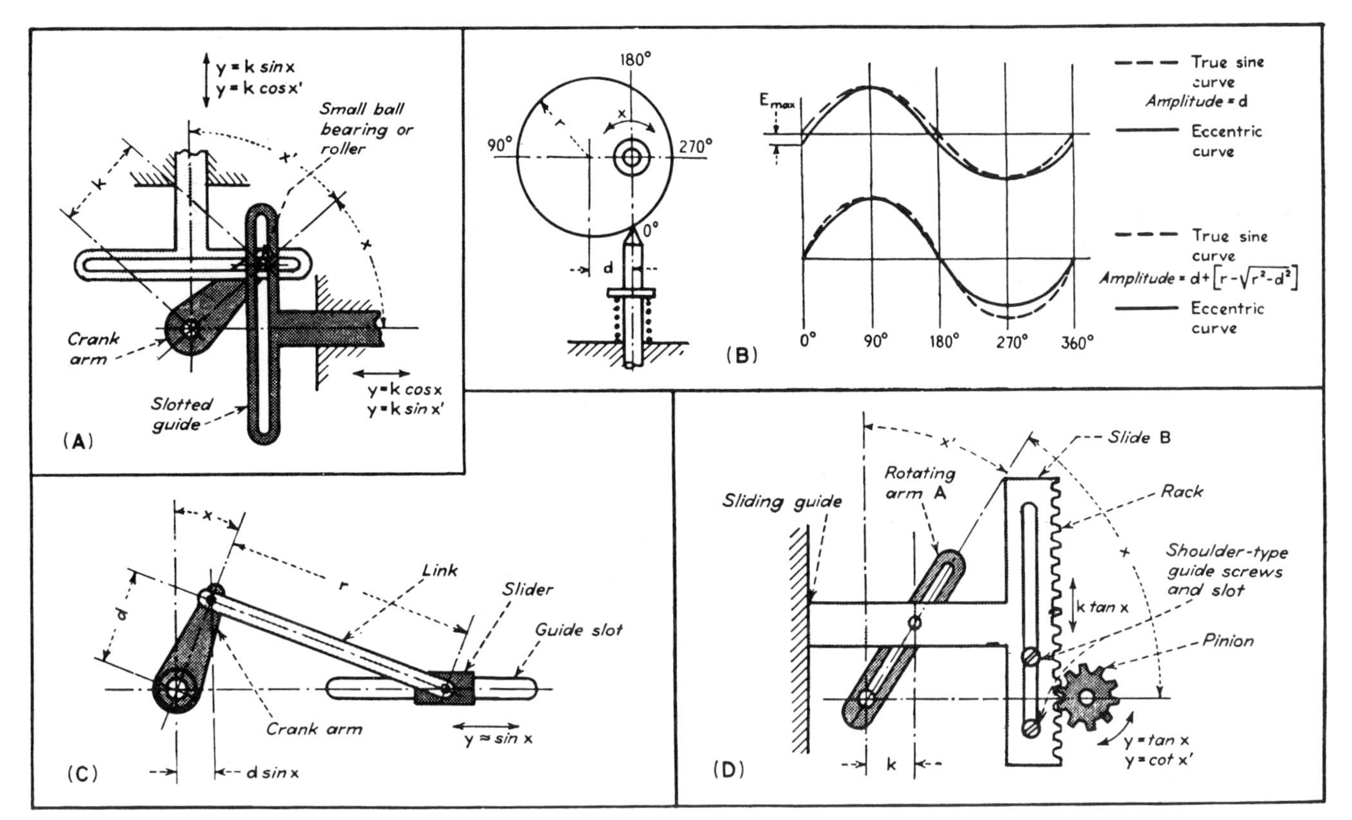

Fig. 5—Mechanisms for reproducing trigonometric functions: (A) Scotch yoke for sine-cosine relationships; (B) Eccentric, (C) Crank mechanism, for approximating sine-cosine functions; (D) Tangent-cotangent mechanism, approximate only.

Generation of Functions

A function generator may be considered the simplest form of computer, since it actually mechanizes an equation. Some of the mechanisms to be described use functions of variables to perform the computation.

A functional relationship,

$$y = f(x)$$

which involves no discontinuities, may be established between any two variables (x) and (y) by using some form of cam mechanism. Certain trigonometric functions, such as

$$y = \sin x$$

may be obtained from devices that mechanize the geometry of the trigonometric relationships. A type of cam mechanism using both rotary and translatory motion (two degrees of freedom) may be used to obtain the relation

$$w = f(x, y)$$

Cams with a single degree of freedom are commonplace in designs of gasoline engines and automatic machinery. The type most commonly employed is the simple rotary plate cam; cylindrical and radial groove cams are often used, and occasionally a translatory or slide cam finds application. Each of these types is used in computer design to mechanize the relation

$$y = f(x)$$

in which $f\ (x)$ may be a series of values in tabular form, or a mathematical expression in powers of (x). In a sense, the surface of the cam may be considered a "mechanical graph" of the function expressed in polar coordinates for rotary cams and in rectangular coordinates for the slide types. This conception must be modified, of course, to account for the finite sizes of hubs, shafts and cam followers.

Each cam mechanism requires a follower, the motion of which is proportional to the generated function. In computers and instruments, which must operate at a low friction level, ball-bearing followers are the most satisfactory. The motion of the follower is constrained by suitable guides to either rotary or translatory motion.

The same principles apply to rotary computing cam design as to ordinary cam design, in regard to undercutting, maximum pressure angle and compensation for rotary follower motion. An originally sliding input motion is often converted into a rotation, by means of a gear and rack, so as to make use of a rotary cam, which is easy to design and fabricate.

The relations of the mathematical quantities and physical displacements for rotary cams are shown in Fig. 4(*A*). The cam rotation from some initial position represents the independent variable (x); the follower radius (defined for ball-bearing or roller followers as the distance from center of cam shaft to center line of follower) at this point, which determines the position of the follower, is y plus a constant k equal to the minimum cam radius. This constant is eliminated by offsetting the y scale, so the relation

$$y = f(x)$$

is satisfied by the mechanism.

Simple plate cams require a spring to insure contact between the cam periphery and the follower regardless of direction of rotation. Radial groove cams, Fig. 4(*B*), are considerably harder to fabricate, but the elimination of the spring load on the mechanism often justifies using this type.

A very useful slide cam, is the cam slot, Fig. 4(*C*), consisting of a function slot milled into a plate, designed empirically in such a way that a point on the independent variable slide proportional to x corresponds to a point of the y slide designated $f(x)$. The design of these devices consists largely of layout work in which each configuration is tested for self-locking conditions. A good rule-of-thumb for avoiding a self-locked mechanism requires that Δx be always larger than the corresponding Δy.

A functional relationship is established between two variables by a "three-dimensional" or barrel cam, Fig. 4(D). Here the cam is rotated by the x input and simultaneously and independently translated axially by the y input. These actions position the follower at a point.

$$w = f(x, y)$$

Theoretically a point follower is correct for this type of cam, but such a follower may be used only with very light loading, and the point tends to wear quickly. A ball-point follower is preferable.

Designing three dimensional cams is quite laborious. The procedure is to consider laminar sections as simple plate cams, varying w with x and holding y constant, as shown in the diagram Fig. 4(D). It is also necessary to consider radial plane contours in which w varies with y as x remains constant. Each radial and axial section must be checked separately for such undesirable cam characteristics as excessive pressure angle or undercutting. These difficulties may be minimized by increasing the minimum cam radius, with the added disadvantage of a larger cam. The fabrication of such a three-dimensional cam is also quite difficult. The usual method follows the laminar design plan in which successive flat plate cams are fastened together and the edges blended by filing.

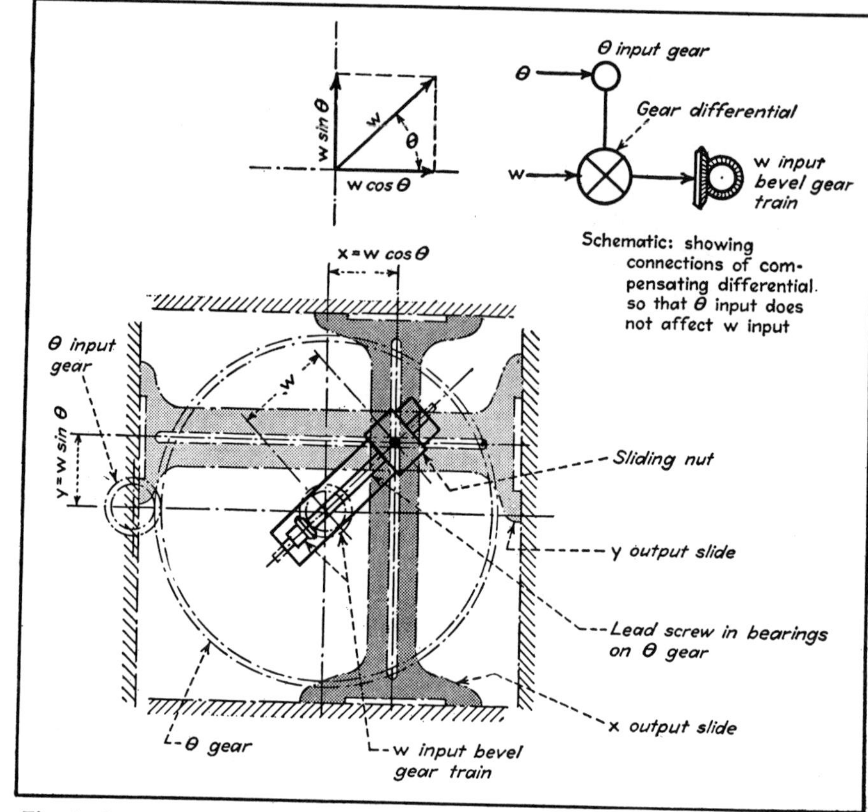

Fig. 6—Component solver, for resolving a vector into x and y components, or for combining two components into a resultant.

Trigonometric Functions

The continuous trigonometric functions

$$y = \sin x \text{ and } y = \cos x$$

may be generated over any portion of the entire range of x from 0 to 360 deg, with any type of cam described, Fig. 5; or a number of geometric devices may be employed to mechanize these functions.

THE SCOTCH YOKE. This familiar mechanism causes exact, simple harmonic motion of the follower slide, and so produces an output function

$$y = f(x) = k \sin x$$
$$y = f(x) = k \cos x$$

depending on the choice of the axis representing the zero value of the angle. The radius, k, of the rotating arm determines the scale factor of the device. Two Scotch yokes at right angles, as in Fig. 5(A), will produce simultaneously as two perpendicular motions, sine and cosine functions of the same angular input variable.

THE ECCENTRIC. An eccentric with a follower constrained to straight-line motion is an excellent device for approximating sine and cosine functions, since it requires a few easily machined parts, and a minimum of space. Considering such a mechanism as a circular cam of radius, r, with the center of rotation offset by an eccentricity d, the output is an approximte sine function of amplitude, d. Comparing this function with a true sine function of the same amplitude, as in Fig. 5(B), the maximum error is

$$\epsilon_{\max} = r - \sqrt{r^2 - d^2}$$

which occurs at 0 deg and 180 deg, the output being correct at 90 deg and 270 deg. However, if the curve generated by the eccentric is raised an amount equal to

$$\frac{\epsilon_{\max}}{2}$$

the error is split and the output is correct at four points; the maximum error is halved, and the curve is a good approximation over the full 360 deg range. This error-splitting scheme is useful in designing approximate computing mechanisms. As a result, with an r/d ratio of 20:1 the maximum error is $1\frac{1}{4}$ percent of d.

Using a true sine function of amplitude,

$$d + [r - \sqrt{r^2 - d^2}]$$

as a basis of comparison, a much closer approximation is obtainable between 0 deg and 180 deg, but the fit is poor between 180 deg and 360 deg, the maximum error being

$$\epsilon_{max} = 2[r - \sqrt{r^2 - d^2}]$$

It is apparent that, if the location of the follower is rotated 90 deg, the device will produce an approximate cosine function with similar error characteristics.

CRANK MECHANISMS. Another approximate sine-cosine mechanism, working on a principle similar to the eccentric just described, is the crank mechanism pictured in Fig. 5(C). With a crank length, d, and a connecting link of length, r, the output is again a sine function of amplitude, d.

The error analysis for the crank is similar to that described for the eccentric where ϵ_{max} is also equal to the same expression, $r - \sqrt{r^2 - d^2}$. The same scheme of error splitting may be applied to the crank mechanism error. Shifting the axis of reference 90 deg will result in the generation of a cosine function.

The trigonometric functions

$$y = \tan x \text{ and } y = \cot x$$

are discontinuous at certain points in the range of x between 0 deg and 360 deg—at 90 deg and 270 deg in the case of the tangent function, and at 0 deg and 180 deg in the case of the cotangent. There is, therefore, no geometric device nor any cam mechanism which will mechanize the full range of these functions. A cam designed to mechanize a portion of the range of the functions must have a very steep slope as it approaches the point of discontinuity. Cams that cover a considerable range, say 0 deg to 80 deg for the tangent, must have a large average radius and a very small total throw to keep the pressure angle within practical limits. The result is a small scale factor output.

The geometric device for tangent and cotangent mechanization, shown in Fig. 5(D), is similar to the Scotch yoke except that the pin is fixed in the output slide, B, and the rotating arm, A, is slotted. The displacement of B is

$$k \tan x$$

and the rotation of the pinion is

$$y = \tan x$$

This mechanism is limited by inherent self-locking to a range of approximately 30 deg of x. Note that, if the complementary angle (x') is considered the independent variable, the

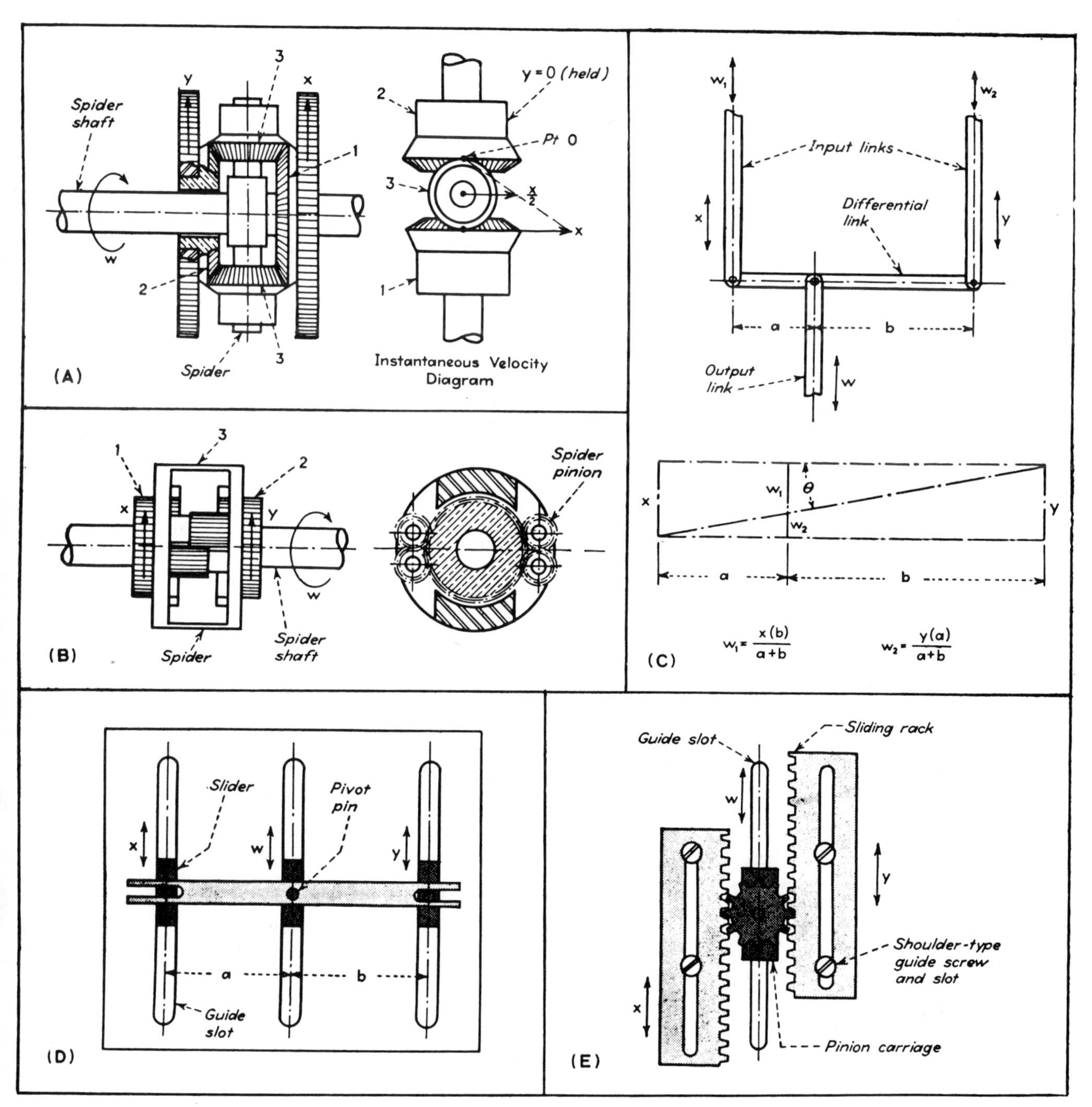

Fig. 7—Addition and subtraction mechanisms: (A) The bevel gear differential and its velocity diagram; (B) The spur gear differentials; (C) Link differential; (D) Slide-link differential; (E) Slide-pinion type of differential (no theoretical error).

output is

$$y = \cot(x')$$

The secant and cosecant functions, which are reciprocals of the cosine and sine functions previously discussed, experience discontinuities similar to tangent and cotangent. The mechanizable portion of their range is best obtained by using cams.

A combination of function generator and multiplier shown in Fig. 6, is a device known as a resolver or component solver. It mechanizes the relations

$$x = w \cos \theta$$
$$y = w \sin \theta$$

If the vector length and angle are inputs, the mechanism is a component solver, and the outputs are the x and y components of the vector. Conversely, if the components are inputs, the resultant length and angle of the vector are outputs.

Addition and Subtraction Devices

Some form of differential mechanism is used universally for adding and subtracting two variables by mechanical means. Four of the most widely applicable types are: bevel gear, spur gear, link and slide. These devices solve the general equation

$$\frac{x \pm y}{2} = w$$

in which w is usually the output and x and y are the inputs. If desired, any two of these variables may be considered inputs, and the third as the output, where x, y, w refer to rotations as in Fig. 7(*A*) below.

Bevel gear differentials are adapted to rotary inputs. Generally the inputs are fed into the differential by spur gearing, attached to the bevel gears. The sum or difference of these inputs appears as the rotation of the spider.

It is convenient to consider the total w rotation as obtained in two steps. In the first step, gear 2 is held while gear 1 is turned through an angle x. Therefore, from the instantaneous velocity diagram, v_2, the velocity of point O on bevel gear 2, is zero, since gear 2 is held and this point is the instantaneous center of rotation (centro) of gear 3. It follows from the velocity diagram that v_s is equal to $v_1/2$, from which it is apparent spider shaft 3 will turn through an angle $x/2$. In the second step, gear 1 is held at this new position, and gear 2 turned through an additional angle, y. From similar considerations, this causes a rotation of spider shaft 3 equal to $y/2$. The total rotation of the spider due to input rotations on gears 1 and 2 of x and y respectively is, therefore

$$w = \frac{x}{2} + \frac{y}{2} = \frac{x + y}{2}$$

If the spider shaft is held so that w equals 0, the rotation of the x input produces an equal rotation of the y gear in an opposite direction, so that

$$x = -y$$

The constant, 2, appearing in the above equation is simply the diameter-radius ratio of gear 3. This constant is inherent in the type of construction in which gears 1 and 2 are of equal size. The size of gear 3 relative to 1 and 2 is immaterial, therefore, and may be chosen at will for particular applications.

A spur gear differential embodying the same characteristics is illustrated in Fig. 7(*B*). This spur gear differential solves the same equation:

$$\frac{x \pm y}{2} = w$$

The advantages of spur gear differentials over the bevel gear are compactness of design and fewer parts.

This mechanism is a special case of planetary gearing, and may be constructed in several ways. Again, the relative size of input gears and spider pinions is independent of the output ratio, as indicated in the equation. Fig. 7(*B*) shows a differential with two pairs of spider pinions. If heavy loads are anticipated, the number of pairs may be increased to decrease gear tooth loading.

The link type of differential, Fig. 7(*C*), is another adding or subtracting mechanism where

$$w = \frac{x(b) + y(a)}{a + b}$$

or, if $a = b$, $w = \frac{x}{2} + \frac{y}{2}$ as before.

Considering the motion of the output w to be composed of two separate components, w_1 due to input x, and w_2 due to input y, then

$$w_1 = \frac{x(b)}{a + b}$$

$$w_2 = \frac{y(a)}{a + b}$$

on the principles of similar triangles. It follows that

$$w = w_1 + w_2 = \frac{x(b)}{a + b} + \frac{y(a)}{a + b} = \frac{x(b) + y(a)}{a + b}$$

As drawn in the schematic form in Fig. 7(*C*), this addition mechanism is an approximate device. The degree of approximation is governed by the limit on the angle θ, determined by the total range of the variables x and y and the length of the differential link $(a + b)$. This follows from the fact that the analysis assumes the x, y and w motions are straight lines, when actually they follow circular paths about moving centers. Since the projected length of a and b along a perpendicular to the input direction equal the actual length of a and b to the desired degree of approximation, the mechanism is large in comparison to a gear type of differential, with a small allowable input and output motion.

Fig. 7 (*D*) shows a more complicated slide-link differential which has no Class B (theoretical) error. Three parallel guides direct the variable motions. The connecting link is pivoted on the w slide, and makes sliding pin-and-slot connections with the x and y input slides.

Another type of slide differential, constructed to eliminate Class B errors, is pictured in Fig. 7(*E*). The pinion meshes with both racks, and rotates about a fixed stud in the w slider. Once again

$$w = \frac{x + y}{2}$$

All these mechanisms also may be used as subtractors when used to add negative numbers.

The Class A errors to be minimized in geared differential mechanisms are (1) backlash, (2) inaccurately cut gears, and (3) incorrectly mounted gears. Precision gears and close tolerances on center distances will minimize these undesirable conditions. If backlash is unavoidable, a take-up spring may be used to keep backlash always in the same direction. But such a spring must be capable of driving the whole gear train on which it operates, so that the forces between teeth are always in one direction regardless of the direction of rotation. A spring sufficiently powerful to accomplish this usually imposes large forces on the computer mechanisms.

Class A errors in slide and link mechanisms arise from lost motion in slides and pin slots. High accuracy in machining parts is the chief remedy.

Mechanical Computing Mechanisms—II

Continuing the analysis of analog mechanisms to perform mathematical operations. Multiplication, division, integration and differentiation devices and their operation are compared.

ROBERT R. REID and DU RAY E. STROMBACK
Research Engineers, The Franklin Institute Laboratories for Research and Development

THE SIMPLEST MULTIPLICATION to mechanize is multiplication of a variable by a constant. That is, $y = cx$, which may be done by a gear ratio equal to the required c for rotary variables, or by a simple lever system for straight line motion.

The problem of multiplying two variables is more complex. The form of this equation is $xy = cw$. Four possible means of solving this equation are: similar triangles method, logarithmic method, slide-link and slide multiplier. A mechanism based on the principle of similar triangles is pictured in Fig. 8(A). In this, the lengths x and y are proportional to input quantities and the length a is fixed, as it was for the mechanism shown in Fig. 5(D) PRODUCT ENGINEERING, From Fig. 8(A), p. 130,

$$\frac{x}{w} = \frac{a}{y} \quad \text{or} \quad xy = aw$$

This mechanism provides a theoretically exact solution to the problem of multiplying two variables and may be constructed to permit both variables to change algebraic sign. Care should be taken to prevent the mechanism from reaching a self-locking condition. The guides should be designed to prevent cocking of the links, which would introduce excessive friction; and the construction should be such that when the linkage is transmitting load there will be no spreading of the slide.

The pins can be replaced by small ball bearings, and theoretically the ball bearing will not roll because the slide is touching the bearing at two points. But in actual construction there is always some clearance, and the ball bearing tends to roll rather than slide.

The space requirements for this type of multiplier are relatively large in two dimensions.

Logarithmic Method

A second, theoretically exact method for multiplying is to add the logarithms of each input variable. The log is taken from a cam; the rotational input represents x and y, and the motions of the two followers ($\log x$) and ($\log y$) respectively. Adding these two functions by one of the methods previously described (the slide type, Fig. 7(D), is usually most appropriate), gives the output

$$(\log x) + (\log y) = \log xy$$

and the output of the differential, $\log xy$, actuates an antilog cam. The motion of the cam follower, Fig. 8(B), represents the value of the answer (xy).

This method has several disadvantages. Neither variable may be permitted to reach zero or negative values, or even to approach zero, since the log of zero is indeterminate. Trouble will ensue if the value of the variable becomes less than 1.0, since beyond this point the log becomes negative. If such a condition is unavoidable, the method still may be used in the design if both sides of the equation can be multiplied by a factor so chosen as to make the minimum value of the variables always greater than 1.0. For example, to multiply w equals xy where

$$1 \leqq y \leqq 5 \text{ and } 0.5 \leqq x \leqq 3$$

if both sides are multiplied by 3

$$3w = 3x\,(y)$$

$\log 3 + \log w = \log 3x + \log y$, and the cam becomes actually a $\log 3x$ cam. The answer taken from the antilog cam is clearly $3w$; but this may be accounted for in choosing the scale factor of w.

Approximate Method of Multiplication

The mechanism shown in Fig. 8(C) is a simple form of an approximate multiplier. It may be constructed from a minimum number of parts, and it finds use in light service and where a limited known error is acceptable. This device mechanizes the equation $w = xy$, where x is the radius of the input arm and y is the input angle.

The slot in the rotating input arm is a circular arc of radius L about the point ($xy = o$) on the output slide. This position of the input arm is taken as $y = o$; at this position the variable x may take on all values within its range without affecting the output. The pivot point of the rotating arm is chosen as $x = o$; similarly, at this point y may take on all values without affecting the result. A layout is required to determine the magnitude of errors inherent in any given mechanism of this type.

This multiplier has the disadvantage of requiring considerable space in two directions for reasonable accuracy; it has the advantage of simple construction.

Another type of linkage which may be used for multiplication is shown in Fig. 8(D). It is somewhat like a similar triangle multiplier except that two of the slides have been replaced by rotary motions. It has the disadvantage of having one of the three scales non-linear. However, the designer may choose which scale is to be non-linear. Also in Fig. 8(D) an equation is given that shows the trigonometric relations between the variables used in determining the y scale.

For example, the linkage is to mechanize the equation w equals xy, where the distance between pivots, c, is equal to 2.000 in. It is desired to find the distance y when variable y is equal to two units. Assume the scales of w and x have the same scale factor of 5 units per in. To find the distance y, for y equals 2 units, assume any value of w, x, that satisfies w equals $2x$; for example, w equals 10, x equals 5; then for

$$w = 10 \text{ units} = 2.000 \text{ in.}$$
$$x = 5 \text{ units} = 1.000 \text{ in.}$$

$$w = \frac{xy}{c - y}$$

$$2.000 = \frac{(1.000)\,y}{(2.000) - y}$$

or $\quad y = 0.750$ in. $\quad (y = 2$ units$)$

Division

While the mechanisms for multiplication are capable of handling zero and negative values of the input vari-

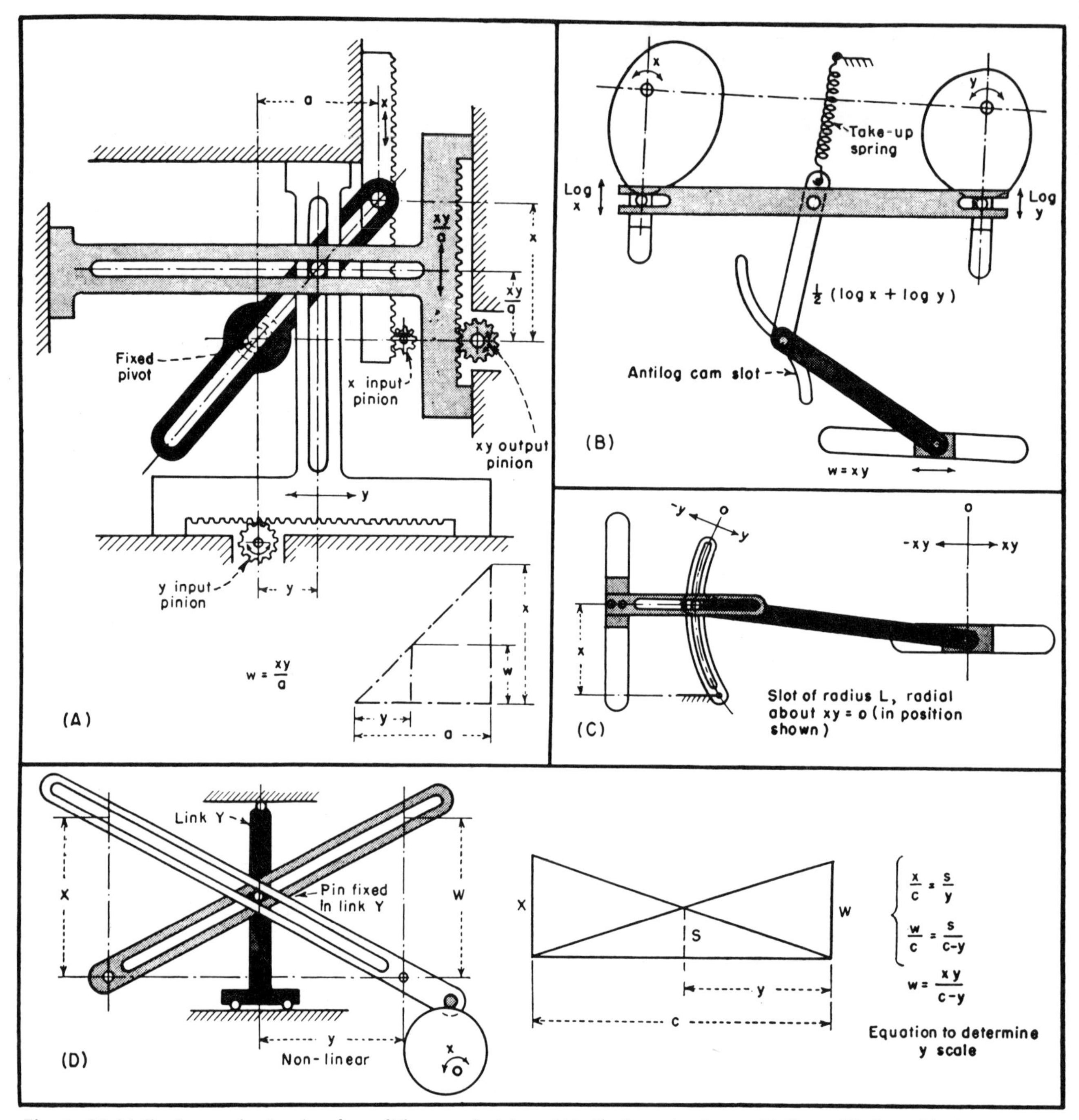

Fig. 8—Multiplication mechanism based on different principles: (A) Similar triangles multiplier. (B) Logarithmic multiplier. (C) Slide-link multiplier, approximate only. (D) Slide multiplier having one non-linear scale.

able, the application of these mechanisms to division are limited by the impossible condition of dividing by zero and the impractical condition of dividing by numbers close to zero. The second condition produces very large values in the answer and results in a small overall scale factor in a computer of reasonable size. With these limitations in mind, and with careful consideration given to scale factors, and possible locking conditions, sometimes it is possible to reverse a multiplying unit and use it for division. For example, consider a line diagram, Fig. 9, of a similar triangle multiplier. From the diagram it can be seen that, as x approaches 0, w increases towards infinity, and when y equals 0, x may vary at will without affecting the answer, w.

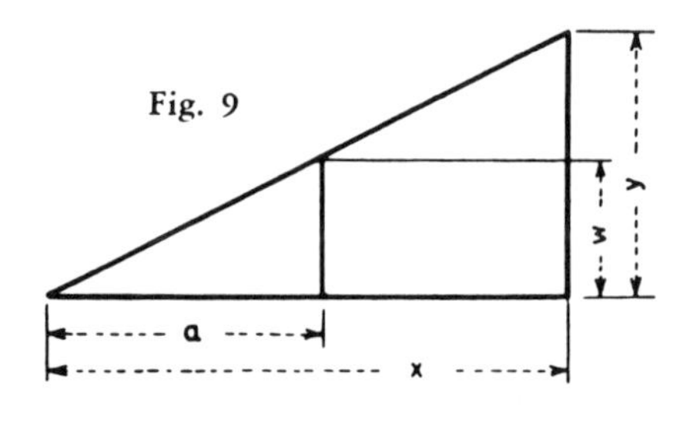

It is possible also to use a log cam multiplier for division. The only change required is that the differential should subtract the logs instead of adding them. In another method, with considerably less limitation, the reciprocal of one input is computed and multiplied by the other variable in accordance with the equation

$$w = (x)\frac{1}{y}$$

The reciprocal function, generated with a cam, experiences the same difficulties of discontinuity at and close to zero. This time it results in a very steep cam rise which is sometimes too steep for good camming action.

Division may be performed by using a multiplier and a servo motor, but this arrangement represents more mechanism and still suffers from the inability to approach zero efficiently. Referring to Fig. 10, a servo motor drives one input w to a multiplier; the other input to the multiplier is the quantity y. The multiplier answer yw is fed to a differential where the other variable x is subtracted from it; that is, the output of the differential is $yw - x$. The follow-up control on the differential causes the servo motor to drive so that yw — *equals* 0 at all times, or w equals x/y.

If y equals 0, the output from the multiplier is zero. If x has a finite value, the differential will give an error signal to the servo motor, but, regardless of the value of w fed into the multiplier, the output is zero. The signal will in effect attempt to drive the value of w to infinity, or until the mechanism drives to the limit stops.

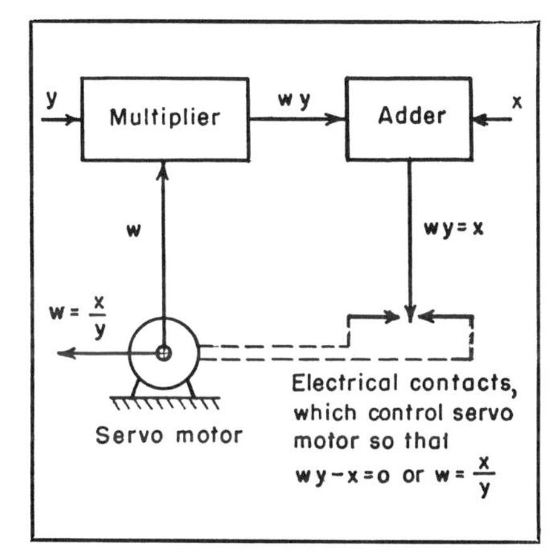

Fig. 10—Division mechanism using a multiplier and servo motor. This device, too, is limited by inability to approach zero efficiently.

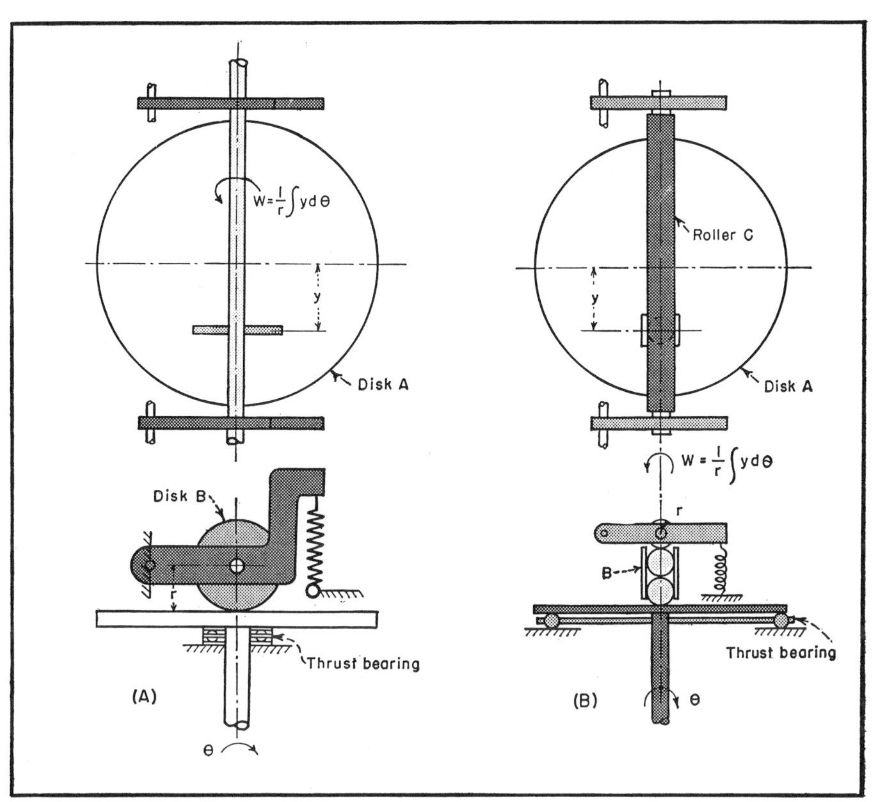

Fig. 11—Two types of integrators. (A) Two-disk integrator. (B) Ball-disk integrator, which overcomes some of the inherent difficuties in two-disk type.

Integration

The classic two-disk integrator, pictured in Fig. 11(A), consists of a smooth disk, A, rotated by an input variable θ, and another output disk, B, perpendicular to, and spring-loaded against disk A. The radial displacement of disk B from the center line of rotation of disk A is the input variable y. Assuming perfect non-slipping contact between the rotations of the disks, and perfect slipping as disk B moves radially along disk A, the arc $y\,d\theta$ on disk A equals the arc $r\,dw$ on disk B.

Then

$$dw = \frac{1}{r} y\,d\theta$$

$$w = \frac{1}{r} \int y\,d\theta$$

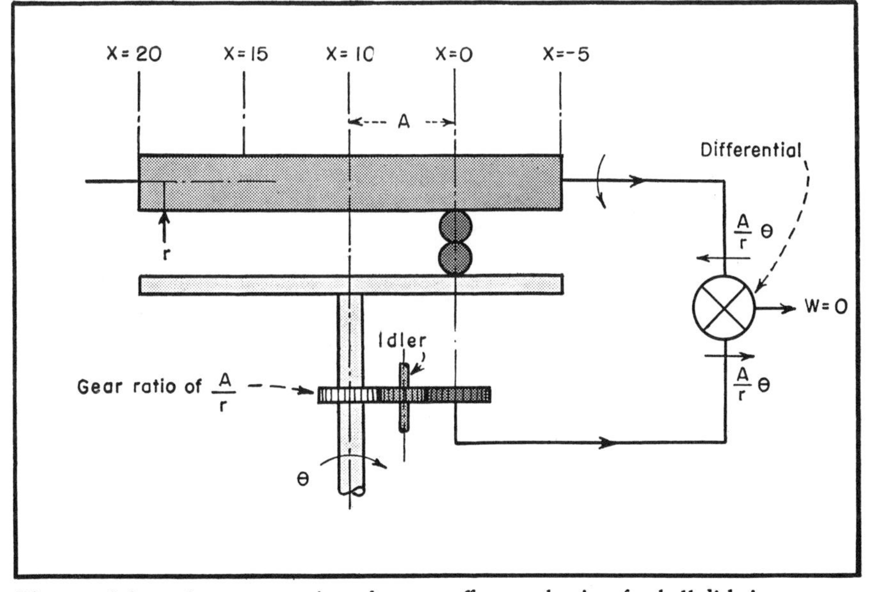

Fig. 12—Schematic representation of a zero offset mechanism for ball-disk integrators. There is less tendency toward pitting in this design, but a loss of scale factor.

The value of $1/r$ is known as the integrator constant.

The disadvantage of the two-disk integrator is that the output torque is limited by the frictional force developed between the two disks. But, the force required to change the second variable y, is directly proportional to the frictional force between disks. The two requirements for good operation are opposed.

This objection may be alleviated in part by using the ball-disk integrator shown in Fig. 11(B). Replacing disk B by two balls (B) and a cylindrical roller (C) produces rolling action radially across the disk, though the basic principle remains the same. Larger output torque may be obtained by increasing the spring force between roller, balls, and disk, with consideration given to a large radius thrust bearing to prevent cocking action of disk A. Note that continued operation at zero radius (y equals 0) will result in pitting disk A at the center. By using a differential the pitting can be partially avoided.

Fig. 12 shows a method whereby the point x equals 0 has been displaced a distance, A, from the centerline of the disk. With this method, the balls are rolling when the variable x is equal to zero, and there is less tendency toward pitting. However, with this method the range of $+x$ and $-x$ are not equal. As seen from this figure, x may vary from -5 to $+20$ units. There is also a loss of scale factor through the differential.

By using the equation

$$d\,(xy) = x\,dy + y\,dx$$

or

$$xy = \int x\,dy + \int y\,dx$$

multiplication may be performed by using two integrators. Fig. 13 shows that the variables x and y must be fed to each integrator subsequent to which the output of both integrators is fed into a differential. The output of the differential is the desired product.

Differentiation

The general problem of differentiating one variable with respect to another is solved by reversing an integrator, but unfortunately this does not yield a theoretically exact solution. The method requires an additional mechanical differential. The mechanism is shown in Fig. 14. The required derivative is

$$\frac{dy}{dx} = w$$

where w is the position of the ball carriage, x is the angular input to the disk, and y is the second input driving one side of the differential. The other side of the differential is driven by the output of the integrator, $\int w\,dx$. The spider of the differential is made to drive the ball carriage, either mechanically or through a servo system, so that $w = k\,(y - \int w\,dx)$ where k is the scale of w in terms of the inputs to the differential.

Rearranging

$$y - \frac{1}{k}w = \int w\,dx$$

and differentiating with respect to x,

$$\frac{dy}{dx} - \frac{1}{k}\frac{dw}{dx} = w$$

The error function of this differentiation may be expressed as

$$\epsilon = -\frac{1}{k}\frac{dw}{dx}$$

This inherent error, a time function, will disappear only when dw/dx equals 0 or when the carriage does not move as x is changed, implying dy/dx equals 1.

Theoretically dy/dx equals 1 will occur only after an infinite time. But it is possible to go in the direction of decreasing error by increasing k, that is, by causing w to move rapidly for small changes of y. Generally the increasing of k is accomplished by a servo follow-up in the w link of the differentiator. Increasing the speed of response too greatly, however, may cause undesirable oscillation of the system.

The preceding mechanism is the general form of differentiator, solving the general equation

$$w = \frac{dy}{dx}$$

In many cases encountered in computer design it is necessary only to find

$$w = \frac{dy}{dt}$$

that is, to differentiate with respect to time. This may be accomplished with a great deal less mechanism, since it amounts to building an accurate tachometer. The input to the device, y, is a shaft rotation, hence dy/dt is the speed of shaft rotation.

The mechanism of Fig. 15 is a force-balance servo system making use of the principle that viscous drag forces are proportional to velocity. The input quantity (y) drives a serrated drum closely fitted inside a housing, the housing being restrained from rotation by a spring. A viscous fluid is contained in the small clearance between housing and the drum.

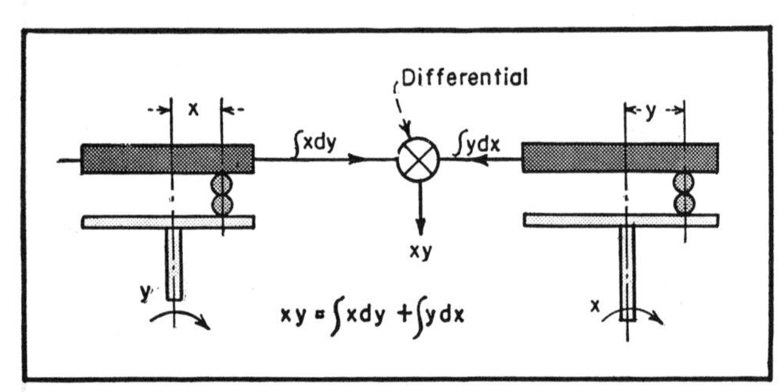

Fig. 13—Showing how two integrators can be combined for multiplication. Differential output represents the product.

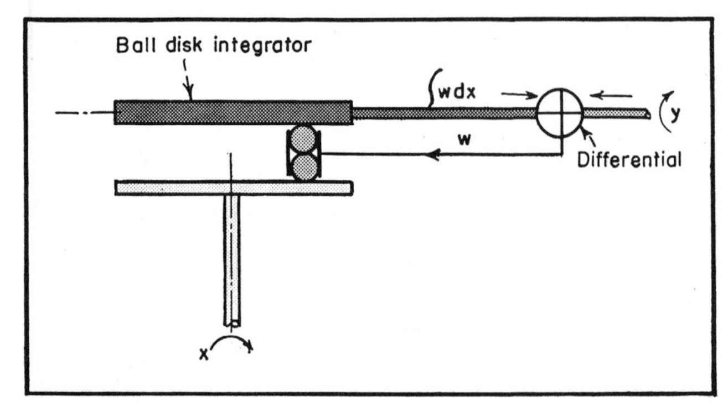

Fig. 14—Differentiating can be done by driving a ball-disk integrator in reverse. An additional differential is required for theoretical accuracy, driving mechanically or through a servo.

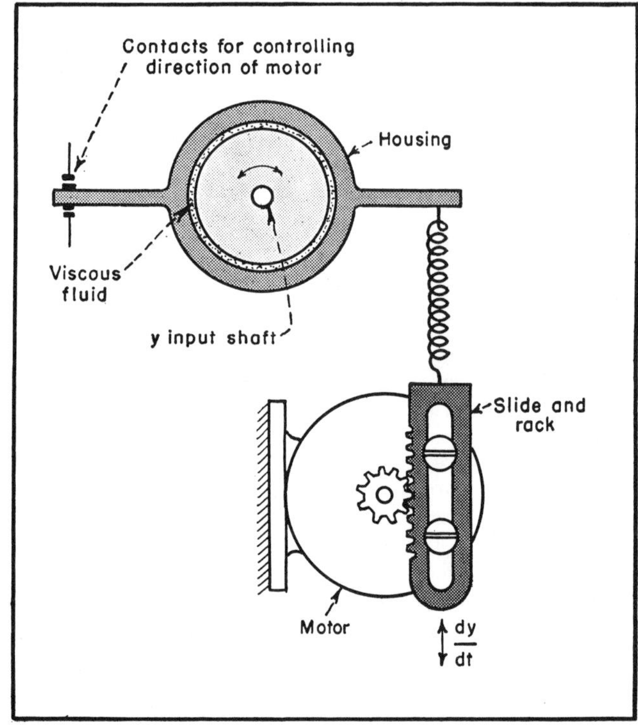

Fig. 15—Viscous drag differentiator has the advantage of sensitivity to slow rates. But, it may have several undesirable conditions, such as friction, hunting, or temperature effects.

Mechanical Computing Mechanisms—III

- How to develop a mechanical analog computer from a mathematical relationship to a complete schematic design.
- Design considerations, such as limit stops, electrical components and vibration and acceleration effects.

ROBERT R. REID and DU RAY E. STROMBACK

Research Engineers, The Franklin Laboratories for Research and Development

The procedure when combining several components into a workable computer now will be demonstrated by analyzing a simple equation such as the well-known Pythagorean theorem, and the principles employed can be extended to more complicated relations involving more variables.

First step is to set up the problem in a mathematical sense and secondly in a mechanical design sense. If a and b are the sides of a right triangle of which c is the hypotenuse, the Pythagorean relation states that $a^2 + b^2 = c^2$. Assume that lengths b and c are to be the inputs to the computer, and the continuous value of the length of side a is the answer required. By modifying the above relation to solve for a.

$$a = \sqrt{c^2 - b^2} \qquad (1)$$

This will serve as a statement of the problem in mathematical terms, at least in a preliminary sense. For the mechanical statement of the problem some facts about the mechanical nature of the input and output analogs must be established.

Assume that two elements sensitive to length are available, from which shaft rotations of linear proportionality to the input quantities b and c are obtained, at a high power level. Suppose the unit of length being measured is miles and length of side b, expected to vary from 1 to 4 miles, may be measured to an accuracy of ± 0.01 miles. The hypotenuse c will range in value from 5 miles to 20 miles, with the same accuracy of measurement. By substituting appropriate maximum and minimum values in Eq 1, the lower limit of side a as 3 miles and the upper limit as 19.98 miles are determined. Thus the ranges of input and output variables are determined.

Before investigating the output accuracy required of the computer, possible accuracy obtainable from these input analogs should be checked as this is often a limiting factor in design.

A very good error analysis may be made quickly if, in general, $a = f(c, b)$ then

$$da = \left(\frac{\delta f}{\delta c}\right) dc + \left(\frac{\delta f}{\delta b}\right) db \,.$$

Applied to Eq (1), this becomes

$$da = \frac{2c\,dc}{2\sqrt{c^2 - b^2}} - \frac{2b\,db}{2\sqrt{c^2 - b^2}} = \frac{c\,dc - b\,db}{\sqrt{c^2 - b^2}}$$

This equation expresses the total error in a due to errors in b and c at a given condition. Maximum errors will result from a minimum value of the radical, implying minimum c and maximum b, in combination with errors of opposite sense.

The problem assumption was

$$|\,db\,| = |\,dc\,| = 0.01 \text{ miles.}$$

To determine maximum error in a, substitute $c = 5.00$ and $b = 4.00$ miles respectively, while $dc = 0.01$ and $db = 0.01$ miles respectively in the error equation. The maximum value of da is approximately 0.03 miles, and occurs when a is at its minimum value of 3 miles. A good average for all values appears to be about 0.015 miles. It is apparent that errors of this order will occur in the output values, even asuming the actual computer built has no error whatsoever, which is, of course, impossible. If the requirement on output accuracy is 1 percent of full scale reading, it means that an error of 0.20 miles in the length of side a will be accepted. The tolerable maximum error in the computer is, therefore, $0.20 - 0.03 = 0.17$ miles.

Requiring the computer to record the output or control another device would add little of value to this schematic design analysis; so, dial indication of the answer will be assumed.

Many times the size and weight limitations are of paramount importance in selecting components, but detailed analysis of such limitations depends upon the individual project. Suffice it to presume that the finished computer will be of desk-computer size and easily portable.

Now, having arrived at some conclusions concerning the overall picture, the statement of the practical problem is as follows: two shafts are rotating according to the values of b and c respectively. A third shaft is required, whose motion represents a, coupled to shafts b and c by suitable devices so that the relation, $a = \sqrt{c^2 - b^2}$, is everywhere maintained. These three shafts, and the mechanical components which relate their motions, comprise the proposed computer.

To attack the problem from several angles, first re-examine the mathematical statement, Eq (1), and sketch out several block-diagram methods of solution.

This relation may be rewritten as

$$a = b\sqrt{c^2/b - 1} \qquad (2)$$

or as

$$a = \sqrt{(c - b)\,(c + b)} \qquad (3)$$

Block-diagram solutions of these three expressions are shown in Fig. 17. At first glance there appears to be little choice among the three methods, since in each, four operational blocks are required to perform the solution. However, as was demonstrated in Parts II and III describing computer components, not all the blocks represent the same number of parts or present the same degree of difficulty. For example, the problem of division is quite difficult, and should be

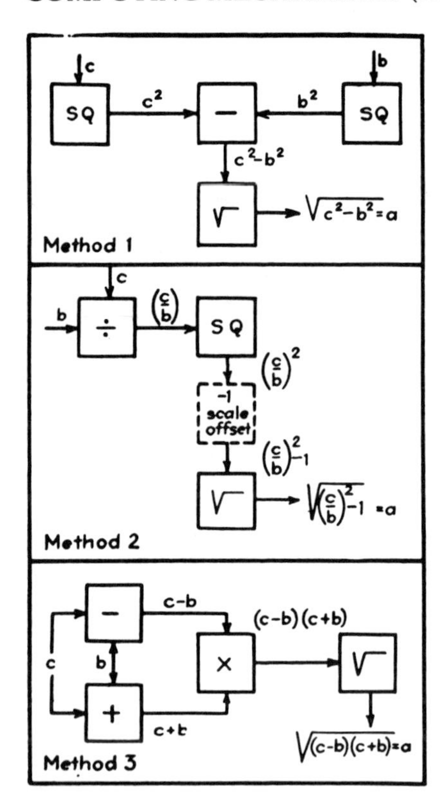

Fig. 17—Block diagram investigation of the different ways of arranging the equation. Method 2 should be discarded because it calls for division. Method 3 is more complex than Method 1, and it is therefore less desirable.

avoided whenever possible; a simple function-generation process such as squaring would certainly be preferred over division; similarly, addition would be preferred over multiplication. Considering the block diagrams on this basis, Method 2 should be discarded first, since it requires both a divider and a multiplier. In choosing between Methods 1 and 3 it can be seen that Method 3 requires a multiplier, which is generally a more complex mechanism than the mechanisms required by Method 1. Therefore, in this particular example nothing is gained by this venture in rewriting the given equation. (Frequently, however, rewriting is definitely justified.) So the original Method 1 can be chosen on the basis of the block-diagram analysis.

Two squaring devices must now be selected, an adder and a square root mechanism, from among the mechanical analog components. A mechanism to obtain the square of an input variable might consist of an integrator in which both inputs are the same quantity and the output is

$$\frac{1}{K}\int b\,db = \frac{1}{K}\frac{b^2}{2} = \frac{b^2}{2K}$$

and similarly $c^2/2K$, from a second integrator. The outputs of the two integrators are then subtracted in a rotary differential to produce a difference $(c^2 - b^2)$, which appears as a shaft rotation. A plate cam on this output shaft, rotating as $(c^2 - b^2)$, may be constructed so the follower motion is $\sqrt{c^2 - b^2} = a$, and a gear train designed so that a is converted from linear to rotary motion, and appears on an output dial. Fig. 18 shows the schematic design of this computer.

Mechanizing the block diagram of Method 1 from another approach, the squaring device is considered to be a function generator to mechanize the relation

$$y = f(x) = x^2 .$$

Here $y_1 = c^2$ and $y_2 = b^2$

Since the shaft rotations are proportional to the quantities c and b already available, two rotary plate cams with roller followers provide the squaring devices shown in Fig. 19. The motions of c^2 and b^2 are converted by these devices into translatory motions of two followers, which suggests a simple link differential connecting the two followers. From this the difference, $(c^2 - b^2)$, is obtained. A function generator is required to extract the square root of the quantity now represented by a translatory motion. A cam is chosen to perform this operation, and a rack and gear train to convert the answer to rotary motion of a dial.

Having sketched each proposed mechanization in the form of a schematic design, Figs. 18 and 19, the next step toward achieving a practical design is the investigation of scale factor of each of the proposed mechanisms.

In Fig. 18 the two integrators and the rotary differential are unbounded components; this means that, since they are capable of rotary motions, any amount of rotation may be assigned as equal to one unit of the variable. The square-root cam is not unbounded however. Design practice usually limits the working range of input rotation to something under 360 deg. Therefore, it is necessary to begin with this square-root cam and work forward and back.

To arrive at a reasonable size cam, the total cam throw, and, consequently, the follower travel, should be arbitrarily limited to approximately ¾ in. The range of variable a represented by the follower travel has been established as 19.98 miles — 3.00 miles = 16.98 miles. Choosing round numbers for ease in computation the scale factor of the a follower is set as ¾ in./15 mile which results in an exact total follower motion of 16.98 miles (1.0 in./20 miles) = 0.849 inches.

The cam-rotation scale factor is similarly determined. The range of $(c^2 - b^2)$ is 399.0 — 9.0 = 390 miles², and, although ordinarily plate cam rotations are limited to less than 360 deg, the rotary constrained follower shown permits somewhat more rotation, at the sacrifice of a return path for possible overtravel. If suitable stops are provided, the rotational cam scale factor may be set as

$$\frac{390 \text{ deg}}{390 \text{ mi}^2} = \frac{1 \text{ deg}}{\text{mi}^2}$$

The differential spider scale factor is the same, since the cam is fastened to the spider shaft. Recalling the characteristics of a bevel gear differential, the input gear scale factors are 2(1 deg/mi²) = 2 deg/mi², and this in turn becomes the output scale factor for the two integrators.

The next step is to assume some reasonable sizes for the integrators. Consider a small integrator having a disk with 2.0 in. working diameter and using 1/4 in. diameter balls and roller. The essential characteristic of an integrator used as a squaring mechanism is the proportionality factor K relating the units of disk rotation to the units of carriage motion, conveniently expressed in in./deg. Assuming the b high power level shaft, having a total rotation of 300 deg corresponding to a variation from 1 to 4 miles, is direct-connected to the integrator disk shaft, the disk scale factor is 300 deg/3 mile = 100 deg/mile. The available working radius of the disk is 1 in., to represent a value of 4 miles, or a scale factor of ¼ in./mile. The proportionality factor between these scale factors is

$$K = \frac{1/4 \text{ in./mile}}{100 \text{ deg/mile}} = 0.0025 \text{ in./deg.}$$

The mathematical derivation of the integrator law, PRODUCT ENGINEERING, October 1949, p 128, where

x = disk rotation
y = carriage displacement = Kx
w = roller rotation
r = roller radius

yields

$$w = \frac{K}{r}\int x\,dx = \frac{K}{r}\left(\frac{x^2}{2}\right) = \frac{Kx^2}{2r}$$

The last expression relates the input and output rotations. Substituting the input values corresponding to one mile, the output rotational scale factor is obtained where K = 0.0025 in./deg, x = 100 deg and r = 0.125 inches.

$$w = \frac{0.0025\,(100)^2}{2\,(0.125)} = 100 \text{ deg/mile}^2$$

The required scale factor was 2 deg/mile^2 for the differential. Therefore, a 50:1 gear reduction must be provided between the output of the integrator and the input to the differential to match the scale factors. The c input, although having the same rotation of 300 deg, represents a variation of 20 — 5 or 15 miles, a scale factor of 300 deg/15 mile = 20 deg/mile. Substituting this value in the expression for the output of the second integrator,

$$w = \frac{0.0025\ (20)^2}{2\ (0.125)} = 4\ \text{deg/mile}^2,$$

and a 2:1 reduction is needed to match the required differential input of 2 deg/mile^2.

All scale factors and essential dimensions from the input shafts to the square root cam shaft have been established. Next, consider the scale factors from the square-root cam to the output dial. If the cam is so designed that the angular rotation of the follower arm is linearly proportional to a, a total rise of 3/4 in. will result in an angular displacement of $2\ [\arcsin\ (1/2 \times 3/4)/L] = 2\alpha$ where L is the length of the follower arm. If α is chosen as 7 1/2 deg. then a value of $L = 2.875$ in. is required to produce a total angle of $2\alpha = 15$ deg, corresponding to 15 miles, and a scale factor of 1 deg/mile. A gear ratio of 20:1 between the follower arm and an output dial will then result in a dial scale factor of 20 deg/mile, and a total rotation of almost 340 deg for variation in a between limits of 3.0 and 19.98 miles.

At this point essential dimensions and scale factors for the entire computer have been established. Though there are no Class B (theoretical) errors in the solution, this design has approximately 75 major parts, including two integrator lead-screws, approximately 9 pairs of spur gears, and a bevel-gear differential, all of which will tend to cause excessive Class A (mechanical) errors. Further analysis might eliminate some of the spur gears, but a tentative conclusion can be made that another mechanization containing fewer parts should be attempted.

The arrangement of Fig. 19 has input scale factors of 100 deg/mile for b and 20 deg/mile for c, as previously stated. Again each input rotates a total of 300 deg to represent the total change in variable. Assume a 1.0 in. cam throw, for the variable c, that is, a scale factor of

$$1.0\ \text{in.}/(400\ \text{mile}^2 - 25\ \text{mile}^2) = 1\ \text{in.}/375\ \text{mile}^2.$$

If the output is located at the center of the differential link, the output scale factor must be equal to 1/2 in./375 mile^2 and the scale factor of b must be identical with that of c. This results in a total throw of the b cam of (16 mile^2 — 1 mile^2) (1 in./375 mile^2) = 0.040 in. The total maximum motion of slide A is calculated as 399 — 9 = 390 mile^2 = 0.52 in. For approximately 300 deg rotation of the square-root cam, the pitch diameter of gear B is approximately

$$D = 360\ (0.52)/300 = 0.1985\ \text{in.}$$

Assuming an 80 diametral pitch pinion, the nearest even number of teeth is 16, with a pitch diameter of 0.200 in., therefore, 298 deg/390 mile^2 = 1 deg/1.31 mile^2 is the rotational scale factor of the pinion. Assume again a dial rotation of 340 deg = 17 miles, and a scale factor of 20 deg/mile on the output dial. For simplicity in fabrication assume the same pinion is used on the output of slide A; therefore the throw of the square root cam must be 0.2(340)/360 = 0.593 in.

All scale factors and essential dimensions are now assigned to both mechanizations. The second method, the cam solution, again has no Class B (theoretical) errors. The primary objection to the other method was the large number of parts, with inherent Class A (mechanical) errors, and the possibility of slippage in the integrators. The cam solution looks very promising on this basis, having only 10 major parts and workable scale-factors throughout. Final schematic design is shown in Fig. 20, with scale factors.

Continued on next page

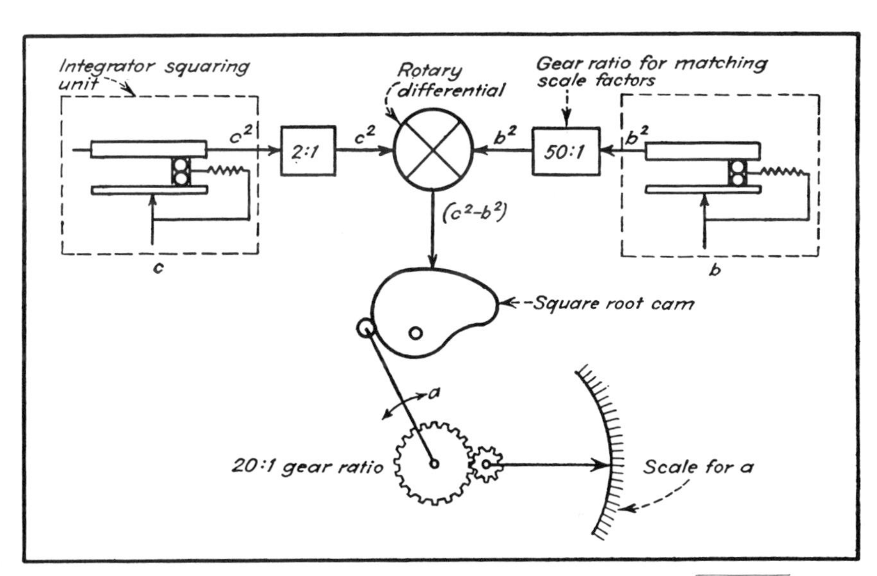

Fig. 18—The integrator method of mechanizing the equation $a = \sqrt{c^2 - b^2}$, shown in schematic form. It requires an excessive number of parts.

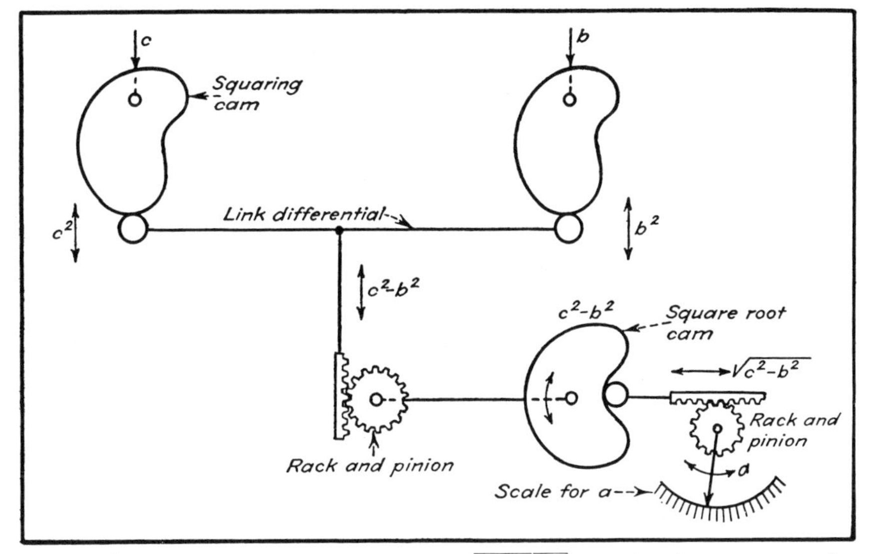

Fig. 19—Cam method of mechanizing $a = \sqrt{c^2 - b^2}$ uses function generators for squaring and a link differential for subtraction. Note the reduction in parts.

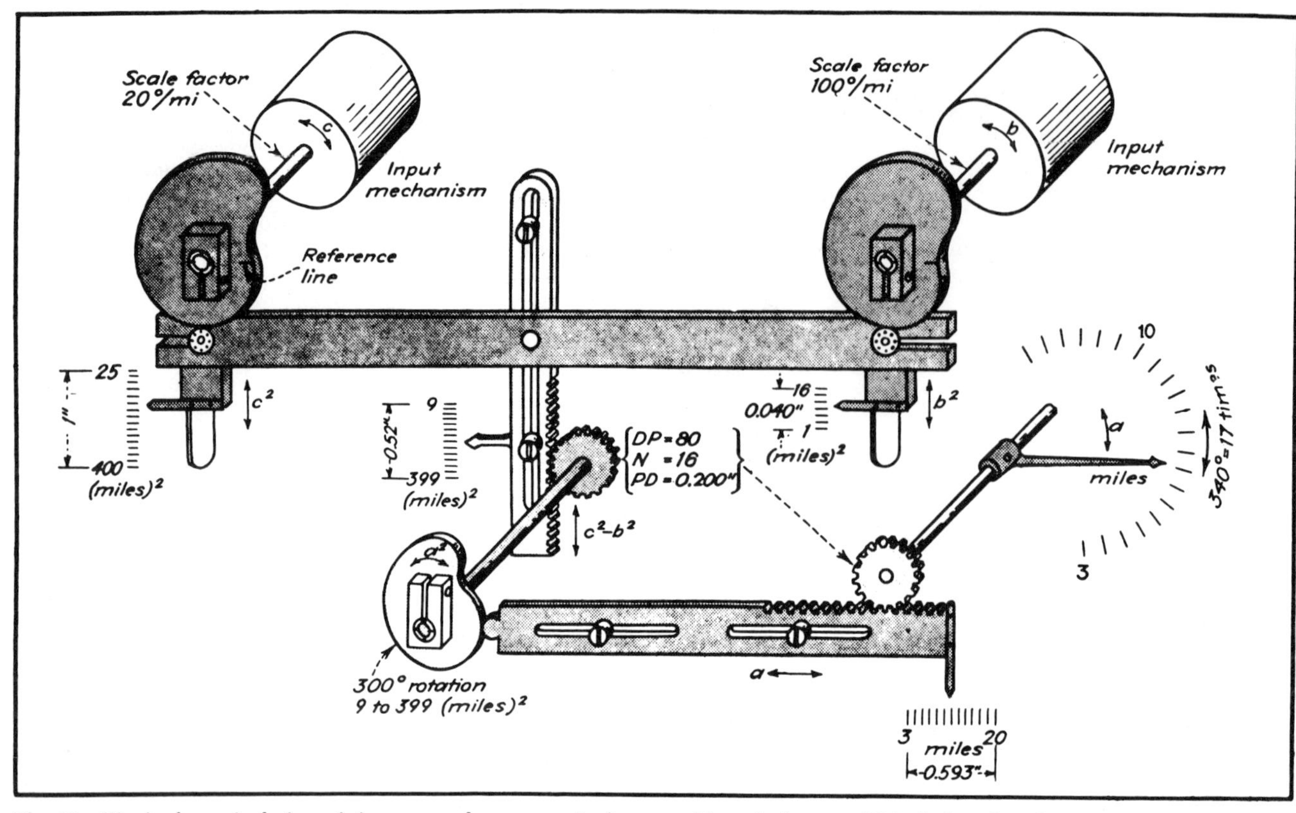

Fig. 20—Final schematic design of the proposed computer is shown, with scale factors. This design, based on the cam method, was selected because far fewer parts (and potential mechanical errors) are included. The value of careful analysis is demonstrated.

FUNCTION MECHANISMS

Precision-function mechanism

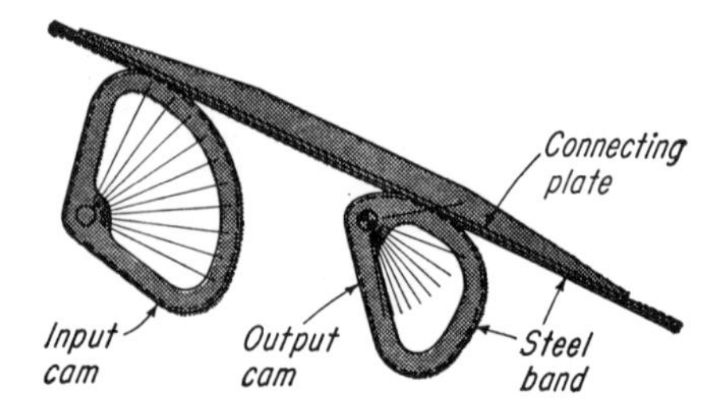

A steel band wrapped around two cams and fastened to a connecting plate can transfer the precise angular rotation of one shaft on to another over long distances. With proper cam design, the mechanism also produces a mathematical-function rotation for a given constant input rotation; for example, it can square, or give a multiple of the input angle. Developed by the author.

Sinusoidal-adding mechanism

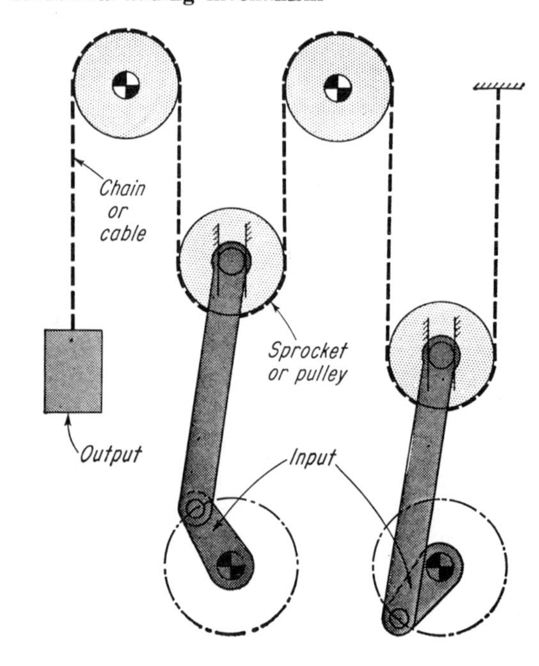

A simple arrangement which can quickly be set up in a lab to superimpose two input motions on to an output. This permits adding sinusoidal motions which have different amplitudes, phase angles, and frequencies.

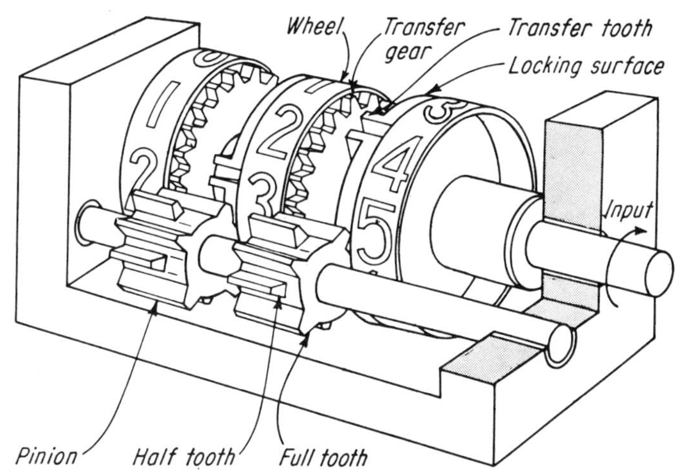

EXTERNAL PINION COUNTER carries the numbered wheels on one shaft, the pinions on a countershaft. The first wheel is direct drive; the rest driven by the pinions. Once per revolution, the transfer tooth (kicker) on the first wheel picks up a full tooth on the first pinion, and turns the second wheel one number. When the next full tooth hits the locking surface, the pinion stops. Until the next transfer the pinion cradles the first wheel between two successive full teeth, locking itself and the second wheel. Because of this locking method, accumulation of tolerances misaligns each number slightly more than the preceding one. If tolerances are too large, the numbers spiral.

The pinion rotates the second wheel one number while the first wheel turns 36°. Thus, if a 2-wheel counter reads 09, it changes to 10 between 9.5 and 10.5 Start and finish of a transfer can be shifted about, though still taking place over 36° rotation. For instance, one manufacturer designs his counter to transfer between 9.0 and 10.0; another has the transfer occur between 9.1 and 10.1.

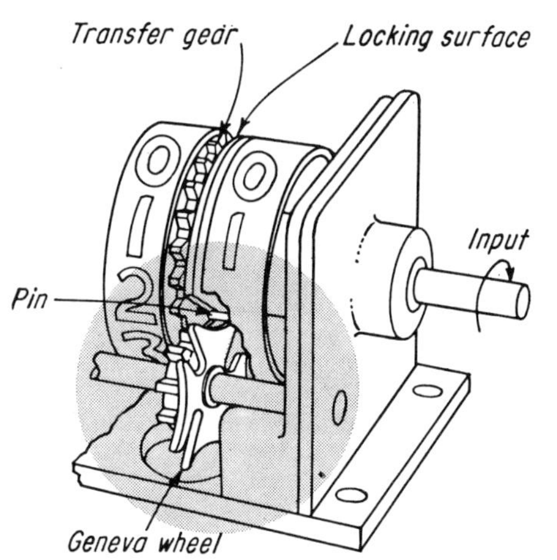

MODIFIED GENEVA TRANSFER turns second wheel one number while the first turns 72° whereas standard 4-slot Geneva wheel turns 90° while the driver turns 90° (with both methods, pin on first wheel slides in and out of slot as it turns the Geneva wheel). Unlike usual Geneva transfer, modified version does not lock with its cutout against the first wheel. Instead, the pinion locks as in the external-pinion type. This way, diameter of the locking surface is larger and positioning more accurate.

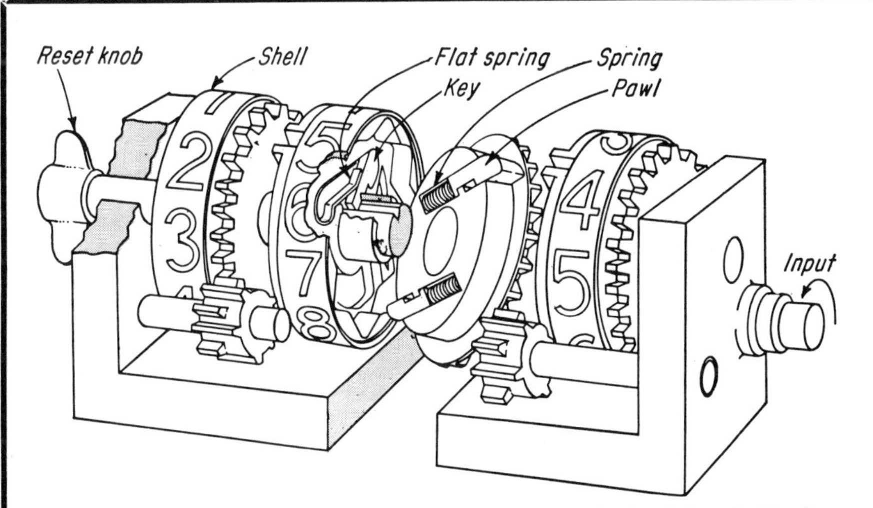

RESET, MOVING SHELL is separate part of wheel. Other half of wheel which carries the transfer gear drives the shell through spring-loaded pawls. Except during reset, wheel-shaft does not turn; therefore, input to first wheel must be through gearing.

At the start of count, with a zero reading, all keys sit in the keyway. Movement of wheels drags them out. During reset, the shells move and transfer gears stand still. A single turn of the wheel-shaft drops all keys back into the keyway; then a slight additional turn brings up all zeros.

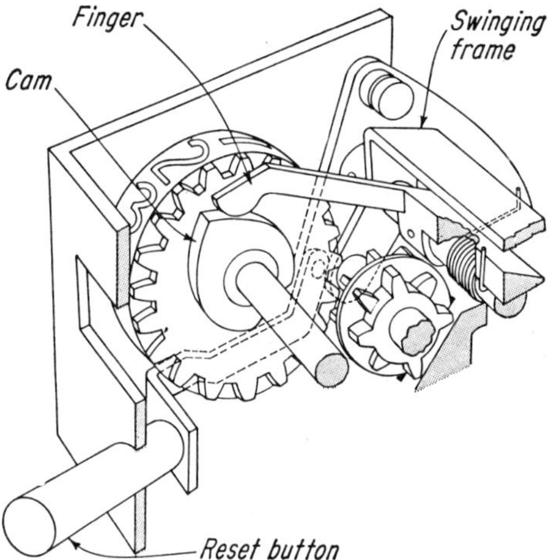

RESET, HEART-SHAPE CAM has steadily decreasing radius from point to cleft. Thus the pressing fingers always turn wheels cleft upward so that zeros face out the window.

The swinging frame that carries pinions is linked to fingers by springs. When a push of the reset button forces down the fingers, springs swing the frame, disengaging the pinions. Center disks keep pinions aligned during the reset cycle. Counting method is same as for an ordinary external pinion counter.

Chapter 6 RATCHETS, DETENTS, TOGGLES, AND DIFFERENTIALS

6 Escapement Mechanisms

FEDERICO STRASSER

Escapements, a familiar part of watches and clocks, can also be put to work in various control devices, and for the same purpose—to interrupt the motion of a gear train at regular intervals.

Each escapement shown here contains an energy absorber. Whether balance wheel or pendulum, it allows the gear train to advance one tooth at a time, locking the wheel momentarily after each half-cycle. Energy from the escape wheel is transferred to the escapement by the advancing teeth, which push against the locking surfaces (pallets).

Seven more escapement systems will be illustrated in a forthcoming issue.

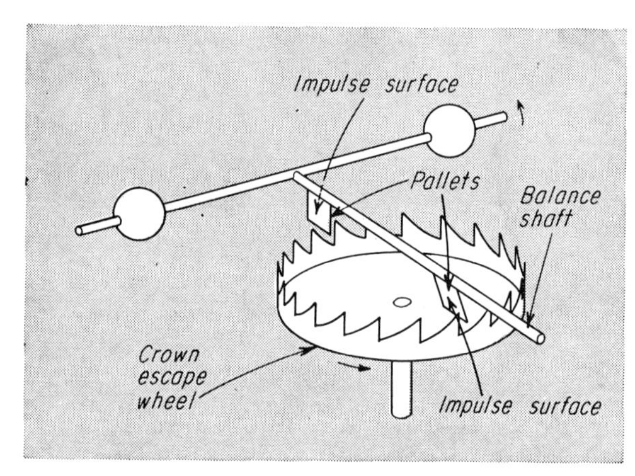

1 VERGE. Pallets are alternately lifted and cleared by diametrically opposed teeth on escape wheel, thus rocking balance shaft. One pallet or the other is always in contact with a tooth of the escape wheel.

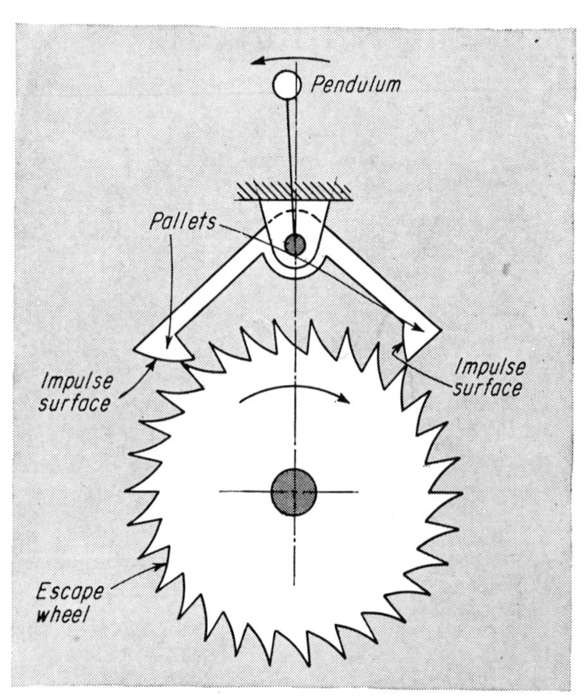

2 ANCHOR—RECOIL. As pendulum swings to one side, pallet engages tooth of escape wheel. When wheel is gripped, it recoils slightly. Pendulum starts in opposite direction by push of wheel transmitted through pallet. As pendulum swings to other side, the other pallet engages escape-wheel tooth, recoils, and repeats cycle.

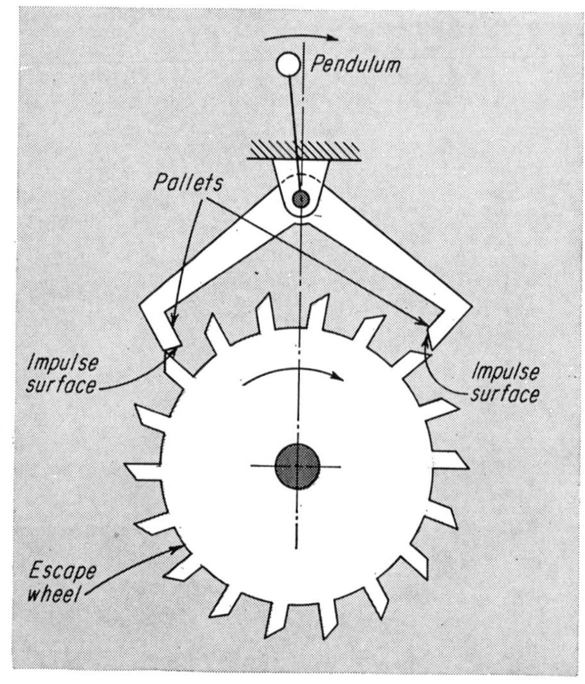

3 ANCHOR—DEAD BEAT. For greater accuracy, recoil escapement is modified by changing angle of impulse surfaces of pallets and teeth so that recoil is eliminated. The teeth of escape wheel fall "dead" upon pallets; as pendulum completes its swing, escape wheel does not turn backward.

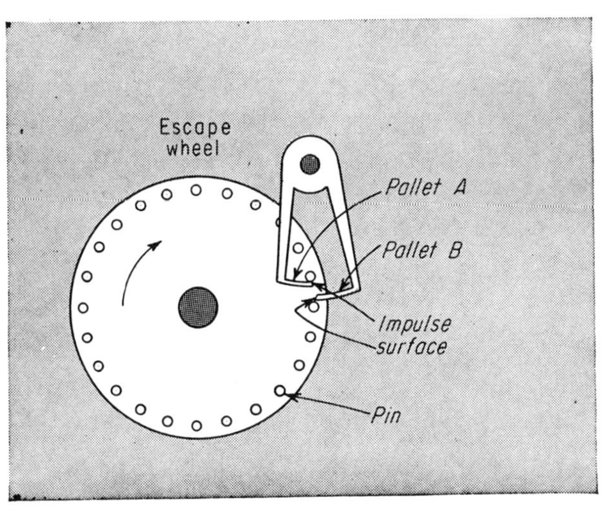

4 PIN WHEEL. Pallet A is about to unlock escape wheel. Pin will give a push to pallet A while sliding down impulse surface. As pendulum continues to swing, pallet B will stop pin, thus locking escape wheel. After pendulum reverses direction and swings toward dead center, pallet B releases wheel and receives an impulse from pin.

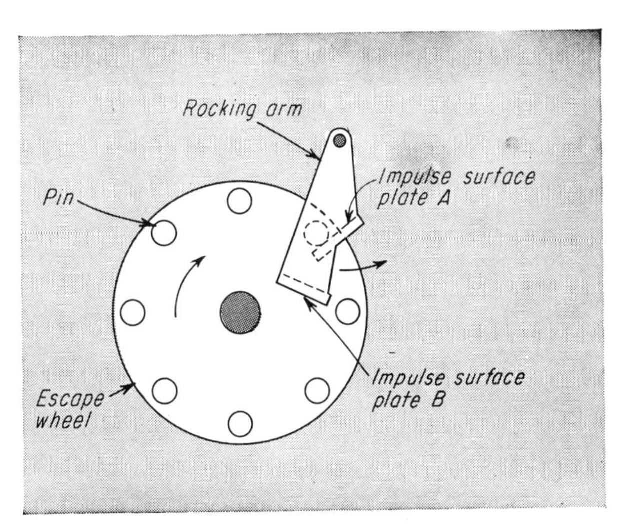

5 LANTERN WHEEL. Escape wheel is stopped when pin hits plate A of pallet, attached to rocking arm. As arm swings to right, pin pushes against impulse surface of plate until pin is clear. Escape wheel then turns freely until pin is stopped by plate B of pallet. Arm starts back toward left, receiving impulse from pin until latter clears plate.

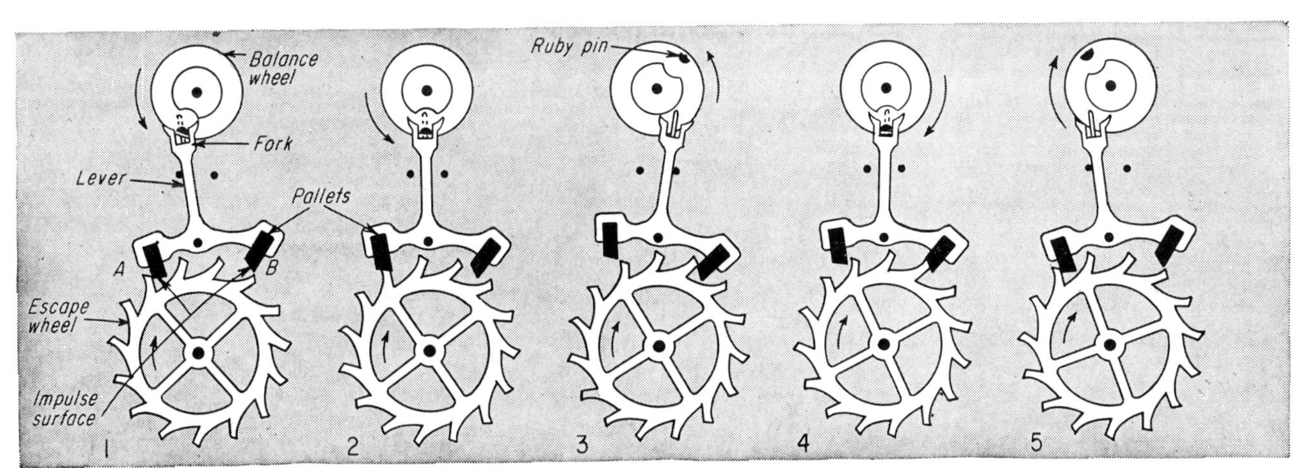

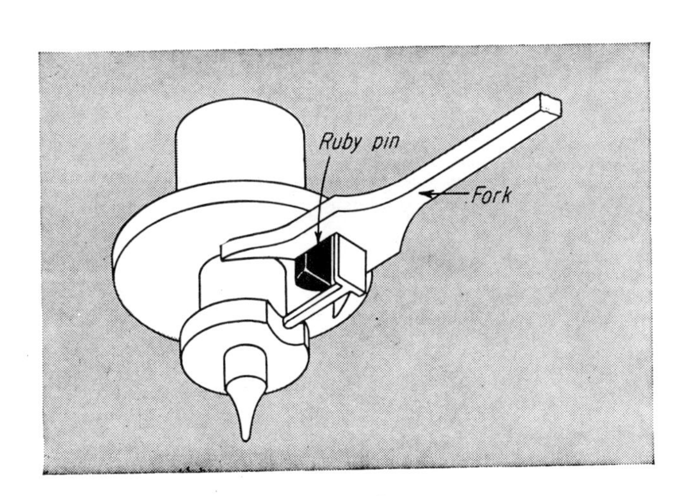

6 MUDGE'S LEVER. As balance wheel swings counterclockwise (1), the ruby pin enters fork of lever and pushes it to the right. The lever releases tooth of escape wheel, which starts turning and gives a tiny push (2) to impulse surface of pallet A. This is transmitted to balance wheel through fork and ruby pin. Balance wheel continues rotating counterclockwise (3) while lever swings to right and pallet B stops the escape wheel by engaging tooth. When balance wheel reaches its hairspring's limit, it reverses direction and re-enters fork (4). Ruby pin pushes lever to left, and receives impulse when escape wheel starts to move. Balance wheel continues clockwise until it reaches opposite limit of its hairspring. Lever continues moving to left until it again stops escape wheel (5) and awaits return of ruby pin. Another view of Mudge's lever is given in (6). This escapement is very precise because of the small portion of each half-cycle that escape wheel and balance wheel are in contact. Friction drag is at a minimum.

7 More Escapements

Another selection of these ingenious mechanical components for use in control systems

FEDERICO STRASSER

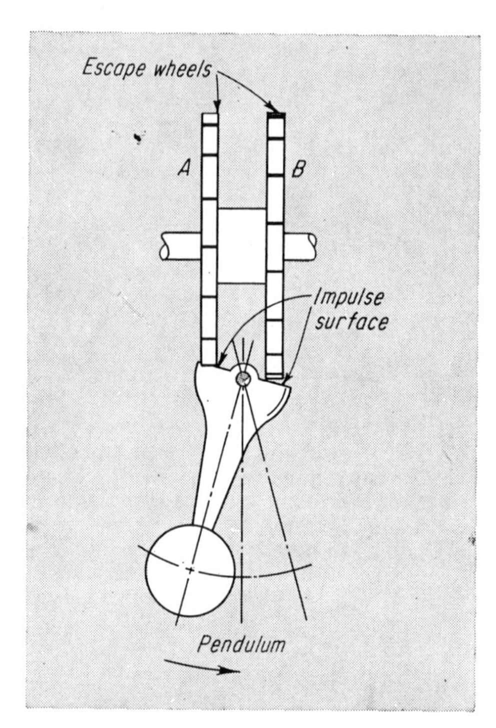

2 DOUBLE RATCHET. Two escape wheels are mounted on a common axis and pinned together so teeth are aligned with a half-pitch distance between them. Drawing shows pendulum in its extreme left position, with its impulse surface locking a tooth of the escape wheel A. As pendulum starts to right, escape wheels start to turn, imparting a push to the pendulum. As it approaches its extreme right position, pendulum again stops the escape wheels, by engaging a tooth on wheel B, then receives a second impulse as it starts toward the left.

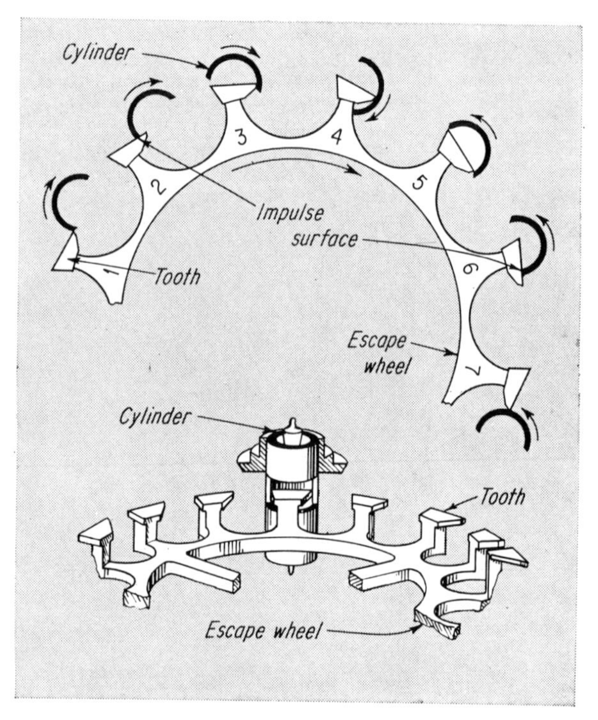

1 CYLINDER. Tooth of escape wheel and cylinder (balance-wheel shaft) are shown in seven positions. Tooth has been locked (1) and is about to enter cycle. Tooth is providing impulse (2) by sliding action along cylinder lip. Escape wheel is locked while cylinder rotates clockwise (3) to its limit (4), then starts back (5). As tooth starts forward, impulse is again imparted (6) to cylinder lip. Second tooth is locked (7) during period of backswing.

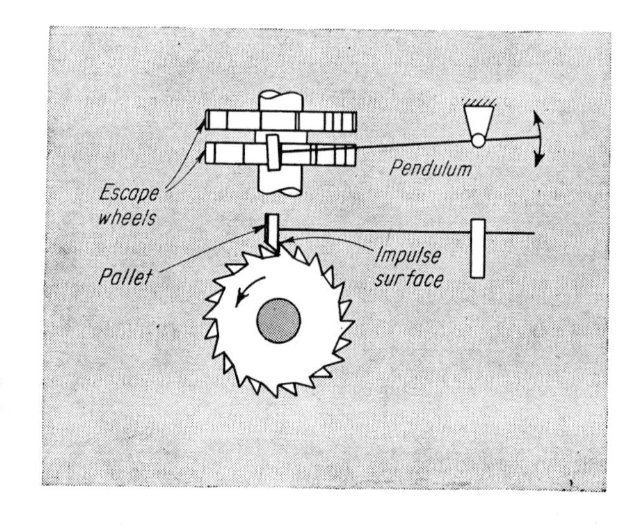

3 HIGH-SPEED DOUBLE RATCHET. Same principle as in 2, but used with a small-mass pendulum. Speeds to 50 beats per second can be obtained.

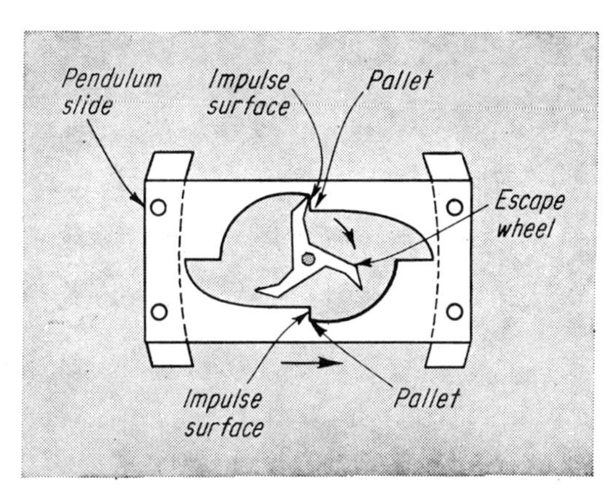

4 THREE-LEGGED. The three teeth of escape wheel work alternately upon top and bottom pallets. Top pallet is shown being driven to right. When tooth clears pallet, escape wheel will turn until tooth engages bottom pallet. When pendulum slide starts back toward left, bottom pallet will receive push from escape wheel, until tooth clears pallet.

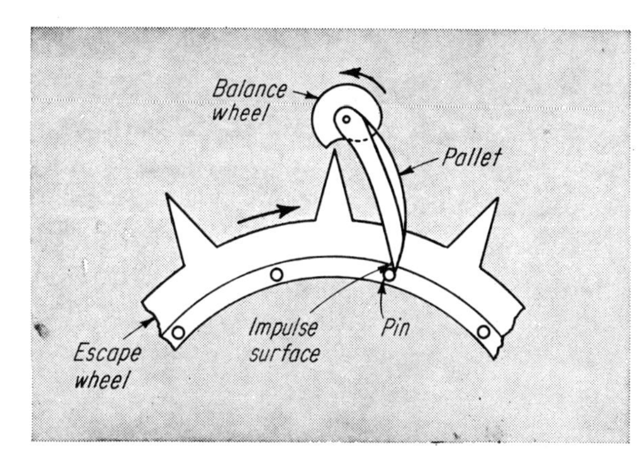

5 DUPLEX. Single pallet of balance wheel is shown receiving impulse from pin of escape wheel. As balance wheel rotates counterclockwise, tooth of escape wheel clears notch, permitting escape wheel to turn clockwise. Next tooth of escape wheel is stopped by balance wheel until balance wheel reaches limit, reverses direction, and pallet swings back to position shown. Escapement receives but one impulse per cycle.

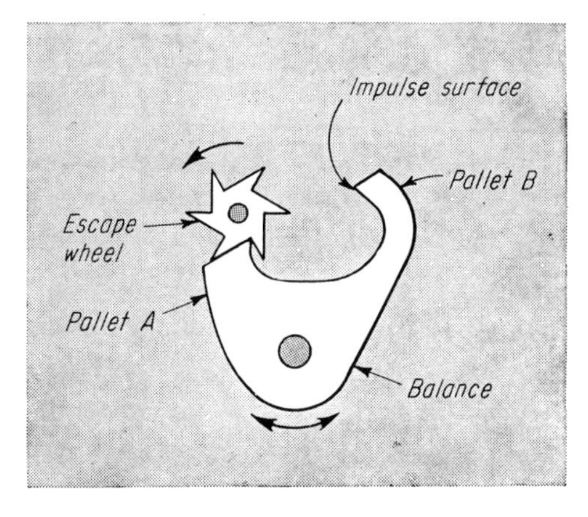

6 STAR WHEEL. Dead-beat escapement in which pallets alternately act upon diametrically opposed teeth. Illustration shows balance in extreme clockwise position with escape wheel locked. Balance starts turning counterclockwise, releasing escape wheel. At other extreme, pallet *B* locks escape wheel for a moment before direction of balance is again reversed and push is imparted.

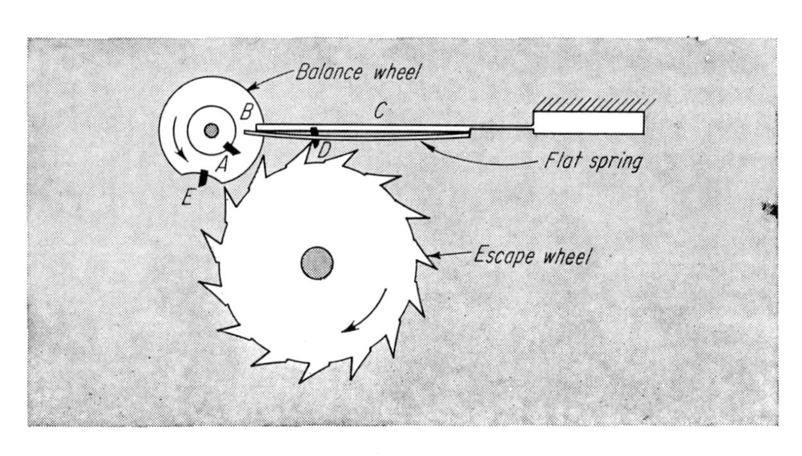

7 CHRONOMETER. As balance wheel turns counterclockwise, jewel *A* pushes flat spring *B* and raises bar *C*. Tooth clears jewel *D* and escape wheel turns. A tooth imparts push to jewel *E*. As jewel *A* clears flat spring, jewel *D* returns to position and catches next tooth. On return of balance wheel (clockwise), jewel *A* passes flat spring with no action. Thus, escapement receives a single impulse per cycle.

RATCHET LAYOUT ANALYZED

EMERY E. ROSSNER
New York, N. Y.

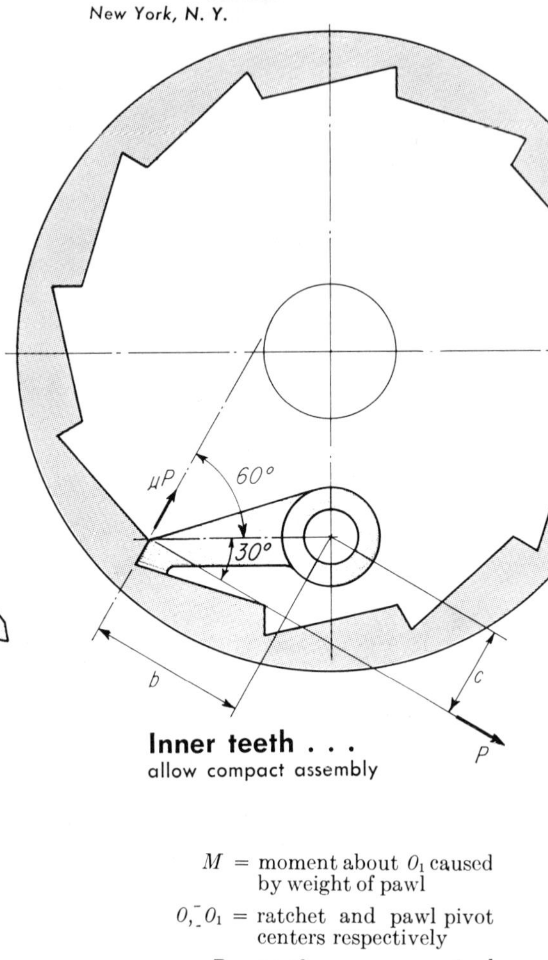

Inner teeth . . .
allow compact assembly

Pawl in compression . . .
has tooth pressure *P* and weight of pawl producing a moment that tends to engage pawl. Friction-force μP and pivot friction tend to oppose pawl engagement.

M = moment about O_1 caused by weight of pawl

O, O_1 = ratchet and pawl pivot centers respectively

P = tooth pressure = wheel torque/a

$P\sqrt{(1+\mu^2)}$ = load on pivot pin

μ, μ_1 = friction coefficients

Other symbols as defined in diagrams

The ratchet wheel is widely used in machinery, mainly to transmit intermittent motion or to allow shaft rotation in one direction only. Ratchet-wheel teeth can be either on the perimeter of a disc or on the inner edge of a ring.

The pawl, which engages the ratchet teeth, is a beam pivoted at one end; the other end is shaped to fit the ratchet-tooth flank. Usually a spring or counterweight maintains constant contact between wheel and pawl.

It is desirable in most designs to keep the spring force low. It should be just enough to overcome the separation forces—inertia, weight and pivot friction. Excess spring force should not be relied on to bring about and maintain pawl engagement against the load.

To insure that the pawl is automatically pulled in and kept in engagement independently of the spring, a properly layed out tooth flank is necessary.

The requirement for self-engagement is

$$Pc + M > \mu Pb + P\sqrt{(1+\mu^2)}\,\mu_1 r_1$$

Neglecting weight and pivot friction

$$Pc > \mu Pb$$

but $c/b = r/a = \tan\phi$, and since $\tan\phi$ is approximately equal to $\sin\phi$

$$c/b = r/R$$

Substituting in term (1)

$$rR > \mu$$

For steel on steel, dry, $\mu = 0.15$. Therefore, using

$$r/R = 0.20 \text{ to } 0.25$$

the margin of safety is large; the pawl will slide into engagement easily. For internal teeth with ϕ of 30°, c/b is $\tan 30°$ or 0.577 which is larger than μ, and the teeth are therefore self engaging.

When laying out the ratchet wheel and pawl, locate points O, A and O_1 on the same circle. AO and AO_1 will then be perpendicular to one another; this will insure that the smallest forces are acting on the system.

Ratchet and pawl dimensions are governed by design sizes and stress. If the tooth, and thus pitch, must be larger than required in order to be strong enough, a multiple pawl arrangement can be used. The pawls can be arranged so that one of them will engage the ratchet after a rotation of less than the pitch.

A fine feed can be obtained by placing a number of pawls side by side, with the corresponding ratchet wheels uniformly displaced and interconnected.

Watch Shows Time to Tenths of Seconds

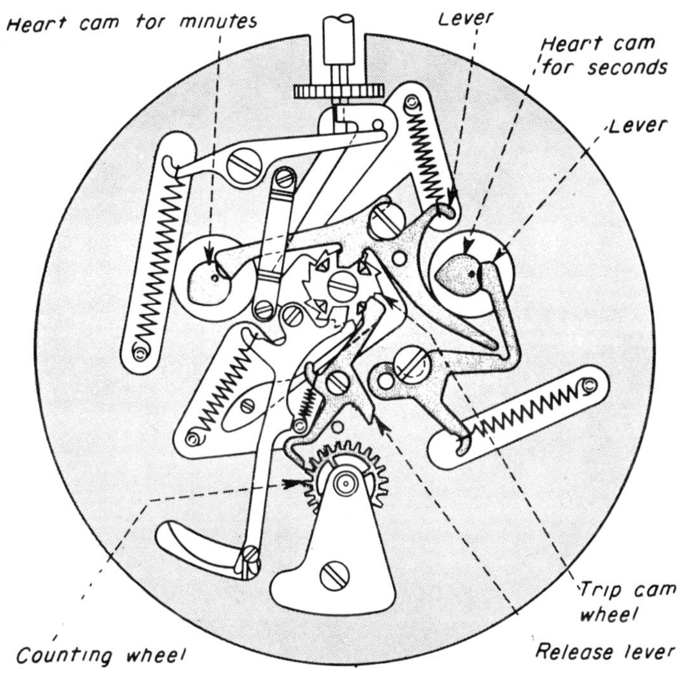

A stop watch readable directly to tenths of a second has been developed by the Junghans Watch Co., Schramberg, Germany. For continuous reading, tenths are indicated on a 120° sector of a three-pointer dial.

Minutes, seconds and tenths . . . are on separate dials so no pointer can hide another. A hood covers all but 120° of the tenths dial and always hides two of its three pointers. The three-pointer hand is on the shaft of the counting wheel and requires no intermediate gearing; its zero position is indexed by one long gear tooth. Zero positions of minute and second hands are indexed by heart cams and spring loaded levers.

Centrifugal Ratchet

For quick positive lock against pump motor backspin, this vertical hollow shaft pump, supplied by ACEC Electric Corp., New York City, has a ball ratchet mounted just above the motor. When pump stops, water column on discharge side tends to run back through the pump, reversing rotation; excessive shaft speed can result, damaging bearings. Earlier solutions included a pin ratchet which was subject to shear, and later a spring coupling which tended to lose tension with age. This new ball ratchet locks against reverse spin within 5 deg rotation after shaft reversal begins.

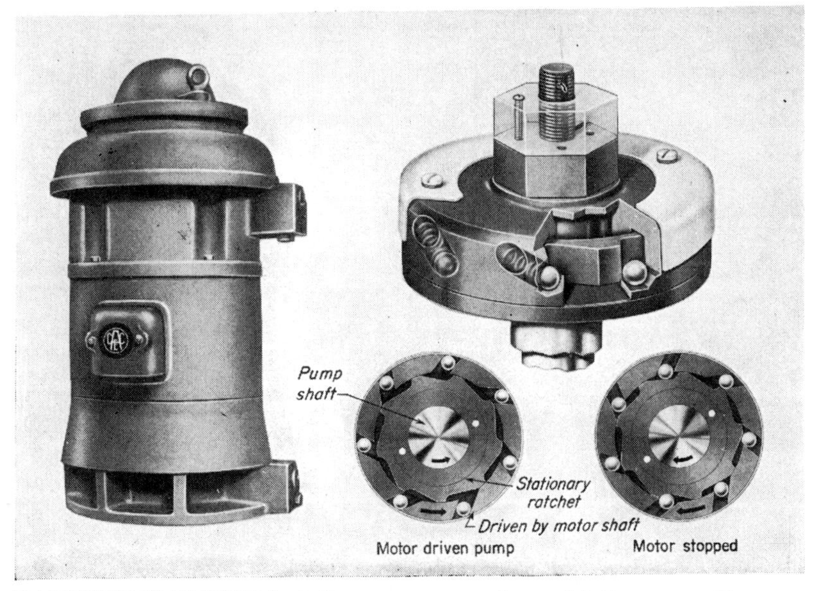

RATCHET ELEMENT is stationary; outer member and balls rotate with motor shaft. When motor drives the pump, steel balls are lifted clear of ratchet into slots by centrifugal force. As motor coasts to a stop, balls roll down slots and come to rest against the stationary ratchet. When pump shaft stops and starts to reverse, one of the balls engages the ratchet. Combination of 7 balls and 9 ratchet teeth assures that pump shaft will be locked with no more than 5 deg reverse rotation.

NO TEETH ON THESE RATCHETS

With springs, rollers and other devices they keep motion going one way.

L KASPER
design consultant
Philadelphia

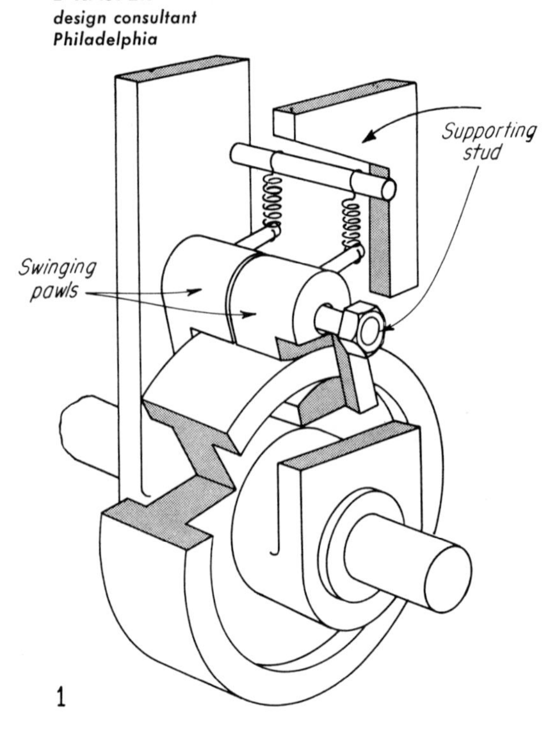

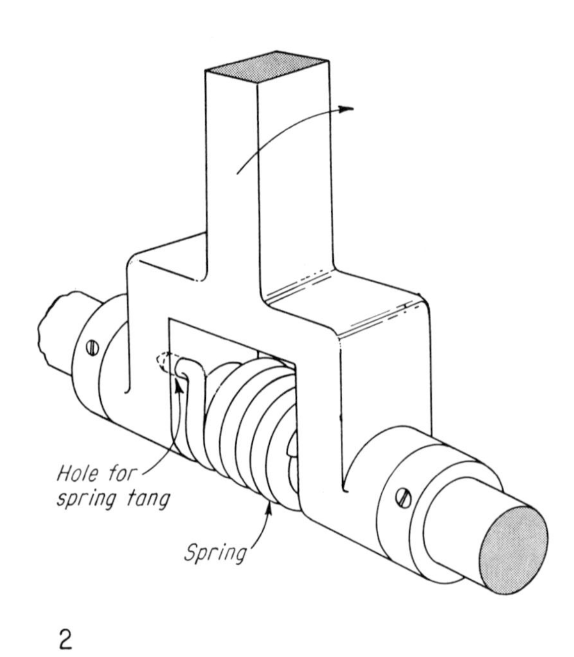

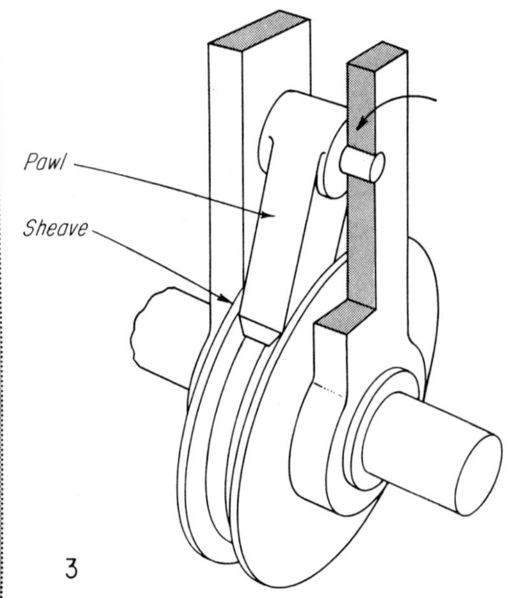

1 SWINGING PAWLS lock on rim when lever swings forward, and release on return stroke. Oversize holes for supporting stud make sure both top and bottom surfaces of pawls make contact.

2 HELICAL SPRING grips shaft because its inner diameter is smaller than the outer diameter of shaft. During forward stroke, spring winds tighter; during return stroke, it expands.

3 V-BELT SHEAVE is pushed around when pawl wedges in groove. For a snug fit, bottom of pawl is tapered like a V-belt.

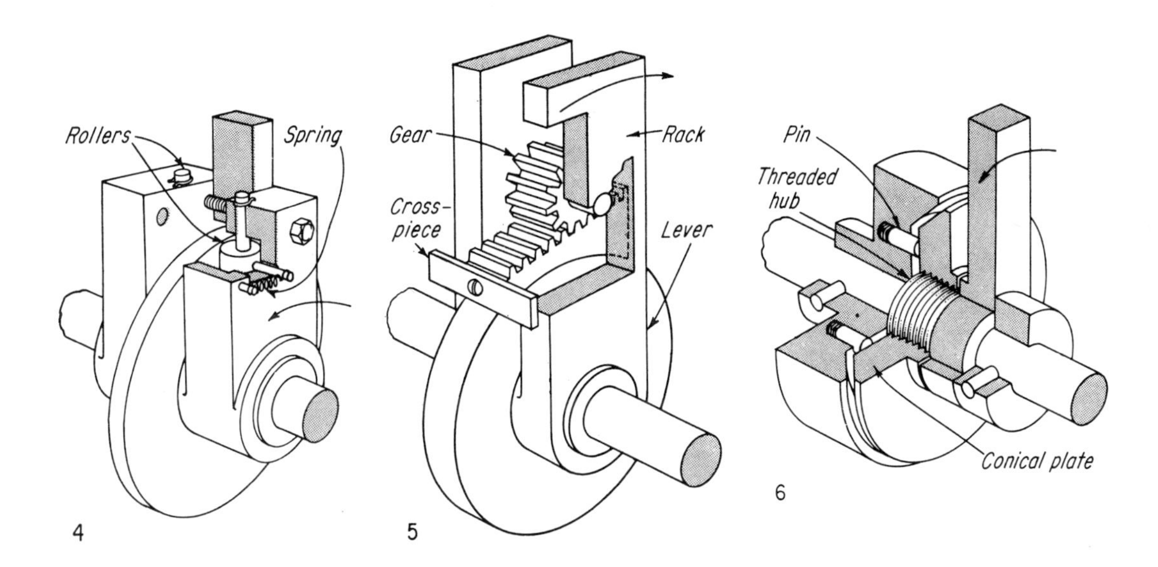

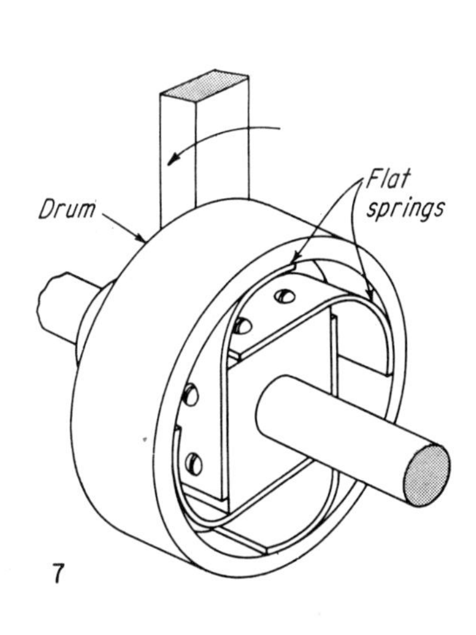

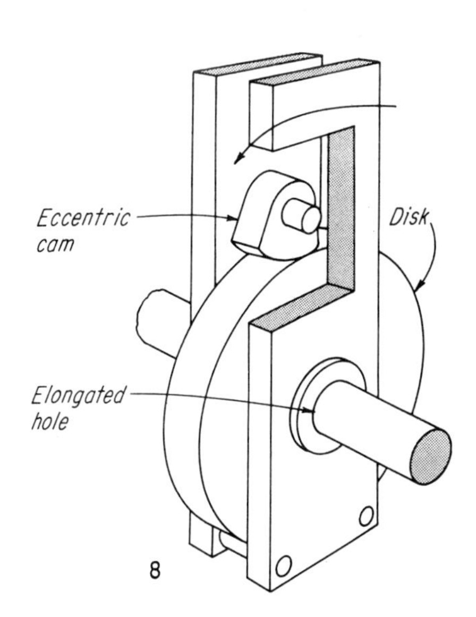

4 ECCENTRIC ROLLERS squeeze disk on forward stroke. On return stroke, rollers rotate backwards and release their grip. Springs keep rollers in contact with disk.

5 RACK is wedge-shape so that it jams between the rolling gear and the disk, pushing the shaft forward. When the driving lever makes its return stroke, it carries along the unattached rack by the cross-piece.

6 CONICAL PLATE moves as a nut back and forth along the threaded center hub of the lever. Light friction of spring-loaded pins keeps the plate from rotating with the hub.

7 FLAT SPRINGS expand against inside of drum when lever moves one way, but drag loosely when lever turns drum in opposite direction.

8 ECCENTRIC CAM jams against disk during motion half of cycle. Elongated holes in the levers allow cam to wedge itself more tightly in place.

Sheet Metal Gears, Sprockets,

When a specified motion must be transmitted at intervals rather than continuously, and the loads are light, these mechanisms are ideal because of their low cost and adaptability to mass production. Although not generally considered precision parts, ratchets and gears can be stamped to tolerances of ±0.007 in. and if necessary, shaved to closer dimensions. Sketches indicate some variations used on toys, household appliances and automobile components.

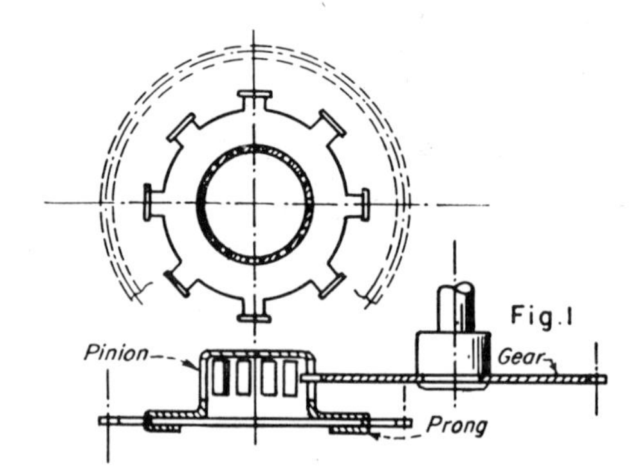

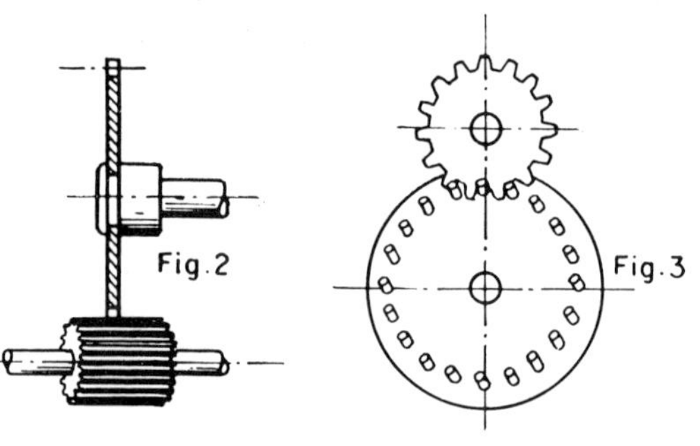

Fig. 4

Fig. 1—Pinion is a sheet metal cup, with rectangular holes serving as teeth. Meshing gear is sheet metal, blanked with specially formed teeth. Pinion can be attached to another sheet metal wheel by prongs, as shown, to form a gear train.

Fig. 2—Sheet metal wheel gear meshes with a wide face pinion, which is either extruded or machined. Wheel is blanked with teeth of conventional form.

Fig. 3—Pinion mates with round pins in circular disk made of metal, plastic or wood. Pins can be attached by staking or with threaded fasteners.

Fig. 4—Two blanked gears, conically formed after blanking, make bevel gears meshing on parallel axis. Both have specially formed teeth.

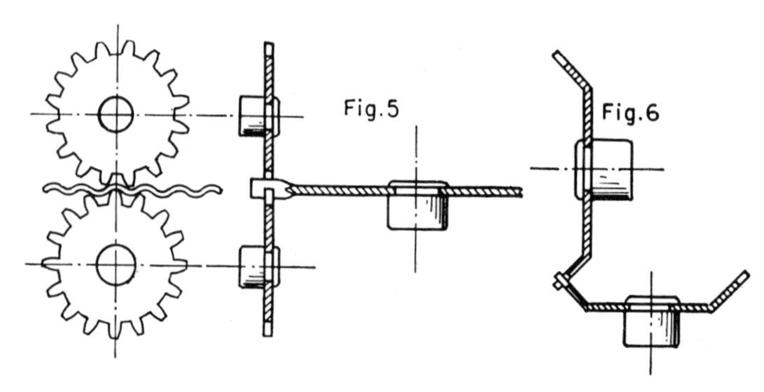

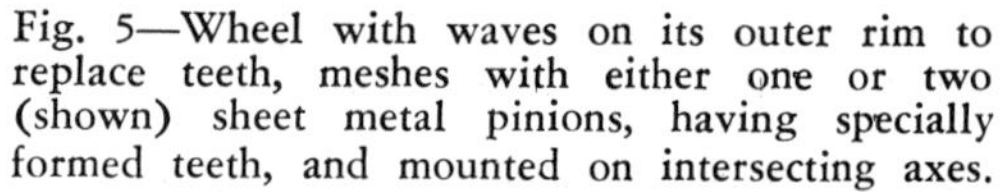

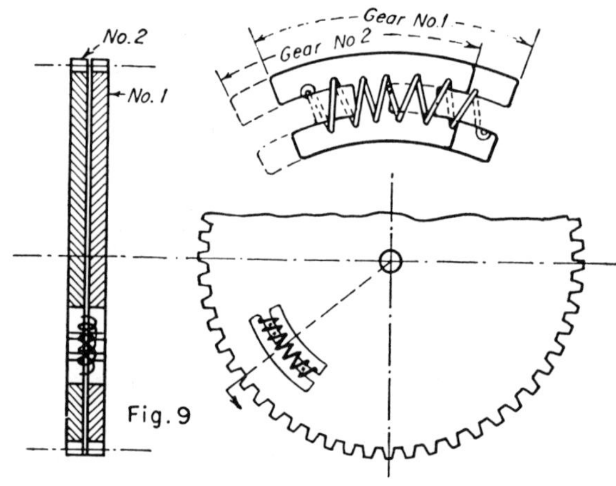

Fig. 5—Wheel with waves on its outer rim to replace teeth, meshes with either one or two (shown) sheet metal pinions, having specially formed teeth, and mounted on intersecting axes.

Fig. 6—Two bevel type gears, with specially formed teeth, mounted for 90 deg intersecting axes. Can be attached economically by staking to hubs.

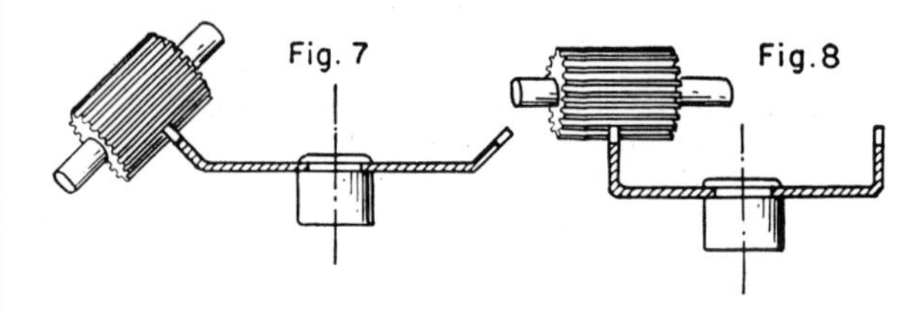

Fig. 7—Blanked and formed bevel type gear meshes with solid machined or extruded pinion. Conventional form of teeth can be used on both gear and pinion.

Fig. 8—Blanked, cup-shaped wheel meshes with solid pinion for 90 deg intersecting axes.

Fig. 9—Backlash can be eliminated from stamped gears by stacking two identical gears and displacing them by one tooth. Spring then bears one projection on each gear taking up lost motion.

Worms, and Ratchets

HAIM MURRO

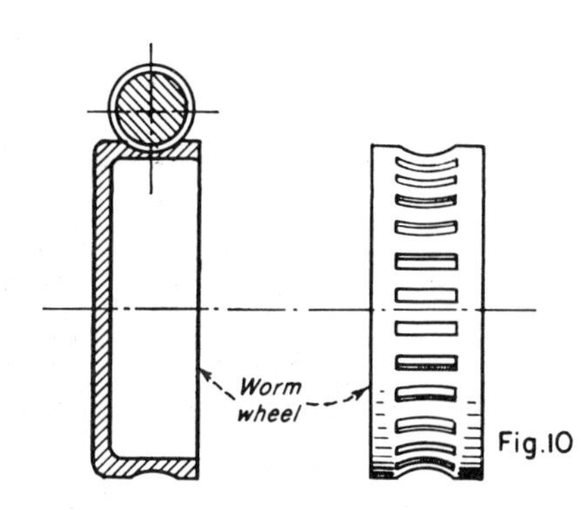

Fig. 10—Sheet metal cup which has indentations that take place of worm wheel teeth, meshes with a standard coarse thread screw.

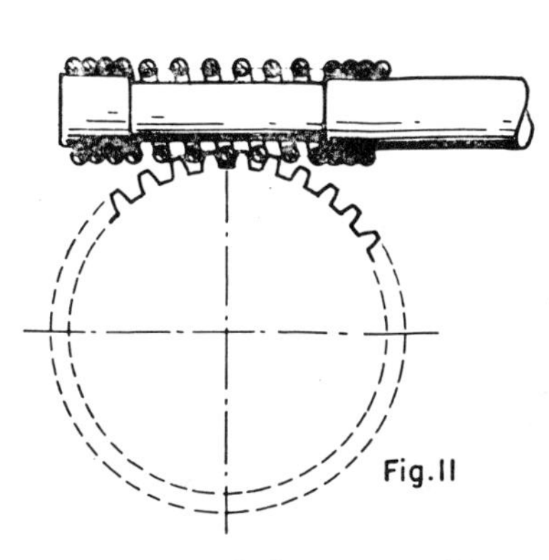

Fig. 11—Blanked wheel, with specially formed teeth, meshes with a helical spring mounted on a shaft, which serves as the worm.

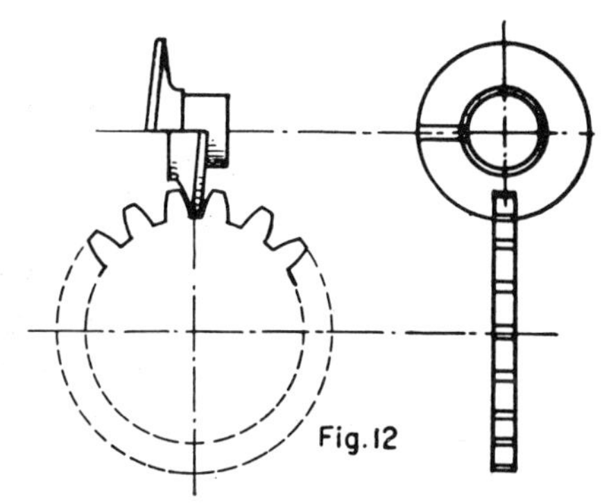

Fig. 12—Worm wheel is sheet metal blanked, with specially formed teeth. Worm is made of sheet metal disk, split and helically formed.

Fig. 13—Blanked ratchets with one sided teeth stacked to fit a wide, sheet metal finger when single thickness is not adequate. Ratchet gears can be spot welded.

Fig. 14—To avoid stacking, single ratchet is used with a U-shaped finger also made of sheet metal.

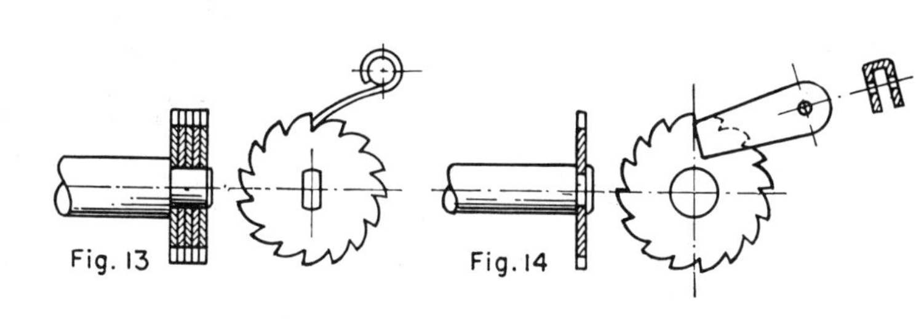

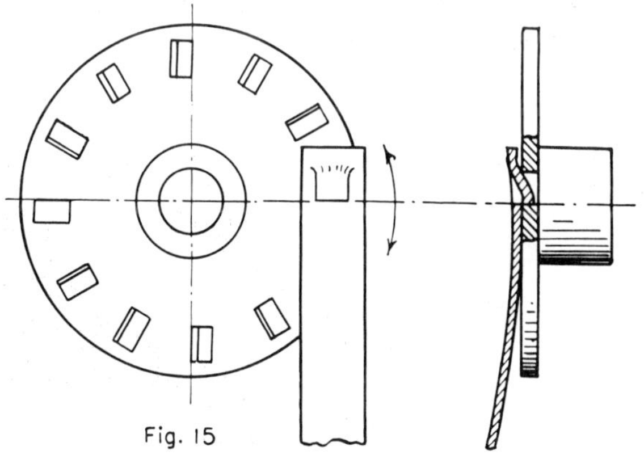

Fig. 15—Wheel is a punched disk with square punched holes to selve as teeth. Pawl is spring steel.

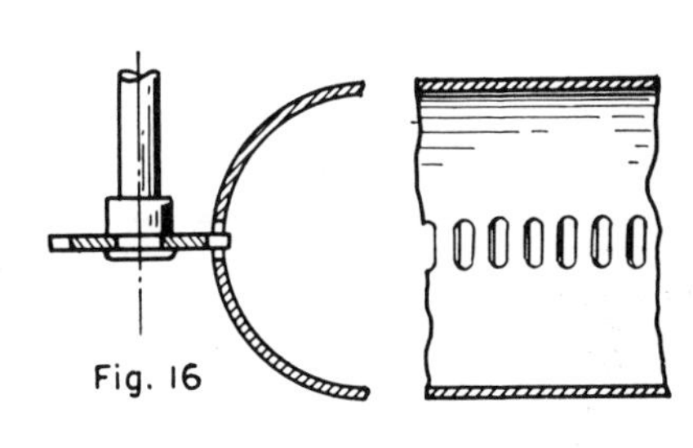

Fig. 16—Sheet metal blanked pinion, with specially formed teeth, meshes with windows blanked in a sheet metal cylinder, to form a pinion and rack assembly.

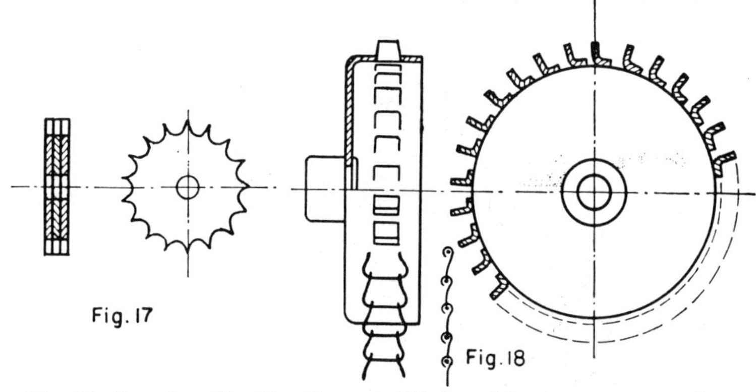

Fig. 17—Sprocket, like Fig. 13, can be fabricated from separate stampings.
Fig. 18—For a wire chain as shown, sprocket is made by bending out punched teeth on a drawn cup.

12 detents for mechanical

Some of the more robust and practical devices for locating or holding mechanical movements are surveyed by the author.

LOUIS DODGE, Consultant, New Richmond, Ohio

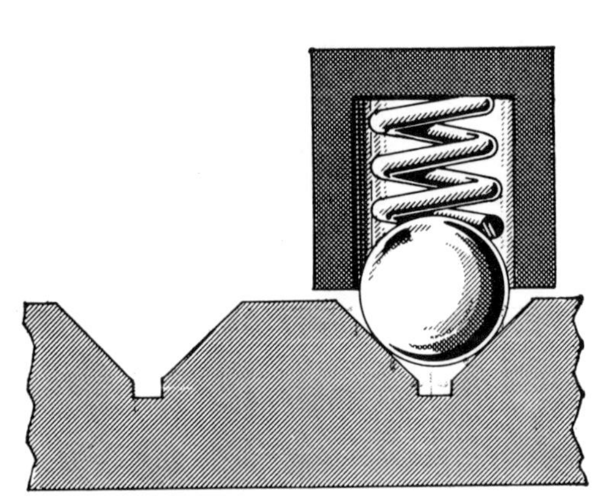

1 FIXED HOLDING POWER IS CONSTANT IN BOTH DIRECTIONS

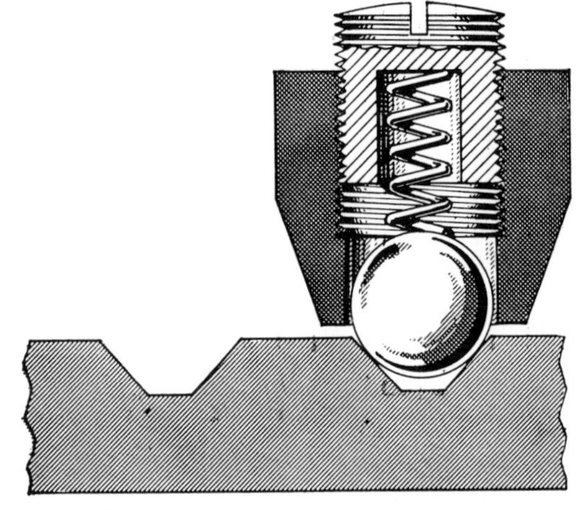

2 ADJUSTABLE HOLDING POWER

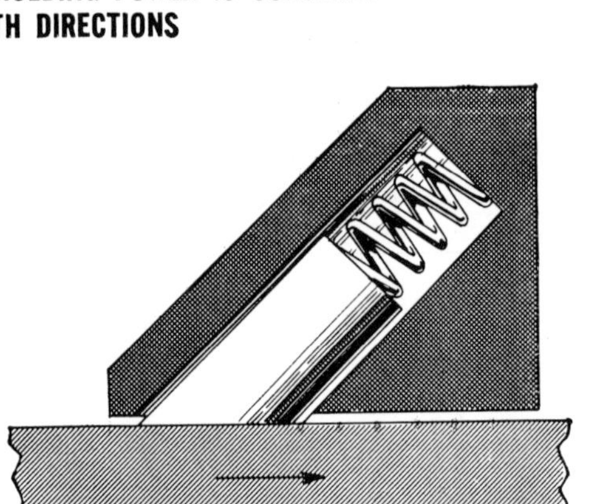

5 WEDGE ACTION LOCKS MOVEMENT IN DIRECTION OF ARROW

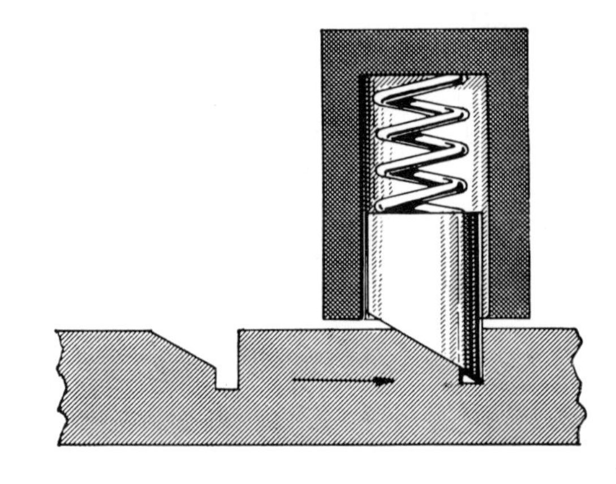

6 NOTCH SHAPE DICTATES DIRECTION OF ROD MOTION

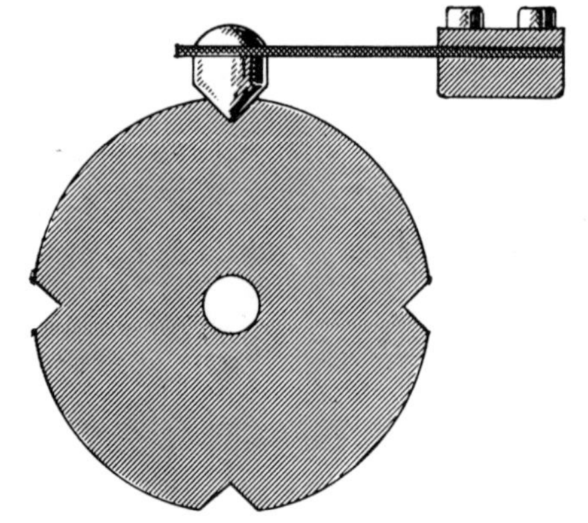

9 LEAF SPRING PROVIDES LIMITED HOLDING POWER

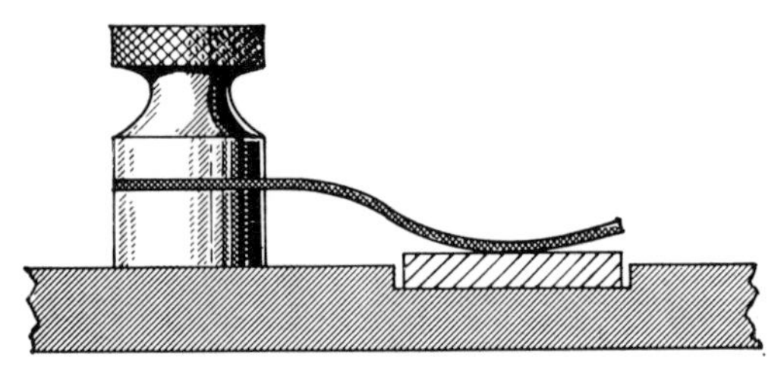

10 LEAF SPRING FOR HOLDING FLAT PIECES

movement

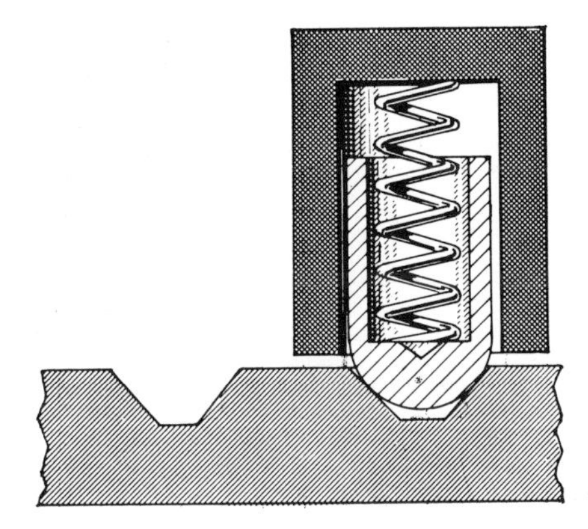

3 DOMED PLUNGER HAS LONG LIFE

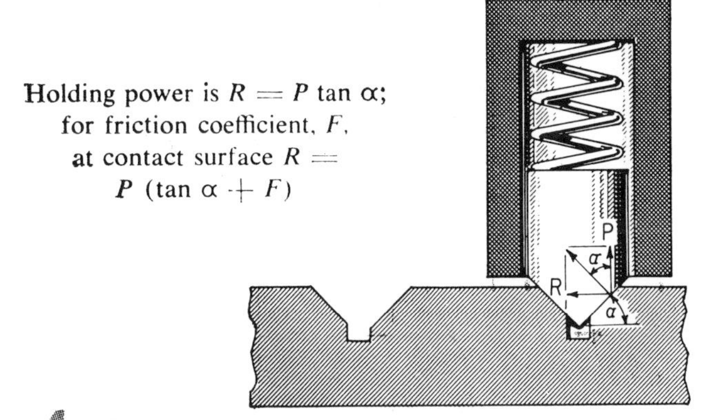

4 CONICAL OR WEDGE-ENDED DETENT

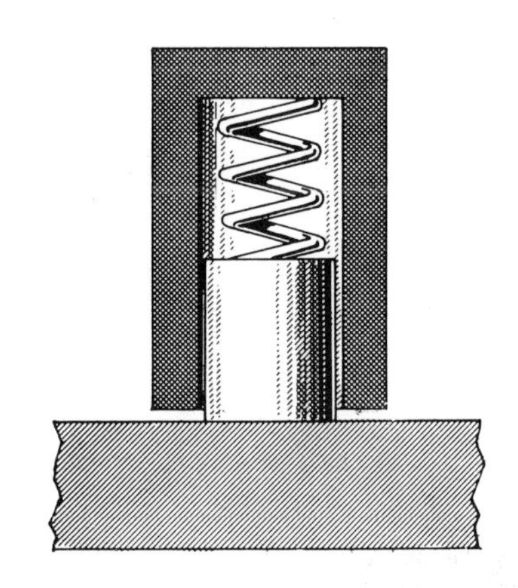

7 FRICTION RESULTS IN HOLDING FORCE

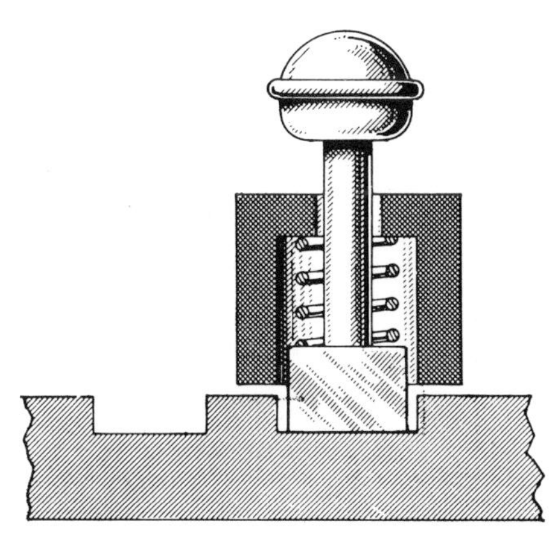

8 POSITIVE DETENT HAS MANUAL RELEASE

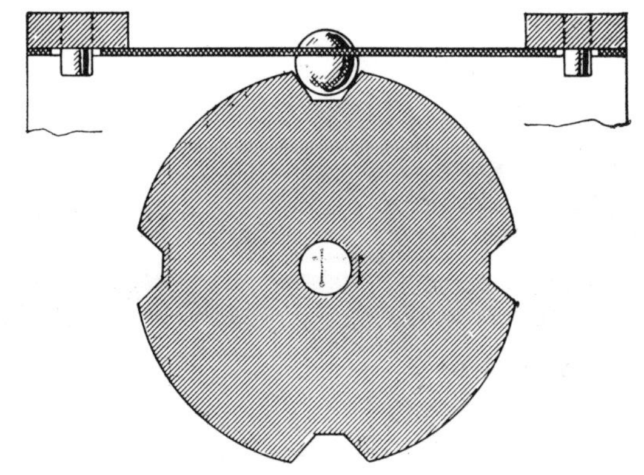

11 LEAF SPRING DETENT CAN BE REMOVED QUICKLY

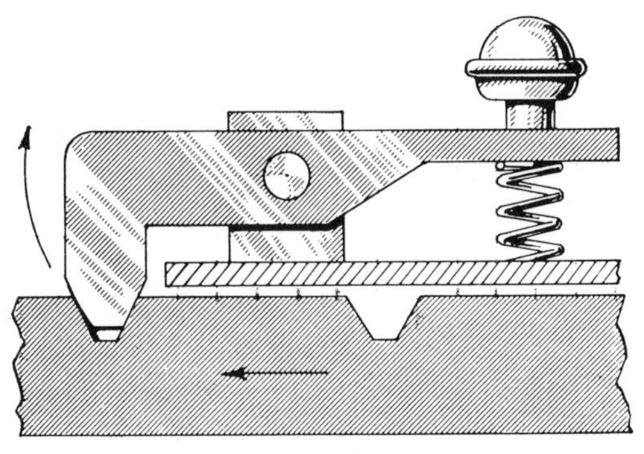

12 AUTOMATIC RELEASE OCCURS IN ONE DIRECTION, MANUAL RELEASE NEEDED IN OTHER DIRECTION

Six more mechanical

Author concludes his survey of devices for indexing or holding mechanical movements.

LOUIS DODGE, Consultant, New Richmond, Ohio

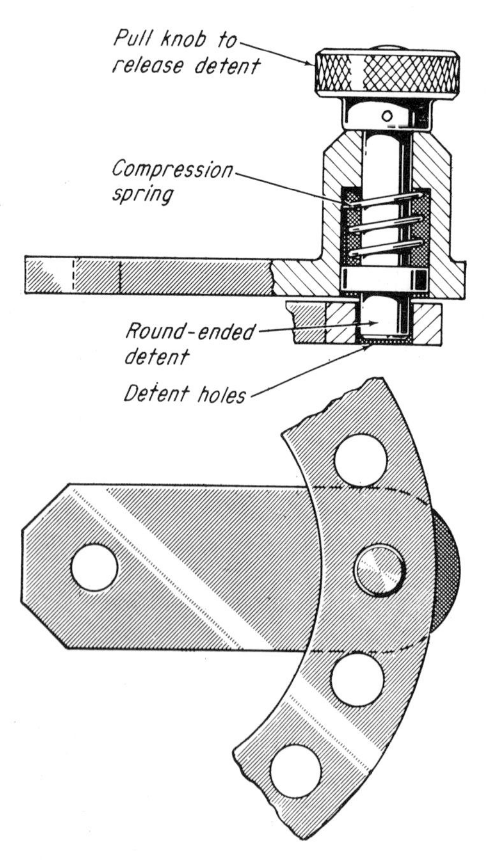

1 AXIAL POSITIONING (INDEXING) BY MEANS OF SPACED HOLES IN INDEX BASE

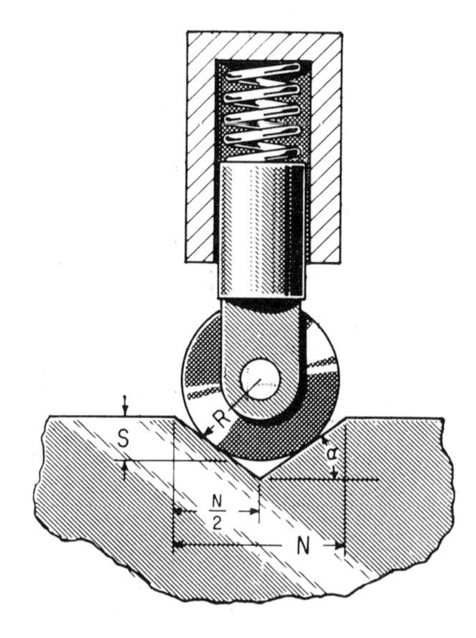

3 ROLLER DETENT POSITIONS IN A NOTCH:

RISE, $S = \frac{N \tan a}{2} - R \times \frac{1 - \cos}{\cos a}$

ROLLER RADIUS, $R = \left(\frac{N \tan a}{2} - S\right)\left(\frac{\cos a}{1 - \cos a}\right)$

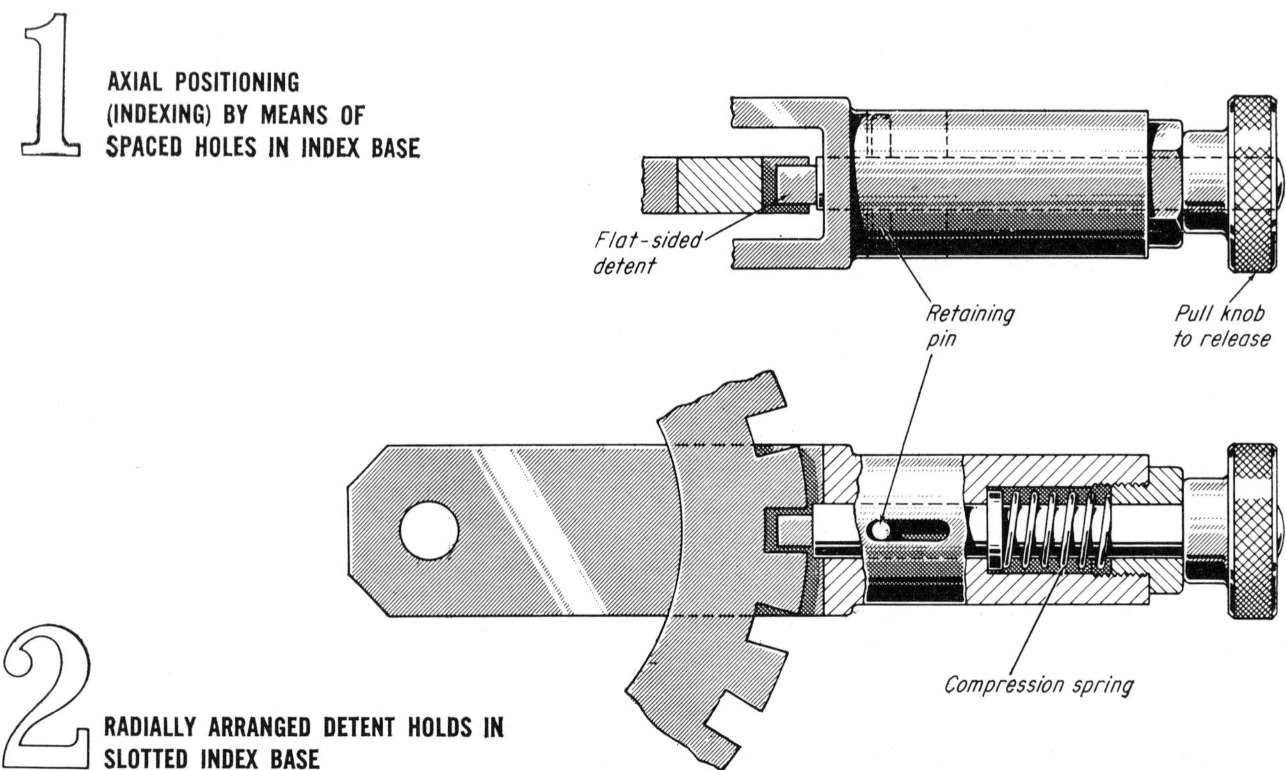

2 RADIALLY ARRANGED DETENT HOLDS IN SLOTTED INDEX BASE

detents

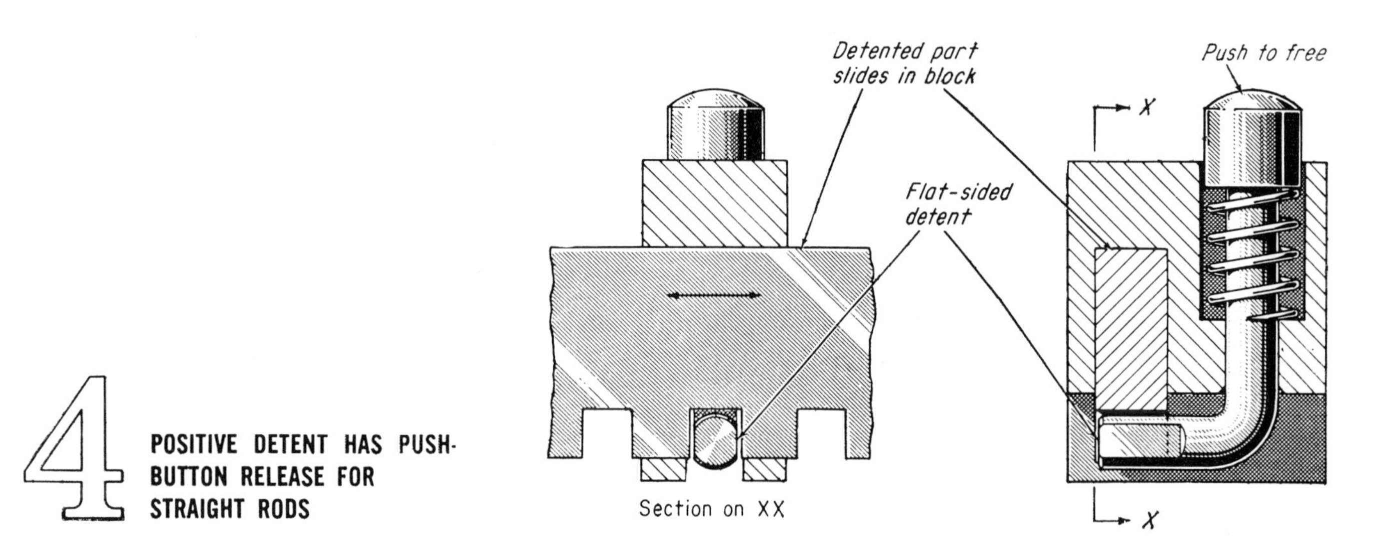

4 POSITIVE DETENT HAS PUSH-BUTTON RELEASE FOR STRAIGHT RODS

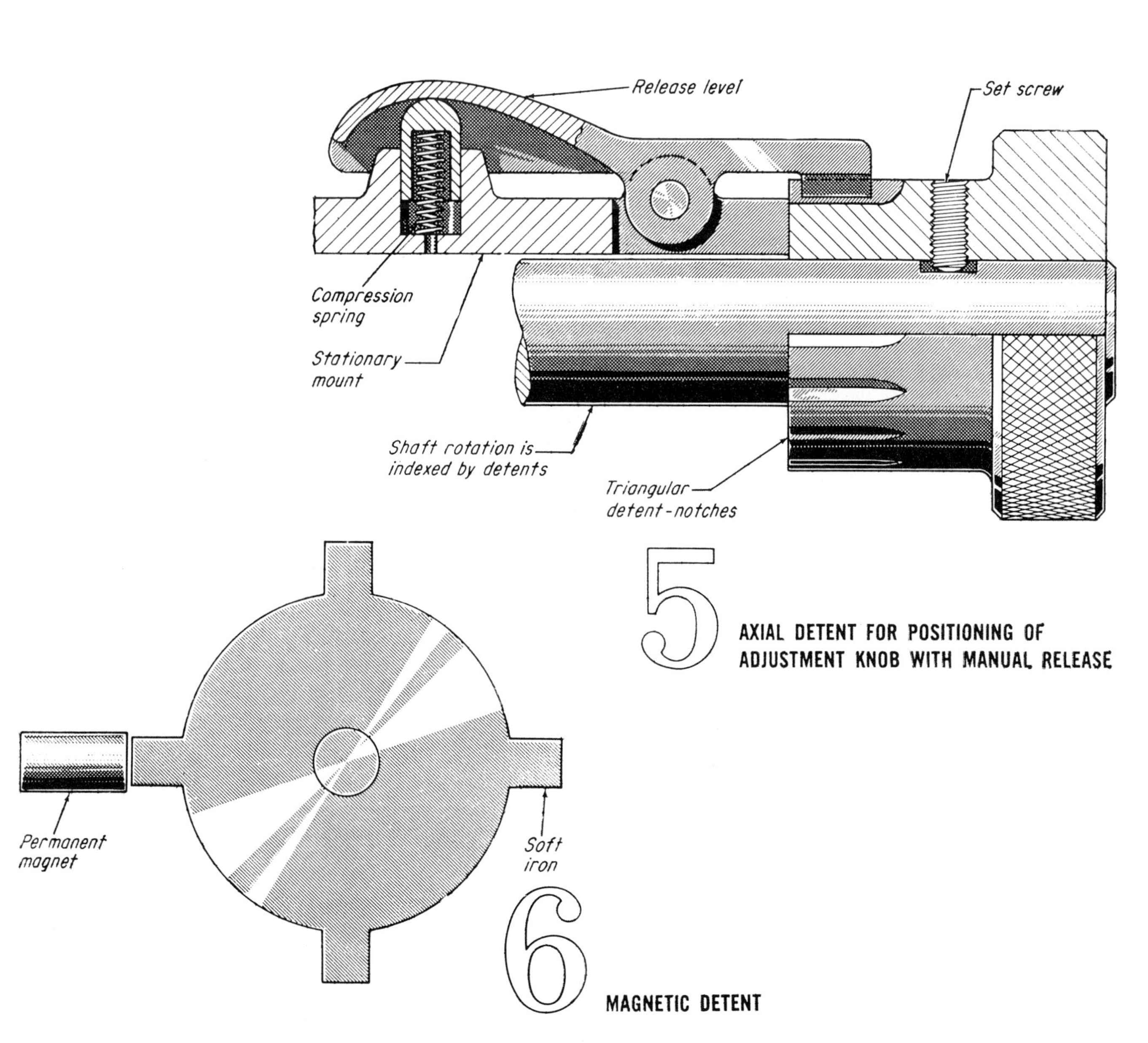

5 AXIAL DETENT FOR POSITIONING OF ADJUSTMENT KNOB WITH MANUAL RELEASE

6 MAGNETIC DETENT

Toggle Linkage Applications in

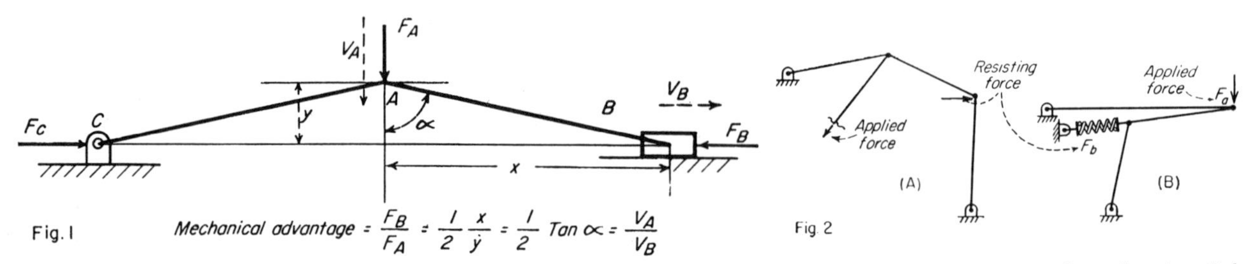

Fig. 1

Fig. 2

MANY MECHANICAL LINKAGES are based on the simple toggle which consists of two links that tend to line-up in a straight line at one point in their motion. The mechanical advantage is the velocity ratio of the input point A to the outpoint point B: or V_A/V_B. As the angle α approaches 90 deg, the links come into toggle and the mechanical advantage and velocity ratio both approach infinity. However, frictional effects reduce the forces to much less than infinity although still quite high.

FORCES CAN BE APPLIED through other links, and need not be perpendicular to each other. (A) One toggle link can be attached to another link rather than to a fixed point or slider. (B) Two toggle links can come into toggle by lining up on top of each other rather than as an extension of each other. Resisting force can be a spring force.

HIGH MECHANICAL ADVANTAGE

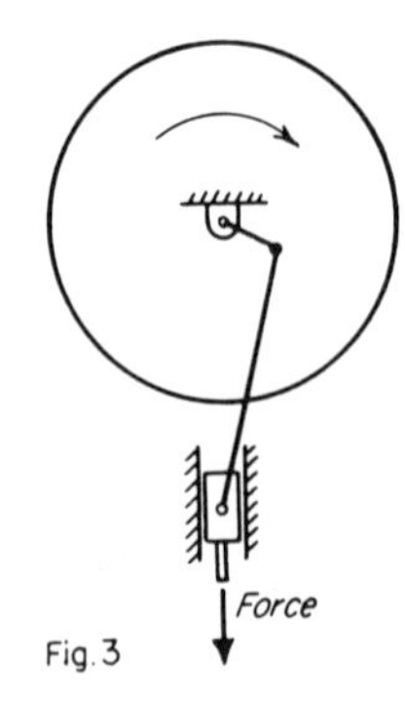

Fig. 3

IN PUNCH PRESSES, large forces are needed at the lower end of the work-stroke, however little force is required during the remainder. Crank and connecting rod come into toggle at the lower end of the punch stroke, giving a high mechcanical advantage at exactly the time it is most needed.

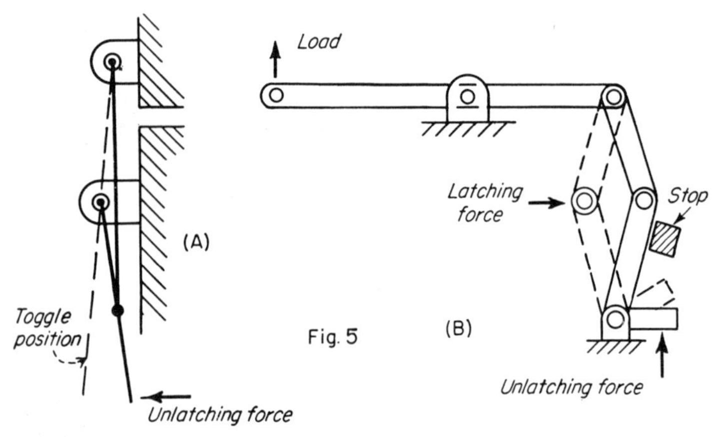

Fig. 5

COLD-HEADING RIVET MACHINE is designed to give each rivet two successive blows. Following the first blow (point 2) the hammer moves upward a short distance (to point 3). Following the second blow (at point 4), the hammer then moves upward a longer distance (to point 1) to provide clearance for moving the workpiece. Both strokes are produced by one revolution of the crank and at the lowest point of each stroke (points 2 and 4) the links are in toggle.

LOCKING LATCHES produce a high mechanical advantage when in the toggle portion of the stroke. (A) Simple latch exerts a large force in the locked position. (B) For positive locking, closed position of latch is slightly beyond toggle position. Small unlatching force opens linkage.

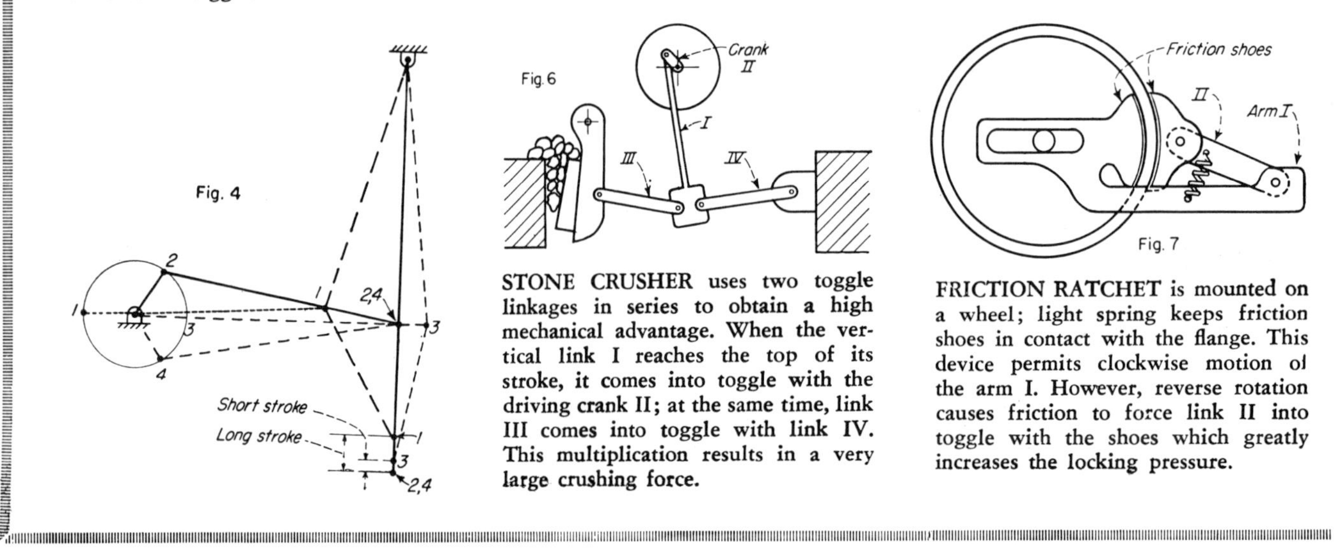

Fig. 4

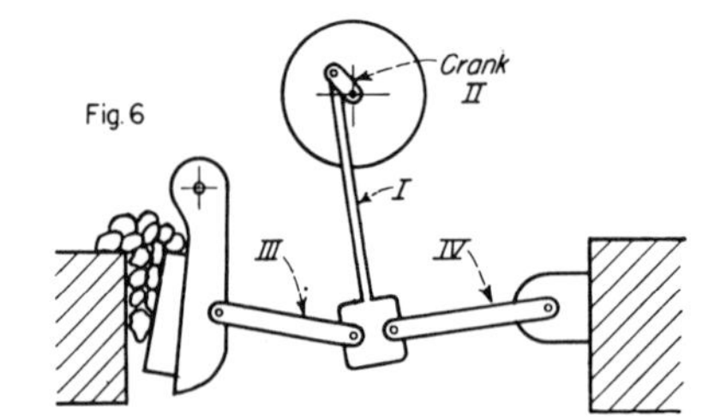

Fig. 6

STONE CRUSHER uses two toggle linkages in series to obtain a high mechanical advantage. When the vertical link I reaches the top of its stroke, it comes into toggle with the driving crank II; at the same time, link III comes into toggle with link IV. This multiplication results in a very large crushing force.

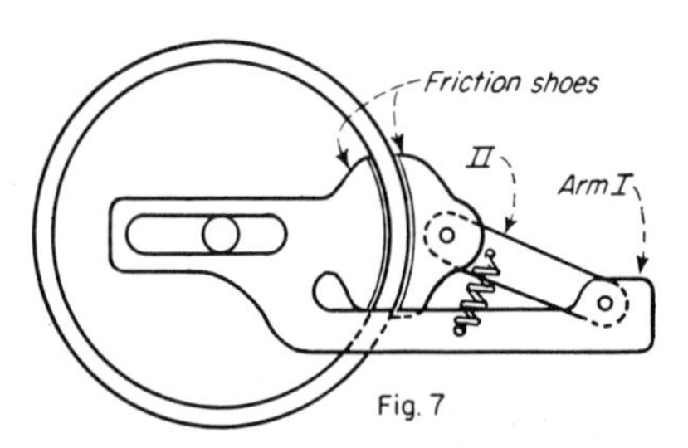

Fig. 7

FRICTION RATCHET is mounted on a wheel; light spring keeps friction shoes in contact with the flange. This device permits clockwise motion of the arm I. However, reverse rotation causes friction to force link II into toggle with the shoes which greatly increases the locking pressure.

Different Mechanisms

THOMAS P. GOODMAN
Westinghouse Electric Corporation

HIGH VELOCITY RATIO

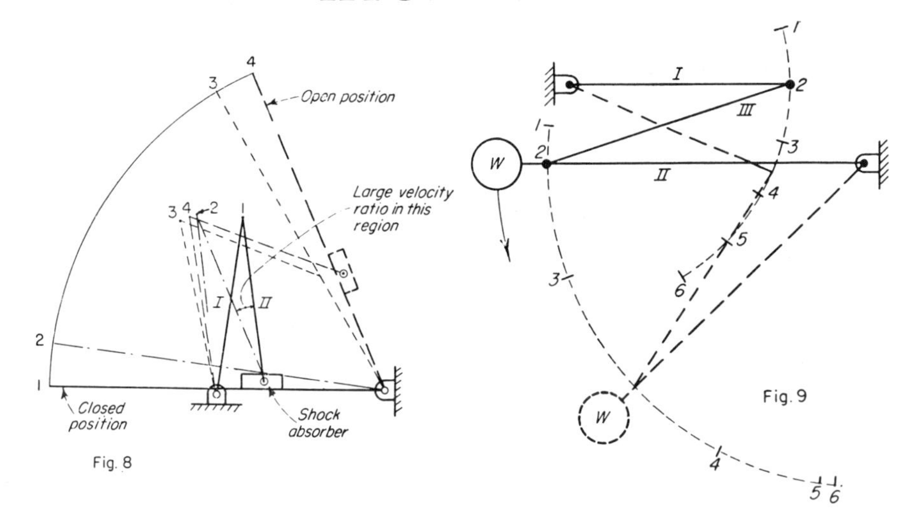

DOOR CHECK LINKAGE gives a high velocity ratio at one point in the stroke. As the door swings closed, connecting link I comes into toggle with the shock absorber arm II, giving it a large angular velocity. Thus, the shock absorber is more effective retarding motion near the closed position.

IMPACT REDUCER used on some large circuit breakers. Crank I rotates at constant velocity while lower crank moves slowly at the beginning and end of the stroke. It moves rapidly at the mid stroke when arm II and link III are in toggle. Accelerated weight absorbs energy and returns it to the system when it slows down.

VARIABLE MECHANICAL ADVANTAGE

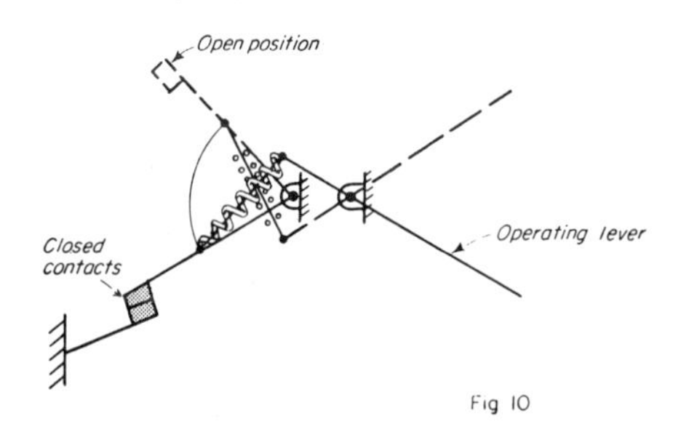

TOASTER SWITCH uses an increasing mechanical advantage to aid in compressing a spring. In the closed position, spring holds contacts closed and the operating lever in the down position. As the lever is moved upward, the spring is compressed and comes into toggle with both the contact arm and the lever. Little effort is required to move the links through the toggle position; beyond this point, the spring snaps the contacts open. A similar action occurs on closing.

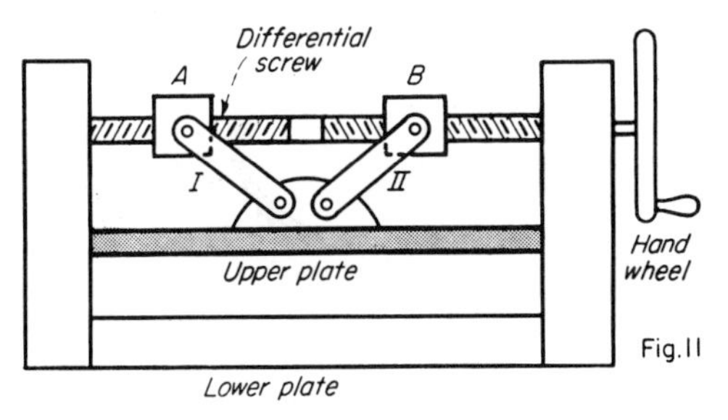

TOGGLE PRESS has an increasing mechanical advantage to counteract the resistance of the material being compressed. Rotating handwheel with differential screw moves nuts *A* and *B* together and links I and II are brought into toggle.

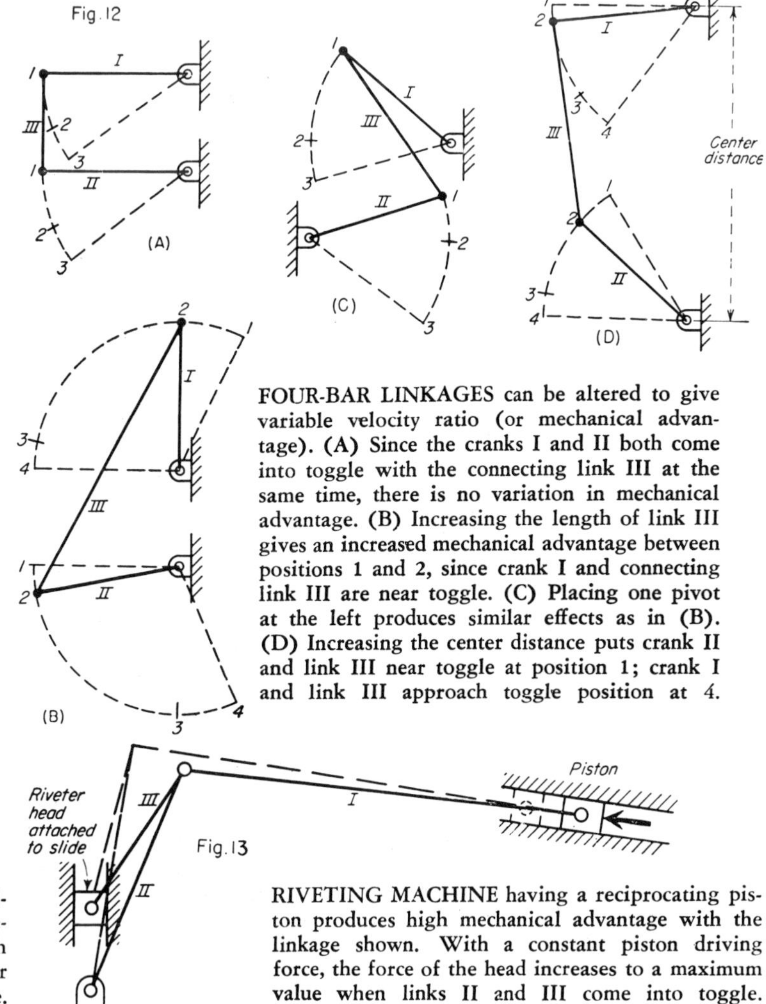

FOUR-BAR LINKAGES can be altered to give variable velocity ratio (or mechanical advantage). (A) Since the cranks I and II both come into toggle with the connecting link III at the same time, there is no variation in mechanical advantage. (B) Increasing the length of link III gives an increased mechanical advantage between positions 1 and 2, since crank I and connecting link III are near toggle. (C) Placing one pivot at the left produces similar effects as in (B). (D) Increasing the center distance puts crank II and link III near toggle at position 1; crank I and link III approach toggle position at 4.

RIVETING MACHINE having a reciprocating piston produces high mechanical advantage with the linkage shown. With a constant piston driving force, the force of the head increases to a maximum value when links II and III come into toggle.

16 Latch, Toggle and

Diagrams of basic latching and quick-release mechanisms.

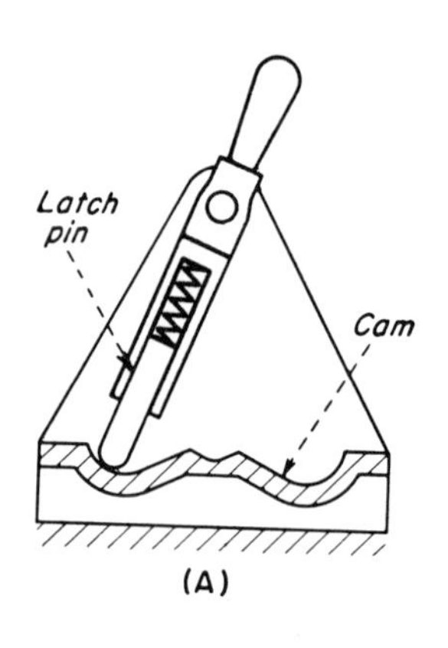

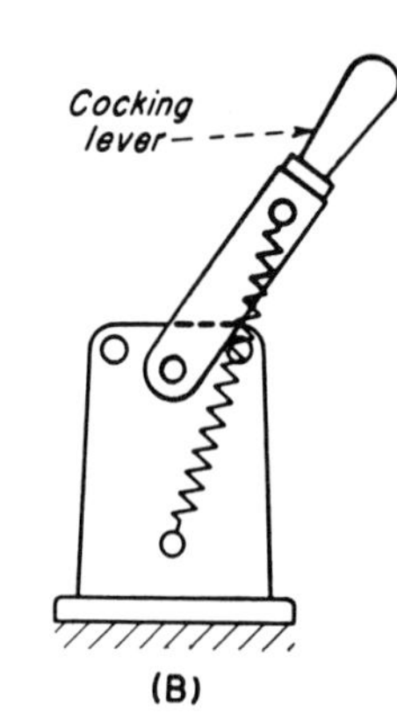

Fig. 1—Cam-guided latch (A) has one cocked, two relaxed positions. (B) Simple over-center toggle action. (C) Over-center toggle with slotted link. (D) Double toggle action is often used in electrical switches.

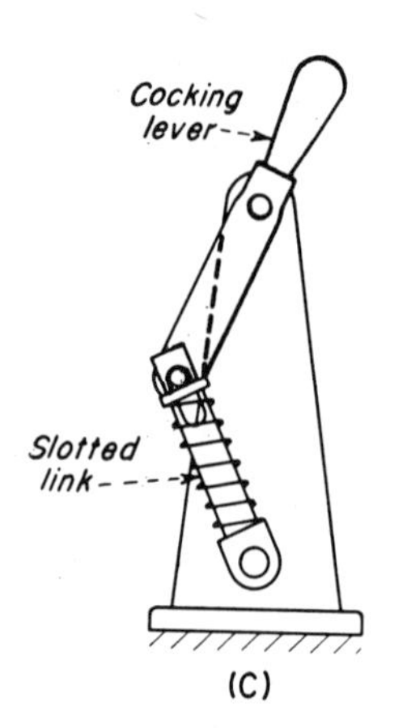

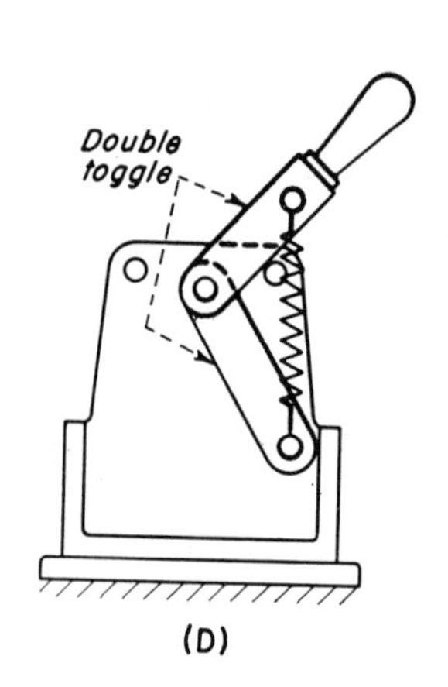

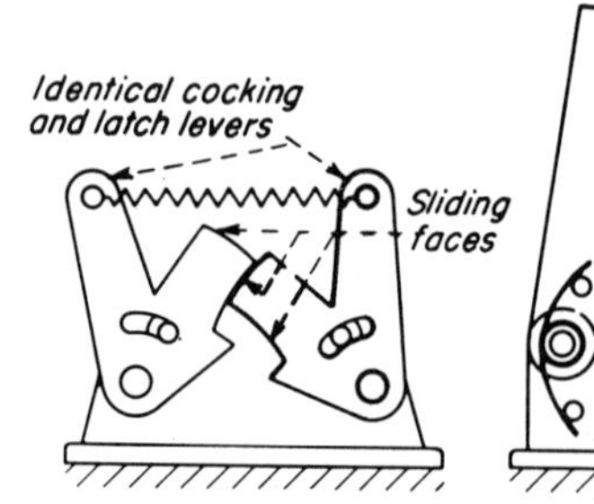

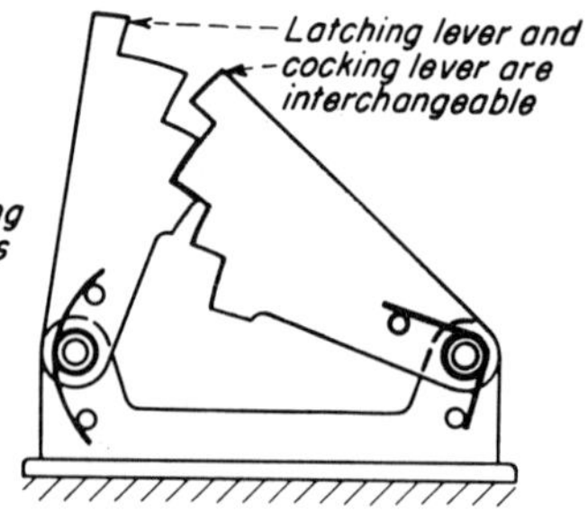

Fig. 2—Identically shaped cocking lever and latch (A) allow functions to be interchangeable. Radii of sliding faces must be dimensioned for mating fit. Stepped latch (B) has several locking positions.

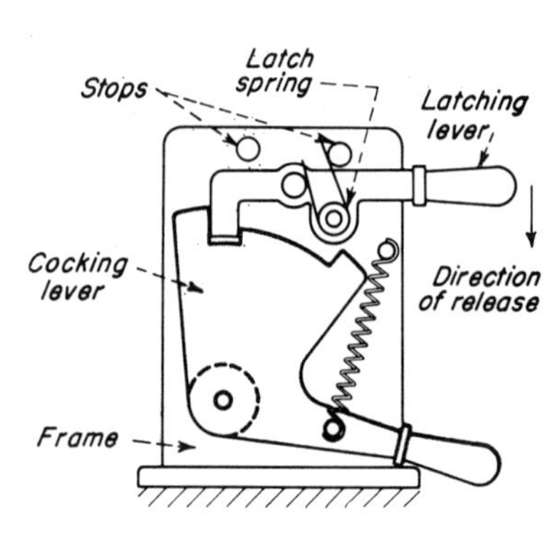

Fig. 3—Latch and cocking lever spring-loaded so latch movement releases cocking lever. Cocked position can be held indefinitely. Studs in frame provide stops, pivots or mounts for springs.

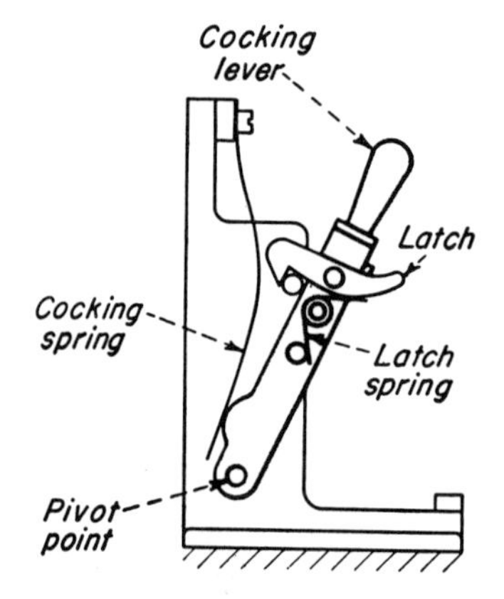

Fig. 4—Latch mounted on cocking lever allows both levers to be reached at same time with one hand. After release, cocking spring initiates clockwise lever movement, then gravity takes over.

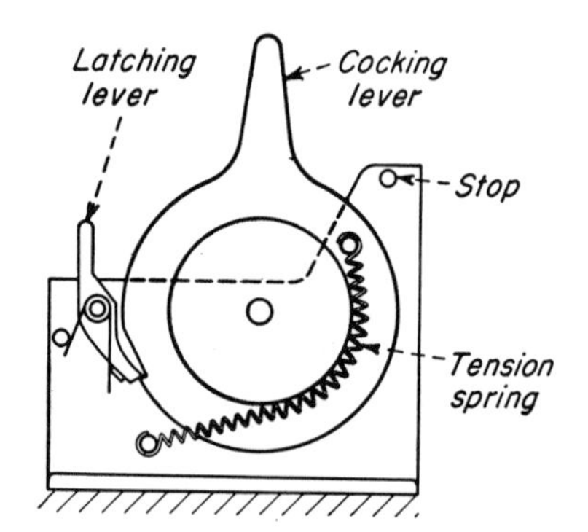

Fig. 5—Disk-shaped cocking lever has tension spring resting against cylindrical hub. Spring force thus alway acts at constant radius from lever pivot point.

Trigger Devices

SIGMUND RAPPAPORT

Project Supervisor, Ford Instrument Company
Adjunct Professor of Kinematics,
Polytechnic Institute of Brooklyn

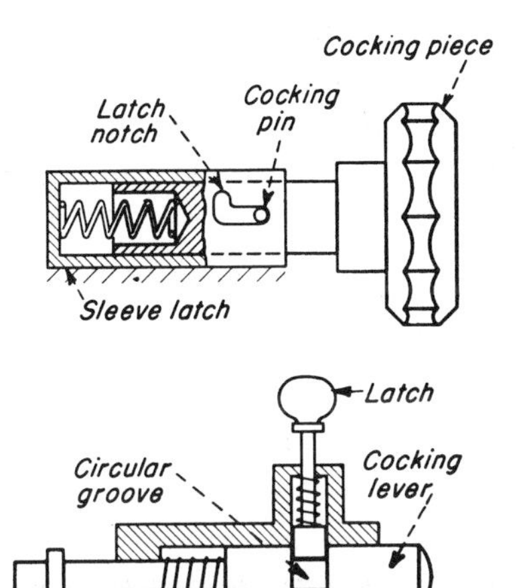

Fig. 6—Sleeve latch (A) has L-shaped notch. Pin in shaft rides in notch. Cocking requires simple push and twist action. (B) Latch and plunger use axial movement for setting and release. Circular groove needed if plunger can rotate.

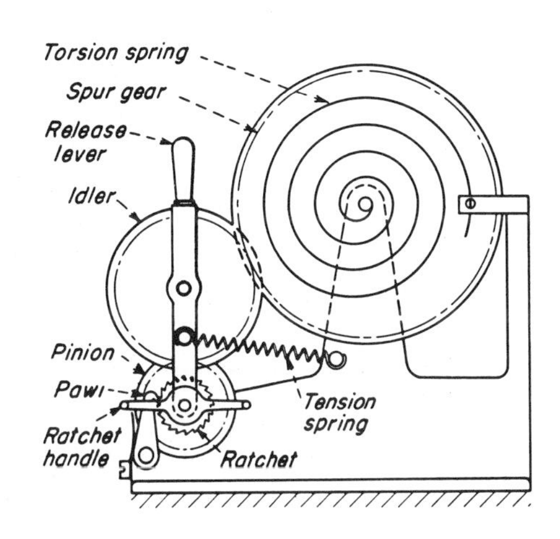

Fig. 7—Geared cocking device has rachet fixed to pinion. Torsion spring exerts clockwise force on spur gear; tension spring holds gear in mesh. Device wound by turning ratchet handle counter-clockwise which in turn winds torsion spring. Moving release-lever permits spur gear to unwind to original position without affecting ratchet handle.

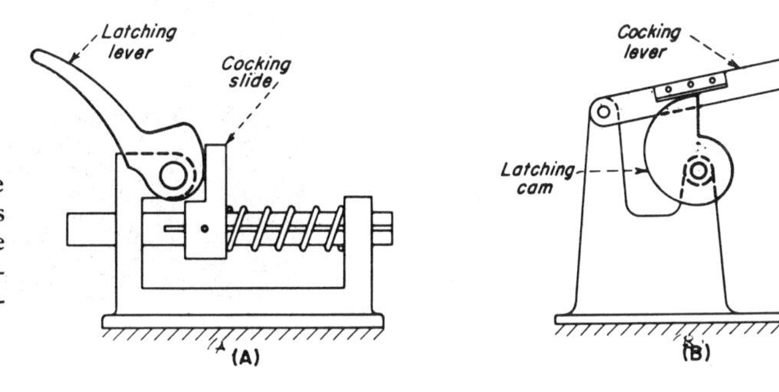

Fig. 8—In over-center lock (A) clockwise movement of latching lever cocks and locks slide. Release requires counter-clockwise movement. (B) Latching-cam cocks and releases cocking lever with same counter-clockwise movement.

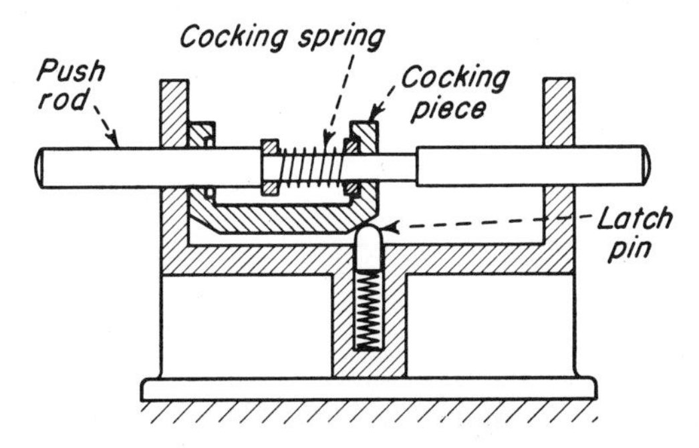

Fig. 9—Spring-loaded cocking piece has chamfered corners. Axial movement of push-rod forces cocking piece against spring-loaded ball or pin set in frame. When cocking builds up enough force to overcome latch-spring, cocking piece snaps over to right. Action can be repeated in either direction.

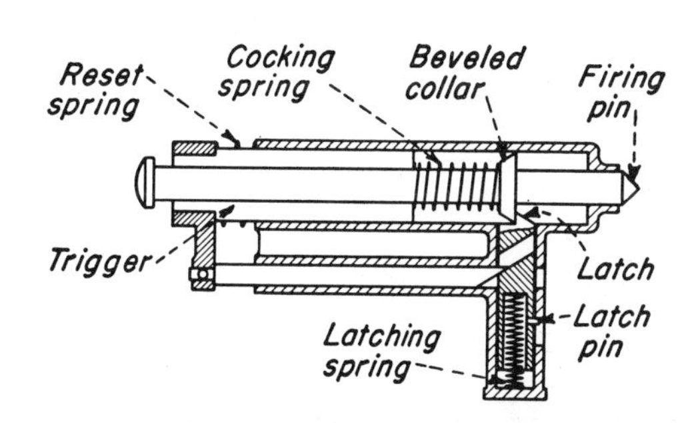

Fig. 10—Firing-pin type mechanism has bevelled collar on pin. Pressure on trigger forces latch down until it releases collar, when pin snaps out under force of cocking spring. Reset spring pulls trigger and pin back. Latch is forced down by bevelled collar on pin until it snaps back after overcoming force of latch spring. (Latch pin retains latch if trigger and firing pin are removed.)

Data based on material and sketches in AWF und VDMA Getrieblaetter, published by Ausschuss fuer Getriebe beim Ausssbuss fuer Wirtschaftiche Fertigung, Leipzig, Germany.

6 SNAP-ACTION

Diagrams show the basic ways to produce mechanical snap action.

PETER C NOY
Production engineer
Canadian General Electric Co
Barrie, Ont

Snap action results when a force is applied to a device over a period of time; buildup of this force to a critical level causes a sudden motion to occur. The ideal snap device would have no motion until the force reached critical level. This, however, is not possible, and the way in which the mechanism approaches this ideal is a measure of its efficiency as a snap device. Some of the designs shown here approach closely to the ideal; others less so, but may nevertheless have other good features.

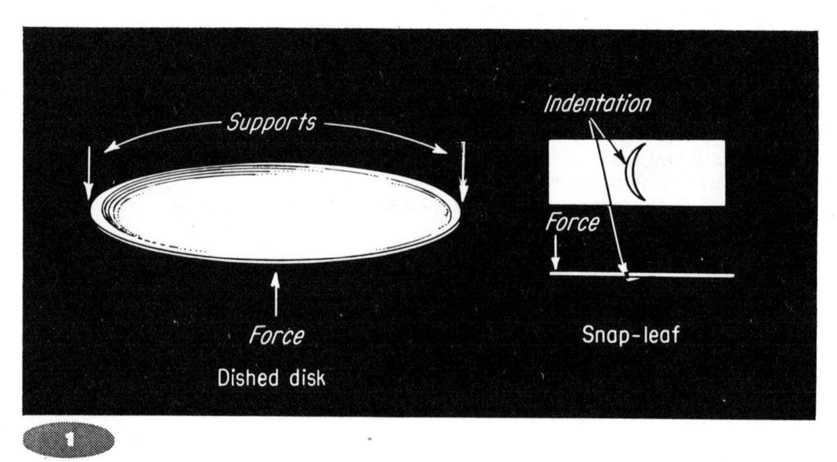

1

Dished disk . . .
is a simple, common method of producing snap action. Snap leaf made from spring material may have various-shaped impressions stamped at the point where overcentering action occurs. Frog clacker is, of course, a typical application. A bimetal made this way will reverse itself at a predetermined temperature.

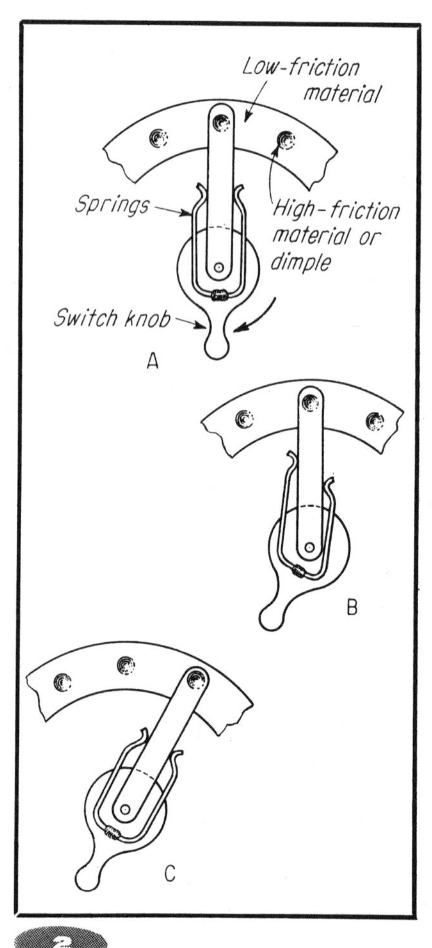

2

Friction override . . .
may be used to hold against an increasing load until friction is suddenly overcome. This is a useful action for small sensitive devices where large forces and movements are undesirable. It is interesting to note that this is the way we snap our fingers, and thus is probably the original snap mechanism. Moisture affects this action.

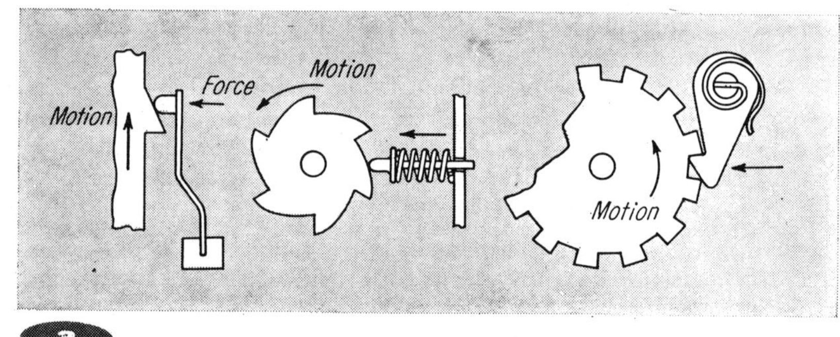

3

Ratchet-and-pawl . . .
principle is perhaps the most widely used type of snap mechanism. Its many variations are an essential feature in practically every complicated mechanical device. By definition, however, this movement is not true snap-action.

MECHANISMS

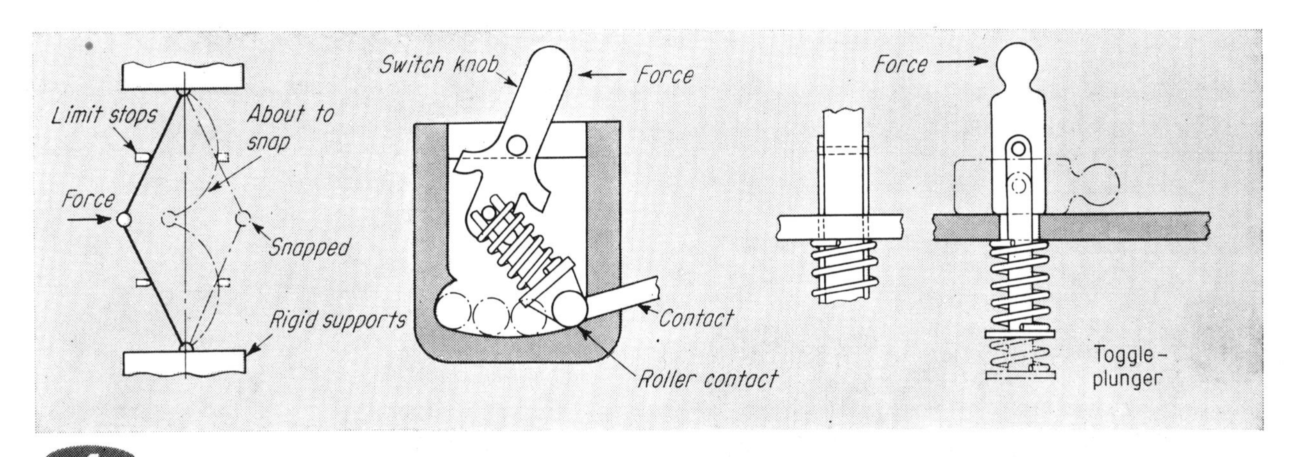

4

Over-centering . . .
devices find their greatest application in electrical switches. Considerable design ingenuity has been used to fit this principle into many different mechanisms. In the final analysis, over-centering is actually the basis of most snap-action devices.

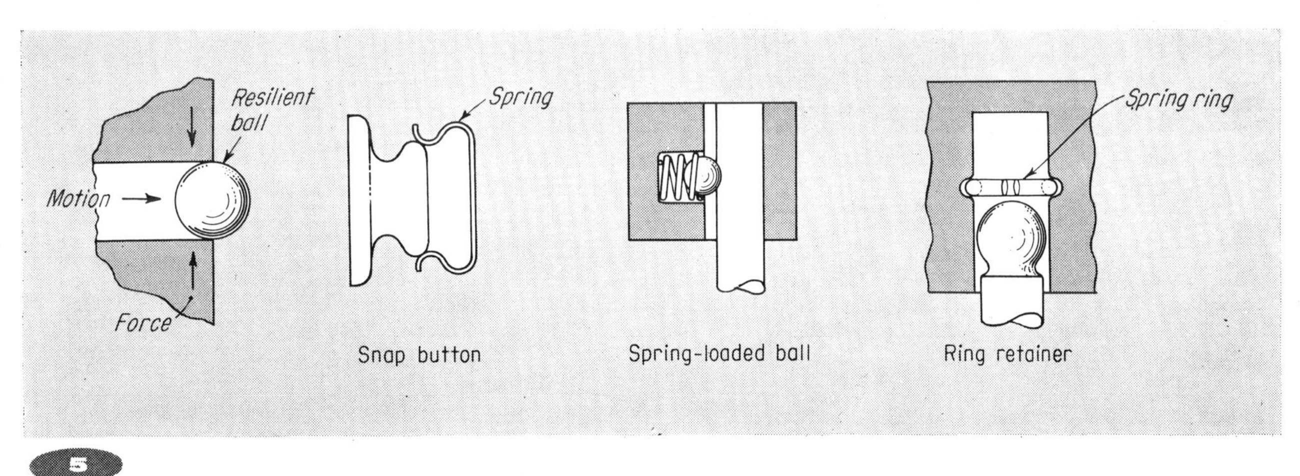

Snap button

Spring-loaded ball

Ring retainer

5

Sphere ejection . . .
principle employs snap buttons, spring-loaded balls and catches, and retaining-rings for fastening that will have to be used repeatedly. Action can be adjusted in design to provide either easy or difficult removal. Wear may change force required.

Pneumatic dump valve . . .
produces snap action by preventing piston movement until air pressure has built up in front end of cylinder to a relatively high pressure. Dump-valve area in low-pressure end is six times larger than its area on high-pressure side, thus the pressure required on the high-pressure side to dislodge the dump valve from its seat is six times that required on the low-pressure side to keep the valve seated.

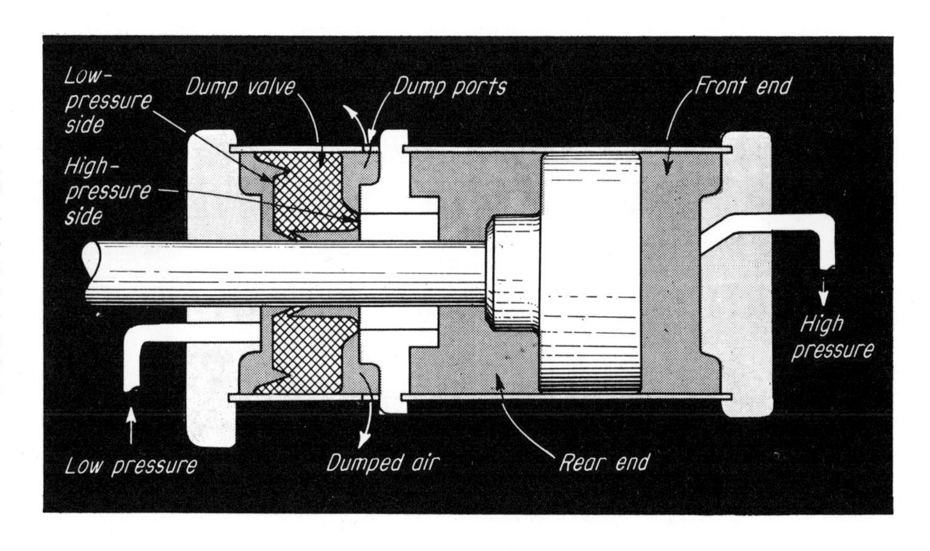

8 SNAP-ACTION

A further selection of basic arrangements for obtaining sudden motion after gradual buildup of force.

PETER C NOY, *production engineer*
Canadian General Electric Co, Barrie, Ont.

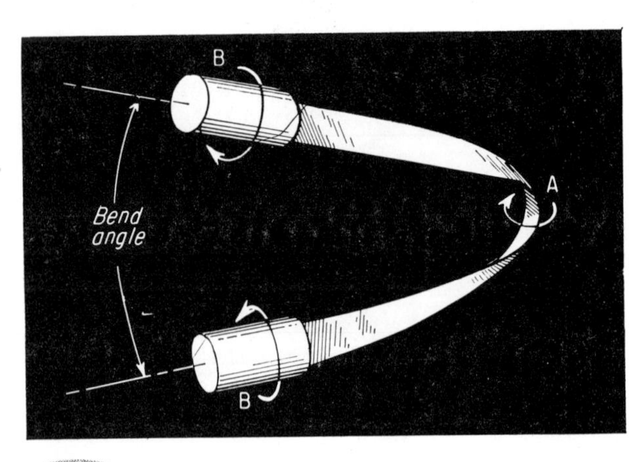

Torsion ribbon . . .
bent as shown will turn "inside out" at A with a snap action when twisted at B. Design factors are ribbon width and thickness, and bend angle.

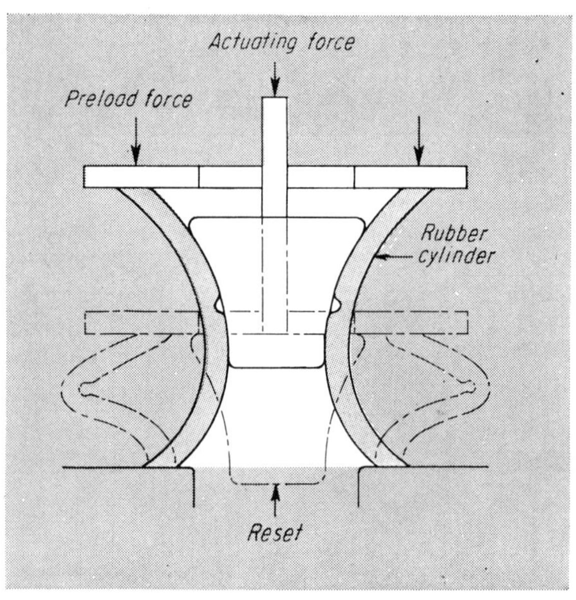

Collapsing cylinder . . .
has elastic walls that may be deformed gradually until their stress changes from compressive to bending with the resultant collapse of the cylinder

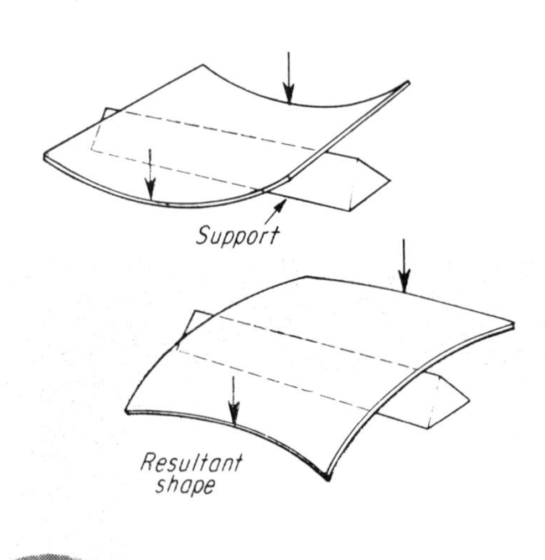

Bowed spring . . .
will collapse into new shape when loaded as shown. "Push-pull" type of steel measuring tape illustrates this action; the curved material stiffens the tape so that it can be held out as a cantilever until excessive weight causes it to collapse suddenly.

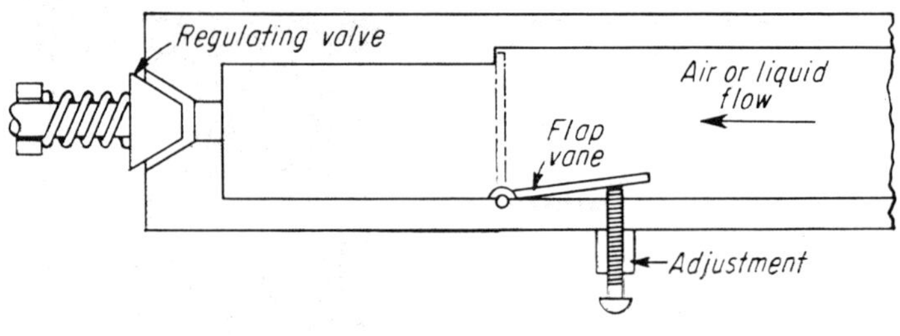

Flap vane . . .
is for air or liquid flow cutoff at a limiting velocity. With a regulating valve, vane will snap shut (because of increased velocity) when pressure is reduced below a certain value.

DEVICES

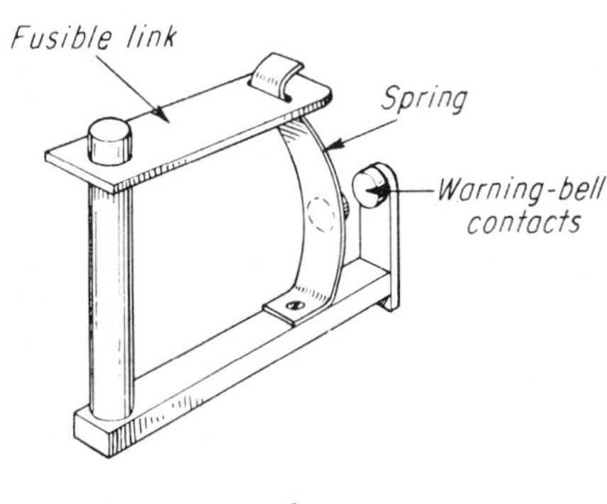

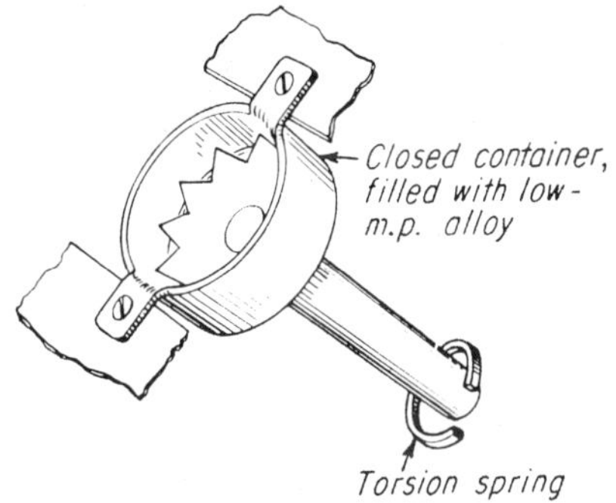

Sacrificing link . . .
is used generally where high heat or chemically corrosive conditions would be hazardous —if temperature becomes too high, or atmosphere too corrosive, link will yield at whatever conditions it is designed for. Usually the device is required to act only once, although a device like the lower one is quickly reset but restricted to temperature control.

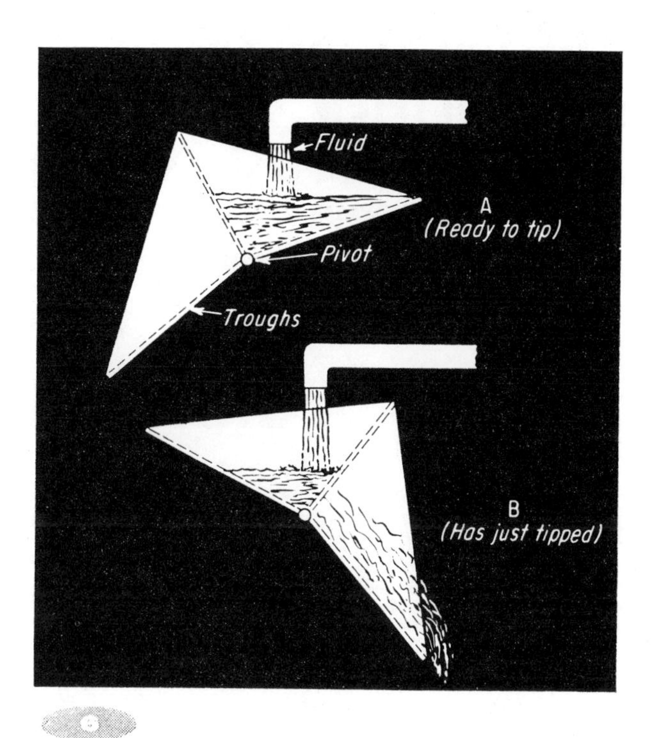

Gravity-tips . . .
although slower acting than most snap mechanisms, can be called snap mechanisms because they require an accumulation of energy to trigger an automatic release. Tipping-troughs used to spread sewage exemplify arrangement shown in A, once overbalanced, action is fast.

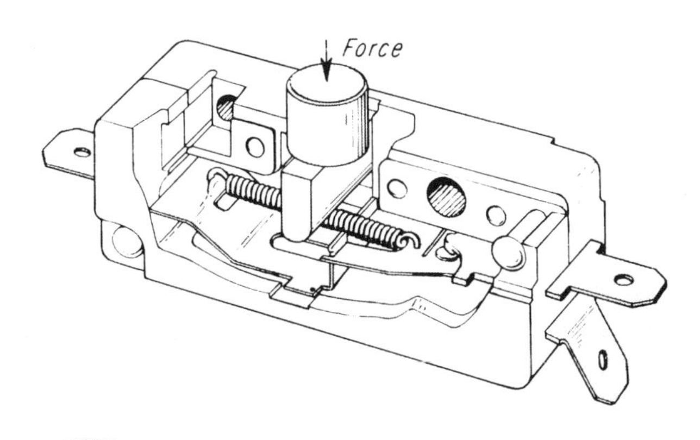

Overcentering tension . . .
spring combined with pivoted contact-strip is one arrangement among many similar ones used in switches. Arrangement shown here is somewhat unusual, since the actuating force bears on the spring itself.

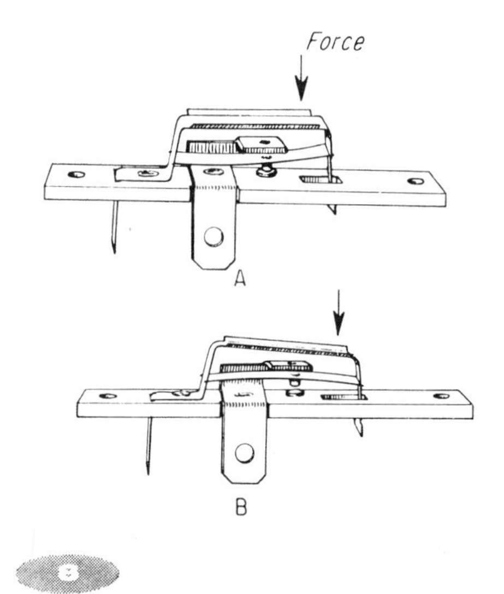

Overcentering leaf-spring . . .
action is also the basis for many ingenious snap-action switches used for electrical control. Sometimes spring action is combined with the thermostatic action of a bimetal strip to make the switch respond to heat or cold either for control purposes or as a safety feature.

6 APPLICATIONS
FOR THE CAPSTAN-TYPE POWER AMPLIFIER

Precise positioning and movement of heavy loads are two basic jobs for this all-mechanical torque booster.

L. A. ZAHORSKY
Universal Match Co.

Capstan principle (above) is basis for mechanical power amplifier (below) that combines two counterrotating drums. Drums are continuously rotating but only transmit torque when input shaft is rotated to tighten band on drum A. Overrun of output is stopped by drum B, when overrun tightens band on this drum.

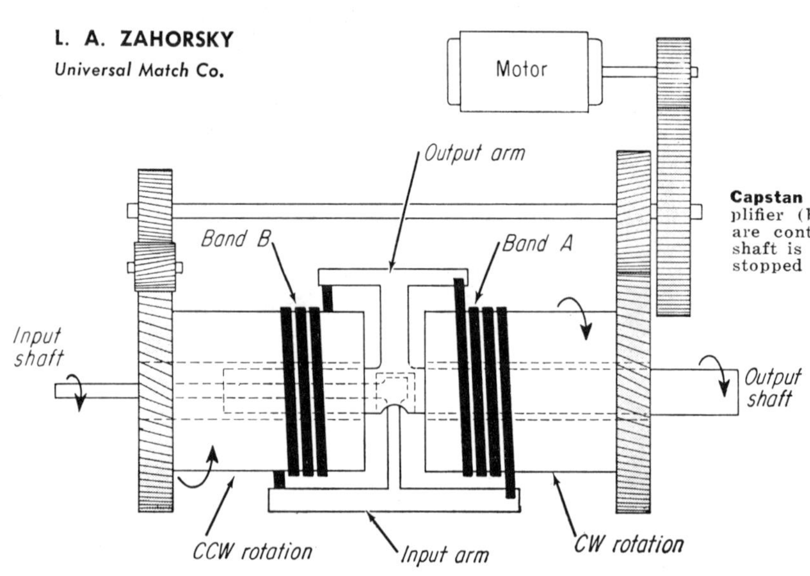

CAPSTAN is simple mechanical amplifier—rope wound on motor-driven drum slips until slack is taken up on the free end. Force needed on free end to lift the load depends on the coefficient of friction and number of turns of rope.

By connecting bands A and B to an input shaft and arm, the power amplifier provides an output in both directions, plus accurate angular positioning. When the input shaft is turned clockwise, the input arm takes up the slack on band A, locking it to its drum. Inasmuch as the load end of locked band A is connected to the output arm, it transmits the CW motion of the driven drum on which it is wound, to the output shaft. Band B therefore slacks off and slips on its drum. When the CW motion of the input shaft stops, tension on band A is released and it slips on its own drum. If the output shaft tries to overrun, the output arm will apply tension to band B, causing it to tighten on the CCW rotating drum and stop the shaft.

From: Control Engineering

This mechanical power amplifier has a fast response. Power from its continuously rotating drums is instantaneously available. When used for position-control applications, such methods as pneumatic, hydraulic, and electrical systems—even with continuously running power sources—require transducers of some kind to change signals from one energy form to another. The mechanical power amplifier, on the other hand, permits direct sensing of the controlled motion.

Four major advantages of this all-mechanical device are:

1. Kinetic energy of the power source is continuously available for rapid response.
2. Motion can be duplicated and power amplified without converting energy forms.
3. Position and rate feedback are inherent design characteristics.
4. Zero slip between input and output eliminates the possibility of cumulative error.

One other important advantage is the ease with which this device can be adapted to perform special functions—jobs for which other types of systems would require the addition of more costly and perhaps less reliable components. The six applications which follow illustrate how these advantages have been put to work in solving widely divergent problems.

1. Nonlinear broaching

Problem: In broaching large bore rifles, the twist given to the lands and grooves represent a nonlinear function of barrel length. Development work on such rifles usually requires some experimentation with this function. At present, rotation of the broaching head is performed by a purely mechanical arrangement consisting of a long heavy wedge-type cam and appropriate gearing. For steep twist angles, however, the forces acting on this mechanism becomes extremely high.

Solution: A suitable mechanical power amplifier, with its inherent position feedback, was added to the existing mechanical arrangement, as shown in sketch. The cam and follower, instead of having to drive the broaching head, simply furnish enough torque to position the input shaft of the amplifier.

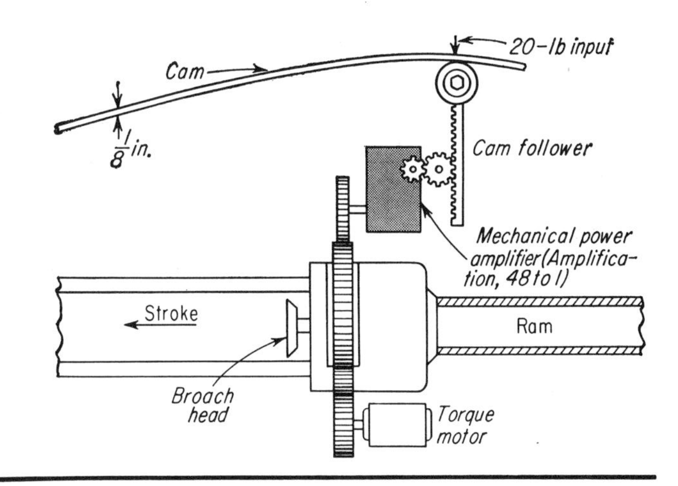

2. Hydraulic winch control

Problem: Hydraulic pump-motor systems represent an excellent method of controlling position and motion at high power levels. In the 10- to 150-hp range, for example, the usual approach is to vary the output of a positive displacement pump in a closed-loop hydraulic circuit. In many of the systems that might be used to control this displacement, however, a force feedback proportional to system pressure can lead to serious errors or even oscillations.

Solution: Sketch shows an external view of the complete package. The output shaft of the mechanical power amplifier controls pump displacement, while its input is controlled by hand. In a more recent development, requiring remote manual control, a servomotor replaces this local handwheel. Approximately 10 lb-in. torque drives a 600 lb-in. load. If this system had to transmit 600 lb-in. the equipment would be more expensive and more dangerous.

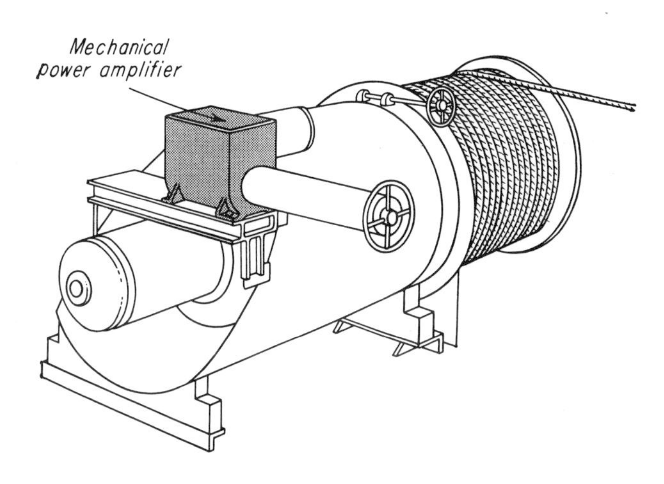

3. Load positioning

Problem: It was necessary for a 750-lb load to be accelerated from standstill in 0.5 sec and brought into speed and position synchronization with a reference linear motion. It was also necessary that the source of control motion be permitted to accelerate more rapidly than the load itself. Torque applied to the load could not be limited by means of a slipping device.

Solution: A system using a single mechanical power amplifier provided the solution, sketch. Here a mechanical memory device preloaded for either rotation is used to drive the input shaft of the amplifier. This permits the input source to accelerate as rapidly as desired. Total control input travel minus the input travel of the amplifier shaft is temporarily stored. After 0.5 sec the load reaches proper speed, and the memory device transmits position information in exact synchronization with the input.

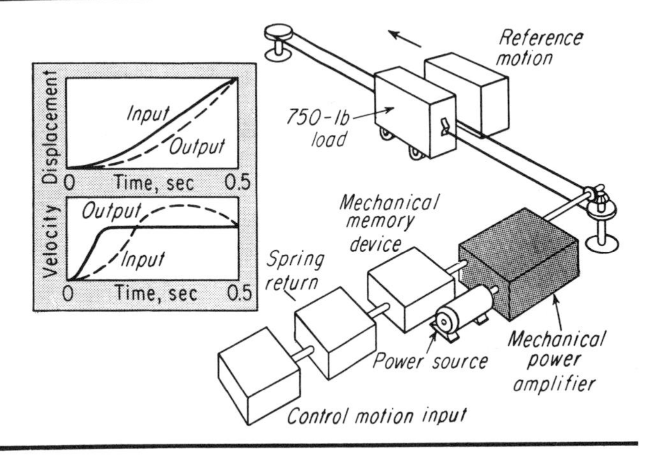

4. Tensile testing machine

Problem: On a hydraulic tensile testing machine, stroke of the power cylinder had to be controlled as a function of two variables: tension in, and extension of, the test specimen. A programming device designed to provide a control signal proportional to these variables, had an output power level of about 0.001 hp—too low to drive the pressure regulator controlling the flow to the cylinder.

Solution: An analysis of the problem revealed three requirements: output of the programmer had to be amplified about 60 times, position accuracy had to be within 2 deg, and acceleration had to be held at a very low value. A mechanical power amplifier satisfied all three requirements. Sketch illustrates the completed system. Its design is based principally on steady-state characteristics.

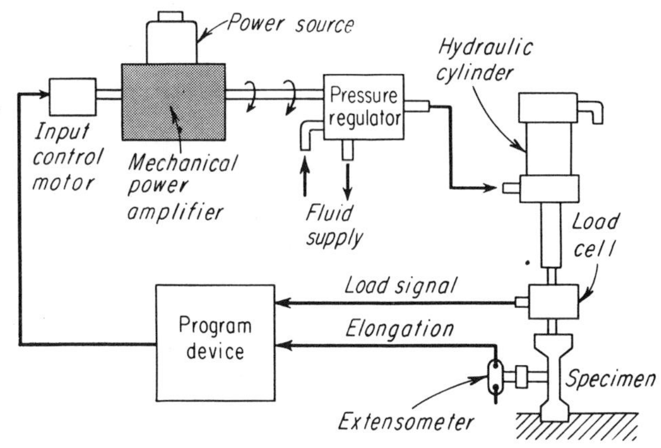

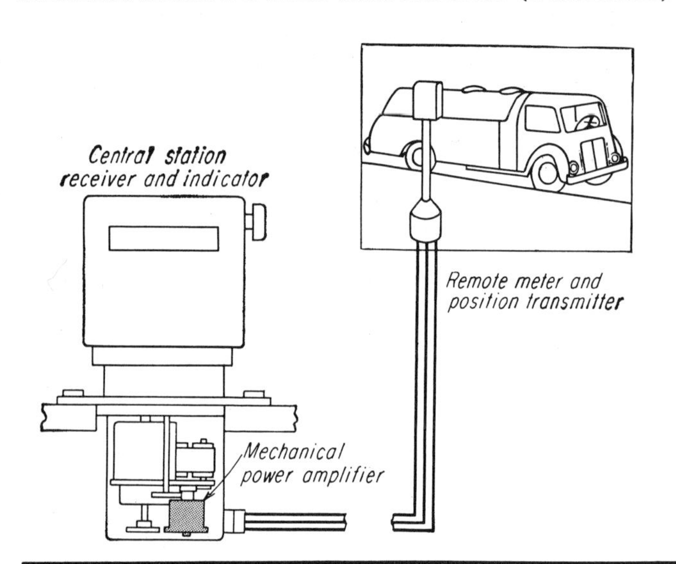

5. Remote metering & counting

Problem: For a remote liquid metering job, synchro systems had been used to transmit remote meter readings to a central station and repeat this information on local indicating counters. The operation involved a large number of meters and indicators. As new devices were added (ticket printers, for instance), the torque requirement also grew.
Solution: Use of mechanical power amplifiers in the central station indicators not only supplied extra output torque but also made it possible to use synchros even smaller than those originally selected to drive the indicators alone.

The synchro transmitters currently used operate at a maximum speed of 600 rpm and produce only about 3 oz-in. of torque. The mechanical power amplifiers furnish up to 100 lb-in. and are designed to fit in the bottom of the registers as shown in sketch. Total accuracy is within 0.25 gallon, error is noncumulative.

6. Irregular routing

Problem: To remotely control the table position of a routing machine from information stored on a film strip. The servoloop developed to interpret this information produced only about 1 oz-in. of torque. About 20 lb-ft was required at the table feedscrew.
Solution: Sketch shows how a mechanical power amplifier supplied the necessary torque at the remote table location. A position transmitter converts the rotary motion output of the servoloop to a proportional electrical signal and sends it to a differential amplifier at the machine location. A position receiver, geared to the output shaft, provides a signal proportional to table position. The differential amplifier compares these, amplifies the difference, and sends a signal to either counterrotating electromagnetic clutch, which drives the input shaft of the mechanical power amplifier.

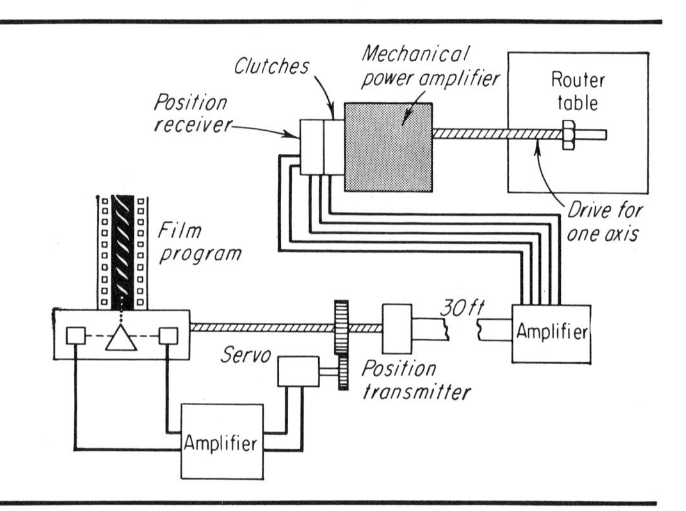

OPERATING DATA FOR MECHANICAL POWER AMPLIFIERS

APPLICATION PARAMETERS		RIFLE BROACH	ROUTER		ACCELERATION SPEED AND POSITION CONTROL		HYDRAULIC WINCH CONTROL	TENSILE TESTER	REMOTE METERING
			SYSTEM	MPA	SYSTEM	MPA			
INPUT TORQUE LB. IN.		75	0.75	2	20+	2–6	5	0.1	0.75
OUTPUT TORQUE LB. IN.		3600	230	230	——	240	600	5	6.25
AMPLIFICATION		48	308	115	——	40	120	50	8.13
NORMAL RPM		——	——	——	50	50	0	50	200
MAXIMUM RPM		INPUT 108 OUTPUT 18	950	190	INF.	50	100	96	200
NUMBER OF REVERSALS PER MINUTE		——	1100	1100	120	120	20	——	——
STROKE FOR ABOVE REVERSALS		——	335°	67°	——	——	720°	2880°	——
ERRORS PERMITTED	MPA	BROACH							
BACKLASH	2°	0°	——	1.5°	——	1.5°	2°	2°	6°
PROPORTIONAL AT FULL LOAD	1°	0.009°	——	1.0°	——	1.0°	10°	1°	3°

Mechanical power amplifier . . .
that drives crossfeed slide uses the principle of the windlass. By varying the control force, all or any part of power to the drum can be utilized.

Two drums mounted back to back supply the bi-directional power needed in servo systems. Replacing the operator with a 2-phase induction servomotor permits using electronic or magnetic signal amplification. A rotating input avoids linear input and output of the simple windlass. Control and output ends of the multiturn bands are both connected to gears mounted concentrically with the drum axis.

When servomotor rotates the control gear it locks the band-drum combination, forcing output gear to rotate with it. Clockwise rotation of the servomotor produces CW power output while second drum idles. Varying the servo speed, by changing servo voltage, varies output speed.

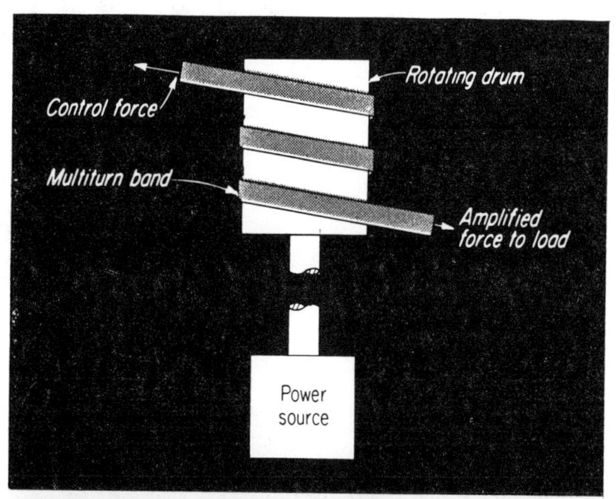

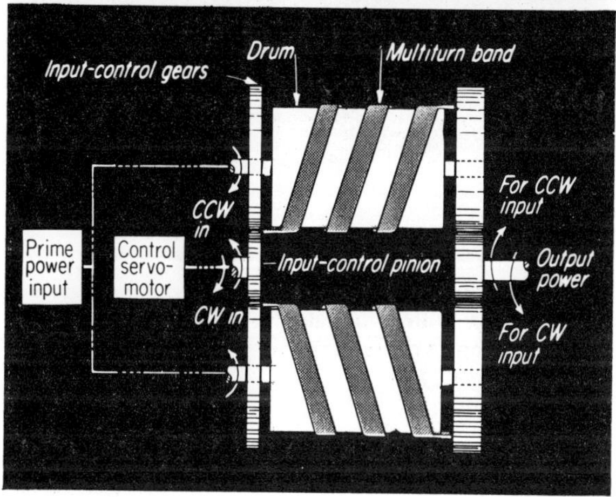

WINCHES

For that let-down feeling

Developed to lower soldiers and equipment from a hovering helicopter, the descent control replaces a bulky winch which can only lower one load at a time. Since the user can control the speed of descent, safer landings are possible. The lowering rope is wrapped around the Genie's shaft several times and the shield slipped over the shaft. An eye on the shaft is attached to the object to be lowered and it is dropped free. Each wrap around the shaft will safely lower 50 lb.

Paratroopers weighing over 200 lb with equipment use a single wrap and break their descent near the ground by manipulating the rope's tension. For multiple use a series of Genies can be slipped over one rope and used in succession.

A single casting of A-360 aluminum is used for the shaft and a piece of tubing completes the assemby. Manual deburring and ball burnishing of the shaft are the only finishing operations required. Produced by LA Die Casting Corp, the casting meets military safety requirements of a 2000-lb tensile test. Possible applications for the Sky Genie include its use for rescue operations, fire escapes, mountain climbing, rigging, or snubbing any moving object.

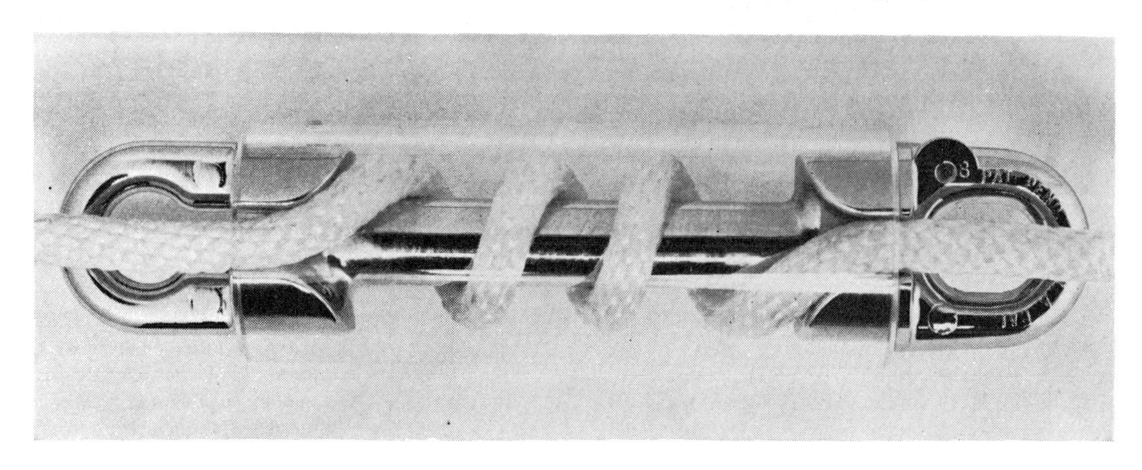

10 Ways to amplify

How levers, membranes, cams, and gears are arranged to measure, weigh, gage, adjust, and govern.

FEDERICO STRASSER

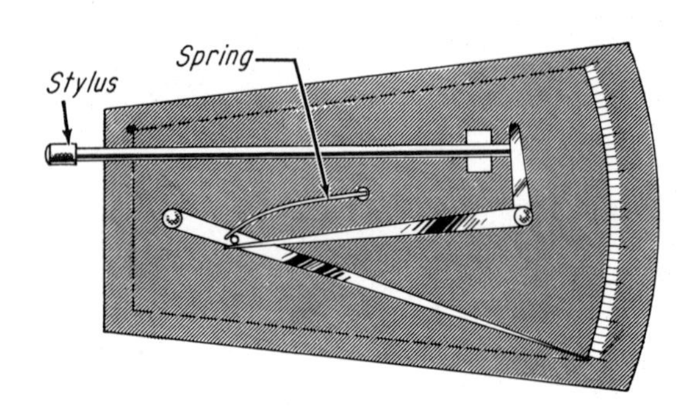

1 **HIGH AMPLIFICATION** for simple measuring instruments is provided by double lever action. Accuracy can be as high as 0.0001 in.

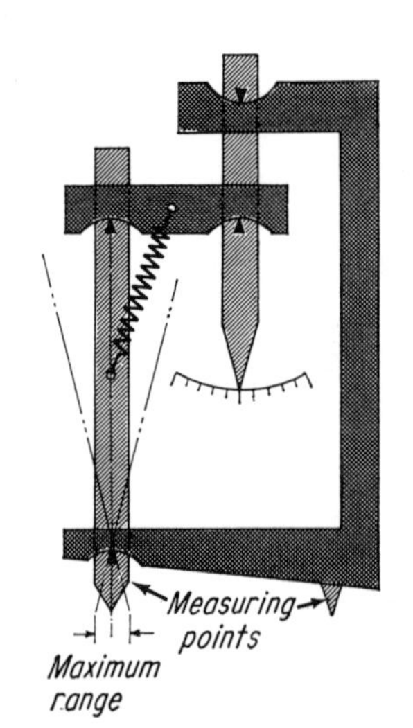

2 **PIVOTED LEVERS** allow extremely sensitive action in comparator-type measuring device shown here. The range, however, is small.

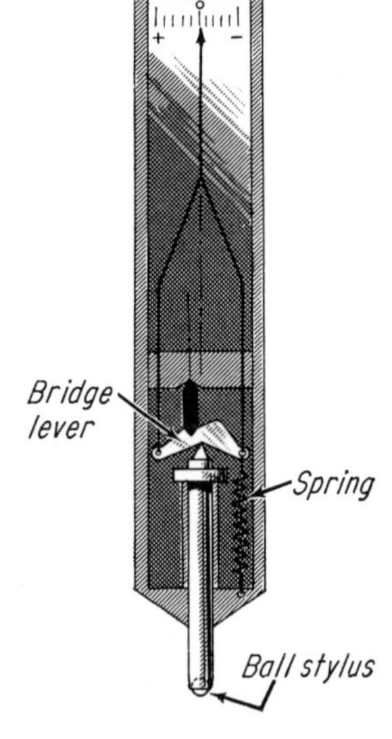

3 **ULTRA-HIGH AMPLIFICATION,** with only one lever, is provided in the Hirth-Minimeter shown here. Again, the range is small.

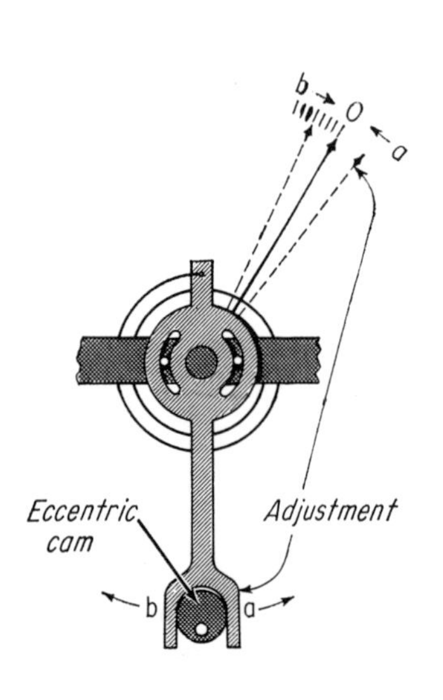

7 **FOR CLOSE ADJUSTMENT,** electrical measuring instruments employ eccentric cams. Here movement is reduced, not amplified.

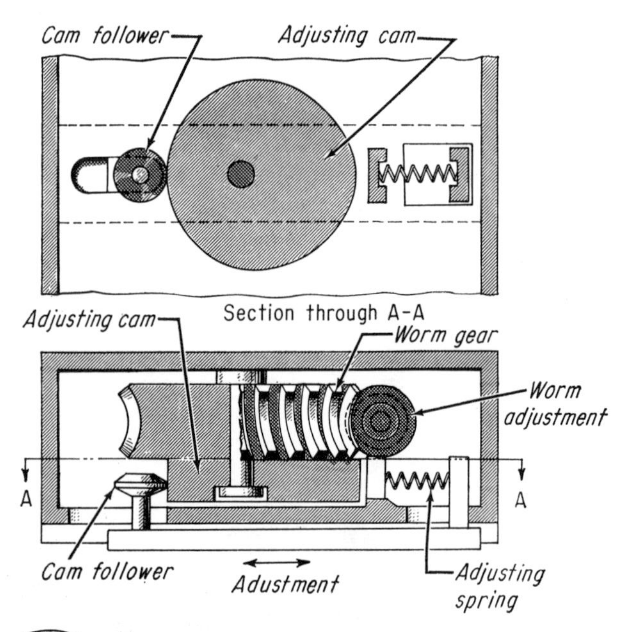

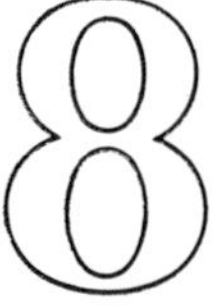

MICROSCOPIC ADJUSTMENT is achieved here by employing a large eccentric-cam coupled to a worm-gear drive. Smooth, fine adjustment result.

mechanical movements

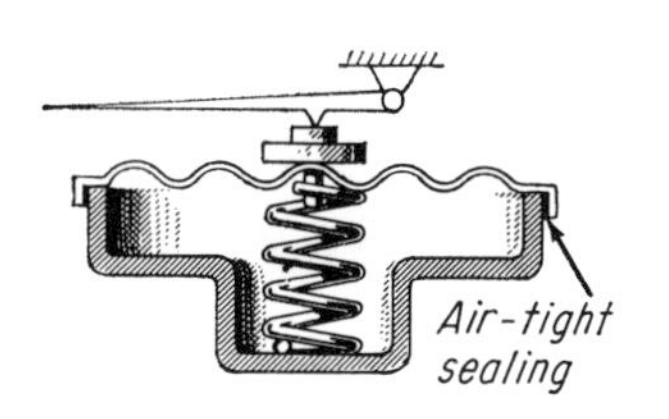

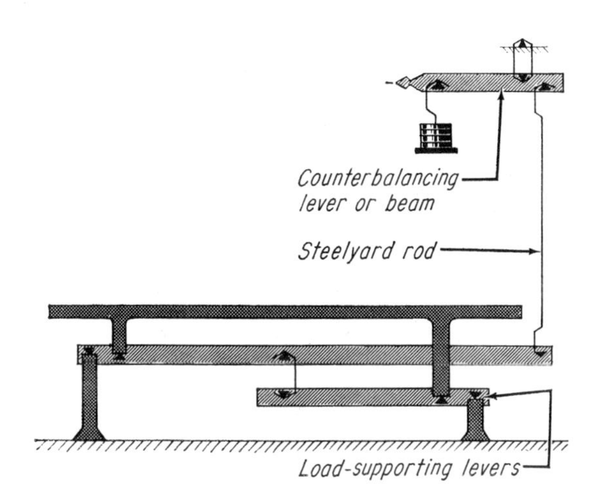

5 **CAPSULE UNIT** for gas-pressure indicators should be provided with a compression spring to preload the membrane for more positive action.

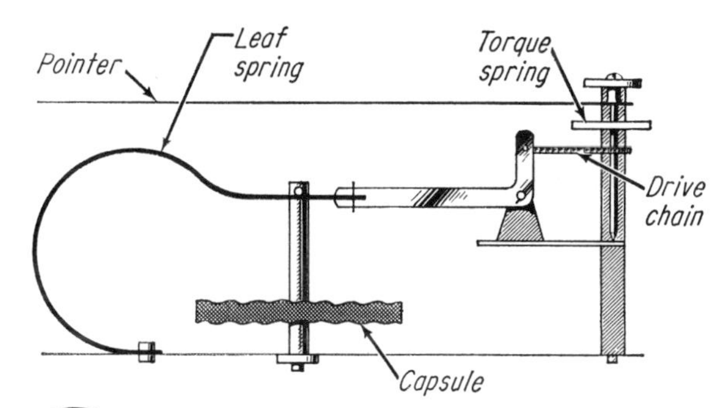

LEVER - ACTUATED weigh-scale needs no springs to maintain balance. The lever system, mounted on knife edges, is extremely sensitive.

6 **AMPLIFIED MEMBRANE MOVEMENT** can be gained by the arrangement shown here. A small chain-driven gear links the lever system.

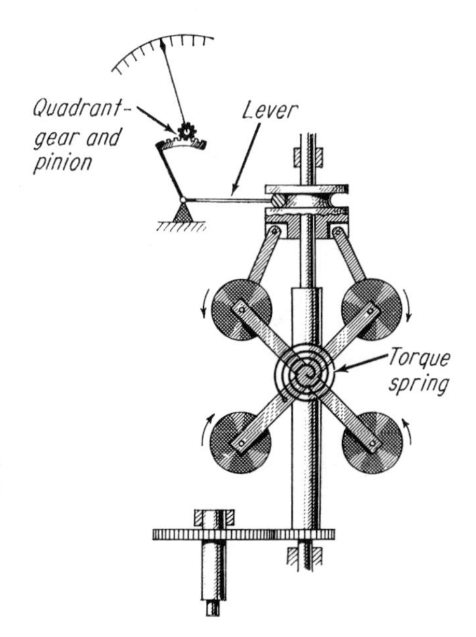

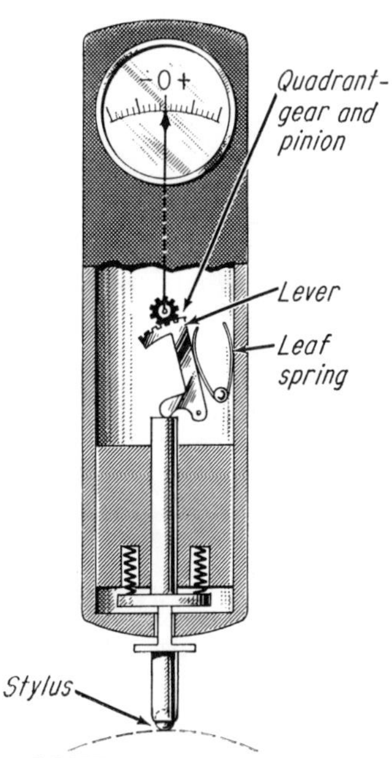

QUADRANT-GEAR AND PINION coupled to an L-lever provide ample movement of indicator needle for small changes in governor speed.

COMBINATION LEVER AND GEARED quadrant are used here to give the comparator maximum sensitivity combined with ruggedness.

10 more ways to amplify

Levers, wires, hair, and metal bands are arranged to give high velocity ratios for adjusting and measuring.

FEDERICO STRASSER

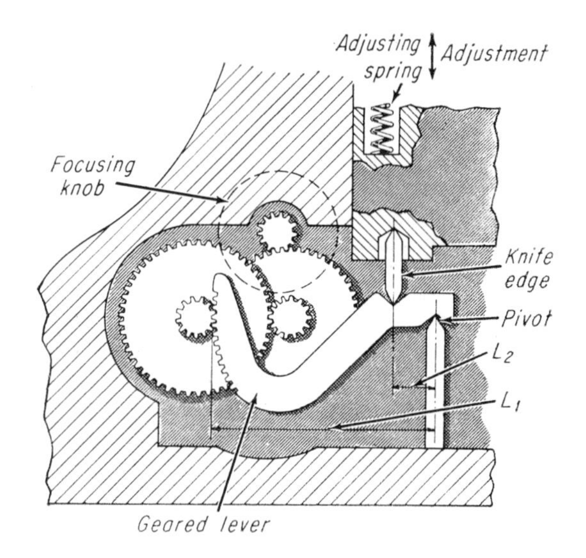

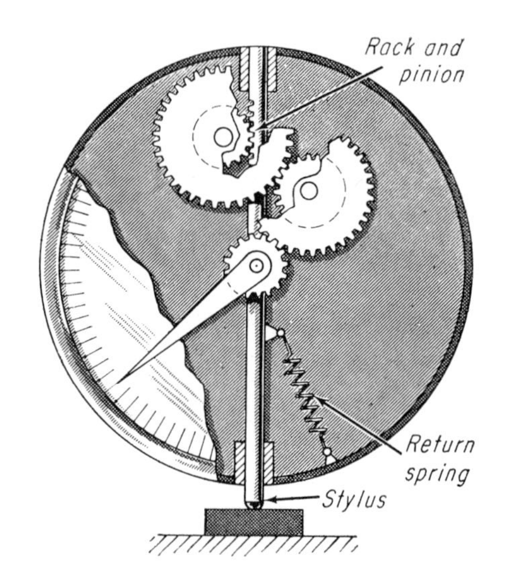

1 **LEVER AND GEAR** train amplify the microscope control-knob movement. Knife edges provide frictionless pivots for lever.

2 **DIAL INDICATOR** starts with rack and pinion amplified by gear train. The return-spring takes out backlash.

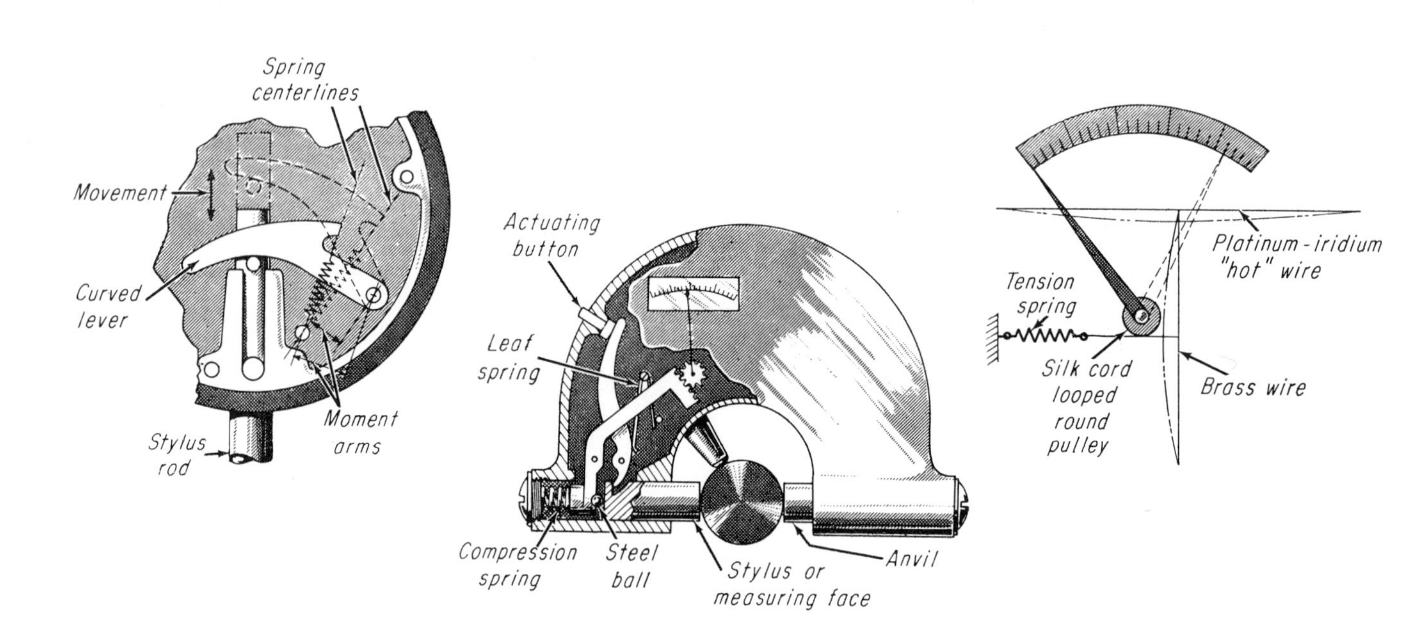

3 **CURVED LEVER** is so shaped and pivoted that the force exerted on the stylus rod, and thus stylus pressure, remains constant.

ZEISS COMPARATOR is provided with a special lever to move the stylus clear of the work. A steel ball greatly reduces friction.

5 **"HOT-WIRE" AMMETER** relies on the thermal expansion of a current-carrying wire. A relatively large needle movement occurs.

mechanical action

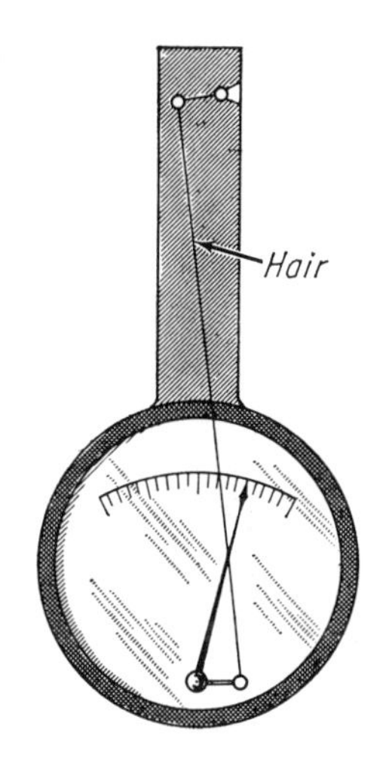

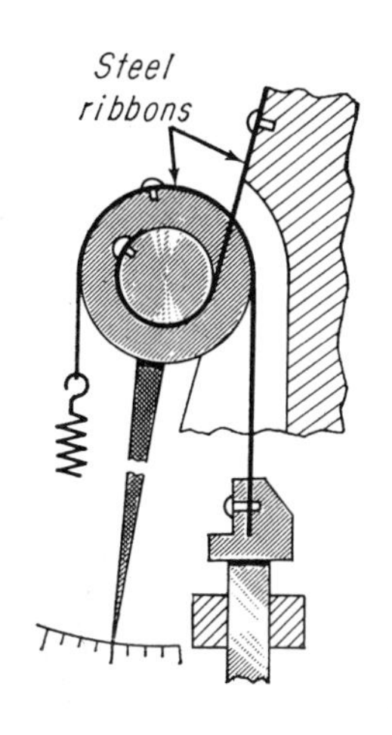

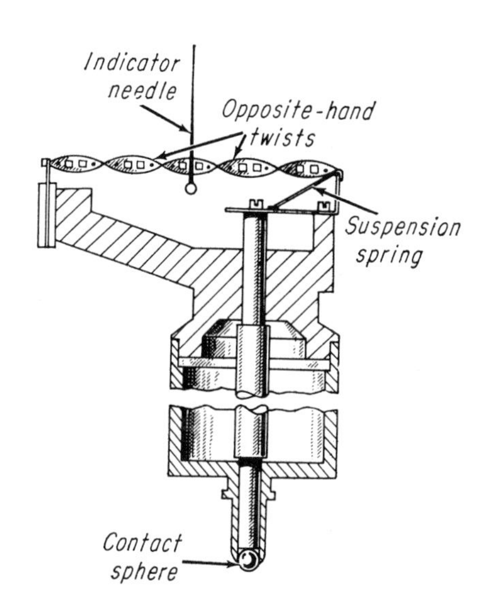

HYGROMETER is actuated by a hair. When humidity causes expansion of the hair, its movement is amplified by a lever.

STEEL RIBBONS transmit movement without the slightest backlash. The movement is amplified by differences in diameter.

METAL BAND is twisted and supported at each end. Small movement of contact sphere produces large needle movement.

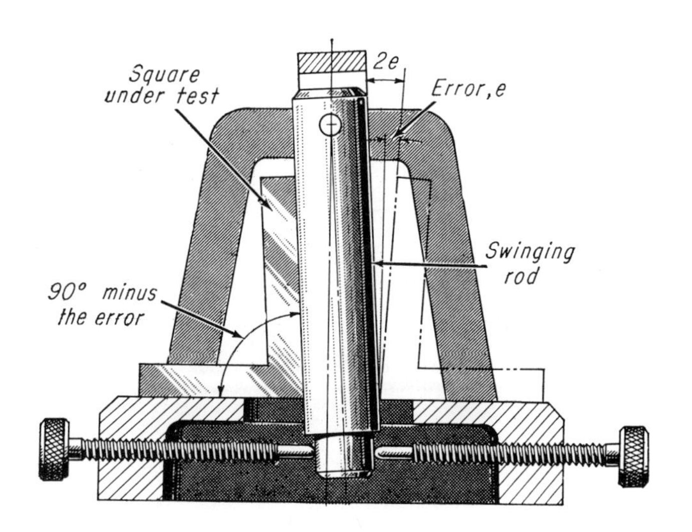

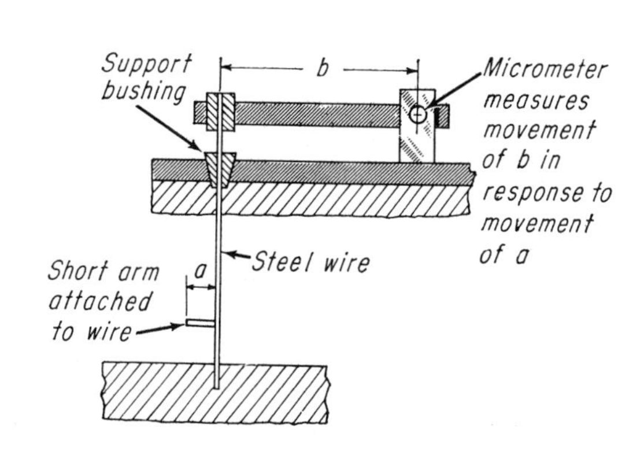

ACCURACY of 90° squares can be checked with a device shown here. The rod makes the error much more apparent.

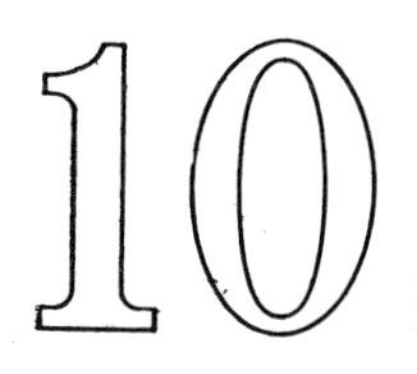

TORSIONAL deflection of the short arm is transmitted with low friction to the longer arm for micrometer measurement.

for self-locking at high efficiency . . .

the TWINWORM GEAR

Developed by an Israeli engineer, this innovation in gearing combines two worm-screws to give self-locking characteristics, or to operate as a fast-acting brake when power is shut off. Model above shows basic components.

NICHOLAS CHIRONIS
Associate editor

The term "self-locking" as applied to gear systems denotes a drive which gives the input gear freedom to rotate the output gear in either direction—but the output gear locks with the input when an outside torque attempts to rotate the output in either direction. This characteristic is often sought by designers who want to be sure that loads on the output side of the system cannot affect position of the gears. Worm gears are one of the few gear systems that can be made self-locking, but at the expense of efficiency—they seldom exceed 40%, when self-locking.

An Israeli engineer displayed a simple dual-worm gear system that not only provided self-locking with over 90% efficiency, but exhibited a new phenomenon which the inventor calls "deceleration-locking."

A point in favor of the inventor—B. Popper, an engineer with the Scientific Department of the Israel Ministry of Defense in Tel Aviv—is that his "Twinworm" drive has been employed in Israel-designed counters and computers for several years and with marked success.

The Twinworm drive is quite simply constructed. Two threaded rods, or "worm" screws, are meshed together. Each worm is wound in a different direction and has a different pitch angle. For proper mesh, the worm axes are not parallel, but slightly skewed. (If both worms had the same pitch angle, a normal, reversible drive would result—similar to helical gears.) But by selecting proper, and different, pitch angles, the drive will exhibit either self-locking, or a combination of self-locking and deceleration-locking characteristics, as desired. Deceleration-locking is a completely new property best described in this way.

When the input gear decelerates (for example, when the power source is shut off, or when an outside force is applied to the output gear in a direction which tends to help the output gear) the entire transmission immediately locks up and comes to an abrupt stop moderated only by any elastic "stretch" in the system.

Almost any type of thread will work with the new drive—standard, 60° screw threads, Acme threads, or any arbitrary shallow-profile thread. Hence, the worms can be produced on standard machine-shop equipment.

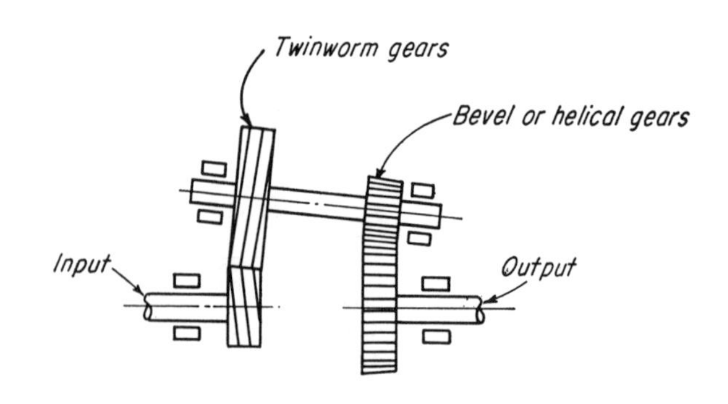

ANGLE BETWEEN SHAFTS IN NEW DRIVE is easily compensated by pairing with bevel or helical gears.

JOBS FOR THE NEW DRIVE

Applications for Twinworm can be divided into two groups:

(1) Those employing self-locking characteristics to prevent the load from affecting the system.

(2) Those employing deceleration-locking characteristics to brake the system to an abrupt stop if the input decelerates.

Self-locking Applications

Mechanical counters. This is the application that led to development of Twinworm gears. Popper was

APPLICATIONS FOR TWINWORM

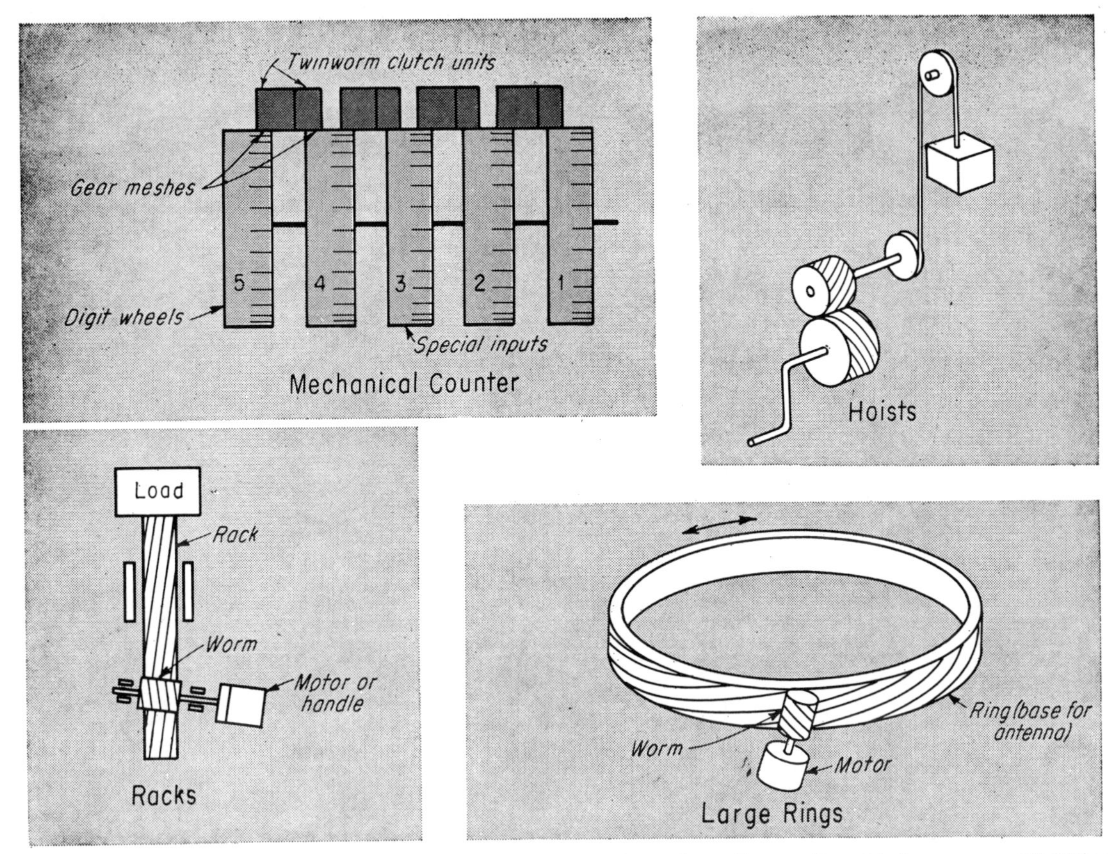

MECHANICAL COUNTER was first application. Sketches show three other Twinworm possibilities.

given the problem of developing a gear system that would permit inputs to be made directly to any one of five digit-wheels. The inputs were to affect higher-digit wheels (see upper left sketch), but not the lower digits.

This was accomplished by coupling Twinworm drives to slip clutches. An impulse to digit-wheel 3 would cause it to also rotate wheel 4, but the Twinworm between wheels 2 and 3 would lock up, causing it to slip against its clutch. Standard worm gears were originally employed but their low efficiency was compounded because of their series-arrangement. If each worm has 40% efficiency, the over-all efficiency, when impulse is to wheel *1*, is 0.4 x 0.4 x 0.4 x 0.4 = 0.0256 or only $2\frac{1}{2}$%.

Hoists and lifts—Popper believes the drive could be employed advantageously wherever loads must be raised, such as in hoists, elevators, lift trucks, mechanisms for adjusting car windows, and so forth. The drive not only prevents the load from rotating the gears but can be so designed that the same input torque is required both for raising and lowering the load (even without a counter-weight). In the illustrations above, for example, one can crank the load to any position with the same force—and remove his hand without fear that the load will fall. Also, the power unit (if a motor is employed) will be the smallest possible.

The principle of the drive could also be applied to large external or internal rings; for example, for rotation of antennas. Wind loads on the antenna cannot shift the ring.

Self-locking occurs as soon as $\tan \phi_1$ is equal to or smaller than μ, or when

$$\tan \phi_1 = \frac{\mu}{S_1}$$

Angles ϕ_1 and ϕ_2 represent the respective pitch angles of the two worms, and $\phi_2 - \phi_1$ is the angle between the two worm shafts (angle of misalignment). Angle ϕ_1 is quite small (usually in the order of 2 to 5°).

Here, S_1 represents a "safety factor" (selected by the designer) which must be somewhat greater than one to make sure that self-locking is maintained even if μ should fall below an assumed value. Neither ϕ_2 nor the angle $(\phi_2 - \phi_1)$ affects self-locking.

Deceleration-locking occurs as soon as $\tan \phi_2$ is also equal to or smaller than μ; or, if a second safety factor S_2 is employed (where $S_2 > 1$), when

$$\tan \phi_2 = \frac{\mu}{S_2}$$

For the equations to hold true, ϕ_2 must always be made greater than ϕ_1. Also, μ refers to the idealized case where the worm threads are square. If the threads are inclined (as with Acme-threads or V-threads) then a modified value of μ must be employed, where

$$\mu_{modified} = \frac{\mu_{true}}{\cos \theta}$$

Relationship between input and output forces during rotation is:

$$\frac{P_1}{P_2} = \frac{\sin \phi_1 + \mu \cos \phi_1}{\sin \phi_2 + \mu \cos \phi_2}$$

Efficiency

$$\eta = \frac{1 + \mu/\tan \phi_2}{1 + \mu/\tan \phi_1}$$

Applications of Differential

Known for its mechanical advantage, the differential winch is a control mechanism that can supplement the gear and rack and four-bar linkage systems in changing

ALEXANDER B. HULSE, JR. and ROBERT AYMAR
Factoring Instruments & Devices, Inc.

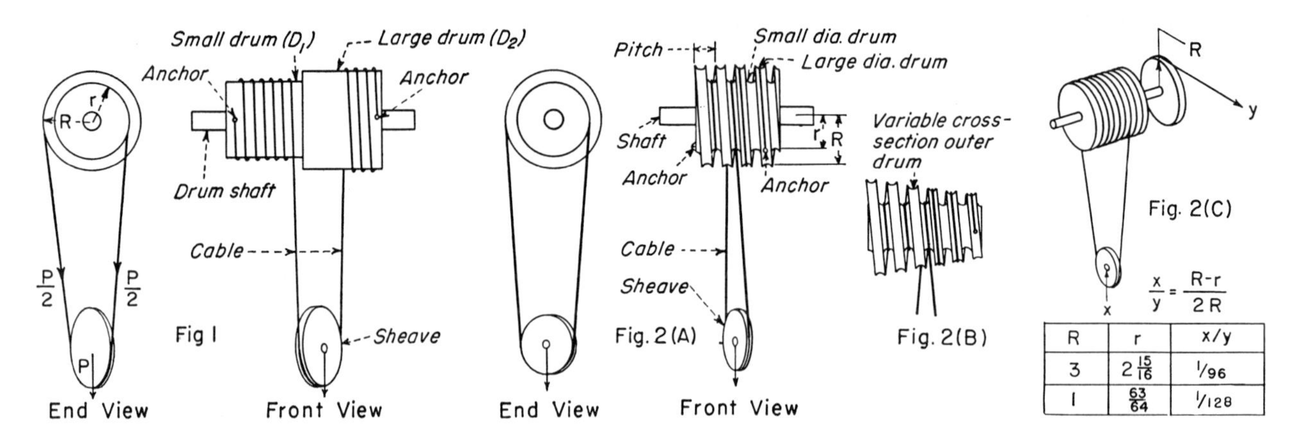

R	r	x/y
3	$2\frac{15}{16}$	1/96
1	$\frac{63}{64}$	1/128

FIG. 1—Standard Differential Winch; consists of two drums D_1 and D_2 and a cable or chain which is anchored on both ends and wound clockwise around one drum and counterclockwise around the other. The cable supports a load carrying sheave and if the shaft is rotated clockwise, the cable, which unwinds from D_1 on to D_2, will raise the sheave a distance

$$\text{Sheave rise/rev} = \frac{2\pi R - 2\pi r}{2} = \pi(R - r)$$

The winch, which is not in equilibrium, exerts a counterclockwise torque.

$$\text{Unbalanced torque} = \frac{P}{2}(R - r)$$

FIG. 2(A)—Hulse Differential Winch*. Two drums, which are in the form of worm threads contoured to guide the cables, concentrically occupy the same longitudinal space. This keeps the cables approximately at right angles to the shaft and eliminates cable shifting and rubbing especially when used with variable cross sections as in Fig. 2(B) where any equation of motion can be satisfied by choosing suitable cross sections for the drums.

Ways for resisting or supporting the axial thrust may have to be considered in some installations. Fig. 2(C) shows typical reductions in displacement.

*Pat. No. 2,590,623

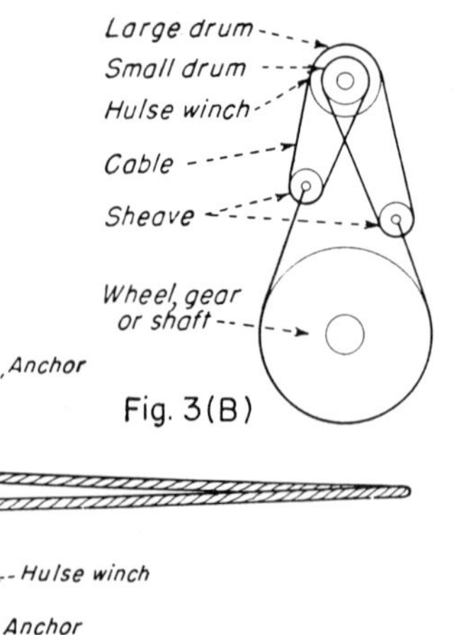

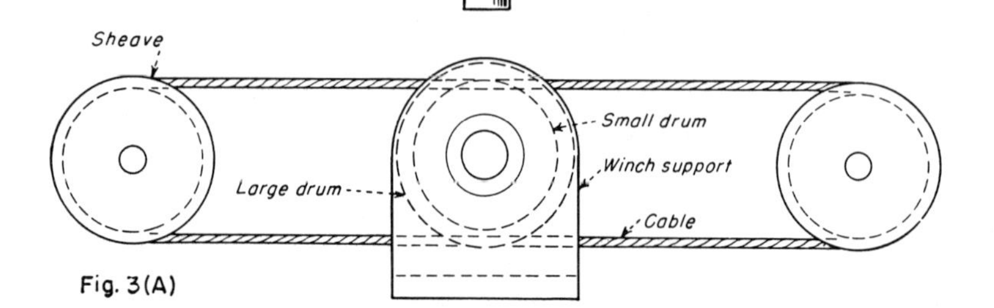

FIG. 3(A)—Hulse Winch with Opposing Sheaves. This arrangement which uses two separate cables and four anchor points can be considered as two winches back-to-back using one common set of drums. Variations in motion can be obtained by: (1) restraining the sheaves so that when the system is rotated the drums will travel toward one of the sheaves; (2) restraining the drums and allowing the sheaves to travel. The distance between the sheaves will remain constant and are usually connected by a bar; (3) permitting the drums to move axially while restraining them transversely. When the system is rotated, drums will travel axially one pitch per revolution, and sheaves remain in same plane perpendicular to drum axis. This

Winch to Control Systems

rotary motion into linear. It can magnify displacement to meet the needs of delicate instruments or be varied almost at will to fulfill uncommon equations of motion.

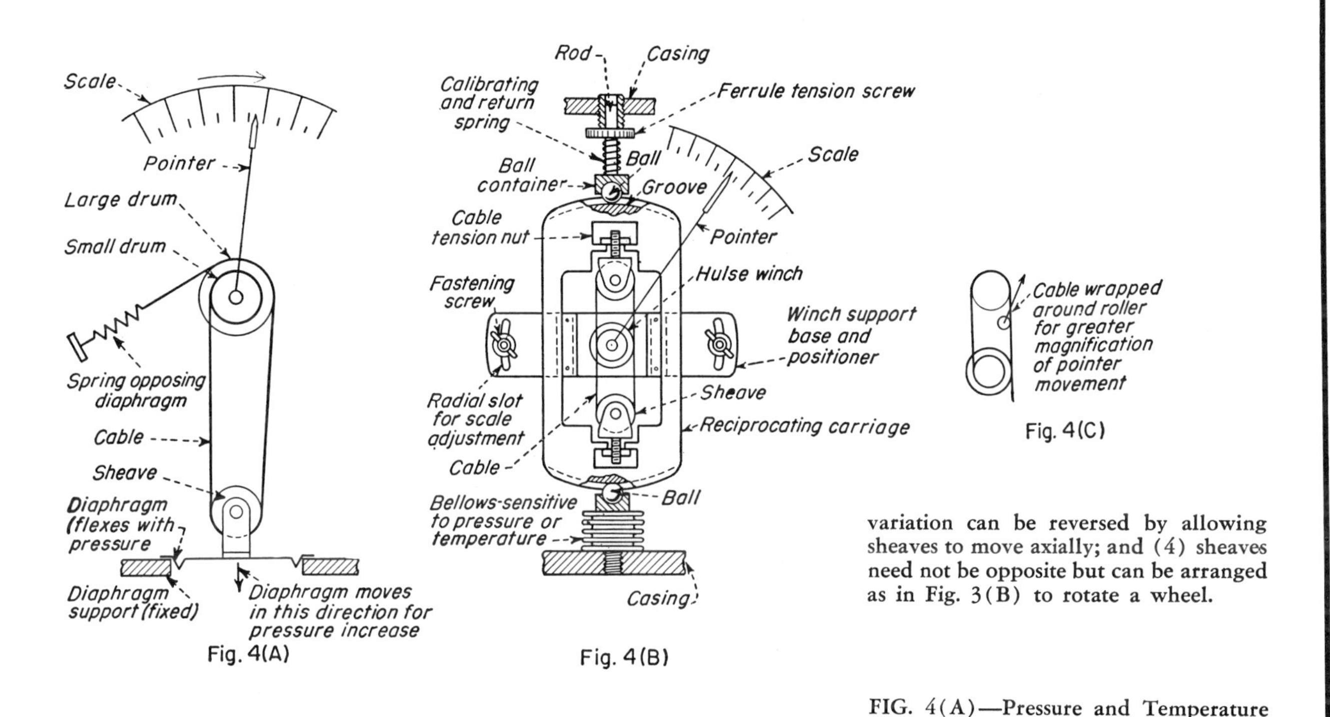

variation can be reversed by allowing sheaves to move axially; and (4) sheaves need not be opposite but can be arranged as in Fig. 3(B) to rotate a wheel.

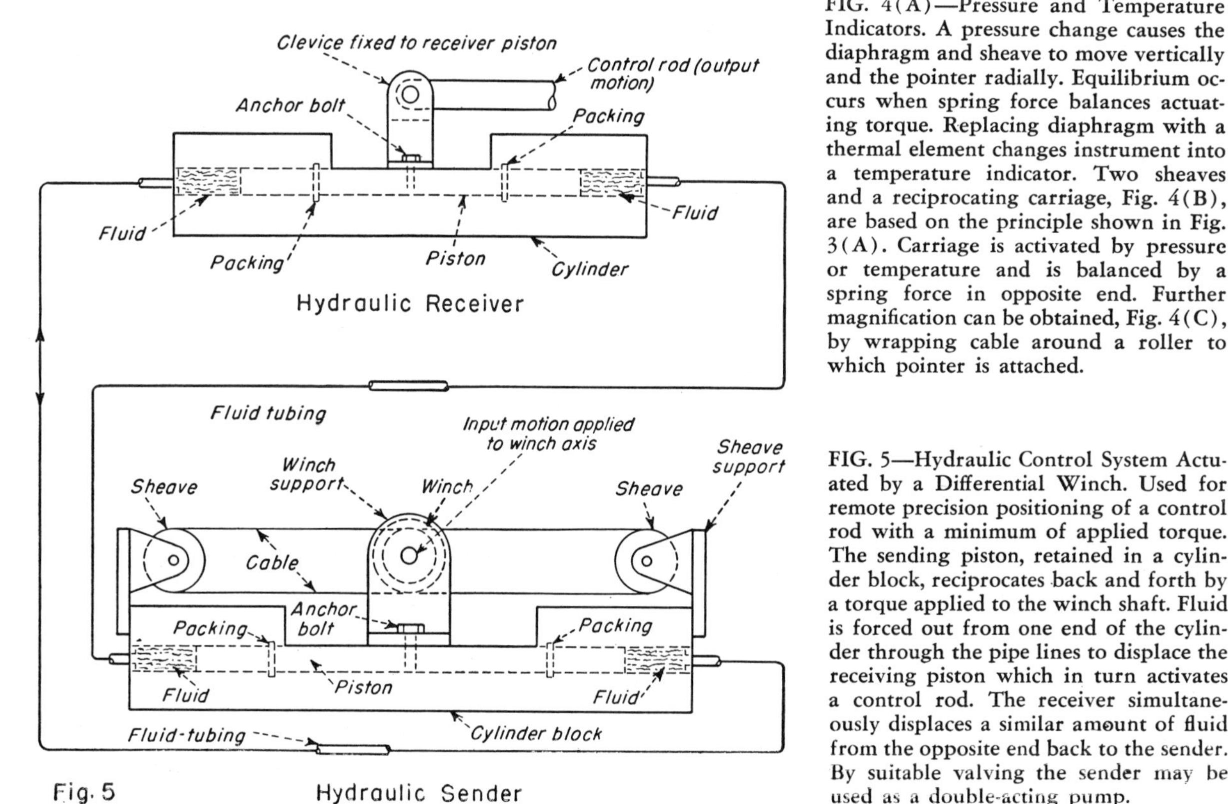

FIG. 4(A)—Pressure and Temperature Indicators. A pressure change causes the diaphragm and sheave to move vertically and the pointer radially. Equilibrium occurs when spring force balances actuating torque. Replacing diaphragm with a thermal element changes instrument into a temperature indicator. Two sheaves and a reciprocating carriage, Fig. 4(B), are based on the principle shown in Fig. 3(A). Carriage is activated by pressure or temperature and is balanced by a spring force in opposite end. Further magnification can be obtained, Fig. 4(C), by wrapping cable around a roller to which pointer is attached.

FIG. 5—Hydraulic Control System Actuated by a Differential Winch. Used for remote precision positioning of a control rod with a minimum of applied torque. The sending piston, retained in a cylinder block, reciprocates back and forth by a torque applied to the winch shaft. Fluid is forced out from one end of the cylinder through the pipe lines to displace the receiving piston which in turn activates a control rod. The receiver simultaneously displaces a similar amount of fluid from the opposite end back to the sender. By suitable valving the sender may be used as a double-acting pump.

18 Variations of the Differential Mechanism

ALFRED KUHLENKAMP

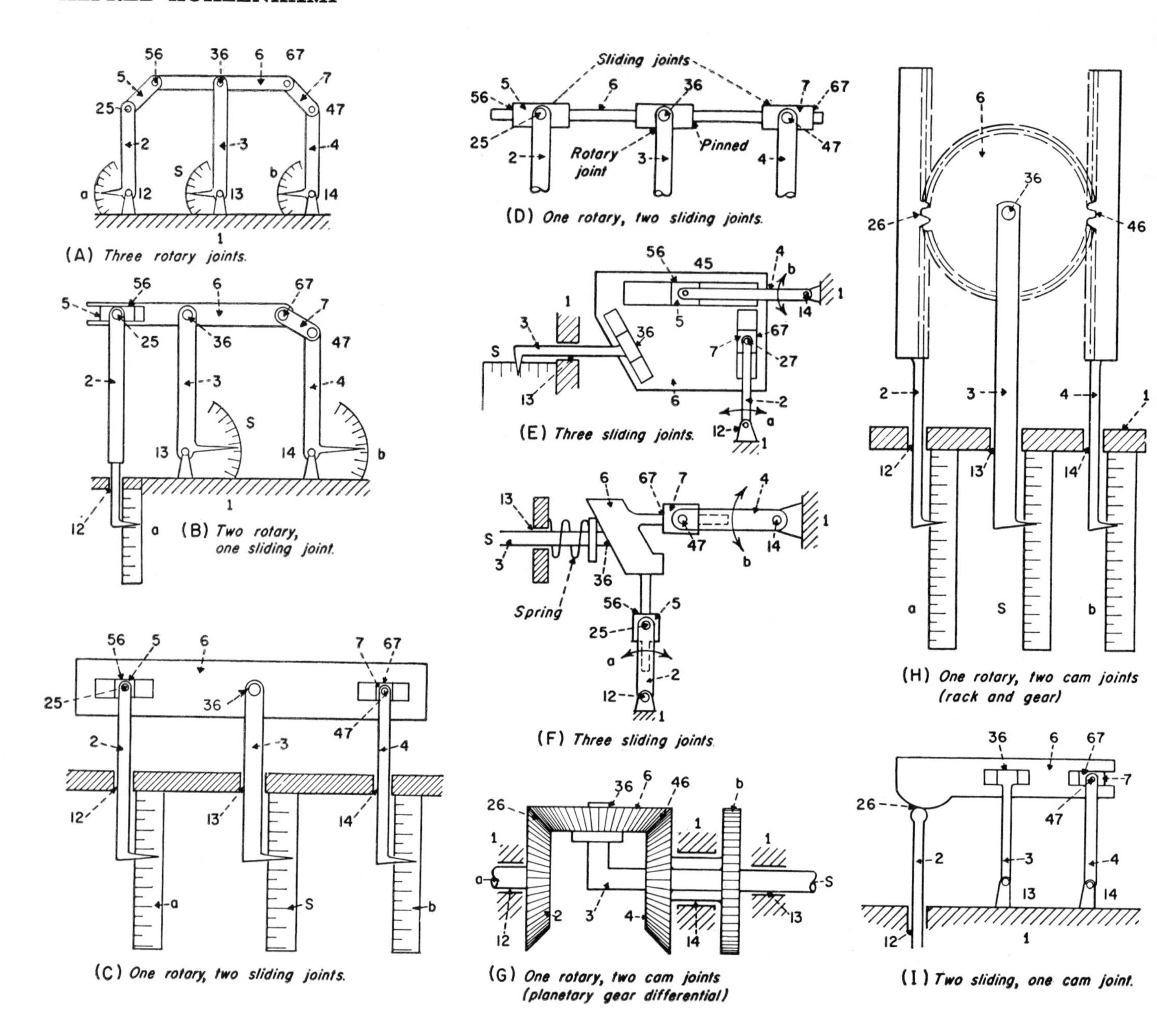

Fig. 1—Modifications of the differential linkage shown in Fig. 2(A) based on variations in the triple-jointed intermediate link 6.

Fig. 2—Two basic arrangements of a differential linkage which converts the independent movements of two input links 2 and 4 into one output motion, link 3. (A) is used as a basis for the family of mechanisms shown in Fig. 3;

The links are designated as follows:

Frame-links: links 2, 3, and 4; two-jointed intermediate links: links 5 and 6; three-jointed intermediate links: link 6.

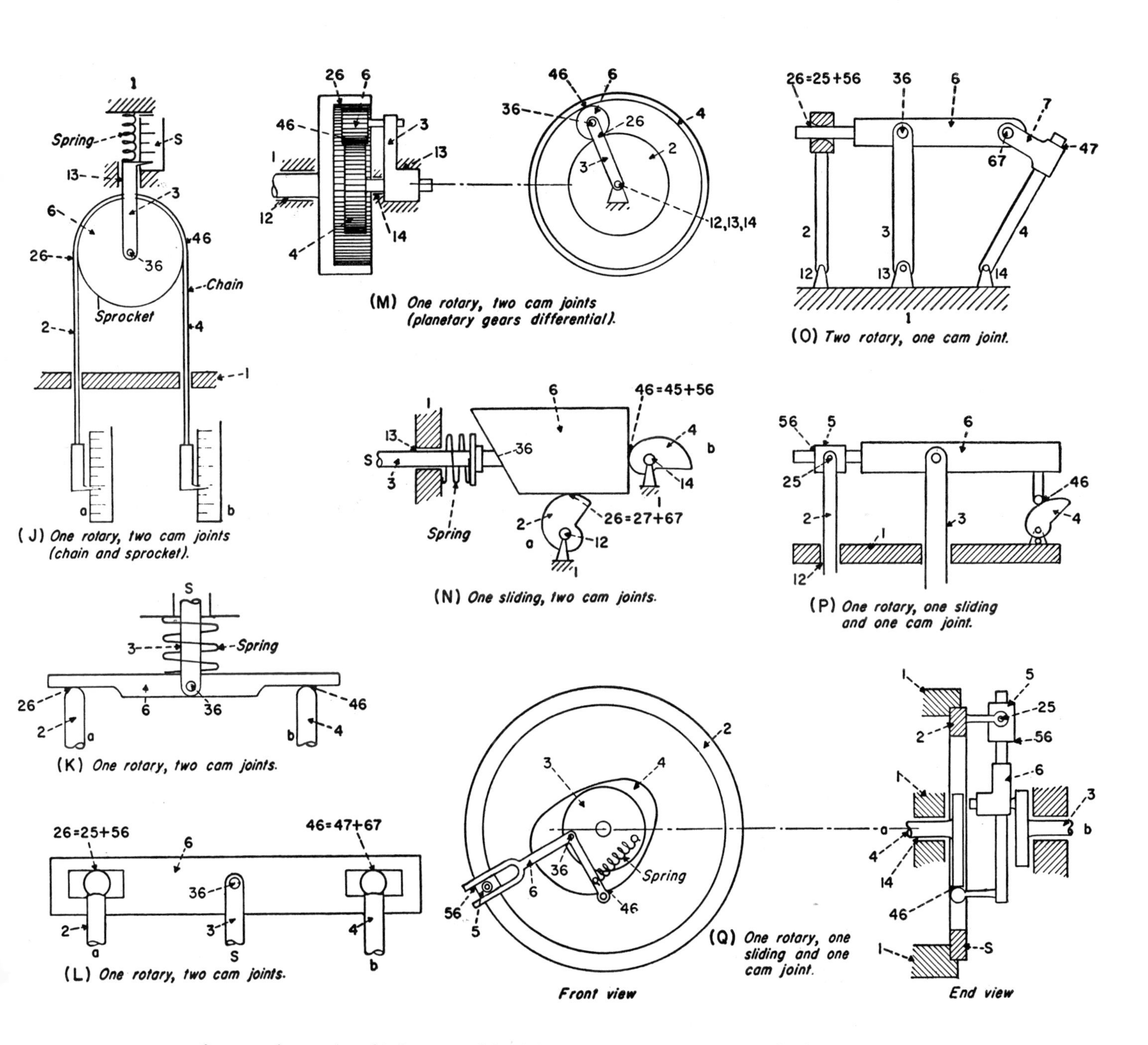

Input motions to be added are a and b; their sum s is equal to $c_1\ a + c_2\ b$ where c_1 and c_2 are scale factors. The links are numbered in the same manner as in Fig. 2(*A*).

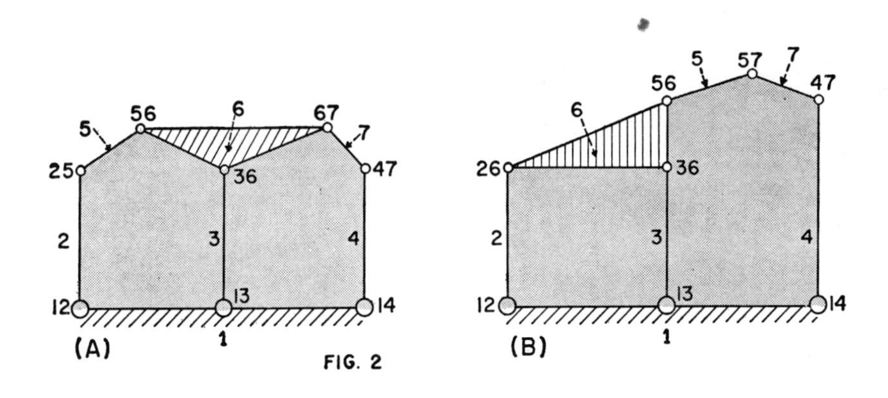

FIG. 2

AUTOMOTIVE DIFFERENTIAL

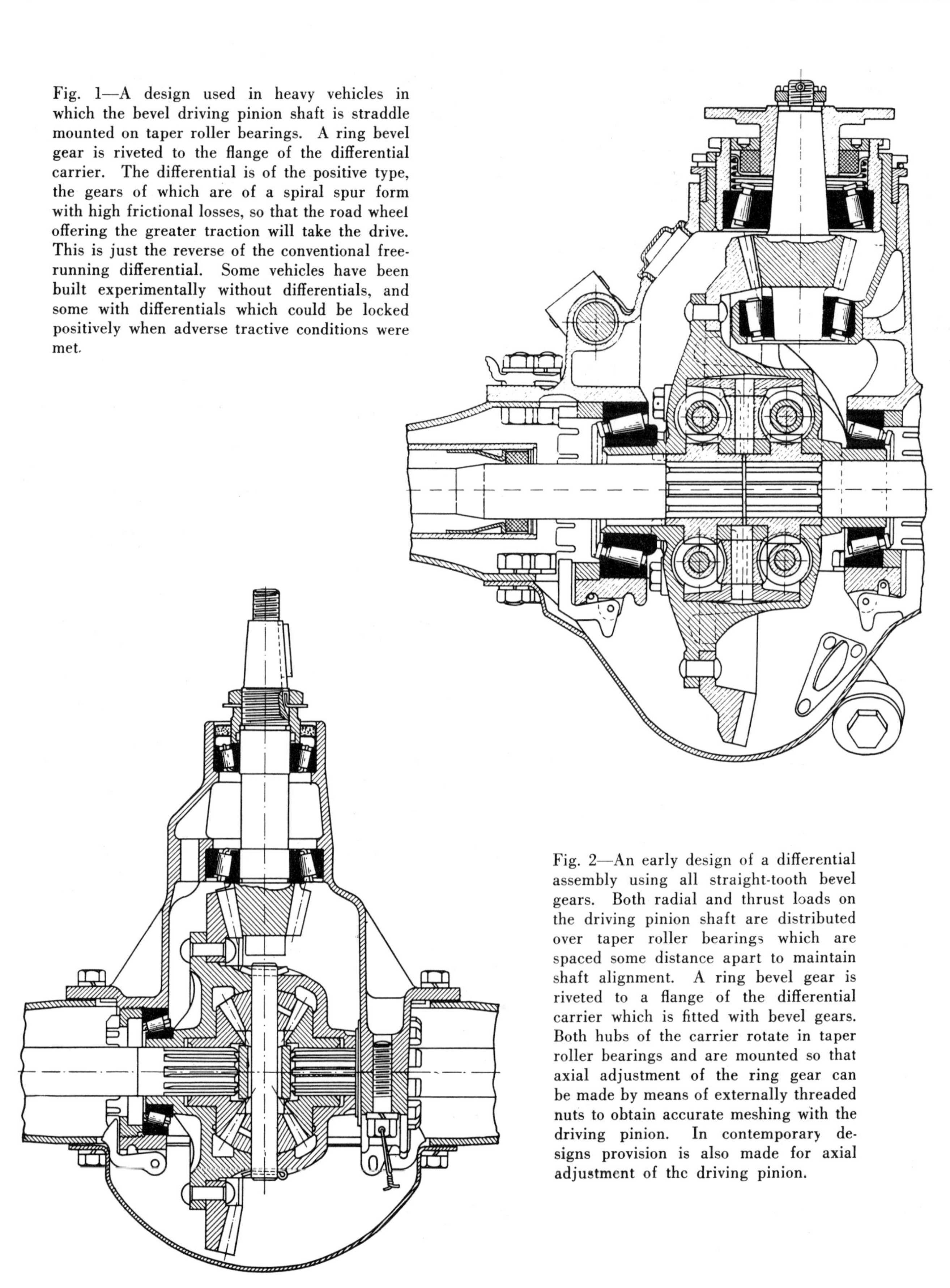

Fig. 1—A design used in heavy vehicles in which the bevel driving pinion shaft is straddle mounted on taper roller bearings. A ring bevel gear is riveted to the flange of the differential carrier. The differential is of the positive type, the gears of which are of a spiral spur form with high frictional losses, so that the road wheel offering the greater traction will take the drive. This is just the reverse of the conventional free-running differential. Some vehicles have been built experimentally without differentials, and some with differentials which could be locked positively when adverse tractive conditions were met.

Fig. 2—An early design of a differential assembly using all straight-tooth bevel gears. Both radial and thrust loads on the driving pinion shaft are distributed over taper roller bearings which are spaced some distance apart to maintain shaft alignment. A ring bevel gear is riveted to a flange of the differential carrier which is fitted with bevel gears. Both hubs of the carrier rotate in taper roller bearings and are mounted so that axial adjustment of the ring gear can be made by means of externally threaded nuts to obtain accurate meshing with the driving pinion. In contemporary designs provision is also made for axial adjustment of the driving pinion.

GEAR MECHANISMS

HERBERT CHASE

Examples of rear axle drives that have been used on passenger cars and trucks to distribute torque equally between the driving wheels and also to permit one wheel to turn faster than the other on curves

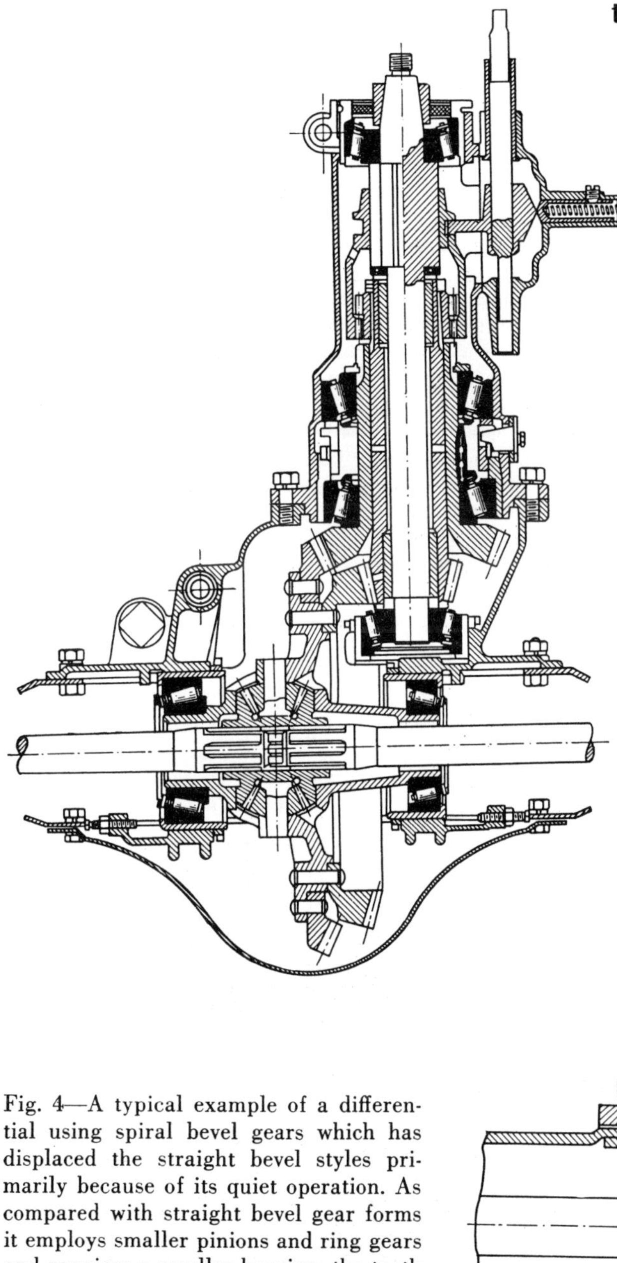

Fig. 3—Two-speed drive combined with differential, consisting of a selective driving clutch, two driving sleeves each integral with a bevel pinion, two ring bevel gears riveted to a flange of a differential carrier, and differential gears. At the forward end of the large pinion sleeve are cut splines, at the forward end of the small pinion sleeve a splined unit is keyed on, these form halves of a positive clutch arranged so as to mesh with a sliding selective driving member which is splined to the driving axle. With the larger pinion engaged the engine speed is lower for a given road speed than with the smaller pinion in engagement, thereby increasing the power and the fuel economy but reducing the ability to accelerate.

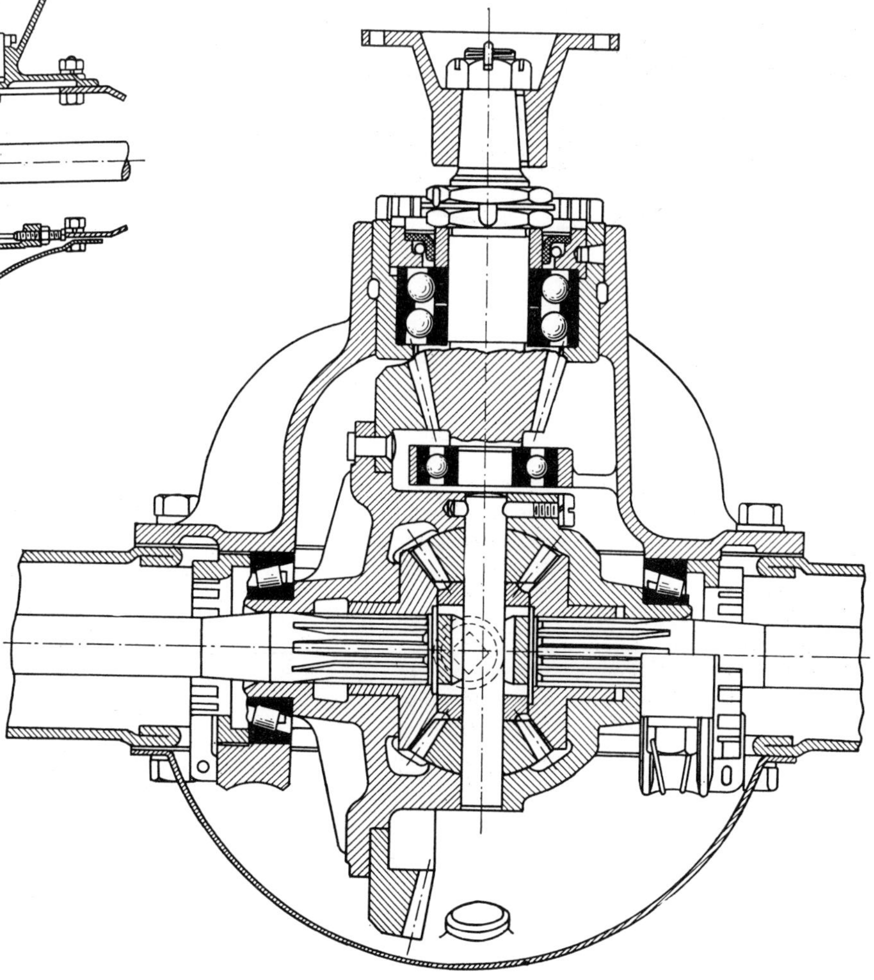

Fig. 4—A typical example of a differential using spiral bevel gears which has displaced the straight bevel styles primarily because of its quiet operation. As compared with straight bevel gear forms it employs smaller pinions and ring gears and requires a smaller housing, the tooth loading is higher but the teeth are stronger and the area of contact is greater. A rigid mounting is essential and is obtained by using an outboard bearing and a double-row ball bearing placed close to the pinion to take both thrust and radial loads. A rigid differential carrier is riveted to the ring gear. The carrier is mounted in tapered roller bearings.

Further examples of rear-axle gearing that has been used on passenger cars and trucks for transmitting power and distributing torque equally to both driving wheels

HERBERT CHASE

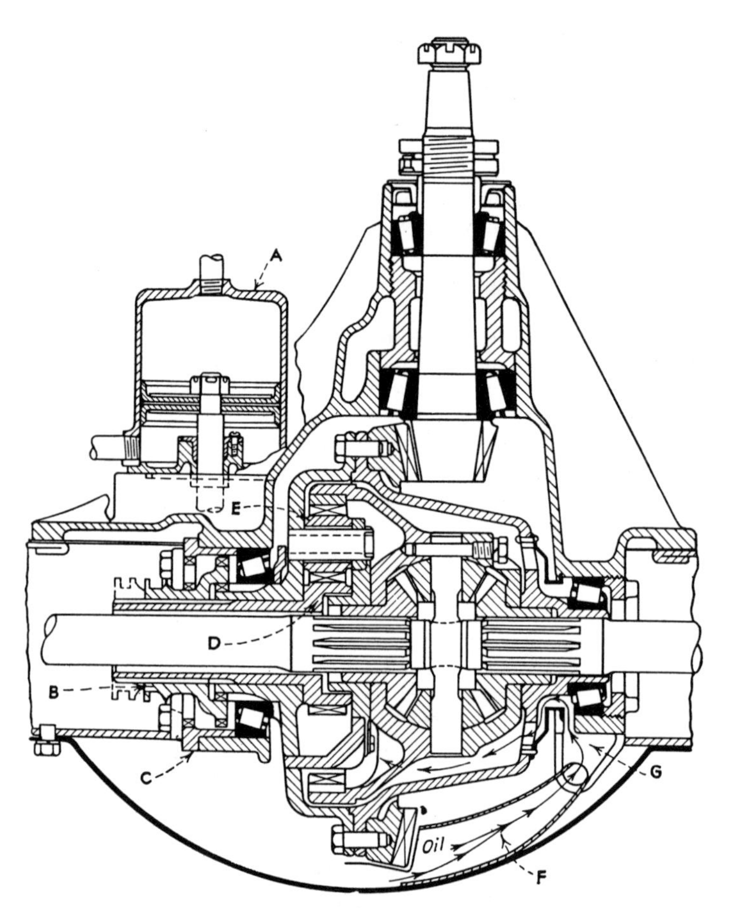

Dual-ratio rear axle as used on 8- and 12-cylinder Auburns. The change in ratio is obtained by using or locking out a planetary gearset built into the unit and is effected by motion of piston in vacuum cylinder *A* which is connected to intake manifold of engine through control valve on instrument panel of car. Piston operates positive clutch *B*. When clutch is at right, inside teeth engage with teeth on differential case, to which crown gear is fastened, and lock overdrive system. When at left, outer teeth engage fixed member *C*, holding control gear *D* stationary. Planet pinions *E* then revolve around *D*, increasing the speed of the differential housing to which the axle shafts are splined. Oil is forced through scoop *F* into space *G* and, after traveling along path indicated by arrows, lubricates planetary gears.

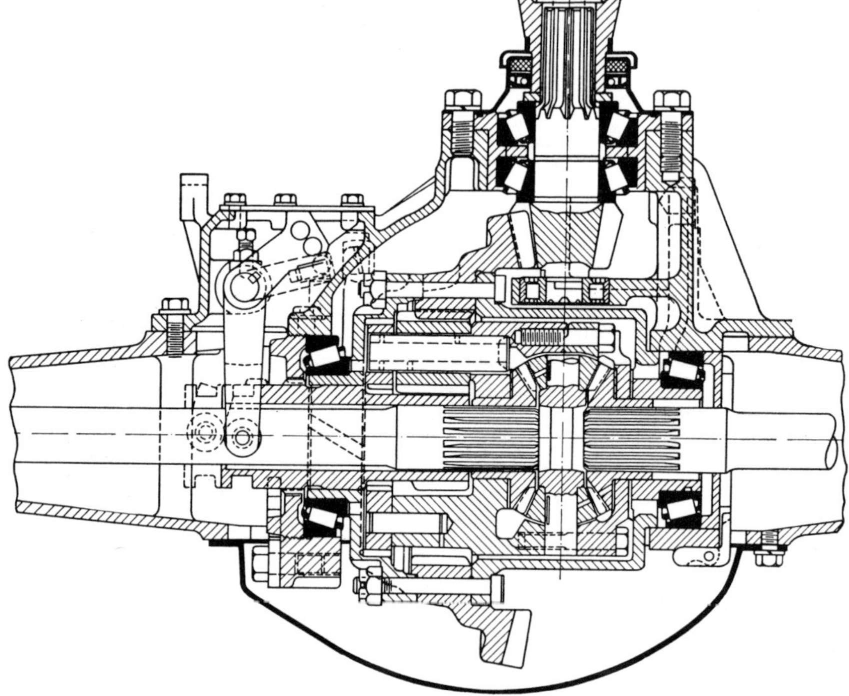

Two-speed bevel-gear axle (Eaton) which provides ratios of 5.14 and 7.15 to 1. The extra speed between the crown gear and the differential is produced by a simple planetary train of which the internal gear is integral with the crown gear and is made of 3.5 per cent nickel steel. Four planet pinions are used. They mesh with a sun gear which slides into or out of mesh with teeth on the planet carrier, locking the epicyclic train and causing it to turn as a solid mass.

Chapter 7 SCREW AND CAM DEVICES

7 Special screw

How differential, duplex, and other types of screws can provide slow and fast feeds, minute adjustments, and strong clamping action.

LOUIS DODGE, Consulting Engineer, New Richmond, Ohio

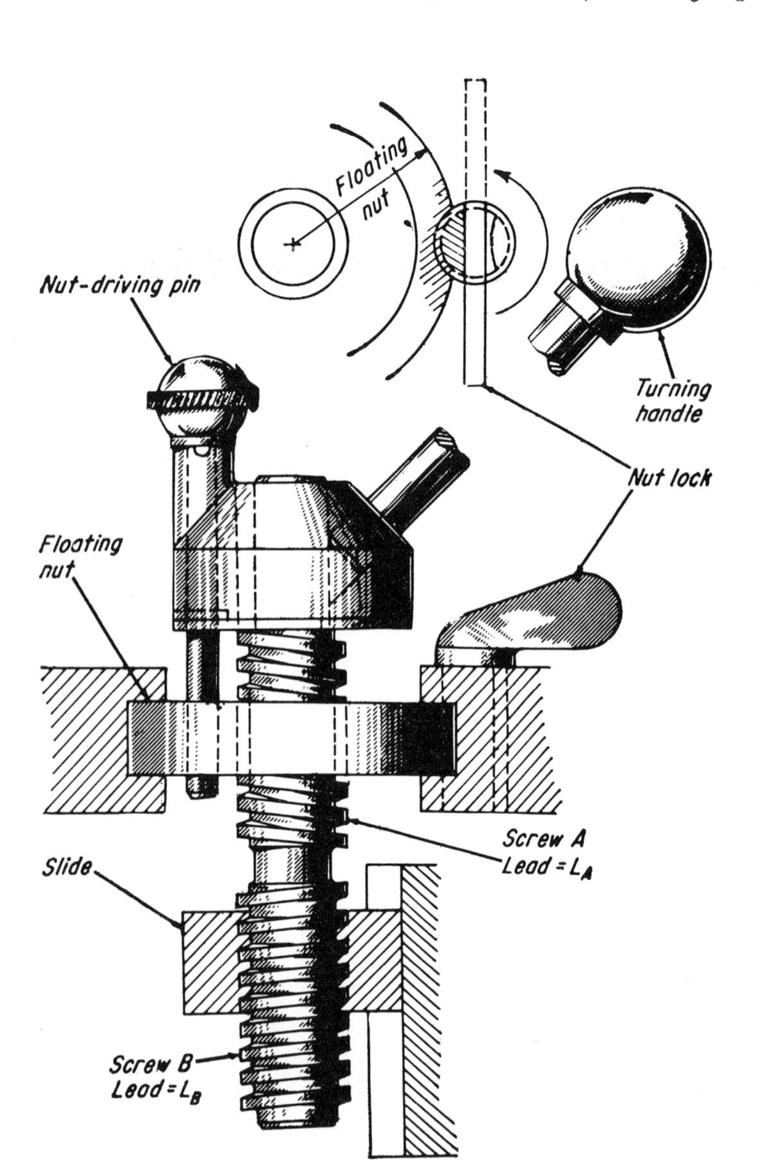

1 **RAPID AND SLOW FEED.** With left- and right-hand threads, slide motion with nut locked equals L_A plus L_B per turn; with nut floating, slide motion per turn equals L_B. Get extremely fine feed with rapid return motion when threads are differential.

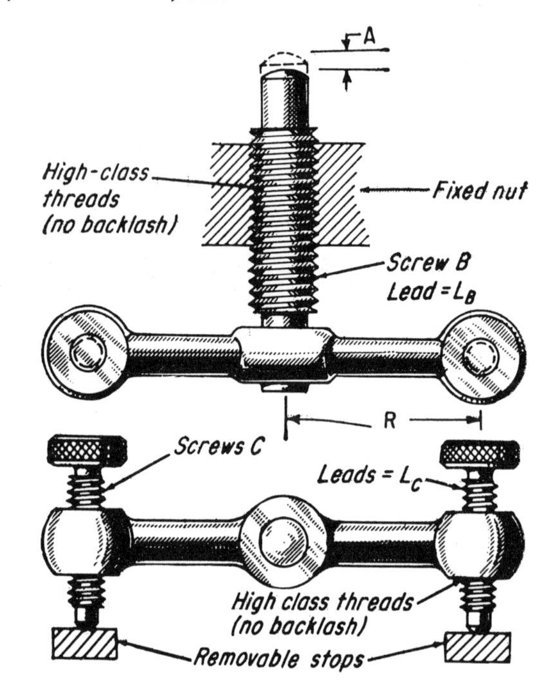

2 **EXTREMELY SMALL MOVEMENTS.** Microscopic measurements, for example, are characteristic of this arrangement. Movement A is equal to $N(L_B \times L_C)/2\pi R$, where N equals number of turns of screw C.

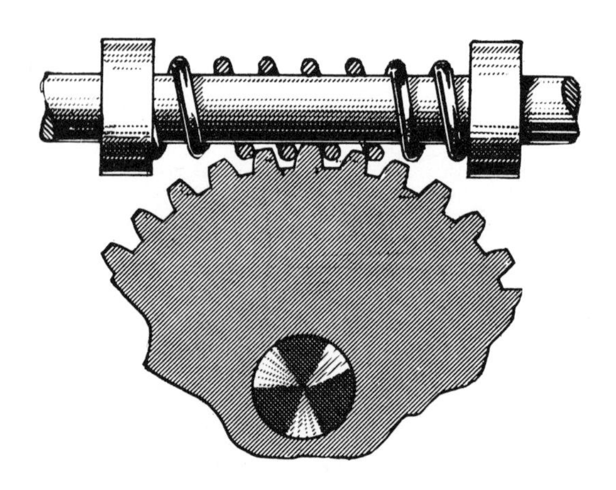

5 **SHOCK ABSORBENT SCREW.** When springs coiled as shown are used as worm drives for light loads, they have the advantage of being able to absorb heavy shocks.

arrangements

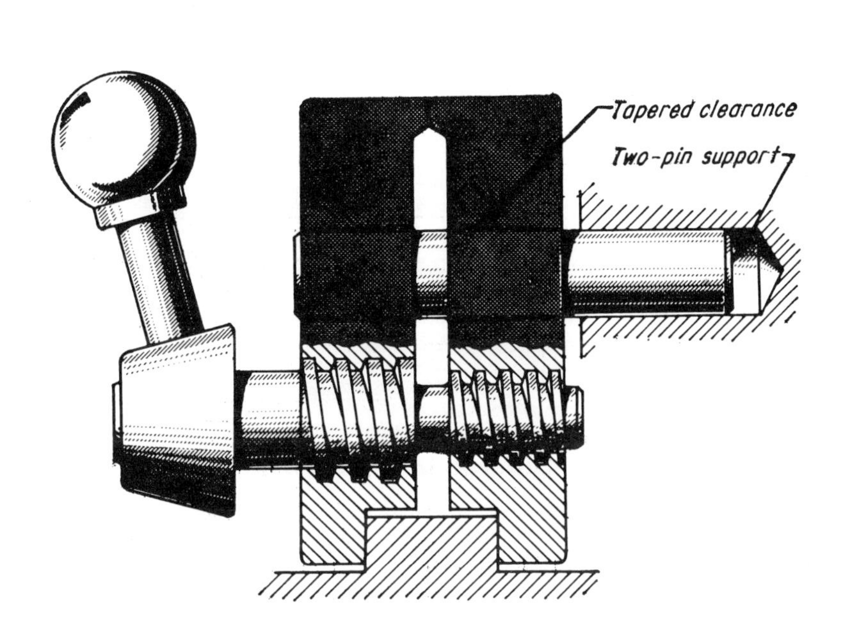

3 **DIFFERENTIAL CLAMP.** This method of using a differential screw to tighten clamp jaws combines rugged threads with high clamping power. Clamping pressure, $P = Te/[R(\tan\phi + \tan\alpha]$, where T = torque at handle, R = mean radius of screw threads, ϕ = angle of friction (approx. 0.1), α = mean pitch angle of screw, and e = efficiency of screw (generally about 0.8).

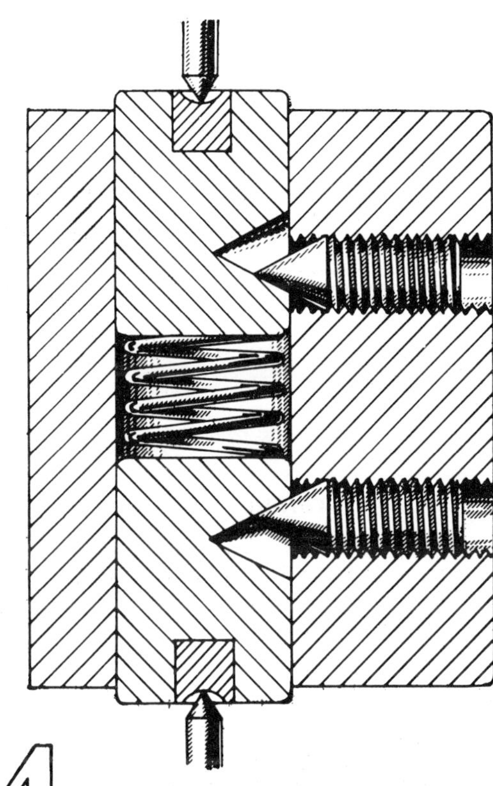

4 **BEARING ADJUSTMENT.** This screw arrangement is a handy way of providing for bearing adjustment and overload protection.

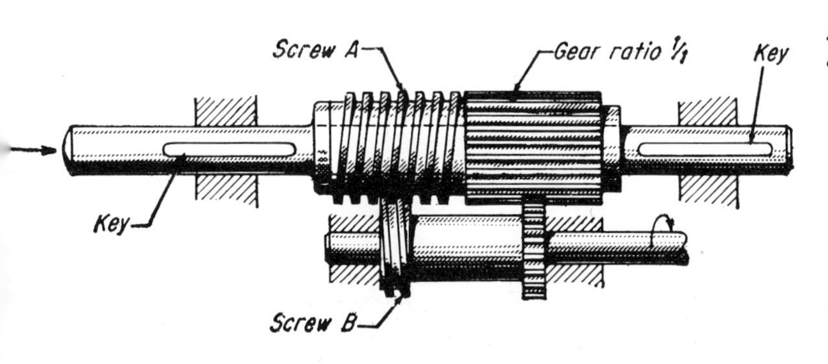

6 **HIGH REDUCTION** of rotary motion to fine linear motion is possible here. Arrangement is for low forces. Screws are left and right hand. $L_A = L_B$ plus or minus a small increment. When $L_B = 1/10$ and $L_A = 1/10.05$ the linear motion of screw A will be 0.05 in. per turn. When screws are the same hand, linear motion equals $L_A + L_B$.

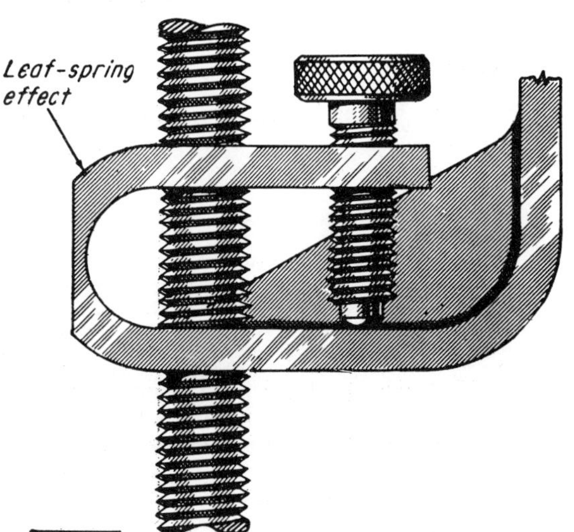

7 **BACKLASH ELIMINATION.** The large screw is locked and all backlash is eliminated when the knurled screw is tightened – finger torque is sufficient.

10 WAYS TO EMPLOY SCREW

Three basic components of screw mechanisms are: actuating member (knob, wheel, handle), threaded device (screw-nut set) and sliding device (plunger-guide set).

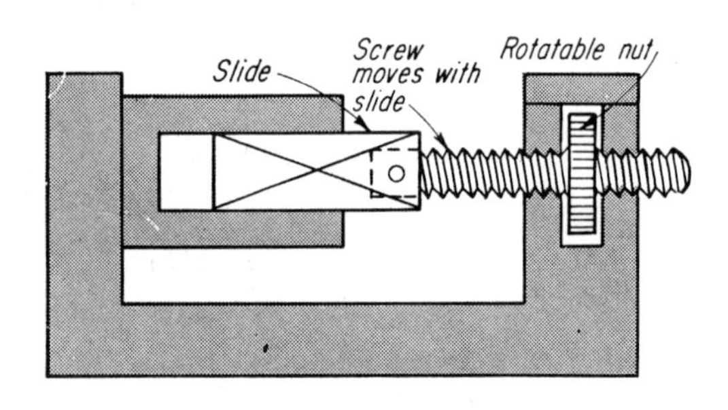

Nut can rotate . . .

but will not move longitudinally. Typical applications: screw jacks, heavy vertically-moved doors; floodgates, opera-glass focusing, vernier gages, Stillson wrenches.

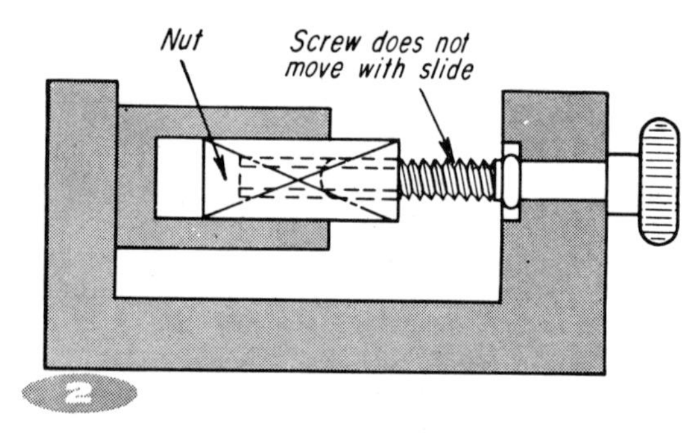

Screw can rotate . . .

but only the nut moves longitudinally. Typical applications: lathe tailstock feed, vises, lathe apron.

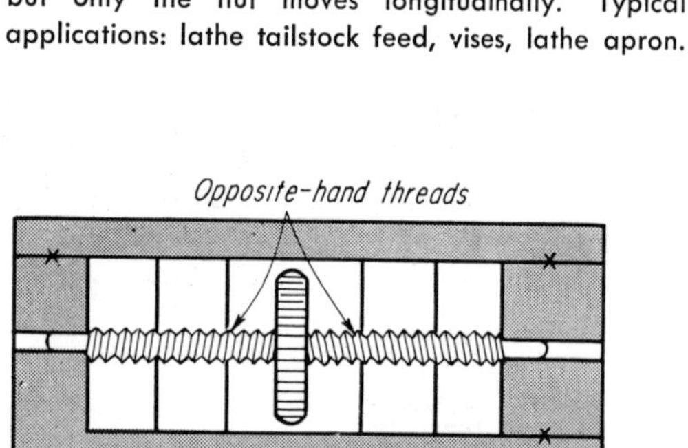

Opposing movement . . .

of lateral slides; adjusting members or other screw-actuated parts can be achieved with opposite-hand threads.

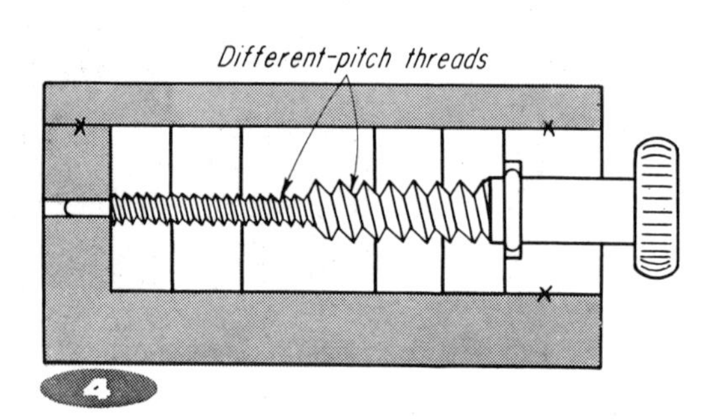

Differential movement . . .

is given by threads of different pitch. When screw is rotated the nuts move in same direction but at different speeds.

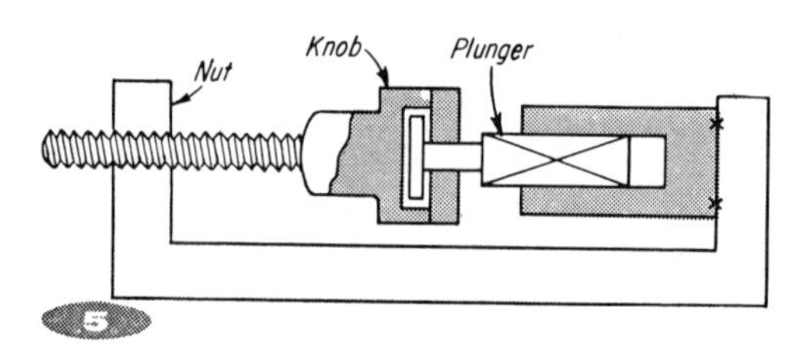

Screw and plunger . . .

are attached to knob. Nut and guide are stationary. Used on: screw presses, lathe steady-rest jaws for adjustment, shaper slide regulation.

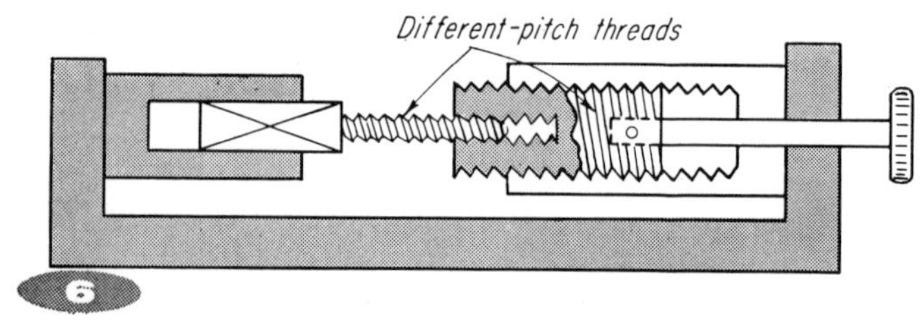

Concentric threading . . .

also gives differential movement. Such movements are useful wherever rotary mechanical action is required. Typical example is a gas-bottle valve, where slow opening is combined with easy control.

MECHANISMS

FEDERICO STRASSER

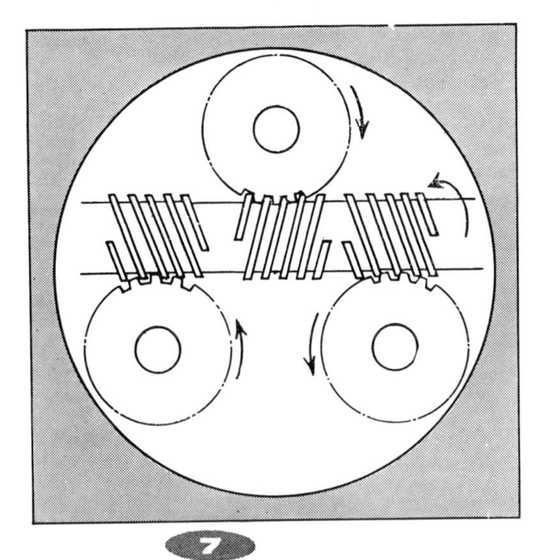

7

One screw actuates three gears . . .

simultaneously. Axes of gears are at right angles to that of screw. This type of mechanism can sometimes replace more expensive gear setups where speed reduction and multi-output from single input is required.

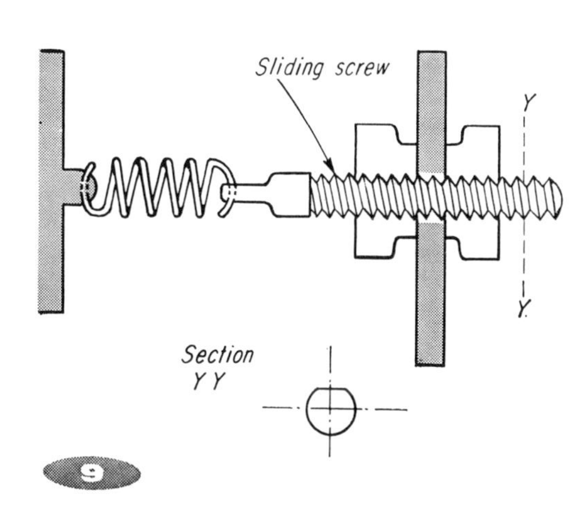

9

Locking nuts . . .

are often placed on opposite sides of a panel, to prevent axial screw movement and simultaneously lock against vibrations. Drill-press depth stops, and adjustable stops for shearing and cutoff dies are some examples.

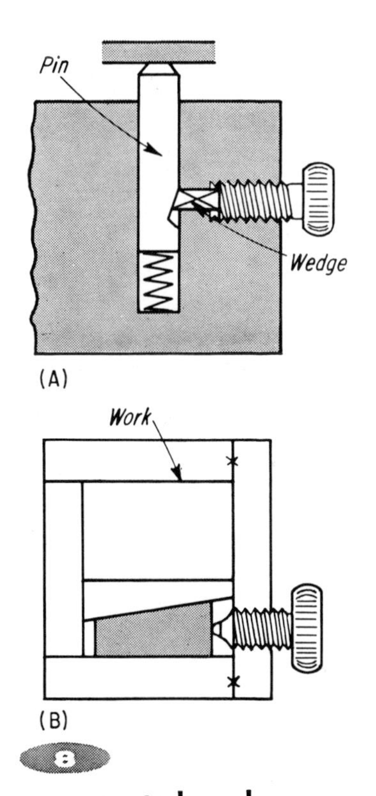

8

Screw-actuated wedges . . .

lock locating pin (A) and hold work in fixture (B). These are just two of many tool and diemaking applications for these screw actions.

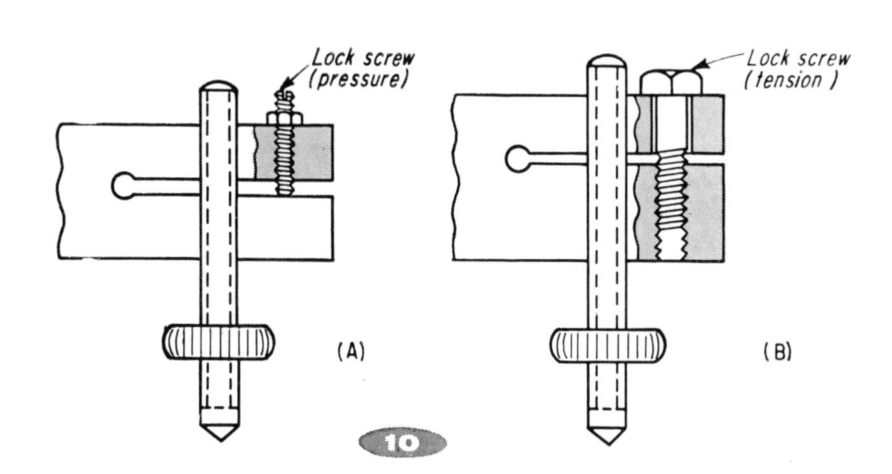

10

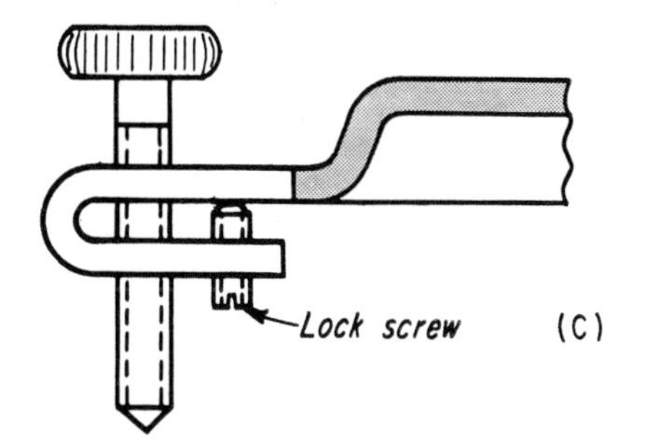

Adjustment screws . . .

are effectively locked by either a pressure screw (A) or tension screw (B). If the adjusting screw is threaded into a formed sheet-metal component (C) a setscrew can be used to lock the adjustment.

20 DYNAMIC APPLICATIONS FOR SCREW THREADS

KURT RABE, *consulting engineer*
Berlin, Germany

You need a threaded shaft, a nut . . . plus some way for one of these members to rotate without translating and the other to translate without rotating. That's all. Yet these simple components can do practically all of the adjusting, setting, or locking used in design.

Most such applications have low-precision requirements. That's why the thread may be a coiled wire or a twisted strip; the nut may be a notched ear on a shaft or a slotted disk. Standard screws and nuts right off your supply shelves can often serve at very low cost.

Here are the basic motion transformations possible with screw threads (Fig 1):

- transform rotation into linear motion or reverse (A),
- transform helical motion into linear motion or reverse (B),
- transform rotation into helical motion or reverse (C).

Of course the screw thread may be combined with other components: in a 4-bar linkage (Fig 2), or with multiple screw elements for force or motion amplification.

continued, next page

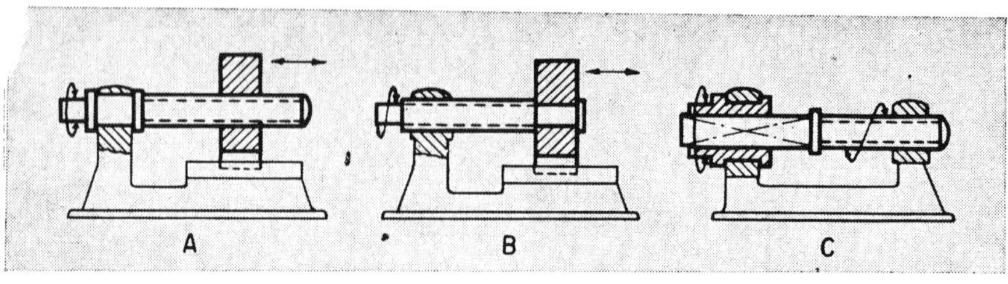

1 MOTION TRANSFORMATIONS of a screw thread include: rotation to translation (A), helical to translation (B), rotation to helical (C). Any of these is reversible if the thread is not self-locking (see screw-thread mathematics on following page—thread is reversible when efficiency is over 50%).

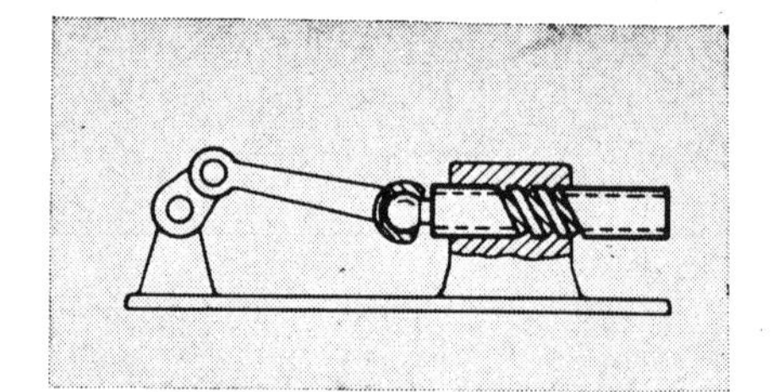

2 STANDARD 4-BAR LINKAGE has screw thread substituted for slider. Output is helical rather than linear.

A REVIEW OF

a — friction angle, $\tan a = f$
r — mean radius of thread
$= \frac{1}{2}$ (root radius + outside radius), in inches
l — lead, thread advance in one revolution, in.
b — lead angle, $\tan b = l/2\pi r$, deg
f — friction coefficient
P — equivalent driving force at radius r from screw axis, lb
L — axial load, lb
e — efficiency
c — half angle between thread faces, deg

SQUARE THREADS:

$$P = L \tan (b \pm a) = L \frac{(1 \pm 2\pi r f)}{(2\pi r \mp f l)}$$

Where upper signs are for motion opposed in direction to L. Screw is self-locking when $b \leqq a$.

$$e = \frac{\tan b}{\tan (b + a)} \quad \text{(motion opposed to } L)$$

$$e = \frac{\tan (b - a)}{\tan b} \quad \text{(motion assisted by } L)$$

V THREADS:

$$P = L \frac{(l \pm 2\pi r f \sec c)}{(2\pi r \mp l f \sec c)}$$

$$e = \frac{\tan b\,(1 - f \tan b \sec c)}{(\tan b + f \sec c)} \quad \text{(motion opposed to } L)$$

$$e = \frac{\tan b - f \sec c}{\tan b\,(1 + f \tan b \sec c)} \quad \text{(motion assisted by } L)$$

For more detailed analysis of screw-thread friction forces, see Marks *Mechanical Engineers' Handbook*, McGraw-Hill Book Co.

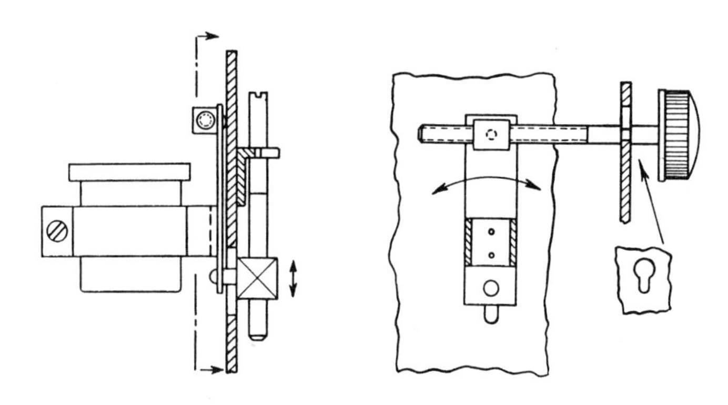

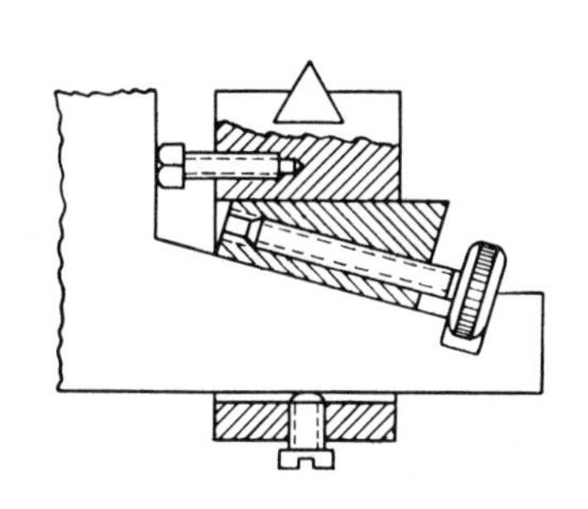

3 TWO-DIRECTIONAL LAMP ADJUSTMENT with screwdriver to move lamp up and down. Knob adjust (right) rotates lamp about pivot.

4 KNIFE-EDGE BEARING is raised or lowered by screw-driven wedge. Two additional screws locate the knife edge laterally and lock it.

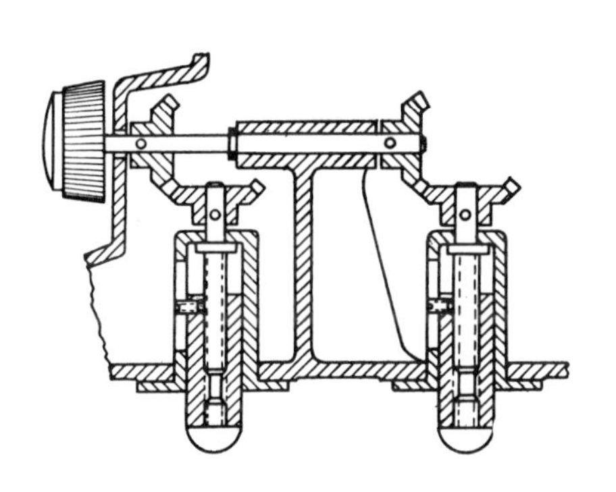

5 SIDE-BY-SIDE ARRANGEMENT of tandem screw threads gives parallel rise in this height adjustment for projector.

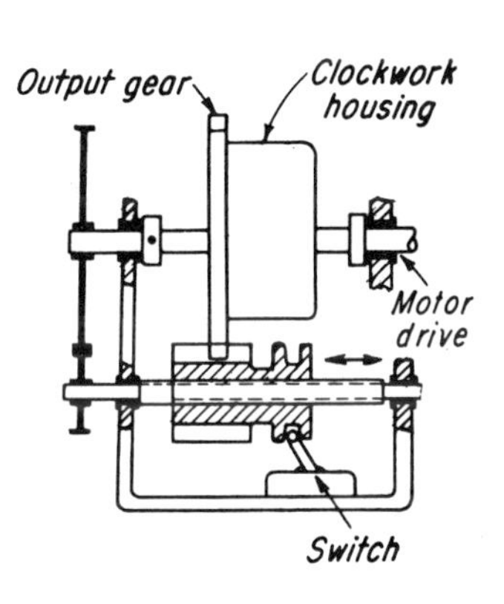

6 AUTOMATIC CLOCKWORK is kept wound tight by electric motor turned on and off by screw thread and nut. Note motor drive must be self-locking or it will permit clock to unwind as soon as switch turns off.

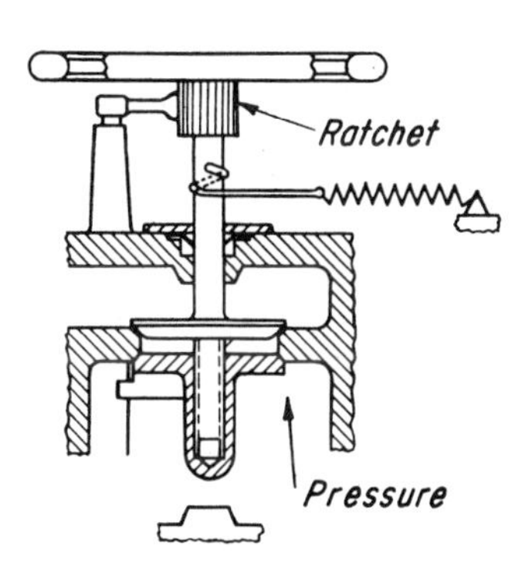

7 VALVE STEM has two oppositely moving valve cones. When opening, the upper cone moves up first, until it contacts its stop. Further turning of the valve wheel forces the lower cone out of its seat. The spring is wound up at the same time. When the ratchet is released, spring pulls both cones into their seats.

SCREW-THREAD MATHEMATICS

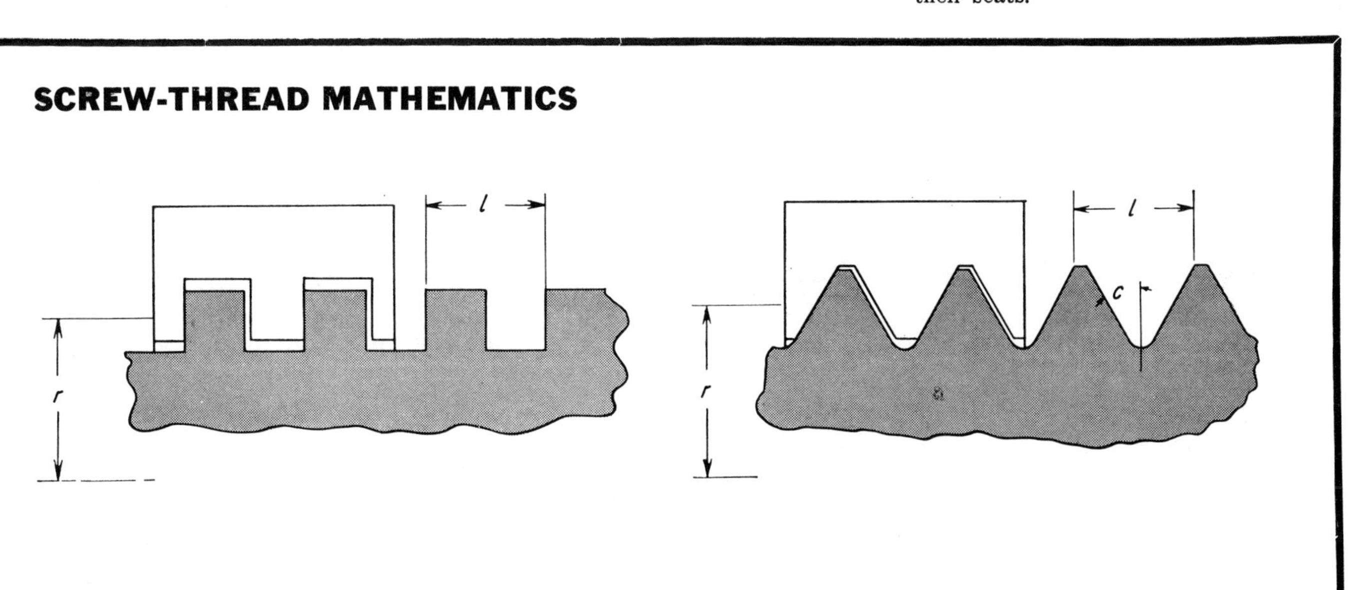

Translation to Rotation

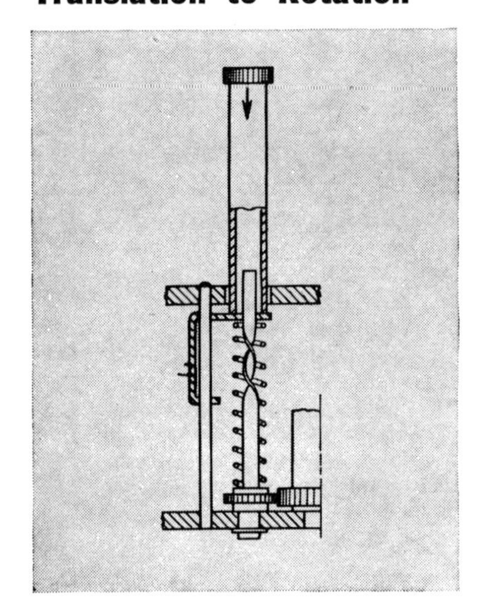

8 A METAL STRIP or square rod may be twisted to make a long-lead thread, ideal for transforming linear into rotary motion. Here a push-button mechanism winds a camera. Note that the number of turns or dwell of output gear is easily altered by changing (or even reversing) twist of the strip.

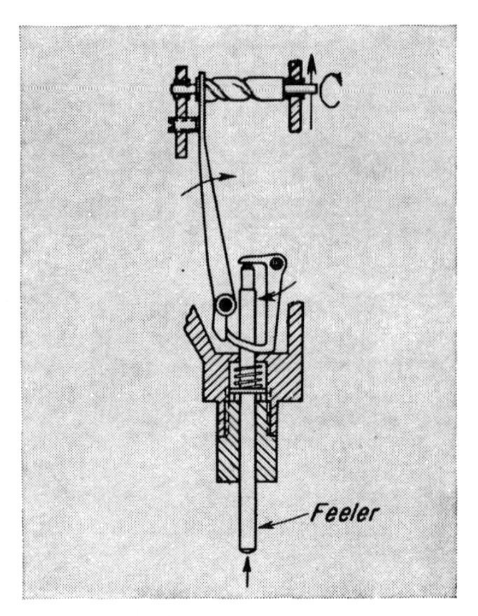

9 FEELER GAGE has its motion amplified through a double linkage and then transformed to rotation for dial indication.

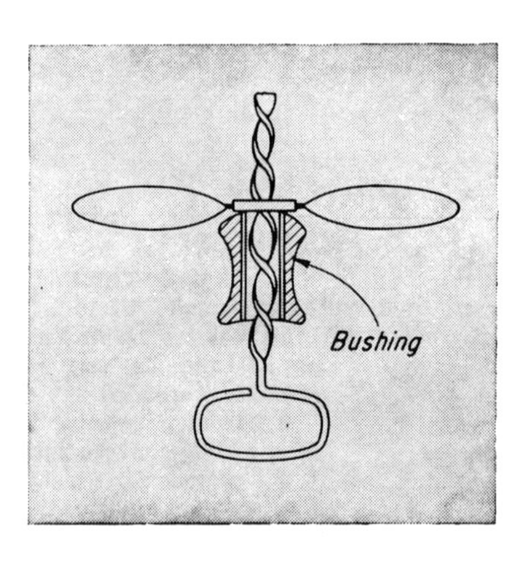

10 THE FAMILIAR flying propeller-toy is operated by pushing the bushing straight up and off the thread.

Self-Locking

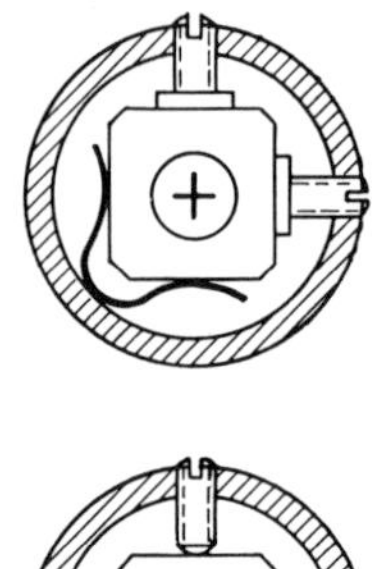

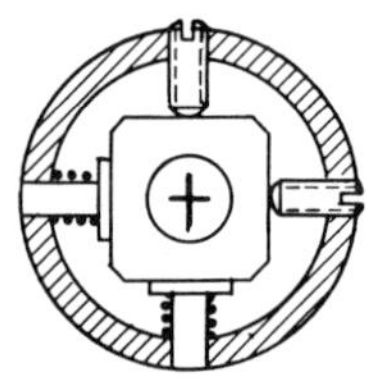

11 HAIRLINE ADJUSTMENT for a telescope, with two alternative methods of drive and spring return.

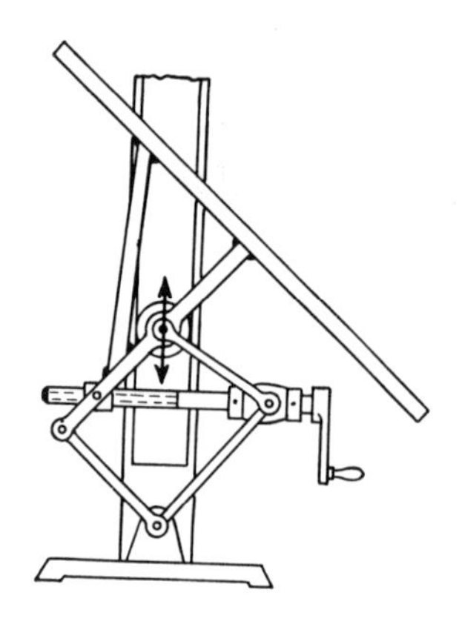

12 SCREW AND NUT provide self-locking drive for a complex linkage.

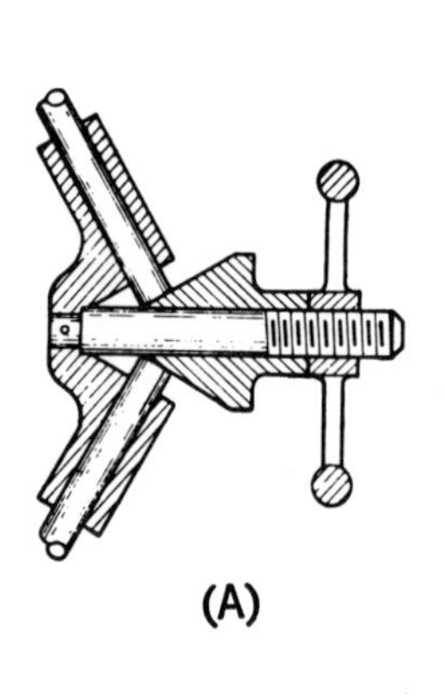

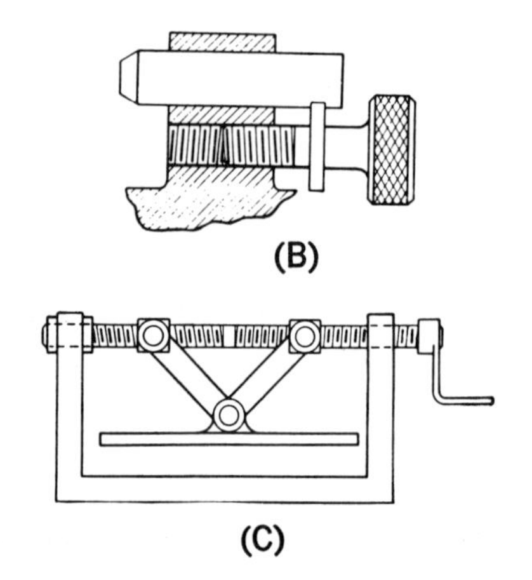

FORCE TRANSLATION. Threaded handle in (A) drives coned bushing which thrusts rods outwardly for balanced pressure. Screw in (B) retains and drives dowel pin for locking applications. Right- and left-handed shaft (C) actuates press.

Double Threading

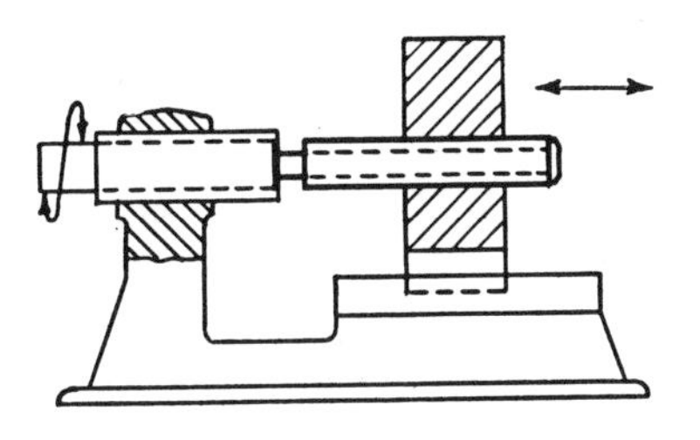

13 DOUBLE-THREADED SCREWS, when used as differentials, provide very fine adjustment for precision equipment at relatively low cost.

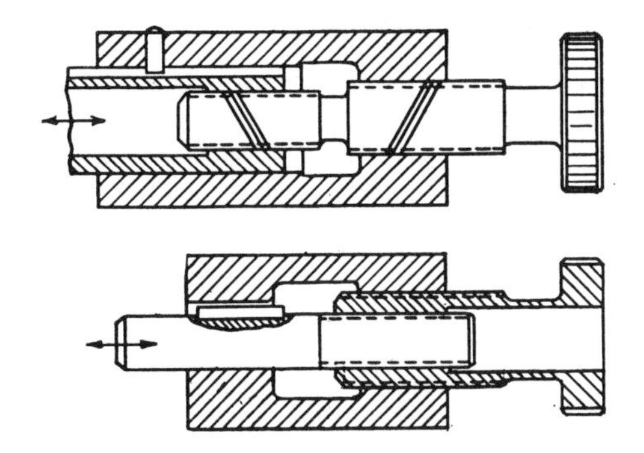

14 DIFFERENTIAL SCREWS can be made in dozens of forms. Here are two methods: above, two opposite-hand threads on a single shaft; below, same hand threads on independent shafts.

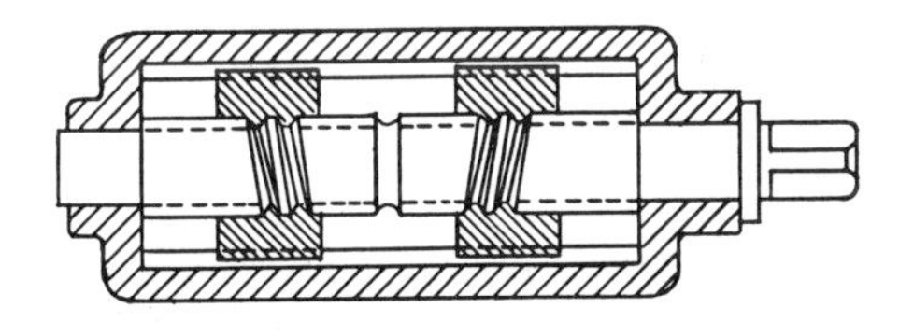

15 OPPOSITE-HAND THREADS make a high-speed centering clamp out of two moving nuts.

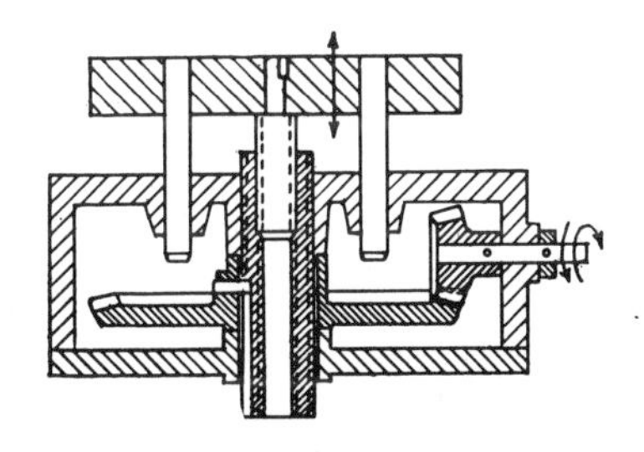

16 MEASURING TABLE rises very slowly for many turns of the input bevel gear. If the two threads are 1½—12 and ¾—16, in the fine-thread series, table will rise approximately 0.004 in. per input-gear revolution.

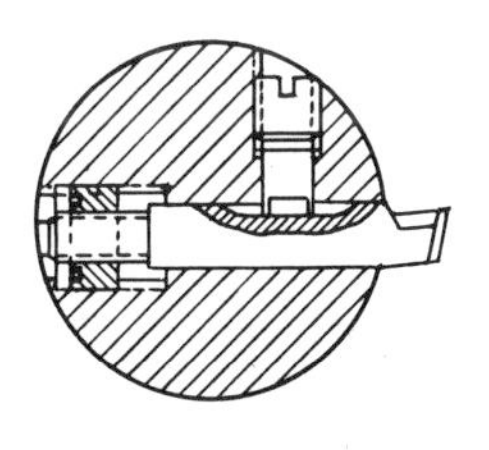

17 LATHE TURNING TOOL in drill rod is adjusted by differential screw. A special double-pin wrench turns the intermediate nut, advancing the nut and retracting the threaded tool simultaneously. Tool is then clamped by setscrew.

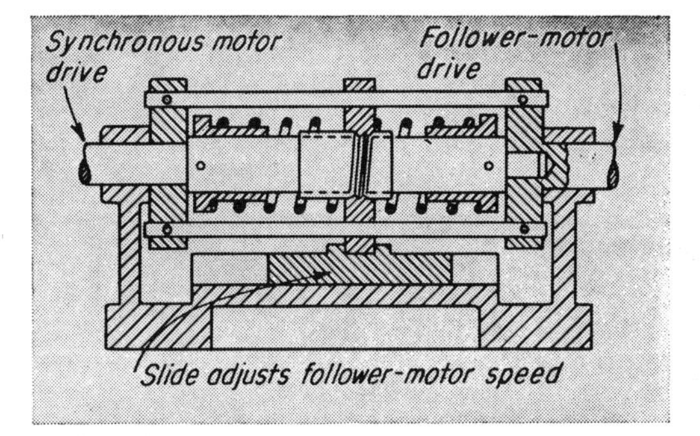

18 ANY VARIABLE-SPEED MOTOR can be made to follow a small synchronous motor by connecting them to the two shafts of this differential screw. Differences in number of revolutions between the two motors appear as motion of the traveling nut and slide so an electrical speed compensation is made.

19 (left) **A WIRE FORK** is the nut in this simple tube-and screw design.

20 (below) **A MECHANICAL PENCIL** includes a spring as the screw thread and a notched ear or a bent wire as the nut.

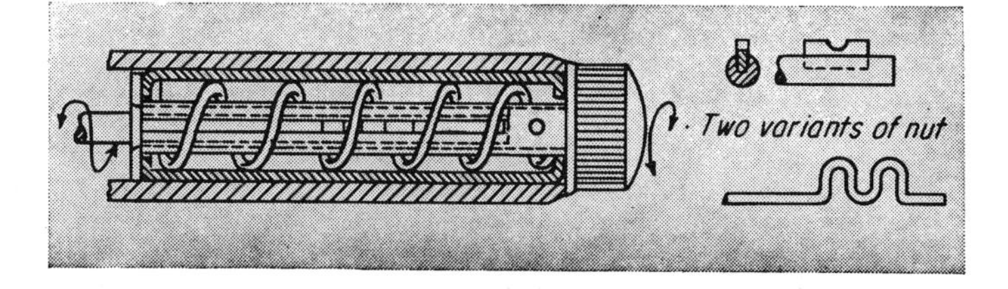

How to Provide for Backlash

These illustrations are based on two general methods of providing for lost motion or backlash. One allows for relative movement of the nut and screw in the plane

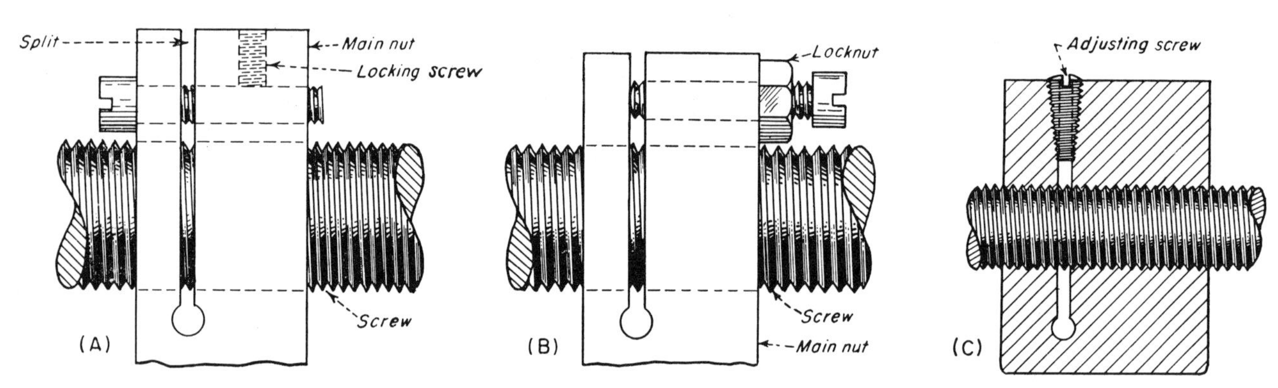

THREE METHODS of using slotted nuts. In *(A)*, nut sections are brought closer together to force left-hand nut flanks to bear on right-hand flanks of screw thread and vice versa. In *(B)*, and *(C)* nut sections are forced apart for same purpose.

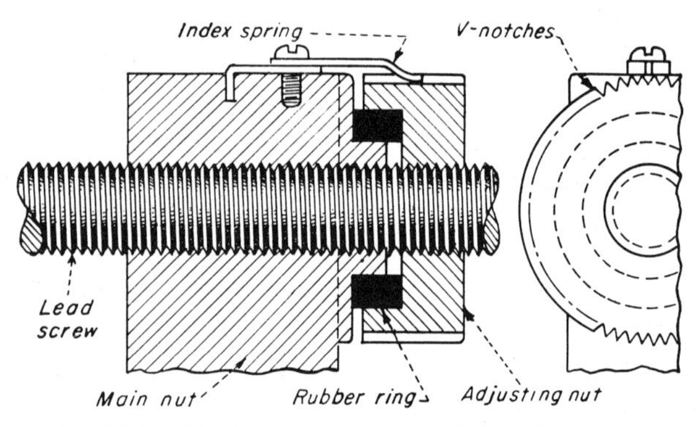

AROUND THE PERIPHERY of the backlash-adjusting nut are "v" notches of small pitch which engage the index spring. To eliminate play in the lead screw, adjusting nut is turned clockwise. Spring and adjusting nut can be calibrated for precise use.

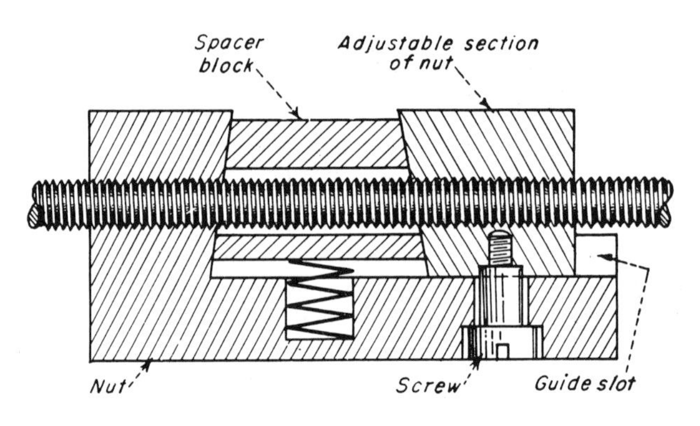

SELF-COMPENSATING MEANS of removing backlash. Slot is milled in nut for an adjustable section which is locked by a screw. Spring presses the tapered spacer block upwards, forcing the nut elements apart, thereby taking up backlash.

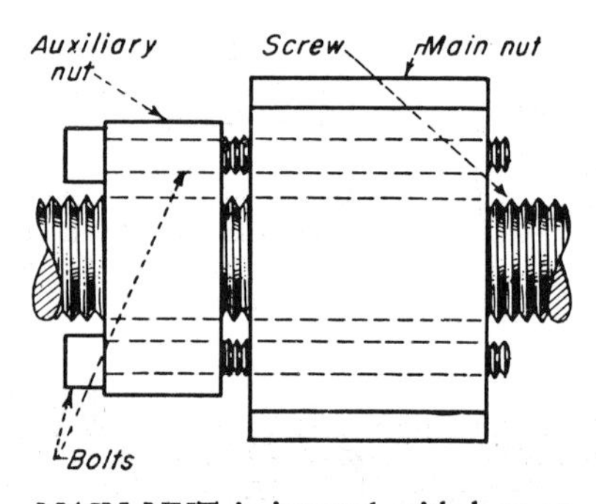

MAIN NUT is integral with base attached to part moved by screw. Auxiliary nut is positioned one or two pitches from main nut. The two are brought closer together by bolts which pass freely through the auxiliary nut.

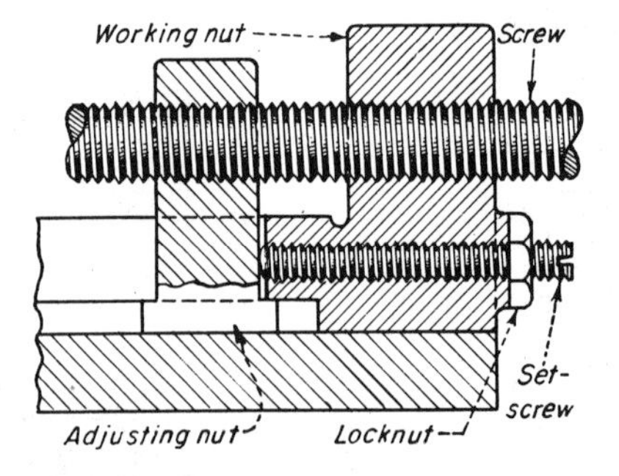

ANOTHER WAY to use an auxiliary or adjusting nut for axial adjustment of backlash. Relative movement between the working and adjusting nuts is obtained manually by the set screw which can be locked in place as shown.

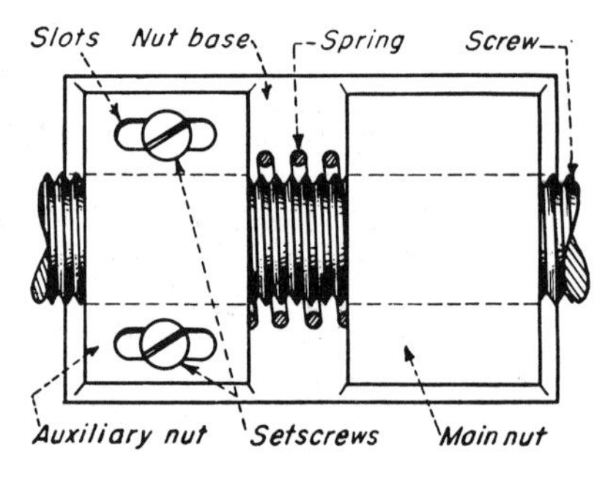

COMPRESSION SPRING placed between main and auxiliary nuts exerts force tending to separate them and thus take up slack. Set screws engage nut base and prevent rotation of auxiliary nut after adjustment is made.

in Threaded Parts

CLIFFORD T. BOWER
London, England

parallel to the thread axis; the other method involves a radial adjustment to compensate for clearance between sloping faces of the threads on each element.

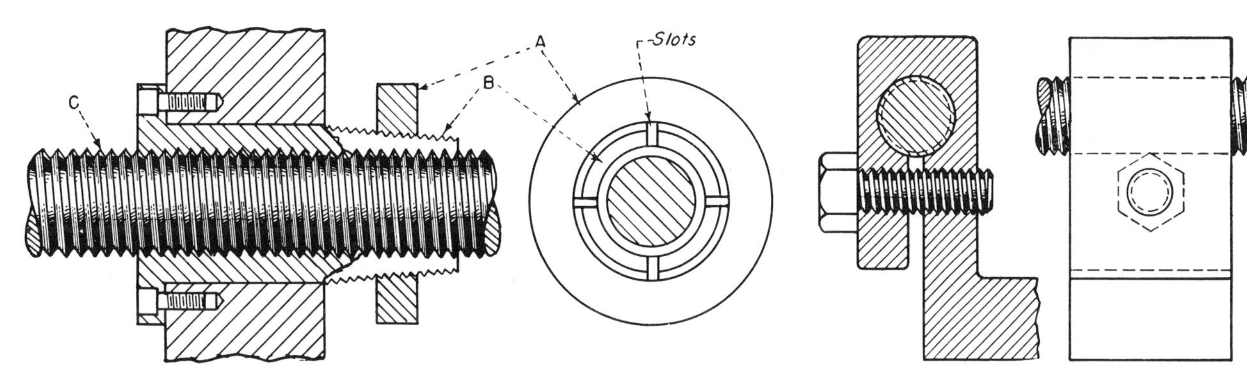

NUT *A* IS SCREWED along the tapered round nut, *B*, to eliminate backlash or wear between *B* and *C*, the main screw, by means of the four slots shown.

ANOTHER METHOD of clamping a nut around a screw to reduce radial clearance.

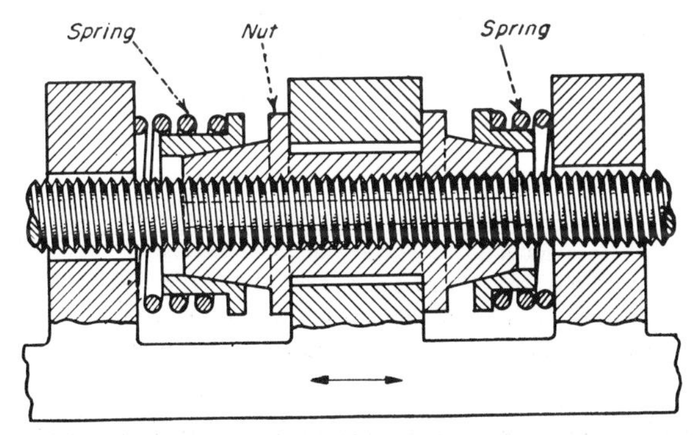

AUTOMATIC ADJUSTMENT for backlash. Nut is flanged on each end, has a square outer section between flanges and slots cut in the tapered sections. Spring forces have components which push slotted sections radially inward.

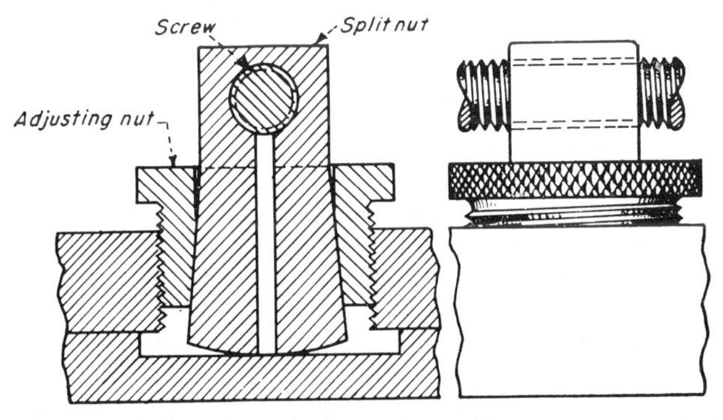

SPLIT NUT is tapered and has a rounded bottom to maintain as near as possible a fixed distance between its seat and the center line of the screw. When the adjusting nut is tightened, the split nut springs inward slightly.

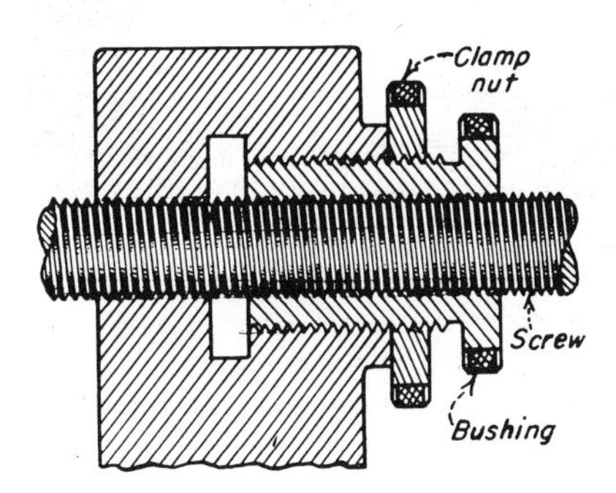

CLAMP NUT holds adjusting bushing rigidly. Bushing must have different pitch on outside thread than on inside thread. If outer thread is the coarser one, a relatively small amount of rotation will take up backlash.

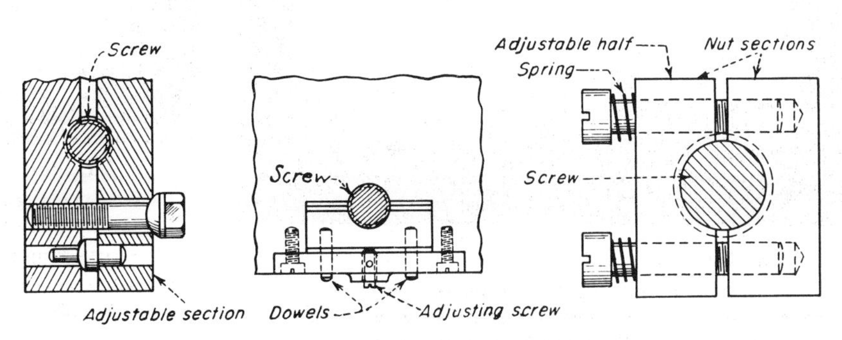

TYPICAL CONSTRUCTIONS based on the half nut principle. In each case, the nut bearing width is equal to the width of the adjustable or inserted slide piece. In the sketch at the extreme left, the cap screw with the spherical seat provides for adjustments. In the center sketch, the adjusting screw bears on the movable nut section. Two dowels insure proper alignment. The third illustration is similar to the first except that two adjusting screws are used instead of only one.

Ways to Control Backlash in Gears

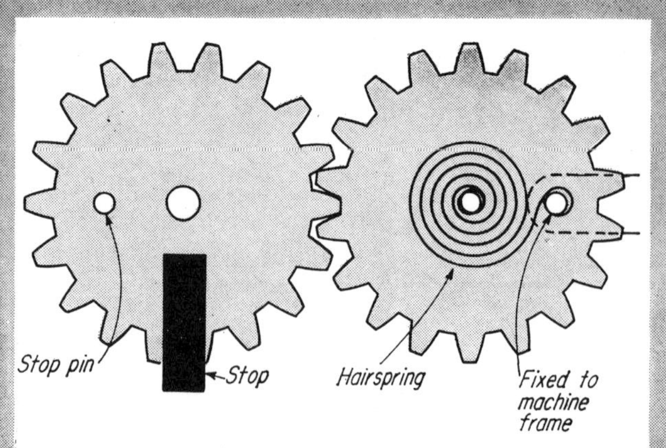

Hairspring type . . .
This device is useful for limited rotation only. One of the gears — usually the output — is loaded by attaching one end of a hairspring to a shaft, other end to a fixed pin.

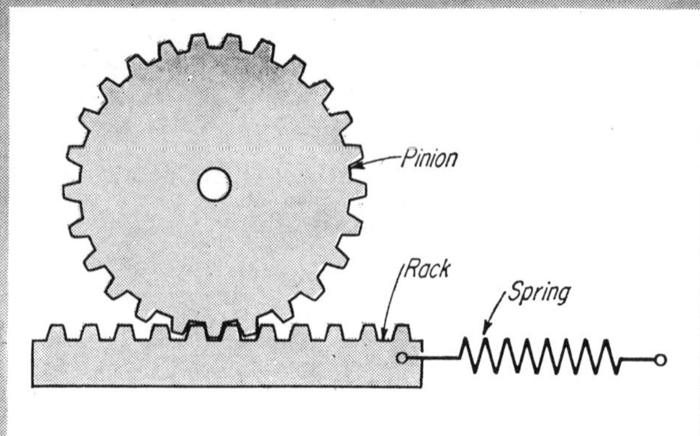

For rack and gears . . .
The loading spring can be a coil spring, or preferably a spring of constant tension. For some designs, counterweights may be substituted, but they add mass and may require undue accelerating and breaking forces.

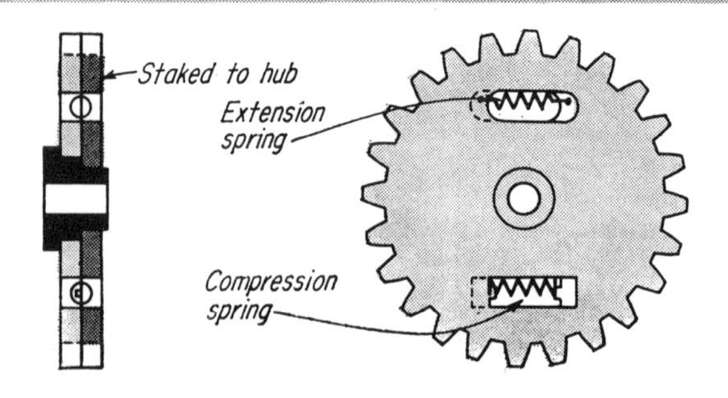

Unlimited rotation . . .
Can be obtained with split spring-loaded gears. One gear is staked to the common hub; the other is free to rotate, but loaded by springs in one direction. The mating gear is solid and should be slightly wider than the split gear. The spring force can be varied by winding the split gear to vary amount of scissors action on mating gear.

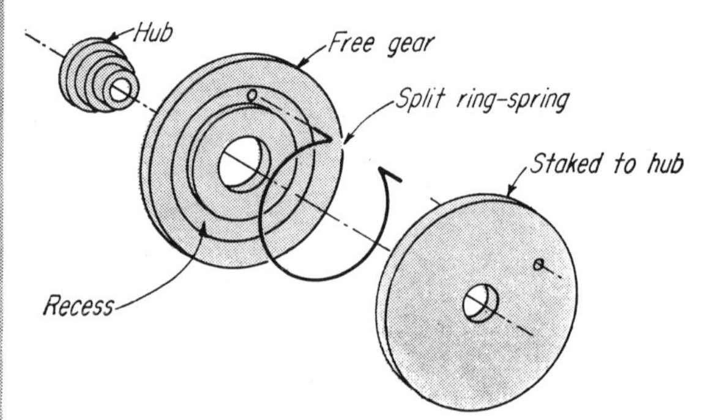

Another variation . . .
Uses a ring-spring. This requires an annular groove which may be simpler to machine than milled slots for extension springs. Spring loading, of course, adds to the tooth load and wear, and requires a higher driving torque as compared with unloaded gearing.

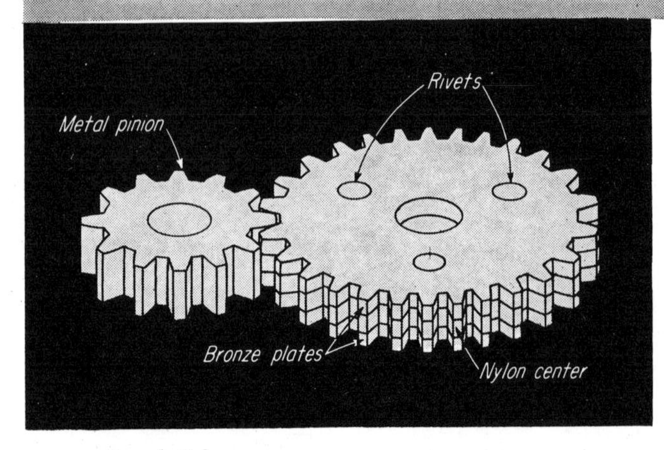

Sandwich gears . . .
having nylon lamination sandwiched between two phosphor-bronze outer plates are employed successfully, shown above, in a drilling machine manufactured by Industrial Technics Ltd., Southampton, England. The three layers are machined to the same dimensions, but the nylon lamination swells slightly and is thus larger than the metal sections. This permits tighter meshing with standard pinions because of nylon's high elasticity combined with good abrasive resistance.

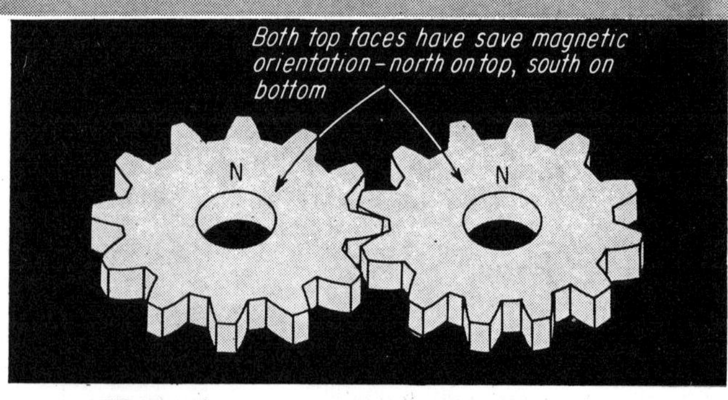

Magnetic gears . . .
Another type reported from England depends on the repelling forces between magnetized gears. Gear faces of the same magnetic orientation face the same way.
This design is useful only for very small torques, in order of 0.01 to 0.02 in.-oz.

New Anti-Backlash Devices

Austrian engineer designs no-backlash drive

To prevent spindle rotation in an extension beam drive, Felix Fritsch of Simmering Graz Pauker, AG, Vienna, has devised the dual-nut system diagramed here.

In the new spindle drive, there are two separate nuts, one driven directly, the other through a freewheeling friction-type clutch. The latter is adjustable from outside the spindle gear housing. Thus it is possible to preselect the pressure between the two spindle nuts, assuring, says SGP, close fit without backlash.

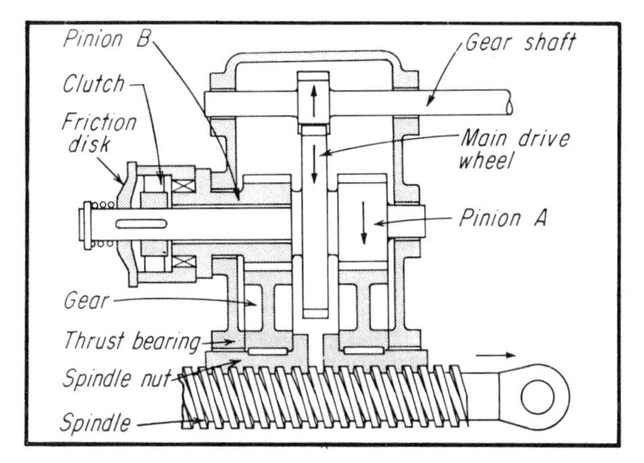

AUSTRIAN DRIVE features twin nuts, one driven directly (by gear), the other through a free-wheeling clutch. In the unit diagramed, pinion A is directly driven; pinion B is driven through the clutch.

Automatic backlash eliminator operates during both directions of table feed screw. Moving rack in or out rotates crown gear, which tends to rotate two independent nuts in opposite directions. Used on Cincinnati Dial Type Milling Machines to eliminate backlash during both down milling and up milling.

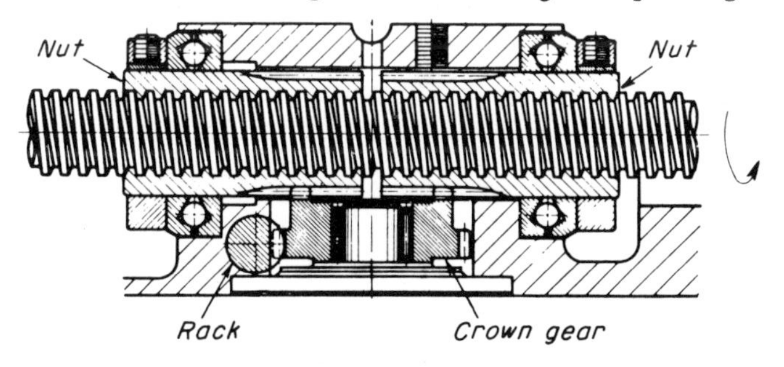

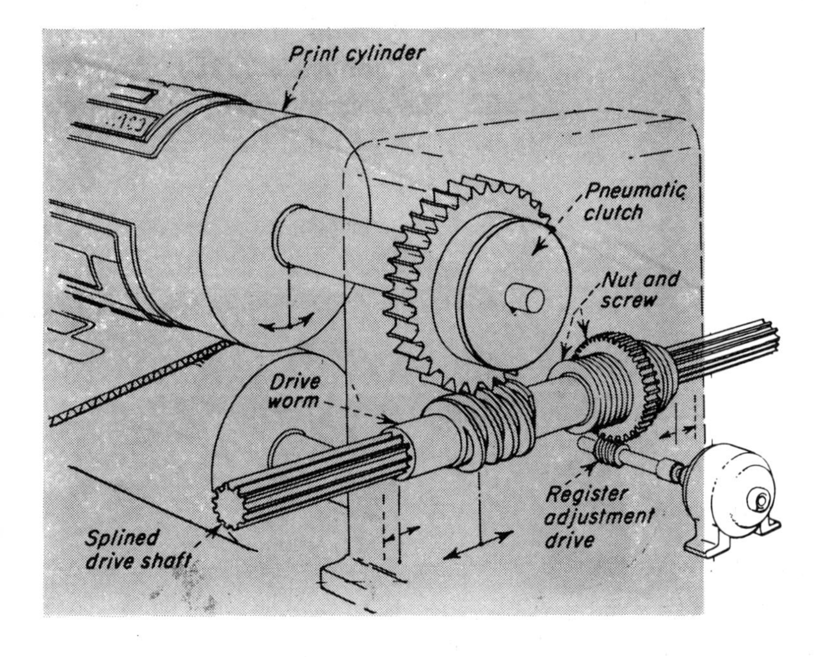

A drive that avoided backlash build-up . . .

inherent in long gear trains was required for the four-section, split-construction design because accumulated backlash would seriously affect register. Solution was a splined drive shaft running full length of the machine, with an independent worm-gear power-takeoff at each feeding, printing and slotting station. The worm drives absorb, as thrust, much of the cyclic load patterns of the kicker and slotter mechanisms. Running register of the print cylinders, and the upper slotting shaft is adjusted by shifting the drive worm axially. Main drive motor and slotting station are in the fixed section. Each worm drive includes a remotely controlled pneumatic clutch. Stations can be driven in any position.

15 IDEAS FOR Cam

This assortment of devices reflects the variety of ways in which cams can be put to work.

F STRASSER

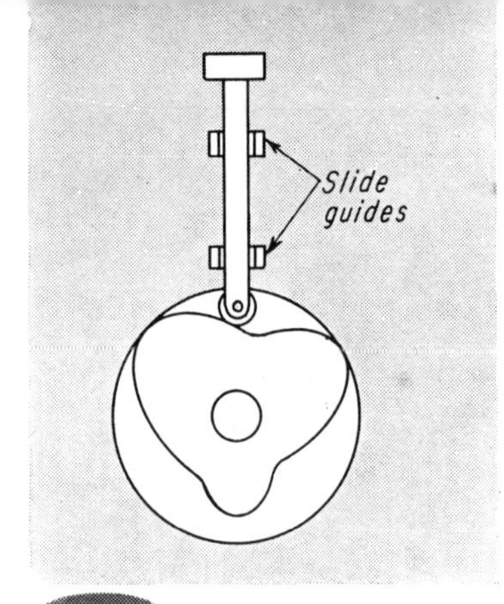

1

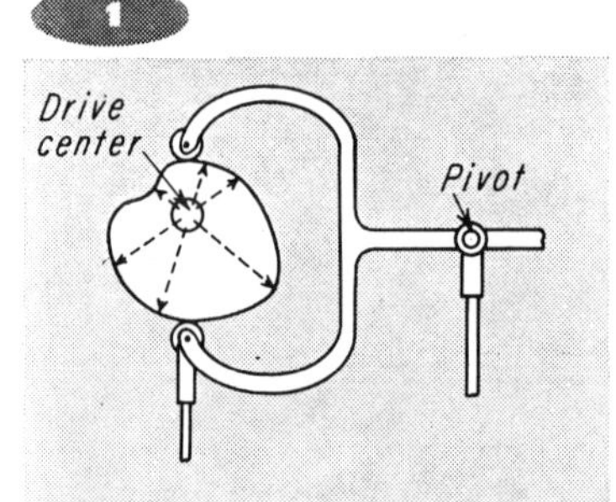

2

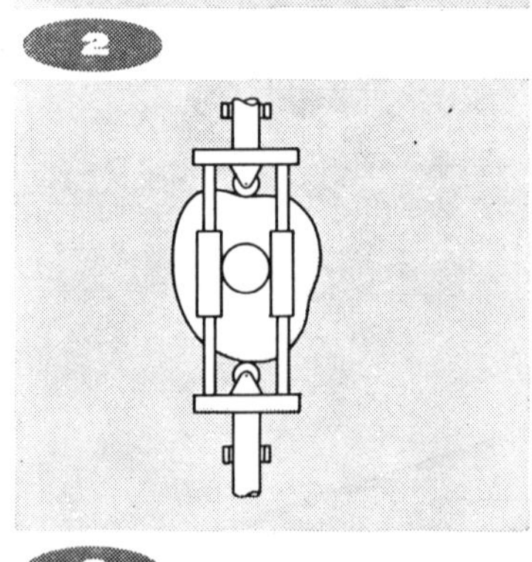

3

Constant-speed rotary . . .
motion is converted (1) into a variable, reciprocating motion; into rocking or vibratory motion of simple forked follower (2); or more robust follower (3), which can provide valve-moving mechanism for steam engines. Vibratory-motion cams must be designed so that their opposite edges are everywhere equidistant when measured through drive-shaft center.

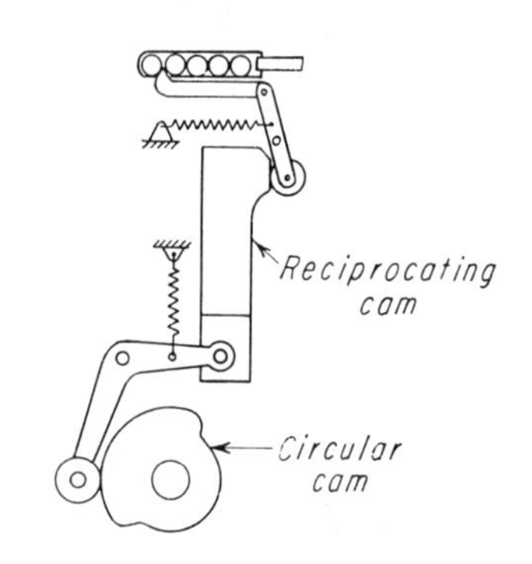

4

Automatic feed . . .
for automatic machines. There are two cams; one with circular motion, the other with reciprocating motion. This combination eliminates the trouble caused by irregularity of feeding and lack of positive control over stock feed.

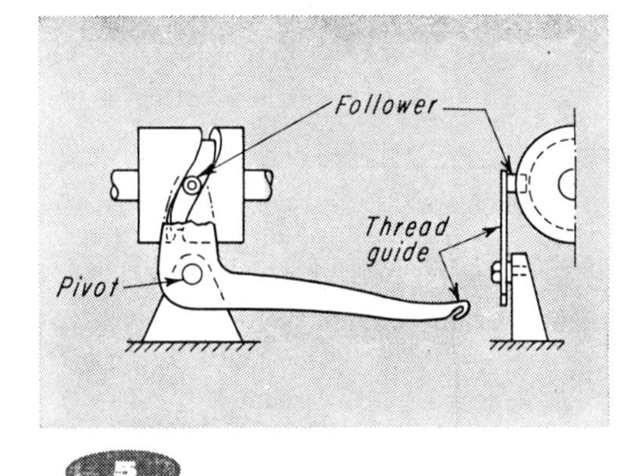

5

Barrel cam . . .
has milled grooves; is used in, for example, sewing machines to guide thread. This type of cam is also used extensively in textile manufacturing machines such as looms and other intricate fabric-making devices.

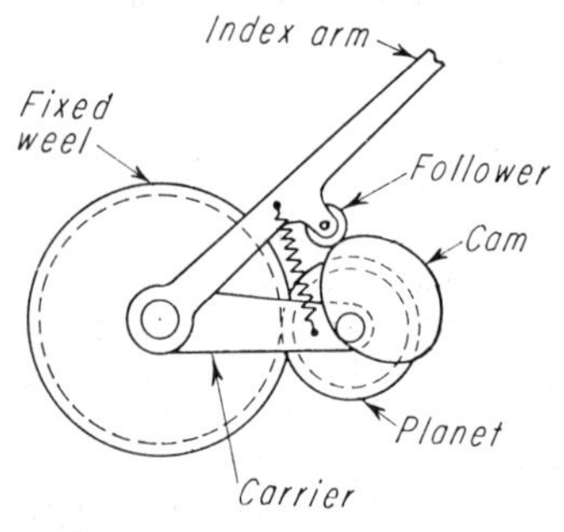

6

Indexing mechanism . . .
combines epicyclic gear and cam. Planetary wheel and cam are fixed relative to one another; the carrier is rotated at uniform speed round the fixed wheel. Index arm has non-uniform motion, with dwell periods.

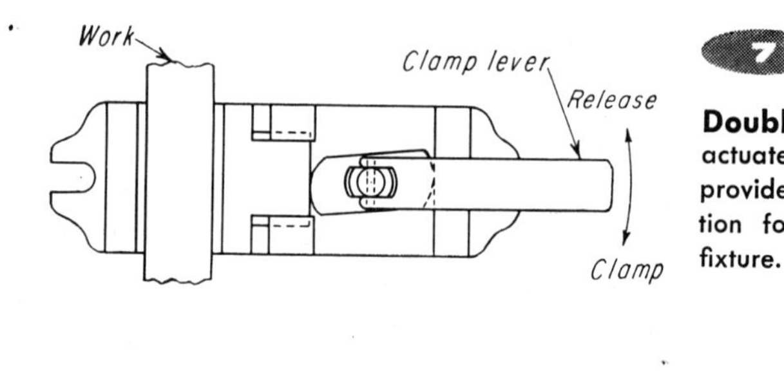

7

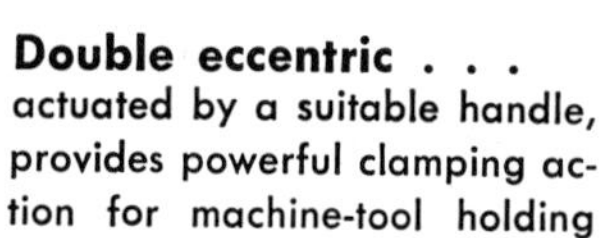

Double eccentric . . .
actuated by a suitable handle, provides powerful clamping action for machine-tool holding fixture.

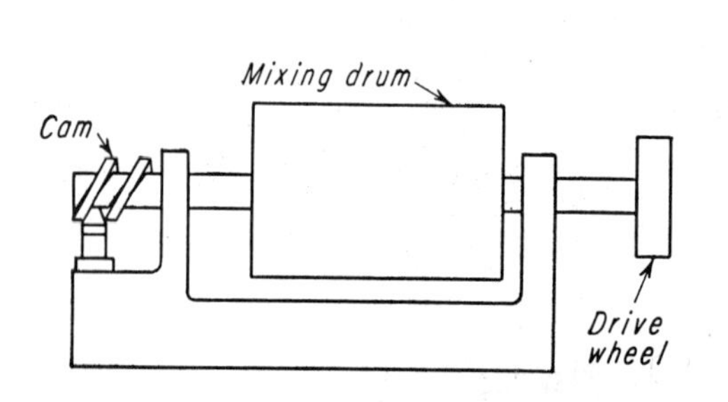

8

Mixing roller . . .
for paint or candy, etc. Mixing drum has small oscillating motion as well as rotation.

Mechanisms

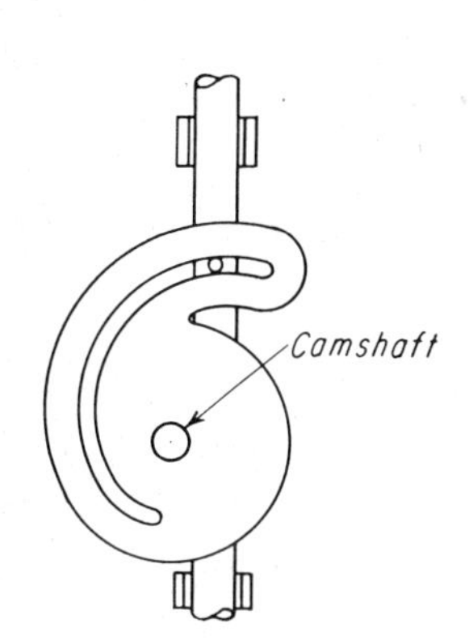

Slot cam . . .
converts oscillating motion of camshaft to variable but straight-line motion of rod. According to slot shape, rod motion can be made to suit specific design requirements such as straight-line and logarithmic motion.

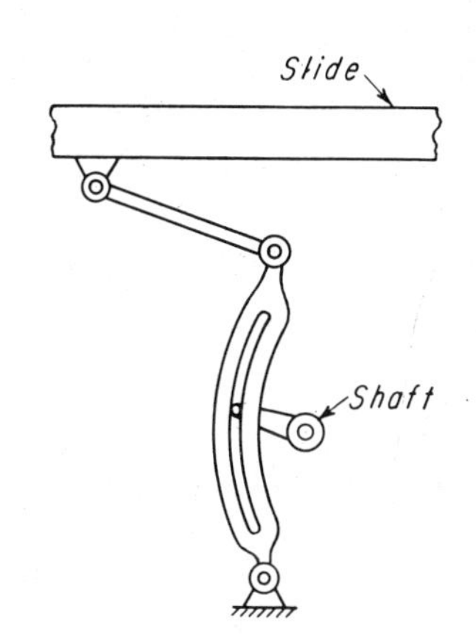

10

Continous rotary motion . . .
of shaft is converted into reciprocating motion of a slide. Device is used on certain types of sewing machines and printing presses.

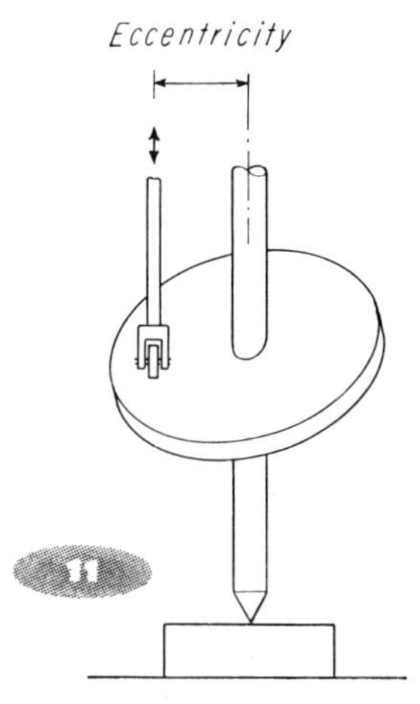

11

Swash-plate cams . . .
are feasible for light loads only, such as in a pump. Eccentricity produces resultant forces that cause excessive loads. Multiple followers may ride on plate, thereby providing smooth pumping action for multipiston pump.

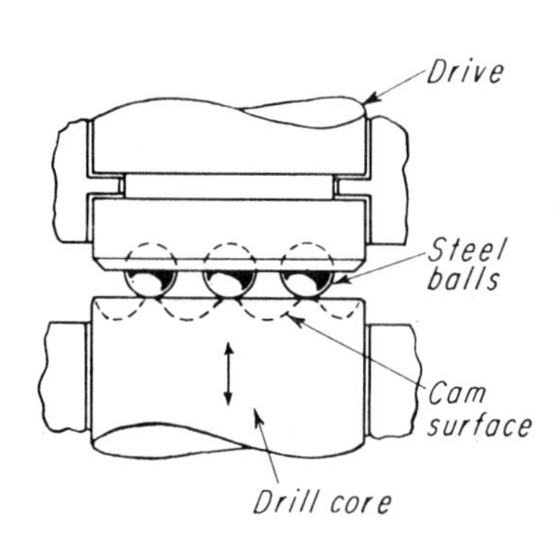

12

Steel-ball cam . . .
converts high-speed rotary motion of, for example, an electric drill into high-frequency vibrations that power the drill core for use as a rotary hammer for masonry, concrete, etc. Attachment can be designed to fit hand drills.

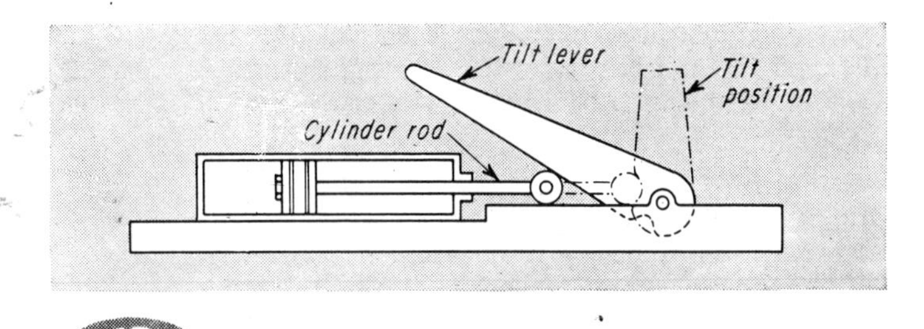

13

Tilting device . . .
can be designed so that lever remains in tilted position when cylinder rod is withdrawn, or can be spring-loaded to return with cylinder rod.

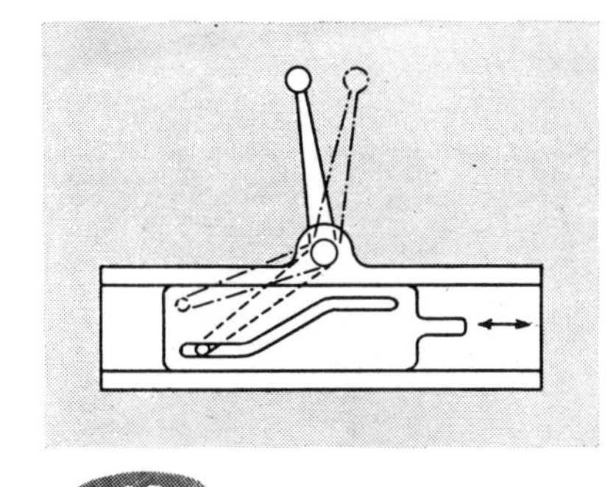

14

Sliding cam . . .
finds employment in remote control to shift gears in position that is otherwise inaccessible on machines.

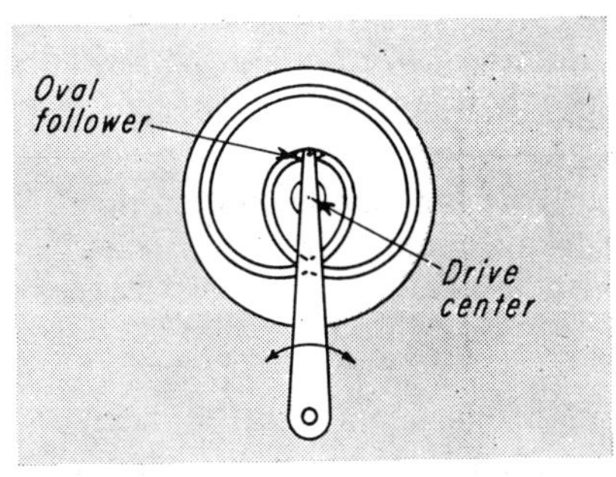

15

Groove and oval . . .
follower provide device that requires two revolutions of cam for one complete follower cycle.

CAMS FOR CONTROLS

HAROLD A ROTHBART
The City College, NYC

There are occasions in control design when it is necessary to develop an electrical output that is some complex function of a nonlinear mechanical input. A very satisfactory solution to this problem is to interpose a cam between the input and a linear potentiometer, which is a dependable and widely available component. More complex devices can be constructed based on this same approach. A good example is the addition of a drive cam to a variable-core transformer. The resulting electrical signals are thereby modified by the configuration of the cam.

The first step in cam design is selection of the scale. The input scale is fixed by the fact that the cam usually rotates less than 360° for the full range. The output scale or slope, on the other hand, offers unlimited opportunities for choice. For accuracy it is desirable to have the scale as large as possible; i.e. a minimum rise per inch of periphery per revolution. Backlash and dimensional accuracy then have less effect.

The second step is to establish cam size and the direction of follower movement, outward or inward. The steepest portion of the function curve should be at the cam outer radius, permitting the use of smaller cams for the same pressure-angle limit. The maximum pressure angle (angle between the cam slope and the follower axis) is 30°, but some designs have used angles as high as 45°. In addition, the control engineer should reduce excessive wear or undercutting by increasing cam size or reducing the output scale. In general, the designer can feel free to specify close manufacturing tolerances, on the order of ±0.0002 in. and 2 min. of arc or better.

The basic function or motion relationship between related parameters X and Y is:

$$Y = F(X)$$

The variable X may be called the input parameter and Y the output parameter. $F(X)$ is any single-valued continuous function whose derivative is held within certain limits to prevent the resulting cam from being impractically large. Among the many mathe-

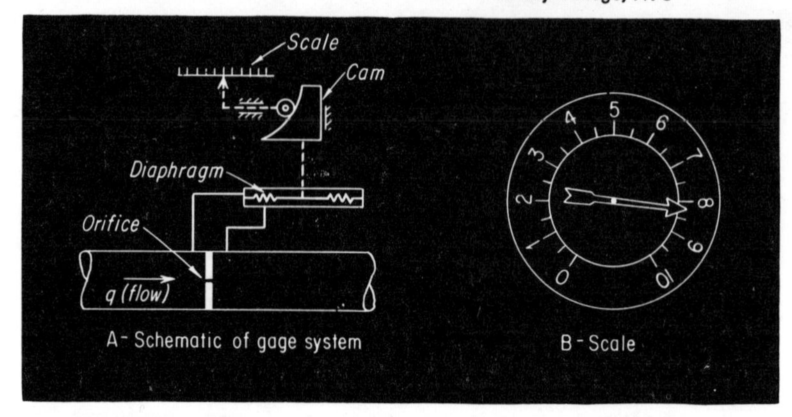

1 . . LINEARIZING CAM converts output for display on uniform scale.

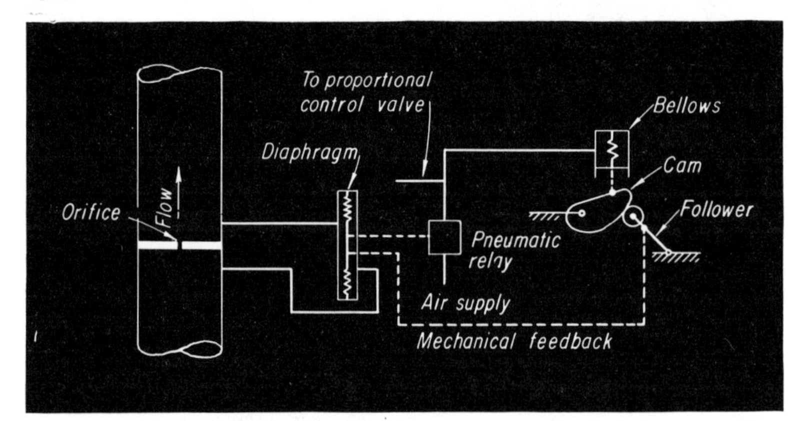

2 . . FEEDBACK balances forces on diaphragm. Since flow is not directly proportional to control-valve opening, cam must compensate.

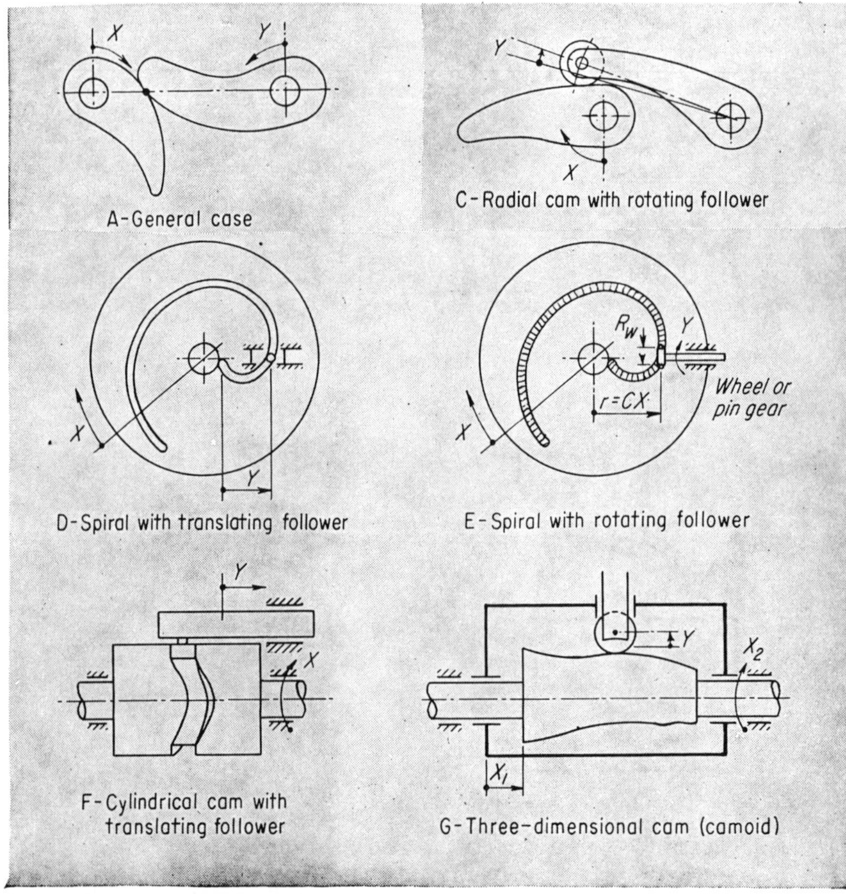

3 . . COMMON TYPES of control cams.

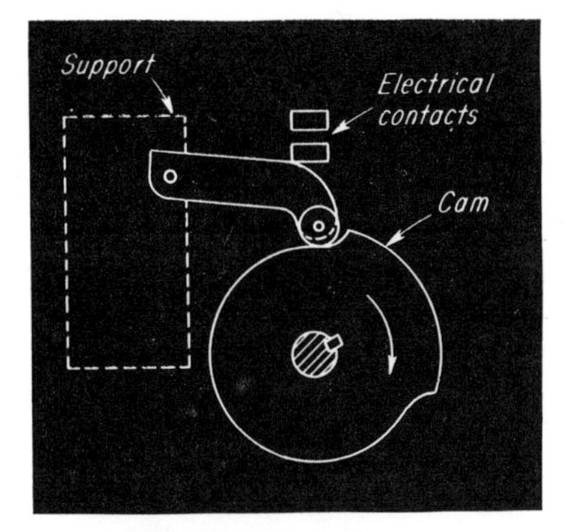

4 . . **PROGRAMMING** cam keeps switch on or off during duration of cam rise.

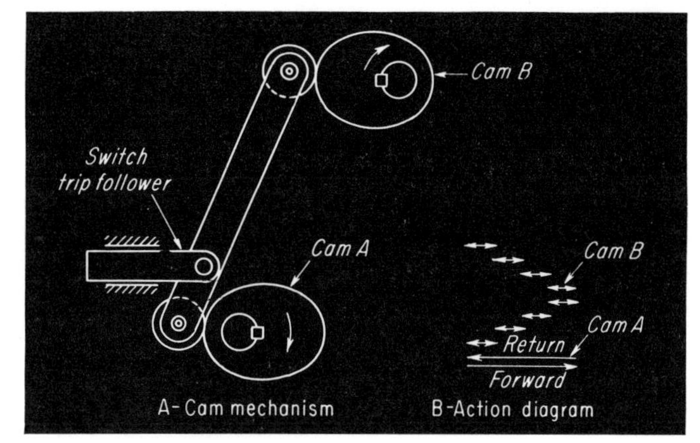

5 . . **TWO CAMS** rotating at different speeds superimpose their motions on trip-follower.

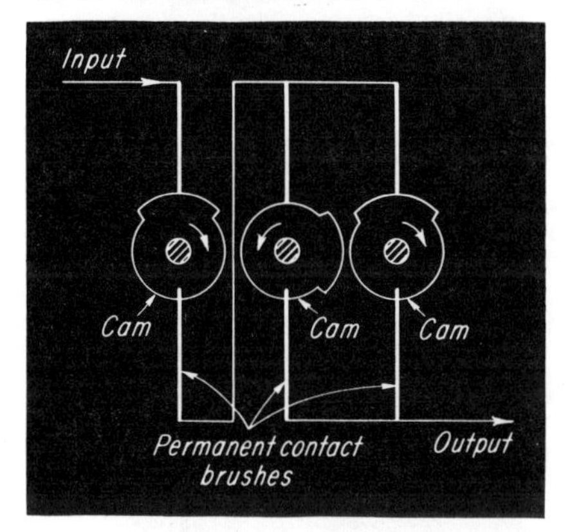

6 . . **CONDUCTING CAMS** achieve two-speed switching in a different way.

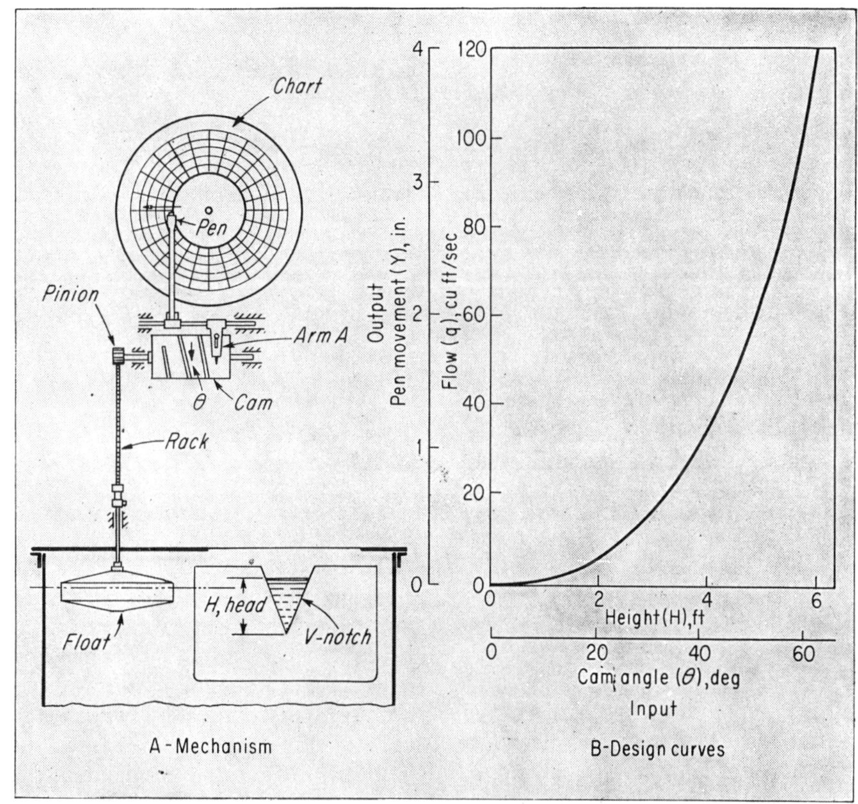

7 . . **FLOWMETER** uses cam because flow rate through V-notch is nonlinear. Cam modifies signal from float so that is can be recorded on uniform scale.

matic operations that can be implemented are squares, roots, reciprocals, empirical functions, and trigonometric functions. A cam, for example, can be designed to produce an output proportional to the tangent of the cam angle. Even non-monotonic functions (those having two possible values of Y for each value of X) can be mechanized by means of two cams rotating in synchronism.

Cylindrical cams

The radial or cylindrical cam employing an oscillating follower is perhaps the simplest computer element employed in measuring systems. Simple polynomial expressions are the most common functions, although any relationship between cam rotation and follower displacement is possible. The polynomials are of the form

$$y = C\,\theta^n$$

The value of n depends upon the system under investigation and is typically $\frac{1}{2}$, 1, or 2. For $n = 1$, the resulting straight-line cam resembles an Archimedes spiral for which a linear input-output relationship exists of the form

$$y = \frac{h\,\theta}{\beta}$$

The nomenclature listed here applies to the above equation and the discussion that follows:

θ = Cam angle of rotation for displacement y
n = Any value
c = Constant
h = Maximum follower output rise
β = Angle of cam rotation for rise h
y = Output

To illustrate, an open channel has a triangular weir for measuring water flow—Fig 7(A). A float moves up and down with the water level. Because flow does not vary linearly with head, a cam is used to make the necessary conversions. A rack on the float rod meshes with a pinion to turn the cam with the recording pen and arm A attached. The translational drive is by means of a pin on arm A riding in the cam groove. The cam follower moves linearly with flow, and so does the recording pen. Assume that the V-notch has a height of 6 ft and an included angle of 60° and that the chart scale is 4 in. long. The flow formula is $q = 1.45H^{2.47}$ where q is the flow (cu ft/sec) and H is the head in feet.

Fig 7(B) shows input H plotted against output q. If the diameter of the pinion meshing with the rack is 2 in., the number of cam revolutions for the total stroke of 6 ft is:

$$\frac{6 \times 12}{2} = 11.4 \text{ radians} = 655°$$

which is an acceptable figure. In Fig 7(B) there is also presented a plot of pen displacement Y against the angle of cam rotation θ. The cam can now be drawn to scale.

Roll-Cam Devices

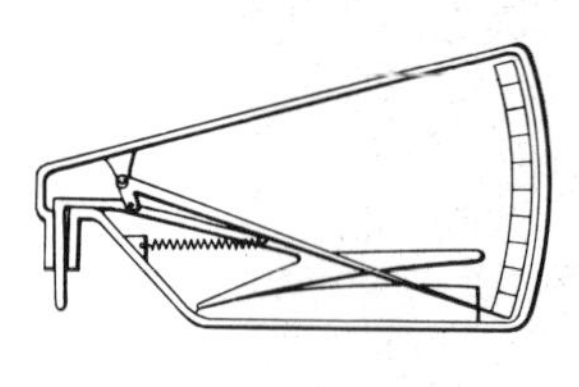

1

Sensitive contact gage uses a rocking pair to decrease the effect of friction and increase the accuracy.

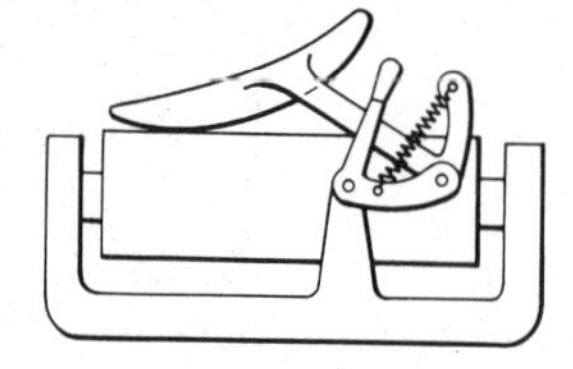

Variable electrical resistor has a rocking surface instead of a sliding brush to reduce wear with smooth operation.

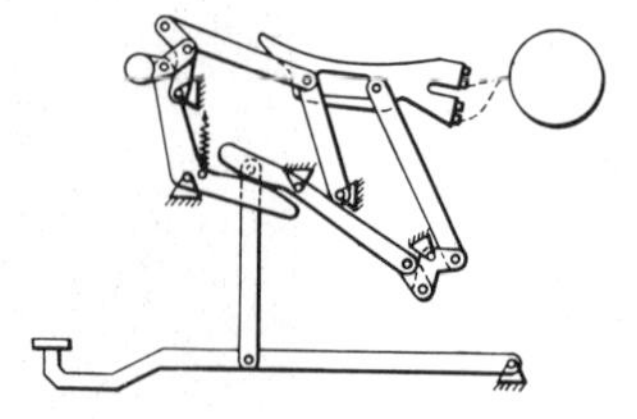

Typewriter linkage has a rocking pair that actuates upper or lower case letters with smoother, quieter action.

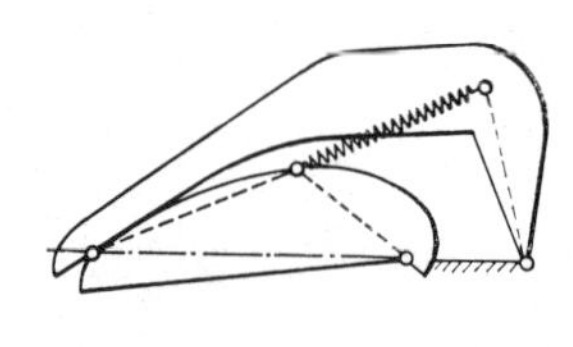

Rocking mechanism derived from 4-bar linkage has constant spring length which transmits no force to bearing.

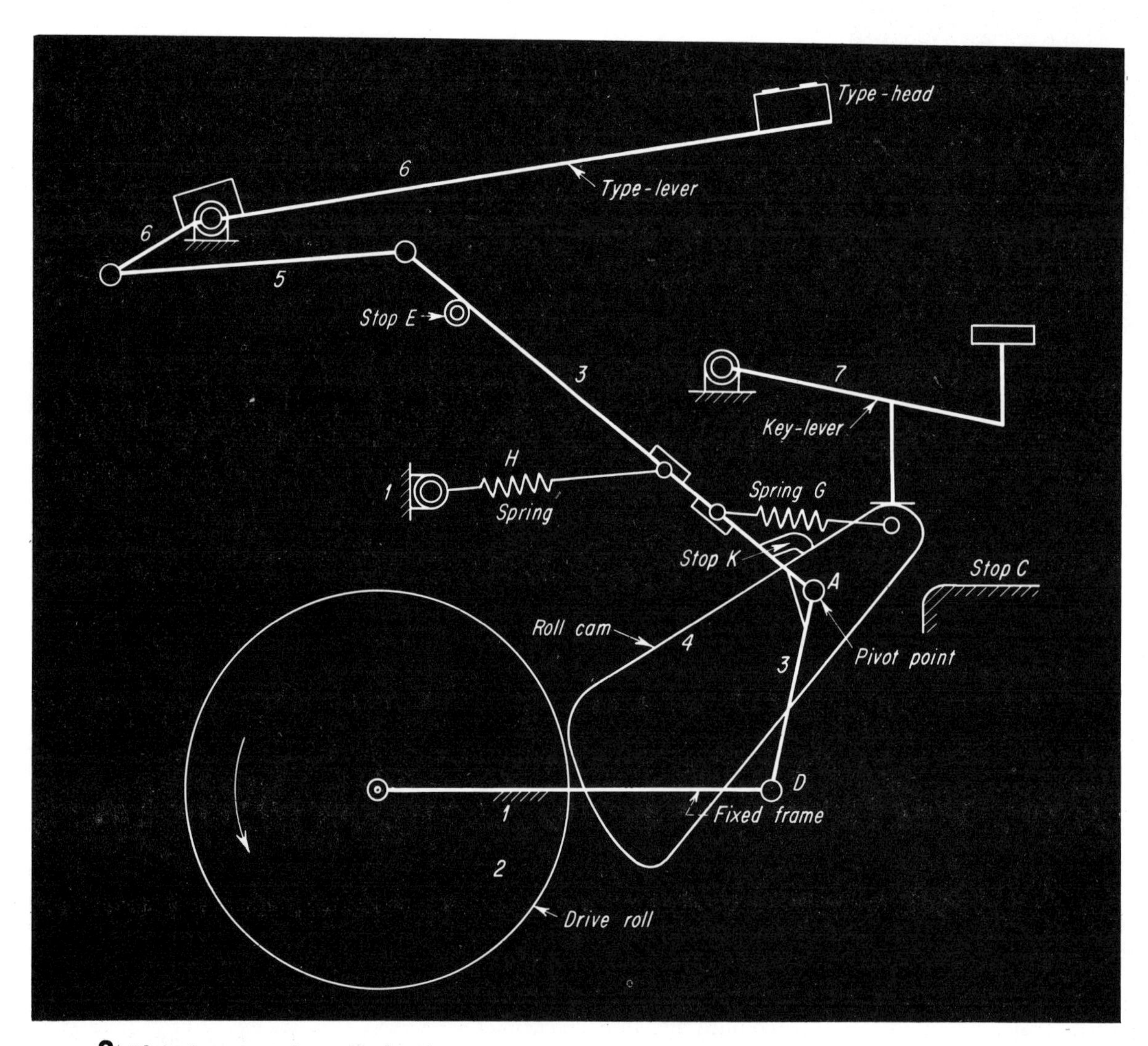

2 Electric-typewriter mechanism . . .
uses roll cam for motion amplification. Here, path of pivot point on roll cam is curvilinear.

Roll cams are also employed in IBM electric typewriters, Fig. 2. Here the cam is triggered by a touch of the typist's finger to power the type heads.

The roll is driven by an electric motor at constant speed. Cycle begins when the typist depresses the key lever which rotates the cam into contact with the drive roll. The cam is connected to link 3 at pin point A. Rotation of drive roll makes link 3 rotate clockwise, causing link 6 to rotate (via link 5) until the type head contacts the platen (not shown).

At end of the cycle, the cam strikes stop C and loses contact with drive roll. Spring G then returns the cam to its position against the stop *K*.

During this time, while cam and drive are disengaged, the type head continues to approach the platen because of kinetic energy stored in type lever (link 6). After type head strikes platen, spring *H* returns the linkage to the home position where link 3 contacts stop *E*.

How a Roll Cam Works

A C DUNK
Assistant Professor
Mechanical Engineering, Purdue U

The roll-cam mechanism shown in Fig. 1 is for a rectilinear type—its job in an RCA 45-rpm record changer is to control tone-arm motion and record changing. Basic components are:

- **Drive roll**—rotates at constant speed about fixed pivot and provides input power.
- **Roll cam**—its cutout allows it to dwell until activated.
- **Slide rod**—is reciprocated by rotation of the roll cam.
- **Frame**—provides a fixed pivot and slot for the rod.

The sketch shows the mechanism in the dwell position --cam and drive not in contact. To start the cycle, finger pressure is employed momentarily (through a lever not shown) to slightly rotate the cam until it comes into contact with the drive roll (which an electric motor is rotating at constant speed). Friction between the surface then rotates the cam, in turn causing the rod to move upward. The cam goes through a complete cycle and returns down to its original position. Here the stop on the frame will arrest the cam, which will again be out of engagement until activated again.

Spring pressure prevents slippage during rotation—especially during the return cycle. However, to make the transfer of motion more positive, one member can be made of rubber and the other of metal with small serrations on its periphery. To prevent any slip whatsoever, conjugate tooth forms could be used.

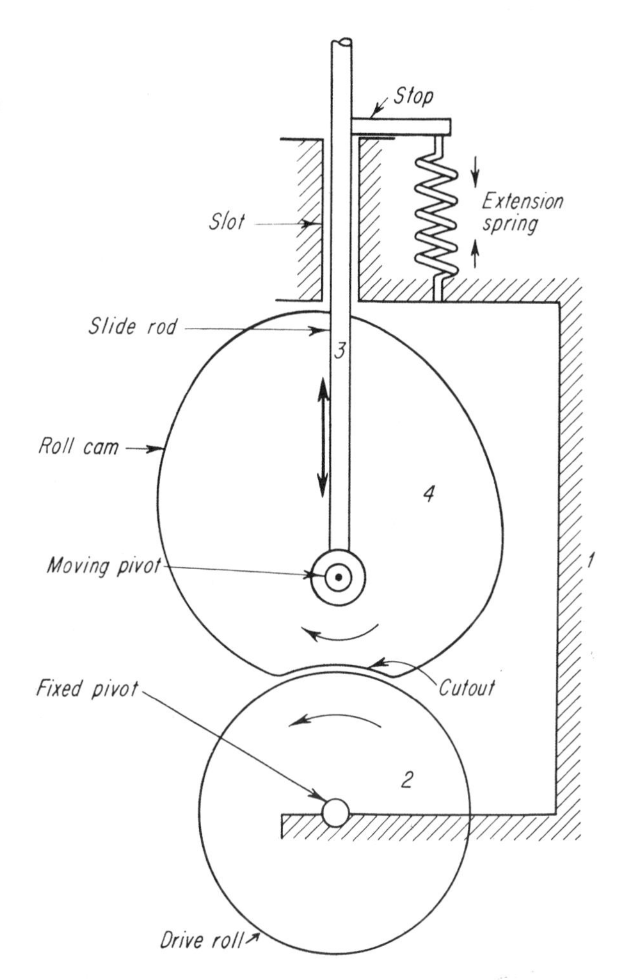

3 Record changer . . . has this roll-cam mechanism. Path of movable pivot is rectilinear. Cam is triggered by rotating cam slightly until it contacts drive.

Cam Drives for Machine Tools

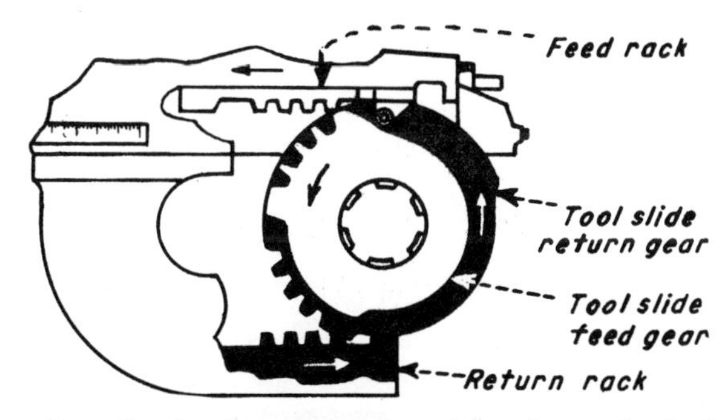

Two-directional rack-and-gear drive for main tool slide combines accuracy, uniform movement and minimum idle time. Mechanism makes a full double stroke per cycle. It approaches fast, shifts smoothly into feed and returns fast. Point of shift is controlled by an adjustable dog on a calibrated gear. Automatic braking action assures smooth shift from approach to feed. Used on Greenlee 4-Spindle Bar Automatics.

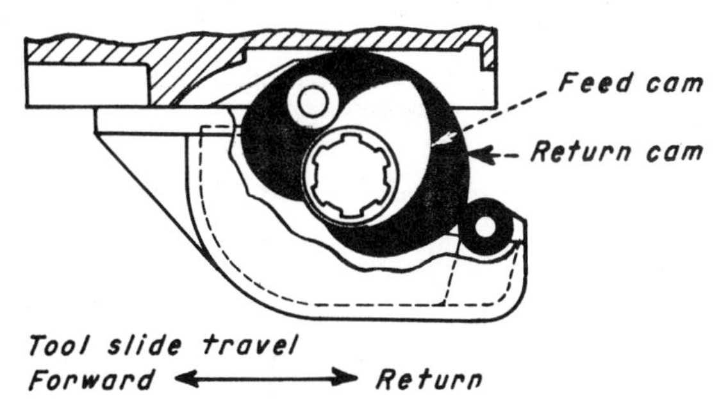

Cam drive for tool slide mechanism is used instead of rack feed when short stroke is required to get fast machining cycle on Greenlee Automatic. Sketch shows cams and rollers with slide in retracted position.

Special-Function Cams

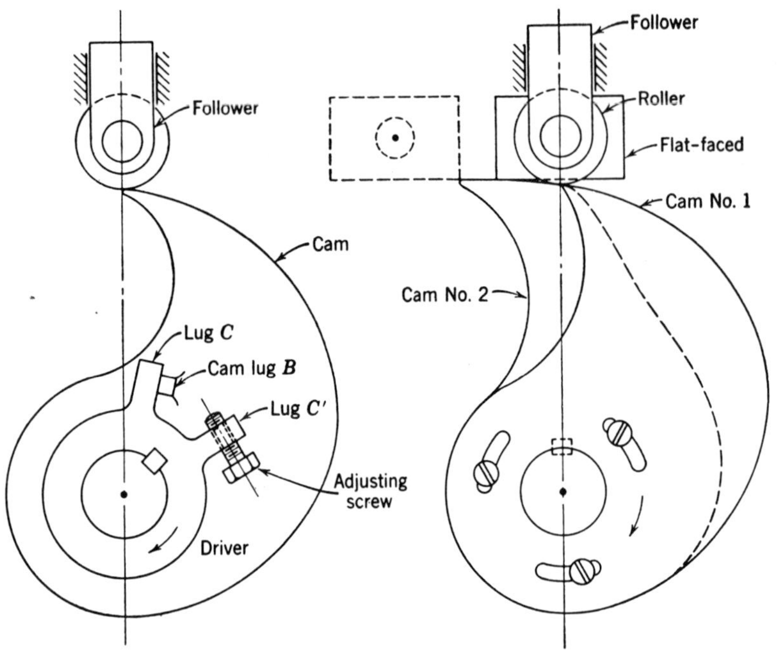

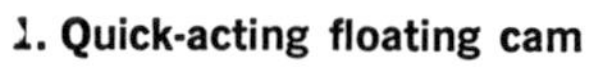

1. Quick-acting floating cam

2. Quick-acting dwell cams

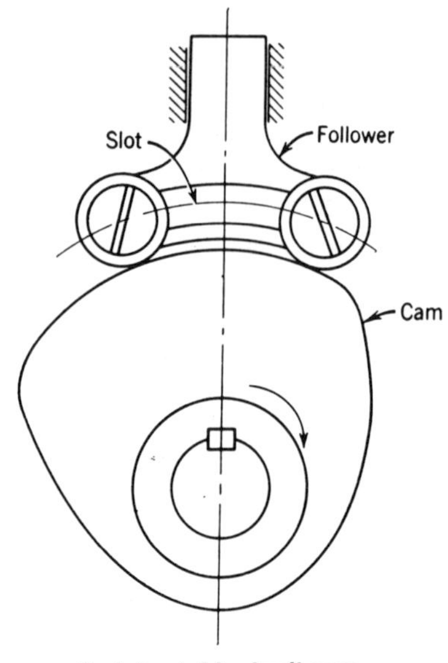

3. Adjustable-dwell cam

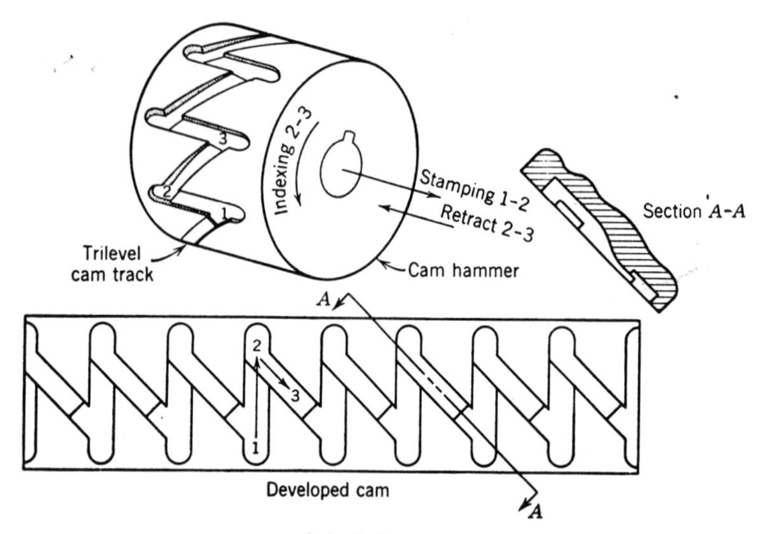

4. Indexing cam

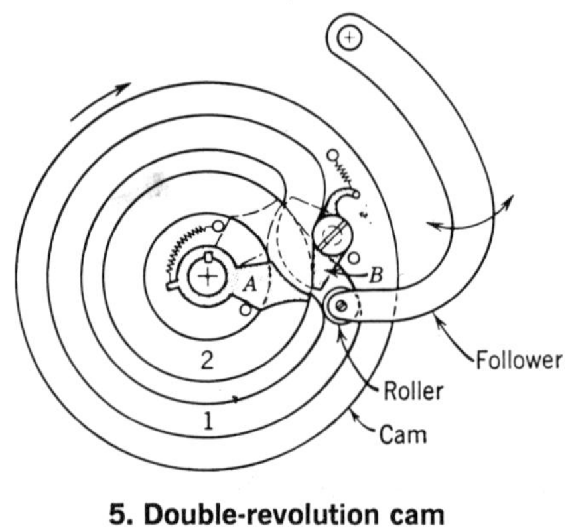

5. Double-revolution cam

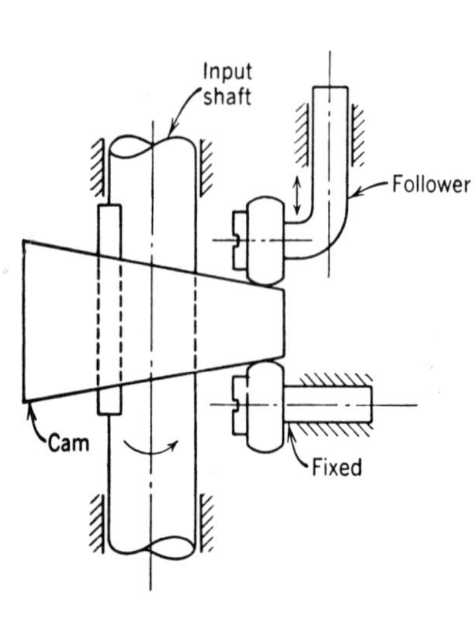

6. Increased-stroke barrel cam

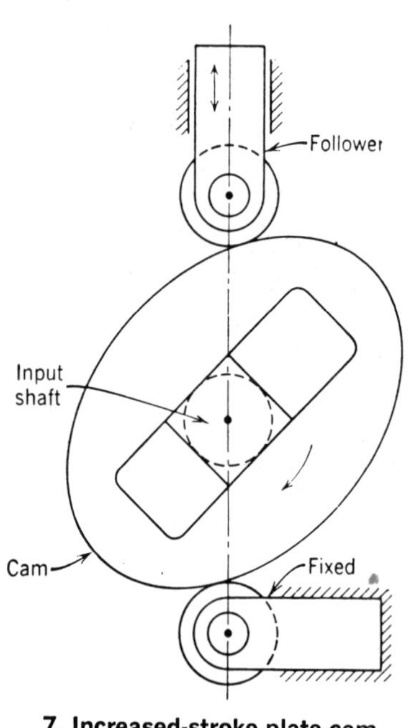

7. Increased-stroke plate cam

Adjustable-Dwell Cams

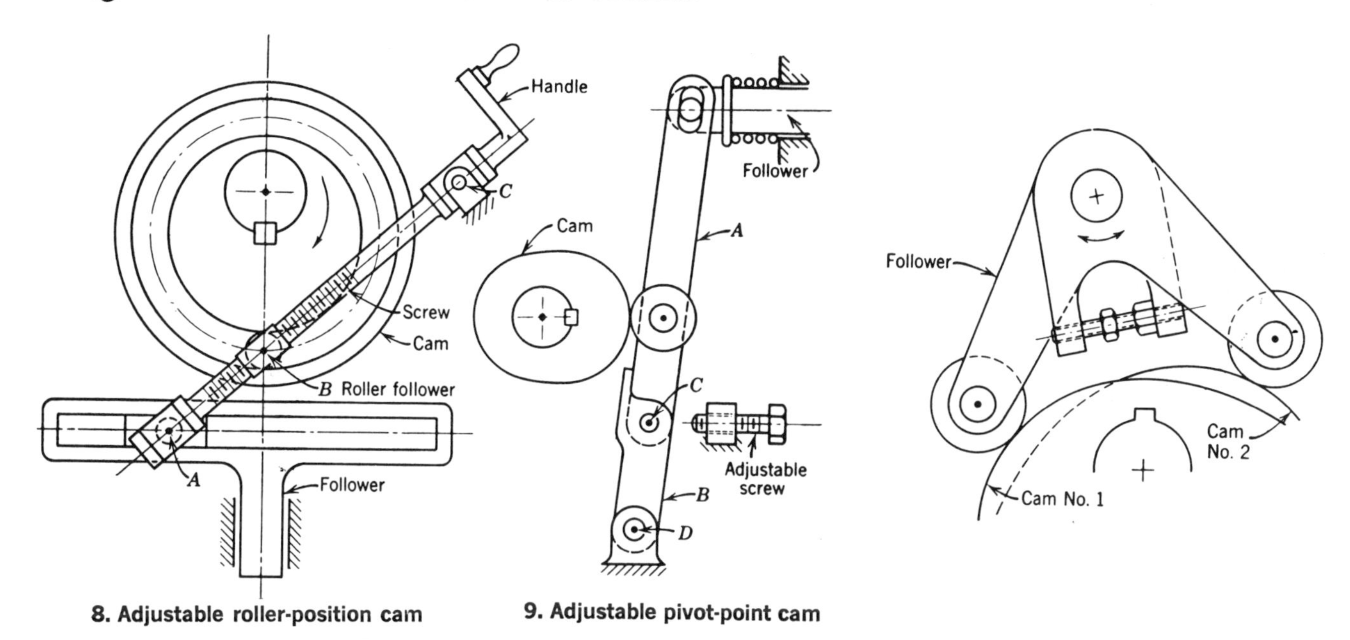

8. Adjustable roller-position cam

9. Adjustable pivot-point cam

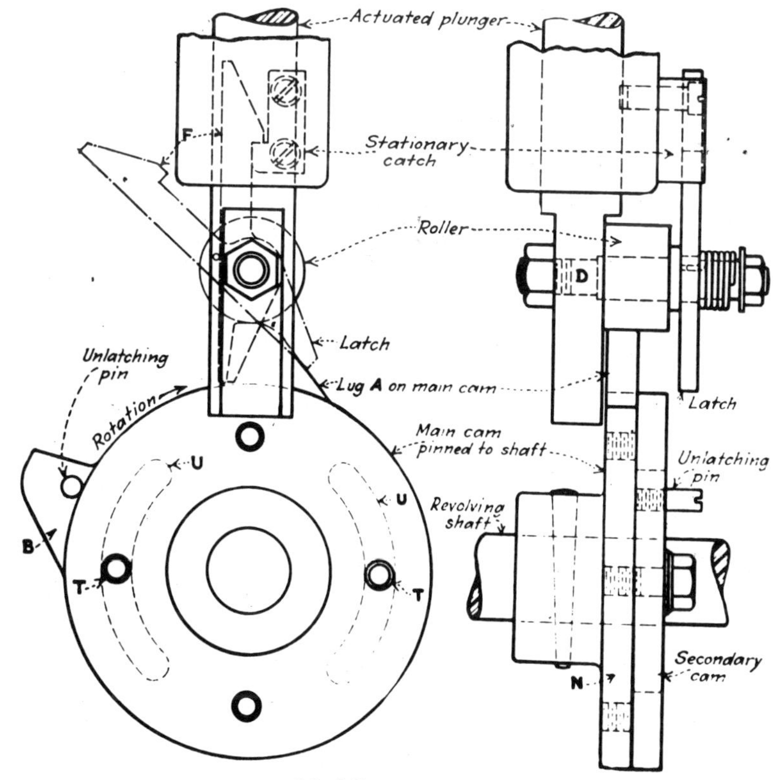

10. Adjustable lug cam

Fig 1 — A quick drop of the follower is obtained by permitting the cam to be pushed out of the way by the follower itself as it reaches the edge of the cam. Lugs *C* and *C′* are fixed to the cam-shaft. The cam is free to turn (float) on the cam-shaft, limited by lug *C* and the adjusting screw. With the cam rotating clockwise, lug *C* drives the cam through lug *B*. At the position shown, the roller will drop off the edge of the cam which is then accelerated clockwise until its cam lug *B* hits the adjusting screw of lug *C′*.

Fig 2 — Instantaneous drop is obtained by the use of two integral cams and followers. The roller follower rides on cam *1*. Continued rotation will transfer contact to the flat-faced follower which drops suddenly off the edge of cam *2*. After the desired dwell, the follower is restored to its initial position by cam *1*.

Fig 3 — Dwell period of cam can be varied by changing the distance between two rollers in the slot.

Fig 4 — Reciprocating pin (not shown) causes the barrel cam to rotate intermittently. Cam is stationary while pin moves from *1* to *2*. Groove *2-3* is at a lower level; thus as the pin retracts it cams the barrel cam, then it climbs the incline from *2* to the new position of *1*.

Fig 5 — Double groove cam makes two revolutions for one complete movement of the follower. Cam has movable switches, *A* and *B*, which direct the follower alternately in each groove. At the instant shown, *B* is ready to guide the roller follower from slot *1* to slot *2*.

Fig 6 and 7 — Increased stroke is obtained by permitting the cam to shift on the input shaft. Total displacement of follower is therefore the sum of the cam displacement on the fixed roller plus the follower displacement relative to the cam.

Fig 8 — Stroke of follower is adjusted by turning the screw handle which changes distance *AB*.

Fig 9 — Pivot point of connecting link to follower is changed from point *D* to point *C* by adjusting the screw.

Fig 10 — Adjustable dwell is obtained by having the main cam, with lug *A*, pinned to the revolving shaft. Lug *A* forces the plunger up into the position shown and allows the latch to hook over the catch, thus holding the plunger in the up position. The plunger is unlatched by lug *B*. The circular slots in the cam plate permit shifting of lug *B*, thereby varying the time that the plunger is held in the latched position.

REFERENCE: Rothbart, H. A. *Cams—Design, Dynamics, and Accuracy,* John Wiley and Sons, Inc., New York.

Quick Release Mechanisms

Quick Release Mechanism

George A. Fries
Philadelphia, Pa.

Quick release mechanisms have numerous applications. Although the design shown here operates as a tripping device for a quick release hook, mechanical principles involved undoubtedly have many other applications. Fundamentally, it is a toggle-type mechanism featured by the fact that the greater the load the more effective the toggle.

The hook is suspended from the shackle, and the load or work is supported by the latch which is machined to fit the fingers *C*. The fingers *C* are pivoted about a pin. Assembled to the fingers are the arms *E*, pinned at one end and joined at the other by the sliding pin *G*. Inclosing the entire unit are the side plates *H* containing the slot *J* for guiding the pin *G* in a vertical movement when the hook release. The helical spring returns the arms to the bottom position after they have been released.

To trip the hook, the tripping lever is pulled by the cable *M* until the arms *E* pass their horizontal center line. The toggle effect is then broken, thereby releasing the load.

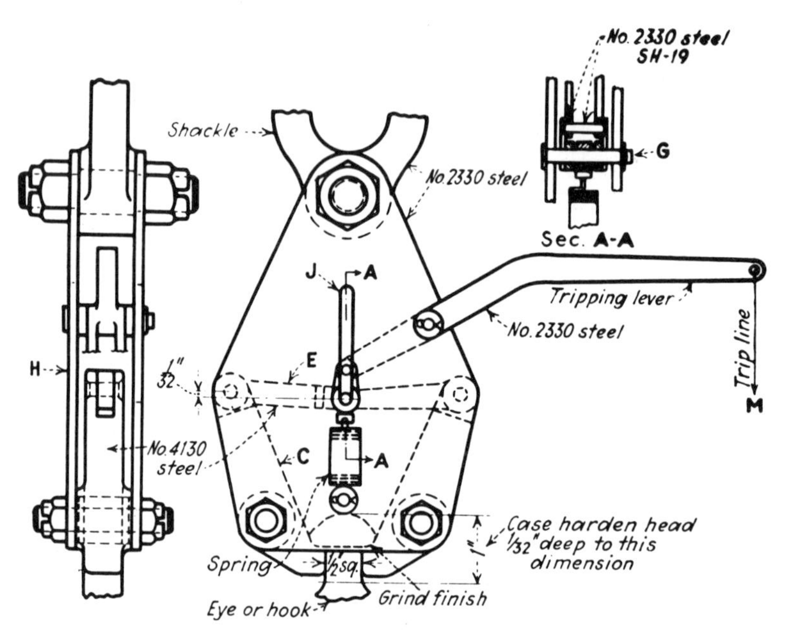

Simple quick-release toggle mechanism as designed for tripping a lifting hook

Positive Locking and Quick Release Mechanism (Brief 63-10420)

The objective was to design a simple device which would hold two objects together securely and quickly release them on demand.

One object, such as a plate, is held to another object, such as a vehicle, by a spring-loaded slotted bolt, which is locked in position by two retainer arms. The retainer arms are constrained from movement by a locking cylinder. To release the plate, a detent is actuated to lift the locking cylinder and rotate the retainer arms free from contact with the slotted bolt head. As a result of this action, the spring-loaded bolt is ejected and the plate is released from the vehicle.

Actuation of the slidable detent can be initiated by a squib, a fluid-pressure device, or a solenoid and the principle of this device can be employed wherever a positive engagement that can be quickly released on demand is required. Some suggested applications of this principle are in coupling devices for load-carrying carts or trucks, hooks or pick-up attachments for cranes, and quick-release mechanisms for remotely controlled manipulators. No patent application is involved in this idea.

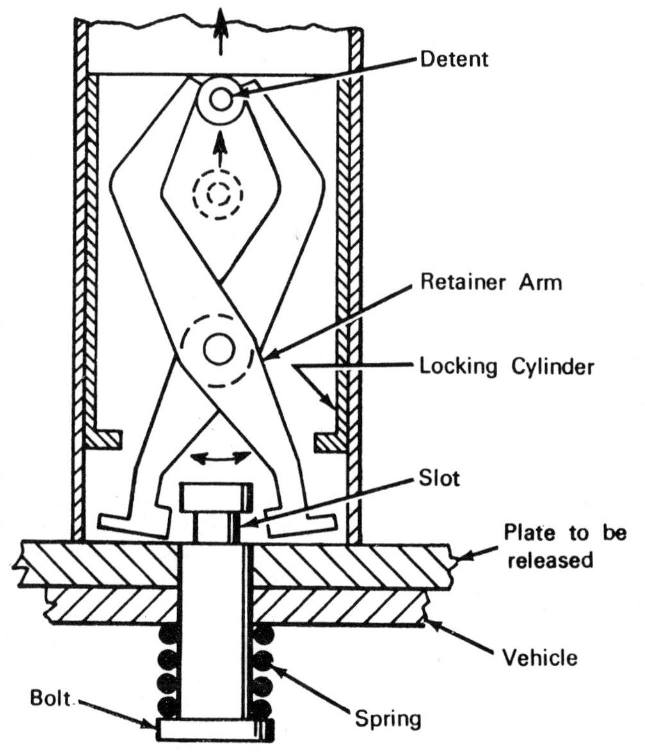

Diagram of the mechanism locking a vehicle and plate.

Chapter 8 SPRING AND BELLOW DEVICES

12 ways to put springs

Variable-rate arrangements, roller positioning, space saving, and other ingenious ways to get the most from springs.

L. KASPER, Philadelphia

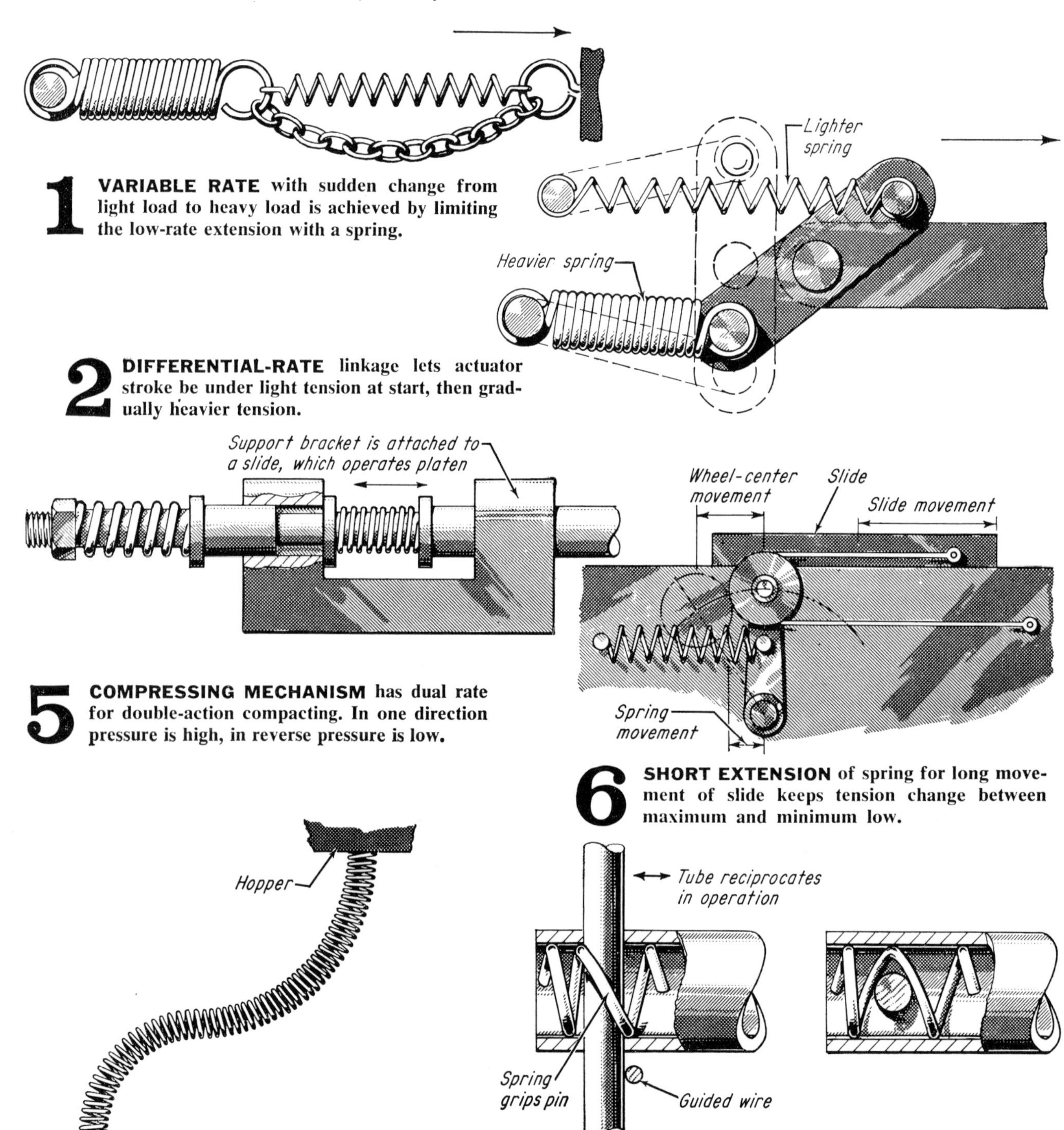

1 **VARIABLE RATE** with sudden change from light load to heavy load is achieved by limiting the low-rate extension with a spring.

2 **DIFFERENTIAL-RATE** linkage lets actuator stroke be under light tension at start, then gradually heavier tension.

5 **COMPRESSING MECHANISM** has dual rate for double-action compacting. In one direction pressure is high, in reverse pressure is low.

6 **SHORT EXTENSION** of spring for long movement of slide keeps tension change between maximum and minimum low.

9 **CLOSE-WOUND SPRING** is attached to a hopper and will not buckle when used as a movable feed-duct for nongranular material.

10 **PIN GRIP** is spring that holds pin by friction against end movement or rotation, but lets pin be repositioned without tools.

to work

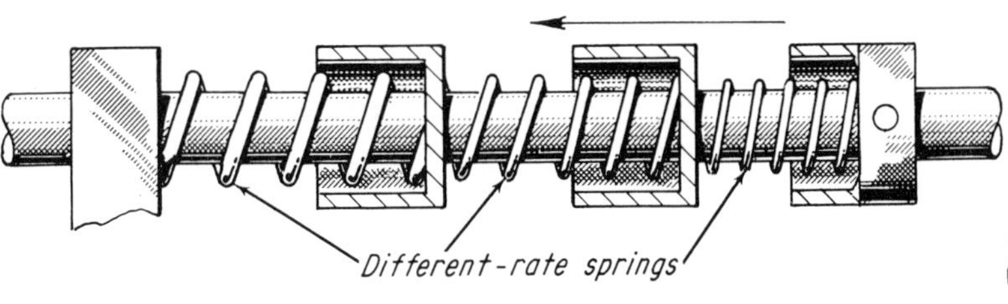

3 **THREE-STEP RATE** change at predetermined positions. The lighter springs will always compress first regardless of their position.

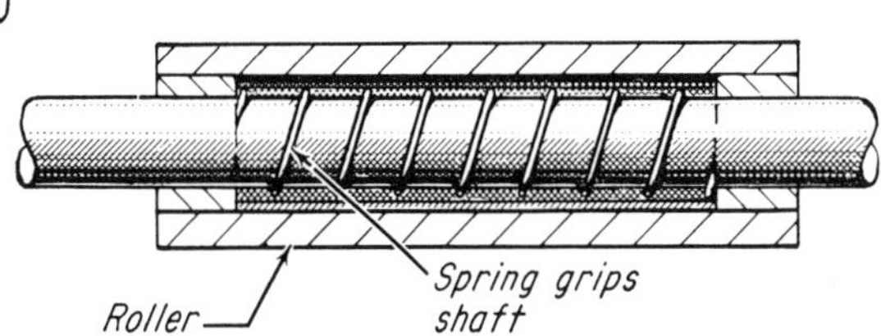

4 **ROLLER POSITIONING** by tight-wound spring on shaft obviates necessity for collars. Roller will slide under excess end thrust.

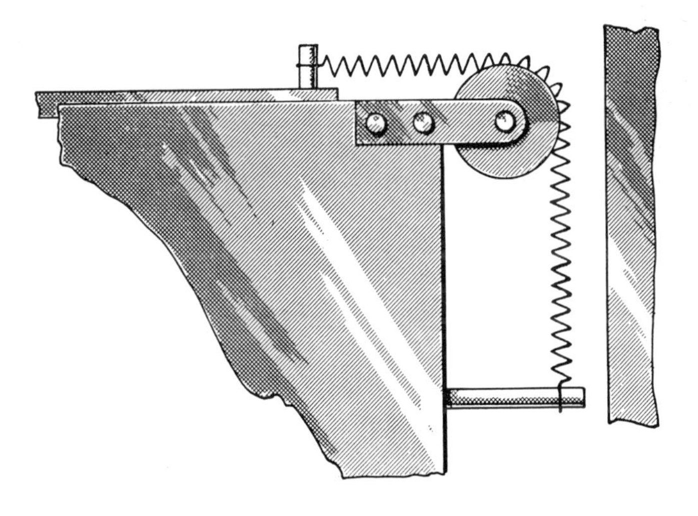

7 **SPRING WHEEL** helps distribute deflection over more coils than if spring rested on corner. Less fatigue and longer life result.

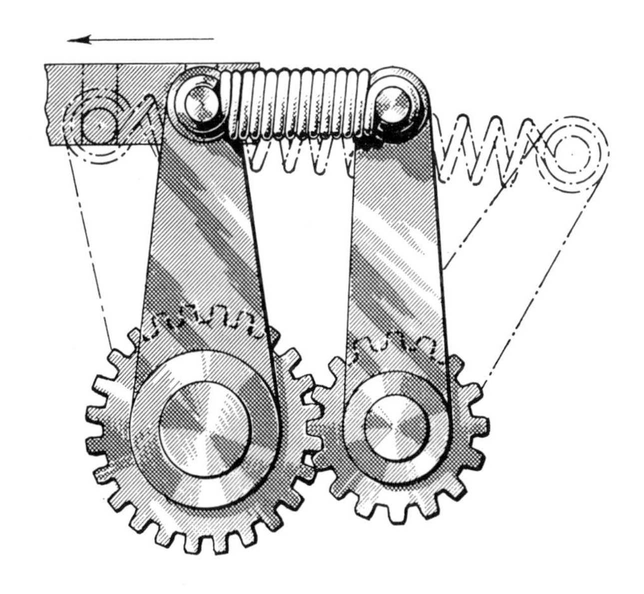

8 **INCREASED TENSION** for same movement is gained by providing a movable spring mount and gearing it to the other movable lever.

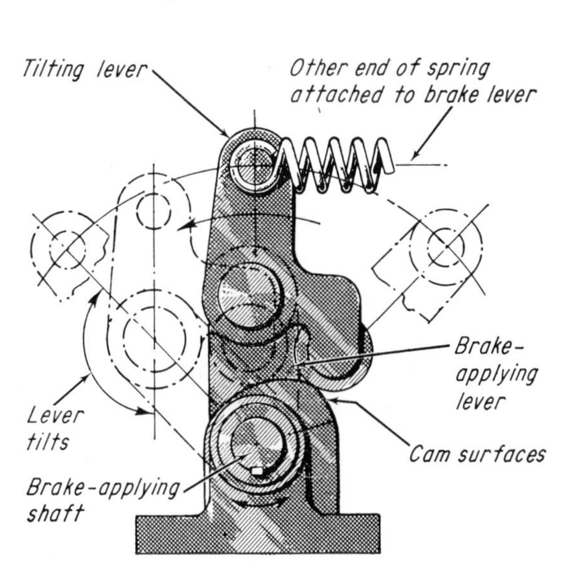

11 **TENSION VARIES** at different rate when brake-applying lever reaches the position shown. Rate is reduced when tilting lever tilts.

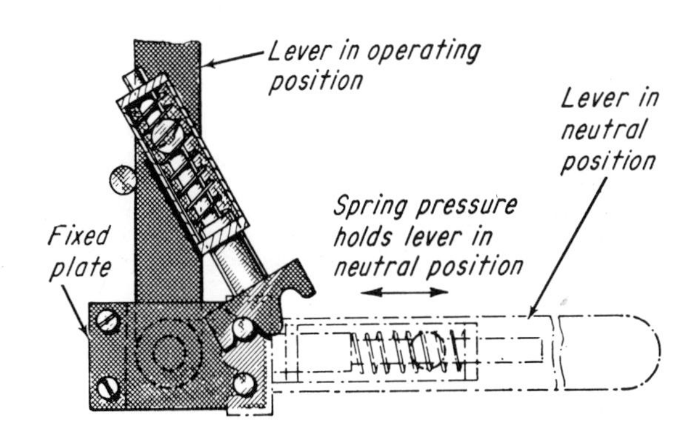

12 **TOGGLE ACTION** here is used to make sure the gear-shift lever will not inadvertently be thrown past neutral.

Flat springs in

These devices all rely on a flat spring for their efficient actions, which would otherwise need more complex configurations.

L. KASPER, Philadelphia

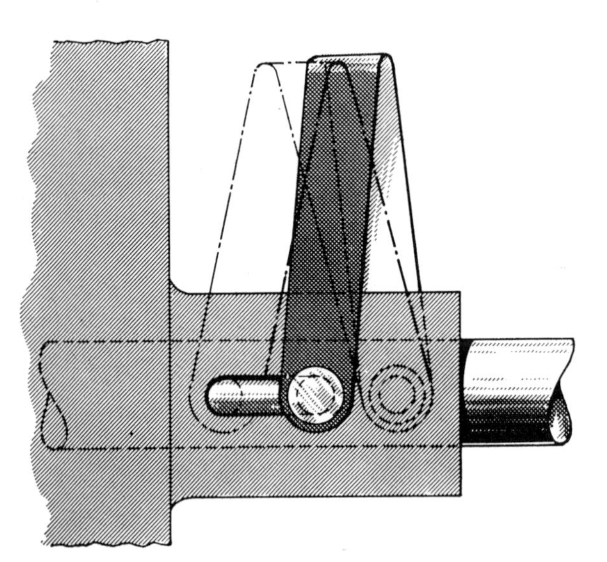

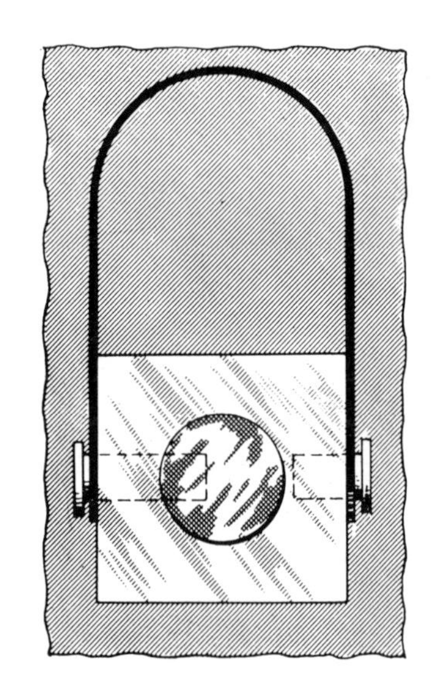

1

CONSTANT FORCE is approached because of the length of this U-spring. Don't align studs or spring will fall.

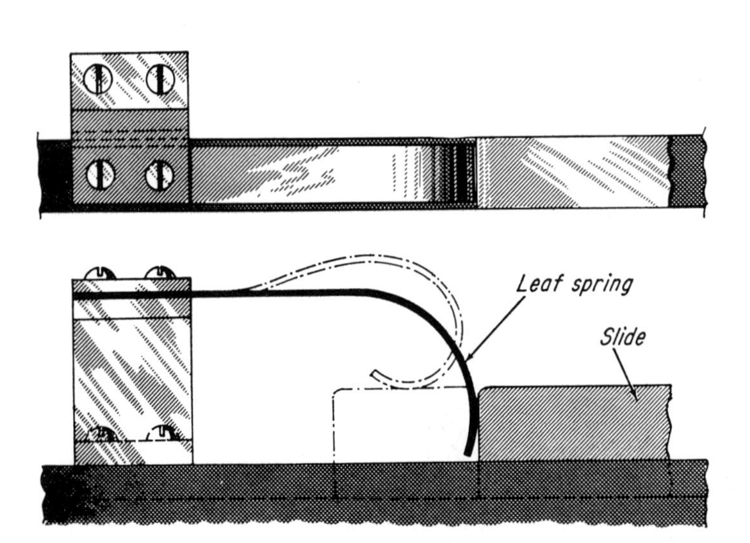

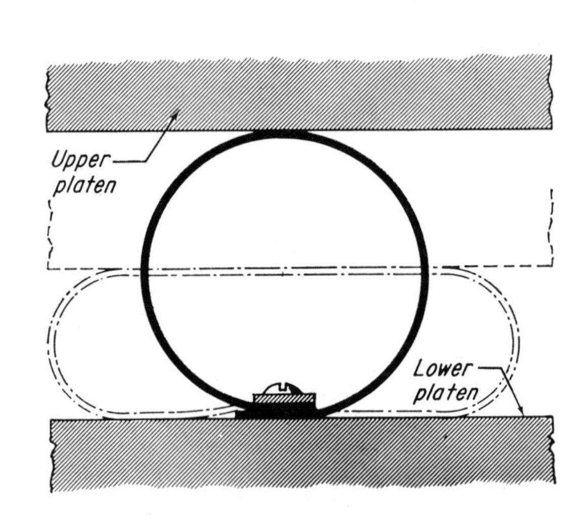

4

SPRING-LOADED SLIDE will always return to its original position unless it is pushed until the spring kicks out.

INCREASING SUPPORT AREA as the load increases on both upper and lower platens is provided by a circular spring.

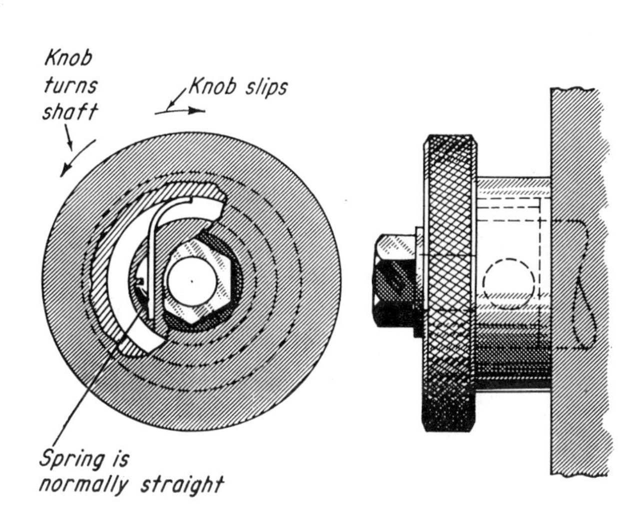

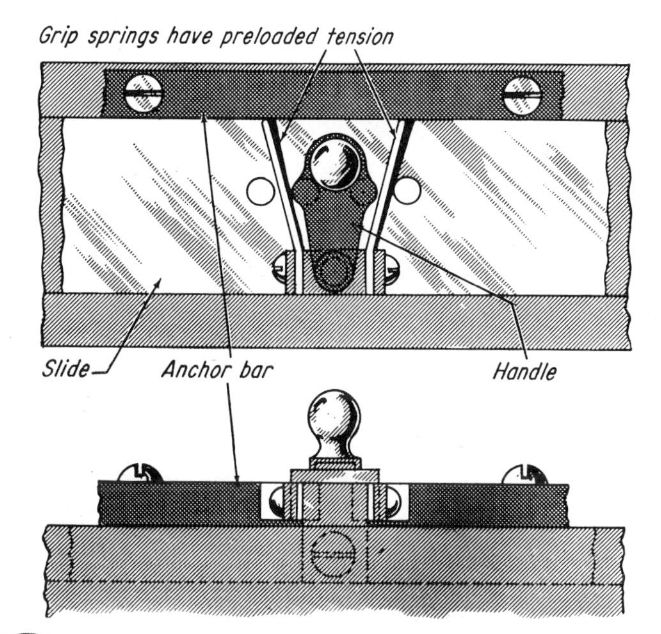

FLAT-WIRE SPRAG is straight until the knob is assembled; thus tension helps the sprag to grip for one-way clutching.

3

EASY POSITIONING of the slide is possible when the handle pins move a grip spring out of contact with the anchor bar.

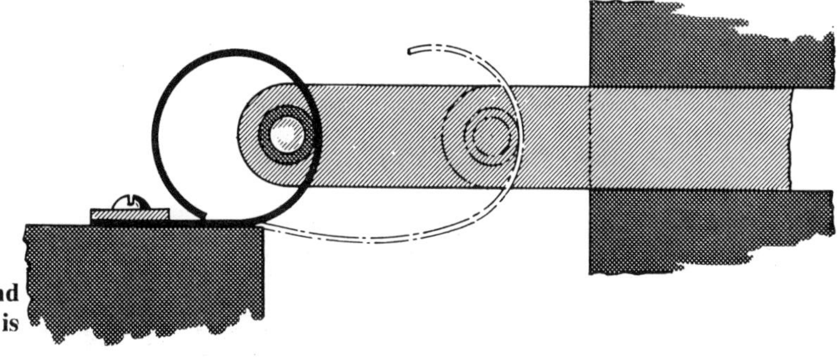

CONSTANT TENSION in the spring, and thus force required to activate slide, is (almost) provided by this single coil.

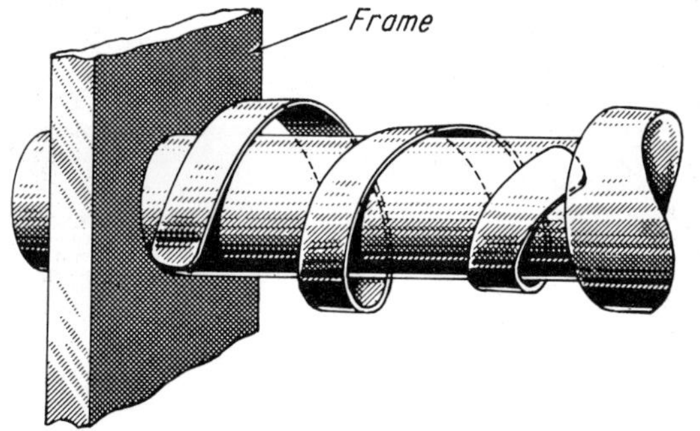

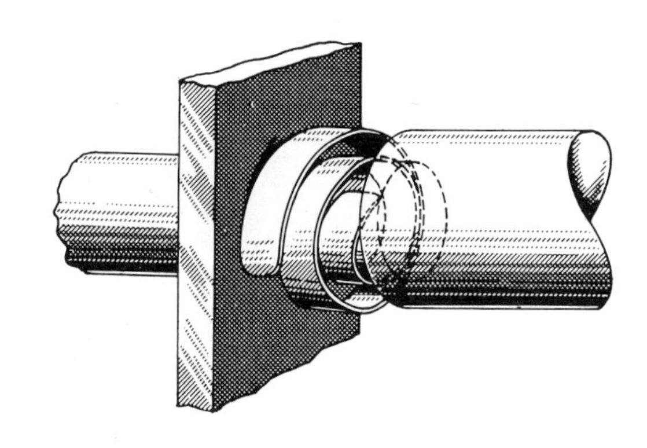

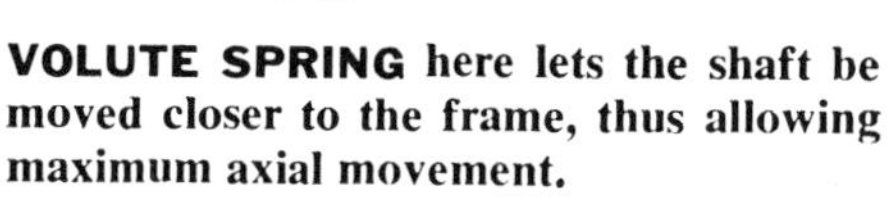

VOLUTE SPRING here lets the shaft be moved closer to the frame, thus allowing maximum axial movement.

Flat springs find

Five additional examples of the way flat springs perform important jobs in mechanical devices.

L. KASPER, *Philadelphia*

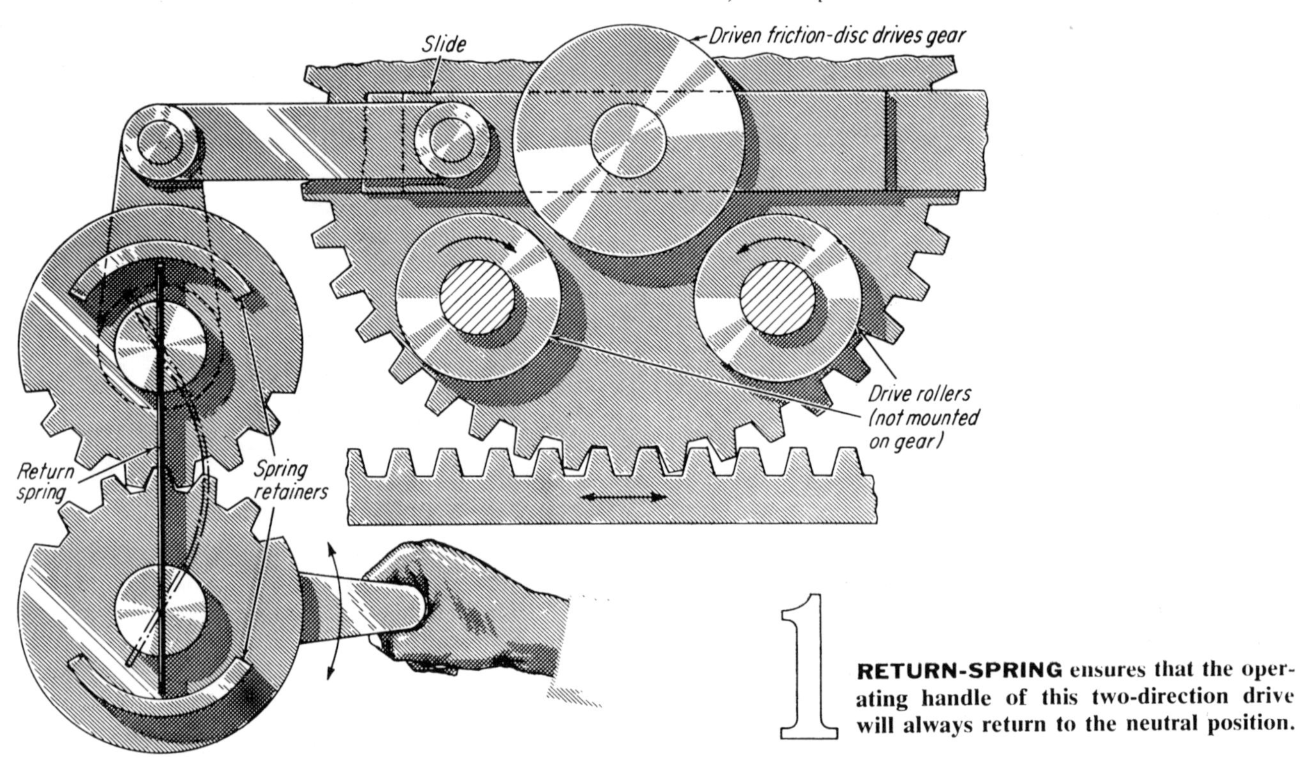

1

RETURN-SPRING ensures that the operating handle of this two-direction drive will always return to the neutral position.

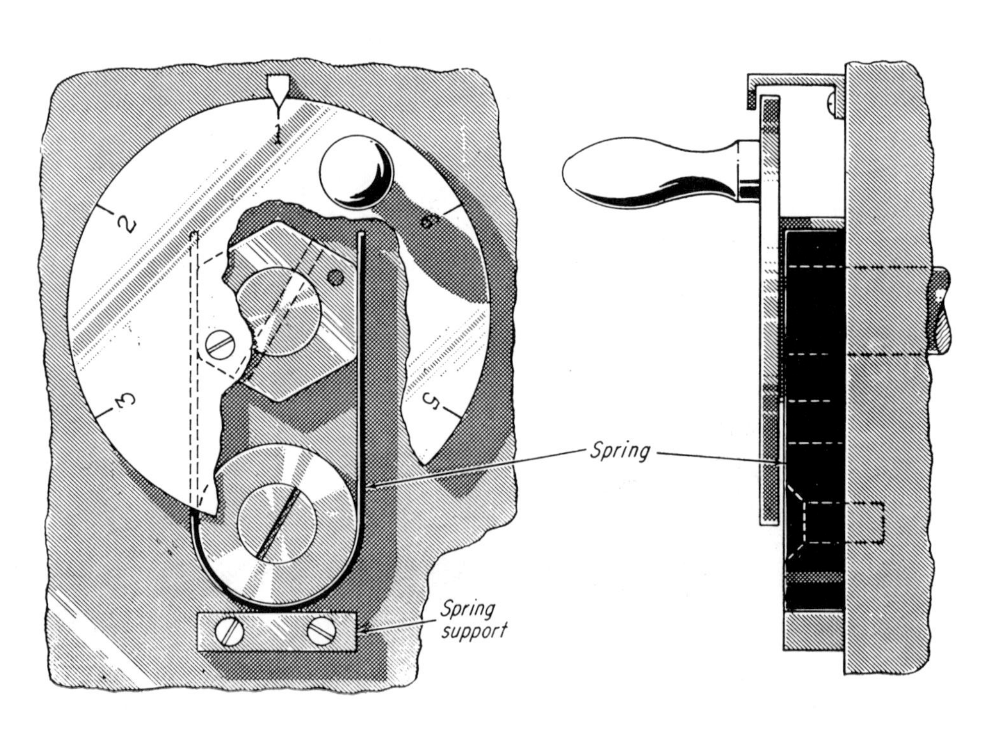

INDEXING is accomplished simply, efficiently, and at low cost by the flat-spring arrangement shown here.

more work

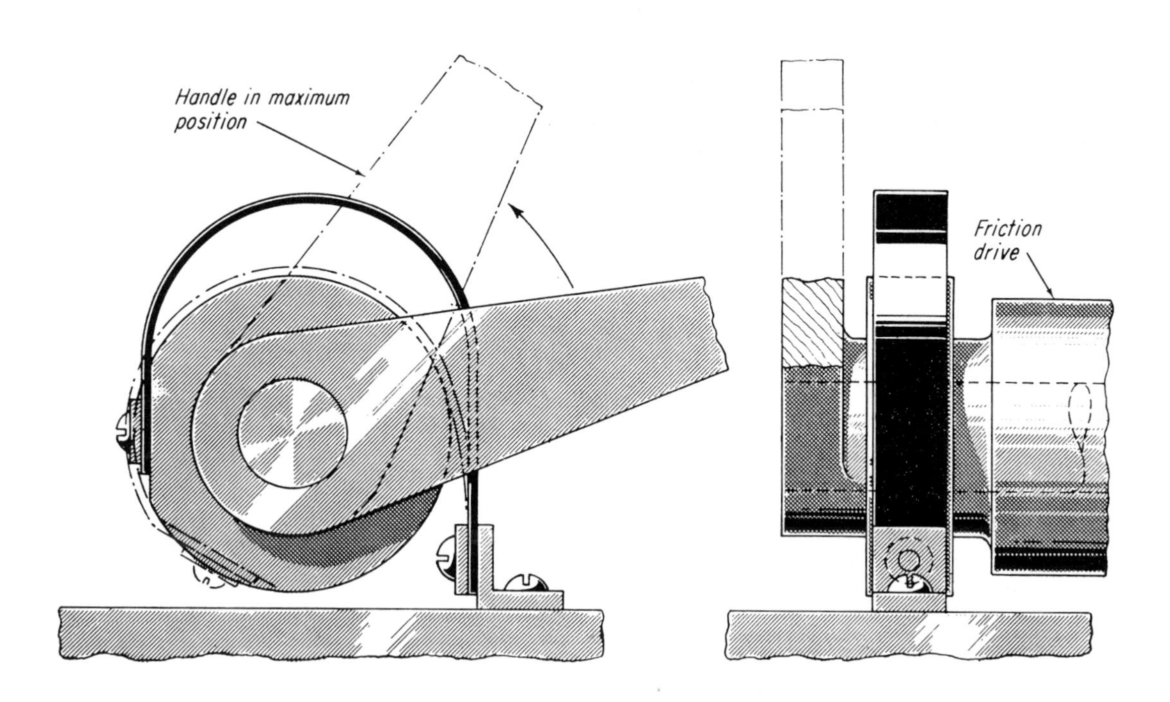

2

SPRING-MOUNTED DISK changes center position as handle is rotated to move friction drive, also acts as built-in limit stop.

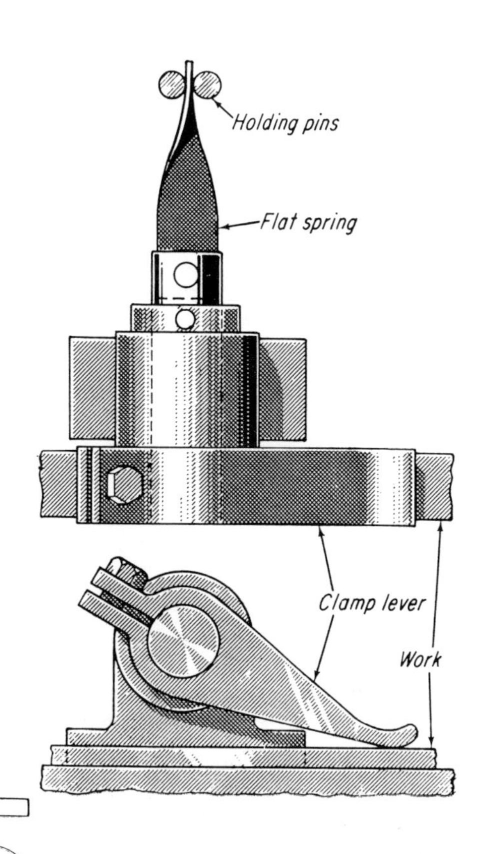

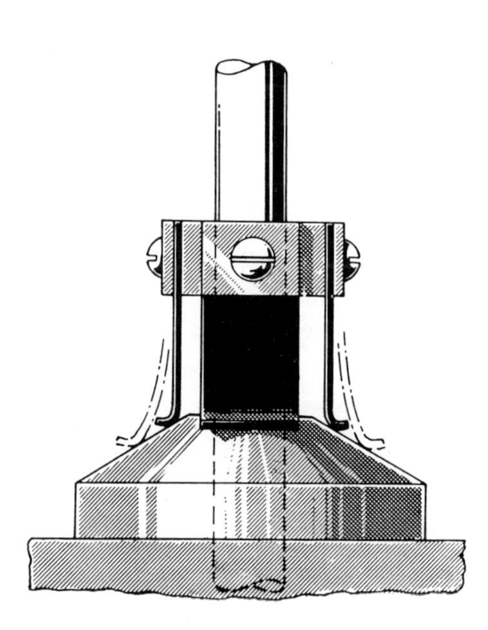

CUSHIONING device features rapid increase of spring tension because of the small pyramid angle. Rebound is minimum, too.

5

HOLD-DOWN CLAMP has flat spring assembled with initial twist to provide clamping force for thin material.

Overriding Spring Mechanisms

Extensive use is made of overriding spring mechanisms in the design of instruments and controls. Anyone of the arrangements illustrated allows an incoming motion to override the outgoing motion whose limit has been reached. In an instrument, for example, the spring device can be placed between

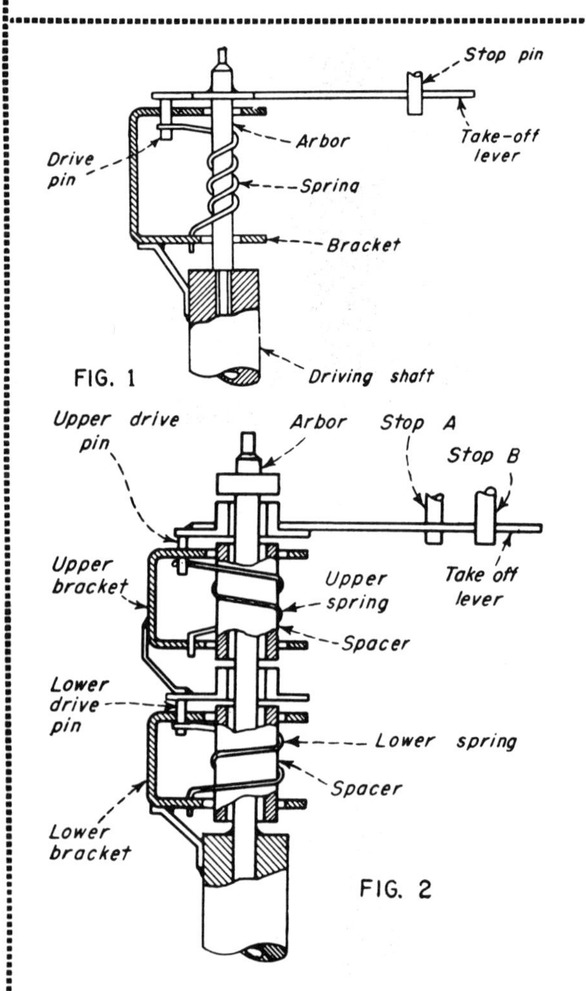

Fig. 1—Unidirectional Override. The take-off lever of this mechanism can rotate nearly 360 deg. It's movement is limited by only one stop pin. In one direction, motion of the driving shaft also is impeded by the stop pin. But in the reverse direction the driving shaft is capable of rotating approximately 270 deg past the stop pin. In operation, as the driving shaft is turned clockwise, motion is transmitted through the bracket to the take-off lever. The spring serves to hold the bracket against the drive pin. When the take-off lever has traveled the desired limit, it strikes the adjustable stop pin. However, the drive pin can continue its rotation by moving the bracket away from the drive pin and winding up the spring. An overriding mechanism is essential in instruments employing powerful driving elements, such as bimetallic elements, to prevent damage in the overrange regions.

Fig. 2—Two-directional Override. This mechanism is similar to that described under Fig. 1, except that two stop pins limit the travel of the take-off lever. Also, the incoming motion can override the outgoing motion in either direction. With this device, only a small part of the total rotation of the driving shaft need be transmitted to the take-off lever and this small part may be anywhere in the range. The motion of the driving shaft is transmitted through the lower bracket to the lower drive pin, which is held against the bracket by means of the spring. In turn, the lower drive pin transfers the motion through the upper bracket to the upper drive pin. A second spring holds this pin against the upper drive bracket. Since the upper drive pin is attached to the take-off lever, any rotation of the drive shaft is transmitted to the lever, provided it is not against either stop *A* or *B*. When the driving shaft turns in a counterclockwise direction, the take-off lever finally strikes against the adjustable stop *A*. The upper bracket then moves away from the upper drive pin and the upper spring starts to wind up. When the driving shaft is rotated in a clockwise direction, the take-off lever hits adjustable stop *B* and the lower bracket moves away from the lower drive pin, winding up the other spring. Although the principal uses for overriding spring arrangements are in the field of instrumentation, it is feasible to apply these devices in the drives of major machines by beefing up the springs and other members.

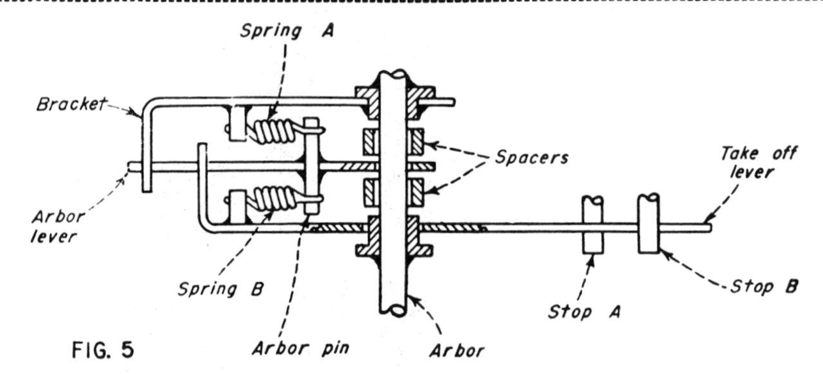

Fig. 5—Two-directional, 90 Degree Override. This double overriding mechanism allows a maximum overtravel of 90 deg in either direction. As the arbor turns, the motion is carried from the bracket to the arbor lever, then to the take-off lever. Both the bracket and the take-off lever are held against the arbor lever by means of springs *A* and *B*. When the arbor is rotated counterclockwise, the take-off lever hits stop *A*. The arbor lever is held stationary in contact with the take-off lever. The bracket, which is soldered to the arbor, rotates away from the arbor lever, putting spring *A* in tension. When the arbor is rotated in a clockwise direction, the take-off lever comes against stop *B* and the bracket picks up the arbor lever, putting spring *B* in tension.

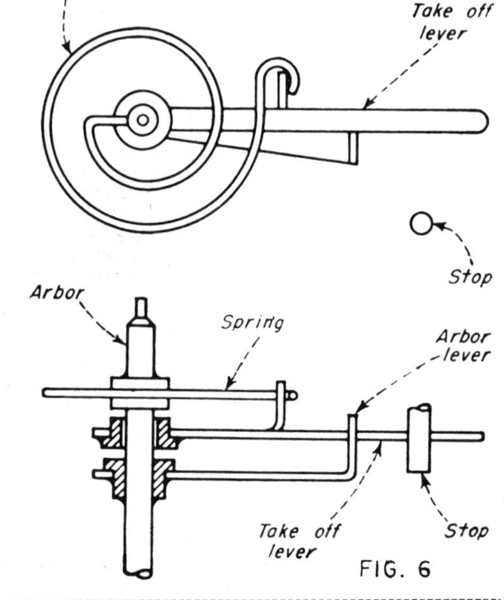

for Low-Torque Drives

HENRY L. MILO, JR.
Division Engineer,
The Foxboro Company

the sensing and indicating elements to provide over-range protection. The dial pointer is driven positively up to its limit, then stops; while the input shaft is free to continue its travel. Six of the mechanisms described here are for rotary motion of varying amounts. The last is for small linear movements.

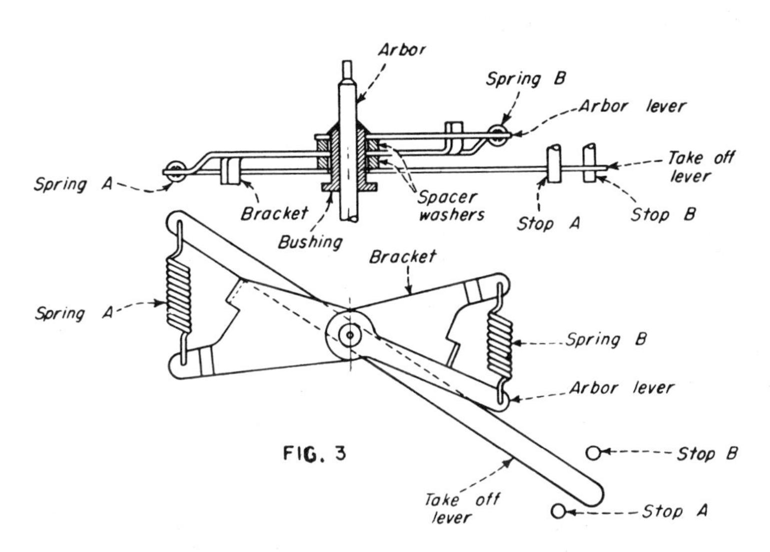

Fig. 3—Two-directional, Limited-Travel Override. This mechanism performs the same function as that shown in Fig. 2, except that the maximum override in either direction is limited to about 40 deg, whereas the unit shown in Fig. 2 is capable of 270 deg movement. This device is suited for uses where most of the incoming motion is to be utilized and only a small amount of travel past the stops in either direction is required. As the arbor is rotated, the motion is transmitted through the arbor lever to the bracket. The arbor lever and the bracket are held in contact by means of spring *B*. The motion of the bracket is then transmitted to the take-off lever in a similar manner, with spring *A* holding the take-off lever and the bracket together. Thus the rotation of the arbor is imparted to the take-off lever until the lever engages either stops *A* or *B*. When the arbor is rotated in a counterclockwise direction, the take-off lever eventually comes up against the stop *B*. If the arbor lever continues to drive the bracket, spring *A* will be put in tension.

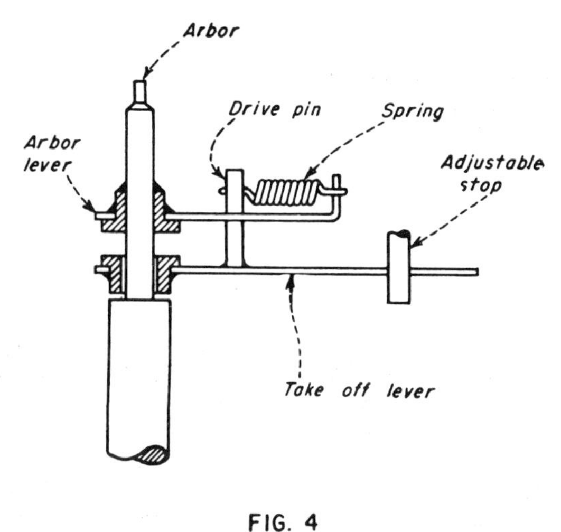

Fig. 4—Unidirectional, 90 Degree Override. This is a single overriding unit, that allows a maximum travel of 90 deg past its stop. The unit as shown is arranged for over-travel in a clockwise direction, but it can also be made for a counterclockwise override. The arbor lever, which is secured to the arbor, transmits the rotation of the arbor to the take-off lever. The spring holds the drive pin against the arbor lever until the take-off lever hits the adjustable stop. Then, if the arbor lever continues to rotate, the spring will be placed in tension. In the counterclockwise direction, the drive pin is in direct contact with the arbor lever so that no overriding is possible.

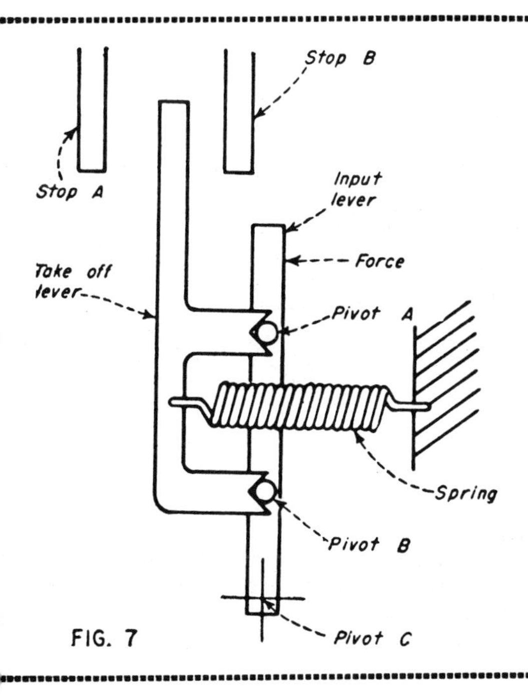

Fig. 6—Unidirectional, 90 Degree Override. This mechanism operates exactly the same as that shown in Fig. 4. However, it is equipped with a flat spiral spring in place of the helical coil spring used in the previous version. The advantage of the flat spiral spring is that it allows for a greater override and minimizes the space required. The spring holds the take-off lever in contact with the arbor lever. When the take-off lever comes in contact with the stop, the arbor lever can continue to rotate and the arbor winds up the spring.

Fig. 7—Two-directional Override, Linear Motion. The previous mechanisms were overrides for rotary motion. The device in Fig. 7 is primarily a double override for small linear travel although it could be used on rotary motion. When a force is applied to the input lever, which pivots about point *C*, the motion is transmitted directly to the take-off lever through the two pivot posts *A* and *B*. The take-off lever is held against these posts by means of the spring. When the travel is such the take-off lever hits the adjustable stop *A*, the take-off lever revolves about pivot post *A*, pulling away from pivot post *B* and putting additional tension in the spring. When the force is diminished, the input lever moves in the opposite direction, until the take-off lever contacts the stop *B*. This causes the take-off lever to rotate about pivot post *B*, and pivot post *A* is moved away from the take-off lever.

7 APPLICATIONS FOR THE

This spring, known as the Neg'ator, finds some up-to-date applications in a magazine feed, one-way brake, motion transfer device, mechanical servo and lifting jack, etc.

HARRY E. NANKONEN
design engineer,
Hunter Spring Co, Lansdale, Pa

1

Strip of spring material . . .
prestressed, formed into a tight coil, and mounted on a freely rotating drum (A) resists withdrawal from the coil with a force that remains constant throughout any extension. In this form the Neg'ator is widely used as a long-deflection spring to perform such functions as counterbalancing, constant-force tensioning and retracting. Spring motor (B) is a second basic form of spring. Here the band is extended and reverse wound about a larger dia output drum. Motor produces almost constant torque throughout entire rundown—50-60 turns or more.

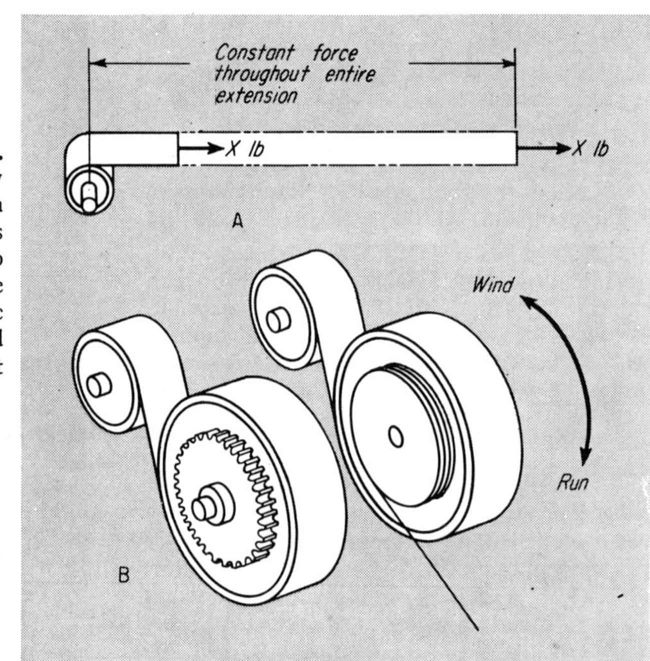

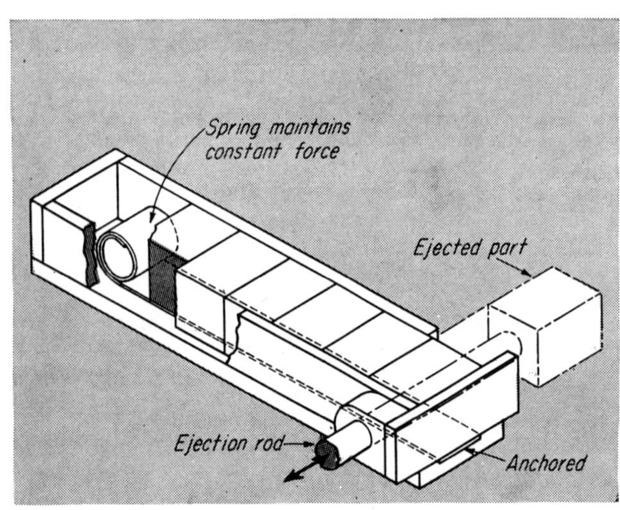

2

Magazine feed . . .
is powered by the self-adjusting coil of extended spring pushing against material being fed. Long feed has no force variation. Device advances work in machine tools and feeds products in vending machines. Assembly stations, where parts must be fed one at a time, can often be made more efficient with these feeds.

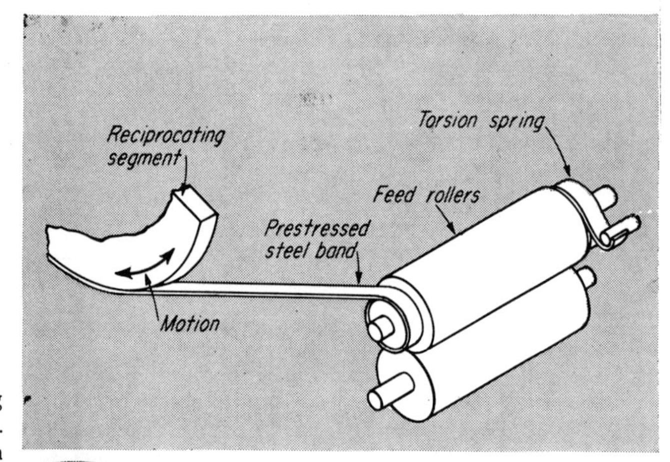

3

Motion transfer . . .
is quiet and accurate; band replaces gears, transmits oscillating motion of cam-operated segment, produces accurately registered rotation of feed rollers. Steel band is prestressed to insure dimensional stability. Torsion spring returns rollers to starting position.

CONSTANT-FORCE SPRING

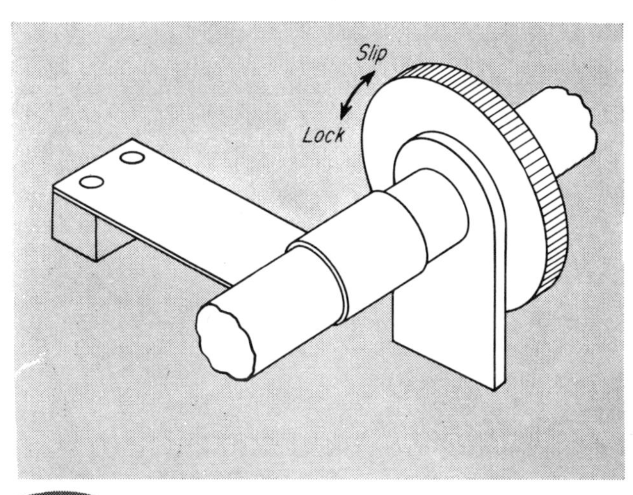

4

One-way brake . . .
prevents counter clockwise rotation of this mechanism, as spring tends to grip more tightly. Clockwise rotation expands the spring coil, lets shaft slip. There is no backlash.

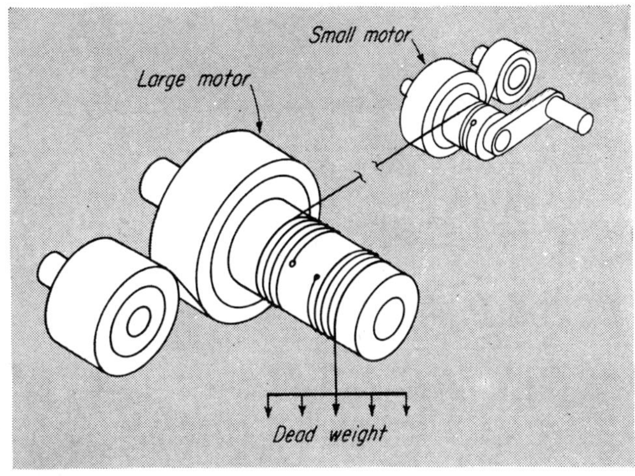

5

Mechanical servo . . .
is formed by two opposed, cable-connected motors. Larger motor provides an output torque sufficient to support both dead weight and force of smaller motor. Friction keeps system static until the remotely located smaller motor is manipulated. Then, the larger motor and load follow.

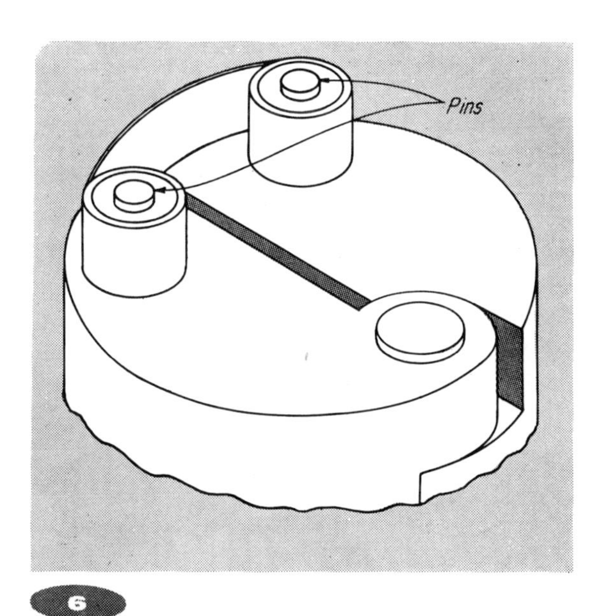

6

Governor spring . . .
provides constant resistance to centrifugal force when hinged governor opens. Pins holding equal coils of spring may either be capped or fitted with retaining rings to prevent spring from working off during rotation.

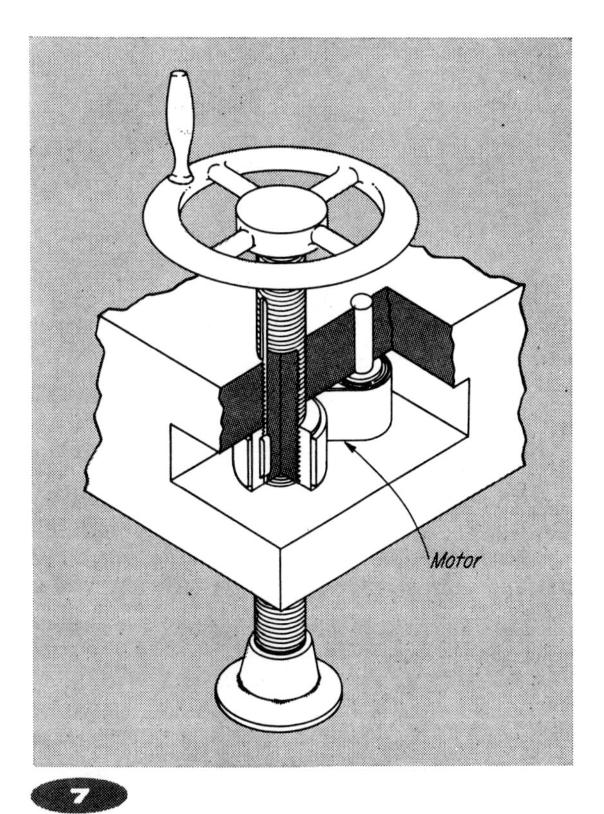

7

Spring-assisted jack . . .
reduces manual effort required to raise heavy equipment. Without added torque supplied by spring motor a larger handwheel is needed for easy operation.

SPRING MOTORS AND TYPICAL

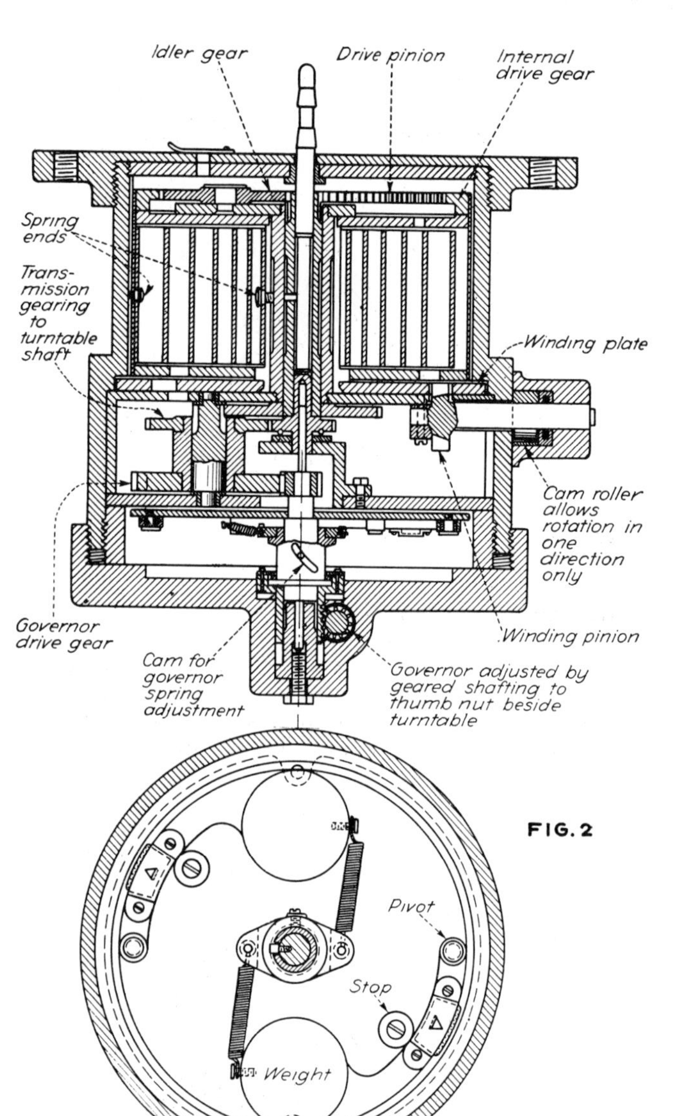

MANY applications of spring motors in clocks, phonographs, motion picture cameras, rotating barber poles, game machines and other mechanisms offer practical ideas for adaptation to any mechanism in which operation for an appreciable length of time is desirable. While spring motors are usually limited to comparatively small power applications where other sources of power are unavailable or impracticable, they may also be useful for intermittent operation requiring

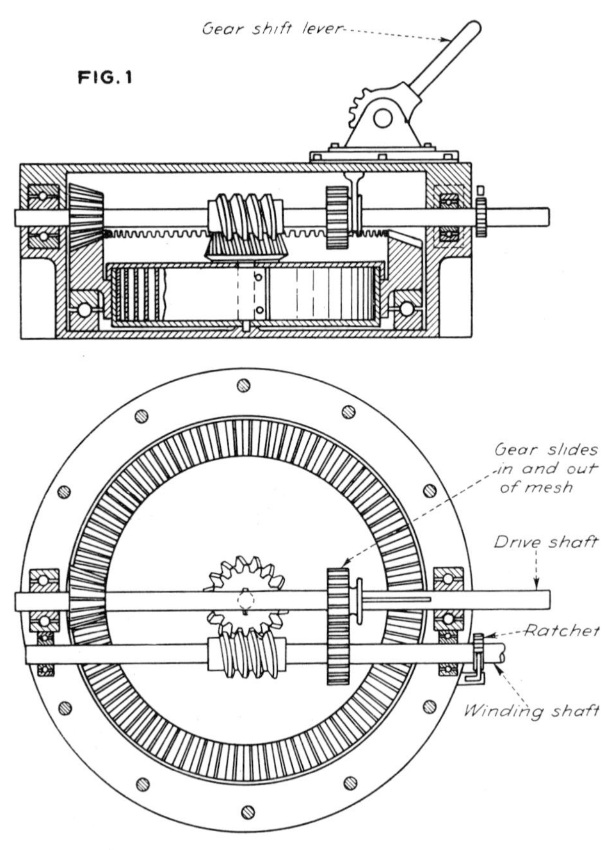

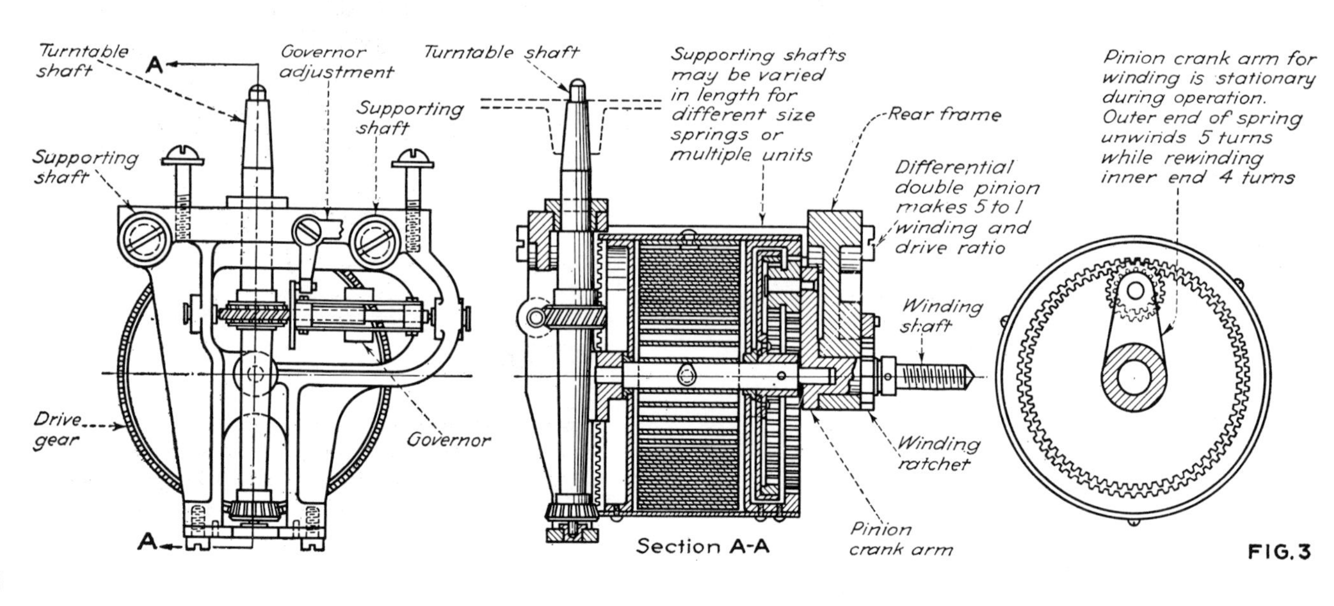

comparatively high torque or high speed, using a low power electric motor or other means for building up energy.

The accompanying patented spring motor designs show various methods of transmission and control of spring motor power. Flat-coil springs, confined in drums, are most widely used because they are compact, produce torque directly, and permit long angular displacement. Gear trains and feed-back mechanisms hold down excess power so that it can be applied for a longer time, and governors are commonly used to regulate speed.

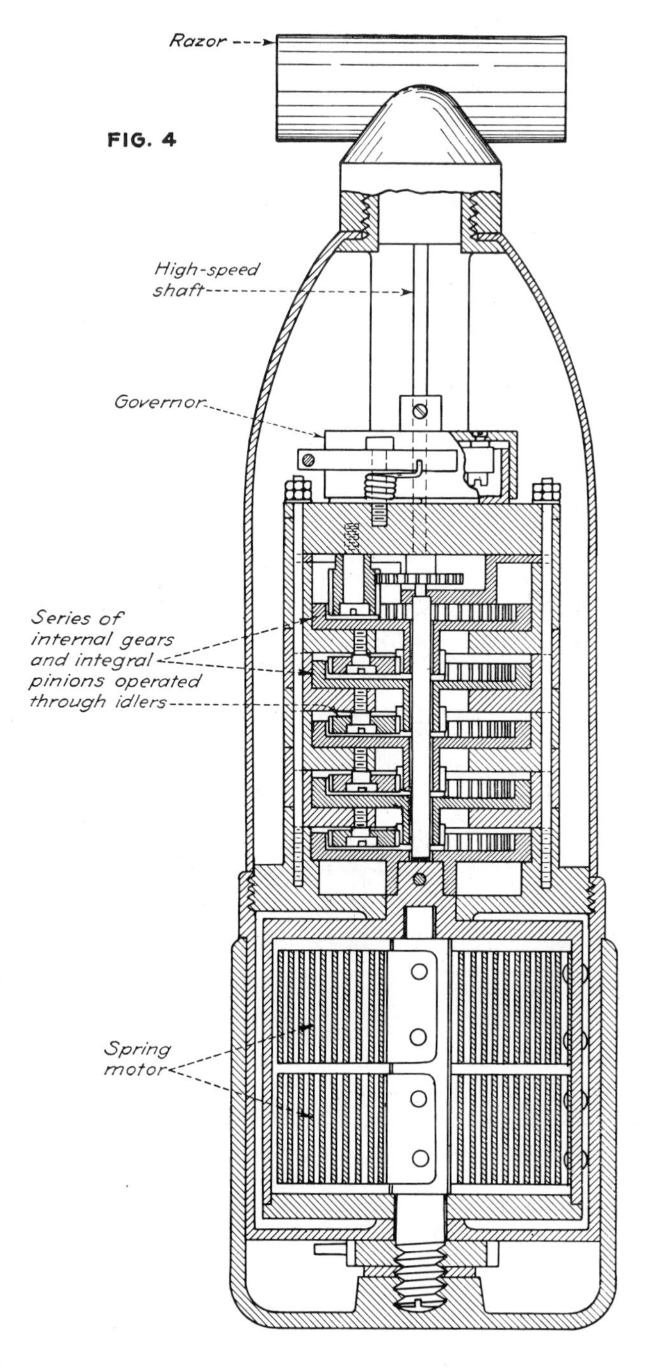

FIG. 4

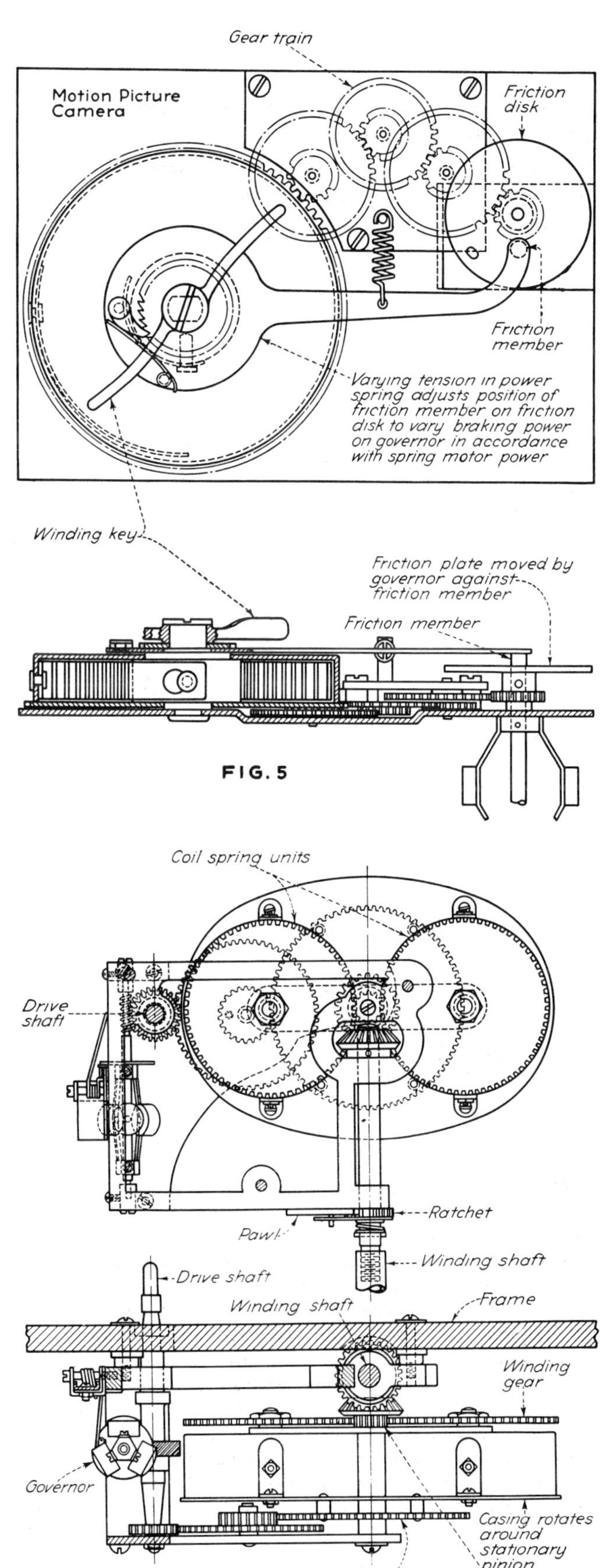

FIG. 5

FIG. 6

Spring and Linkage Arrangements for Vibration Control

Need a buffer between vibrating machinery and the surrounding structure? These isolators, like a capable fighter, absorb the light jabs and stand firm against the forces that are haymakers.

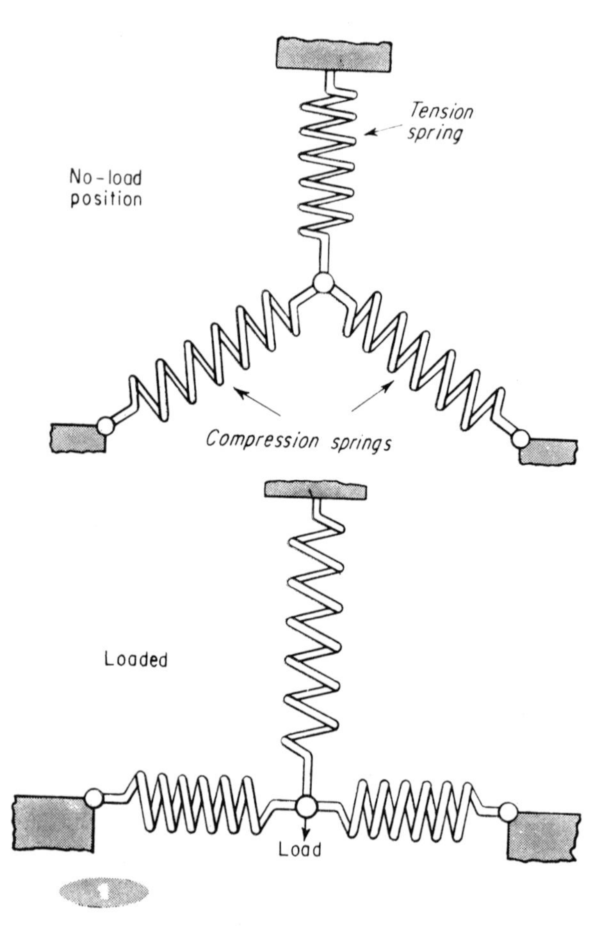

Basic spring arrangement . . .
has zero stiffness, is "soft as a cloud" when compression springs are in line, as illustrated in the loaded position. But change the weight or compression-spring alignment, and stiffness increases greatly. Such a support is adequate for vibration isolation because zero stiffness gives a greater range of movement than the vibration amplitude—generally in the hundredths-of-an-inch range. Arrangements shown here are highly absorbent when required, yet provide a firm support when large force changes occur. By contrast, isolators depending upon very "soft" springs, such as the sine spring, are unsatisfactory in many applications—they allow large movement of the supported load with any slight weight change or large-amplitude displacing force.

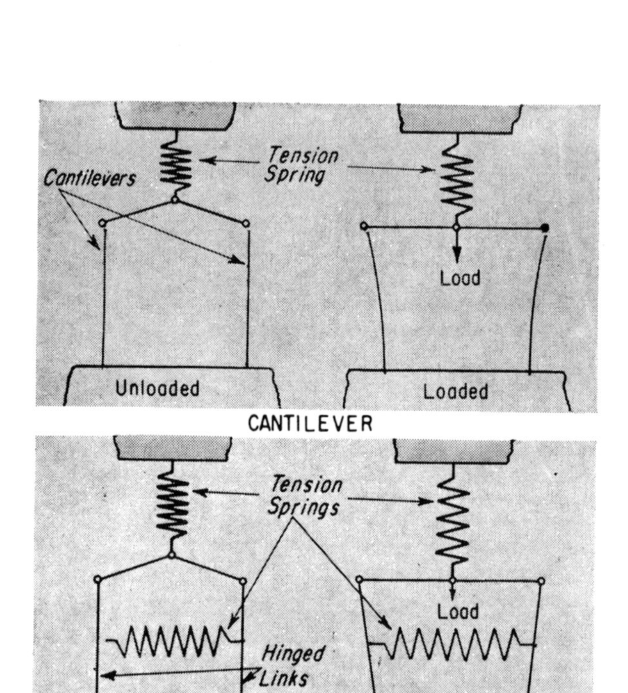

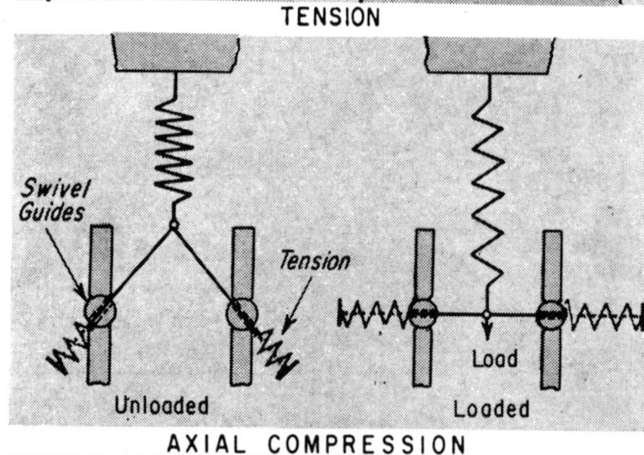

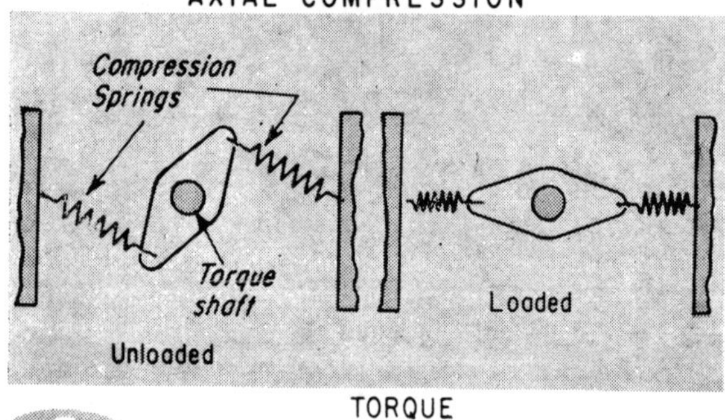

Alternative arrangements . . .
illustrate adaptability of basic design. Here, instead of the inclined, helical compression springs, either tension or cantilever springs can serve. Similarly, different types of springs can replace the axial, tension spring. Zero torsional stiffness can also be provided.

FROM ENGLAND

This new form of spring arrangement is described in "Supports for Vibration Isolation," a British Aeronautical Research Council paper by W. G. Molyneux. He explains and illustrates the basic arrangement, as well as alternatives. From his paper come the applications shown here, together with a brief description of the principles involved.

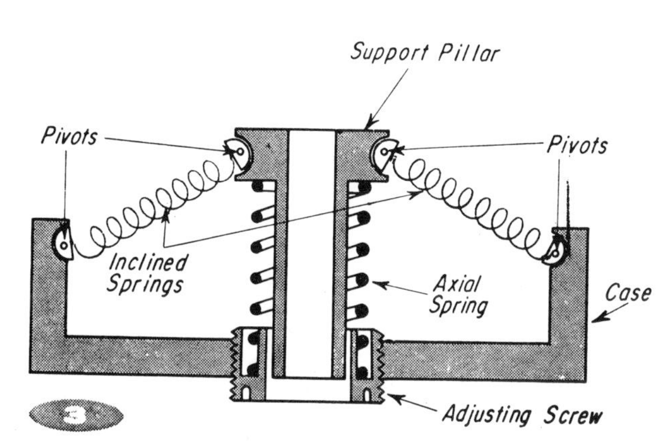

General-purpose support . . .
is based on basic spring arrangement except that axial compression spring is substituted for tension spring. Inclined compression springs, spaced around a central pillar, carry the component to be isolated. When load is applied, adjustment may be necessary to bring inclined springs to zero inclination. Load range that can be supported with zero stiffness on a specific support is determined by the adjustment range and physical limitations of the axial spring.

Various applications . . .
of the principle to vibration isolation show how versatile the design is. Coil spring (4), as well as cantilever and torsion-bar suspension of automobiles can be all reduced in stiffness by adding an inclined spring; stiffness of tractor seat (5) and, consequently, transmitted shocks can be similarly reduced. Mechanical tension meter (6) provides sensitive indication of small variations in tension: as a weighing device, for example, it could detect small variations in nominally identical objects. Nonlinear torque meter (7) provides sensitive indication of torque variations about a predetermined level.

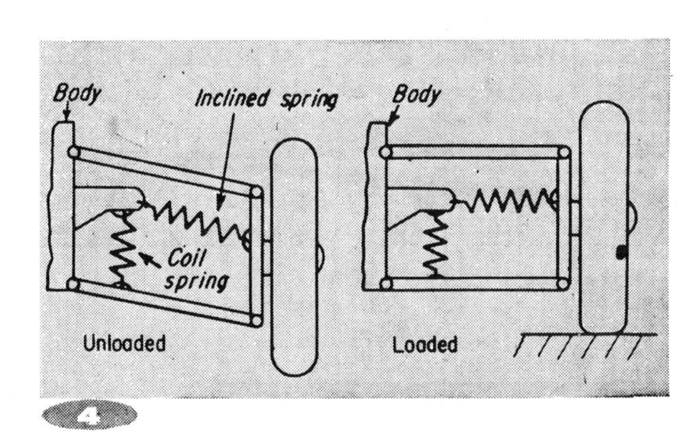

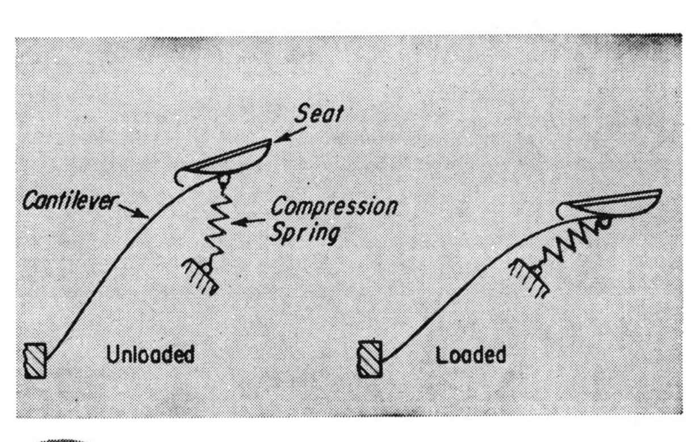

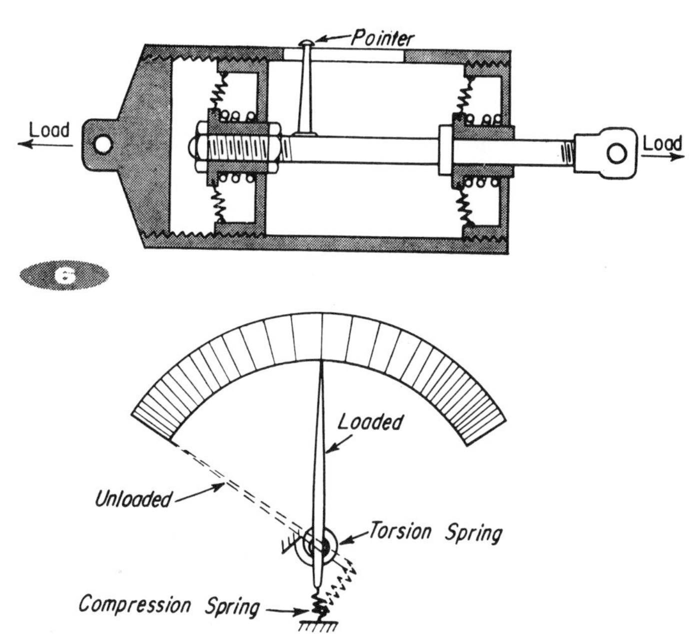

Air Spring Mechanisms

Eight ways to actuate mechanisms with

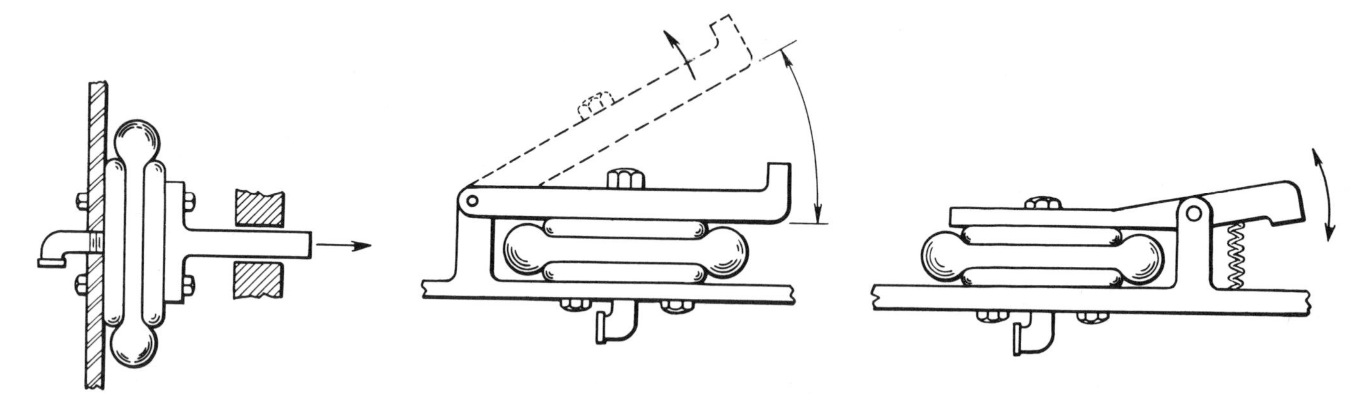

Linear force link—One- or two-convolution air spring drives guide rod. Rod returned by gravity, opposing force, metal spring or, at times, internal stiffness of air spring.

Rotary force link—Pivoted plate can be driven by one-convolution or two-convolution spring to 30 deg of rotation. Limitation on angle is based on permissible spring misalignment.

Clamp—Jaw is normally held open by means of metal spring. Actuation of air spring then closes clamp. Amount of opening in jaws of clamp can be up to 30 deg of arc.

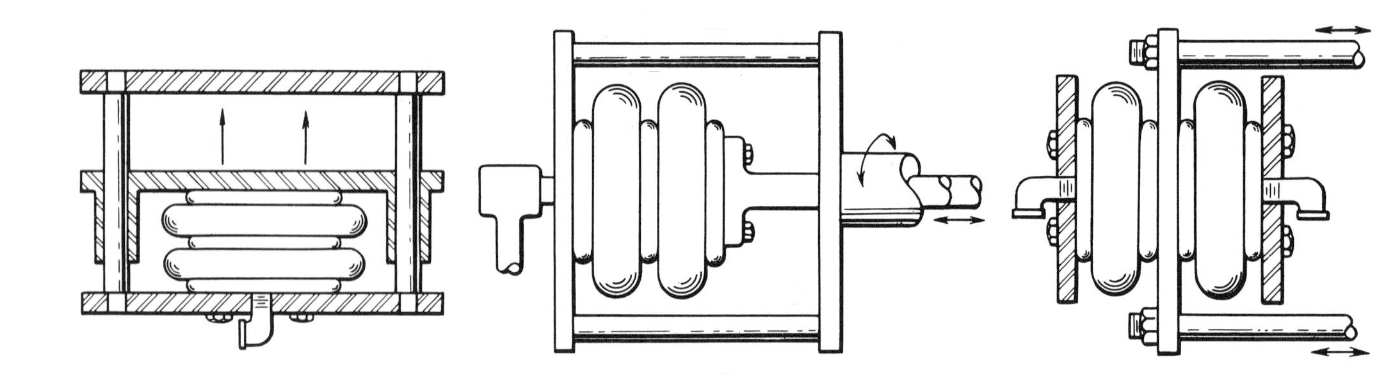

Direct-acting press—One-, two-, or three-convolution air springs used singly or in gangs. Naturally stable when used in groups. Gravity returns platform to starting position.

Rotary shaft actuator — Shifts shaft longitudinally while the shaft is rotating. Air springs with one, two, or three convolutions can be used. Standard rotating air fitting is required.

Reciprocating linear force link—With one-, two-, or three-convolution air springs in back-to-back arrangement. Two- and three-convolution springs may need guides for force rods.

Popular types of air springs

AIR is an ideal load-carrying medium. It is highly elastic, its spring rate can be easily varied, it is not subject to permanent set.

Air springs are elastic devices that employ compressed air as the spring. They maintain a soft ride and a constant vehicle height under varying load. In industrial applications they control vibration (isolate or amplify it) and actuate linkages to provide either rotary or linear movement. Three kinds of air springs (bellows, rolling sleeve, rolling diaphragm) are illustrated.

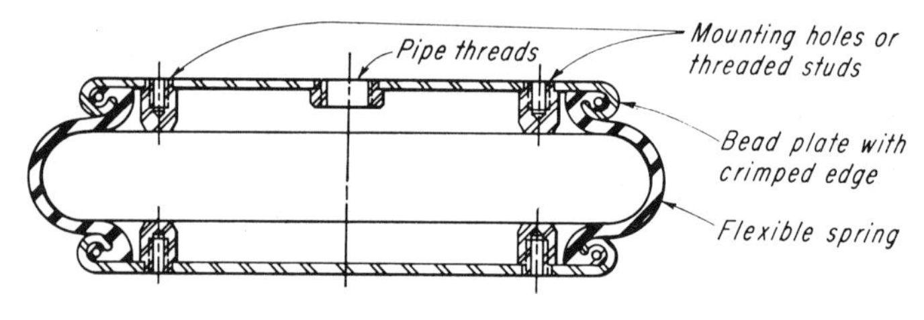

ONE-CONVOLUTION BELLOW

Bellows type

A single-convolution spring looks like a tire lying on its side. It has limited stroke and relatively high spring rate. Natural frequency is about 150 cpm without auxiliary volume for most sizes, and as high as 240 cpm on smallest size. Lateral stiffness is high (about half the vertical rate); therefore the spring is quite stable laterally when used for industrial vibration isolation. It can be filled manually or

air springs

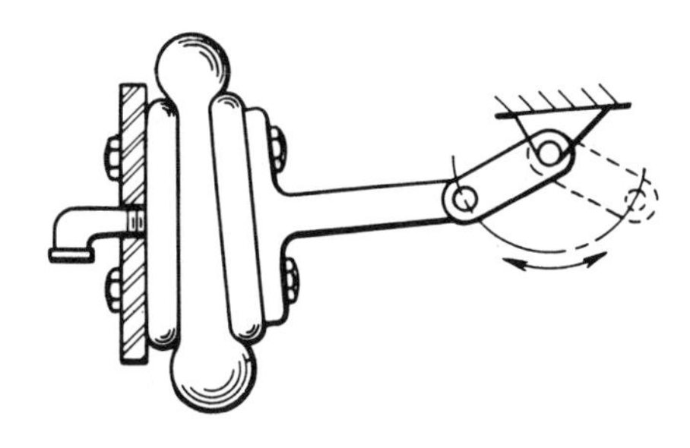

Pivot mechanism—Rotates rod through 145 deg of rotation. Can take 30-deg misalignment owing to circular path of connecting-link pin. Metal spring or opposing force retracts link.

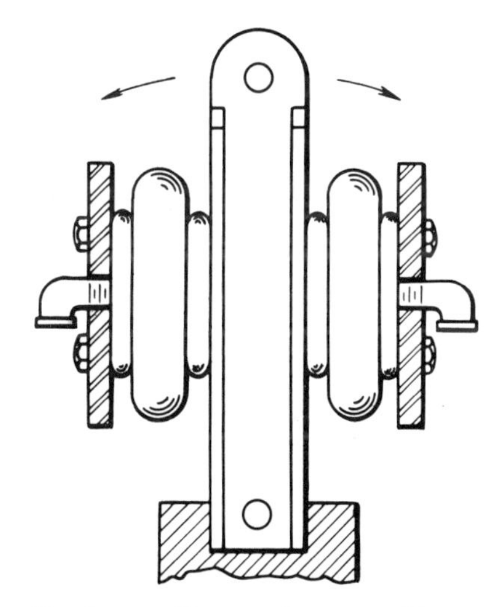

Reciprocating rotary motion—With one-convolution and two-convolution springs. Arc up to 30 deg is possible. Can pair large air spring with smaller one or lengthen lever.

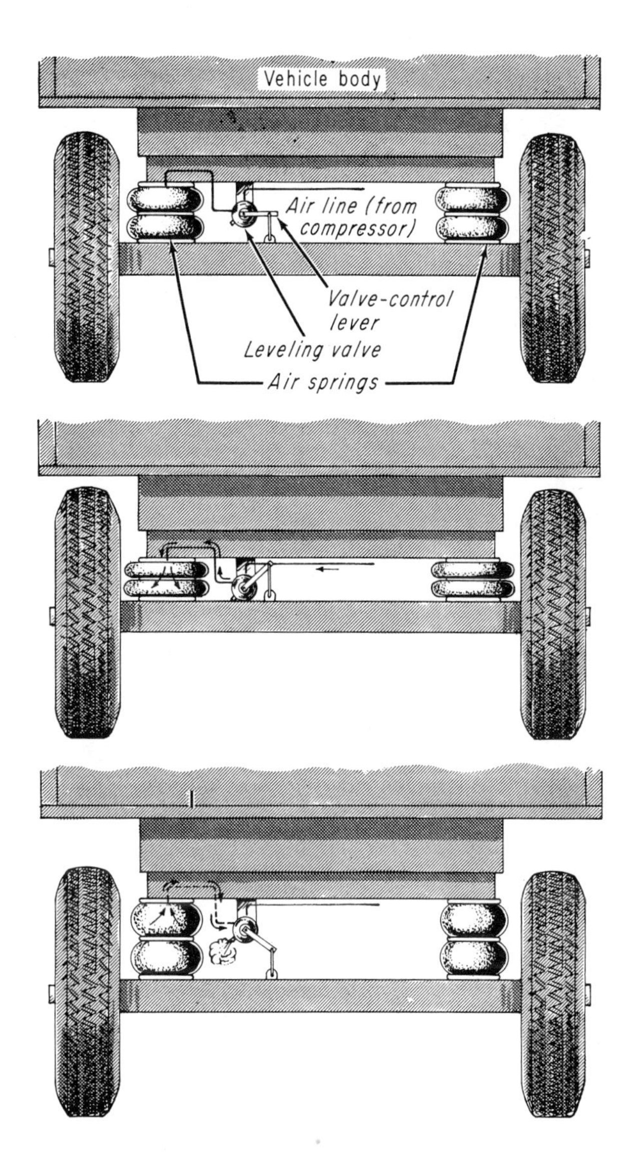

AIR SUSPENSION ON VEHICLE: (a) Normal static conditions —air springs at desired height, height-control valve closed. (b) Load added to vehicle—valve opens to admit air to springs and restore height, but at higher pressure. (c) Load removed from vehicle—valve permits bleeding off excess air pressure to atmosphere and restores design height.

kept inflated to constant height if connected to factory air supply through a pressure regulator. This spring will also actuate linkages where short axial length is desirable. It is seldom used in vehicle suspensions.

Rolling-sleeve type

This is sometimes called the reversible-sleeve or rolling-lobe type. It has telescoping action—the lobe at the bottom of the air spring rolls up and down along the piston. The spring is used primarily in vehicle suspensions because lateral stiffness is almost zero.

Rolling-diaphragm type They are laterally stable and can be used as vibration isolators, actuators, or constant-force springs. But because of the negative effective-area curve, they are not generally supplied by pressure regulators

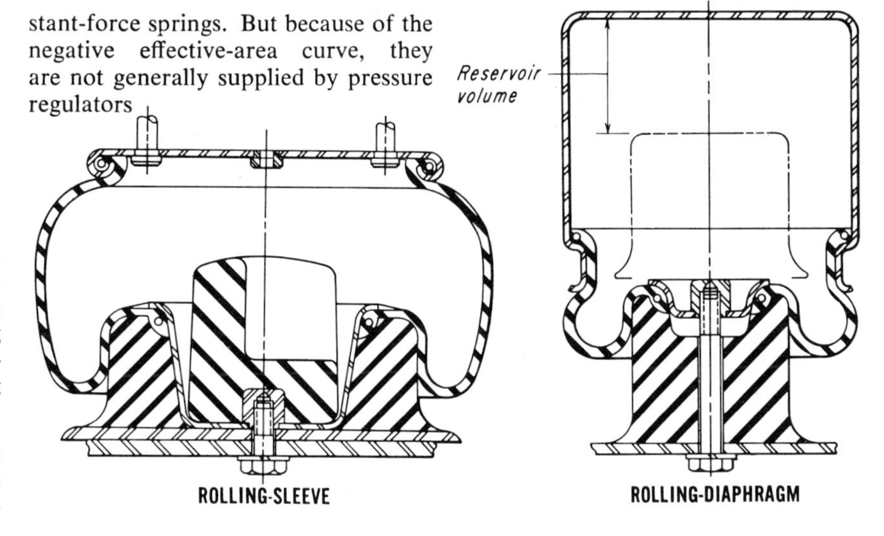

Selecting Metallic Bellows

Common Types of Bellows and Methods of Attaching End Fittings

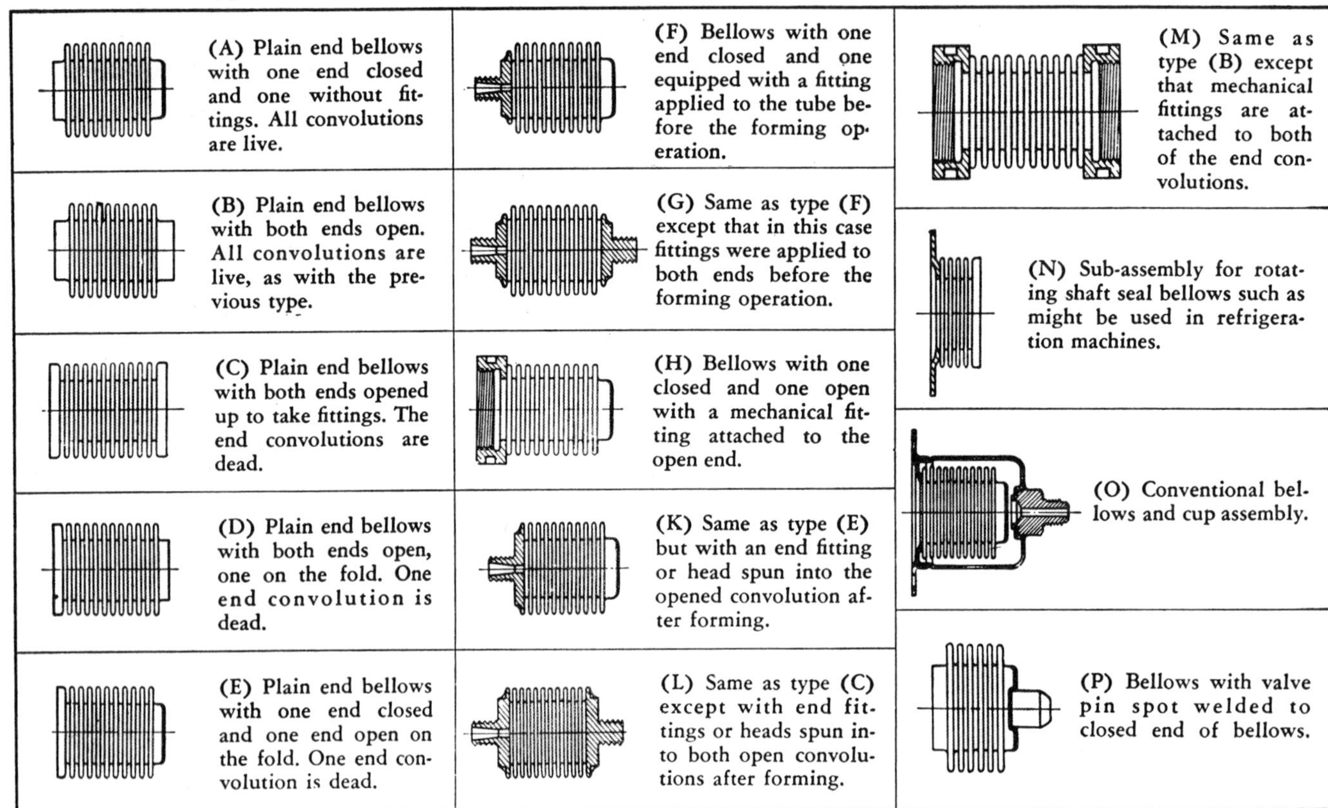

CONSTRUCTION OF BELLOWS

A bellows usually is formed in one continuous operation from a thin seamless tube into the finished form, but the final construction may be varied considerably. Both ends may be opened or closed. If open, they may be open on the folds or between the folds. The ends may be fitted with different attachments: An internal or external screw head attachment; a special fitting having a valve pin on the end; a special shaft seal fitting; or an enclosing envelope. These modifications are shown in the accompanying table. The reasons for these various constructions is to facilitate assembly or to improve adaptation of the bellows to a particular control device, as will be discussed.

FLEXIBILITY

The flexibility of a bellows is directly proportional to number of convolutions in a given length, materials and wall thickness remaining the same. Doubling the number of convolutions doubles the flexibility and halves the spring rate. While this may be done when space is limited, the simpler procedure is to use a longer bellows. However, the length to diameter ratio should not be much greater than 1 to 1 to avoid buckling in service. This tendency can be alleviated to some extent by using an internal or external guide.

The smaller the diameter, the less the flexibility and the smaller the deflection with a given load. Flexibility in terms of stroke per unit load varies directly as the square of the outside diameter of the bellows. The stiffness of a small bellows therefore may require that the length be increased to $1\frac{1}{2}$ or even twice the diameter to obtain the desired length of travel.

FILLING MEDIUM

A bellows may be a thermostatic assembly, in which case it is operated by the fluid with which it is filled. This fluid may be a liquid, gas or vapor.

Bellows that depend on gaseous expansion are filled with an inert gas. Changes in temperature cause corresponding changes in volume. This type of control requires a bulb sub-assembly whose volume is many times greater than that of the bellows. Even with such an arrangement, however, the movement of the bellows per unit change in temperature is small, and the switch or valve to be operated must be capable of functioning with a bellows movement of 0.006 to 0.007 in. or less. This class of bellows usually is applied at low temperatures—on the order of 300 F maximum.

Vapor pressure bellows are widely used for temperature control of either heated or refrigerated spaces. To generate the necessary pressure at temperatures below normal and yet keep the pressures at higher temperatures from becoming excessive, the bellows is filled with a minute amount of fluid whose vapor pressure characteristics are suitable for the temperature range in which the bellows assembly is to operate. By controlling the amount of fluid introduced into the bellows, the total pressure generated at a specified temperature can be limited. Such bellows have greater sensitivity and more power than units filled with gas. Also, good regulation is obtained even with a small bulb and a large bellows.

Liquid filled bellows are usually small and compact, can exert a relatively large force, and, while expansion per unit change in temperature is small, they are usually less affected by

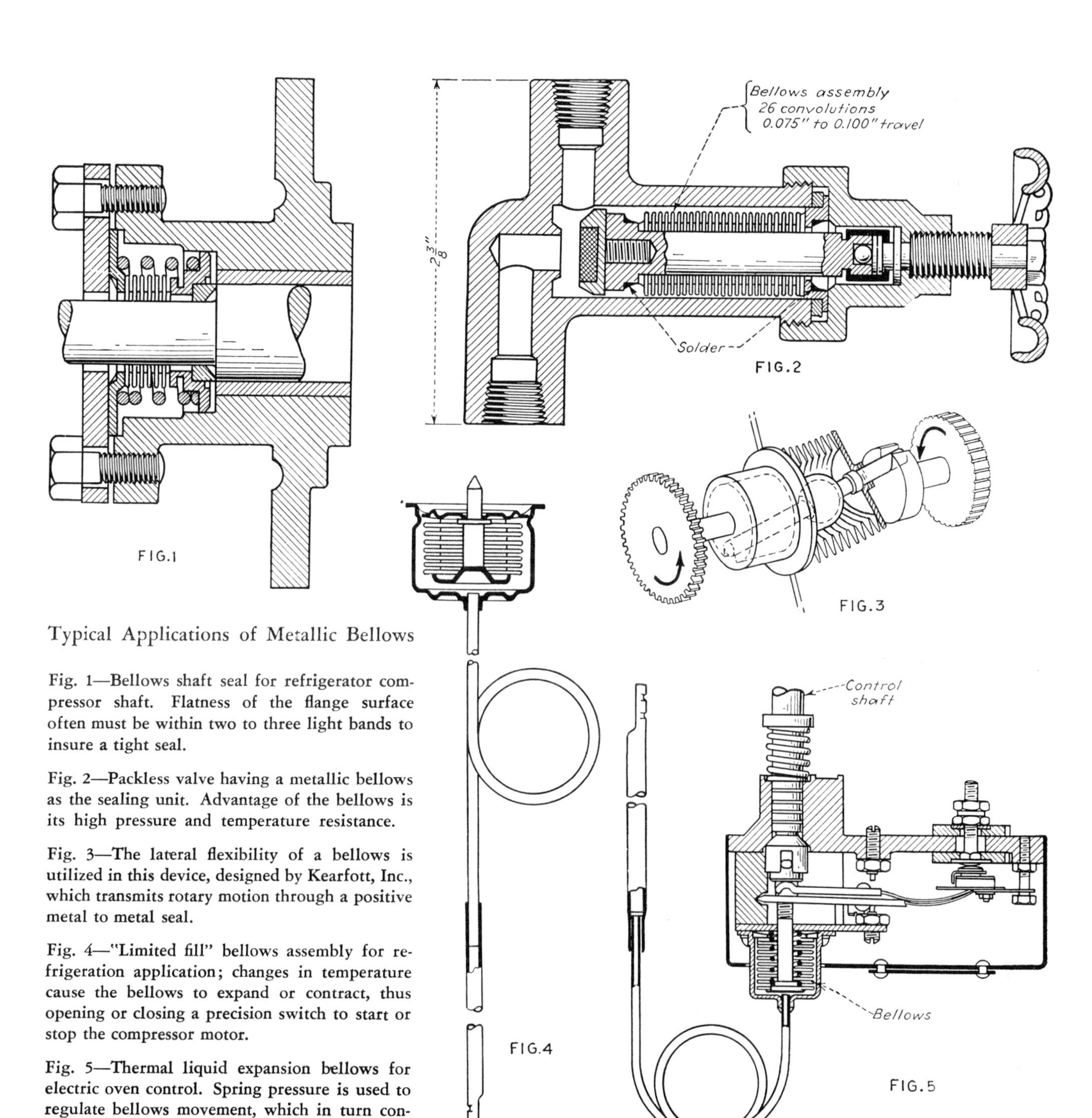

Typical Applications of Metallic Bellows

Fig. 1—Bellows shaft seal for refrigerator compressor shaft. Flatness of the flange surface often must be within two to three light bands to insure a tight seal.

Fig. 2—Packless valve having a metallic bellows as the sealing unit. Advantage of the bellows is its high pressure and temperature resistance.

Fig. 3—The lateral flexibility of a bellows is utilized in this device, designed by Kearfott, Inc., which transmits rotary motion through a positive metal to metal seal.

Fig. 4—"Limited fill" bellows assembly for refrigeration application; changes in temperature cause the bellows to expand or contract, thus opening or closing a precision switch to start or stop the compressor motor.

Fig. 5—Thermal liquid expansion bellows for electric oven control. Spring pressure is used to regulate bellows movement, which in turn controls the on and off switch.

variations in flexibility than bellows of other types. The change in length per degree F change in temperature is on the order of 0.001 in. maximum. However, smaller changes in length per degree F change in temperature make regulation possible over a range from 100 to 650 F; as a general rule, the expansion of the liquid per degree change in temperature is greater for liquids having a low boiling point.

When used for control purposes, liquid filled bellows are extremely accurate.. They are generally applied for actuation of switches that must operate over a wide temperature range at moderate sensitivity. Many bellows used on aircraft are of the liquid filled type, but for another reason: Their accuracy is only slightly affected by changes in barometric pressure.

There is one other type of bellows, the aneroid bellows, that also is extensively used in aircraft. This is an evacuated unit that is extremely sensitive to changes in atmospheric pressure; as the atmospheric pressure becomes less, the bellows expands. One use is for oxygen-air regulators. Another is to regulate the fuel-air ratio to the engine. While there are some uses other than for aircraft for this type of bellows, they are not extensive.

LIFE

The actual life of a bellows is dependent upon many factors, but for any given bellows is dependent on the operating stroke and pressure. The further the bellows operates away from the maximum stroke and pressure rating , the longer the life rating,

Throttle linkage includes bellows . . .

that override accelerator pedal while the clutch remains disengaged. It is disengaged by the diaphragm motor when driver squeezes switch on gearshift lever and energizes shift valve in the control unit. Simultaneously, vacuum is applied to the throttle-closing bellows, which moves down and overrides accelerator linkage. When shift lever and switch are released, clutch starts to re-engage but is retarded by an air-flow regulating valve positioned by the accelerator; depressing the accelerator speeds up re-engagement. A synchronizer switch in the clutch energizes second valve in control unit if engine speed is slower than that of the driven clutch plate. This valve applies vacuum to the throttle-opening bellows, which overrides throttle-closing bellows and accelerator linkage, and also holds clutch disengaged by blocking atmospheric vent of the diaphragm motor. When engine and clutch plate speeds are synchronized, the accelerator linkage again controls engine speed.

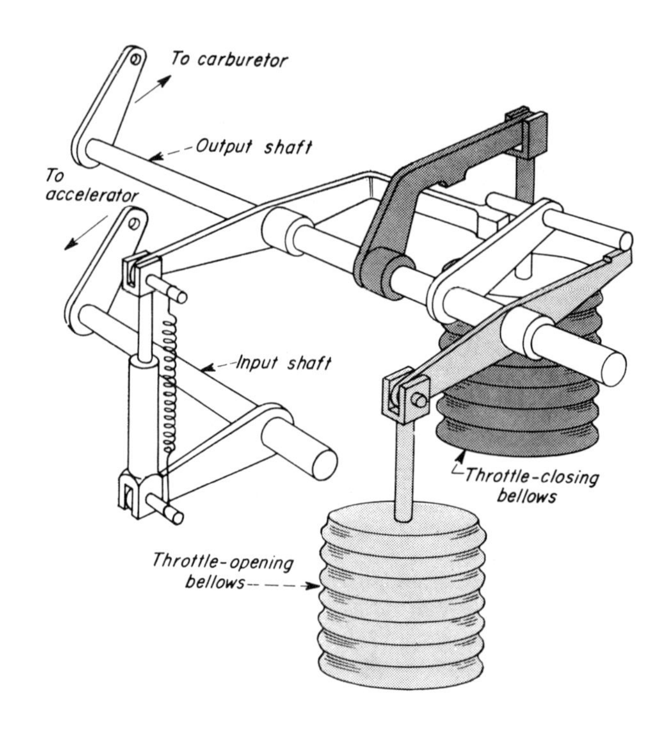

Diaphragms Actuate Sandblast Valve

A SINGLE VALVE controls air supply, abrasive feed and blast in the sandblasting machine developed by the O. Granowski Co., Melbourne, Australia. Since the three functions are regulated simultaneously, there is no waste of sand or air after blast shut-off as occurs when separate valves are employed. When the inlet valve is turned to the "on" position, compressed air is admitted to the mixing chamber of the valve mechanism. Pressure causes the diaphragm to bulge, moving spindle to the left and opening sand gate and outlet valve. When blasting operation is completed, the inlet valve is turned to divert air supply from mixing chamber into sealed pressure chamber. Then, the large diaphragm pushes spindle to right, closing sand gate and blast outlet.

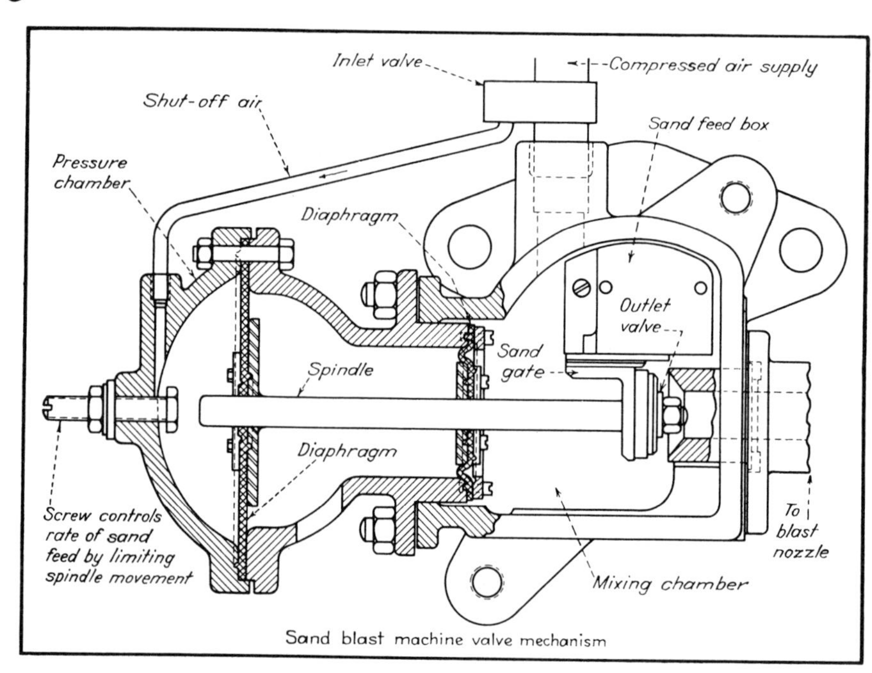

Sand blast machine valve mechanism

Bellows Controls Camera Exposure

Air flowing through holes in a bellows dashpot sets exposure time in an automatic camera. The pointer of a meter movement energized by a photocell controls flow rate by covering more holes to increase exposure time.

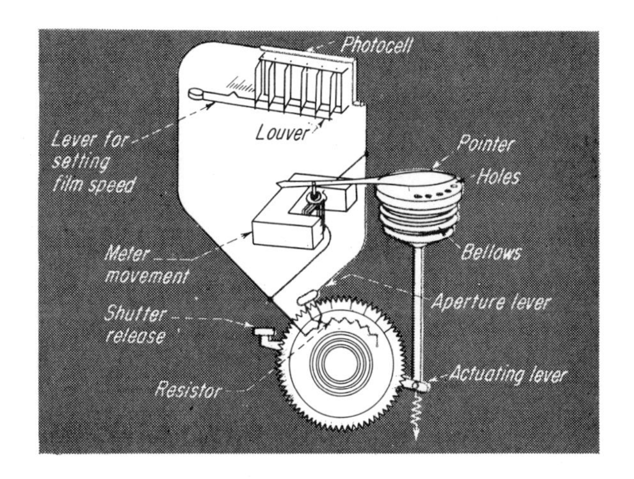

Exposure time . . . is automatically controlled, after aperture is set for desired depth of focus, and louvers on photocell are set for film speed. Shunt resistance across meter element, adjusted by aperture lever, compensates for lens stop. When shutter is released, the spring-loaded actuating lever moves down against restraint of dashpot. Increasing the number of holes covered increases the exposure time; with all holes open, shutter speed is 1/250 sec; all holes covered, 1/15 sec. Design is by Agfa Camera Works, Munich, Germany.

Bellows Expansion Changes Leverage

THE TRUE AIRSPEED at which an airplane is flying is instantly shown by the True Airspeed Indicator, a wartime development of Kollsman Instrument Division of the Square D Company, Elmhurst, N. Y. The instrument consists of three basic, interacting diaphragms—airspeed indicator, altimeter, and air thermometer—that automatically correct for changes in air density and the effect of the compressibility of air at high speeds on the Pitot tube and temperature bulb.

Impact air pressure, picked up by Pitot tube, causes airspeed diaphragm to expand against a lever, successively moving the intermediate rocking shaft, main rocking shaft, sector, hand staff pinion and hand. The altitude diaphragm and temperature diaphragms insert corrective factors in this train. All elements in the case are surrounded by static pressure brought into the otherwise airtight case by the static pressure line. As the pressure decreases with increase in altitude, the altitude diaphragm expands, moving the altitude slide so as to shorten the length of the lever arm that turns the intermediate rocking arm. The change in length of lever arm is calibrated to the altitude. Atmospheric temperature similarly affects the temperature diaphragm.

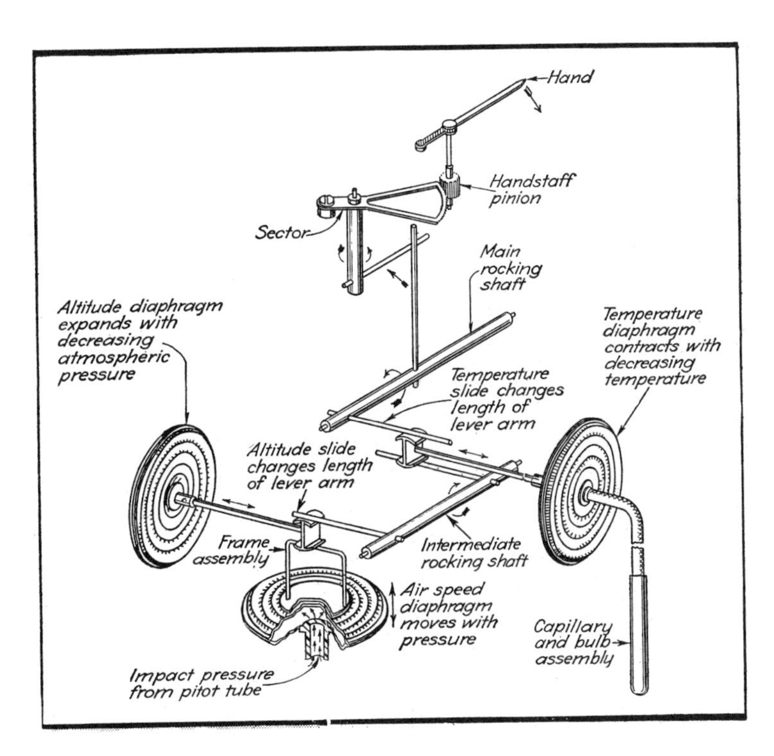

10 ways to use METAL DIAPHRAGMS

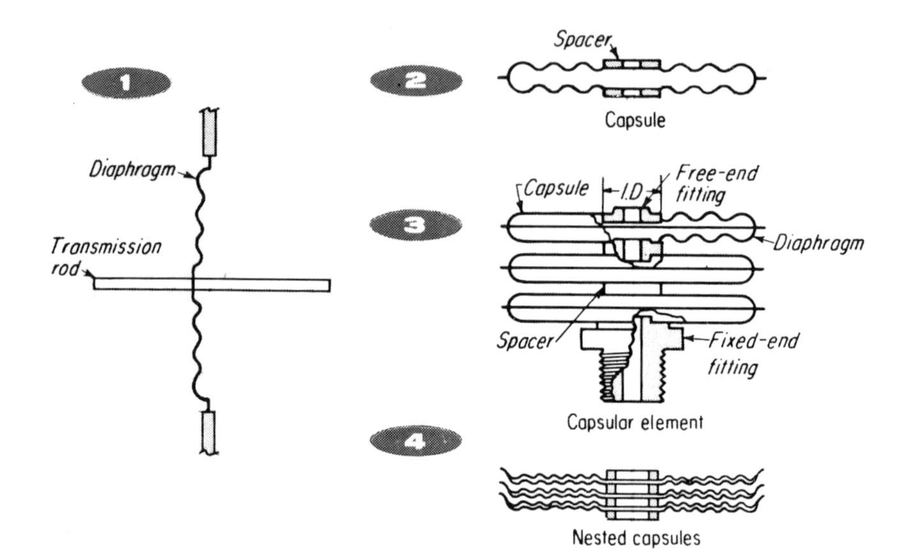

Metal diaphragm . . .

is usually corrugated (1) or formed to some irregular profile. It can be used as a flexible seal for actuating rod. Capsule (2) is an assembly of two diaphragms sealed together at their outer edges, usually by soldering, bronzing or welding. Two or more capsules assembled together are known as a capsular element (3). End fittings for capsules vary according to their function; the "fixed end" is fixed to the equipment. The "free end" moves the related components and linkages. Nested capsule (4) requires less space and can be designed to withstand large external overpressures without damage.

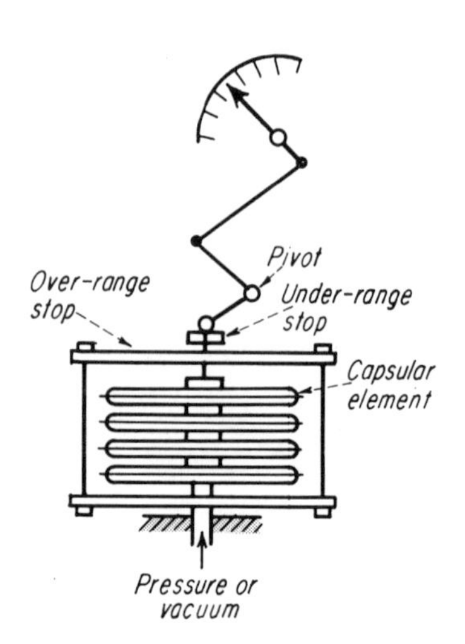

Pressure gage . . .

has capsular element linked to dial indicator by three-bar linkage. Such a gage measures pressure or vacuum relative to prevailing atmospheric pressure. If greater angular motion of indicator is required than can be obtained from three-bar linkage, a quadrant and gear can be used.

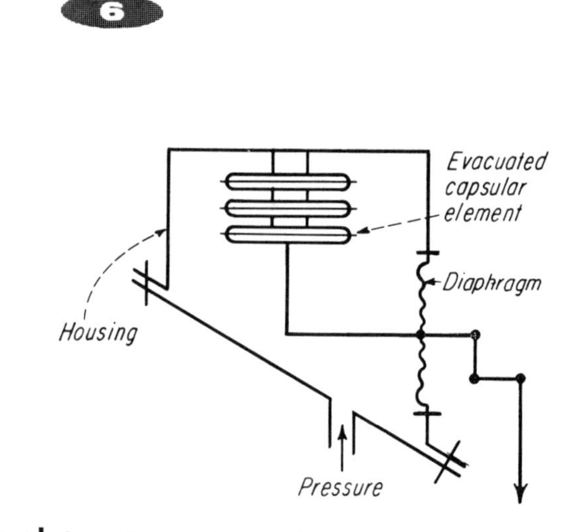

Absolute pressure gage . . .

has an evacuated capsular element inside an enclosure that is connected to pressure source only. Diaphragm allows linkage movement from capsule to pass through sealed chamber. This arrangement can also be used as a differential pressure gage by making a second pressure connection to the interior of the element.

and CAPSULES

D. C. WHITTEN
Development Engineer
Bristol Co., Waterbury, Conn.

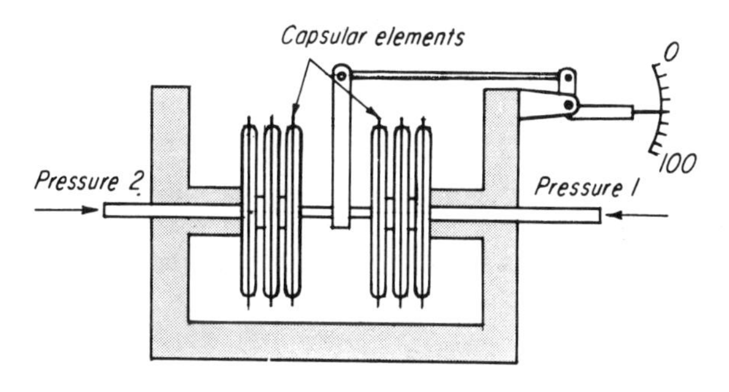

7

Differential pressure gage . . . with opposing capsules can have either single or multicapsular elements. The multicapsular type gives greater movement to indicator. Capsules give improved linearity over bellows for such applications of pressure measuring devices. Force exerted by any capsule is equal to the total effective area of the diaphragms (about 40% actual area) multiplied by the pressure exerted on it. Safe pressure is the max pressure that can be applied to a diaphragm before hysteresis or set become objectionable.

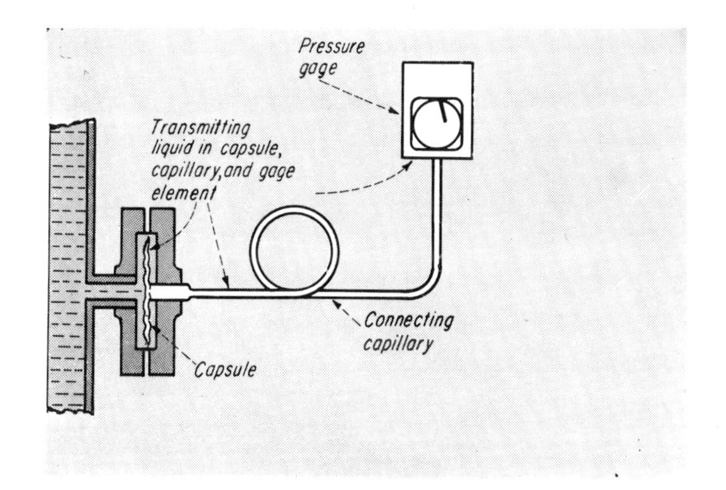

8

Capsule pressure-seal . . . works like a thermometer system except that the bulb is replaced by a pressure-sensitive capsule. The capsule system is filled with liquid such as silicone-oil and is self-compensating for ambient and operating temperatures. When subjected to external pressure changes, the capsule expands or contracts until the internal system pressure just balances the external pressure surrounding the capsule.

9

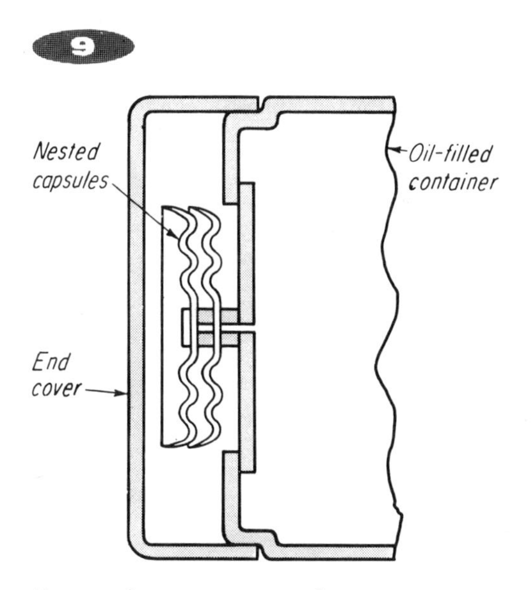

Expansion compensator . . . for oil-filled systems takes up less space when capsules are nested. In this application, one end of capsule is open and connected to oil in system; other end is sealed. Capsule expansion prevents internal oil pressure from increasing dangerously from thermal expansion. Capsule is protected by end cover.

10

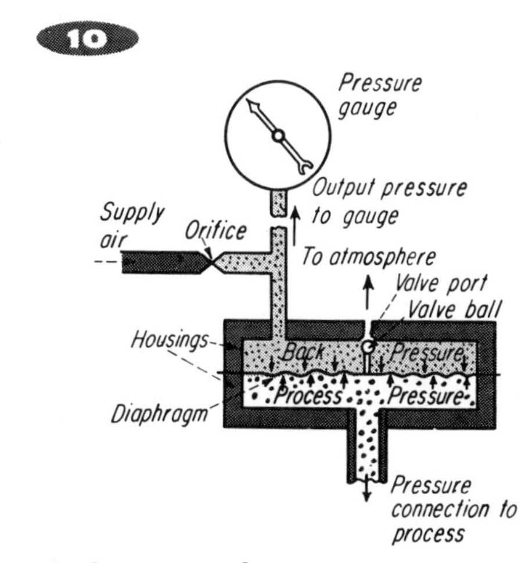

Force-balance seal . . . solves the problem, as in seal 8, of keeping corrosive, viscous or solids-bearing fluids out of the pressure gage. The air pressure on one side of a diaphragm is controlled so as to exactly balance the other side of the diaphragm. The pressure gage is connected to measure this balancing air pressure. Gage, therefore, reads an air pressure that is always exactly equal to the process pressure.

How Bellows Simplify Design of Instruments and Devices

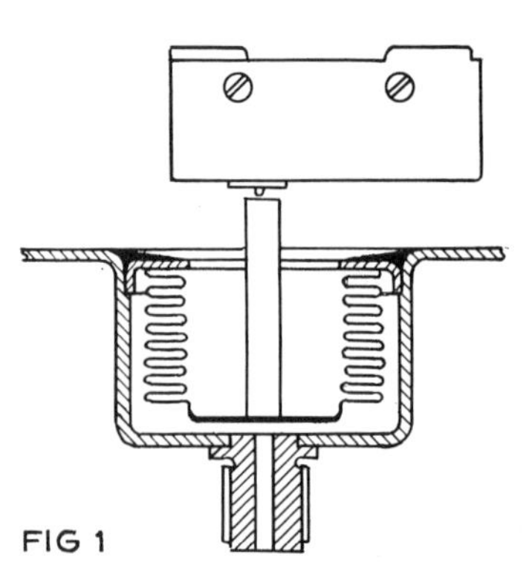

ACTUATE GAGES and switches. Pressures can be as high as 2,000 psi. Maximum value should exist when the bellows is near free-length.

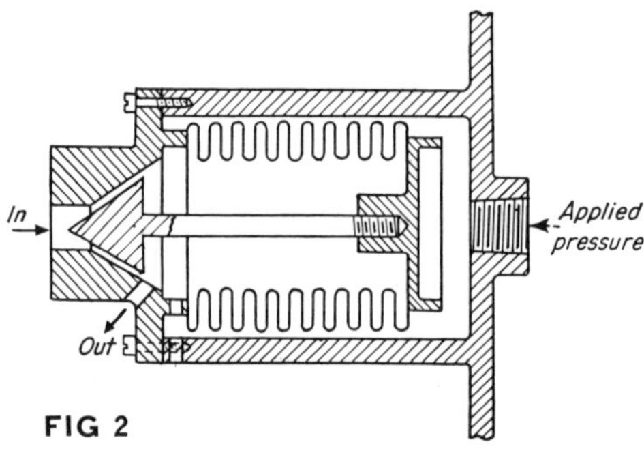

FLOW CONTROL. Variations in pressure adjust needle in flow valve. Also shows how bellows can be packless seals for valve stems and shafts.

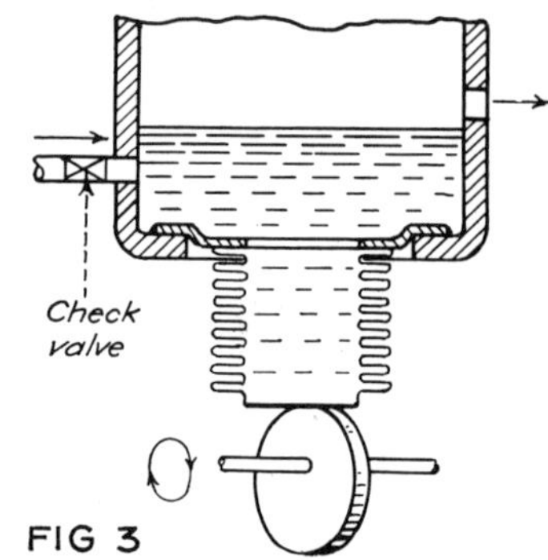

METERING DEVICE. Dispensing machines can use bellows as constant or variable displacement pump to measure and deliver predetermined amounts of liquids.

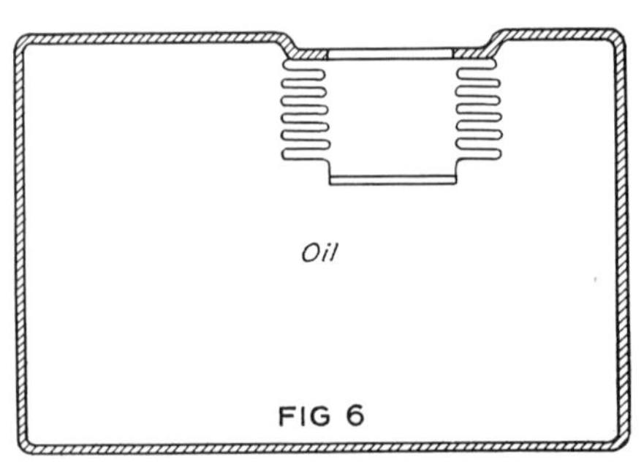

ABSORB EXPANSION OF FLUIDS OR GASES. Transformer (above) uses bellows to absorb increases in volume of oil caused by thermal expansion. Single controls of this type can operate from —70 to 250 F or from 0 to 650 F.

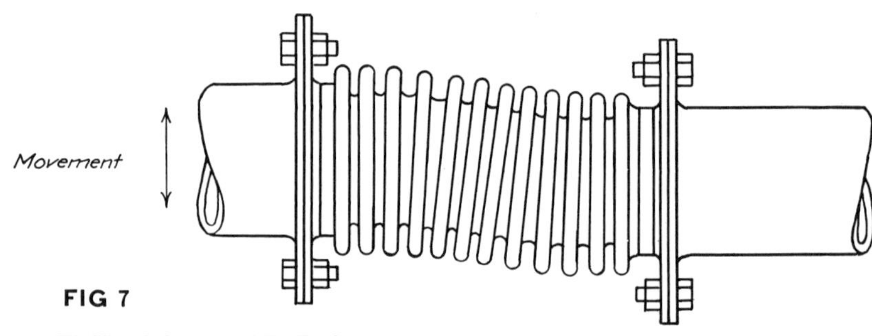

FLEXIBLE CONNECTOR. Suitable for wide range of applications from instruments to jet engines and large piping. Bellows absorb movement caused by thermal expansion, isolate vibration and noise as well as permit misalignment of mating elements. Wide variety of sizes and materials are now possible. Units are now in use from ¼ to 72 inches in diameter, made from such materials as brass, phosphor bronze, beryllium copper and stainless steel.

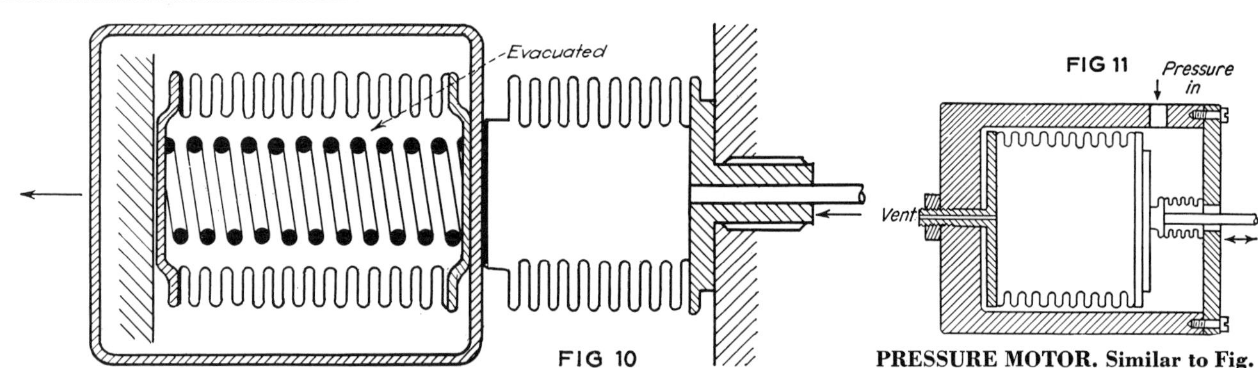

PRESSURE COMPENSATOR. Effect of ambient pressure can be eliminated in a pressure measuring system by matching the area of a pressure bellows with that of an aneroid and combining the two into a single assembly. Errors caused by ambient pressure can be held to a max of one percent. Present materials permit aneroid operation from —70 to 450 F.

PRESSURE MOTOR. Similar to Fig. 2. Bellows used instead of piston and cylinder arrangement. Eliminates effects of leakage and friction. Long stroke can be provided with sensitive response.

E. PERRY CUMMING
Bridgeport Thermostat Division
Robertshaw-Fulton Controls Company

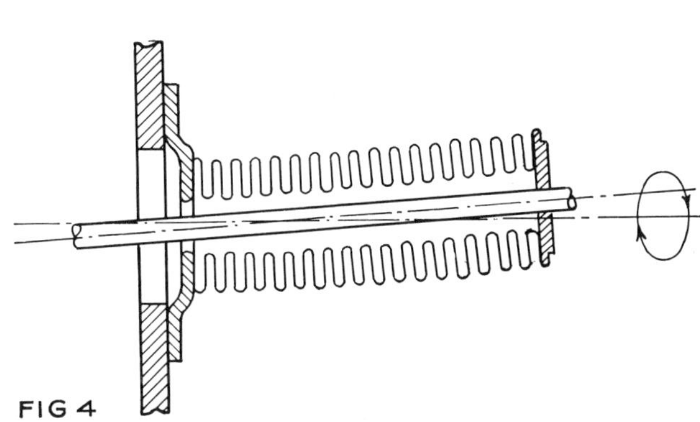

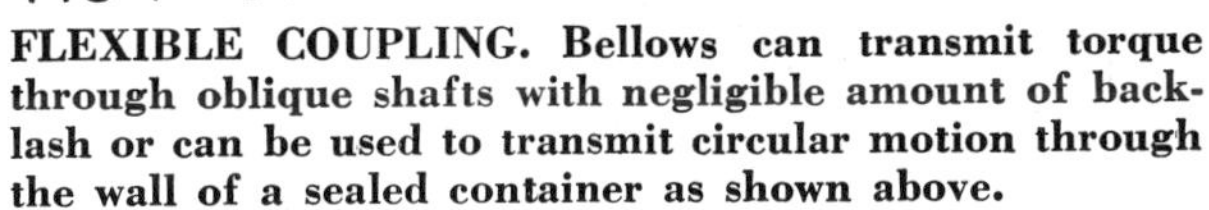
FLEXIBLE COUPLING. Bellows can transmit torque through oblique shafts with negligible amount of backlash or can be used to transmit circular motion through the wall of a sealed container as shown above.

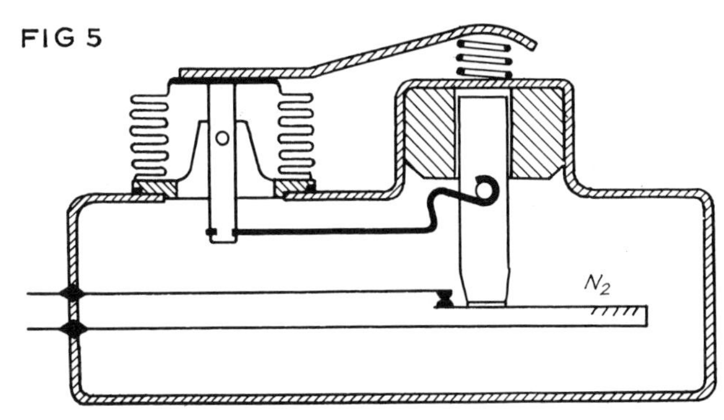

HERMETIC SWITCH. Bellows provide a gas tight flexible member through which motion can be transmitted into a sealed assembly. Flexibility and long life are important characteristics of these elements.

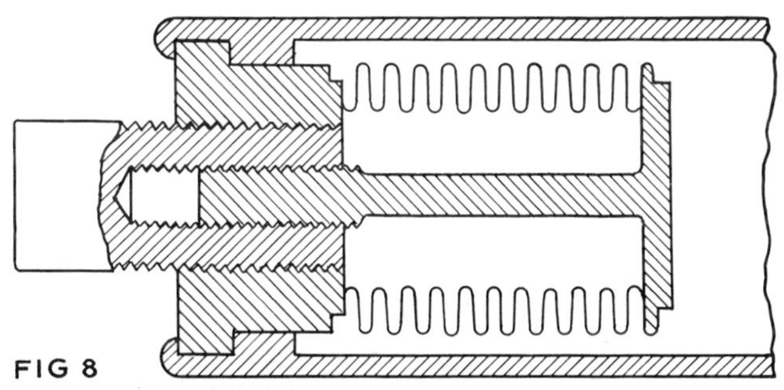

SEALED ADJUSTMENT. Accurately calibrated adjustments inside sealed instruments are possible by means of single or compound threads. To meet varied installation requirements, bellows are available with ends prepared for ring or disk end plates of standard or special design. These plates are fastened by brazing or welding techniques.

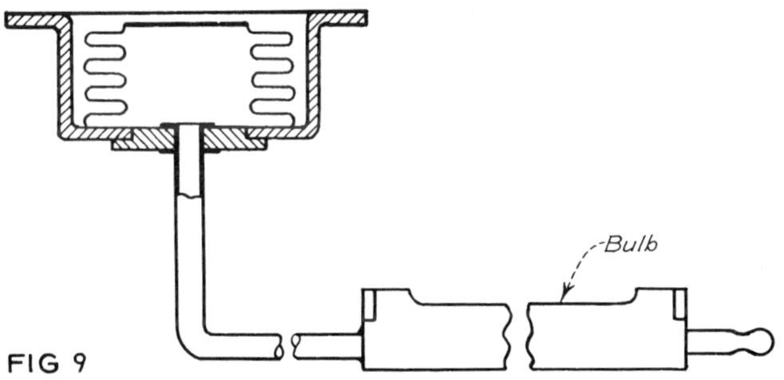

VAPOR PRESSURE THERMOSTAT. Small dia bellows offer large movement over a relatively small, adjustable temperature range. Can be filled so as to be unaffected by over-runs in temperature. Compensation for changes in ambient temperature is unnecessary whether this temperature is above or below the value selected for control purposes.

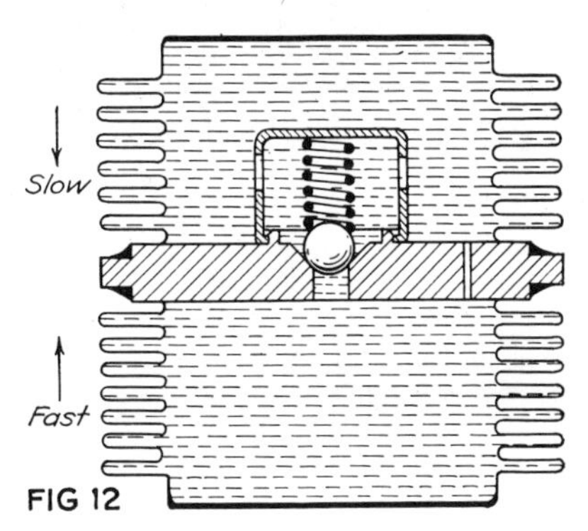

TIME DELAY MECHANISM. A check valve and proper size bleed hole between two liquid filled bellows allows fast motion in one direction and slow motion in the other direction.

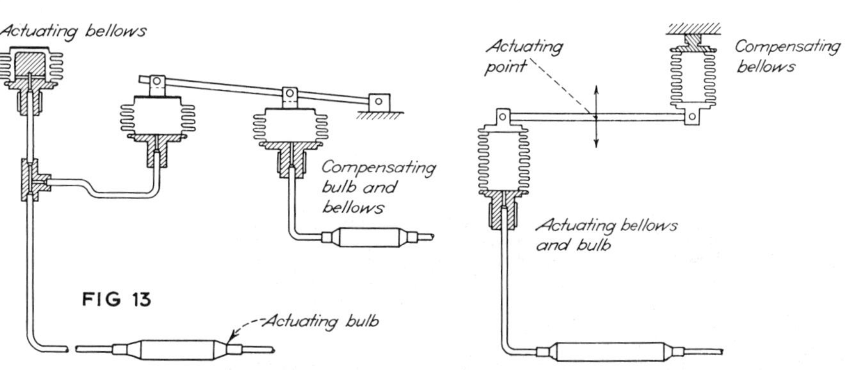

AMBIENT TEMPERATURE COMPENSATION. Two methods are shown. The left one uses two bellows in the actuating system. One is driven by the compensating assembly and correctly positions the actuating bellows as ambient temperature changes. The other method uses a floating lever whose mid-point is positioned by both the actuating and compensating bellows.

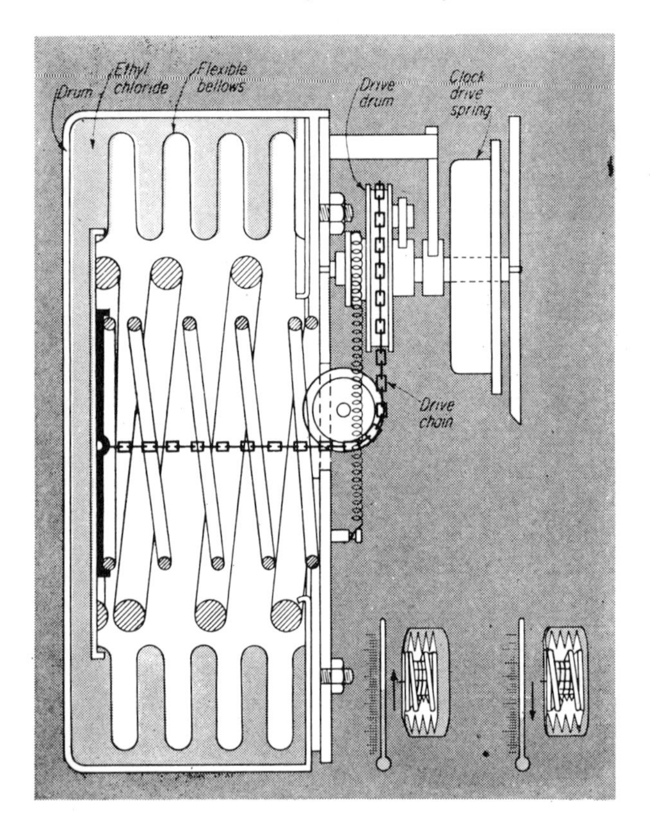

Rising temperature . . .
expands ethyl chloride gas contained in space between flexible bellows and metal drum. Gas compresses the bellows and springs, and drive chain is wound on spring-loaded drive drum. Dropping temperature lowers gas pressure; springs expand the bellows, pulling chain to wind clock drive-spring.

Clocks now wound by HEAT

Gas responds to temperature variations and winds new Swiss timepiece pictured above. Like two random-motion methods also shown on this page (at right and below), it suggests ways to simplify operation and maintenance of unattended equipment.

The self-winding timepiece movements show what the Swiss are doing in this field with unconventional energy-storing methods. Clock is by Le Coultre Watches, Le Sentier, Switzerland.

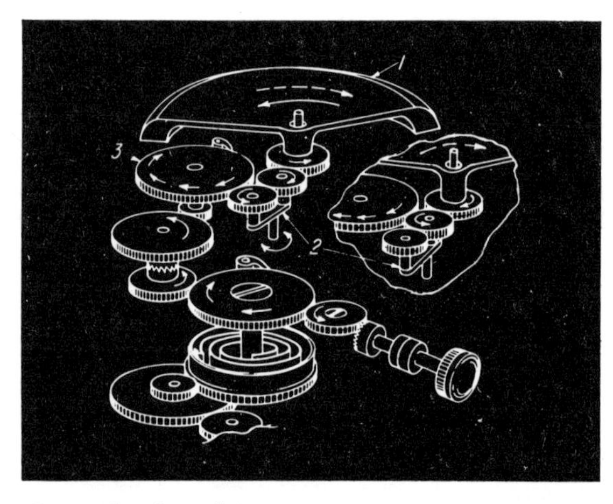

Gearwheel rocker . . .
Alternating movement of swing mass (1) is "rectified" by a gearwheel rocker (2) and transmitted to the drive gear (3). Gearwheel rocker shaft is free to rotate, bringing either gearwheel into mesh according to the direction of rotation of the swing mass. Reduction gearing is used to wind spring.

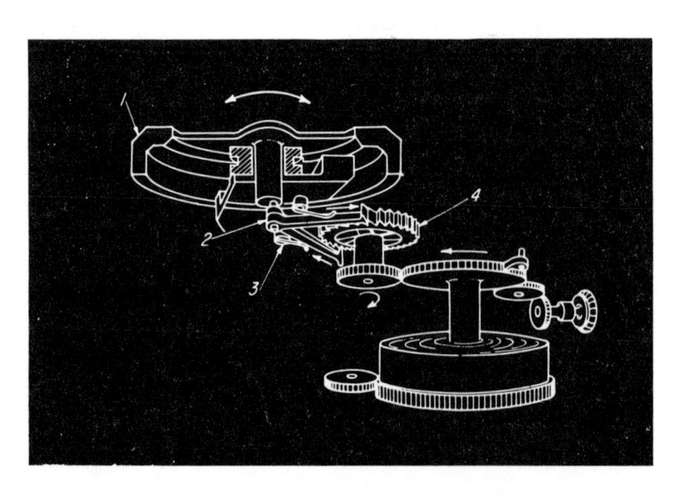

Eccentric drive . . .
Shaft of the free rotor (1) carries two eccentrically mounted actuating pawls (2). Springs (3) keep the pawls and pawl wheel (4) meshed. Pawl wheel is driven in one direction to wind spring, regardless of direction the rotor rotates. Large mechanical advantage of eccentric drive simplifies reduction gearing that winds the mainspring. Knob can be used for manual winding.

Chapter 9 BELT, CHAIN, GEAR, AND FRICTION DEVICES

Belts and Chains for Instrument Drives

Two-dimensional chain

A new type of chain link gives two-dimensional freedom for added flexibility in conveying and indexing. The chain links have simple twists which position alternate pins at right angles to each other. Idlers can be horizontal or vertical. This eliminates transfer points as one chain can do the job of several. With attachments such as rods, flights, and trays, the chain can convey lightweight products or trip off timing mechanisms. From Atlas Chain.

Two-dimensional chain (Atlas Chain)

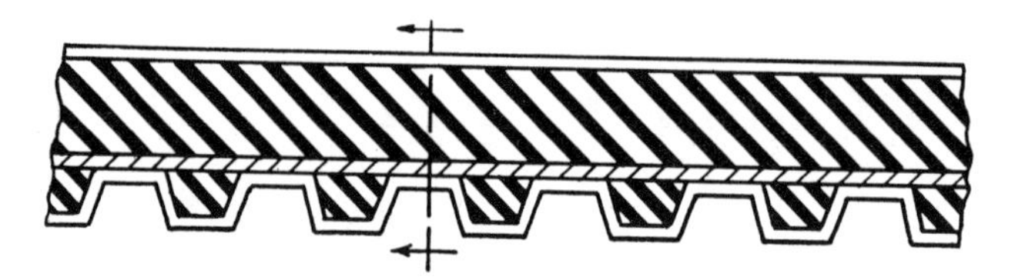

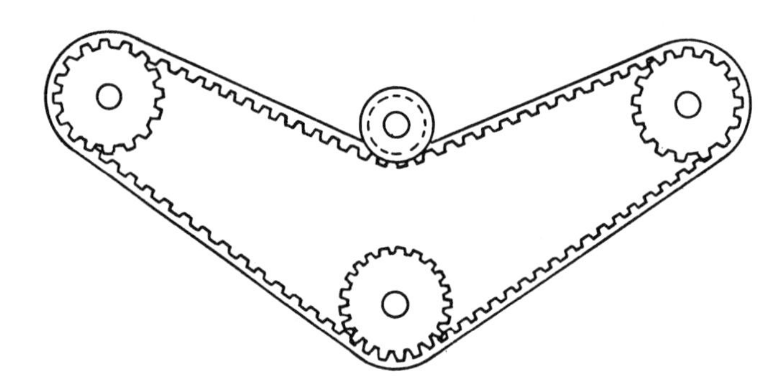

A combination . . . timing and V-belt, just patented by A. E. Carle, Detroit, has the advantages of both, permitting combination of friction and timing drives with the same belt. Basically, the belt consists of a "strain resisting element" or cord along the belt's neutral axis, with a time belt construction on one side and a V-Belt construction on the other. The V may be pointed toward the center of the belt or away from it. In one construction the timing belt element is recessed into the belt, permitting the belt to act as its own guide over unflanged timing wheels.

Miniature chain for low-cost drives

Tiny chains and sprockets in shaft sizes from $\frac{1}{8}$ in. to $\frac{3}{8}$ in., (see illustration) are being molded of acetal plastic by Bohannon Industries, Colorado Springs. In the chain, "unit-link" design eliminates need for master link and makes length adjustable in $\frac{1}{8}$-in. increments just by snapping links together. The chain runs on either side, weighs about $\frac{1}{5}$ as much as steel, is prelubricated and resistant to chemicals, stands temperatures to 250 F, will operate running loads up to 2-lb tension. Sustained static tension is not recommended, however. Prices, in quantity, will be less than $2 a foot for the chain, about 30¢ for an 8-tooth sprocket; 50¢ for one with 32 teeth.

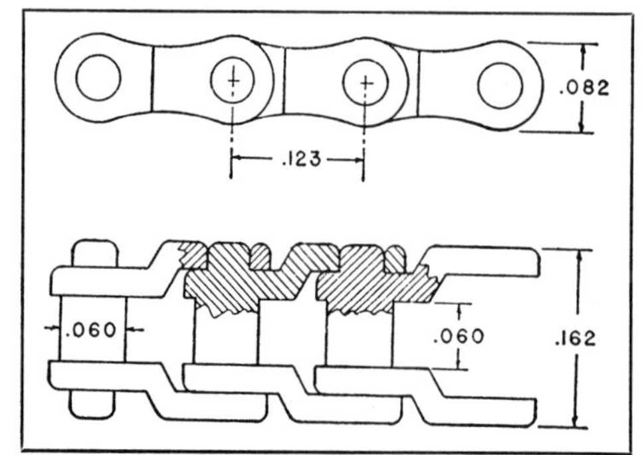

Belt drive

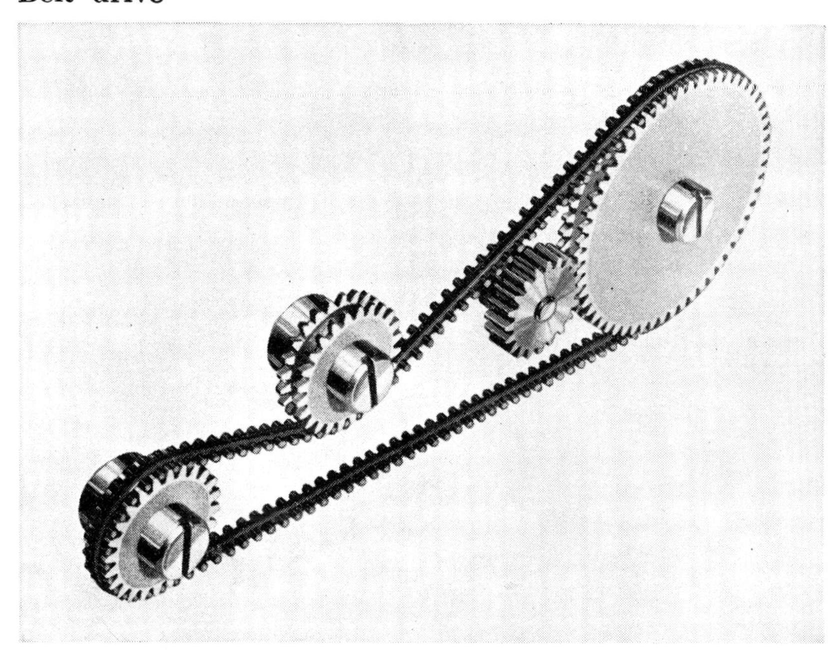

Self-lubricating thermoplastic belt with nylon core has lateral projections which engage pulley, which is essentially a grooved spur gear, and can be arranged to mesh with conventional gears. Lateral walk-off tendency of most types of belt drives is said to be eliminated. It is presently suggested for servo systems, data recorders, chart drives, and other relatively light-duty applications, but heavy-duty versions are under development. Belts available from stock in 3 to 15 in. pitch dia; pulleys in stainless, aluminum, and nylon in 14 to 128-teeth sizes. PIC Design Corp.

TIMING BELT with idler drives counter in step with angular rotation of micrometer screw in the Carson Dice Electronic Micrometer. Electronic sensing circuit stops drive at instant micrometer tip contacts work.

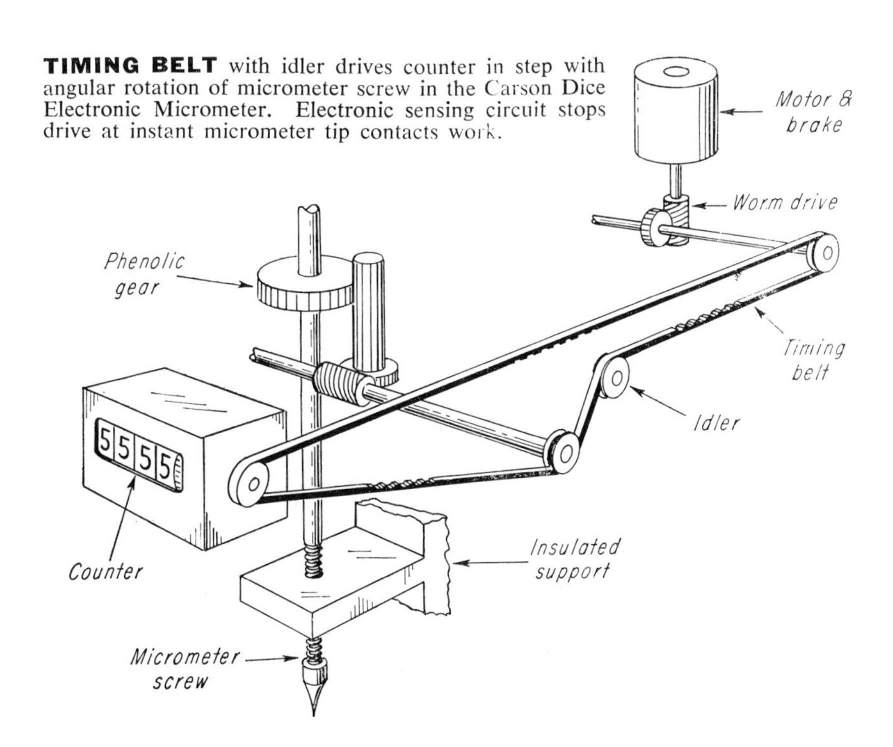

Bead chains for light service

BERNARD WASKO, chief engineer, Voland and Sons, Inc.

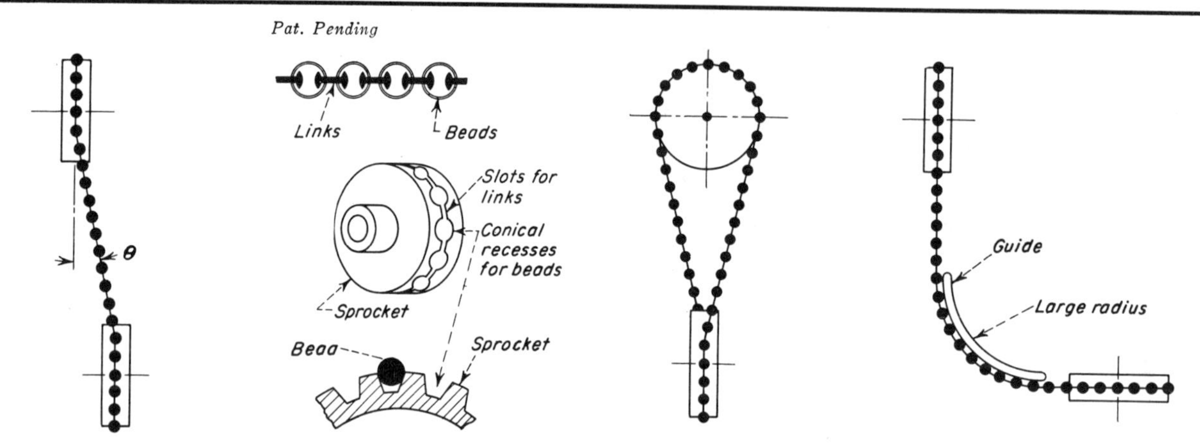

Fig. 1—Misaligned sprockets. Nonparallel planes usually occur when alignment is too expensive to maintain. Bead chain can operate at angles up to θ=20 degrees.

Fig. 2—Details of bead chain and sprocket. Beads of chain seat themselves firmly in conical recesses in the face of sprocket. Links ride freely in slots between recesses in sprocket.

Fig. 3—Skewed shafts normally acquire two sets of spiral gears to bridge space between shafts. Angle misalignment does not interfere with qualified bead chain operation on sprockets.

Fig. 4—Right angle drive does not require idler sprockets to go around corner. Suitable only for very low torque application because of friction drag of bead chain against guide.

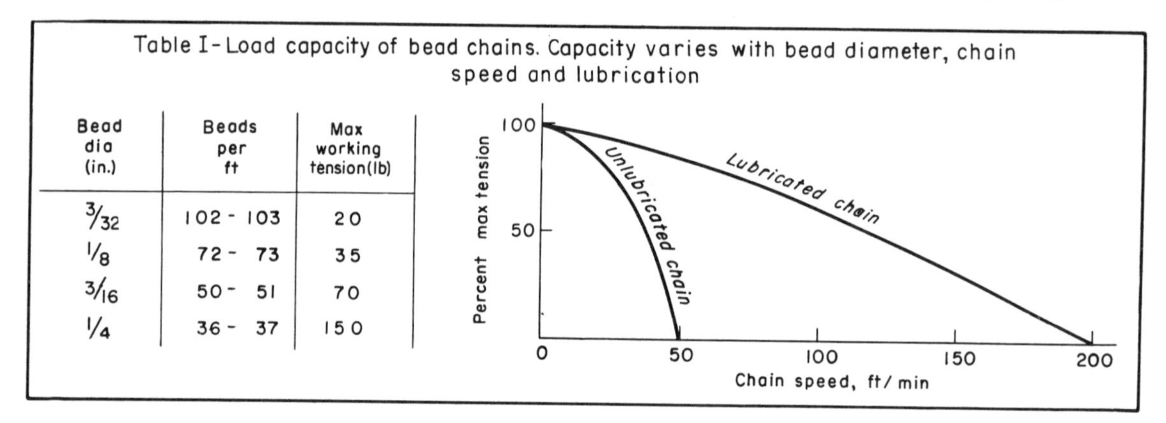

Table I - Load capacity of bead chains. Capacity varies with bead diameter, chain speed and lubrication

Bead dia (in.)	Beads per ft	Max working tension (lb)
3/32	102 - 103	20
1/8	72 - 73	35
3/16	50 - 51	70
1/4	36 - 37	150

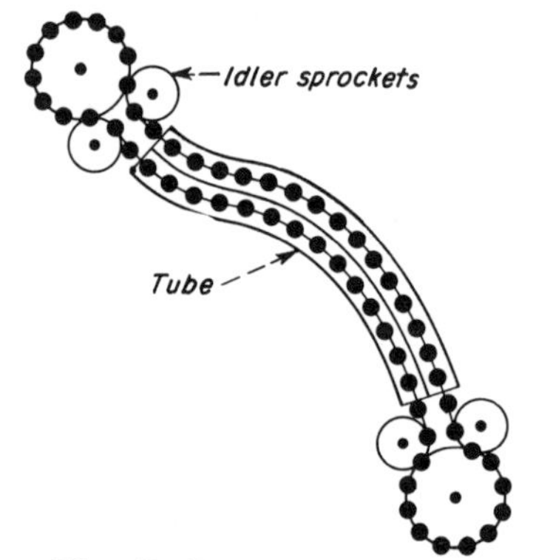

Fig. 5—Remote control through rigid or flexible tube has almost no backlash and can keep input and output shafts synchronized.

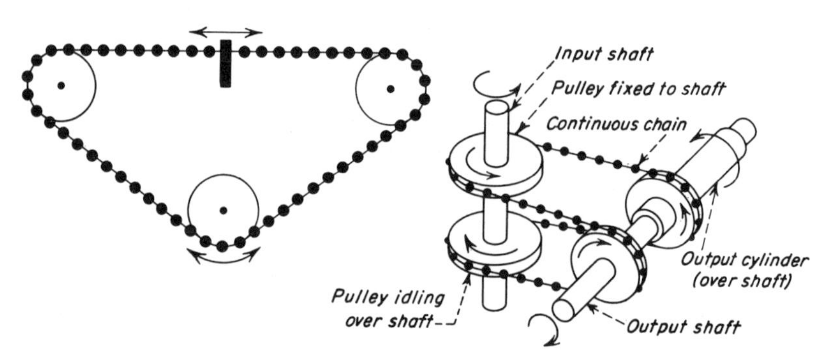

Fig. 6—Linear output from rotary input. Beads prevent slippage and maintain accurate ratio between the input and output displacements.

Fig. 7—Counter-rotating shafts. Input shaft drives two counter-rotating outputs (shaft and cylinder) through a continuous chain.

Where torque requirements and operating speeds are low, qualified bead chains offer a quick and economical way to: Couple misaligned shafts; convert from one type of motion to another; counter-rotate shafts; obtain high ratio drives and overload protection; control switches and serve as mechanical counters.

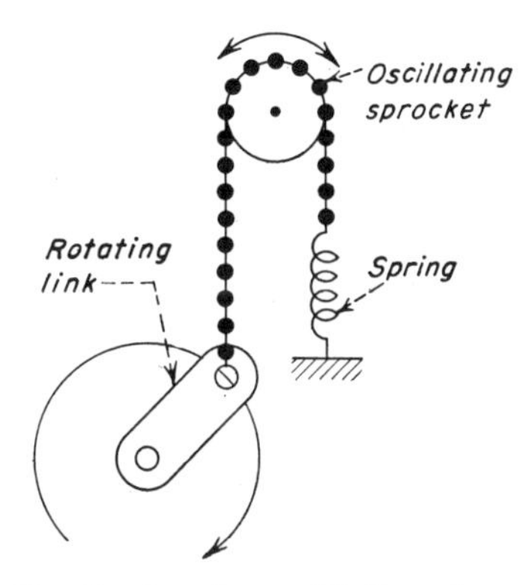

Fig. 8—Angular oscillations from rotary input. Link makes complete revolutions causing sprocket to oscillate. Spring maintains chain tension.

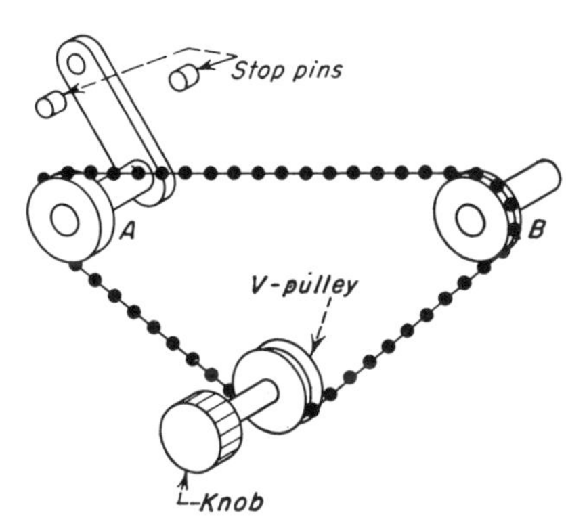

Fig. 9—Restricted angular motion. Pulley, rotated by knob, slips when limit stop is reached; shafts A and B remain stationary and synchronous.

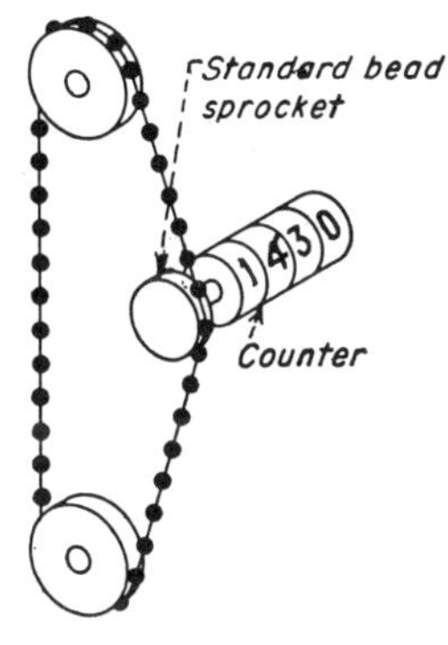

Fig. 10—Remote control of counter. For applications where counter cannot be coupled directly to shaft, bead chain and sprockets can be used.

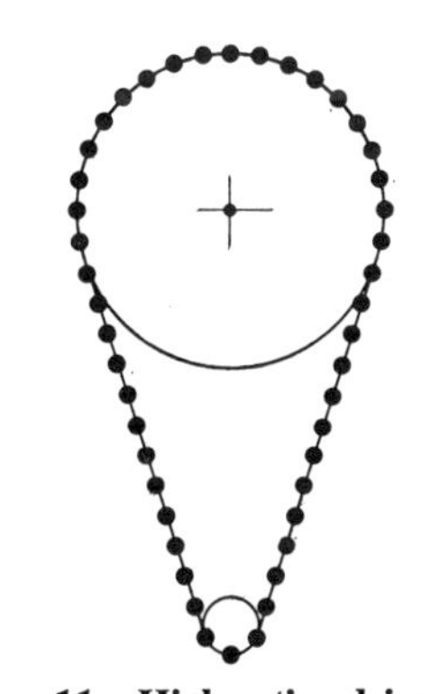

Fig. 11—High-ratio drive less expensive than gear trains. Qualified bead chains and sprockets will transmit power without slippage.

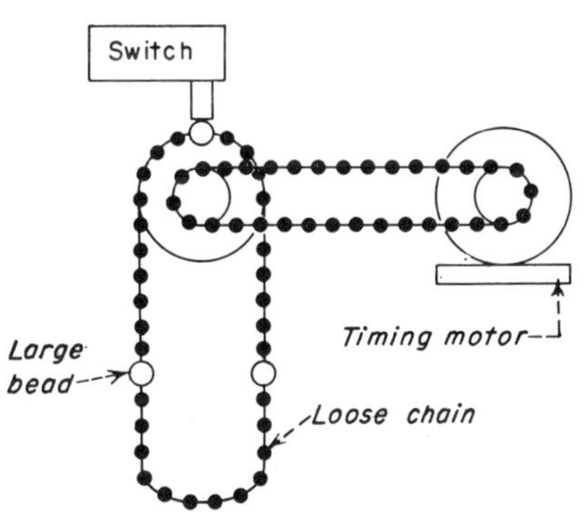

Fig. 12—Timing chain containing large beads at desired intervals operates microswitch. Chain can be lengthened to contain thousands of intervals for complex timing.

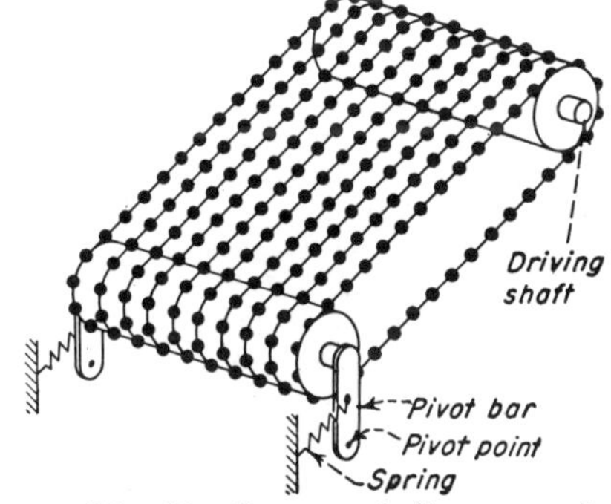

Fig. 13—Conveyor belt composed of multiple chains and sprockets. Tension maintained by pivot bar and spring. Width of belt easily changed.

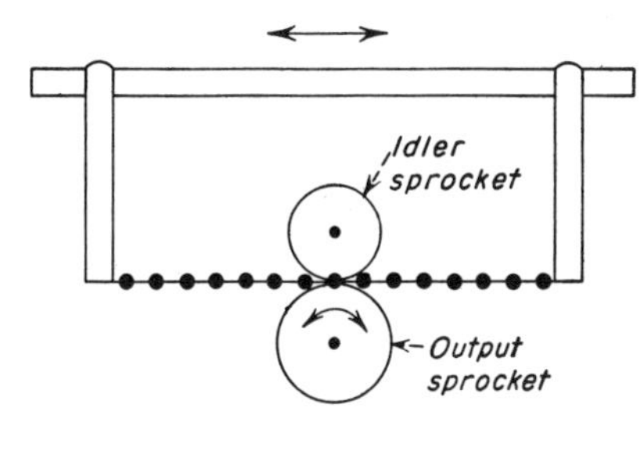

Fig. 14—Gear and rack duplicated by chain and two sprockets. Converts linear motion into rotary motion.

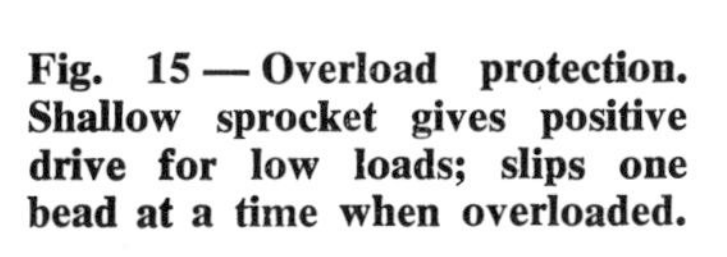

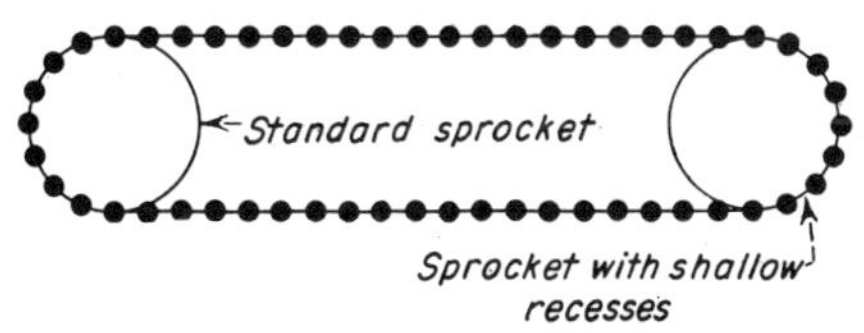

Fig. 15 — Overload protection. Shallow sprocket gives positive drive for low loads; slips one bead at a time when overloaded.

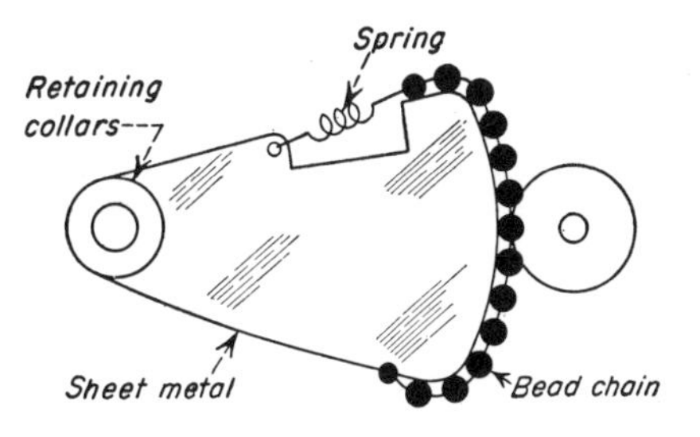

Fig. 16—Gear segment inexpensively made with bead chain and spring wrapped around edge of sheet metal. Retaining collars keep sheet metal sector from twisting on the shaft.

6 INGENIOUS JOBS for ROLLER CHAIN

How this low-cost industrial workhorse can be harnessed in a variety of ways to perform tasks other than simply transmitting power.

PETER C NOY, *manufacturing engineer*
Canadian General Electric Co Ltd, Barrie, Ont

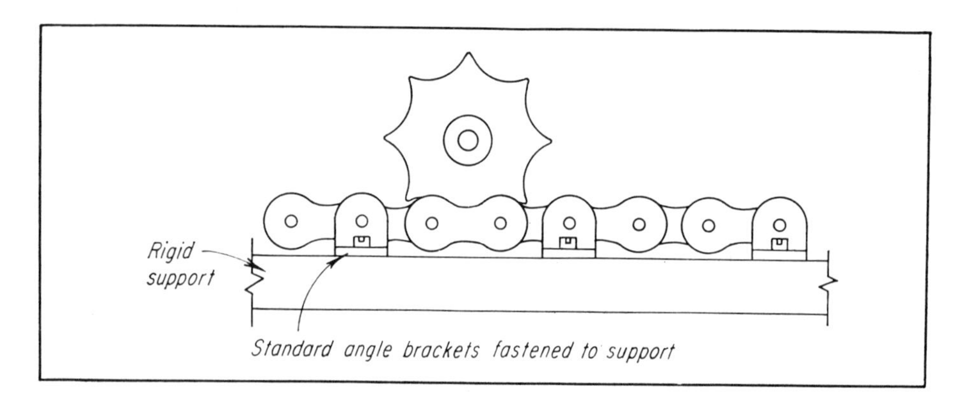

1 LOW-COST RACK-AND-PINION device is easily assembled from standard parts.

2 AN EXTENSION OF RACK-AND-PINION PRINCIPLE— soldering fixture for noncircular shells. Positive-action cams can be similarly designed. Standard angle brackets attach chain to cam or fixture plate.

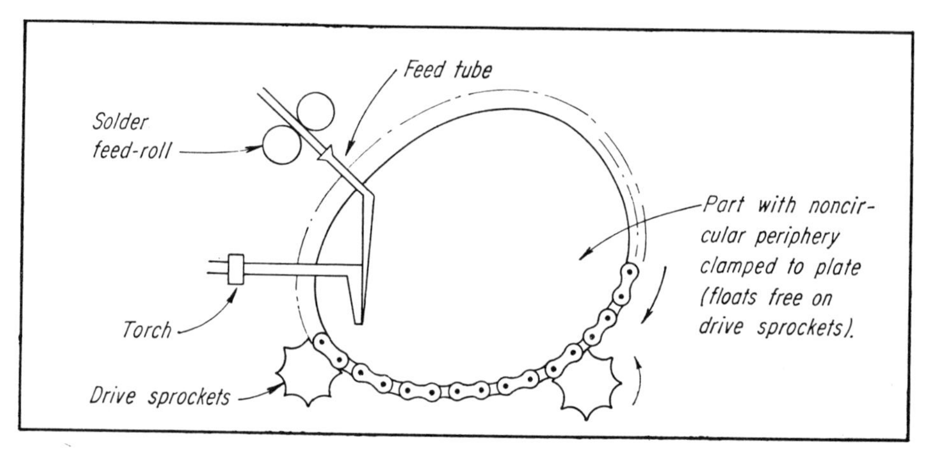

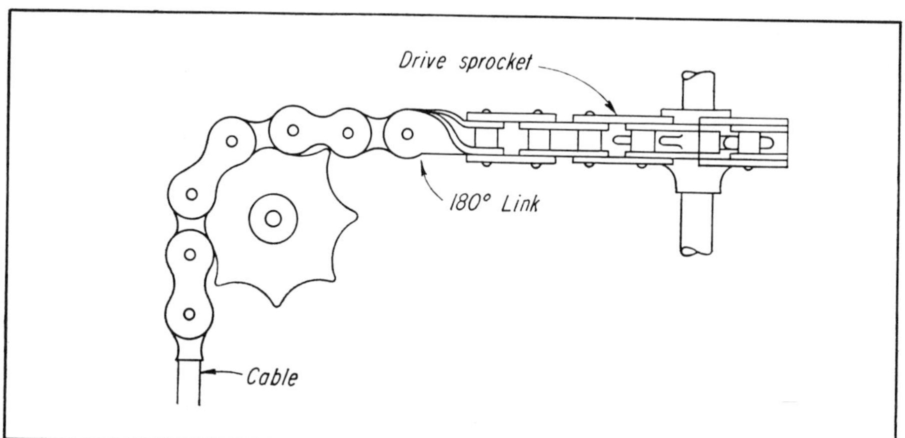

3 CONTROL-CABLE DIRECTION-CHANGER extensively used in aircraft.

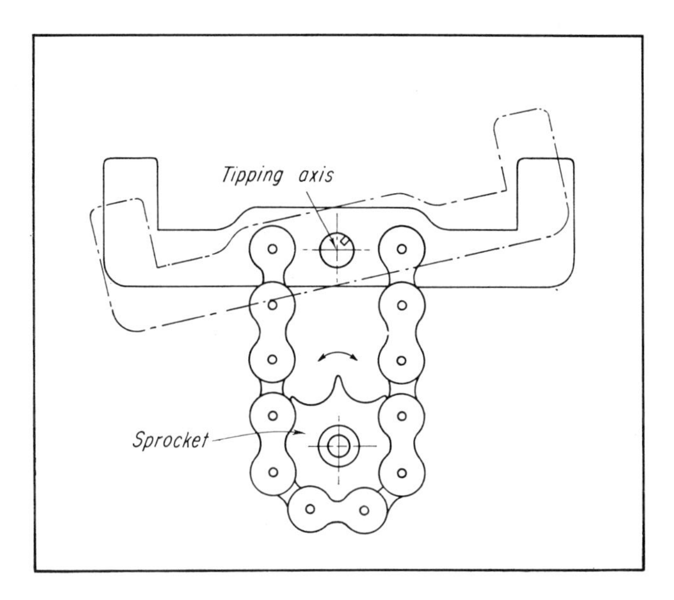

4 TRANSMISSION OF TIPPING OR ROCKING MOTION. Can be combined with previous example (3) to transmit this type of motion to a remote location and around obstructions. Tipping angle should not exceed 40° approx.

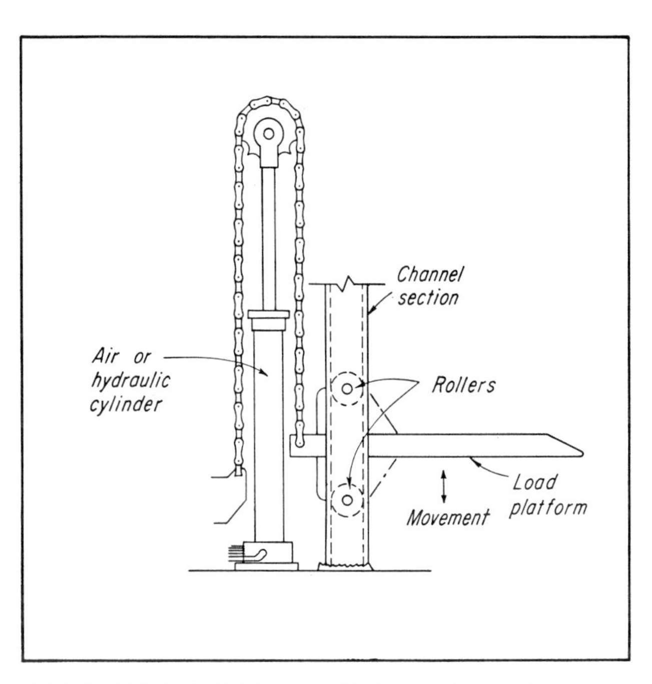

5 LIFTING DEVICE is simplified by roller chain.

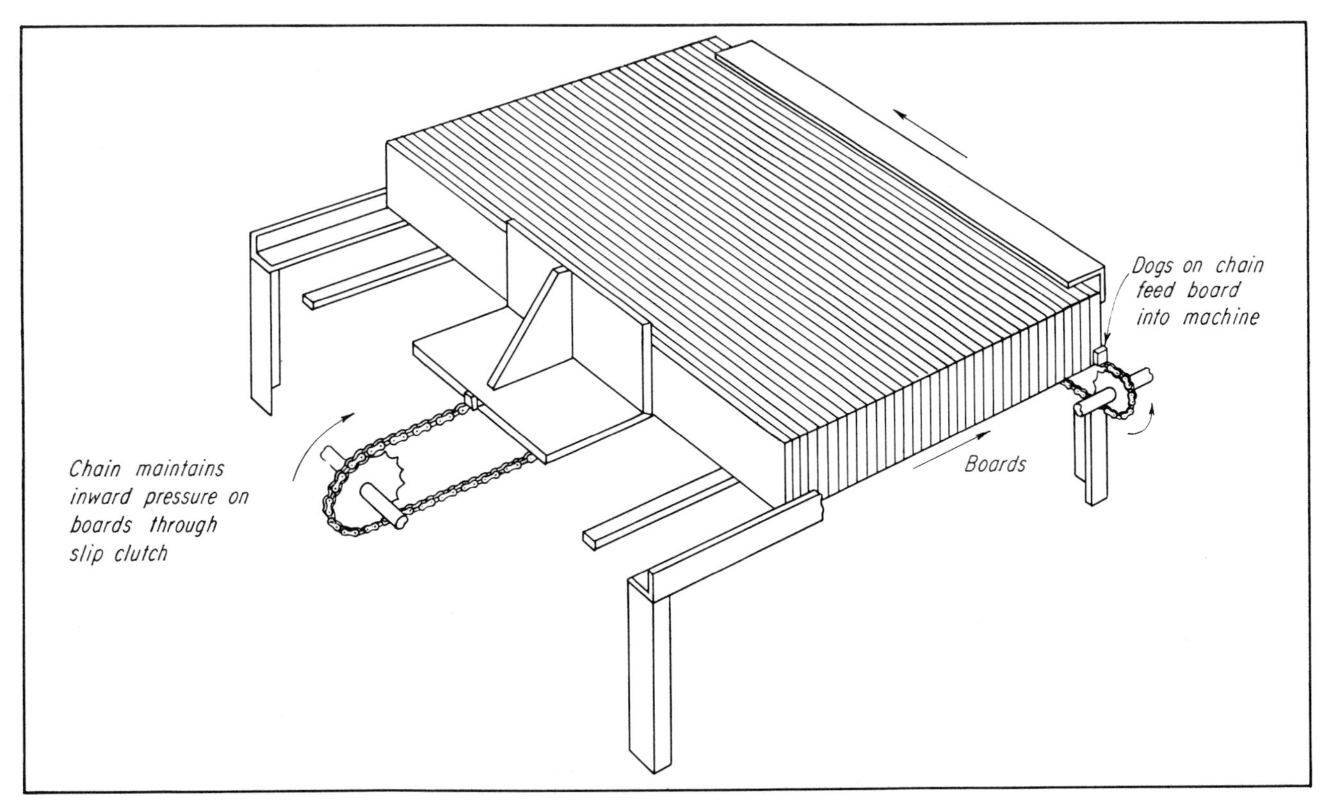

6 TWO EXAMPLES OF INDEXING AND FEEDING uses of roller chain are shown here in a setup that feeds plywood strips into a brush-making machine. Advantages of roller chain as used here are flexibility and long feed.

Mechanisms for Reducing

Pulsations in chain motion created by the chordal action of chain and sprockets can be minimized or avoided by introducing a compensating cyclic motion in driving sprocket.

EUGENE I. RADZIMOVSKY
Ass't. Prof. of Mechanical Engineering, University of Illinois

Fig. 1—The large cast-tooth non-circular gear, mounted on the chain sprocket shaft, has wavy outline in which number of waves equals number of teeth on sprocket. Pinion has a corresponding noncircular shape. Although requiring special-shaped gears, drive completely equalizes chain pulsations.

Fig. 2—This drive has two eccentrically mounted spur pinions (1 and 2). Input power is through belt pulley keyed to same shaft as pinion 1. Pinion 3 (not shown), keyed to shaft of pinion 2, drives large gear and sprocket. However, mechanism does not completely equalize chain velocity unless the pitch lines of pinions 1 and 2 are noncircular instead of eccentric.

Fig. 3—Additional sprocket 2 drives noncircular sprocket 3 through fine-pitch chain 1. This imparts pulsating velocity to shaft 6 and to long-pitch conveyor sprocket 5 through pinion 7 and gear 4. Ratio of the gear pair is made same as number of teeth of spocket 5. Spring-

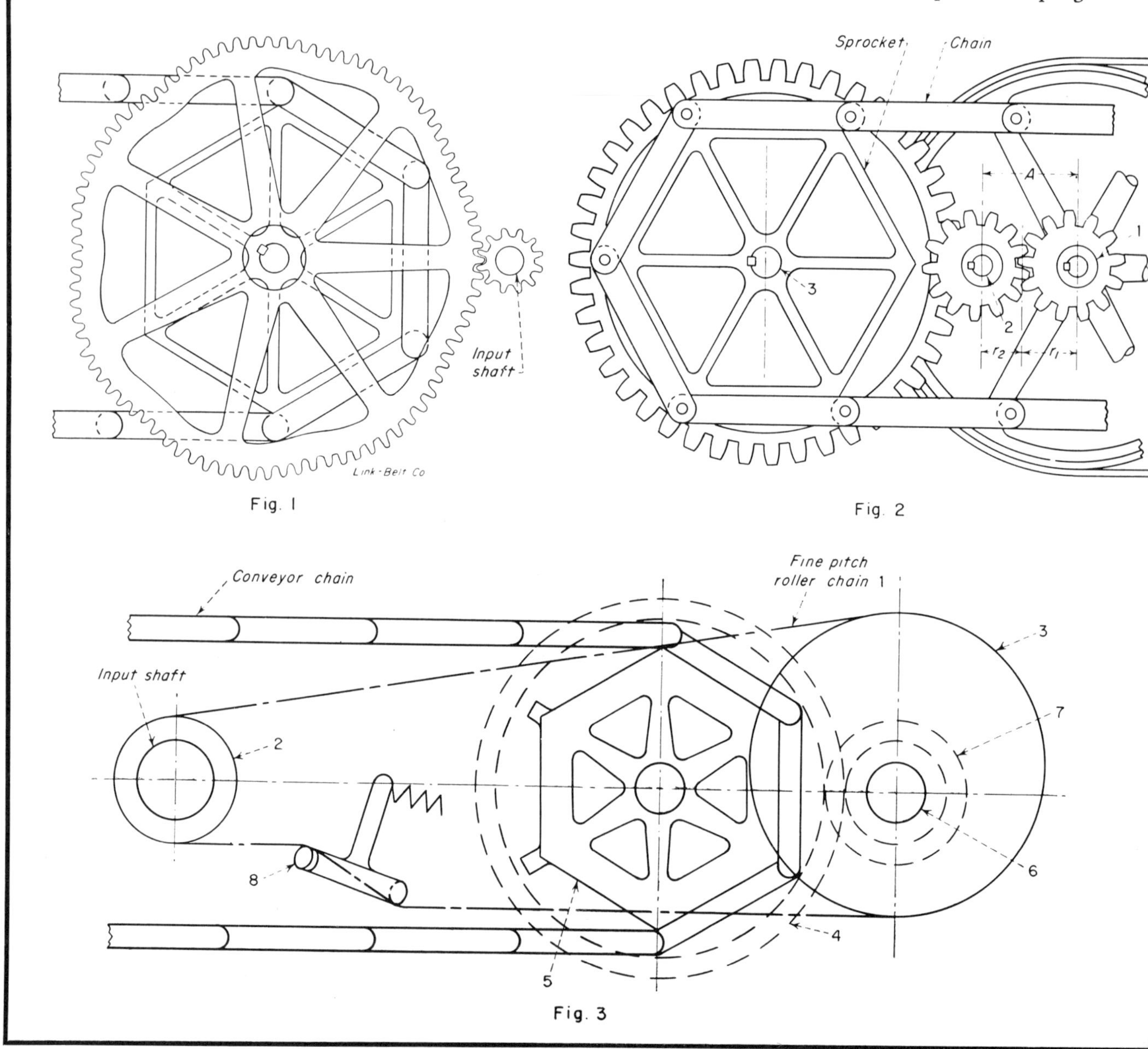

Fig. 1

Fig. 2

Fig. 3

Pulsations in Chain Drives

Mechanisms for reducing fluctuating dynamic loads in chain drives and the pulsations resulting therefrom include non-circular gears, eccentric gears, and cam activated intermediate shafts.

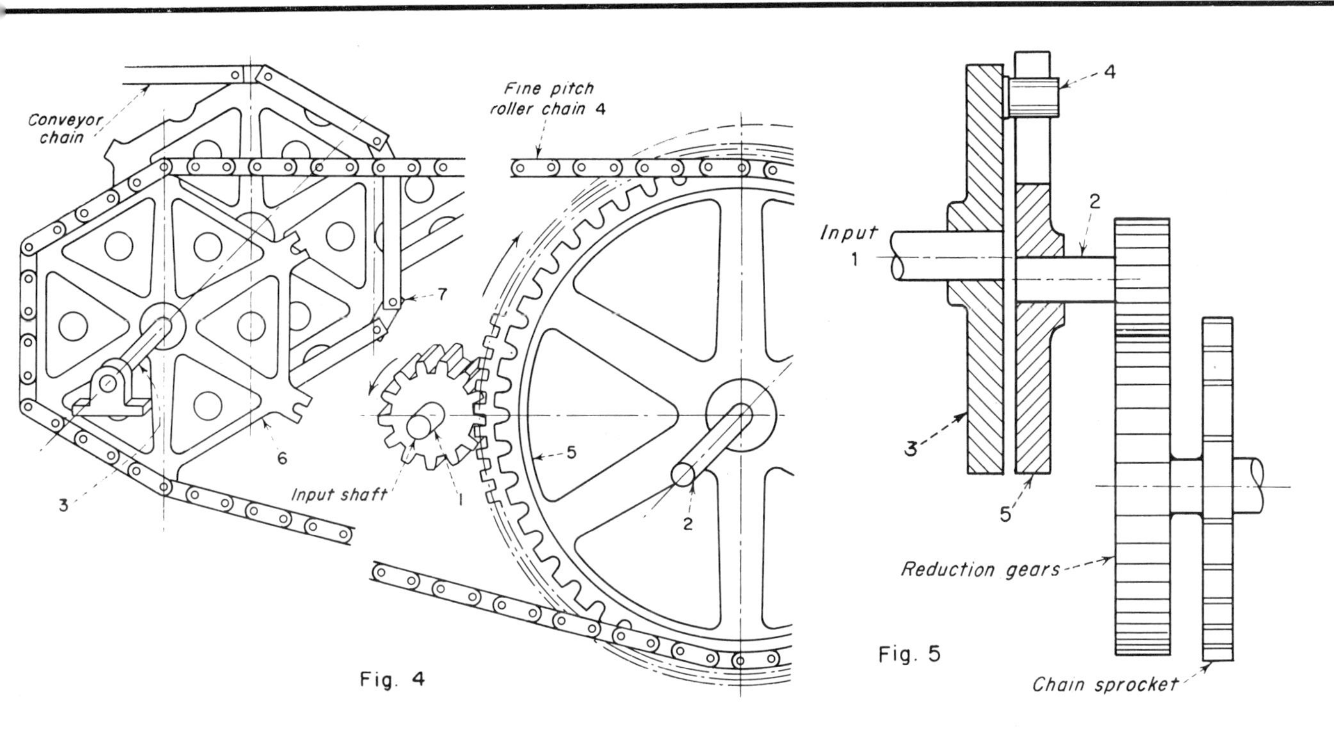

actuated lever and rollers 8 take up slack. Conveyor motion is equalized but mechanism has limited power capacity because pitch of chain 1 must be kept small. Capacity can be increased by using multiple strands of fine-pitch chain.

Fig. 4—Power is transmitted from shaft 2 to sprocket 6 through chain 4, thus imparting a variable velocity to shaft 3, and through it, to the conveyor sprocket 7. Since chain 4 has small pitch and sprocket 5 is relatively large, velocity of 4 is almost constant which induces an almost constant conveyor velocity. Mechanism requires rollers to tighten slack side of chain and has limited power capacity.

Fig. 5—Variable motion to sprocket is produced by disk 3 which supports pin and roller 4, and disk 5 which has a radial slot and is eccentrically mounted on shaft 2. Ratio of rpm of shaft 2 to sprocket equals number of teeth in sprocket. Chain velocity is not completely equalized.

Fig. 6—Integrated "planetary gear" system (gears 4, 5, 6 and 7) is activated by cam 10 and transmits through shaft 2 a variable velocity to sprocket synchronized with chain pulsations thus completely equalizing chain velocity. The cam 10 rides on a circular idler roller 11; because of the equilibrium of the forces the cam maintains positive contact with the roller. Unit uses standard gears, acts simultaneously as a speed reducer, and can transmit high horsepower.

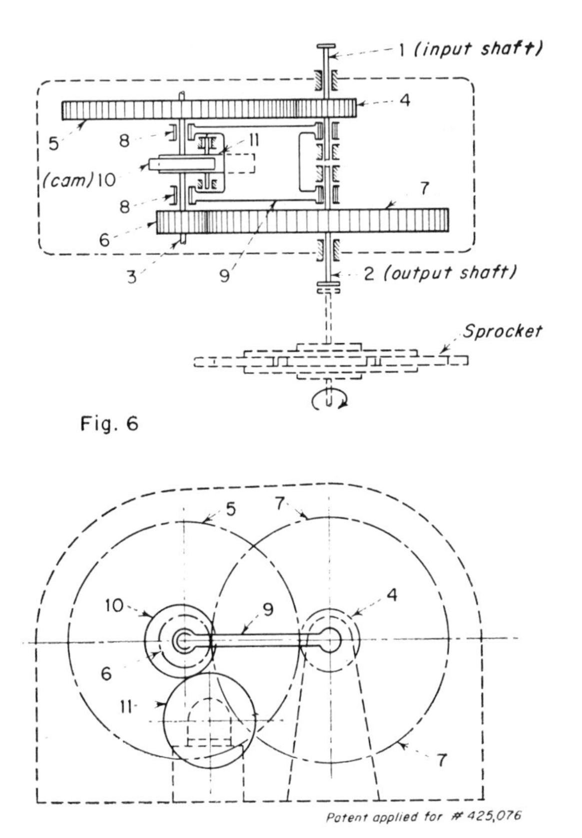

Self-Centering Conveyor Roller

A two-section roller automatically recenters a wide belt used in the American Machine and Foundry Co. pinspotter to carry swept pins and ball into the spotting mechanism. Roller is made of two hollow half-cylinders; one segment is fixed on the shaft and the other segment shifts with the belt. Links which connect segments cause roller diameter to increase when belt pushes movable segment to either side. Belt tension forces, which increase as enlarged roller tends to stretch belt, then recenter both roller segment and belt.

CENTERING ACTION is derived from force amplification produced by elastic characteristics of belt. Belt is wider than it is long, has a pitched idler in the center which deflects ball into return chute, and carries an unevenly-distributed load. When belt moves to one side, it pushes flange of movable roller segment outward and rotates links about pivots in fixed segment. This action increases roller diameter. Belt tension force increases as larger diameter roller tends to stretch belt. Higher belt tension force tends to collapse roller to original diameter, and as roller collapses, movable section swings belt toward central position on pivoted links. Roller will function either as driver or idler. It will operate up to 150 rpm without dynamic balancing, and is suitable for operation over 500 rpm when dynamically balanced.

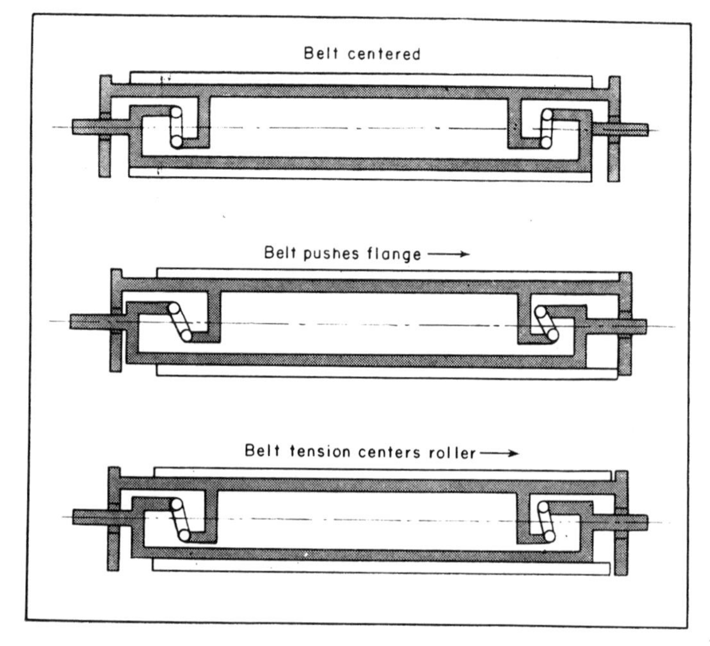

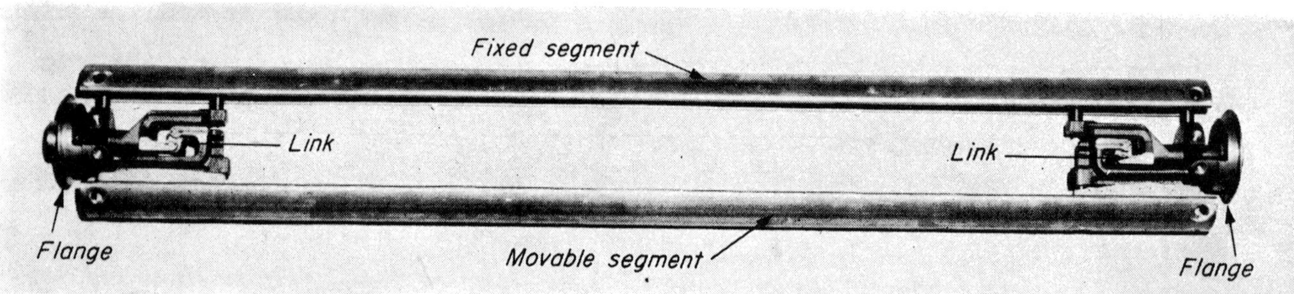

Very thin belting has proved its worth in high-speed transmissions for small devices.

MYLAR belting, engineers at Sandia Corp say, may be the answer to design of transmission systems operating at high rotational speeds. They have demonstrated their ideas with a prototype unit that produces 1000:1 speed reduction in a five-stage device measuring only about 1 cu in. In each stage a 0.0937-in. shaft serves as drive pulley. The driven pulley has major diameter of 0.3735 in. with a ¾-in. radius crown.

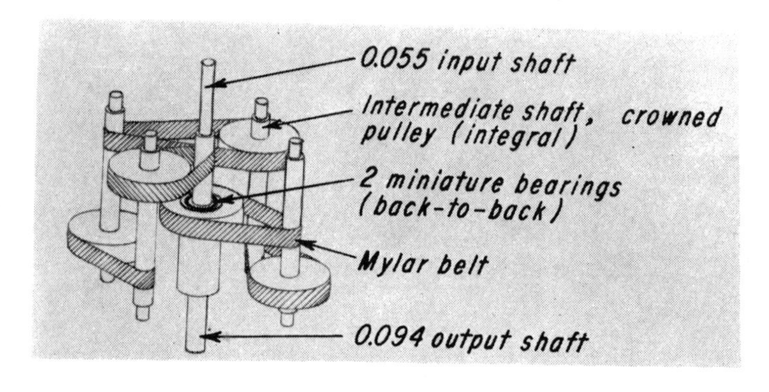

How to change center distance without affecting speed ratio

Increasing the gap between the roller and knife changes chain lengths from F to E. Since the idler moves with the roller sprocket, length G changes to H. The changes in chain length are similar in value but opposite in direction. Chain lengths E minus F closely approximate G minus H. Variations in required chain length occur because the chains do not run parallel. Sprocket offset is required to avoid interference. Slack produced is too minute to affect the drive because it is proportional to changes in the cosine of a small angle (2° to 5°). For the 72-in. chain, variation is 0.020 in.

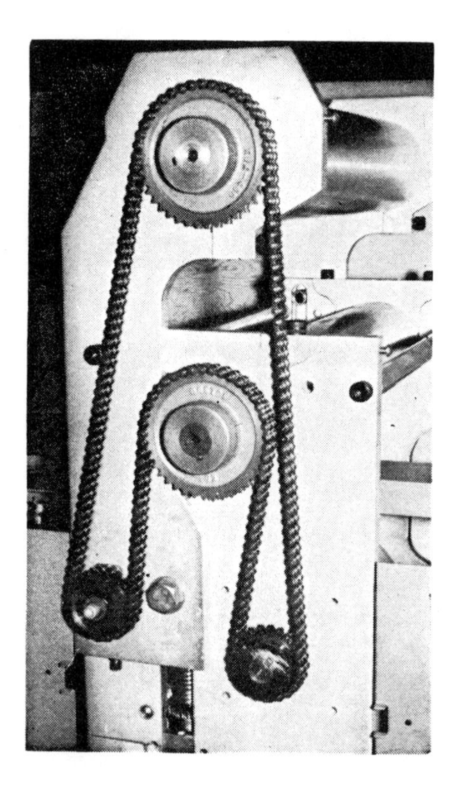

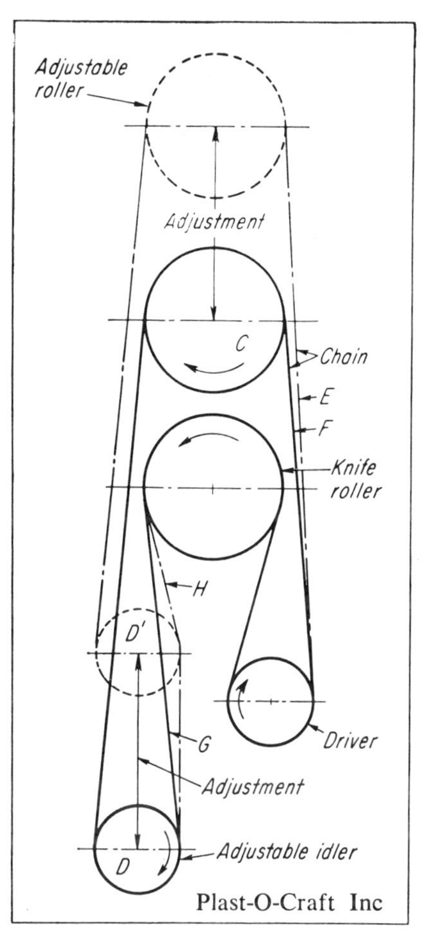

Plast-O-Craft Inc

Motor mount pivots for

Controlled tension

Belt tensioning proportional to load.

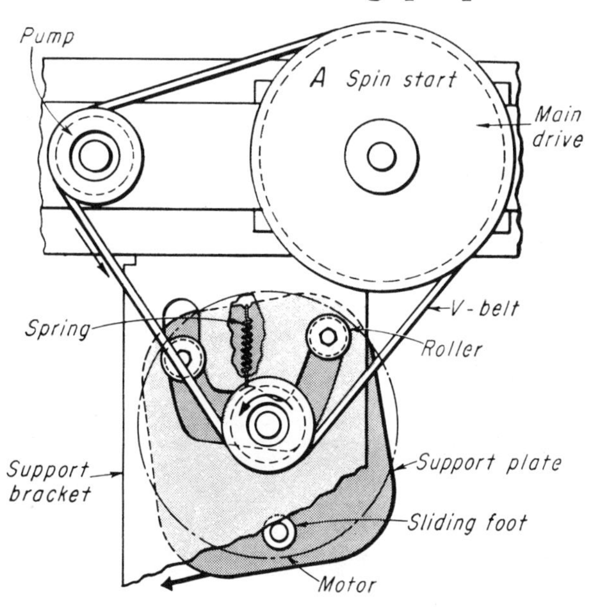

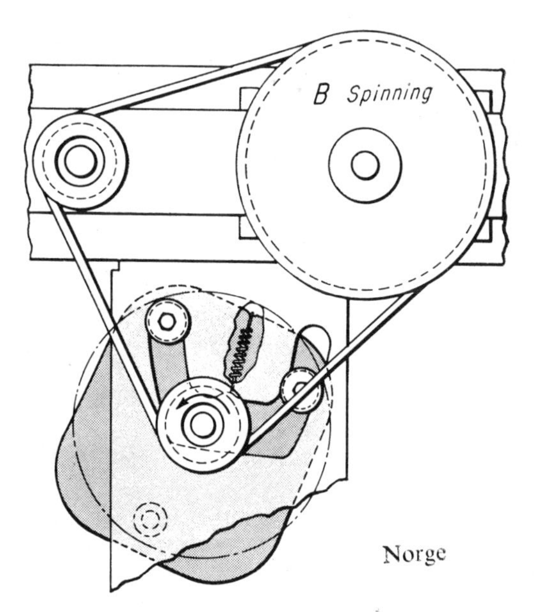

Norge

When the agitation cycle is completed the motor is momentarily idle with the right roller bottomed in the right-hand slot. When spin-dry starts (A) the starting torque produces a reaction at the stator pivoting the motor on the bottomed roller. The motor pivots until the opposite roller bottoms in the left-hand slot. The motor now swings out until restrained by the V-belt, which drives the pump and basket.

The motor, momentarily at zero rpm, develops maximum torque and begins to accelerate the load of basket, water and wash. The motor pivots (B) about the left roller increasing belt tension in proportion to the output torque. When the basket reaches maximum speed the load is reduced and belt tension relaxes. The agitation cycle produces an identical reaction in the reverse direction.

SPECIAL GEARING

A mixed bag of unusual gear arrangements to answer various tricky design problems. Federico Strasser of Santiago, Chile, is credited for much of this material.

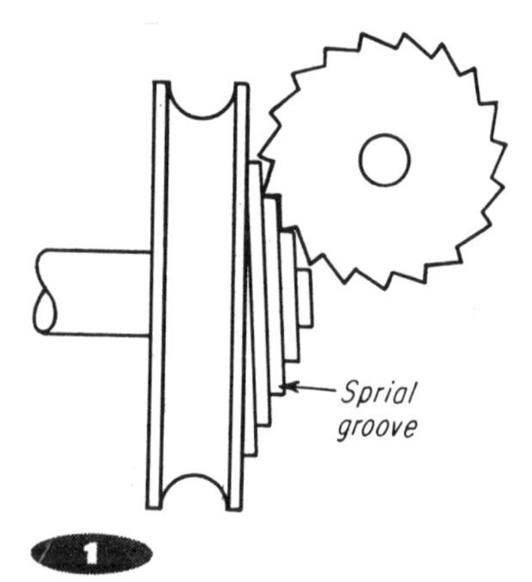

1

Worm gear . . .
has spiral groove. At least two or three pinion teeth are always in mesh.

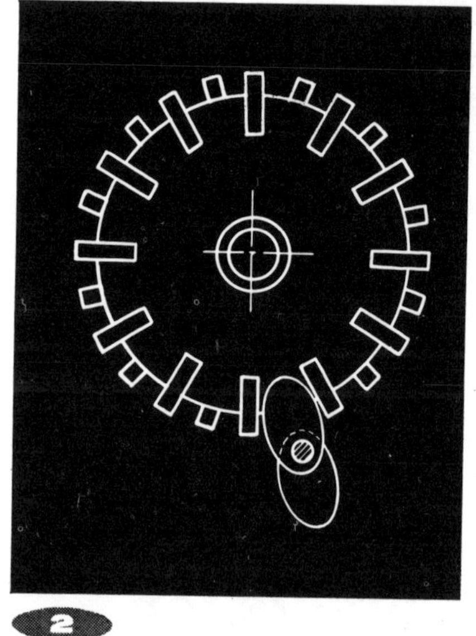

2

Two-tooth pinion . . .
consists of two diametrically opposed, specially shaped teeth. Gear is locked through part of the pinion revolution: by friction type (above), or by cardioid-shape pinion tooth.

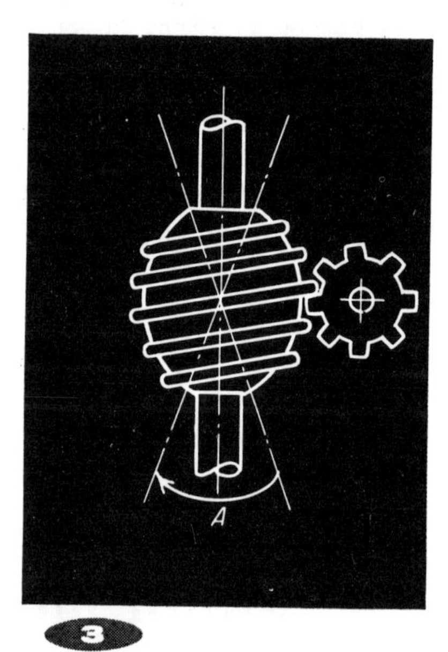

3

Globoid gear . . .
allows shaft to be swung through angle A without varying rotation speed.

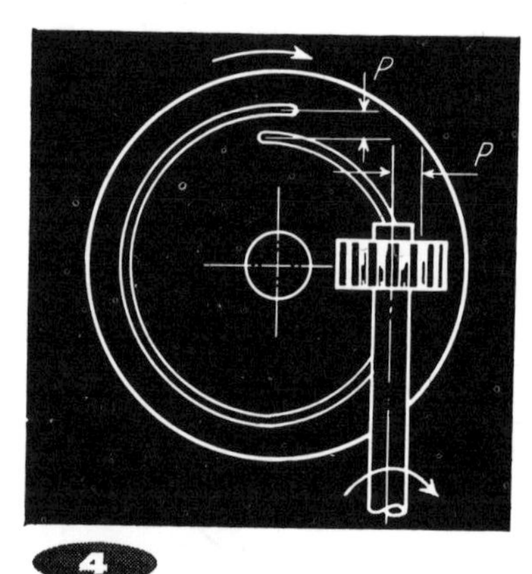

4

Faceplate worm . . .
has spiral ridge on flat face, meshing with worm wheel that turns forward one tooth per faceplate revolution.

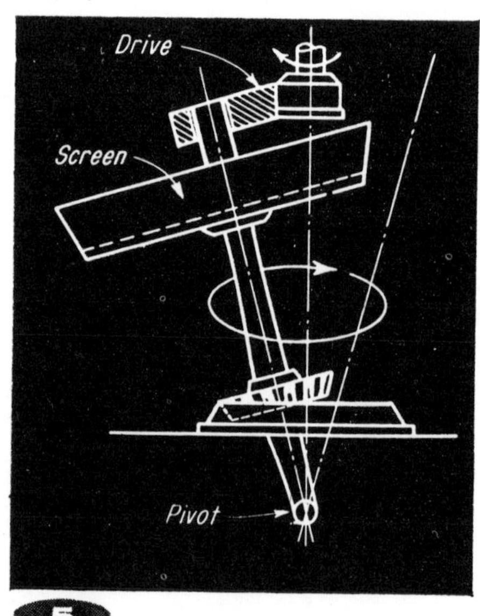

5

Conical-rotary gear . . .
causes shaft to rotate as it is swung about its pivot. Arrangement is used in reaping machines and other applications where screening action is required.

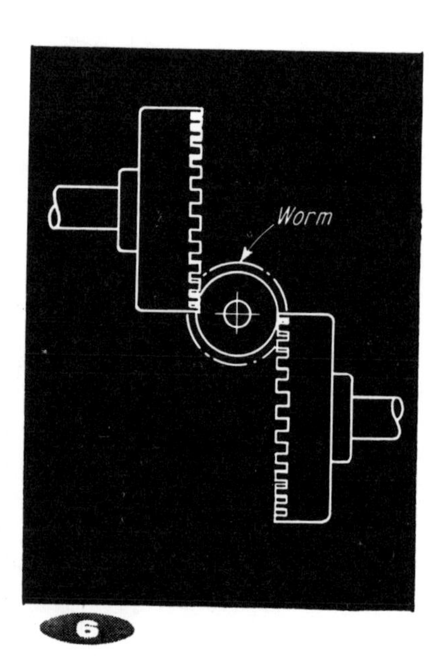

6

Worm and crown-gear . . .
gives slow, simultaneous feed to two shafts, which rotate in opposite directions. An application for this device is in chaffing machines.

DEVICES

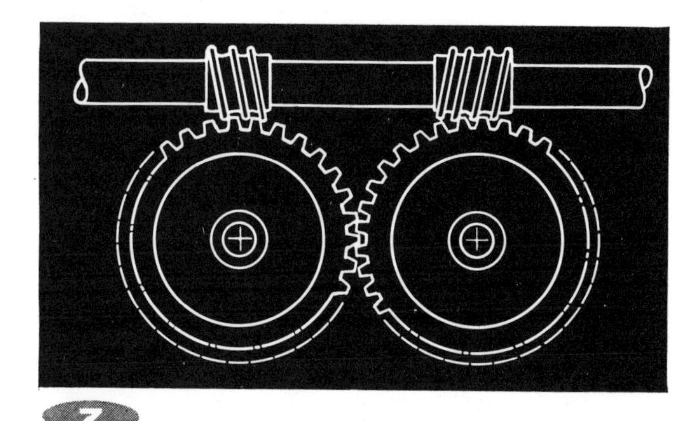

7

Double worms . . .
with opposite-hand threads neutralize end-thrust. Meshing the two gear-wheels gives greater stability to the setup.

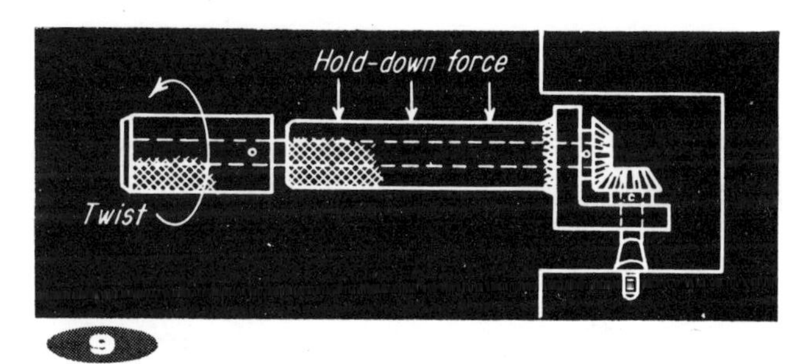

9

Right-angle screwdriver . . .
uses two small bevel-gears to transmit torque through 90°.

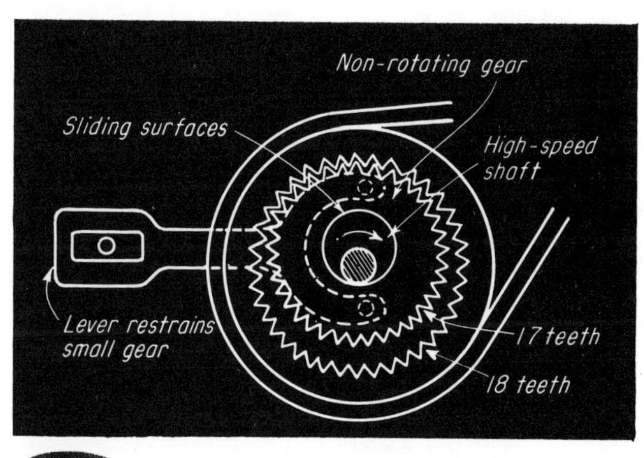

11

"Wabble" gear . . .
provides large speed reduction, depending on numbers of teeth. E.g. for arrangement shown, speed ratio will be 18 to 1—for one shaft revolution the large gear will rotate one tooth space. A 16-to-19 tooth ratio would give a 19-to-3 speed ratio

Bevel-gear differential . . .
in analog computors solves the equation: $z = c(x \pm y)$, where c is scale factor, x and y are inputs and z is output. Motion of x and y in same direction results in addition; opposite direction gives subtraction.

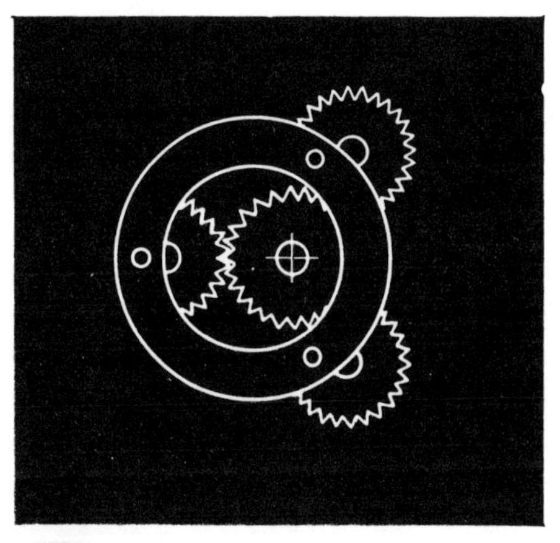

Planetary gears . . .
have ring pinned to them at eccentric points. As the planets rotate, the ring-center rotates about a circle with a radius equal to eccentricity of the planet ring-mounting pin.

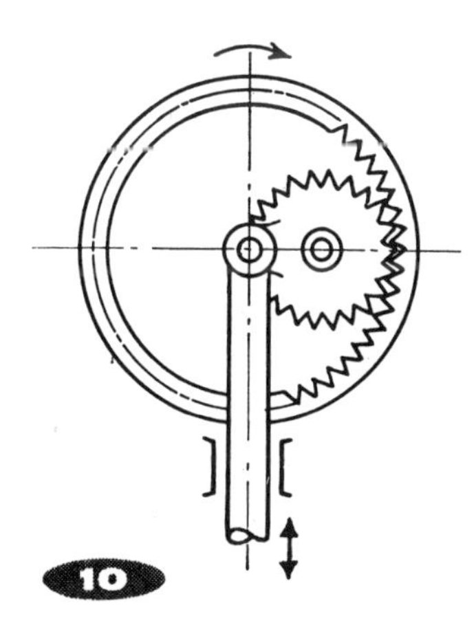

10

For straight-line motion . . .
epicyclic gear has pinion with pitch diameter equal to pitch radius of sun gear. Pivot-point on pinion, at pitch line, generates straight-line motion as sun gear rotates.

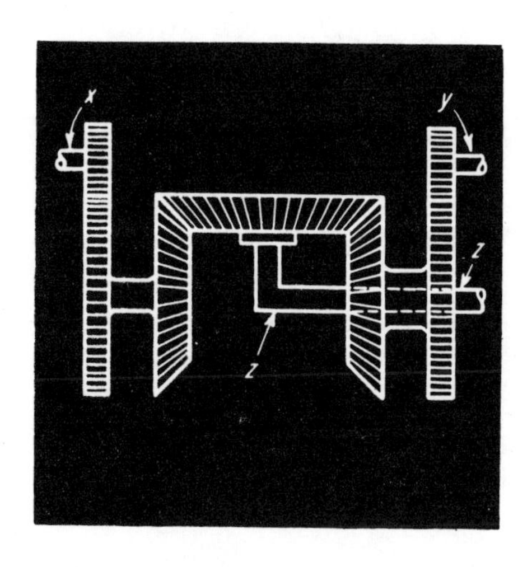

Unusual Gearing Devices

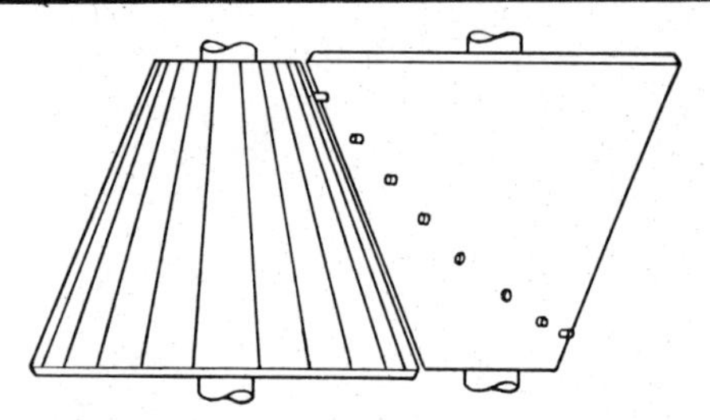

Conical tooth-and-pin gear . . . produces varying speeds in pin-gear shaft, depending on how the pins are located or spaced.

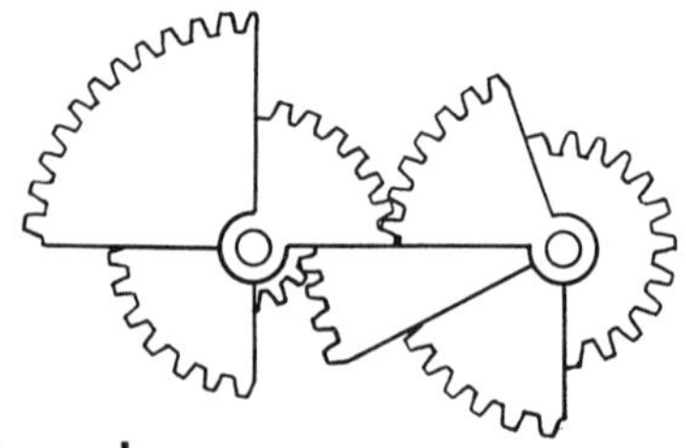

Irregular gear . . . gives sudden speed-variations of the driven-gear shaft as the different sections mesh. The sections, on different planes, must have the same diametral pitch.

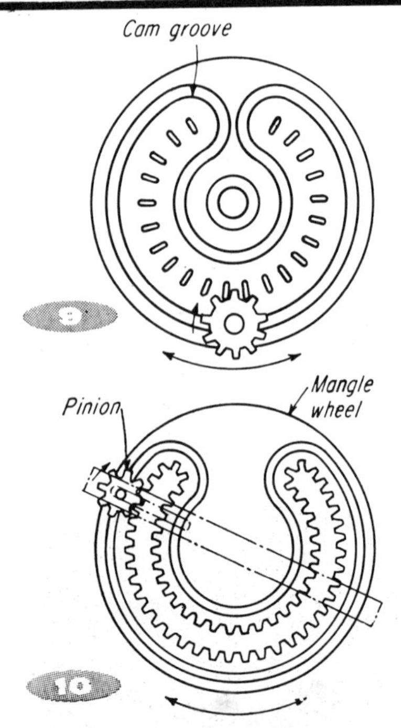

"Mangle" gears . . . cyclicly reverse mangle wheel rotation. Pinion (9) follows cam groove, which keeps teeth in mesh. Convoluted gear-track (10) has more teeth but is far more efficient. Pinion moves in radial slot to maintain mesh with mangle-wheel teeth.

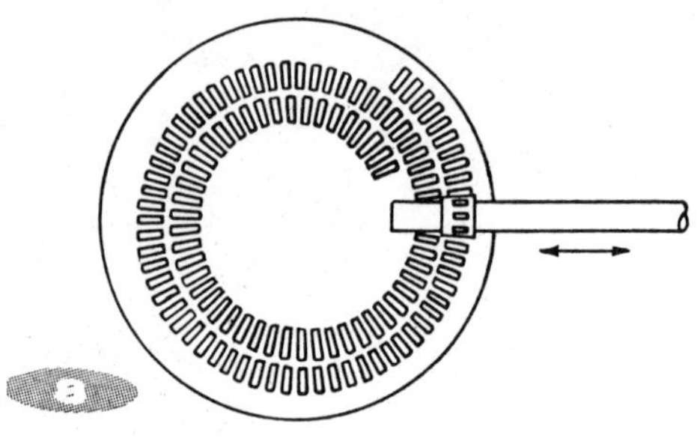

Slotted spiral . . . gear for toy application is not intended for continuous rotation. As it rotates, the pinion shaft is fed in or out and also accelerated or retarded. Gear must be reversed at track end.

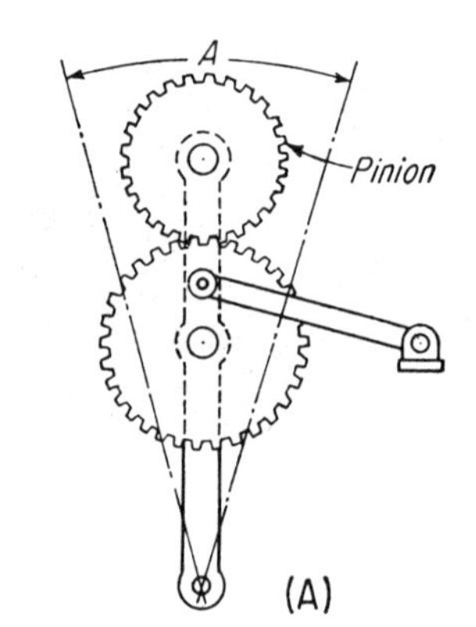

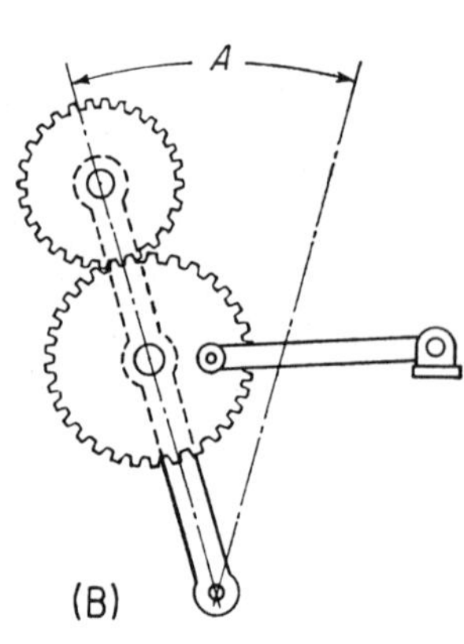

Tied crank and gear . . . give rocking motion to crank through arc A. Pinion drive-shaft has universal joint to permit angular movement as pinion oscillates through arc. Tie mounting-position governs arc length.

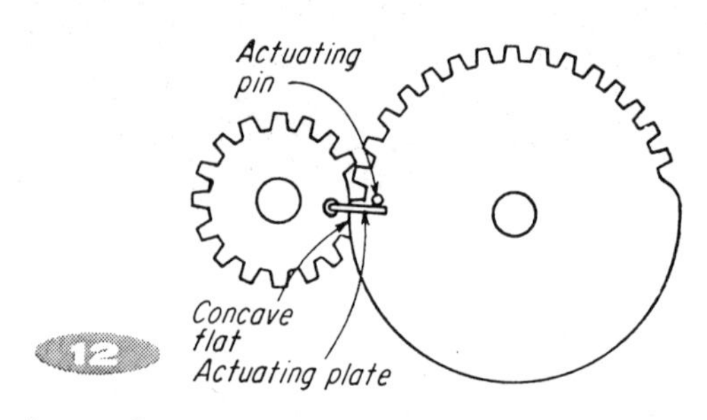

Intermittent rotary motion . . . from continuous rotation is obtained by providing concave flat on pinion, which only meshes with drive wheel when actuating pin strikes actuating plate. The drive wheel and pinion must have same number of teeth.

Types of Noncircular Gears

Noncircular gears generally cost more than competitive devices such as linkages and cams. But with the development of modern production methods, such as the tape-controlled gear shaper, cost has gone down considerably. Also, in comparison with linkages, noncircular gears are more compact and balanced—and can be more easily balanced. These are important considerations in high-speed machinery. Further, the gears can produce continuous, unidirectional cyclic motion—a point in their favor when compared with cams. The disadvantage of cams is that they offer only reciprocating motion.

Applications can be classed into two groups:

- Where only an over-all change in angular velocity of the driven member is required: quick-return drives, intermittent mechanisms as in printing presses, planers, shears, winding machines, automatic-feed machines.
- Where precise, nonlinear functions must be generated, as in computing machines for extracting roots of numbers, raising numbers to any power, generating trigonometric and logarithmic functions.

TYPES OF NONCIRCULAR GEARS

It is always possible to design a special-shaped gear to roll and mesh properly with a gear of any shape—sole requirement is that distance between the two axes must be constant. However, the pitch line of the mating gear may turn out to be an open curve, and the gears can be rotated only for a portion of a revolution—as with two logarithmic-spiral gears (illustrated in Fig 1).

True elliptical gears can only be made to mesh properly if they are twins, and if they are rotated about their focal points. However, gears resembling ellipses can be generated from a basic ellipse. These "higher-order" ellipses (see Fig 2) can be meshed in various interesting combinations to rotate about centers *A*, *B*, *C* or *D*. For example, two 2nd-order elliptical gears can be meshed to rotate about their geometric center; however, they will produce two complete speed cycles per revolution. Difference in contour between a basic ellipse and a 2nd-order ellipse is usually very slight. Note also that the 4th-order "ellipses"

continued, next page

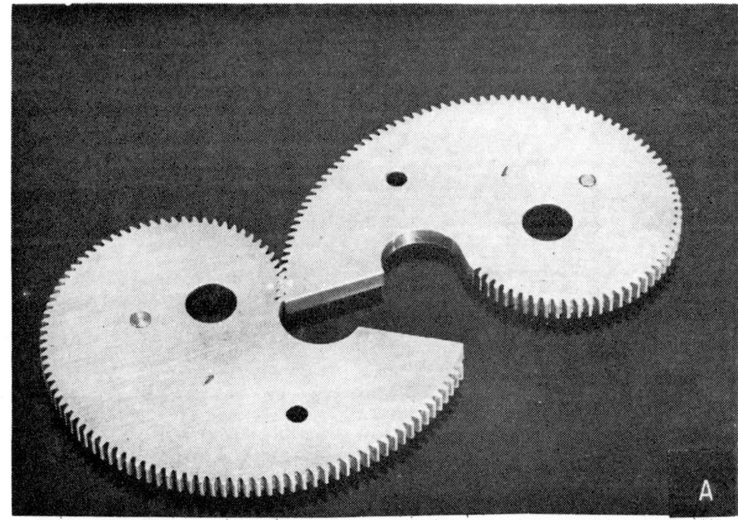

1 LOGARITHMIC SPIRAL GEARS in (A) are open curved, usually employed in computing devices. Elliptical-shape gears (B) are closed curved, frequently found in automatic machinery. Special-shape gears (C) offer wider range of velocity and acceleration characteristics.

resemble square gears (this explains why the square gears, sometimes found as ornaments on tie clasps, illustrated in Fig 3, actually work).

A circular gear, mounted eccentrically, can roll properly only with specially derived curves (shown in Fig 4). One of the curves, however, closely resembles an ellipse. For proper mesh, it must have twice as many teeth as the eccentric gear. When the radiis r, and eccentricity, e, are known, the major semi-axis of the elliptical-shape gear becomes $2r + e$, and the minor $2r - e$. Note also that one of the gears in this group must have internal teeth to roll with the eccentric gear. Actually, it is possible to generate internal-tooth shapes to rotate with noncircular gears of any shape (but, again, the curves may be of the open type).

Noncircular gears can also be designed to roll with special-shaped racks (shown in Fig 5). Combinations include: an elliptical gear and a sinusoid-like rack (a 3rd order ellipse is illustrated but any of the elliptical rolling curves can be used in its place—main advantage is that when the ellipse rolls, its axis of rotation moves along a straight line); and a logarithmic spiral and straight rack (the rack, however, must be inclined to its direction of motion by the angle of the spiral).

DESIGN EQUATIONS

Equations for noncircular gears are shown here in functional form for three common design requirements. They are valid for any noncircular gear pair. Symbols are defined in the box on the next page:

CASE I—Polar equation of one curve and center distance

2 Basic and High-order Elliptical Gear Combinations

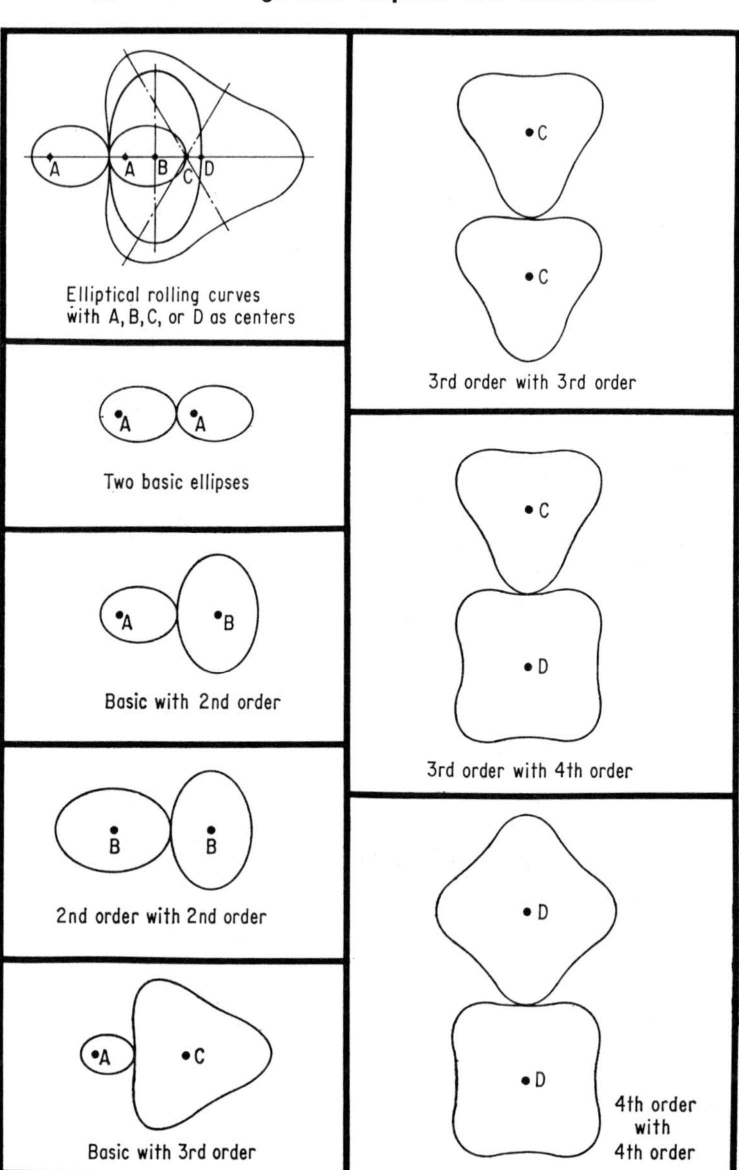

3 SQUARE GEARS ON TIE CLASP seem to defy basic kinematic laws, are actually a takeoff on a pair of 4th order ellipses.

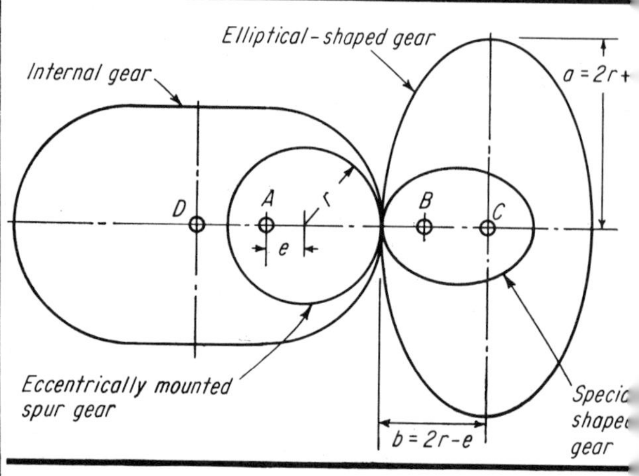

4 ECCENTRIC SPUR GEAR rotating about point **A**, will mesh properly with any of the three gears shown with centers at points **B, C** and **D.**

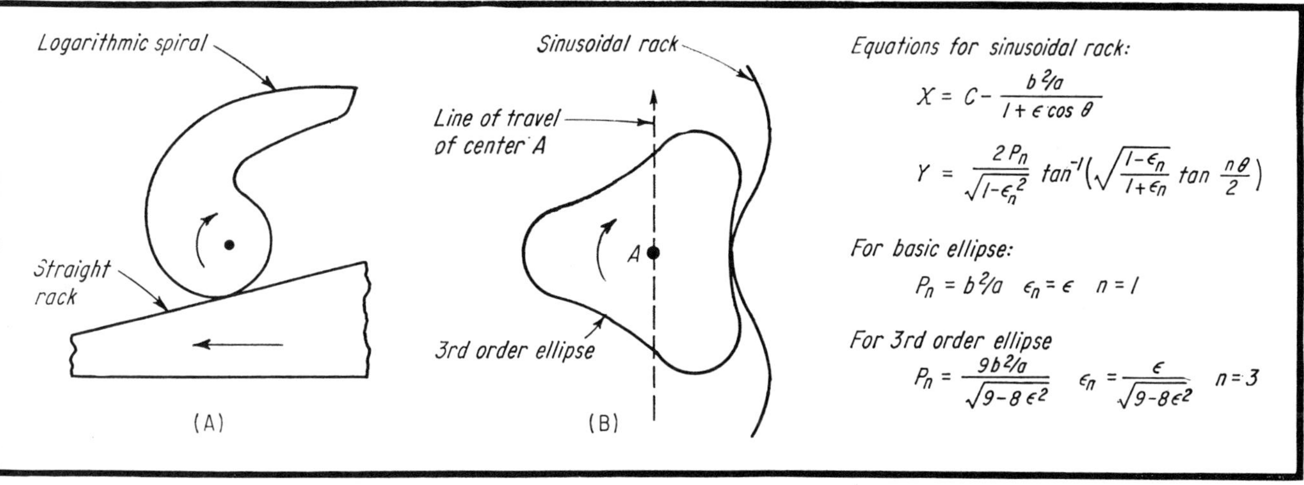

5 RACK AND GEAR COMBINATIONS are possible with noncircular gears. Straight rack for logarithmic spiral (A) must move obliquely; center of 3rd order ellipse (B) follows straight line.

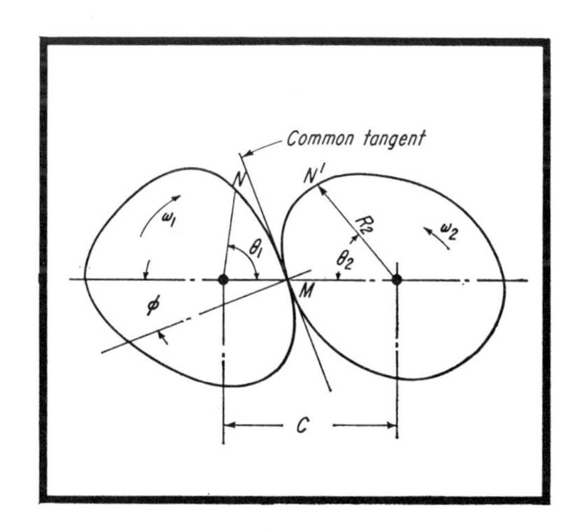

are known; to find the polar equation of the mating gear:

$$R_1 = f(\theta_1)$$
$$R_2 = C - f(\theta_1)$$
$$\theta_2 = -\theta_1 + C\int \frac{d\theta_1}{C - f(\theta_1)}$$

CASE II—Relationship between angular rotation of two members and center distance are known; to find polar equations of both members:

$$\theta_2 = F(\theta_1)$$
$$R_1 = \frac{C\,F'(\theta_1)}{1 + F'(\theta_1)}$$
$$R_2 = C - R_1 = \frac{C}{1 + F'(\theta_1)}$$

CASE III—Relationship between angular velocities of two members and center distance are known; to find polar equations of both members:

$$\omega_2 = \omega_1 G(\theta_1)$$
$$R_1 = \frac{C\,G(\theta_1)}{1 + G(\theta_1)}$$
$$R_2 = C - R_1$$
$$\theta_2 = \int G(\theta_1)\,d\theta_1$$

Velocity equations and characteristics of five types of noncircular gears are listed in the table on the next page.

CHECKING FOR CLOSED CURVES

Gears can be quickly analyzed to determine whether their pitch curves are open or closed by means of the following equations:

In Case I, if $R = f(\theta) = f(\theta + 2N\pi)$, the pitch curve is closed.

In Case II, if $\theta_1 = F(\theta_2)$ and $F(\theta_0) = 0$, the curve is closed when the equation $F(\theta_0 + 2\pi/N_1) = 2\pi/N_2$ can be satisfied by substituting integers or rational fractions for N_1 and N_2. If fractions must be used to solve this

Symbols

- a = semi-major axis of ellipse
- b = semi-minor axis of ellipse
- C = center distance (see above sketch)
- ϵ = eccentricity of an ellipse = $\sqrt{1-(b/a)^2}$
- e = eccentricity of an eccentrically mounted spur gear
- N = number of teeth
- P = diametral pitch
- r_c = radius of curvature
- R = active pitch radius
- S = length of periphery of pitch circle
- X, Y = rectangular coordinates
- θ = polar angle to R
- ϕ = angle of obliquity
- ω = angular velocity
- $f(\theta), F(\theta), G(\theta)$ = various functions of θ
- $f'(\theta), F'(\theta), G'(\theta)$ = first derivatives of functions of θ

Characteristics of Five Noncircular Gear Systems

Type	Comments	Basic equations	Velocity equations ω_1 = constant
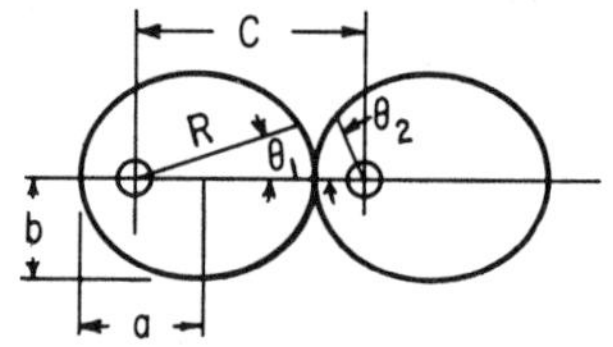 C, R, θ_1, θ_2, b, a Two ellipses rotating about foci	Gears are identical. Comparatively easy to manufacture. Used for quick-return mechanisms, printing presses, automatic machinery	$R = \frac{b^2}{a[1 + \epsilon \cos\theta]}$ ϵ = eccentricity $= \sqrt{1 - \left(\frac{b}{a}\right)^2}$ $a = \frac{1}{2}$ major axis $b = \frac{1}{2}$ minor axis	$\omega_2 = \omega_1 \left[\frac{r^2 + 1 + (r^2 - 1)\cos\theta_2}{2r}\right]$ where $r = \frac{R\ max}{R\ min}$ 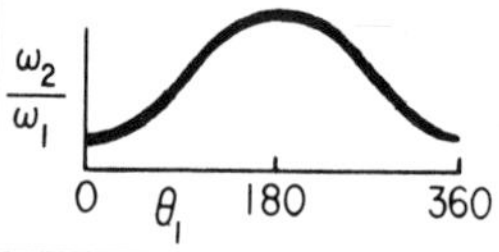 $\frac{\omega_2}{\omega_1}$; 0, θ_1, 180, 360
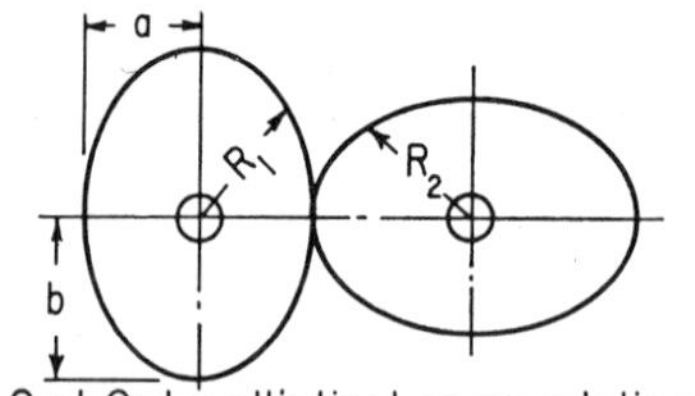 a, R_1, R_2, b 2nd Order elliptical gears rotating about their geometric centers	Gears are identical. Geometric properties well known. Better balanced than true elliptical gears. Used where two complete speed cycles are required for one revolution	$R = \frac{2ab}{(a+b) - (a-b)\cos 2\theta}$ $C = a + b$ a = maximum radius b = minimum radius	$\omega_2 = \omega_1 \left[\frac{r + 1 + (r^2 - 1)\cos 2\theta_2}{2r}\right]$ where $r = \frac{a}{b}$ 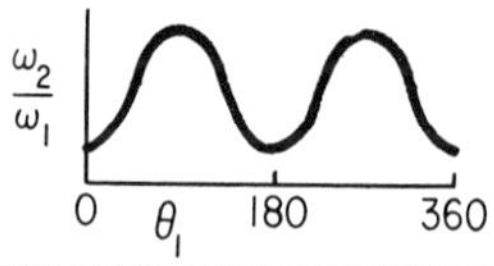 $\frac{\omega_2}{\omega_1}$; 0, θ_1, 180, 360
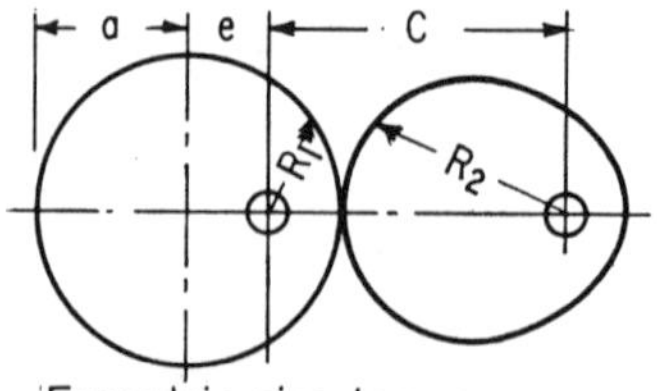 a, e, C, R_1, R_2 Eccentric circular gear rotating with its conjugate	Standard spur gear can be employed as the eccentric. Mating gear has special shape	$R_1 = \sqrt{a^2 + e^2 + 2ae\cos\theta_1}$ $\theta_2 = \theta_1 + C\int \frac{d\theta_1}{C - R_1}$ $C = R_1 + R_2$	$\frac{\omega_2}{\omega_1} = \frac{\sqrt{a^2 + e^2 + 2ae\cos\theta_1}}{C - \sqrt{a^2 + e^2 + 2ae\cos\theta_1}}$ 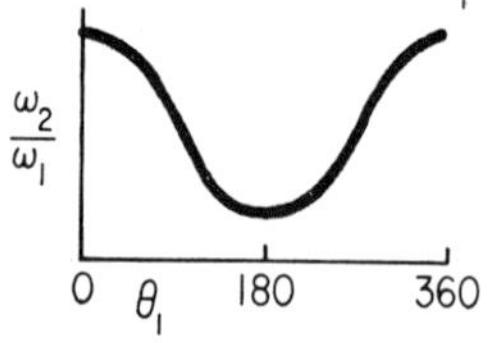 $\frac{\omega_2}{\omega_1}$; 0, θ_1, 180, 360
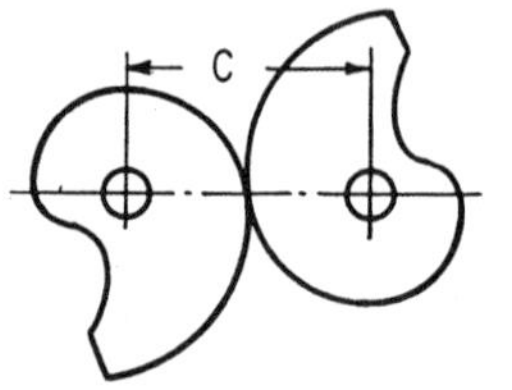 C Logarithmic spiral gears	Gears can be identical although can be used in conbinations to give variety of functions. Must be open gears	$R_1 = Ae^{k\theta_1}$ $R_2 = C - R_1$ $= Ae^{k\theta_2}$ $\theta_2 = \frac{1}{k}\log(C - Ae^{k\theta_1})$ e = natural log base	$\frac{\omega_2}{\omega_1} = \frac{Ae^{k\theta_1}}{C - Ae^{k\theta_1}}$ 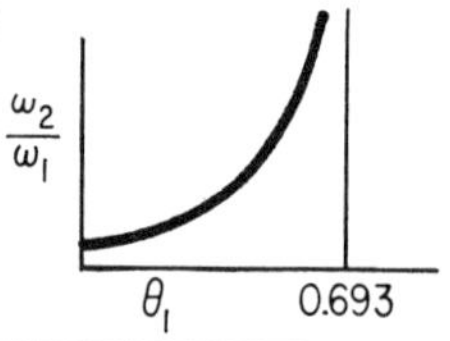 $\frac{\omega_2}{\omega_1}$; θ_1, 0.693
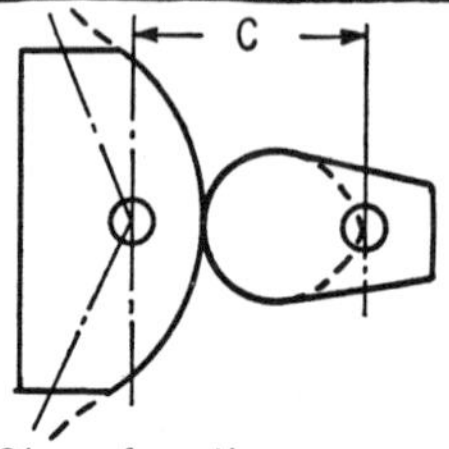 C Sine-function gears	For producing angular displacement proportional to sine of input angle. Must be open gears	$\theta_2 = \sin^{-1}(k\theta_1)$ $R_2 = \frac{C}{1 + k\cos\theta_1}$ $R_1 = C - R_2$ $= \frac{Ck\cos\theta_1}{1 + k\cos\theta_1}$	$\frac{\omega_2}{\omega_1} = k\cos\theta_1$ $\frac{\omega_2}{\omega_1}$; θ_1 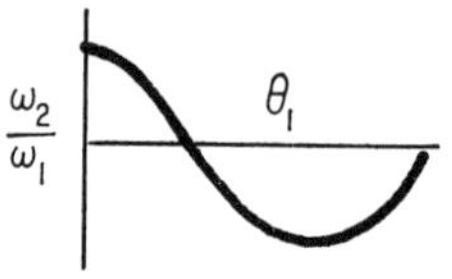

equation, the curve will have double points (intersect itself), which is, of course, an undesirable condition.

In Case III, if $\theta_2 = \int G(\theta_1)\, d\theta_1$, let $G(\theta_1)\, d\theta_1 = F(\theta_1)$, and use the same method as for Case II, with the subscripts reversed.

With some gear sets, the mating gear will be a closed curve only if the correct center distance is employed. This distance can be found from the equation:

$$4\pi = \int_0^{2\pi} \frac{d\theta_1}{C - f(\theta_1)}$$

ELLIPTICAL GEARS for CYCLIC SPEED VARIATIONS

SIGMUND RAPPAPORT, *kinematician*
Ford Instrument Co, Div of Sperry Rand Corp, and
Adjunct professor, Polytechnic Institute of Brooklyn

The twin principle simplifies design. It is illustrated with the pair of identical ellipses drawn at right. They rotate in opposite directions: one around its focus F_1; the other around corresponding focus F'_1. When the gears rotate through angles θ_1 and θ_2, respectively, all points on arc MN come in successive contact with points of the equally long and identically shaped arc MP. Basic rule is that $R_1 + R_2$ must be constant. From symmetry, $F_1'P$ (which is R_2) equals F_2N; therefore distance $F_1N + F_2N$ is constant, which is the definition of an ellipse.

From this same condition the instantaneous angular velocity $d\theta_2dt$ of the output gear easily be found if the constant input velocity ω is given. The instantaneous velocity ratio $(d\theta_2/dt)$: (ω) is determined by the inverse ratio of the instantaneous radii vectors:

$$\frac{d\theta_2}{dt} = \omega \frac{R_1}{R_2}$$

From the polar equation of the ellipse,

$$R_1 = \frac{a(1-\epsilon^2)}{1+\epsilon\cos\theta_1}$$

where ϵ (the numerical eccentricity of the ellipse) is defined as

$$\epsilon = \frac{\sqrt{a^2-b^2}}{a}$$

with a as major, and b as minor semi-axis. This equation can also be written as

$$\epsilon = \frac{e}{a}$$

where e is distance from focus to center of the ellipse.

Symbols

a = major semi-axis
b = minor semi-axis
ϵ = numerical eccentricity
e = distance from focal point to geometric center of ellipse
F_1 = focal point and center of rotation of ellipse
F_2 = focal point of ellipse
K = ratio of velocity change of output gear
N = number of teeth
R = instantaneous radius
θ = instantaneous angle of rotation
ρ = radius for approximating the contour of ellipse
ω = constant angular velocity of input
Subscripts *1* and *2* relate to input and output elliptical gear, respectively, unless otherwise stated.

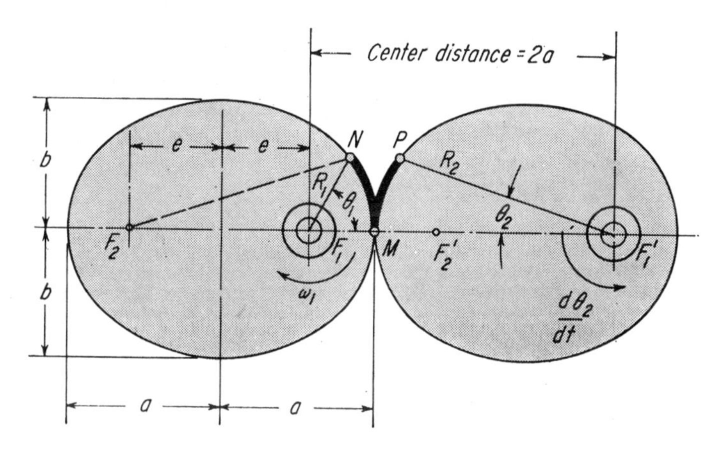

TWO IDENTICAL ELLIPSES rotating around foci F_1 and F_1', have equal arcs MN and MP in successive contact.

From $R_1 + R_2 = 2a$, is it found that

$$R_2 = 2a - \frac{a(1-\epsilon^2)}{1+\epsilon\cos\theta_1}$$

$$R_2 = a\left[\frac{1+2\epsilon\cos\theta_1+\epsilon^2}{1+\epsilon\cos\theta_1}\right]$$

or

$$\frac{d\theta_2}{dt} = \omega\left[\frac{1-\epsilon^2}{1+\epsilon^2+2\epsilon\cos\theta_1}\right]$$

The extremes occur at $\theta_1 = 0$ and $180°$, with $\cos\theta_1 = +1$ and $\cos\theta_1 = -1$. Thus, by substituting for ϵ, the minimum velocity at $\theta_1 = 0$ is:

$$\frac{d\theta_2}{dt_{(\min)}} = \omega\left[\frac{a-e}{a+e}\right]$$

and, for $\theta_1 = 180°$:

$$\frac{d\theta_2}{dt_{(\max)}} = \omega\left[\frac{a+e}{a-e}\right]$$

Maximum and minimum angular-output velocities are the reciprocals of each other. If K is the ratio of maximum to minimum, then

$$K = \left(\frac{a+e}{a-e}\right)^2$$

It is good practice to keep K under 5 to insure smooth running without "whip." Design usually starts with the choice of center distance 2a; in other words, with the size of the gears. If the major semi-axis a is given, and a certain over-all ratio K is desired, the minor semi-axis b can be found from

$$b = \frac{2aK^{1/4}}{1+K^{1/2}}$$

3-GEAR DRIVES

DR J HIRSCHHORN,
Senior lecturer in Mechanical Engineering
University of New South Wales, Australia

THREE PATTERNS OF MOTION

The drive is built around a gear mounted eccentrically on the input shaft, plus an idler gear and an output gear. Two links keep the idler in mesh with the output and input gears, but the gears are free to turn with respect to the links.

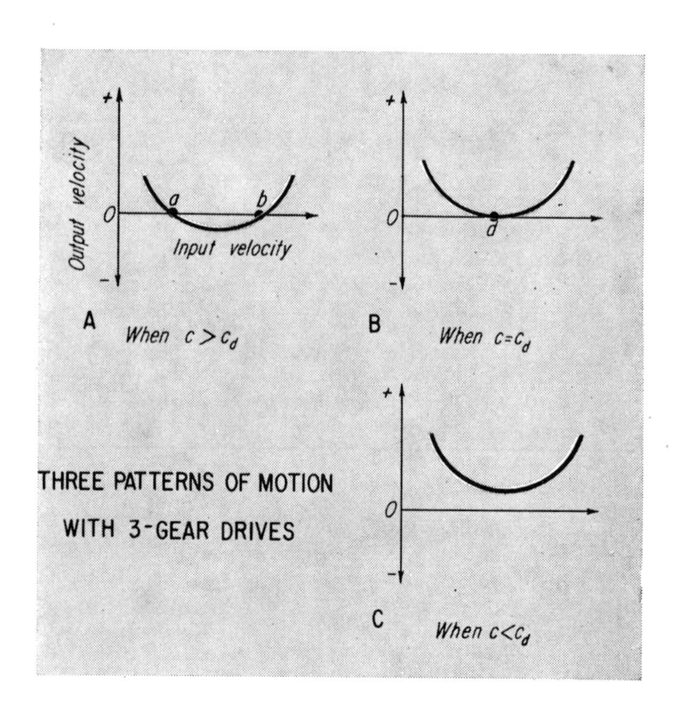

THREE PATTERNS OF MOTION WITH 3-GEAR DRIVES

The above comparison of motion patterns shows that when the distance c between input and output shafts is considered adjustable, three patterns of motion are available:

- If c is made larger than a critical value, c_d, the output gear stops for an instant, reverses for a finite period, stops again, and then resumes its original sense of rotation, while the driving gear is completing one revolution at constant speed.
- If c is made equal to c_d, the output gear dwells for an instant, and then continues to rotate in its original sense. Although, theoretically, it pauses only an instant, in and actual mechanism this may last about 45° of the cycle—and is the motion pattern which is usually required.
- If c is made smaller than c_d, the output gear slows down and then accelerates, but does not actually stop.

The output gear comes to rest—positions *a*, *b* in (A) and *d* in (B)—when the instantaneous center of its rotation relative to the driver coincides with the input

BASIC COMPONENTS OF 3-GEAR DRIVE

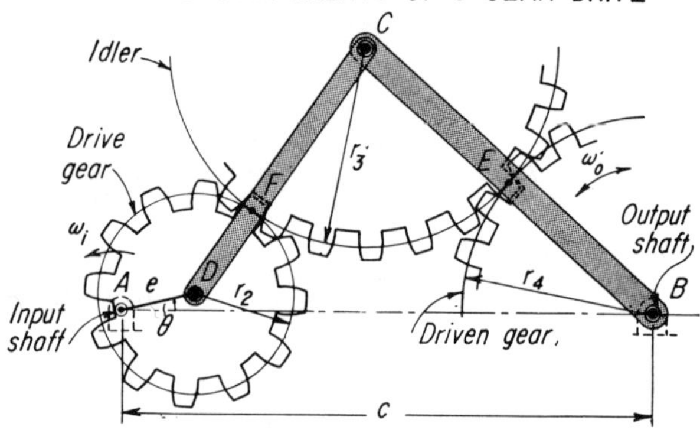

shaft center. This is the case when points A, *F* and *E* are in line, as shown in the dwell-position diagram below. This diagram shows one of the two dwell conditions that occur when c is larger than c_d. As c is made smaller the

IN DWELL POSITION

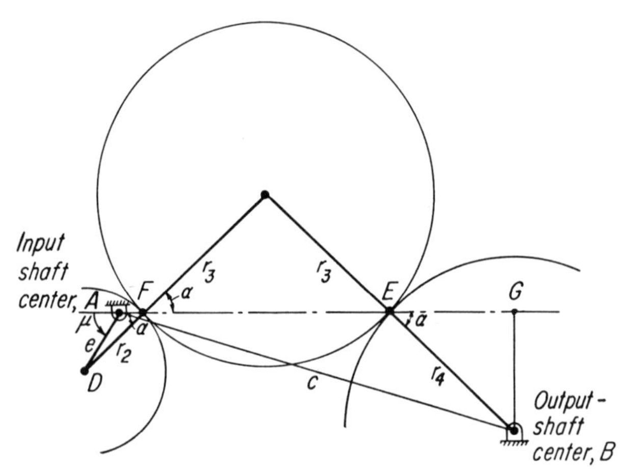

two dwell positions occur closer together; therefore, the critical value c_d can be defined as being the smallest center distance which, for a given mechanism, allows points A, F and E to be in line.

THE CENTER-DISTANCE EQUATION

The center distance c can be expressed in terms of trigonometric quantities:

$$c^2 = (AF + FE + EG)^2 + (GB)^2 \quad (1)$$

with

$$AF = r_2 \cos\gamma - e\cos\mu, \quad (2)$$

$$FE = 2r_3 \cos\gamma, \quad (3)$$

$$EG = r_4 \cos\gamma, \quad (4)$$

$$GB = r_4 \sin\gamma, \quad (5)$$

Eq (1) becomes:

$$c^2 = [(r_2 + 2r_3 + r_4)\cos\gamma - e\cos\mu]^2 + r_4^2 \sin^2\gamma \quad (6)$$

but,

$$e\sin\mu = r_2 \sin\gamma \quad (7)$$

hence,

$$e\cos\mu = \sqrt{r_2^2\cos^2\gamma - (r_2^2 - e^2)} \quad (8)$$

Substituting Eq (8) into Eq (6) gives:

$$c^2 = [(r_2 + 2r_3 + r_4)\cos\gamma - \sqrt{r_2^2\cos^2\gamma - (r_2^2 - e^2)}]^2 + r_4^2\sin^2\gamma \quad (9)$$

To obtain $c_d = c_{min}$, Eq (9) is differentiated with respect to γ, and equated to zero. After some simple but tedious algebraic operations, the following quadratic equation can be obtained involving the angle for the dwell position γ_d:

$$K \cos^4 \gamma_d - L \cos^2 \gamma_d - M = 0 \quad (10)$$

where $K = [(r_4^2 - r_2^2)^2 - 2(r_4^2 + r_2^2)(r_2 + 2r_3 + r_4)^2 + (r_2 + 2r_3 + r_4)^4]r_2^2 \quad (11)$

$L = [(r_4^2 - r_2^2)^2 - 2(r_4^2 + r_2^2)(r_2 + 2r_3 + r_4)^2 + (r_2 + 2r_3 + r_4)^4](r_2^2 - e^2) \quad (12)$

$M = (r_2 + 2r_3 + r_4)^2(r_2^2 - e^2)^2 \quad (13)$

Knowing γ_d, all other quantities, in particular c_d, can be determined.

SAMPLE PROBLEM

Determine the dwell center distance for the three-gear drive with dimensions $r_2 = 1$ in. $e = 0.8$ in., $r_3 = 2$ in., $r_4 = 2$ in.

Substituting these values into Eq (10) gives:

$$1920(\cos^4 \gamma_d) - 691(\cos^2 \gamma_d) - 6.35 = 0$$

which gives

$$\cos^2 \gamma_d = \frac{1417}{3840} \quad \text{and} \quad \gamma_d = 52°36'$$

from Eq (7) $\mu_d = 83°14'$ from Eq (4) $EG = 1.21$ in.

from Eq (2) $AF = 0.51$ in. from Eq (5) $GB = 1.59$ in.

from Eq (3) $FE = 2.43$ in. from Eq (1) $c_d = 4.45$ in.

Two-Tooth Gear Systems

S. RAPPAPORT
Ford Instrument Company

THE PROFILE OF A GEAR TOOTH is determined by the laws of kinematics under the assumption that two coplanar pitch circles roll, without sliding, around each other in opposite directions at equal circumferential speed. But what happens to the gear system if this assumption is modified by the condition that the two planes must rotate at equal speed in the same direction?

Applying the laws of relative motion in the diagram, Fig. 1, the point P is located at distance r from the rotating center A of one plane. Relative to the second plane, which rotates around center B, point P describes a circle having the radius $r\sqrt{2}$.

But the two gears generated using radius $r\sqrt{2}$ do not make up a constrained kinematic chain because the movement of one gear does not completely control the movement of the other. To make one gear precisely define the movement of the other at all times, each gear must consist of a set of at least three two-tooth gear laminations. The latter are spaced at equal angles around the gear center as shown in Fig. 2.

Four of these laminated two-tooth gears mounted in mesh with their centers at the corners of a square fulfill the original condition as stated above. If one of the four gears is rotated, the other three will turn in the same direction. Similarly, five or more of these gears arranged, in mesh, around a circle would increase the number of centers.

Although there is an inherent high friction loss due to the constant sliding of the tooth tip of one gear along the face of the other, which results in very low coefficiency of the gear system, there are undoubtedly possible applications for this type of gear mechanism. It is not suitable for use as a power transmission element, but the pocket formed between the teeth as they rotate could conceivably be utilized for metering semi-solid materials.

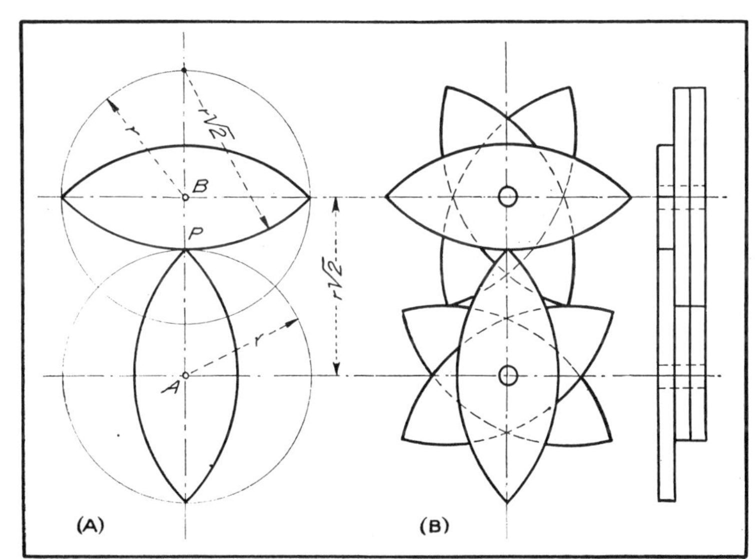

Fig. 1—Diagram (A) shows location of contact point P in two coplanar pitch circles. Three laminations (B) are used for constrained motion.

Wright Machinery Co.

Fig. 2—Four or more gears equally spaced on a circle have same rotation.

Planetary gear systems

Designers keep finding new and useful planetaries. Forty-eight popular types are given here with their speed-ratio equations

JOHN H. GLOVER, Product Design Engineer, Transmission and Chassis Div, Ford Motor Co, Detroit, Mich

1—Minuteman cover drive (American Electric Co.)

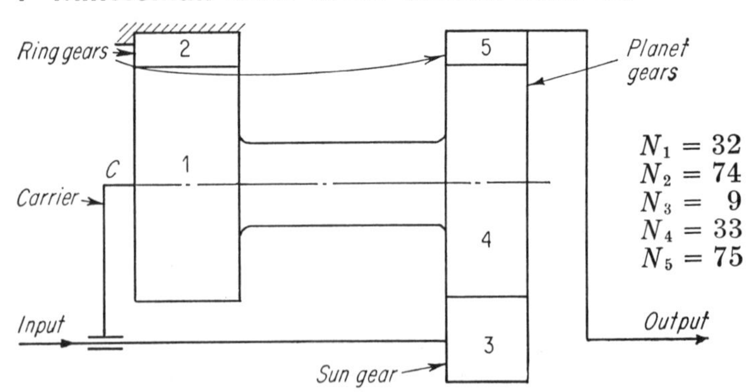

Ring gear 2 fixed; ring gear 5 output

Speed-ratio equation

$$R = \frac{1 + \dfrac{N_4 N_2}{N_3 N_1}}{1 - \dfrac{N_4 N_2}{N_5 N_1}} = \frac{1 + \dfrac{(33)(74)}{(9)(32)}}{1 - \dfrac{(33)(74)}{(75)(32)}} = -541\tfrac{2}{3}$$

Planetary gear for pulling 95-ton blast-resistant lid to cover and uncover underground Minuteman missiles. Schematic at left. Author's equations lead directly to the speed-ratio equation for the system, boxed at left.

Symbols

C = carrier (also called "spider")—a non-gear member of a gear train whose rotation affects gear ratio

N = number of teeth

R = overall speed reduction ratio

1, 2, 3, etc = gears in a train (corresponding to labels on schematic diagram)

Double-eccentric drives (Ref. 4)

Input is through double-throw crank (carrier). Gear 1

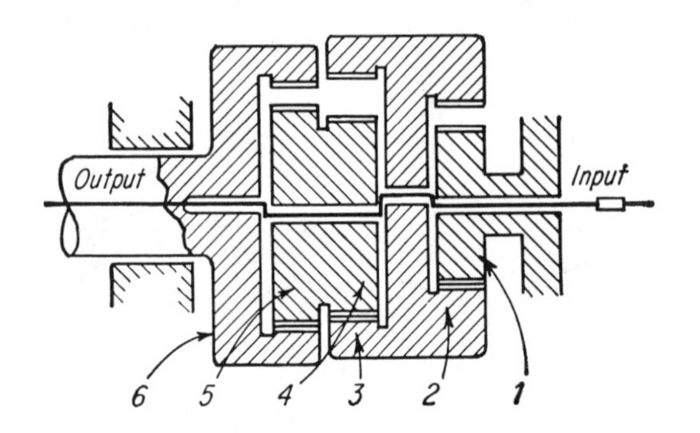

$$R = \frac{1}{1 - \dfrac{N_5 N_3 N_1}{N_6 N_4 N_2}}$$

When $N_1 = 103$, $N_2 = 110$, $N_3 = 109$, $N_4 = 100$, $N_5 = 94$, $N_6 = 96$

$$R = \frac{1}{1 - \dfrac{(94)(109)(103)}{(96)(100)(110)}} = 1505$$

Coupled planetary drives (Ref. 1)

(A)

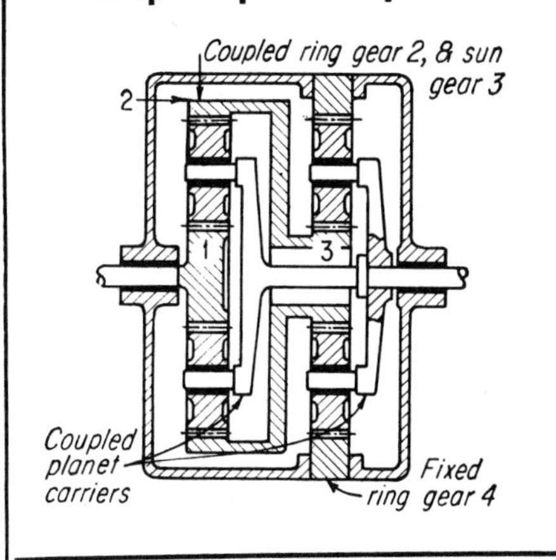

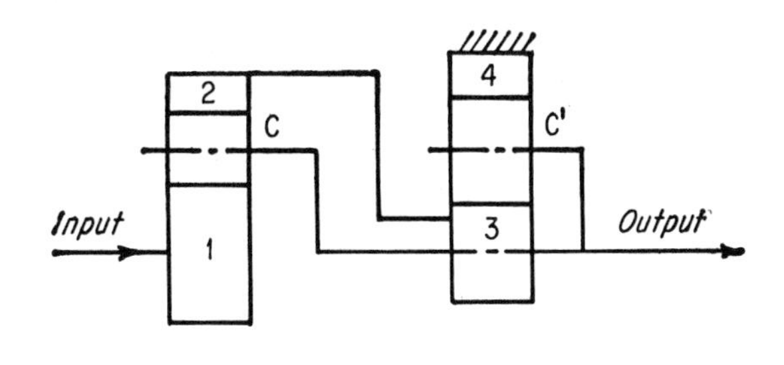

$$R = 1 - \frac{N_2 N_4}{N_1 N_3}$$

(B)

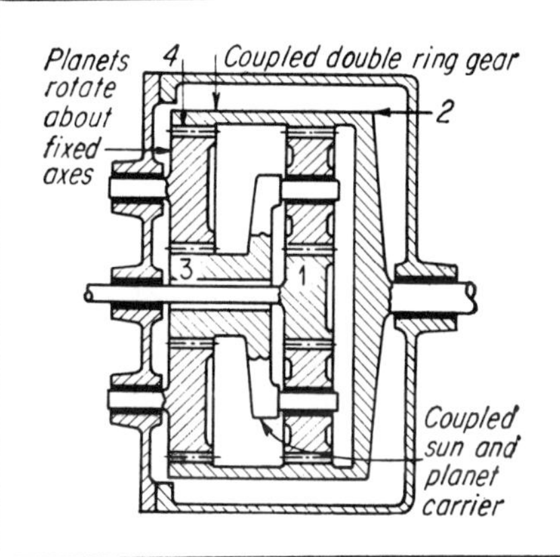

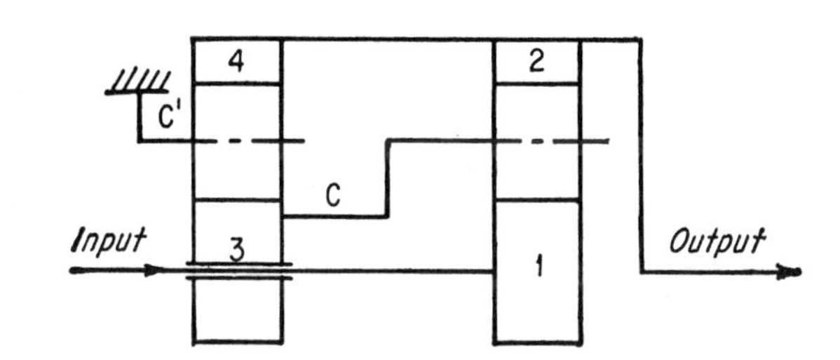

$$R = \left(1 + \frac{N_2}{N_1}\right)\left(-\frac{N_4}{N_3}\right) - \frac{N_2}{N_1}$$

(C)

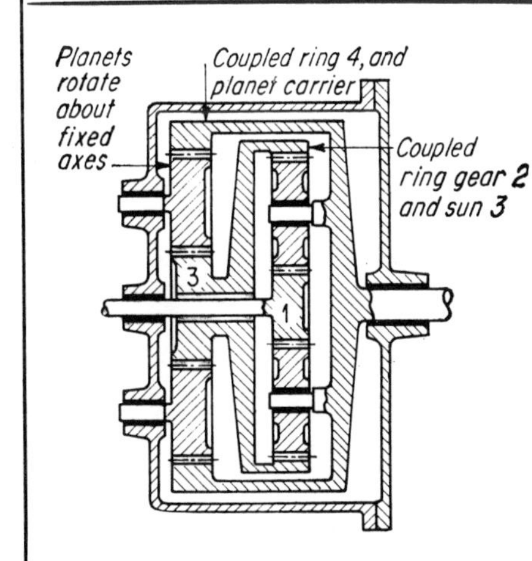

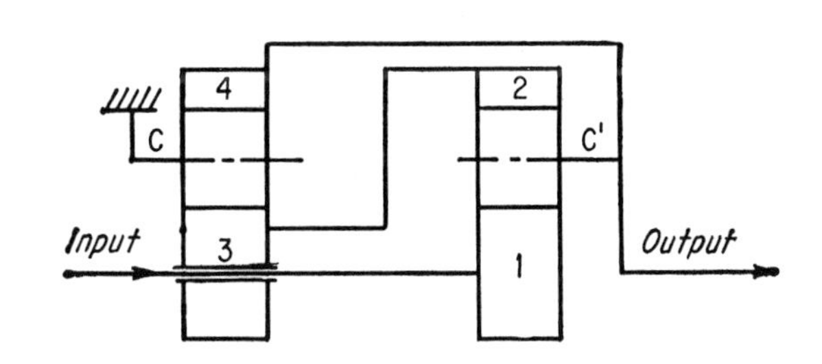

$$R = 1 + \frac{N_2}{N_1}\left(1 + \frac{N_4}{N_3}\right)$$

(D)

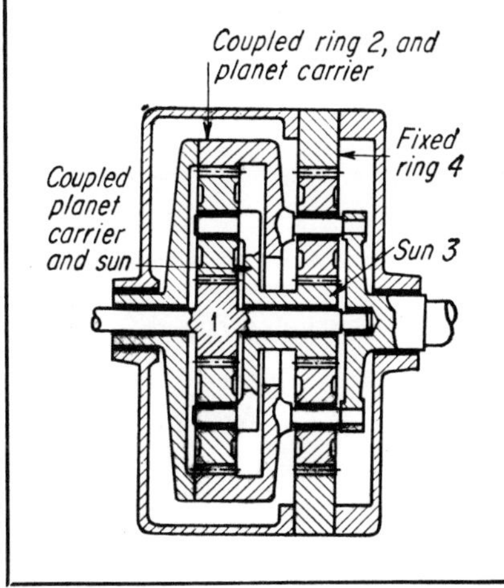

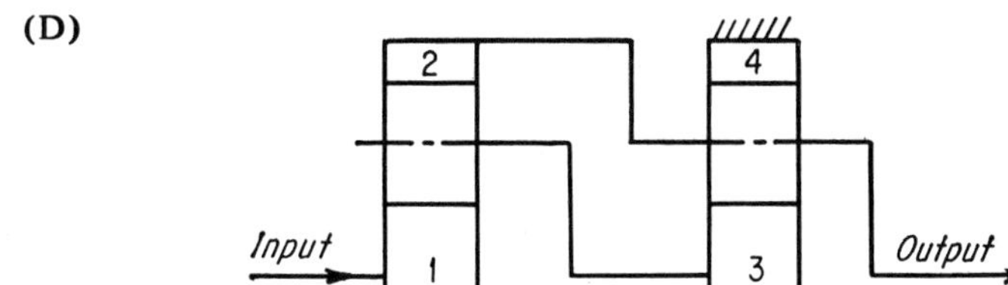

$$R = 1 + \frac{N_4}{N_3}\left(1 + \frac{N_2}{N_1}\right)$$

Fixed-differential drives

Output is difference between speeds of two parts leading to high reduction ratios

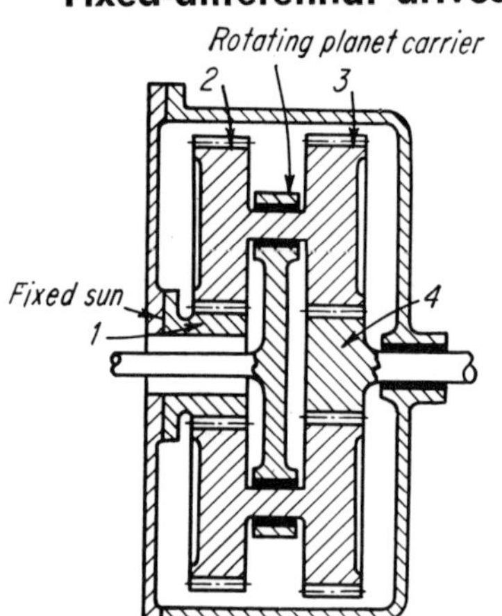

(A)

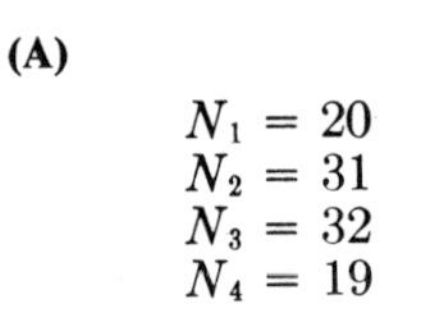

$N_1 = 20$
$N_2 = 31$
$N_3 = 32$
$N_4 = 19$

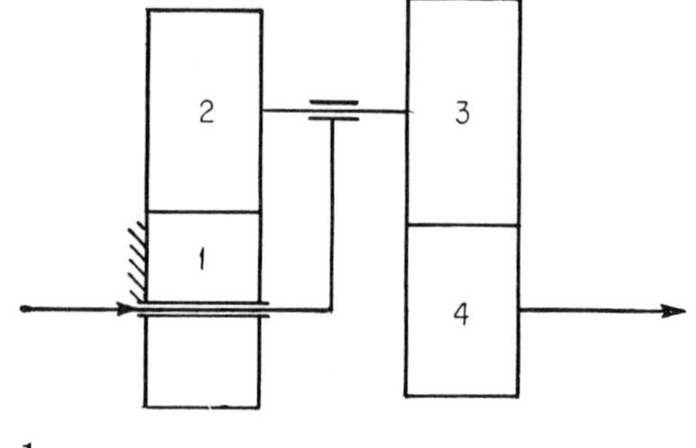

$$R = \frac{1}{1 - \dfrac{N_3 N_1}{N_4 N_2}} = \frac{1}{1 - \dfrac{(32)\,(20)}{(19)\,(31)}} = -11.549$$

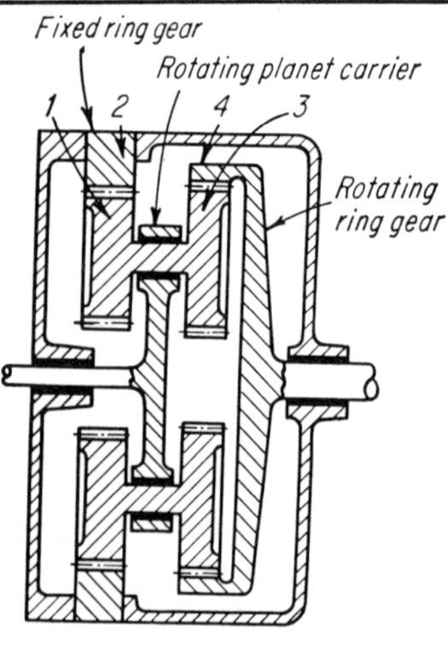

(B)

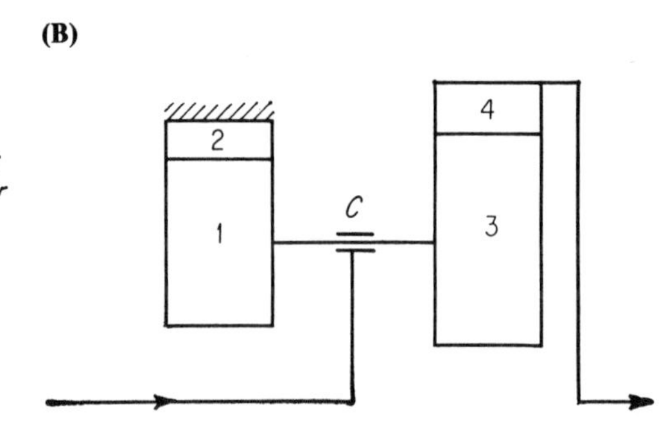

$$R = \frac{1}{1 - \dfrac{N_3 N_2}{N_4 N_1}}$$

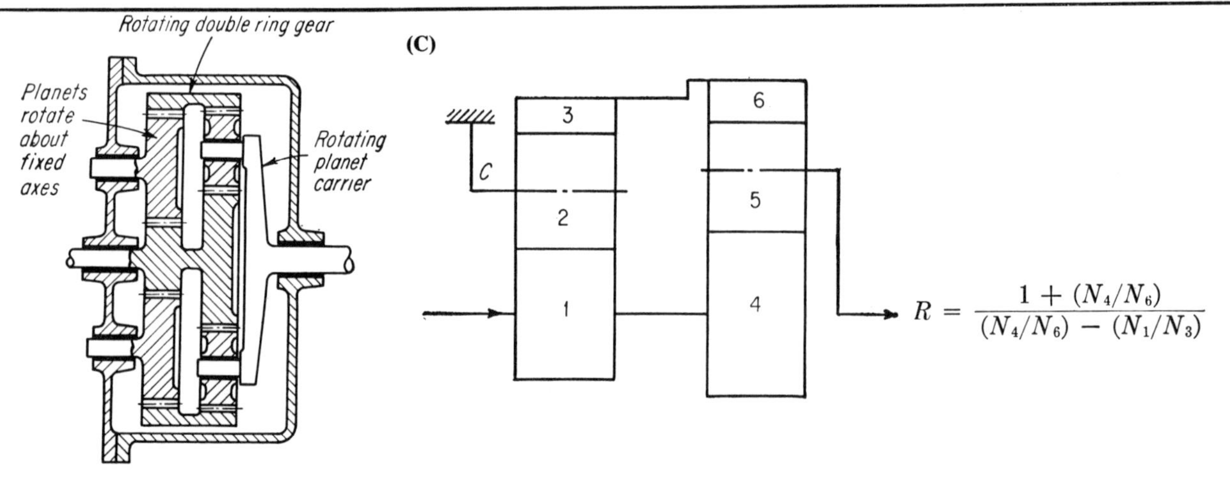

$$R = \frac{1 + (N_4/N_6)}{(N_4/N_6) - (N_1/N_3)}$$

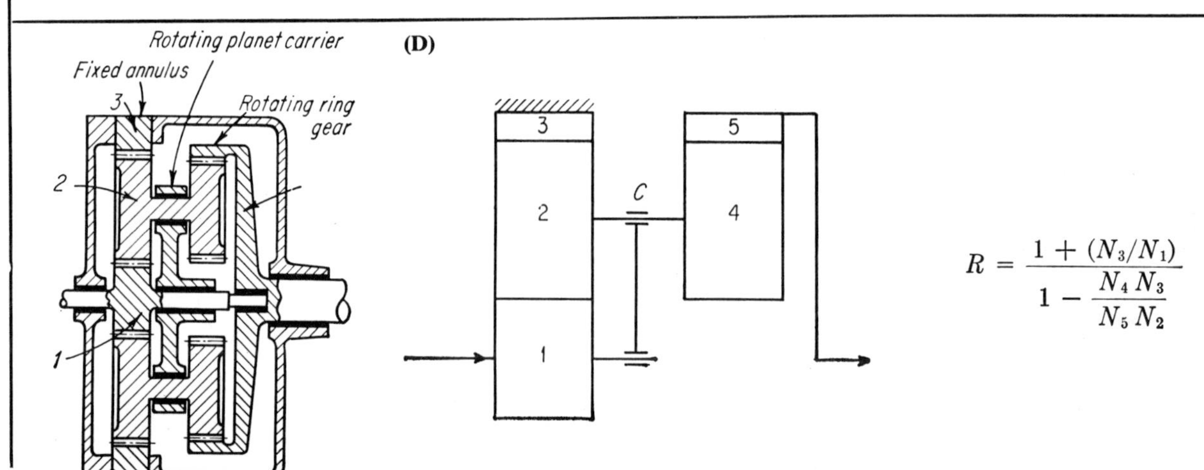

$$R = \frac{1 + (N_3/N_1)}{1 - \dfrac{N_4 N_3}{N_5 N_2}}$$

Simple planetaries and inversions

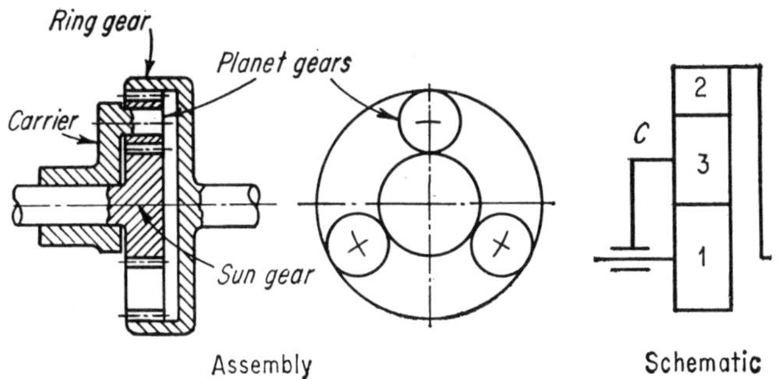

Input member	Fixed member	Output member	Speed-ratio equation
1	C	2	$R = -N_2/N_1$
2	C	1	$R = -N_1/N_2$
1	2	C	$R = 1 + (N_2/N_1)$
2	1	C	$R = 1 + (N_1/N_2)$
C	2	1	$R = \dfrac{1}{1 + (N_2/N_1)}$
C	1	2	$R = \dfrac{1}{1 + (N_1/N_2)}$

Input member	Fixed member	Output member	Speed-ratio equation
1	C	3	$R = \dfrac{N_2 N_3}{N_1 N_4}$
1	3	C	$R = 1 - \dfrac{N_2 N_3}{N_1 N_4}$
3	1	C	$R = 1 - \dfrac{N_1 N_4}{N_2 N_3}$
3	C	1	$R = \dfrac{N_4 N_1}{N_3 N_2}$
C	1	3	$R = 1 \Big/ \left(1 - \dfrac{N_1 N_4}{N_2 N_3}\right)$
C	3	1	$R = 1 \Big/ \left(1 - \dfrac{N_2 N_3}{N_1 N_4}\right)$

Continued on next page

Humpage's bevel gears

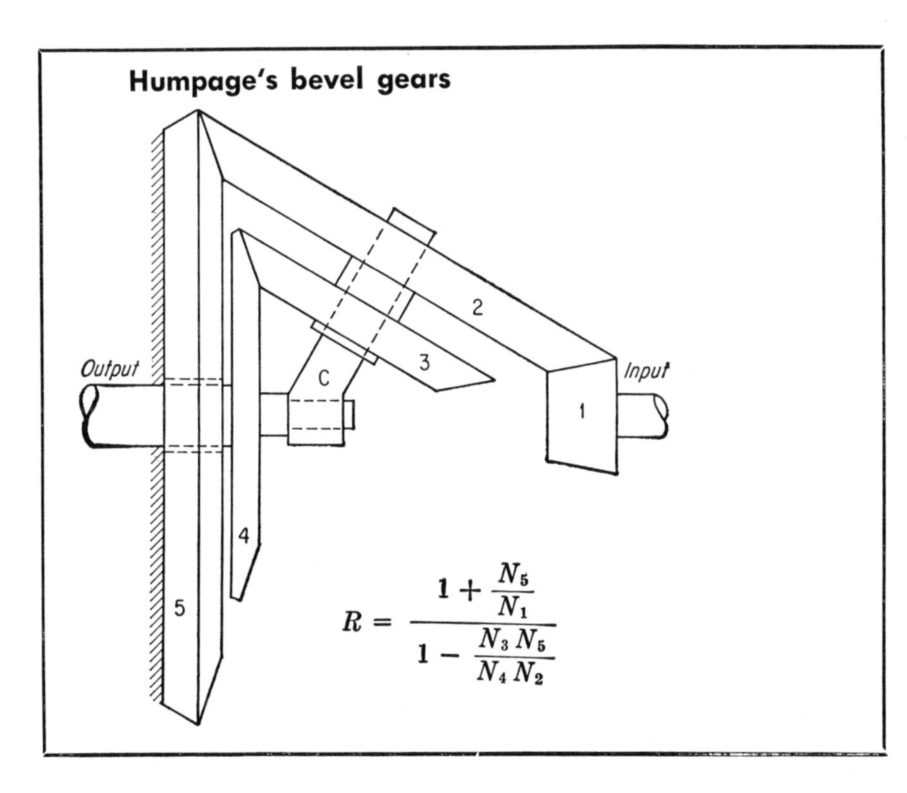

$$R = \frac{1 + \dfrac{N_5}{N_1}}{1 - \dfrac{N_3 N_5}{N_4 N_2}}$$

References:

1. D. W. Dudley, ed, *Gear Handbook,* pp 3-19 to 3-25, McGraw-Hill, New York, 1962.
2. E. F. Obert, "Speed Ratios and Torque Ratios in Epicyclic Gear Trains," pp 270-271, *Product Engineering,* Apr 1945.
3. J. W. Edgemond, Jr, "Epicyclic Gears for Control Mechanisms," pp 194-198, *Product Engineering,* Feb 1957.
4. J. H. Barnwell, "The Double-Eccentric Speed Reducer," pp 135-137, *Machine Design,* Aug 17, 1961.

Hydramatic 4-speed transmission (General Motors)

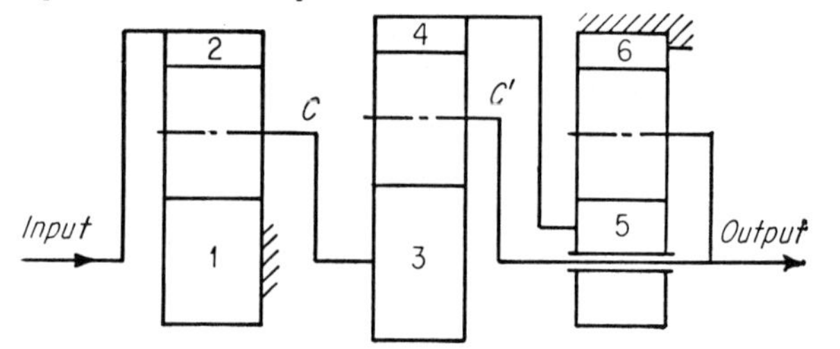

Reverse gear

$$R = \left(1 + \frac{N_1}{N_2}\right)\left(1 - \frac{N_4 N_6}{N_3 N_5}\right)$$

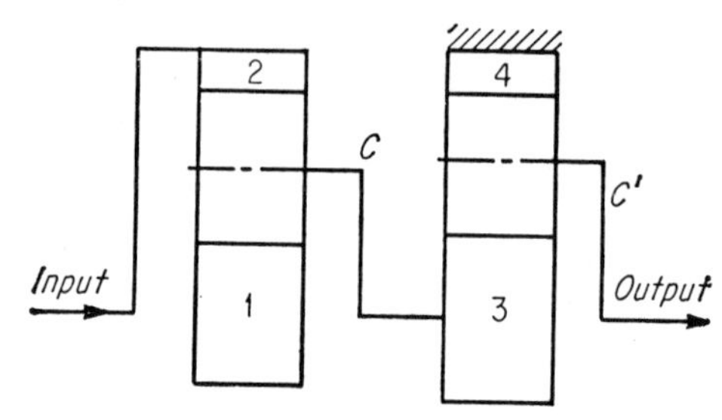

For low gear, uncouple connection between gears 4 and 5 and fix gears 1 and 4

$$R = \left(1 + \frac{N_1}{N_2}\right)\left(1 + \frac{N_4}{N_3}\right)$$

Daimler preselective drive

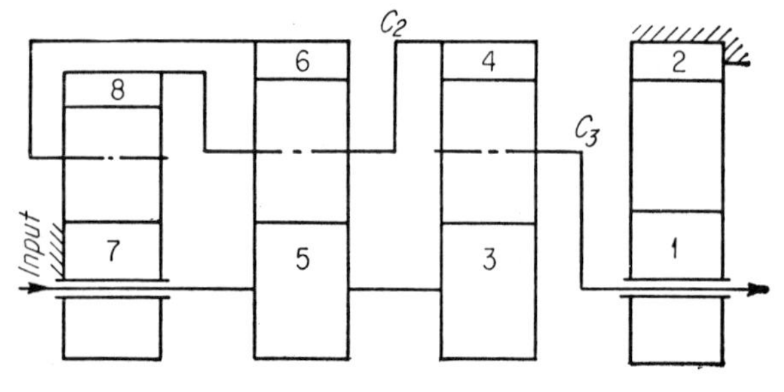

3rd gear (high speed)—gear 1 and 2 inoperative

$$R = \frac{1 + \dfrac{N_6/N_5}{1 + (N_8/N_7)}}{1 + \dfrac{N_4/N_3}{[1 + (N_4/N_3)]\,[1 + N_8/N_7]}}$$

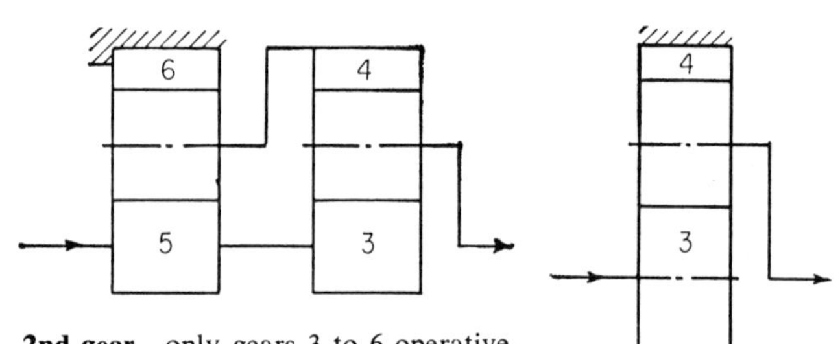

2nd gear—only gears 3 to 6 operative

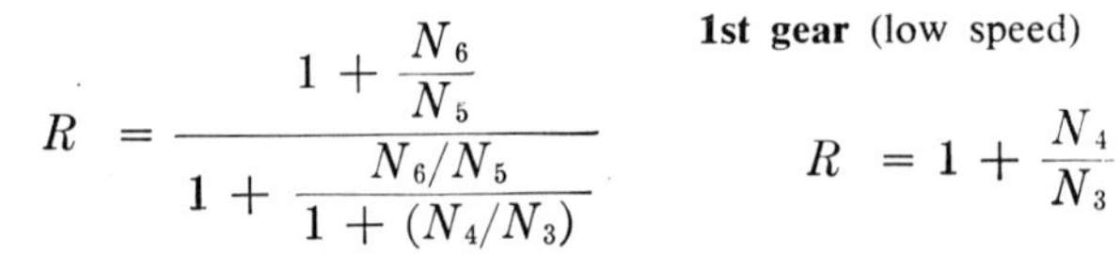

$$R = \frac{1 + \dfrac{N_6}{N_5}}{1 + \dfrac{N_6/N_5}{1 + (N_4/N_3)}}$$

1st gear (low speed)

$$R = 1 + \frac{N_4}{N_3}$$

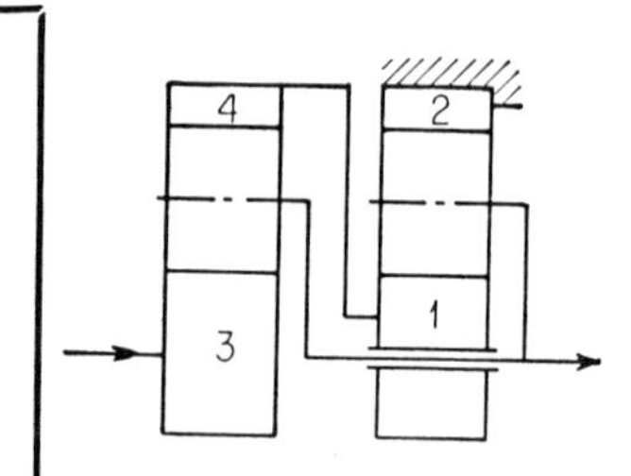

Reverse gear—
only gears 1 to 4 operative

$$R = 1 - \frac{N_4 N_2}{N_3 N_1}$$

Torque-flite transmission (Chrysler Corp.)

Gears 1 and 3 locked together

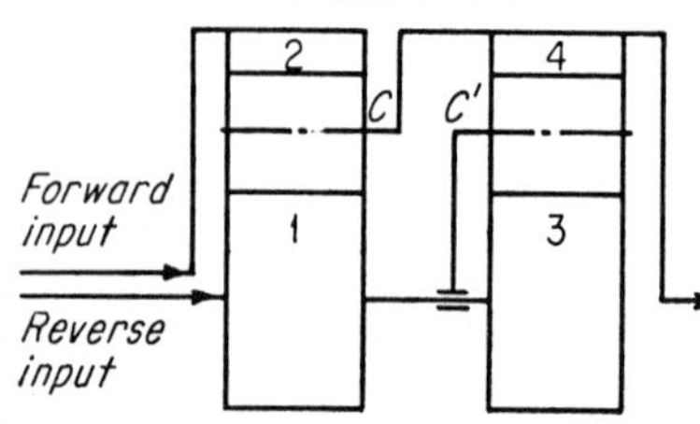

Reverse gear—
input to 3, C′ fixed
Gear set 1C2 not used

$$R = -N_4/N_3$$

Low gear— Input to 2, C′ fixed

$$R = 1 - \frac{N_1}{N_2}\left(1 + \frac{N_4}{N_3}\right)$$

Intermediate gear— input to 2; 1 and 3 fixed

$$R = 1 + \frac{N_1}{N_2}$$

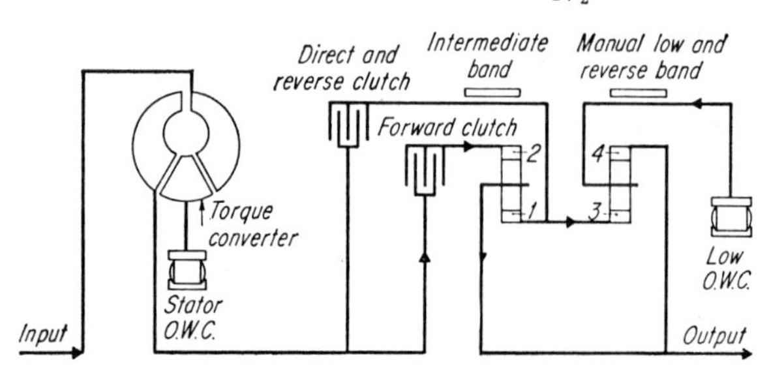

Power-flite transmission (Chrysler Corp.)

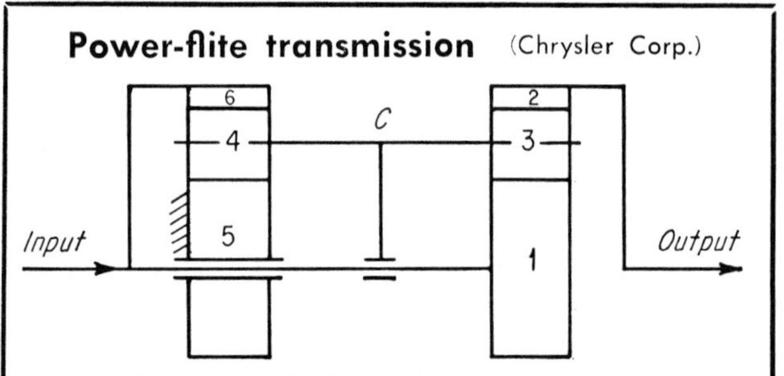

Sun gear 5 fixed
Split input to ring gear 6 and sun gear 1

$$R = \frac{1 + \dfrac{N_5}{N_6}}{1 - \dfrac{N_1 N_5}{N_2 N_6}}$$

Two-speed Fordomatic (Ford Motor Co.)

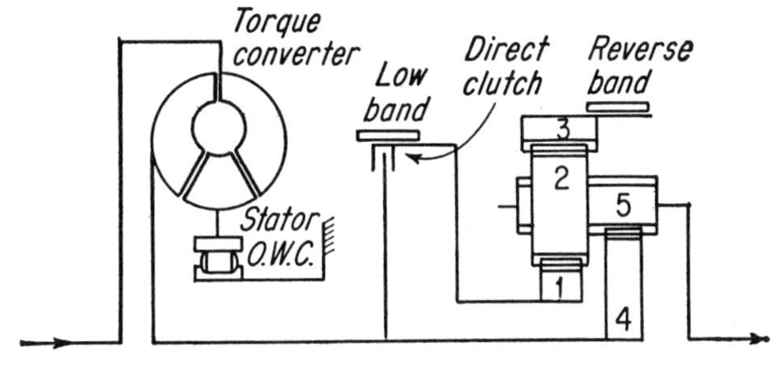

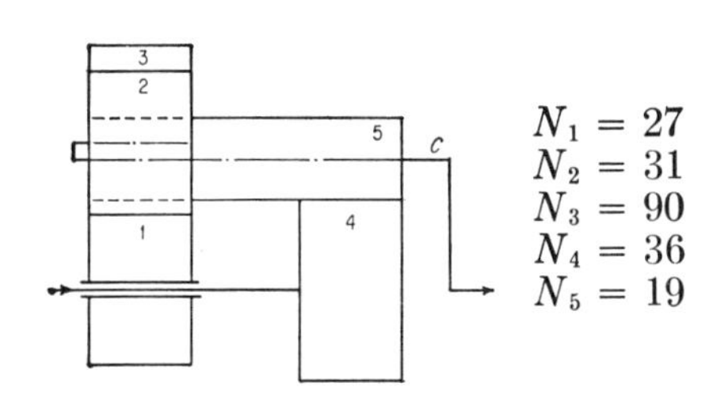

$N_1 = 27$
$N_2 = 31$
$N_3 = 90$
$N_4 = 36$
$N_5 = 19$

Low gear—
gear 1 fixed

$$R = 1 + \frac{N_1}{N_4} = 1.75$$

Reverse gear—
gear 3 fixed

$$R = 1 - \frac{N_3}{N_4} = -1.50$$

Note: Power-Glide Transmission is similar to above, but with $N_1 = 23$, $N_2 = 28$, $N_3 = 79$, $N_4 = 28$, $N_5 = 18$. This produces identical ratios in low and reverse.

$$R = 1 + \frac{23}{28} = 1.82 \qquad R = 1 - \frac{79}{28} = -1.82$$

Cruise-O-Matic 3-speed transmission (Ford Motor Co.)

5
3
C
4
2
Input
1
Output

Long planet, $N_3 = 18$
Short planet, $N_2 = 18$
Sun gears, $N_4 = 36$, $N_1 = 30$
Ring gears, $N_5 = 72$

Low gear—Input to 1 C fixed

$$R = \frac{N_5}{N_1} = 2.4$$

Intermediate gear—
Input to 1, gear 4 fixed

$$R = \frac{1 + \dfrac{N_4}{N_1}}{1 + \dfrac{N_4}{N_5}} = 1.467$$

Reverse gear—
Input to 4, C fixed

$$R = \frac{N_5}{N_4} = -2.0$$

Hydramatic 3-speed transmission (General Motors)

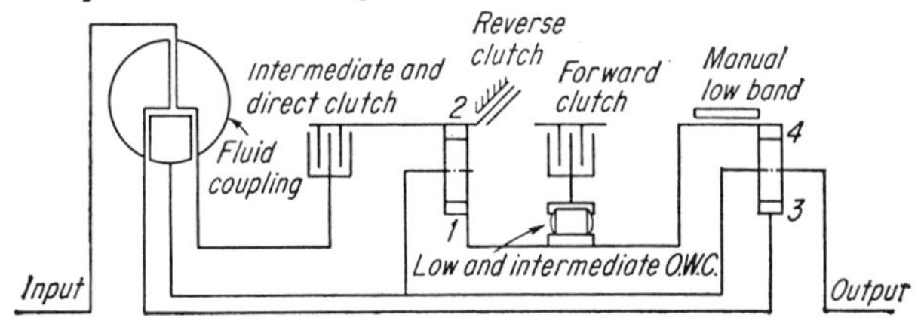

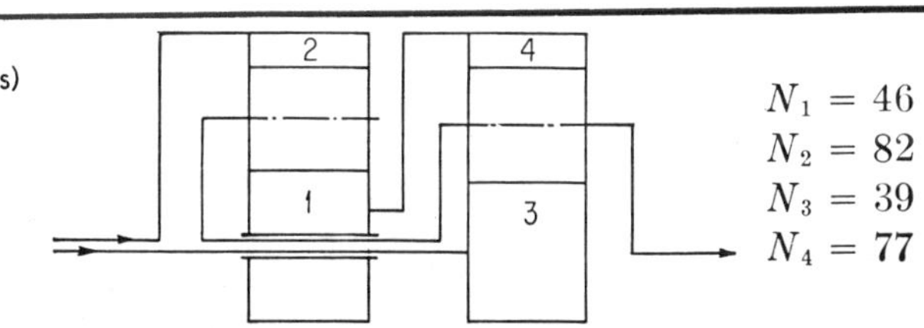

$N_1 = 46$
$N_2 = 82$
$N_3 = 39$
$N_4 = 77$

Low gear—
Input to 3, 4 fixed

$$R = 1 + \frac{N_4}{N_3} = 2.97$$

Intermediate gear—Input to 2, 1 fixed

$$= 1 + \frac{N_1}{N_2} = 1.56$$

Reverse gear—Input to 3, 2 fixed

$$R = 1 - \frac{N_4 N_2}{N_3 N_1} = -2.52$$

Triple planetary drives

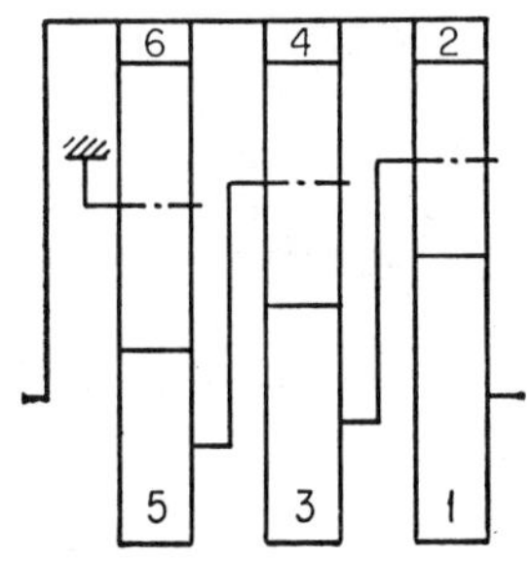

(Ref. V.Ya. Sukhina, U.S.S.R.)

Input to gear 1, output from gear 6

$$R = \left(1 + \frac{N_2}{N_1}\right)\left[\left(1 + \frac{N_4}{N_3}\right)\left(-\frac{N_6}{N_5}\right) - \frac{N_4}{N_3}\right] - \frac{N_2}{N_1}$$

(B) **(C)**

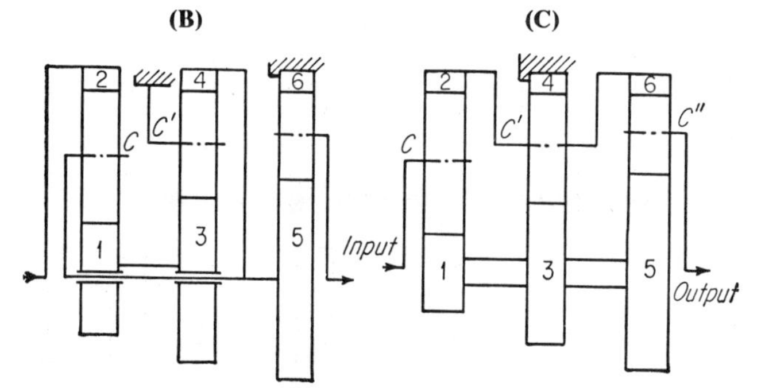

(B) $R = \left[1 + \dfrac{N_1}{N_2}\left(1 + \dfrac{N_4}{N_3}\right)\right]\left(1 + \dfrac{N_6}{N_5}\right)$

(C) $R = \left[1 + \dfrac{N_4/N_3}{1 + (N_2/N_1)}\right] \bigg/ \left[1 + \dfrac{N_4/N_3}{1 + (N_6/N_5)}\right]$

Ford tractor drives

Ring gear 3 coupled to sun gear 1; split output.

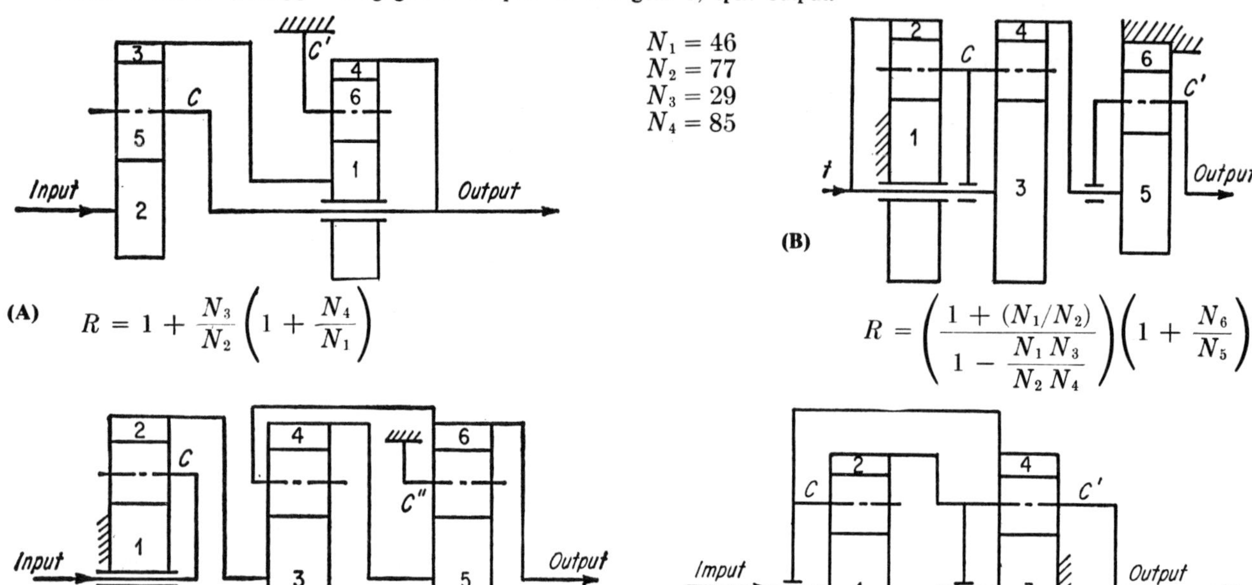

$N_1 = 46$
$N_2 = 77$
$N_3 = 29$
$N_4 = 85$

(A) $$R = 1 + \frac{N_3}{N_2}\left(1 + \frac{N_4}{N_1}\right)$$

(B) $$R = \left(\frac{1 + (N_1/N_2)}{1 - \dfrac{N_1 N_3}{N_2 N_4}}\right)\left(1 + \frac{N_6}{N_5}\right)$$

(C) $$R = \frac{1}{1 + \dfrac{N_1}{N_2}}\left[1 + \frac{N_4}{N_3}\left(1 + \frac{N_6}{N_5}\right)\right]$$

(D) $$R = \frac{N_3}{N_4}\left(1 - \frac{N_4}{N_3} + \frac{N_2}{N_1}\right)$$

Lycoming turbine drive

$$R = \left(1 + \frac{N_3}{N_2}\right) \times \left(1 + \frac{N_4}{N_1}\right)$$

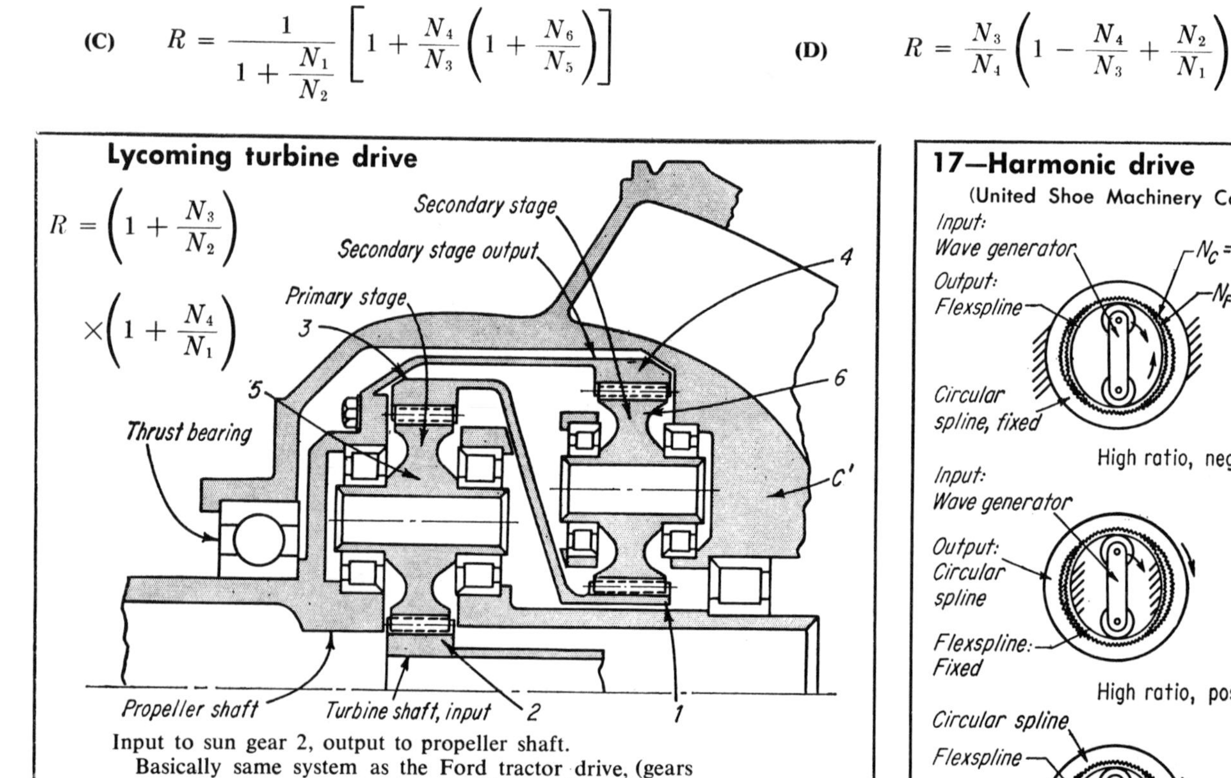

Input to sun gear 2, output to propeller shaft.
Basically same system as the Ford tractor drive, (gears are numbered the same way) and will have the same speed-ratio.

Compound spur-bevel gear drive (Ref. 2)

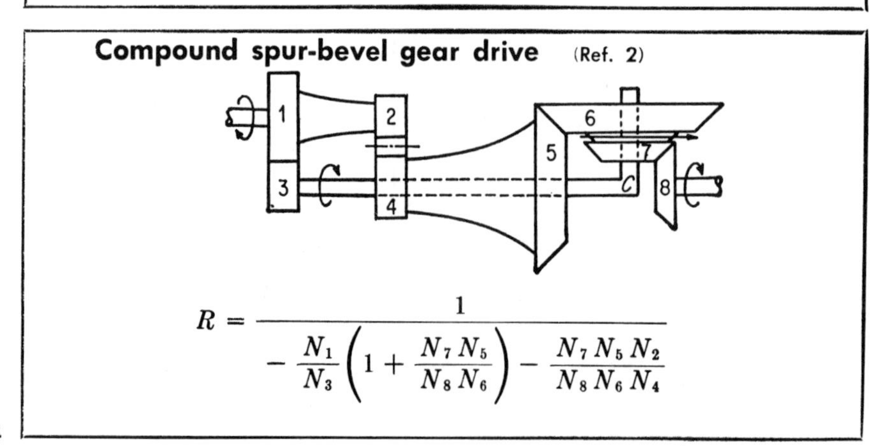

$$R = \frac{1}{-\dfrac{N_1}{N_3}\left(1 + \dfrac{N_7 N_5}{N_8 N_6}\right) - \dfrac{N_7 N_5 N_2}{N_8 N_6 N_4}}$$

17—Harmonic drive

(United Shoe Machinery Corp.)

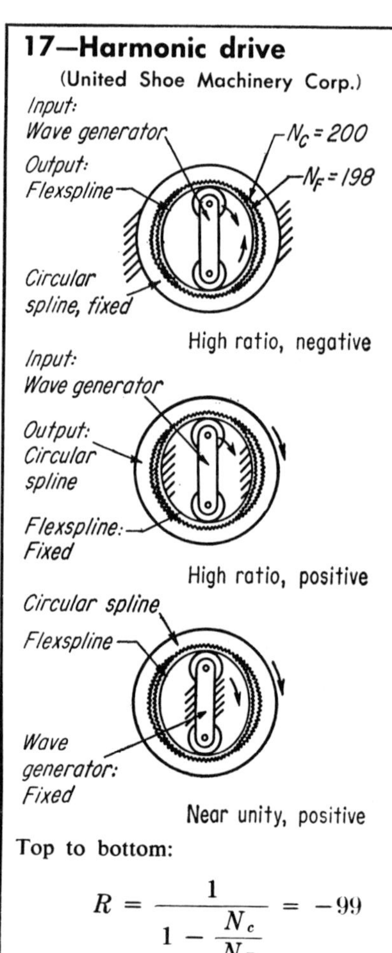

Top to bottom:

$$R = \frac{1}{1 - \dfrac{N_c}{N_F}} = -99$$

$$R = R_i = \frac{1}{1 - \dfrac{N_F}{N_c}} = 100$$

$$R = N_C/N_F = 100/99$$

Two-gear planetary drives (Ref. 3)

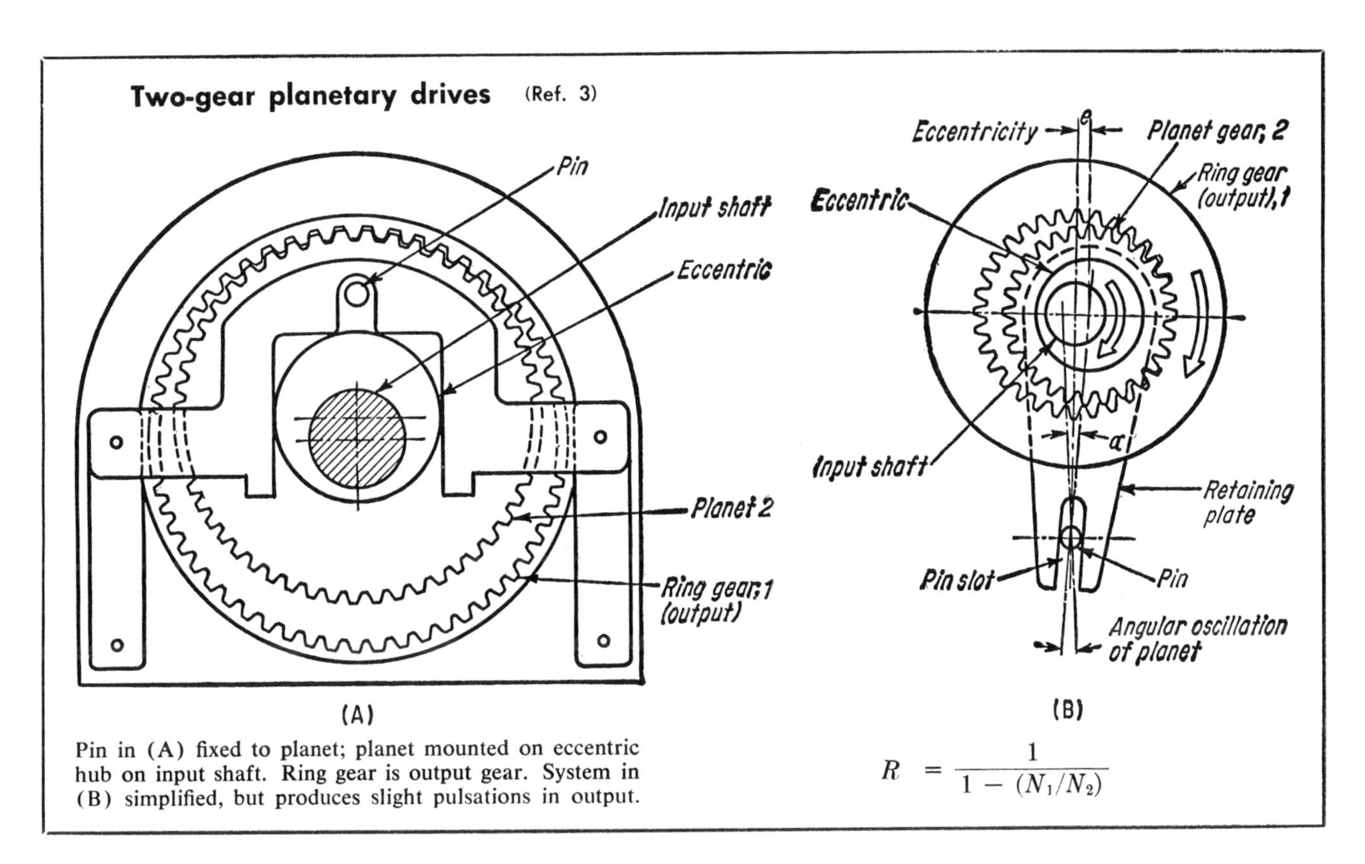

Pin in (A) fixed to planet; planet mounted on eccentric hub on input shaft. Ring gear is output gear. System in (B) simplified, but produces slight pulsations in output.

$$R = \frac{1}{1 - (N_1/N_2)}$$

Planocentric drive (General Electric Co.)

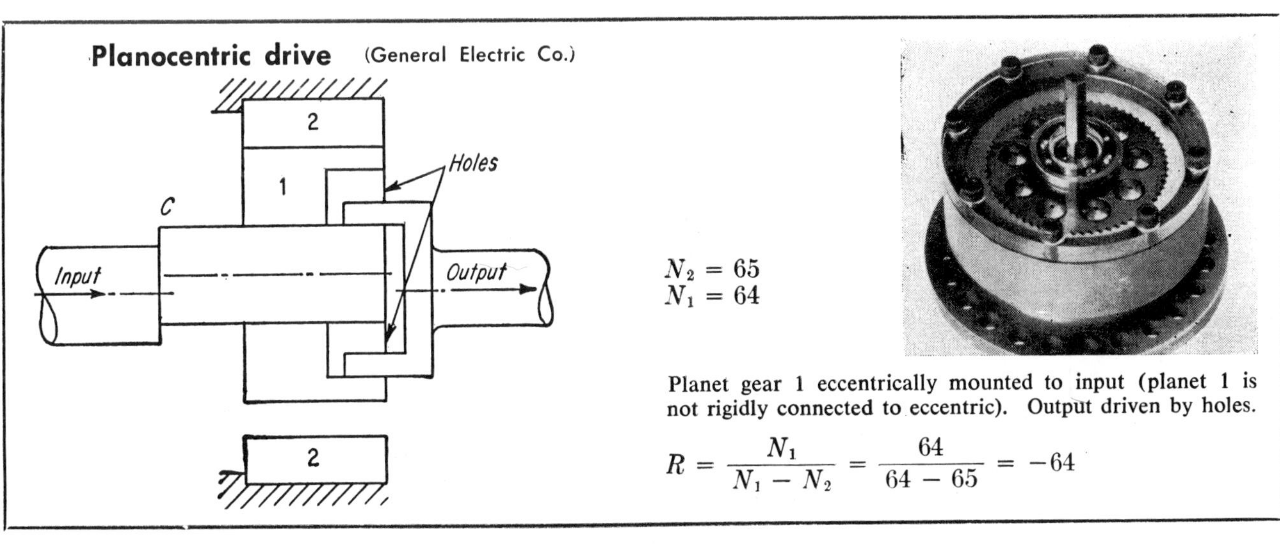

$N_2 = 65$
$N_1 = 64$

Planet gear 1 eccentrically mounted to input (planet 1 is not rigidly connected to eccentric). Output driven by holes.

$$R = \frac{N_1}{N_1 - N_2} = \frac{64}{64 - 65} = -64$$

Wobble-gear drive

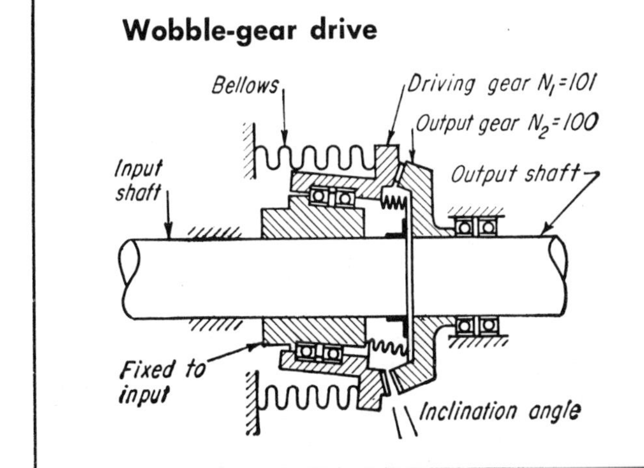

A close relative of the Harmonic Drive. The bevel "wobble" gears mesh at only one point on the circumference because of slight angle of inclination of driving gear, N_1, which has one tooth more than output gear, N_2. The driving gear, N_1, does not rotate but yaws and pitches only.

$$R = R_i = \frac{1}{1 - m_{or}}$$

$$R = \frac{1}{1 - \dfrac{N_1}{N_2}} = \frac{1}{1 - \dfrac{101}{100}} = -100$$

Speed Change Systems

Drive with coarse-fine gearing devised in West Germany

An in-line shaft drive, with reduction ratios of 1:1 and 1:16 or 1:28, combined in a single element, has been designed by Telefunken of West Germany. It consists basically of friction wheels which grip each other elastically.

Crown wheels with a gear ratio of 1:1 are used for the coarse adjustment; and friction spur gearing, with a ratio of 1:16 or 1:28 for the fine or vernier adjustment.

A spring (see diagram) applies pressure to the fine-adjustment pinion, preventing backlash while the coarse adjustment is in use; and uncouples the coarse adjustment when the vernier is brought into play by forward movement of the front shaft. The spring also makes sure that the front shaft is always in gear.

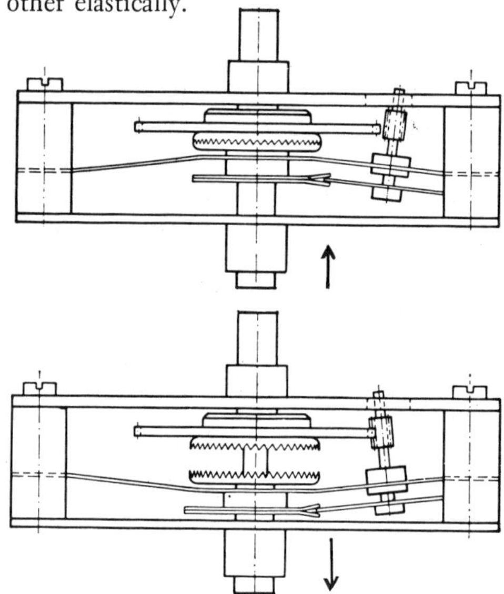

speed-change mechanism

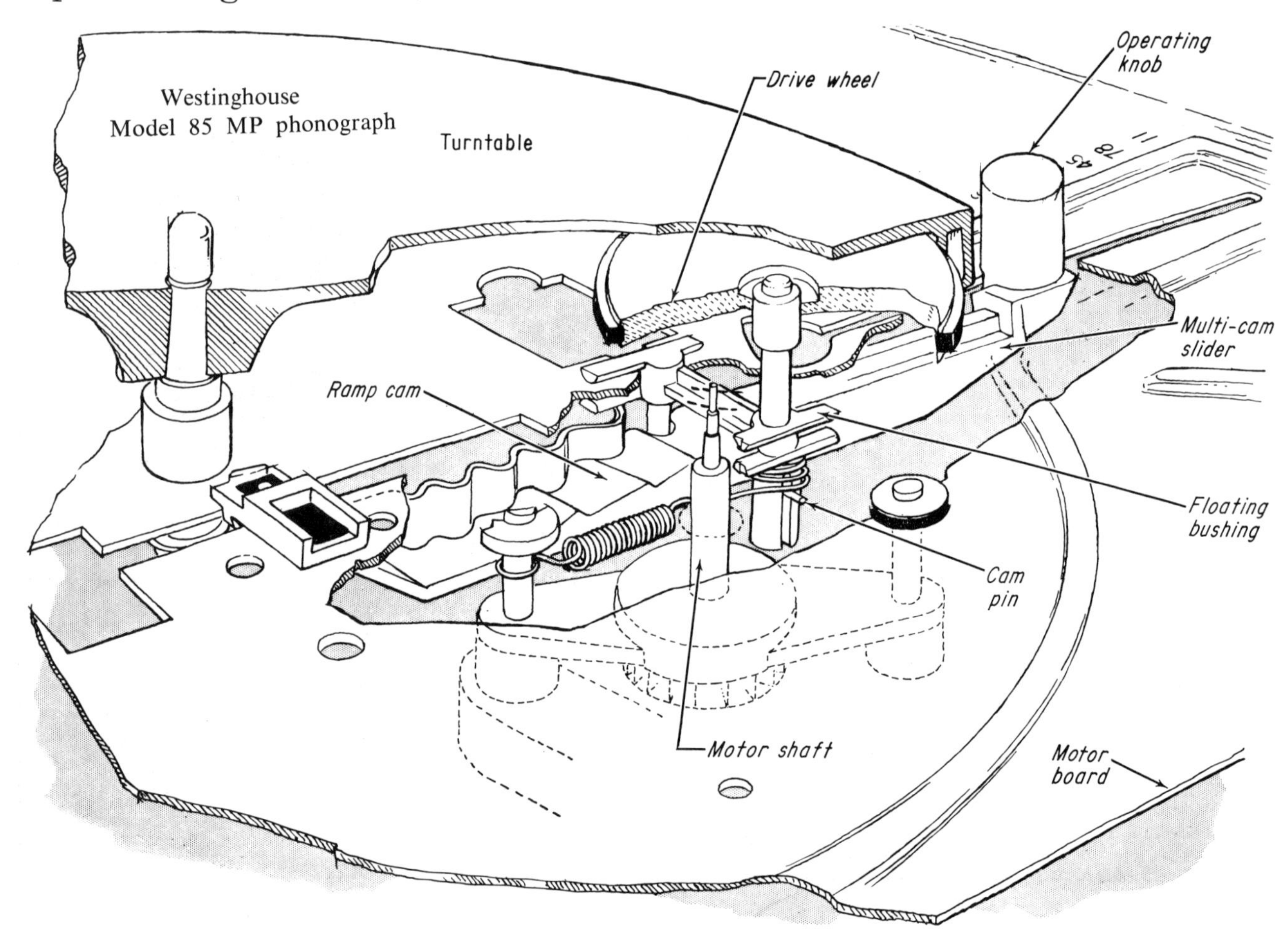

SPEED-CHANGE MECHANISM incorporates a multi-cam slider to move the drive wheel in two planes along the motor shaft. The ramp cam portion of the slider moves the drive-wheel shaft vertically while the serrated portion shifts the floating bushing and wheel away from the shaft during each vertical step. A pin through the spring-loaded drive wheel shaft contacts the ramp cam on the slider. A post on the die-cast floating bushing contacts the serrated cam portion of the slider. An operating knob on the slider projects through the motor board and is moved to change speeds. A pointer on the knob indicates the rpm of the speed selected, while a spring-loaded ball detent on the opposite end of the slider engages holes in the motor board. The entire mechanism snaps together without threaded fasteners.

the Harmonic Drive . . . HIGH-RATIO GEARING

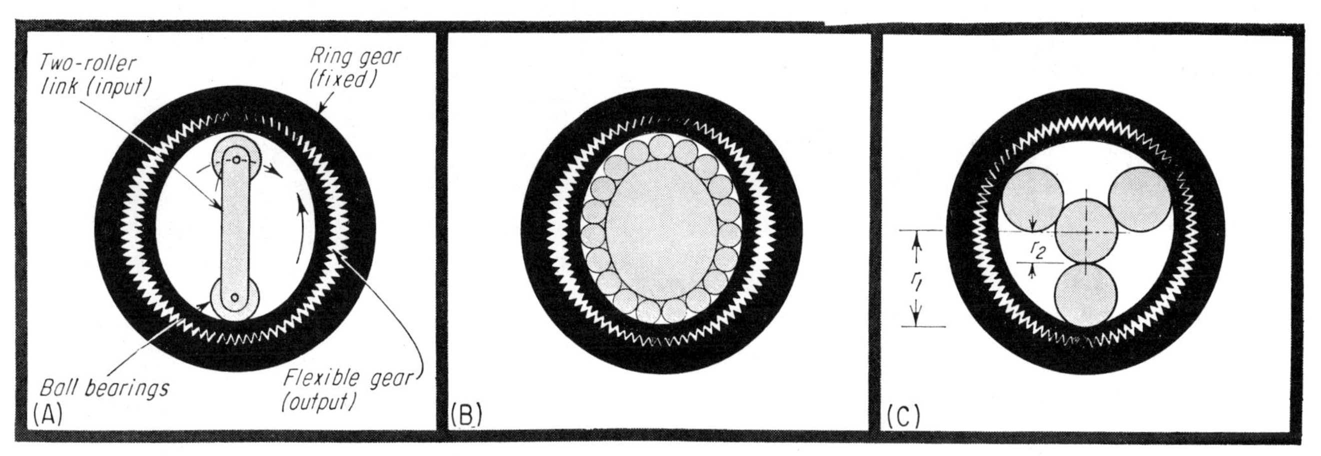

THREE VERSIONS OF DRIVE. Flexible gear is deflected in (A) by two-roller link; (B) by elliptical cam rolling within ballbearings; (C) by planetary-gear system for still-higher speed change.

How the Drive Works

The rotary version has a ring gear with internal teeth mating with a flexible gear with external teeth—see (A) in previous illustration. These teeth are straight-sided, and both gears have the same circular pitch, hence the areas of engagement are in full mesh. But the flexible inner gear has fewer teeth than the outside gear and therefore its pitch circle is smaller. Third member of the drive is a link with two rollers which rotates within the flexible gear, causing it to mesh with the ring gear progressively at diametrically opposite points. This propagates a traveling strain, or deflection wave in the flexible gear—hence, United Shoe's tradename, Harmonic Drive. If motion of the center link or "wave generator" is clockwise, and the ring gear is held fixed, the flexible gear will rotate (or "roll") counterclockwise at a slower rate, with constant angular velocity.

Teeth are stationary where in mesh, thus acting as splines in full contact. Movement of the flexible, driven member is confined to that area where teeth are disengaged. Each rotation of the center link moves the flexible gear a distance equal to the tooth differential between the two gears. Thus, speed ratio between center link (input) and flexible gear (output) is

$$\frac{V_o}{V_i} = \frac{N_f - N_r}{N_f}$$

where V_o = output velocity, rpm; V_i = input velocity, rpm; N_r = number of teeth (or pitch dia) of the ring gear; N_f = number of teeth (or pitch dia if permitted to take its full circular form) of the flexible gear.

For a drive with, say, 180 teeth in the ring gear and 178 teeth in the flexible gear (for the drive illustrated, difference in the number of teeth must be an even integer), the speed ratio will be

$$\frac{V_o}{V_i} = \frac{178 - 180}{178} = -\frac{1}{89}$$

Negative sign indicates that the input and output move in opposite directions. Actually, any one of the three basic parts can be held fixed and the other two used interchangeably as input and output.

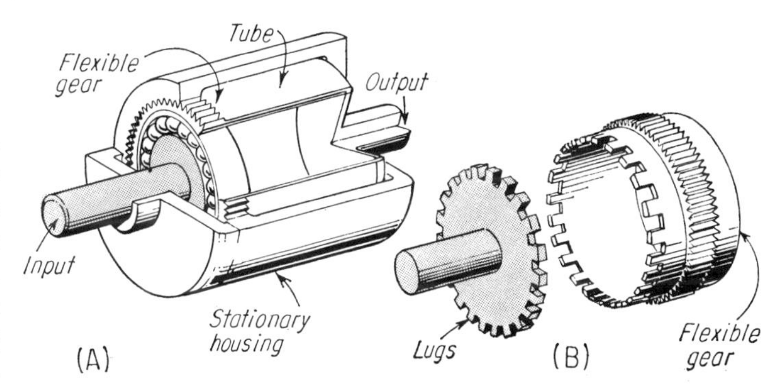

TWO METHODS OF COUPLING flexible gear to shafts: (A) by means of tubing; (B) with lugs.

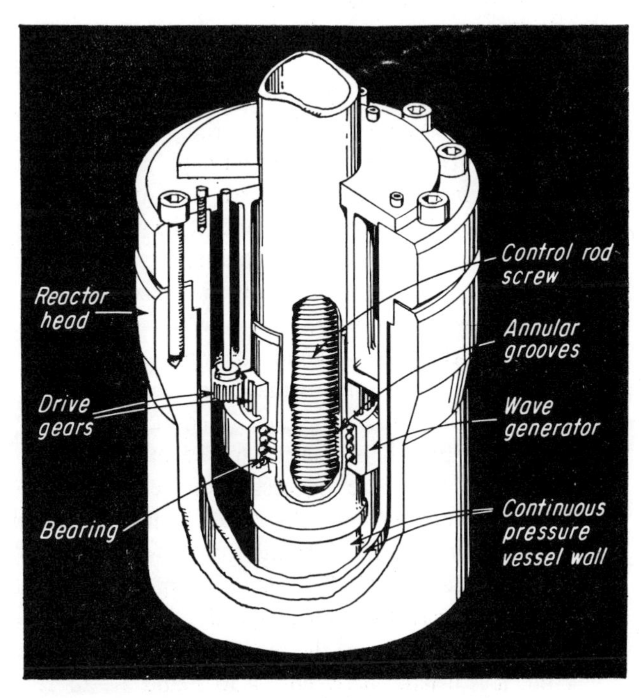

ROTARY-TO-LINEAR VERSION moves the control-rod linearly in this nuclear reactor head without need for mechanical contact through the sealed inner tube.

Friction Drives

Single Wheel for Several Jobs

Elastic-surfaced wheel provides speed reduction, friction drive and right-angle takeoff for dictating machine. Magnetic recording disk combines advantages of record and magnetic tape.

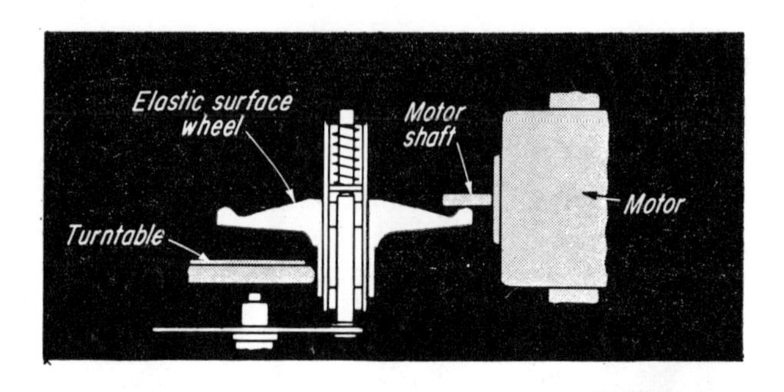

Rapid stop-start . . . ▶ control is essential for accurate dictation. A pivoted elastic wheel, operated by electromagnet, engages and disengages the motor as required. Light weight, slow-speed (9.4 rpm) turntable is stopped by friction of the magnetic head in 1/20 second and reaches operating speed in 1/14 second.

For playback or correction, turntable is reversed by a reversing wheel between the elastic wheel and turntable.

Telefunken GmbH, Ulm/Donau, W Germany.

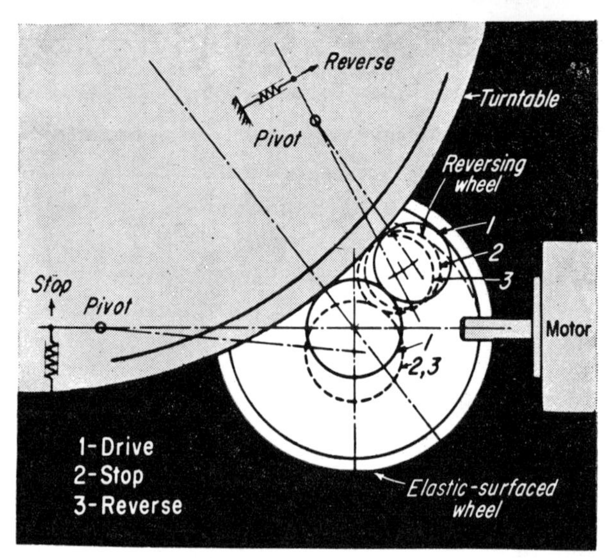

Bearings take the place of gears in a speed-reduction system designed in Britain. The system is based on the planetary action of balls rolling between paired races.

The bearings may be connected in several ways. In the setup illustrated, designed for installation on an electric motor, three similar ball bearings are connected in series. Their split outer races are held stationary in housing rings which are axially clamped to an end cap.

Inner race
Cage
Split outer races
Output shaft
Driving flange
Driving tongue
Slot
Housing rings
Compression springs

The inner race of the first bearing is driven by the motor, and the first bearing cage has driving tongues which engage in slots in the inner race of the second bearing.

Similarly, the second bearing cage drives the inner race of the third bearing, and the third bearing cage is connected to a ring which is secured to the output shaft. Thus three successive stages of speed reduction are achieved. Compression springs load the split outer races to reduce slip and increase the torque which can be transmitted by the assembly.

Designers at the Gear Division of George Angus & Co Ltd, Hebburn-on-Tyne, England, say the actual speed reduction that can be achieved depends on the relative diameters of the rolling elements and the tracks of the inner and outer races, but it is usually between 2.4 to 1 and 3.2 to 1.

The system is now in production for only one product, a roller shutter, and this is an intermittent-duty application. But other applications are planned.

Chapter 10 TORQUE LIMITING, GOVERNING, AND TENSIONING DEVICES

WHICH TYPE OF

R. A. BAREISS, *Senior Project Engineer*
P. A. BRAND, *Project Engineer*
Lessells and Associates, Inc., Boston

Hydraulic . . . magnetic . . . mechanical . . . electromechanical? In text and table the authors analyze 17 conventional and newer ways to prevent motor overloading and strain on equipment.

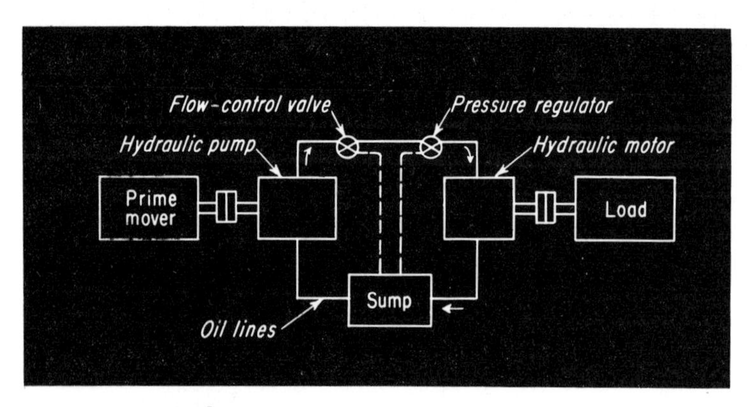

Hydrostatic transmission . . .
limits torque by converting mechanical energy to hydraulic energy, then regulating the hydraulic pressure and flow.

This analysis of torque-limiting devices was made in connection with research undertaken for the US Navy's Bureau of Ships. In general, such devices prevent overloading of motors and act as safety devices by preventing strain on the mechanism, tool or equipment being powered. In this particular project, the search was for a satisfactory unit to limit torque transmitted to boat winches, and thus prevent a cable from snapping when overloaded.

Some of the devices analyzed are standard off-the-shelf items with inherent characteristics that permit direct installation without modification. Others require modification. And a few were specially designed by the authors and their associates.

Specifications called for (1) limiting torque to a preset maximum between 120 to 200 ft-lb, (2) giving such limitation that would be effective for both directions of rotation, (3) resetting without disconnecting the drive. Comparison of cost and sizes, therefore, are made on this basis, but can be extrapolated to meet other specifications.

HYDRAULIC DEVICES

Hydrostatic transmission employs a variable-displacement hydraulic pump, constant-displacement hydraulic motor, and pressure-compensating or relief control. This circuit limits torque by limiting hydraulic pressure and flow, but needs high internal pressures, up to 3000 psi, for efficient operation. Power load on the electric drive motor can be controlled by torque requirements of hydraulic motor; therefore, there is little necessity for a large heat sink. At speeds approaching zero, internal leakage reduces torque of hydraulic motor so that torque is no longer a linear function of oil pressure.

Hydraulic pumps and motors are sized and priced mainly by flow capacity. In this application, however, machinery must be sized to provide the required flow at high speeds and required torque at lower speeds.

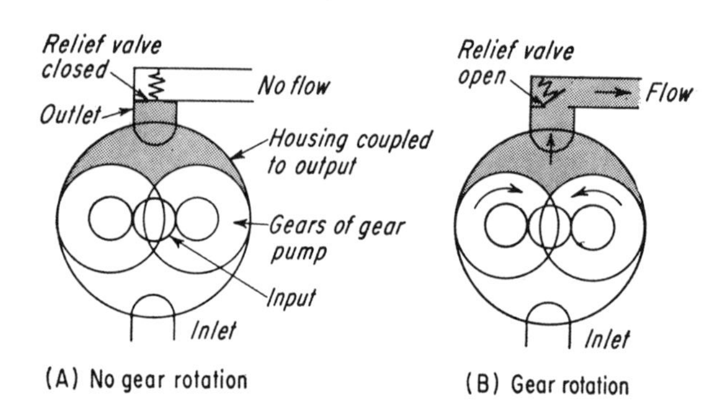

(A) No gear rotation (B) Gear rotation

Hydrostatic coupling . . .
employs gear pump between prime mover and load. Input is coupled to gears of pump; output to housing. In normal operation (A), spring keeps relief valve closed and gears cannot rotate—coupling is "locked up" and there is no slippage between input and output. Under overload (B), relief valve opens, gears rotate relative to the housing and slippage occurs. Oil is scooped into inlet side during rotation.

Hydrostatic coupling—here a torque-sensitive coupling is obtained by connecting the load shaft to the casing of a hydraulic gear pump, the input shaft to the gear train, and using a pressure-relief valve on the outlet side of the pump. Pump gears cannot turn as long as relief valve is closed; overload builds up pressure to open relief valve which permits slippage. Torque limitation is valid

TORQUE-LIMITING DEVICE?

in both directions of rotation. Centrifugally operated valves will prohibit the load from turning a dead motor. Some leakage will occur within the gear motor, resulting in a limited amount of slip—estimated considerably below 5% at 1800 rpm.

For heat dissipation, the relief valve will vent against an outer stationary, finned housing; the bottom of this housing will act as oil sump. A constant-viscosity oil will minimize temperature sensitivity.

Hydrokinetic couplings are of fluid-clutch type using very high fluid flow rates at low internal pressures. Nominal slip of 3% to 5% is required at rated conditions and the torque transmitted increases with slip speed.

Heat buildup is relatively severe at high slip speeds. Torque-transmitting characteristics are dependent upon fluid viscosity. A simple hydrokinetic coupling does not appear to be satisfactory for this application.

Variable hydrokinetic coupling. By permitting one element of a fluid clutch to move in an axial direction at the command of a torque-sensitive member of the drive train, the effective torque transmitted can be held essentially constant. Cam positioning of the coupling element in response to reaction against a torsional spring would be the torque-limiting action. A relatively constant-viscosity oil is recommended. Size and cost data on this type of coupling have not yet been evaluated.

MAGNETIC DEVICES

Electrostatic couplings—manufacturing clearances and design ingenuity required to limit torque in electrostatic type couplings are extremely rigorous and unsuitable to this application.

Eddy-current couplings have an abruptly increasing and then slowly decreasing torque as a function of slip speed—but this is not too severe for the application considered here. Torque limitation is by either constant or speed-regulated rotor dc excitation. Slip capacity and ability to dissipate heat is very high. This is an off-the-shelf item but a special casting is necessary to enclose the coupling between drive motor and gearing, and dc electrical connection to the coupling is required. Salt-air atmosphere may effect the slip-ring performance.

Eddy-current, self-excited couplings require no electrical connections. Radially mounted Alnico permanent-type magnets provide the necessary field flux. Cost is about half that of conventional eddy-current couplings.

Dry magnetic powder clutches have constant torque

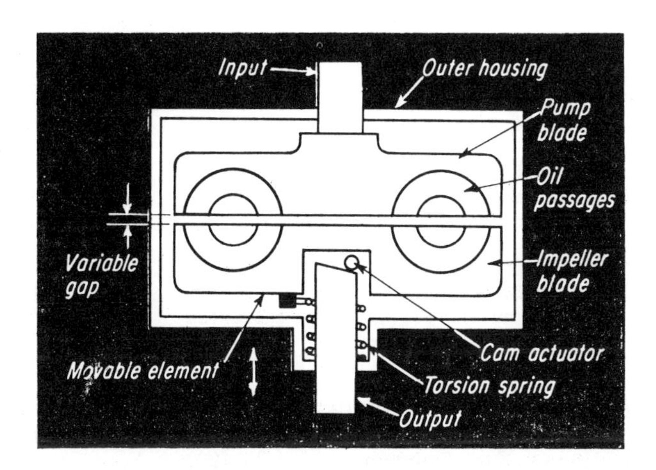

Hydrokinetic coupling . . .
employs a torsion spring to vary gap between blades of a standard fluid clutch. Torsion spring wraps up under load, pulling impeller blade away from pump blade to increase slip.

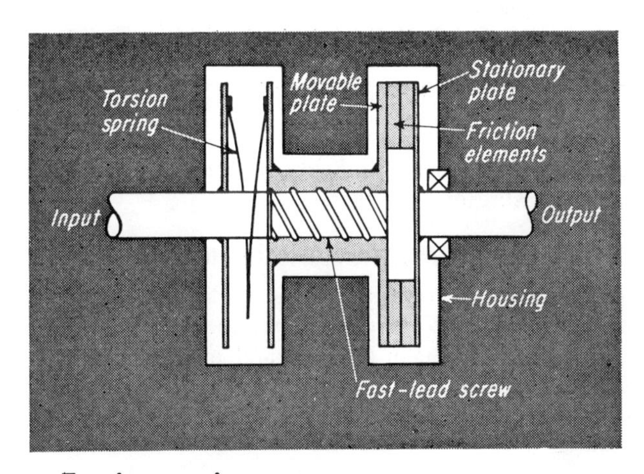

Torsion spring . . .
is key element in constant-torque clutch. Spring winds up under load and pulls back movable friction plate, causing slippage.

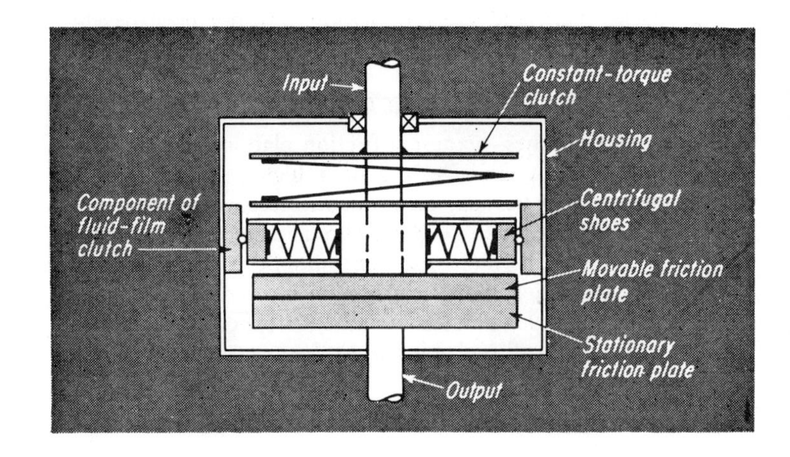

Dual-torque clutch . . .
connects the constant-torque clutch and fluid-film clutch in parallel. This reduces heat buildup during slippage and increases maximum load capacity.

WHICH TYPE OF TORQUE-LIMITING DEVICE?

TORQUE-LIMITING DEVICES—A COMPARISON . . .

METHOD	Torque Limits, max/stalled	Self-resetting	Bi-directional Operation	Self-contained	Reliability	Length & Dia, in.	Weight, lb	Cost, $ (estimated)
HYDRAULIC DEVICES								
Hydrostatic Transmission	200/200	Yes	Yes	Yes	Good	30 x 15	1800	4000
Hydrostatic Coupling	200/200	Yes	Yes	Yes	Good	15 x 11	100	1500
Hydrokinetic Coupling	75/200	Yes	Yes	Yes	Good	....	....	300
Variable Hydrokinetic Coupling	75/200	Yes	Yes	Yes	Good	....	....	....
MAGNETIC DEVICES								
Electrostatic	....	Yes	No	No; dc	Excellent	....	....	....
Eddy-current	200/120	Yes	Yes	No; dc	Good	18 x 19	475	1100
Eddy-current, self-excited	145/93	Yes	Yes	Yes	Excellent	27 x 16	250	600
Dry Magnetic Powder	200/200	Yes	Yes	No	Good	20 x 12	230	1000
ELECTROMECHANICAL DEVICES								
Electromagnetic Friction	200/140	Yes	Yes	No	Good	3¼ x 6½	40	400
Friction Current-limitation	200/200	Yes	Yes	No	Fair	....	....	....
MECHANICAL COUPLINGS								
Slip Clutch	200/140	Yes	Yes	Yes	Fair	8½ x 10¼	40	150
Constant-torque Clutch	125/125	Yes	Yes	Yes	Excellent	4½ x 6½	30	250
Fluid-film Clutch	200/4	No	Yes	Yes	Excellent	3 x 9½	....	....
Dual-torque Clutch	200/80	Yes	Yes	Yes	Excellent	6 x 9½	....	....
Cam and Roller Couplings	200/0	No	Yes	Yes	Good	8 x 7½	....	....
Centrifugal Friction	200/120	Yes	Yes	Yes	Good	4 x 7½	30	85
Centrifugal Dry-fluid Clutch	165/120	Yes	Yes	Yes	Excellent	4 x 9½	30	90

capability at a constant dc field excitation but cannot take 100% slip for more than 5 sec.

ELECTROMECHANICAL DEVICES

Electromagnetic friction clutches can limit torque by employing low power dc for field excitation. Slip capacity is relatively low; slip accuracy and predictability is somewhat questionable under severe storage and environmental conditions because the friction surfaces can corrode.

Current-limitation is another electromechanical torque-limiting method. It employs a mechanical torque-sensing device (perhaps a flexible member in the drive train) to operate a drive-motor current limiter. Calibration and maintenance of such a system are the major drawbacks. Cost and size information have not been developed for this device.

MECHANICAL DEVICES

Slip clutches—because a friction plate slip clutch has the inherent characteristic of a higher static friction than sliding friction, variations in breakaway torque are large—usually ±20%. Some slip clutches can withstand 100% slip for more than one minute without serious damage. Friction elements are of molded asbestos and the effect of long disuse in a salt atmosphere is unknown—though it will probably add to the inaccuracy of breakaway torque.

Constant-torque clutches are available with a torsion spring and fast-lead screw to load the movable element of a friction plate clutch in an axial direction. A "dry" clutch is employed because the friction-surface protection and heat-dissipation advantages of "wet" operation are unnecessary. Torque is limited in only one direction of rotation. Clutch acts as a solid coupling in the other direction, which is satisfactory for applications where jamming during the reverse operation will not damage the equipment or endanger human life.

Fluid-film clutch has radially mounted metal shoes that pivot so that when slip occurs a relatively friction-free wedge of fluid is built up rearwards from the leading edge. When shoe and drum are again synchronized, the oil wedge is squeezed out and the coupling will carry its full design torque.

This type clutch was developed by the Polaroid Corp., which also holds a patent on it. It is not available on the market.

Dual-torque clutch combines constant-torque clutch with fluid-film clutch to give a coupling that has two values of torque transmission. This idea has not yet reached production stage. Clutches are connected in "parallel"—so that input side of each clutch is directly connected to the motor shaft and output side of each to the load. In operation, the constant torque clutch would be sized to handle independently the pick-up load. This would bring two members of the fluid film clutch into synchronization and full lock-up or torque transmission. The unit would then transmit a torque

100% Slip Duration, sec	Adjustable Torque Limit	Development Cost	Availability
60+	Yes	Low	Composed of standard components
60+	Yes	Low	New idea—specially designed
....	Yes	None	Standard item
60+	Yes	Medium	Standard Item—modified
60+	Yes	None	Standard item
60+	Yes	None	Standard item—modified
60+	Yes	None	Standard item
5	Yes	None	Standard item
2	Yes	None	Standard item
60+	Yes	None	Standard item—modified
60	Yes	None	Standard item
60+	Yes	None	Standard item
60+	Yes	Low	Patented item, but never produced
60+	Yes	Low	New idea—not in production yet
60+	Yes	None	Standard item
60	No	None	Standard item
45	Yes	None	Standard item

equal to sum of maximum torques of each of them separately. The lower torque capacity of the constant torque clutch will be sufficient to pick-up the load, and in combination with the fluid film clutch will provide overload torque capacity.

If this maximum torque is exceeded, the constant torque clutch will begin to slip. The fluid film clutch would slip simultaneously and its torque capacity would drop to zero. Full lock-up of the unit would occur again when the load torque dropped below the rating of the constant torque clutch alone. In this way the power (heat) to be dissipated during a jammed condition would be reduced to a minimum while still providing for picking up the load without stopping the motor or manually resetting the coupling.

Centrifugal friction-type clutches are speed sensitive. The maximum torque the clutch can transmit is reduced as the drive motor is slowed down under load. Heat dissipation and heat capacity is higher than for the comparable disk clutch. Corrosion of the friction surfaces can seriously affect performance. Torque capacity is not readily adjustable.

Centrifugal dry-fluid clutches use a charge of steel shot in a rotating housing to clamp and hold a mounted disk rotor. The clutch is speed-sensitive in that the breakaway torque decreases with decreasing input shaft speed. Further, at a given input shaft speed the full slip torque is approximately 75% of breakaway capacity.

How the clutch responds as a function of load characteristics is shown in operation curve above. Assume the drive operating at a torque speed relationship defined by point A. With a sudden increase in torque (or load) output torque of the transmission will rise instantaneously with virtually no drop in speed to point B, the breakaway-torque value. The clutch will then begin to slip, torque capacity will drop to the full slip line at the point of intersection with the motor curve, point C. The clutch will again transmit power when the load torque is relieved to this value or lower.

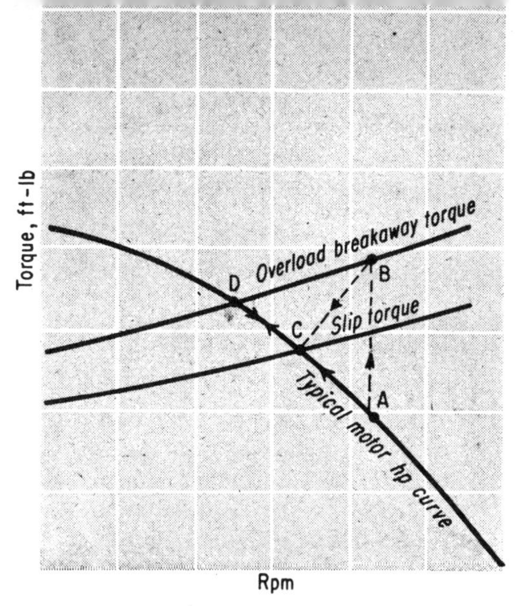

Operation curve . . .
of a centrifugal dry-fluid clutch, showing speed variations under overload conditions.

However, if the load is increased gradually, the torque-speed relationship will proceed from point A to point D. Torque transmitted will then drop to the full-slip value with little change in speed, and equilibrium will be reached at the intersection of the full-slip curve with the motor curve at point C as before—and re-engagement will occur.

If the unit can be designed or sized such that the breakaway torque at impact load does not exceed the maximum permissible load, the unit will be "fail-safe" in that all other torque values at breakaway are lower.

This clutch is adjustable, can be set and sealed at the time of manufacture. A thermal shutoff device is available which can stop the drive motor if there is critically severe overheating of the clutch. This does not seem required for the usual abnormal operation, which involves 100%-slip durations of less than 45 sec followed by 10 to 15 minutes of cooling period.

TORQUE-LIMITERS PROTECT LIGHT-DUTY DRIVES

In such drives the light parts break easily when overloaded. These eight devices disconnect them from dangerous torque surges.

L KASPER,
design consultant
Philadelphia

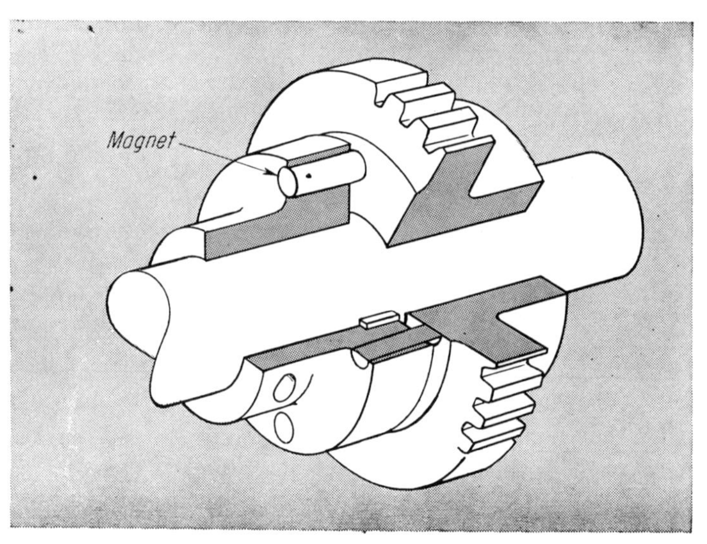

1
MAGNETS transmit torque according to their number and size. In-place control is limited to lowering torque capacity by removing magnets.

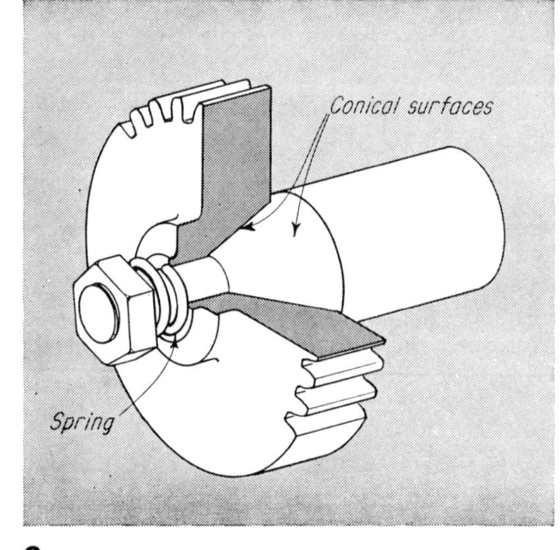

2
CONE CLUTCH is formed by mating taper on shaft to beveled hole through gear. Tightening down on nut increases torque capacity.

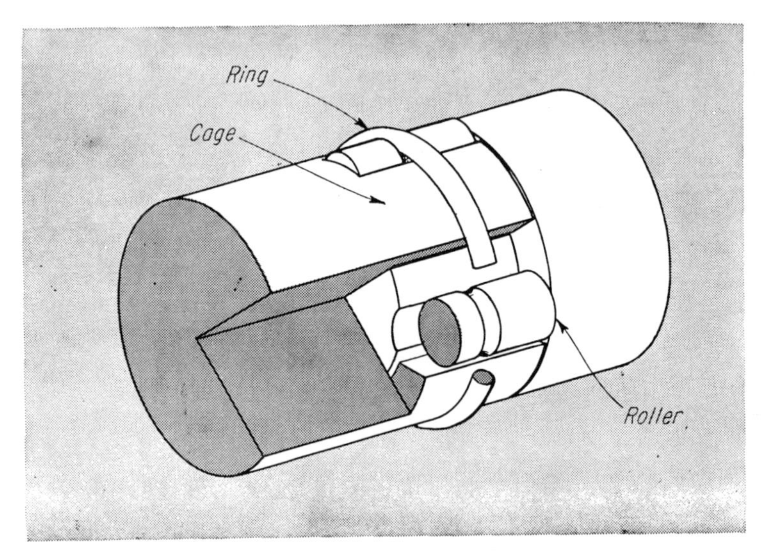

3
RING fights natural tendency of rollers to jump out of grooves cut in reduced end of one shaft. Slotted end of hollow shaft is like a cage.

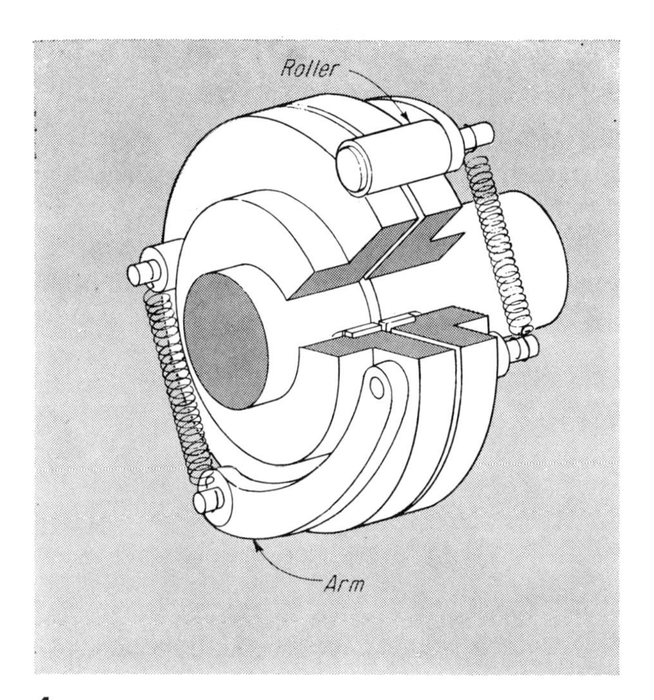

4
ARMS hold rollers in slots which are cut across disks mounted on ends of butting shafts. Springs keep rollers in slots; over-torque forces them out.

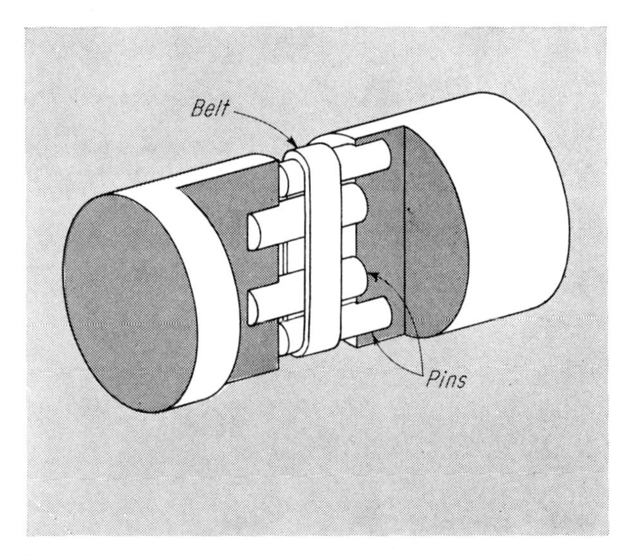

5
FLEXIBLE BELT wrapped around four pins transmits only lightest loads. Outer pins are smaller than inner pins to ensure contact.

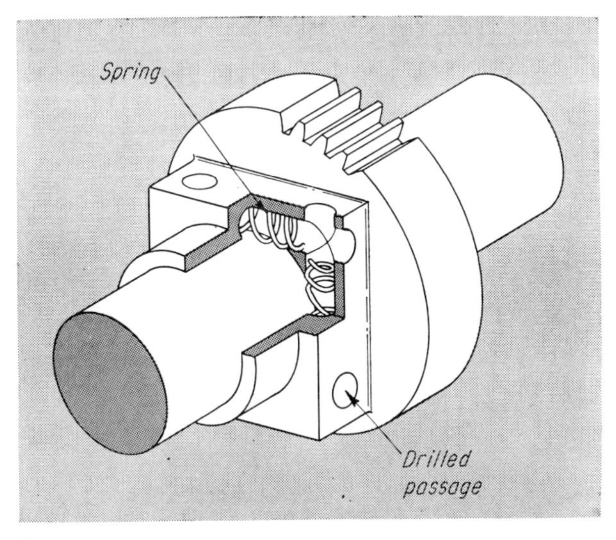

6
SPRINGS inside drilled block grip the shaft because they distort during mounting of gear.

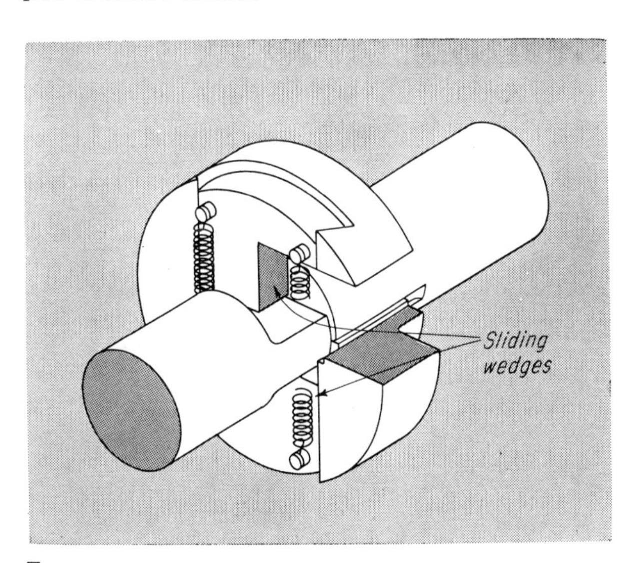

7
SLIDING WEDGES clamp down on flattened end of shaft; spread apart when torque gets too high. Strength of springs which hold wedges together sets torque limit.

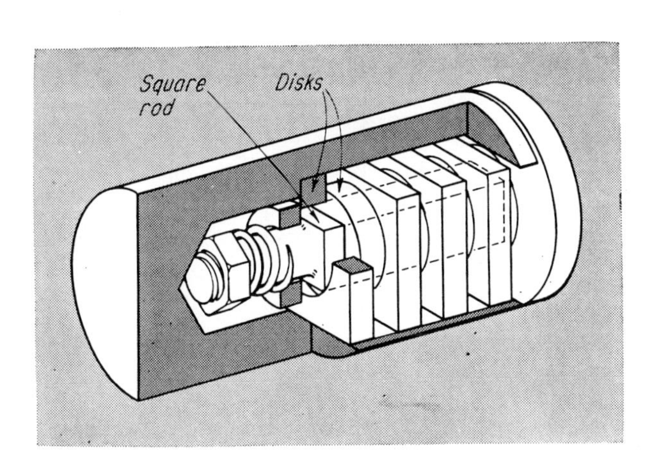

8
FRICTION DISKS are compressed by adjustable spring. Square disks lock into square hole in left shaft; round ones lock onto square rod on right shaft.

ways to PREVENT OVERLOADING

These "safety valves" give way if machinery jams, thus preventing serious damage.

PETER C NOY, *technical representative,*
Nicromatic Ltd
Toronto, Ont

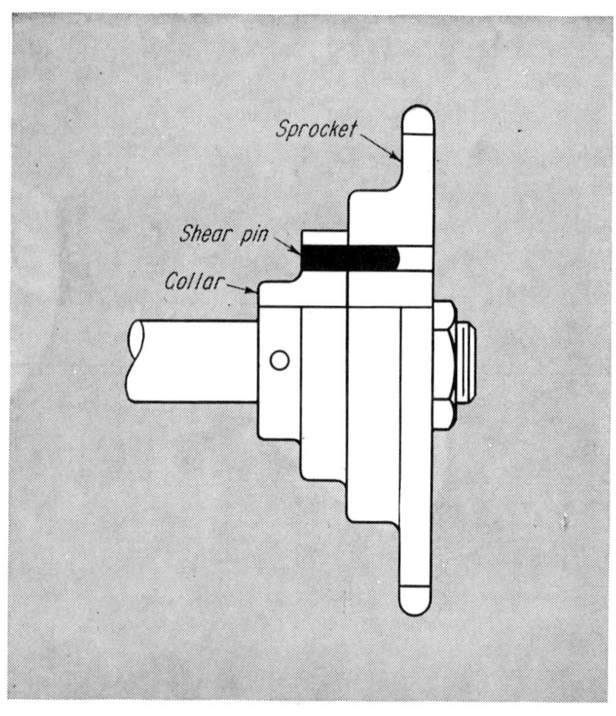

1
SHEAR PIN is simple to design and reliable in service. However, after an overload, replacing the pin takes a relatively long time; and new pins aren't always available.

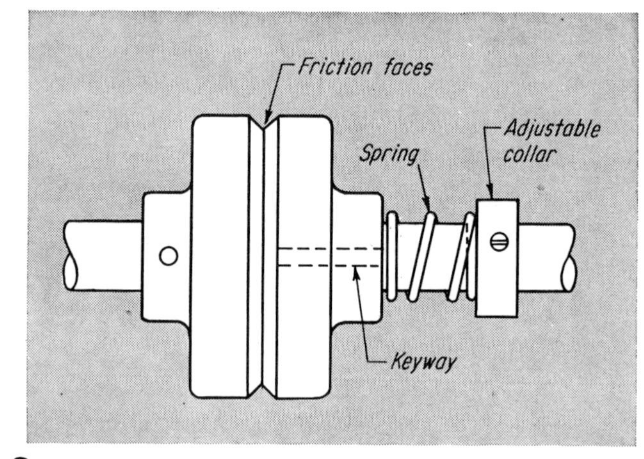

2
FRICTION CLUTCH. Adjustable spring tension that holds the two friction surfaces together sets overload limit. As soon as overload is removed the clutch reengages. One drawback is that a slipping clutch can destroy itself if unnoticed.

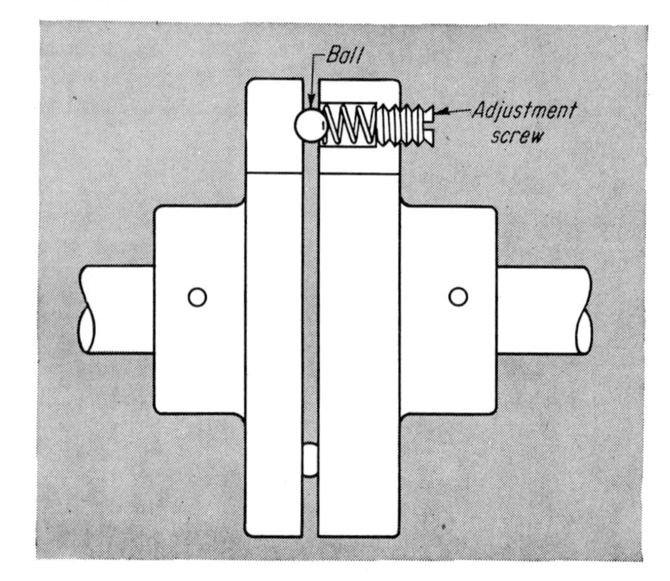

3
MECHANICAL KEYS. Spring holds ball in dimple in opposite face until overload forces the ball out. Once slip begins, wear is rapid, so device is poor when overload is common.

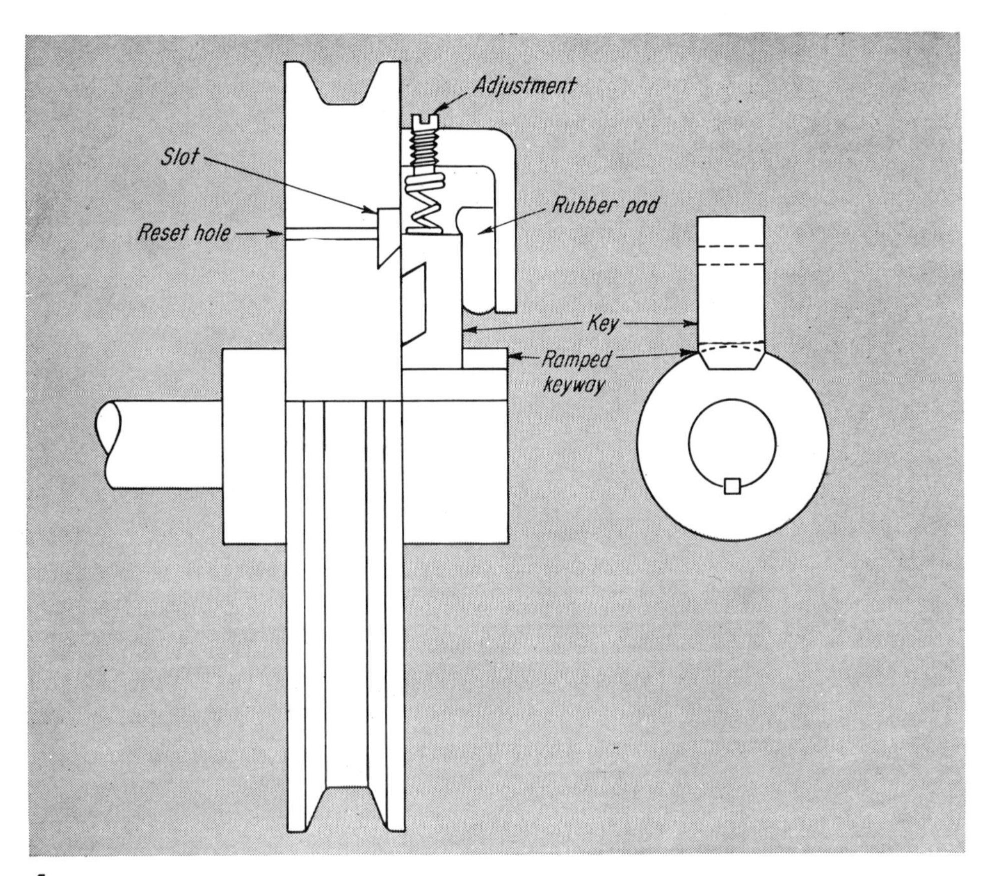

4
RETRACTING KEY. Ramped sides of keyway force key outward against adjustabe spring. As key moves outward, a rubber pad—or another spring—forces the key into a slot in the sheave. This holds the key out of engagement and prevents wear. To reset, push key out of slot by using hole in sheave.

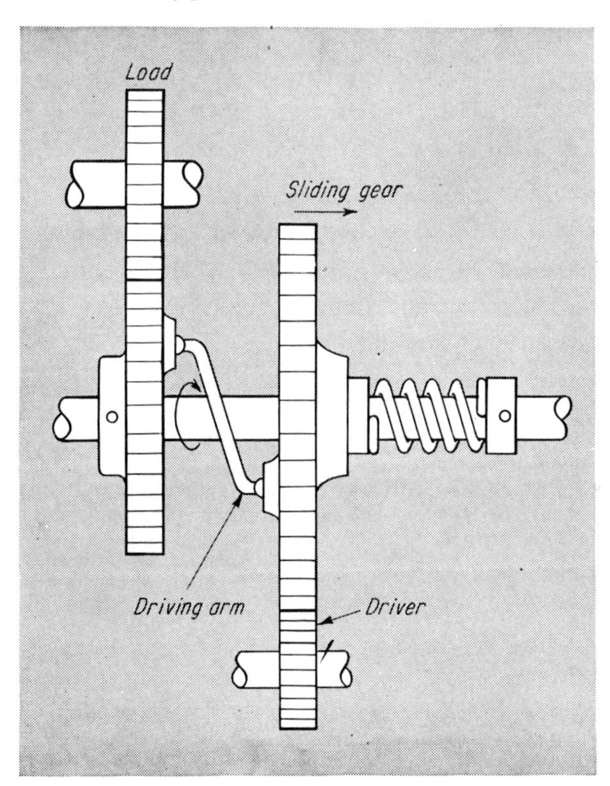

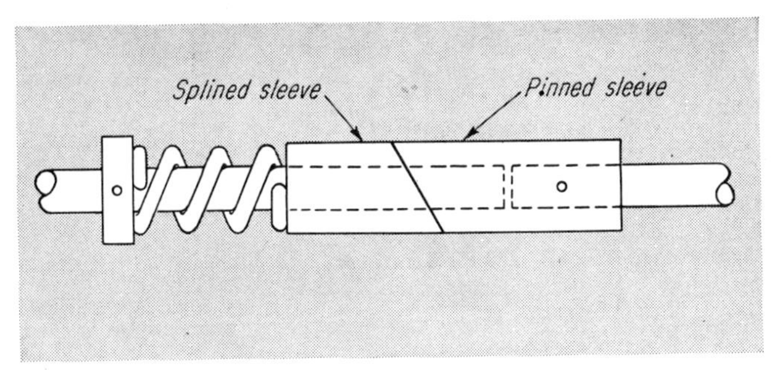

5
ANGLE-CUT CYLINDER. With just one tooth, this is a simplified version of the jaw clutch. Spring tension sets load limit.

6
DISENGAGING GEARS. Axial forces of spring and driving arm balance. Overload overcomes spring force to slide gears out of engagement. Gears can strip once overloading is removed, unless a stop holds gears out of engagement.

more ways to
PREVENT OVERLOADING

For the designer who must anticipate the unexpected, here are ways to guard machinery against carelessness or accident.

PETER C NOY, *technical representative*
Nicromatic Ltd
Toronto, Ont

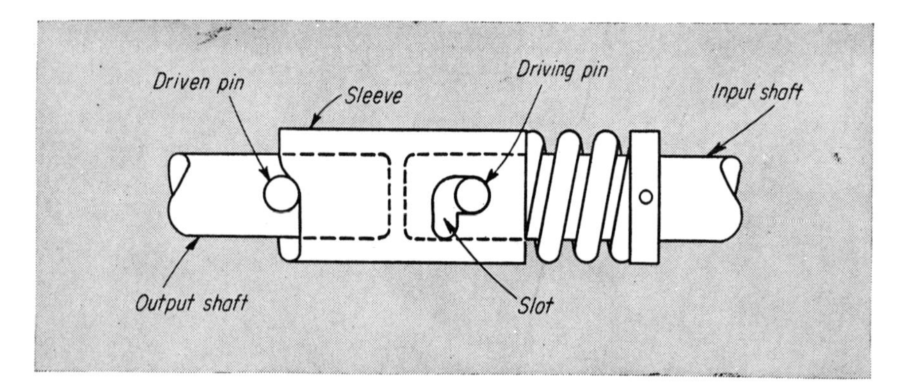

1
CAMMED SLEEVE connects input and output shafts. Driven pin pushes sleeve to right against spring. When overload occurs, driving pin drops into slot to keep shaft disengaged. Turning shaft backwards resets.

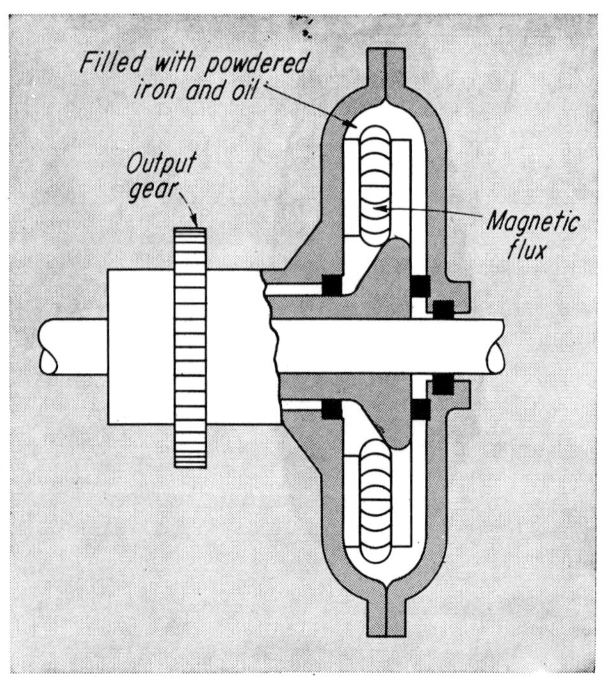

2
MAGNETIC FLUID COUPLING is filled with slurry made of iron or nickel powder in oil. Controlled magnetic flux that passes through fluid varies slurry viscosity, and thus maximum load over a wide range. Slip ring carries field current to vanes.

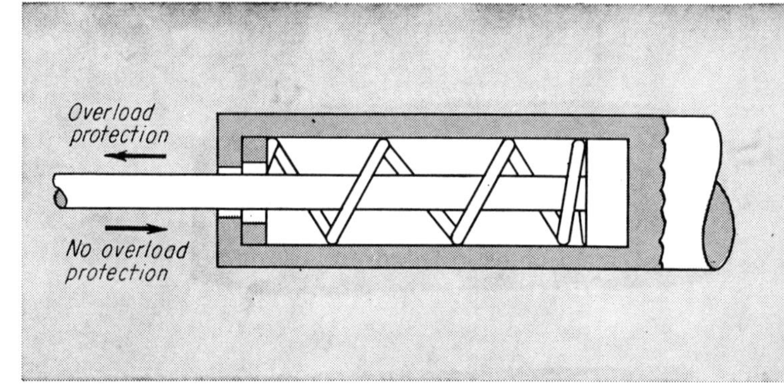

3
SPRING PLUNGER is for reciprocating motion with possible overload only when rod is moving left. Spring compresses under overload.

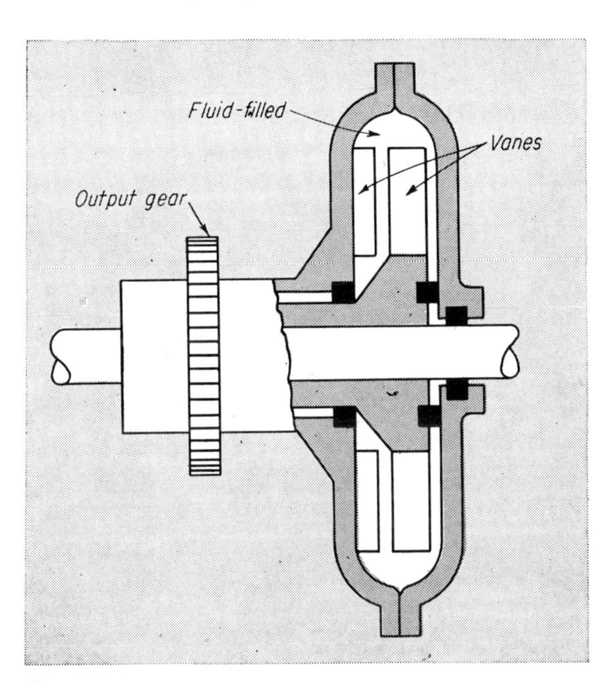

4
FLUID COUPLING. Maximum load can be closely controlled by varying viscosity and level of fluid. Other advantages are smooth transmission and low heat rise during slip.

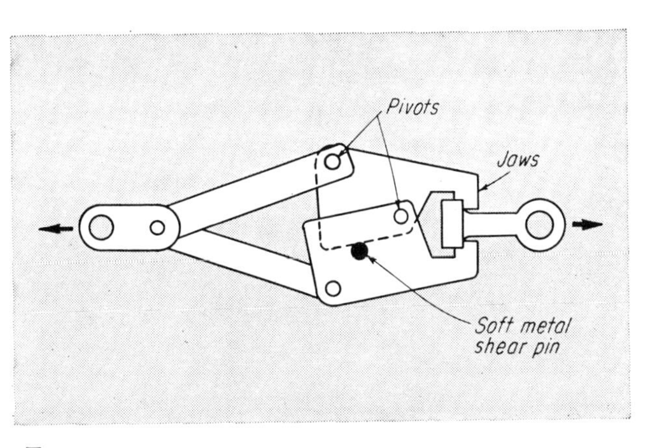

5
TENSION RELEASE. When toggle-operated blade shears soft pin, jaws open to release eye. A spring that opposes the spreading jaws can replace the shear pin.

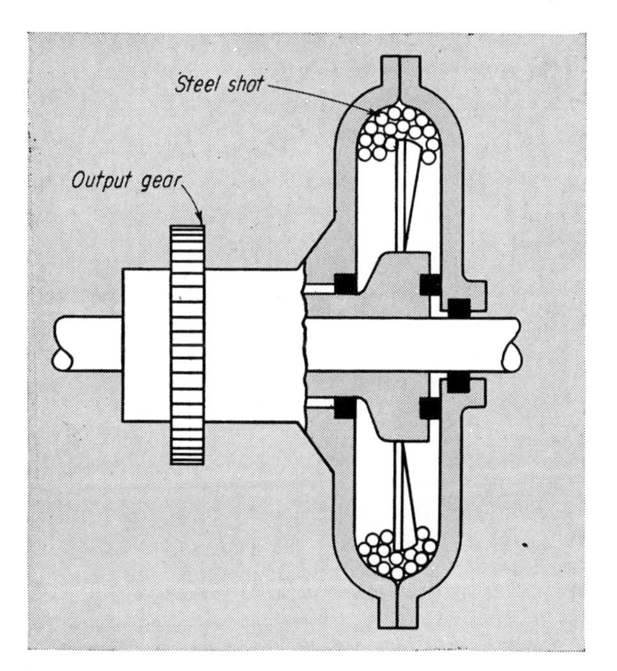

6
STEEL-SHOT COUPLING transmits more torque as speed increases. Centrifugal force compresses steel shot against case, increasing resistance to slip. Adding more steel shot also increases resistance to slip.

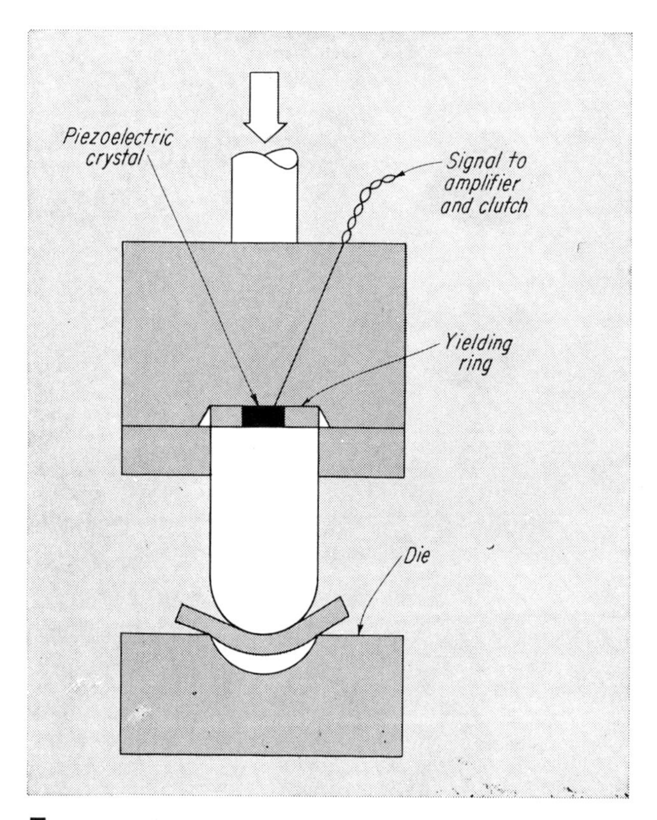

7
PIEZOELECTRIC CRYSTAL sends output signal that varies with pressure. Clutch at receiving end of signal disengages when pressure on the crystal reaches preset limit. Yielding ring controls compression of crystal.

Overspeed Control Devices

Overspeed control for machines uses frangible element

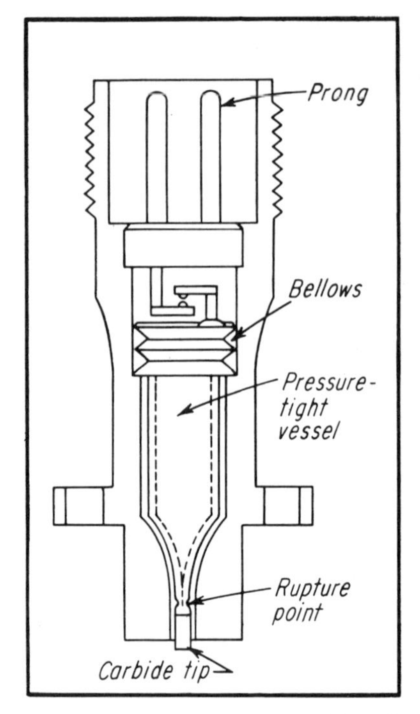

FRANGIBLE FUSE for machines has pressure-tight housing comprising a bellows and carbide tip. The two parts can be combined into a single unit, as shown, with the tip attached directly to the bellows; or the two sections may be mounted independently, with a connection being made by capillary tube. A standard plug may be inserted in the end of the bellows housing to make contact with the prongs.

A nonelectrical "safety fuse" for machines, a device that operates mechanically to close an electrical circuit in the event of overspeeding which might, for example, cause a blade to strike a housing, has been developed by Englehard Industries, Newark, NJ.

Basically, the fuse is a pressure-tight bellows with a tungsten carbide tip (see diagram). Pressure on the unit expands the bellows; but under overspeed conditions, the carbide tip breaks off, relieving the pressure and allowing the bellows to contract, thus closing the circuit. One promising future application: aircraft turbines.

In a turbine application, the unit would be mounted so that there would be a slight, but only very slight, clearance between the turbine blades and the tip. Overspeeding and bearing failure would immediately cause the blade to strike the tip and rupture the pressure vessel.

Englehard has patented the device; but has not yet announced plans for commercial production.

When a Turbine Fails

The fuses here are mechanical ones. They cause gas valves to close if the turbine overspeeds or bearings fail. An added safety feature backing up this protective system is the double-walled exhaust shroud.

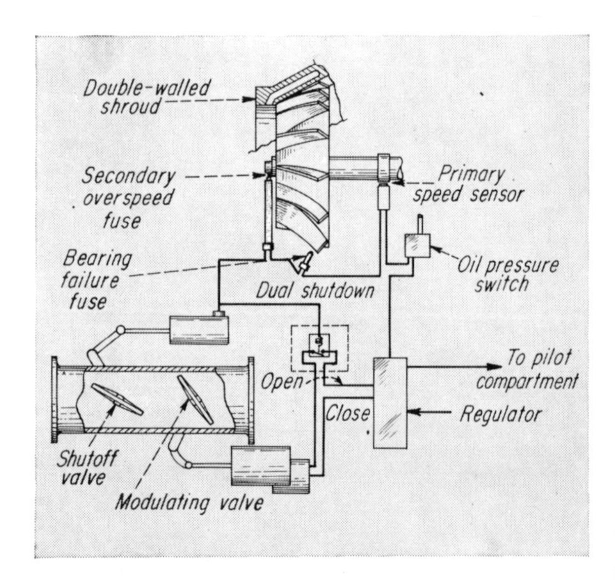

Protective system . . .

takes no chances. Any of the failure-detecting devices will open an electrical circuit that closes both the shut-off and control valves in gas inlet. This results from designing the system to override the gas flow regulator that holds generator voltage constant.

The primary overspeed detector is sensitive to voltage generated at a preset speed by a magnet mounted on the shaft. Secondary overspeed detector works as a fuse. It is a loop of platinum wire held in a ceramic insulator and steel sleeve. At a preset speed, it is broken by a throwout pin in the shaft.

Exhaust shroud holds three bearing-failure fuses. Their tungsten-carbide tips carry a painted electrical circuit on their inside surfaces. If a bearing fails, rotor blades will snap off the tips and break the circuit. The system also has a switch to detect loss of oil pressure.

The method protects pneumatic turbines that drive alternators for aircraft electrical systems. Fuses are made by Charles Englehard, Inc., East Newark, N. J. The alternator drives are built by Thompson Products, Inc., of Cleveland.

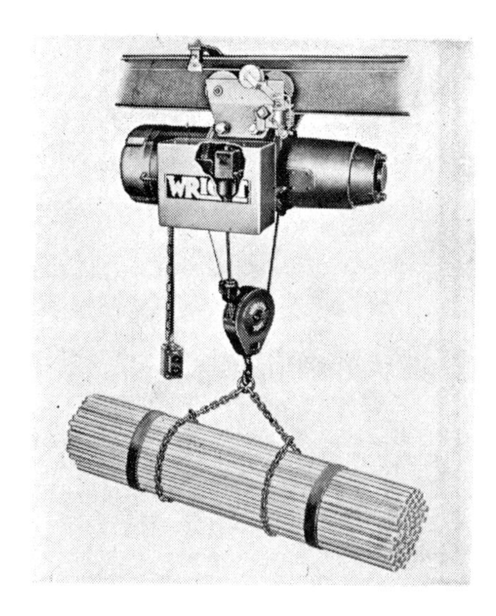

Cutoff Prevents Overloading of Hoist

Fail-safe switch deactivates lifting circuit if load exceeds preset value. Split coupling permits quick attachment of cable.

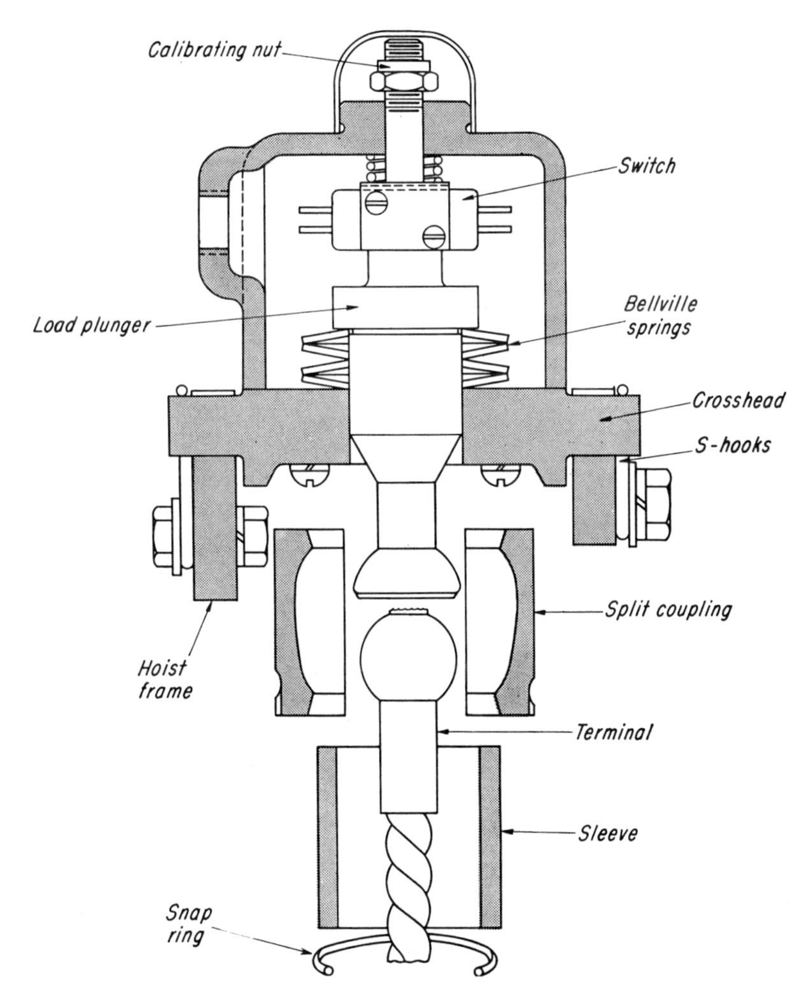

LOAD PLUNGER is inserted through the Belleville springs, which are supported on a swiveling crosshead. The crosshead is mounted on the hoist frame and retained by two S-hooks and bolts. Under load, the Belleville springs deflect and permit the load plunger to move axially. The end of the load plunger is connected to a normally closed switch. When the springs deflect beyond a preset value, the load plunger trips the switch opening the raising-coil circuit of the magnetic hoist-controller. The raising circuit becomes inoperative, but the lowering circuit is not affected. A second contact, normally open, is included in the switch to permit the inclusion of visual or audible overload signal devices.

The load plunger and the swaged-on cable termination have ball-and-socket seat sections to permit maximum free cable movement, reducing the possibility of fatigue failure. A split-coupling and sleeve permits quick attachment of the cable-ball terminal to the load plunger.

SWITCH AND SPRING elements are sealed for mechanical protection and to prevent tampering. The overload-cutoff is calibrated at the factory and sealed. Unless the switch wires are bypassed deliberately, normal operation is insured. A sudden, accidental overloading—for example, if the load is dropped—could cause the switch to fail or fracture. If it does, switch is fail-safe; lifting circuit can't operate until switch is repaired or replaced.

PS . . . Primarily intended to protect the operator and secondarily to prevent damage to equipment, overload-cutoff prevents the lifting of loads greater than the capacity of the hoist. It's standard equipment on new frame-2 and frame-3 electric hoists; can be fitted to all older models. Produced by Wright Hoist Div, American Chain & Cable Co Inc, York, Penna.

7 ways to LIMIT SHAFT

Traveling nuts, clutch plates, gear fingers, and pinned members are the bases of these ingenious mechanisms.

Mechanical stops are often required in automatic machinery and servomechanisms to limit shaft rotation to a given number of turns. Two problems to guard against, however, are: Excessive forces caused by abrupt stops; large torque requirements when rotation is reversed after being stopped.

I M ABELES, *design engineer, Ordnance Dept General Electric Co, Pittsfield, Mass*

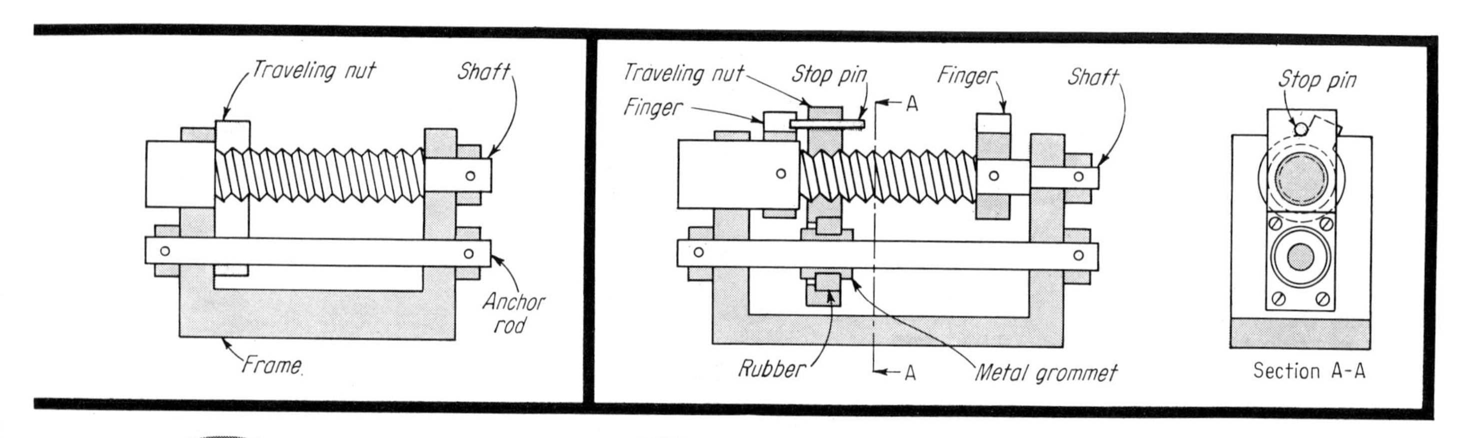

TRAVELING NUT moves (1) along threaded shaft until frame prevents further rotation. A simple device, but nut jams so tight that a large torque is required to move the shaft from its stopped position. This fault is overcome at the expense of increased length by providing a stop pin in the traveling nut (2). Engagement between pin and rotating finger must be shorter than the thread pitch so pin can clear finger on the first reverse-turn. The rubber ring and grommet lessen impact, provide a sliding surface. The grommet can be oil-impregnated metal.

CLUTCH PLATES tighten and stop rotation as the rotating shaft moves the nut against the washer. When rotation is reversed, the clutch plates can turn with the shaft from A to B. During this movement comparatively low torque is required to free the nut from the clutch plates. Thereafter, subsequent movement is free of clutch friction until the action is repeated at other end of the shaft. Device is recommended for large torques because clutch plates absorb energy well.

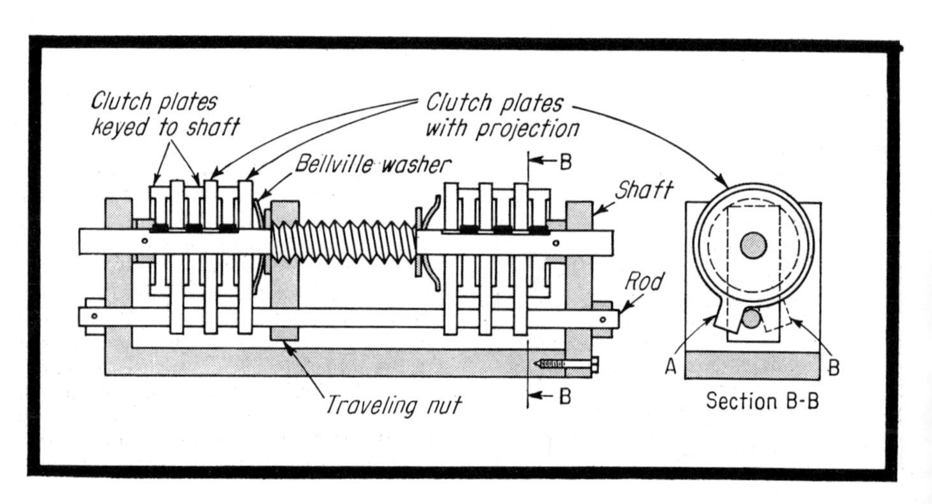

ROTATION

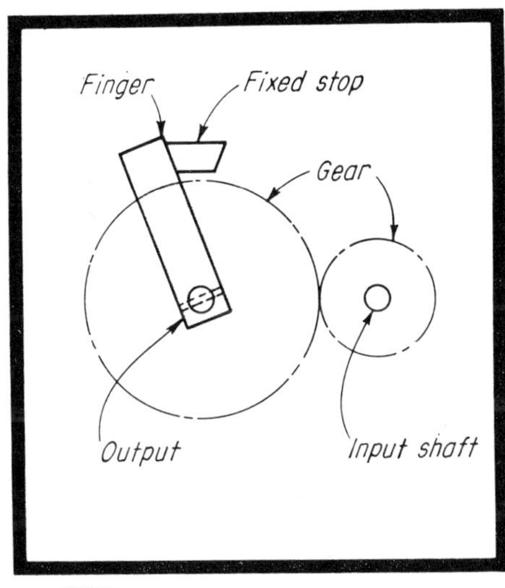

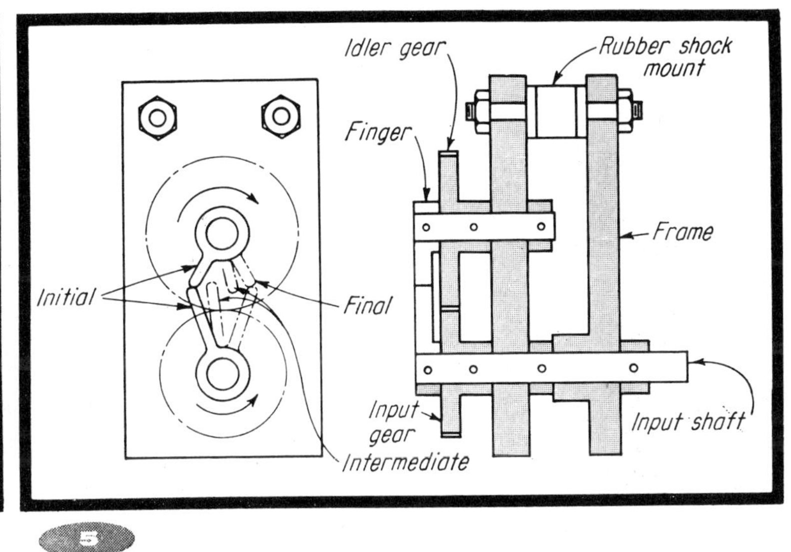

4

SHAFT FINGER on output shaft hits resilient stop after making less than one revolution. Force on stop depends upon gear ratio. Device is, therefore, limited to low ratios and few turns unless a worm-gear setup is used.

5

TWO FINGERS butt together at initial and final positions, prevent rotation beyond these limits. Rubber shock-mount absorbs impact load. Gear ratio of almost 1:1 ensures that fingers will be out of phase with one another until they meet on the final turn. Example: Gears with 30 to 32 teeth limit shaft rotation to 25 turns. Space is saved here but gears are costly.

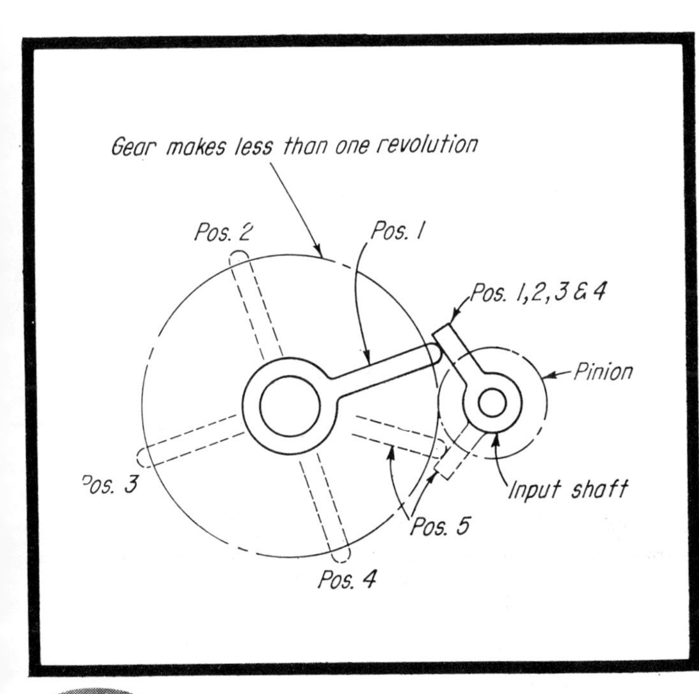

6

LARGE GEAR RATIO limits idler gear to less than one turn. Sometimes stop fingers can be added to already existing gears in a train, making this design simplest of all. Input gear, however, is limited to a maximum of about 5 turns.

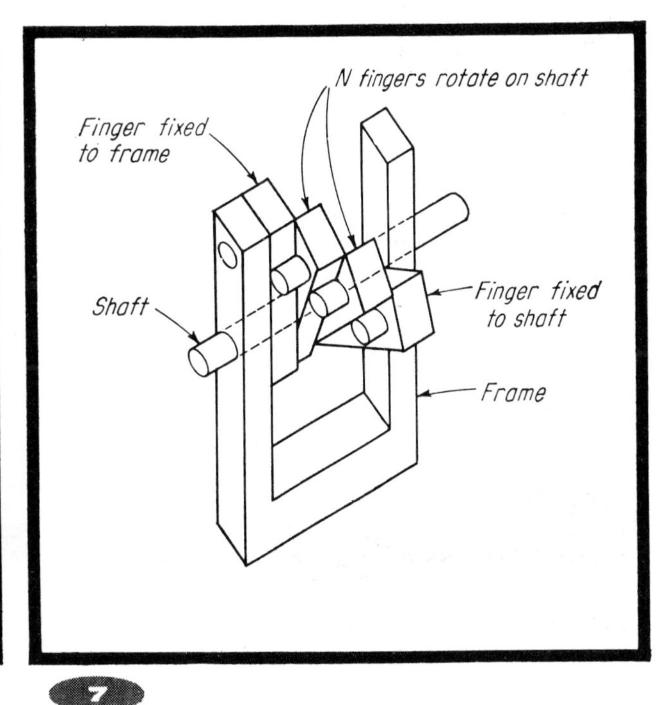

7

PINNED FINGERS limit shaft turns to approximately N + 1 revolutions in either direction. Resilient pin-bushings would help reduce impact force.

Speed control for

Friction devices, actuated by centrifugal force, automatically keep speed constant regardless of variation of load or driving force.

FEDERICO STRASSER

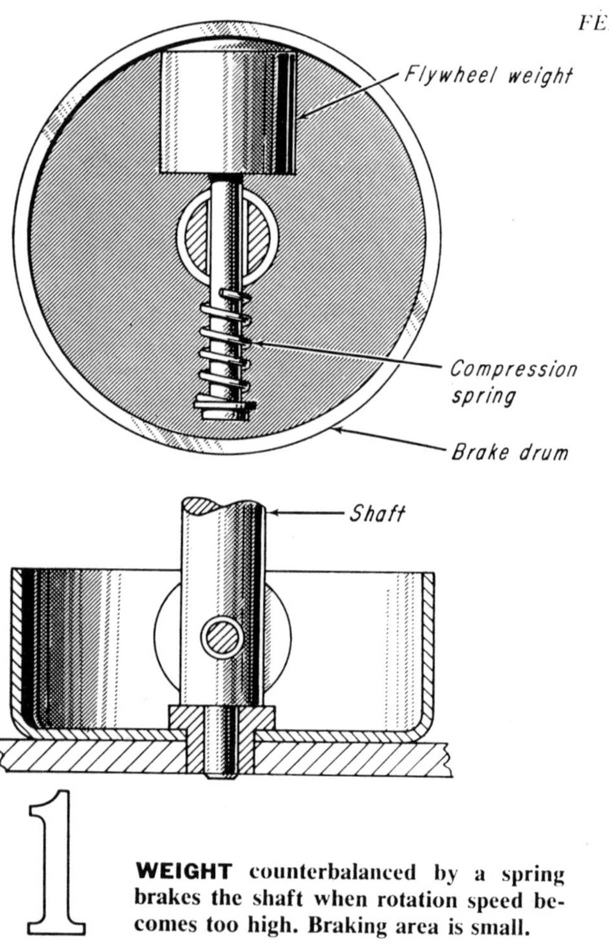

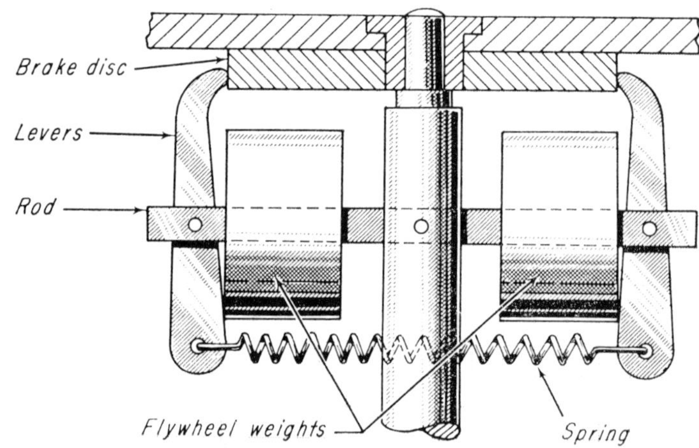

1

WEIGHT counterbalanced by a spring brakes the shaft when rotation speed becomes too high. Braking area is small.

WEIGHT-ACTUATED LEVERS make this arrangement suitable where high braking moments are required.

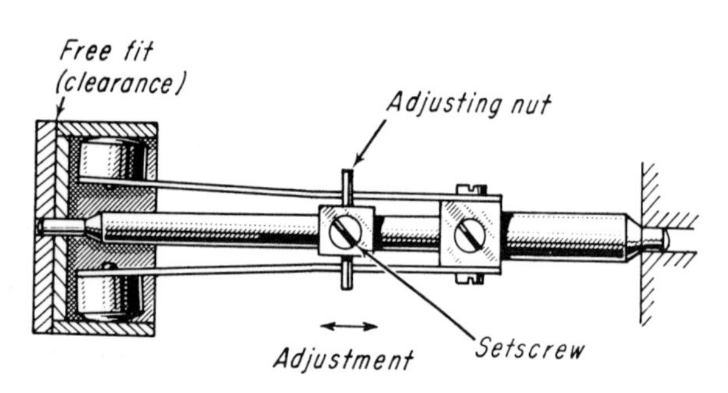

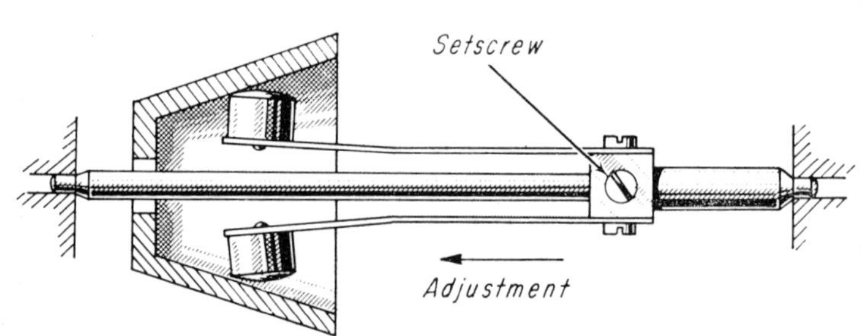

ADJUSTMENT of speed at which this device starts to brake is quick and easy. Adjusting nut is locked in place with setscrew.

TAPERED BRAKE DRUM is another way of providing for varying speed-control. The adjustment is again locked.

small mechanisms

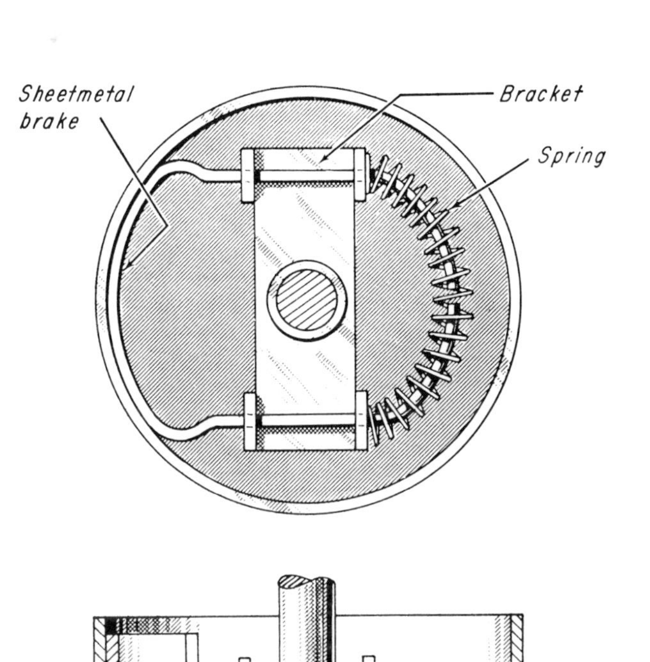

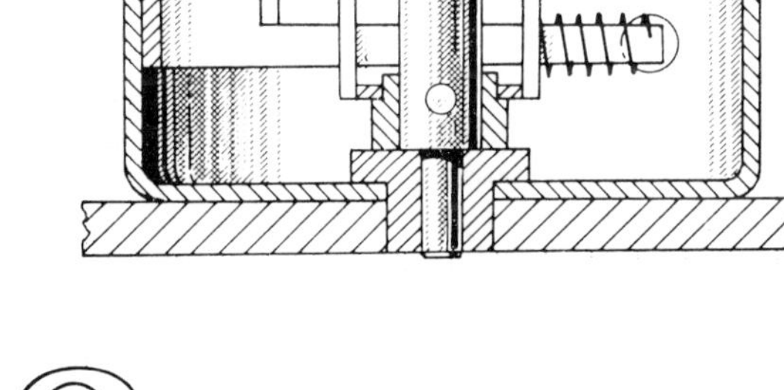

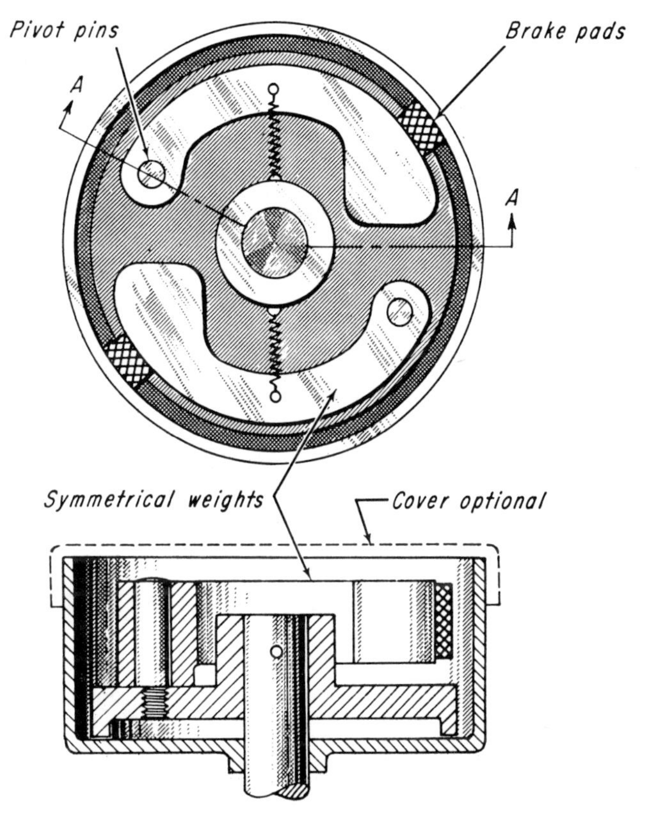

3

SHEETMETAL BRAKE provides larger braking area than previous design. Operation is thus more even and cooler.

SYMMETRICAL WEIGHTS give even braking action when they pivot outward. Entire action can be enclosed.

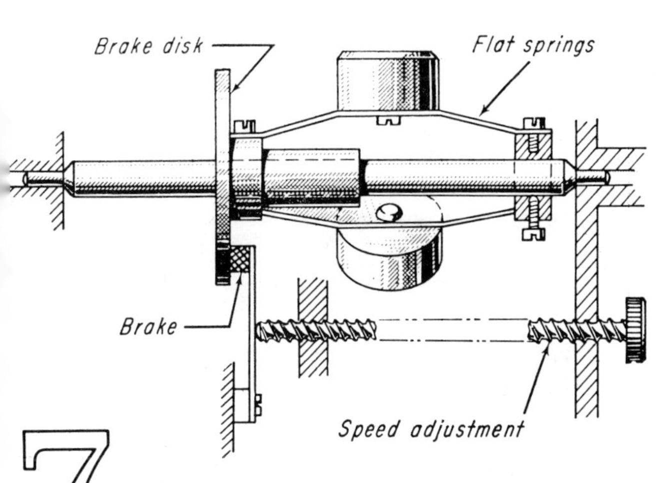

7

THREE FLAT SPRINGS carry weights that provide brake force upon rotation. Device can be provided with adjustment.

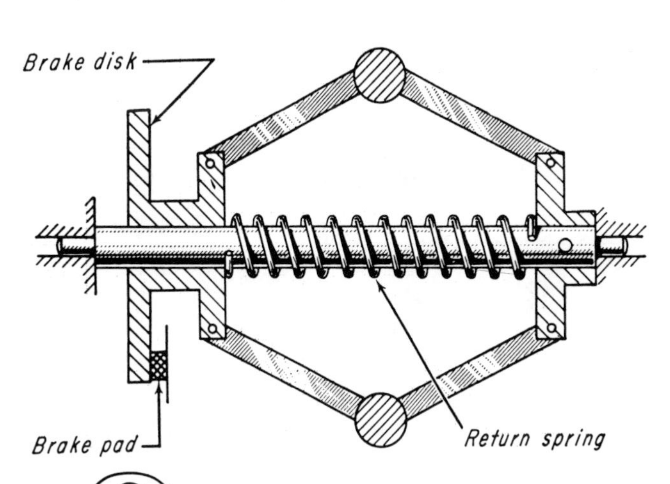

8

TYPICAL GOVERNOR action of swinging weights is utilized here. As in the previous device, adjustment is optional.

Governing Systems for Turbines and Engines

from POWER

Basically, turbine governing systems control throttle steam flow to hold speed constant as load changes, or to hold pressure constant as demand for process steam varies, or to hold both constant. Let's look at speed governing, the most common problem, first.

The simplest and most familiar speed governor is the direct-acting flyball arrangement of Fig 1. In the usual form, flyballs or weights hang on arms from a central shaft driven at speed proportional to turbine speed. Arms connect the weights to a sliding collar acted against by a spring. A lever connects the collar to a governing valve controlling steam flow. As weights rotate, centrifugal force tends to lift them against spring force, raising the collar.

The curves of Fig. 7 show this governor's essential characteristics. To understand them, let's look first at the forces acting on the collar. Centrifugal force depends on the diameter of the weight's circle of rotation and on speed. Greater diameter or higher speed means more force. The spring force depends on dimensions of the spring and how much it is compressed. Let's assume constant speed, say 98% of rated speed, and a spring of known dimensions. With weights at "in" position, a certain centrifugal force results and this will just balance the spring force when the sliding collar is all the way down and the valve wide open. Still holding speed at 98%, let's move the collar all the way up and close the valve. The spring, being compressed fully, now exerts greater force, but the weights also exert greater force, because they swing in a wider circle. Again the forces balance. Similar balances will result at points between these extremes and it is possible to plot *centrifugal force* against *governor travel,* obtaining a sloping line like that labeled 98% in Fig. 7. Shape of this curve depends on governor design and may deviate considerably from a straight line.

At any speed above 98%, the weights exert greater centrifugal force at each position and a stronger spring is needed to balance these forces. Fig. 7 shows force-travel curves for speeds from 98% to 102%. Obviously it is impractical to use a different spring for each speed as assumed for illustration purposes here. An actual governor would have a spring of such dimensions that its force would just balance the centrifugal force at 98% speed and the weights-in position and at 102% speed and the weights-out position. Lines *X* and *Y* of Fig. 7 show limits of governor travel and line *A* is the force line of the actual spring.

SIX BASIC SPEED-GOVERNING ARRANGEMENTS

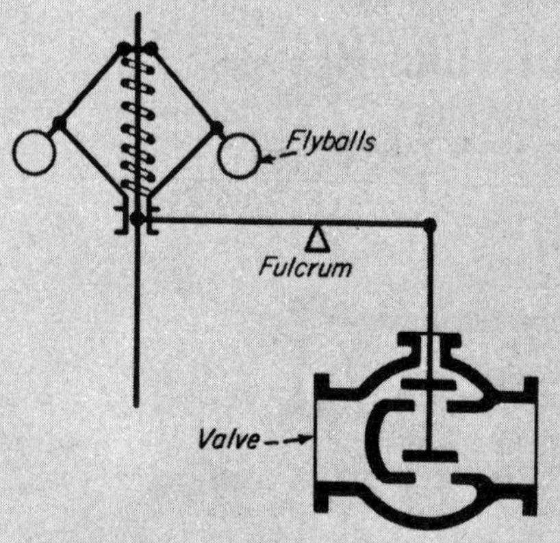

1 Simple, rugged direct-acting governor is limited by its low power and need for speed to change to meet changes in load

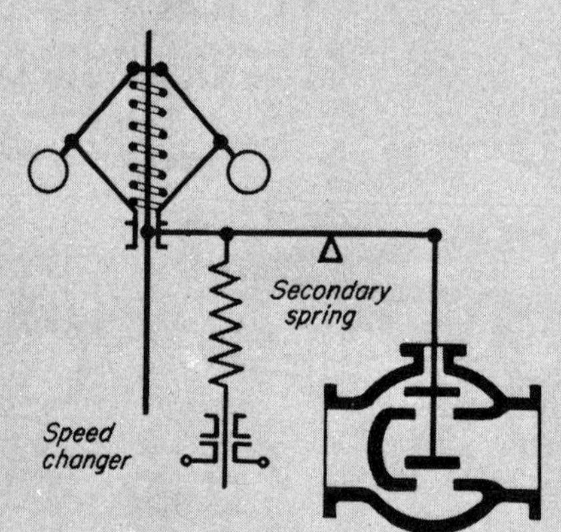

2 Addition of secondary spring permits restoring rated speed after load change but governor-travel curve is distorted

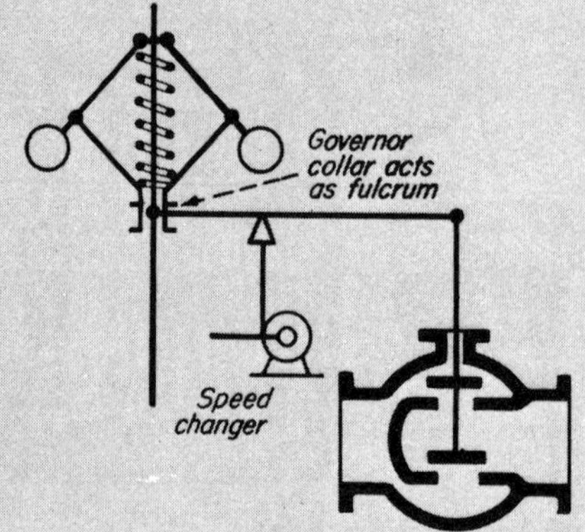

3 Restoring rated speed by moving valve lever about collar as fulcrum gives same speed-travel relationship at all loads

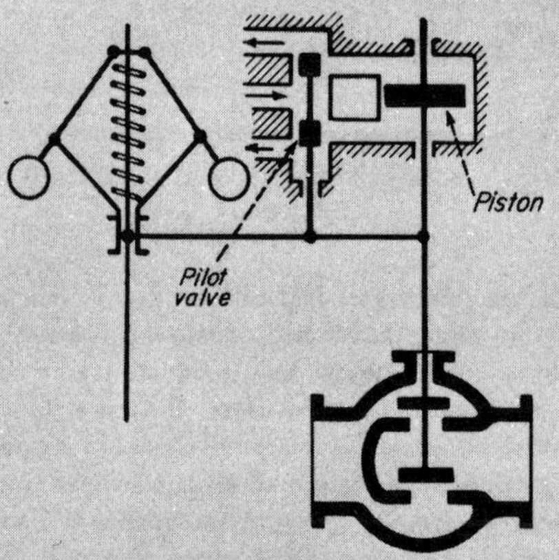

4 Pilot valve and hydraulically powered operating piston amplify governor force

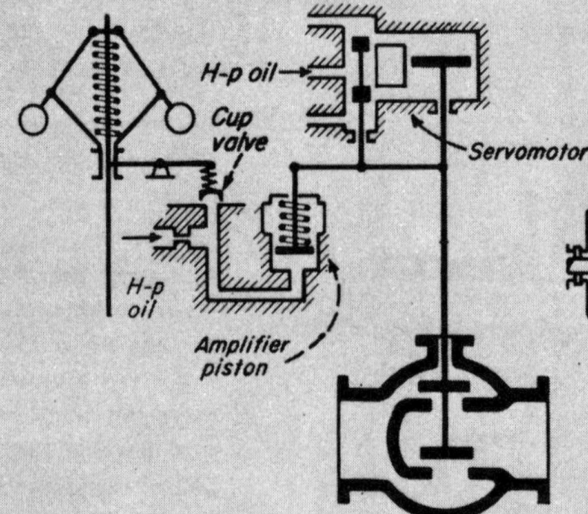

5 Double relaying (two stages of amplification) further boosts governor force

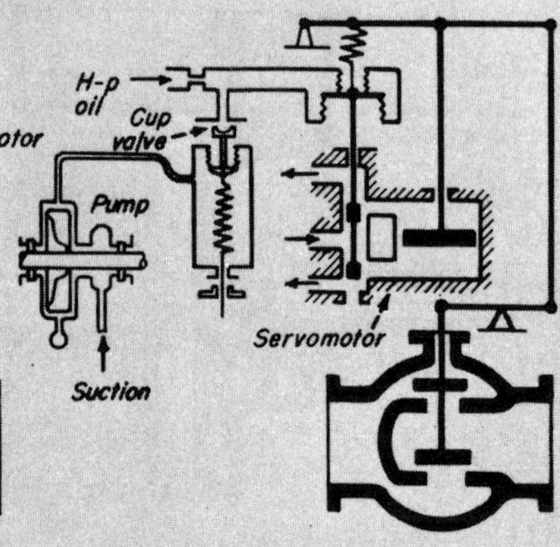

6 Speed-responsive element may be shaft-driven impeller instead of flyballs

With this spring in place, governor opens valve far enough to carry full load at 98% speed and closes it enough to hold speed to 102% at no load. This speed change of 4% is the governor's *regulation* or *speed droop*. To change this characteristic requires changing the spring. For example, to improve regulation to 3%, the new spring must match centrifugal force at 98% as before, but need only match force at 101%, less than before. Since travel and hence spring compression stays the same, this means a lighter spring.

Thus far we have neglected friction. To see its effect, let's assume that difference in centrifugal force at weights-in position between 98 and 99% speed is 10 lb. Now if, because of friction, it takes 10 lb over and above spring force to move the collar, speed must increase from 98 to 99% after a load change before the valve starts to close. As load drops further, governor positions at various speeds are shown along the dotted line above *A*. Note that at the weights-out position, only about ⅔ of 1% speed change produces the needed 10-lb force. Thus in speaking of a governor's *power,* defined as the change in governor force for 1% speed change without motion, it is necessary to refer always to position.

If the process described above is reversed and load applied to the unit, governor positions for various speeds are shown by the dotted line below *A*, Fig. 7. Vertical distance between lines above and below *A* represents the governor's *dead band,* the speed change in which no governor motion occurs. This is a measure of the governor's *sensitivity,* which may be defined as the percentage speed change required to produce a corrective movement.

Before discussing limitations of the simple governor of Fig. 1, and ways of overcoming them, let's see how governor characteristics affect parallel operation of turbine generators. Paralleled synchronous machines operate at the same speed, just as if they were mechanically connected. When load increases, speed tends to fall until *total* output increases sufficiently to meet demand. Percentage speed change is the same for all units and if they are of the same size and have governors with the same regulation, they divide load equally between them, Fig. 9.

Still assuming that units are equal in size, consider what happens if one has greater speed droop than the other. The percentage speed change is the same for both. It can be seen that the lower the speed droop, the more load change the unit will go through before reaching a given percentage speed change. Fig. 10 illustrates a typical example. If the units are of different capacities, the larger ones will take more of the added load. This may be put into a rule: *Division of added load between units varies inversely as their speed droops and directly as their ratings.*

A governor that has zero speed droop is called an *isochronous* governor. If a unit fitted with such a governor is paralleled with another having speed droop, the machine with the isochronous governor tends to take all added load and speed of both units remains constant within the capacity of the isochronous-governed unit to absorb added load, Fig. 11. The need to take all load changes on one machine necessarily limits such an arrangement.

GOVERNOR CHARACTERISTICS AND PARALLEL-OPERATION PRINCIPLES

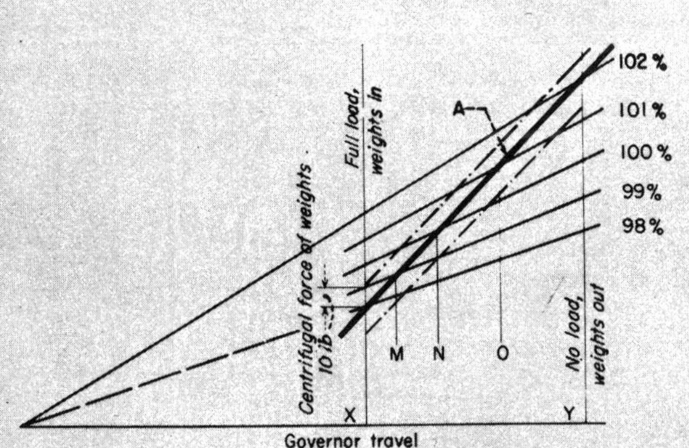

7 Curves show centrifugal force exerted by the weights at different speeds and positions. Line *A* indicates spring-force characteristic needed for this governor. Vertical distance between dotted lines above and below *A* is *dead band*

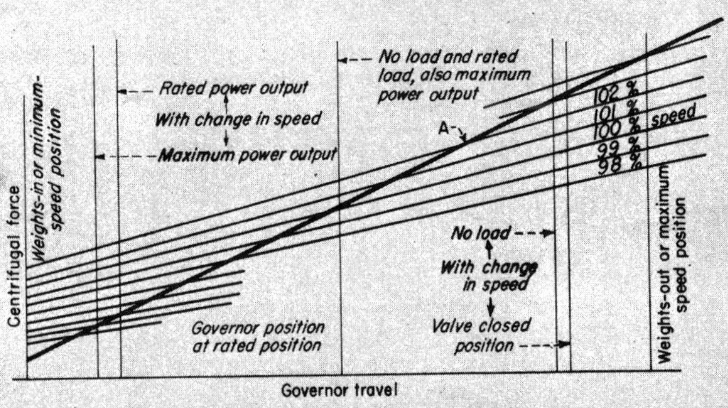

8 When speed changer of Fig. 3 is used, governor weight position doesn't change as long as unit runs at normal speed and speed-travel relationship stays the same at all loads. Governor meets requirement that it open valve fully from no-load position without changing setting of speed-changer, and also drop full load and close valve from maximum-load rated-speed position

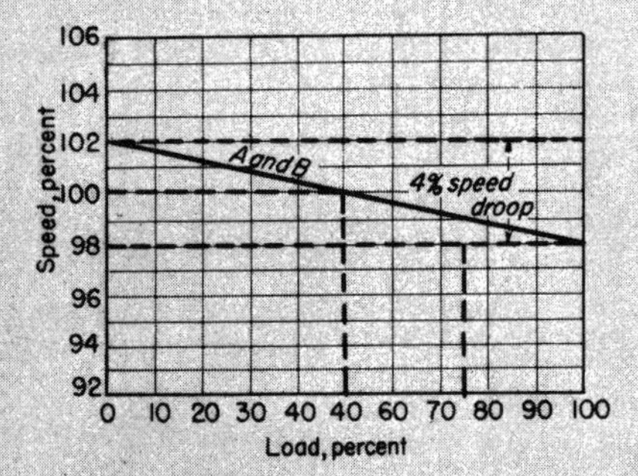

9 Identical turbines *A, B* have governors with 4% speed droop; each carries 50% load at 100% speed. Load increase divides equally between them, bringing them to 75% load at 99% of rated speed

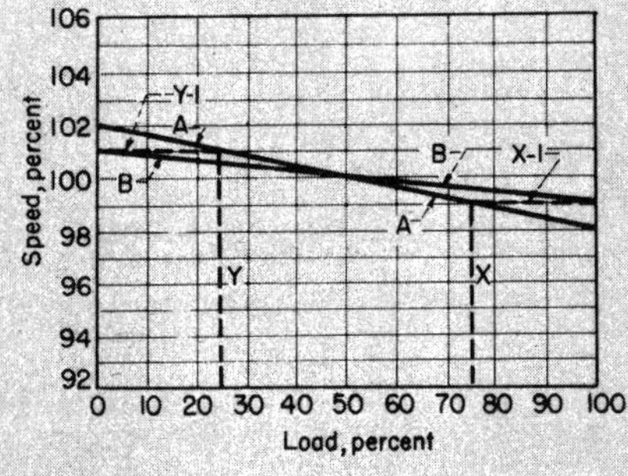

10 Identical units with governors of different speed droops divide load in inverse ratio to speed droop. *B* becomes fully loaded; *A* becomes 75% loaded (*X*); resultant speed is 99% as shown at *X-1*

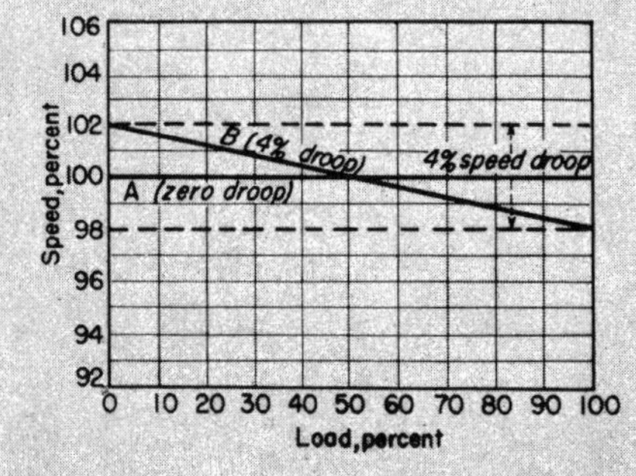

11 Of two identical turbines, unit with isochronous governor takes all new load within its capacity. In example, *A* goes from 50% to full load while *B* takes no added load; speed stays constant at 100%

TYPICAL GOVERNORS WITH CENTRIFUGAL SPEED-RESPONSIVE ELEMENTS

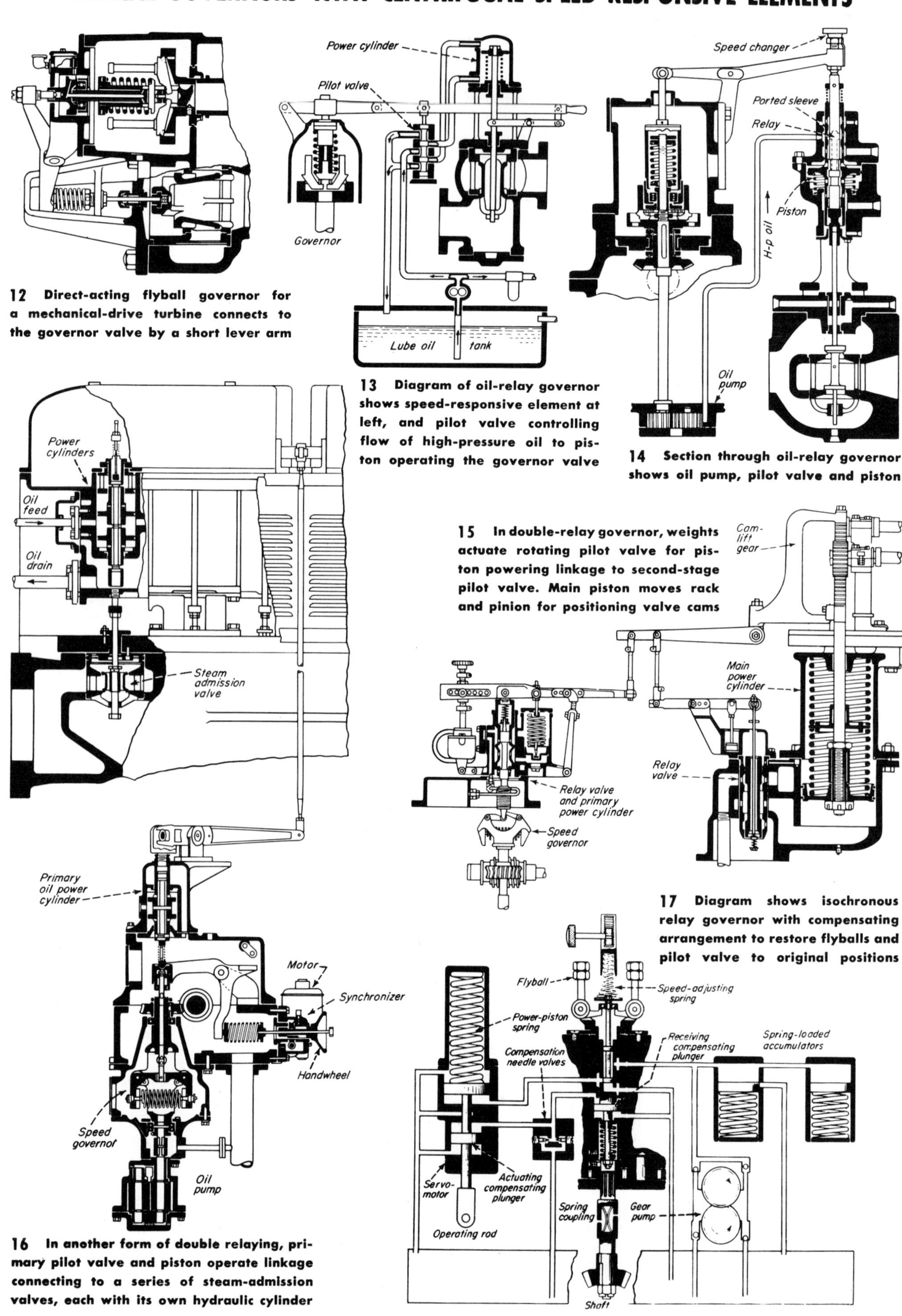

12 **Direct-acting flyball governor for a mechanical-drive turbine connects to the governor valve by a short lever arm**

13 **Diagram of oil-relay governor shows speed-responsive element at left, and pilot valve controlling flow of high-pressure oil to piston operating the governor valve**

14 **Section through oil-relay governor shows oil pump, pilot valve and piston**

15 **In double-relay governor, weights actuate rotating pilot valve for piston powering linkage to second-stage pilot valve. Main piston moves rack and pinion for positioning valve cams**

16 **In another form of double relaying, primary pilot valve and piston operate linkage connecting to a series of steam-admission valves, each with its own hydraulic cylinder**

17 **Diagram shows isochronous relay governor with compensating arrangement to restore flyballs and pilot valve to original positions**

Returning to the direct-acting flyball governor of Fig. 1, we can see it offers simplicity and ruggedness and thus proves suitable for many applications not requiring close speed regulation. However, it has certain definite limitations: (1) change in valve position and thus in turbine output only follows a speed change (2) it is relatively insensitive and (3) the force it can exert can only be increased by boosting governor size and this is definitely limited.

The disadvantage of speed change with load change, serious where close frequency control is desired, may be overcome by various forms of *speed changers,* Fig. 2 and 3. In Fig. 2, a secondary spring, manually adjusted, acts to change spring force and thus restore original speed after a load change has been met. For a given speed the weights assume different positions for different loads and this is called a *travel* governor. The governor-travel curve is distorted and the regulation curve is affected. In a *no-travel* governor, weight position does not change as long as the unit runs at normal speed and the speed-travel relationship stays the same at all loads, Fig. 8. In this type, Fig. 3, the governor sliding collar acts as a fulcrum while the valve position is changed by moving the lever center mechanically.

To increase sensitivity, designers use ball bearings at all joints and mechanical oscillators to keep the system moving slightly at all times, thus substituting lesser kinetic friction for static friction. However, the power available still remains insufficient for moving the extensive linkages and large valves required on many turbines and some form of amplification is necessary.

The *relay* governor of Fig. 4 obtains such amplification by using the governor force to move a light and relatively frictionless pilot valve which controls flow of high-pressure oil to a piston for operating the governor valves and attendant mechanism. By this means the governor's sensitivity may be increased and the force available for valve movement is no longer dependent on the governor's centrifugal force.

In larger units the linkage between the governor proper and the hydraulic cylinder or *servomotor,* or between the servomotor and the admission valves may be extensive and double-relaying may be employed. In Fig. 5, a short linkage connects the governor to the first-stage of amplification, the operating piston of which links to a second stage of amplification driving the admission valves. The system shown uses a cup valve controlling pressure under the first-stage operating piston by regulating high-pressure oil leakoff. In many designs, however, a pilot valve similar to that in the second stage of Fig. 5 is employed and the first and second amplification stages are essentially similar.

In the designs thus far discussed, the *primary element* or *speed responsive element* consists of flyballs generating centrifugal force. A completely hydraulic governor, Fig. 6, may be obtained by using a shaft-driven pump as the primary element. Oil pressure on the bellows, varying with pump speed and thus with shaft speed, moves a cup valve controlling leakoff of high-pressure oil from a system containing a spring-loaded bellows actuating a pilot valve for the main piston.

Fig. 12 through 17 show typical actual governors having flyball primary elements. The unit of Fig. 12 is a direct-acting design as used on a mechanical drive turbine. Fig. 13 shows an oil-relay governor diagrammatically and Fig. 14 is an actual unit of the same type. The use of two stages of amplification, or double-relaying, appears in Fig. 15; the primary element and its hydraulic cylinder connects through a linkage to the pilot valve of the main operating piston powering the cam-lift mechanism for the steam-admission valves. Fig. 16 illustrates another form of double relaying; the primary power cylinder operates a linkage controlling individual steam-admission valves. Each steam-admission valve is operated by a separate hydraulic cylinder and pilot valve. The valves are arranged to open in sequence, the second valve opening when steam-ring pressure in nozzles supplied by the first

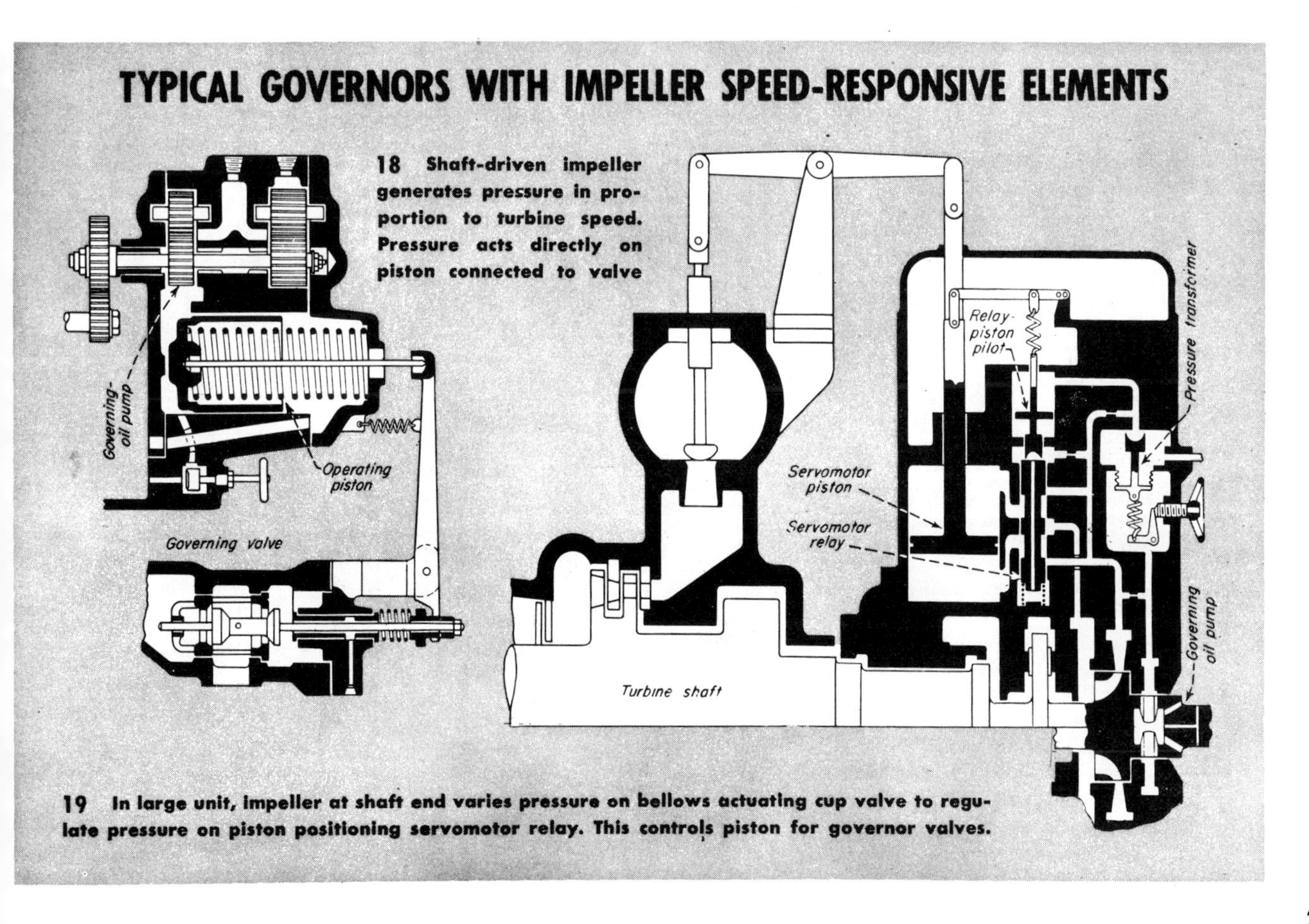

18 **Shaft-driven impeller generates pressure in proportion to turbine speed. Pressure acts directly on piston connected to valve**

19 **In large unit, impeller at shaft end varies pressure on bellows actuating cup valve to regulate pressure on piston positioning servomotor relay. This controls piston for governor valves.**

valve has built up nearly to throttle pressure, and so on.

Fig. 17 shows, schematically, an oil-relay governor designed for isochronous operation with provision for speed adjustment and speed-droop control. Flyball action moves pilot-valve plunger, uncovering regulating port in pilot-valve bushing and releasing oil from under power piston. Motion of power piston causes compensating piston to move and change position of pilot-valve bushing, compressing compensating spring. Movement of power piston and pilot-valve bushing continues until regulating port bushing is covered by land on plunger, at which point power piston stops in the position for the new load. With rated speed restored, flyballs and pilot-valve plunger return to normal position. Compensating spring returns pilot-valve bushing to normal, also. With original conditions fully restored, governor is ready for next load change.

Fig. 18 and 19 show pump or impeller-type governors. In Fig. 18, the pressure from the pump acts directly on a piston connecting to the governor linkage. In Fig. 19, relaying is employed, and the unit operates in the same way as that shown diagrammatically in Fig. 6.

In addition to *operating* speed governors like those previously described, turbines are fitted with various forms of *protective* or *emergency* governors. Fig. 20 and 21 show typical *overspeed trips*. In most designs, a pin or weight, in the main shaft or in a unit attached to it, is acted on by centrifugal force. When shaft speed exceeds a desired safe level, say 10% overspeed, centrifugal action throws the pin or weight against a latching device which releases a spring to close either the governing valve or a special emergency stop valve. In some designs the overspeed trip is incorporated with the operating governor. Fig. 22 shows an emergency valve held open by oil pressure acting on a bellows. When oil pressure fails, spring closes valve, stopping turbine.

PROTECTIVE GOVERNORS

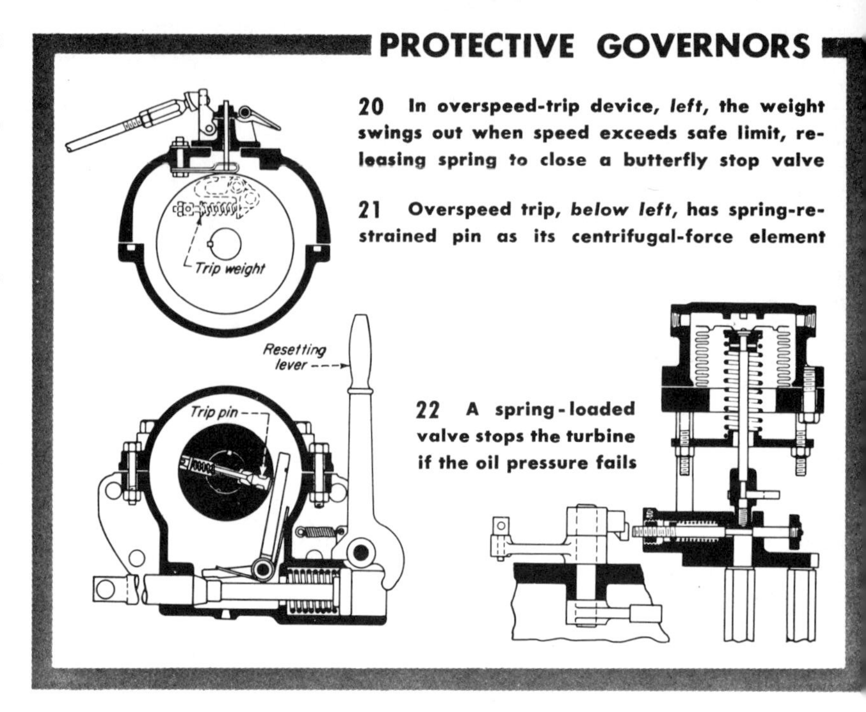

20 In overspeed-trip device, *left*, the weight swings out when speed exceeds safe limit, releasing spring to close a butterfly stop valve

21 Overspeed trip, *below left*, has spring-restrained pin as its centrifugal-force element

22 A spring-loaded valve stops the turbine if the oil pressure fails

TYPICAL GOVERNORS WITH PRESSURE-RESPONSIVE ELEMENTS

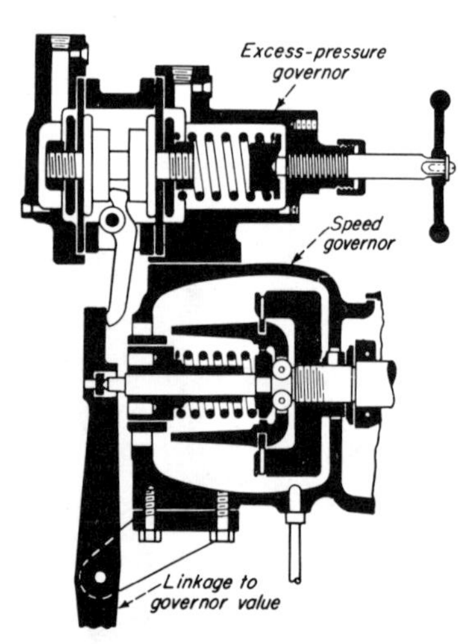

23 Excess-pressure governor acts on the governor valve to vary speed of turbine and driven pump thus maintaining differential pressure needed for boiler feed

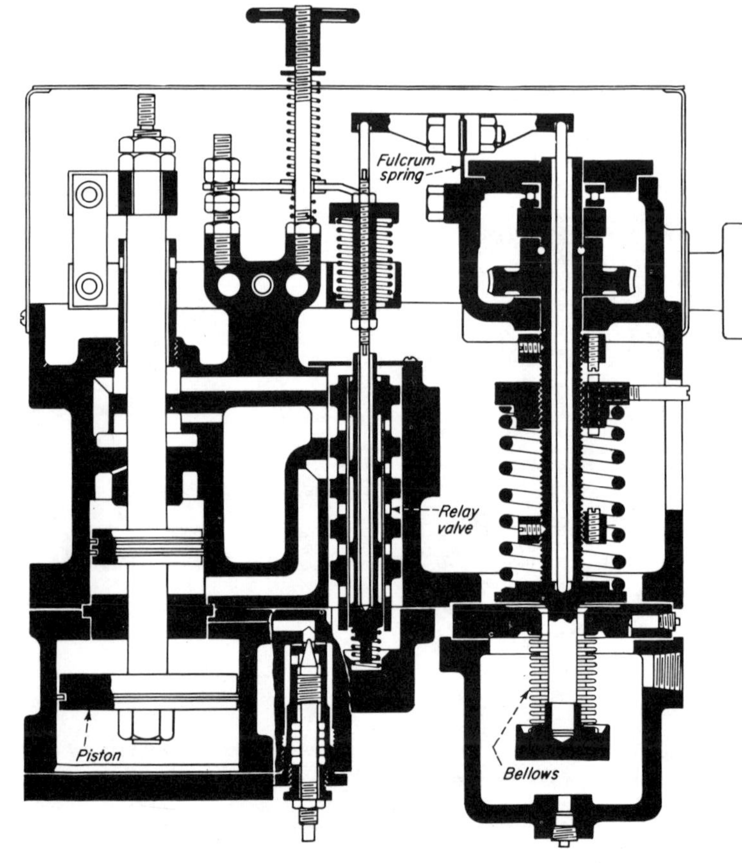

24 In this extraction-pressure regulator, steam pressure from the extraction stage acts on a spring-loaded bellows connecting through a strap fulcrum to a pilot valve controlling high-pressure oil to piston powering valve in turbine

For control of extraction pressure and for other hookups where pressure-control is essential, the primary element is *pressure responsive.* Fig. 23 shows an excess-pressure governor used in controlling a turbine driving a boiler-feed pump. Regulated by differential pressure, the pump governor acts on the governor valve, changing turbine speed as required. In the extraction pressure regulator of Fig. 24, steam pressure at the extraction point acts on a spring-loaded sylphon bellows. Motion is transmitted through a strap fulcrum to an oil-relay pilot valve controlling high-pressure oil to the piston operating the valve in the extraction stage of the turbine.

The various governing elements for speed and pressure control may be used independently, as in the case of a condensing turbine-generator fitted only for speed control. In the case of noncondensing turbines and extraction machines, and units for variable-speed drive, the elements are combined in governing systems.

When a noncondensing turbine must operate at constant backpressure, the pressure governor forms the primary control and the speed governor acts only as an emergency device, taking control of speed when it rises above, say, 3% more than normal. Such a turbine must be operated in parallel with speed-governed turbines, which set its speed. The electrical output varies in fixed relation to the exhaust flow and cannot be controlled independently. In a variation of this *back-pressure governing,* known as *waste-heat governing,* throttle steam pressure is held constant instead of exhaust pressure. The turbine thus uses all steam available up to its capacity limit. Such a system is commonly applied to low-pressure condensing turbines.

For condensing turbines designed for single or double extraction it is possible to provide control of extraction pressure and electrical output, since the relation between steam load and electrical load can be varied within limits. Each extraction point has its pressure governor and the throttle flow is under speed-governor control. These governing elements may be connected mechanically as in Fig. 26, or hydraulically as in Fig. 27.

Discussion of governor characteristics at beginning of this section is based on articles by A F Schwendner (POWER).

VARIABLE-SPEED AND EXTRACTION GOVERNING SYSTEMS

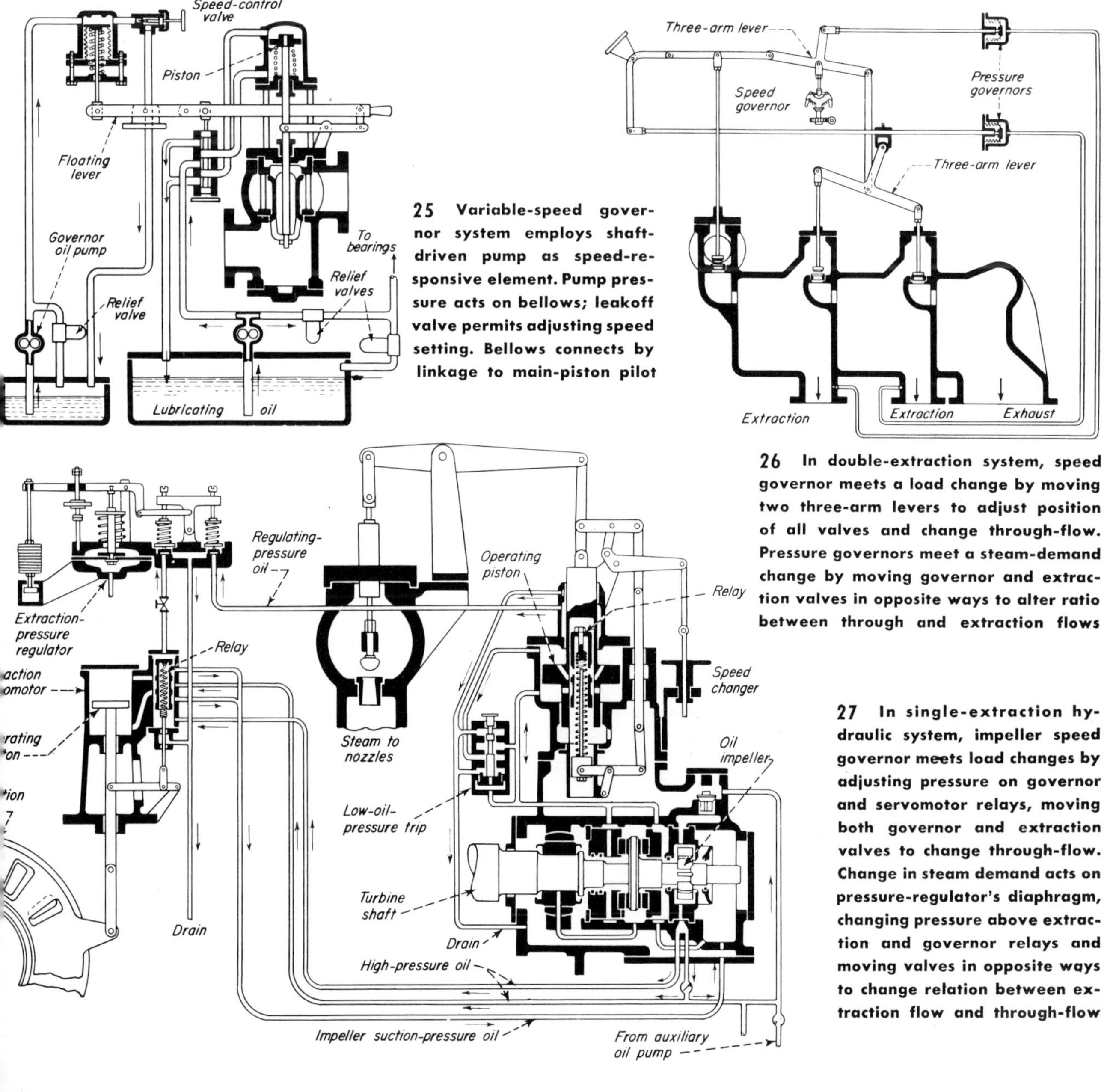

25 Variable-speed governor system employs shaft-driven pump as speed-responsive element. Pump pressure acts on bellows; leakoff valve permits adjusting speed setting. Bellows connects by linkage to main-piston pilot

26 In double-extraction system, speed governor meets a load change by moving two three-arm levers to adjust position of all valves and change through-flow. Pressure governors meet a steam-demand change by moving governor and extraction valves in opposite ways to alter ratio between through and extraction flows

27 In single-extraction hydraulic system, impeller speed governor meets load changes by adjusting pressure on governor and servomotor relays, moving both governor and extraction valves to change through-flow. Change in steam demand acts on pressure-regulator's diaphragm, changing pressure above extraction and governor relays and moving valves in opposite ways to change relation between extraction flow and through-flow

Centrifugal governors

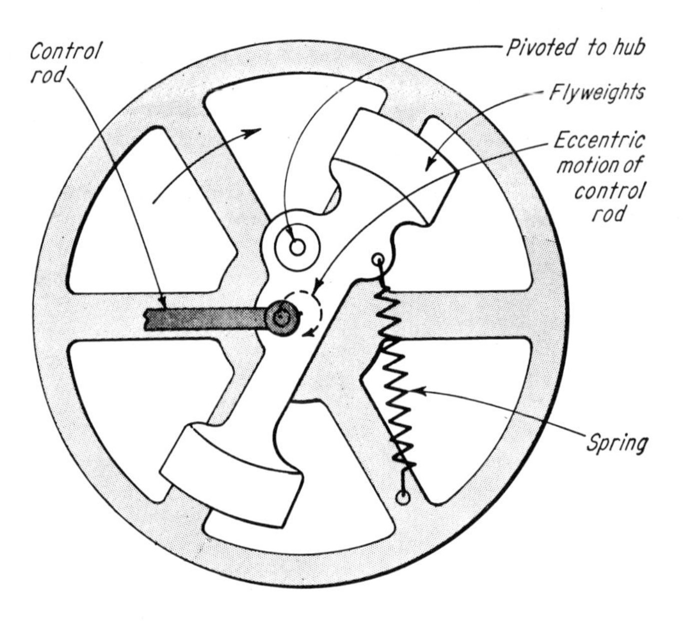

ACCELERATION GOVERNOR
(steam engine)

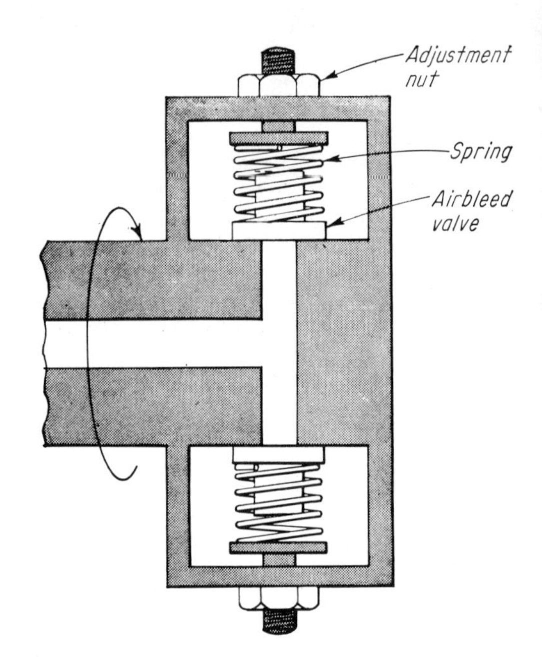

CENTRIFUGAL VALVE

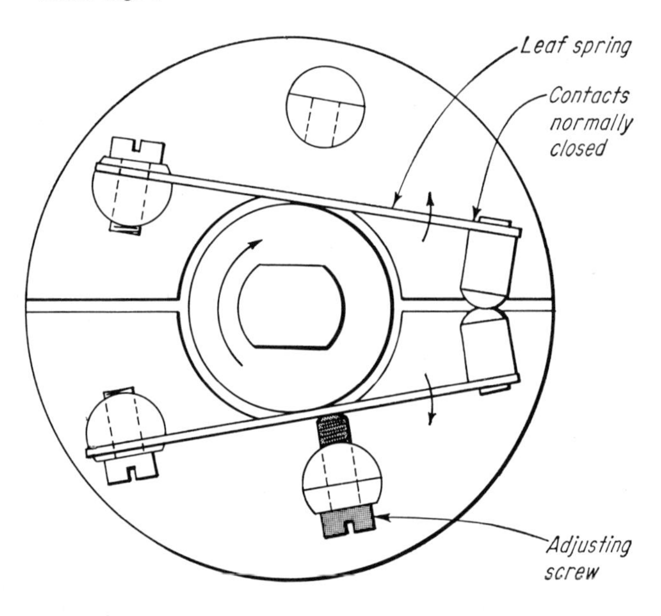

CENTRIFUGAL CONTACTS

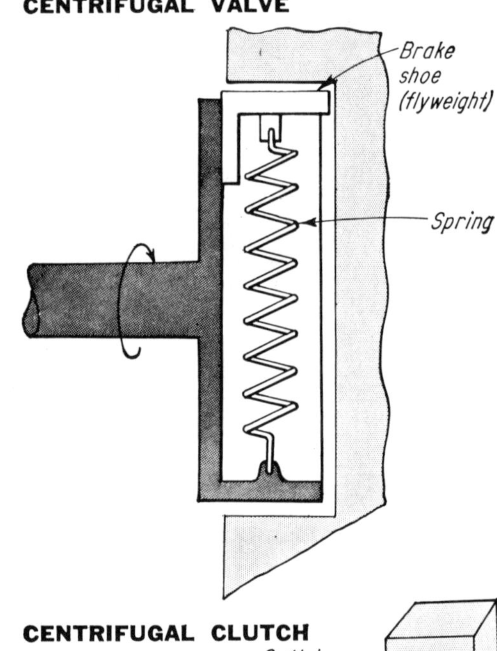

CENTRIFUGAL CLUTCH

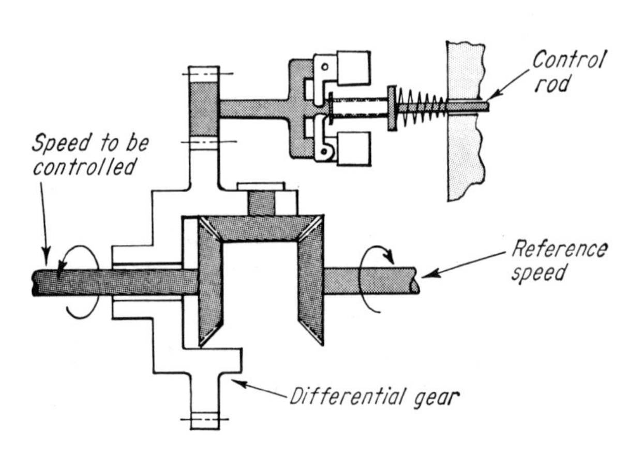

DIFFERENTIAL CENTRIFUGAL

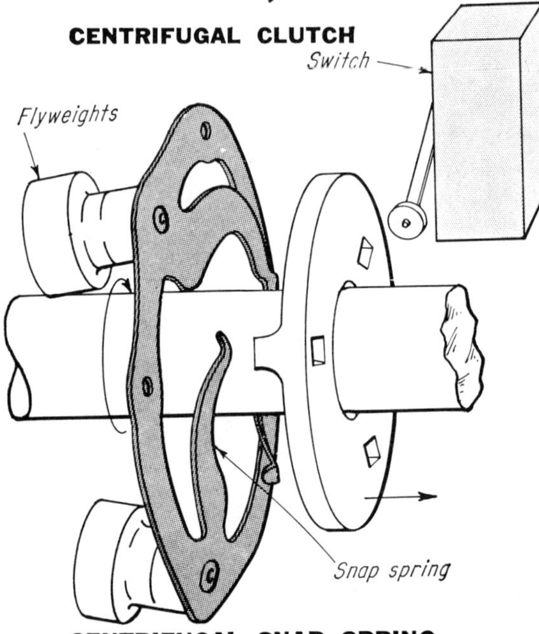

CENTRIFUGAL SNAP SPRING

Performance

BERYL A. BOGGS, Goodyear Atomic Corp,

SIMPLE GOVERNORS

	CENTRIFUGAL	PNEUMATIC	HYDRAULIC	ELECTRIC
Regulation, % full speed	1½ to 10	5 to 20	5 to 15	2 to 10
Repeatability, rpm (based on 5000 rpm)	5 to 20	20 to 50	15 to 40	10 to 30
Response time, sec	½	2	1½	½
Adjustability of operating speed	Engine stopped or running	With engine running	With engine running	Engine running
Speed range, ratio	2:1	4:1	4:1	3:1
Cost (rank: 1 to 4)	2	1 (lowest)	3	4

WITH HYDRAULIC SERVO

	CENTRIFUGAL	PNEUMATIC	HYDRAULIC	ELECTRIC
Regulation, % full speed (adjustable)	0 to 5	1 to 6	1 to 5	0 to 5
Repeatability, rpm (based on 5000 rpm)	2	10	8	2
Response time, sec (adjustable)	⅛ to 2	¼ to 2	¼ to 2	⅛ to 2
Adjustability of operating speed	With engine running			
Speed range, ratio	2:1	4:1	4:1	2:1
Cost (rank: 1 to 4)	3	1 (lowest)	2	4

Note: This table is based on simple governors; any given value can be improved. See discussion below.

Using performance table

The table above is very conservative. For instance, Woodward Governor Co offers 15% or higher speed regulation for one of its standard centrifugal governors. And if you want it, you can get speed ranges to 10:1 for the centrifugals, 15:1 for electric governors.

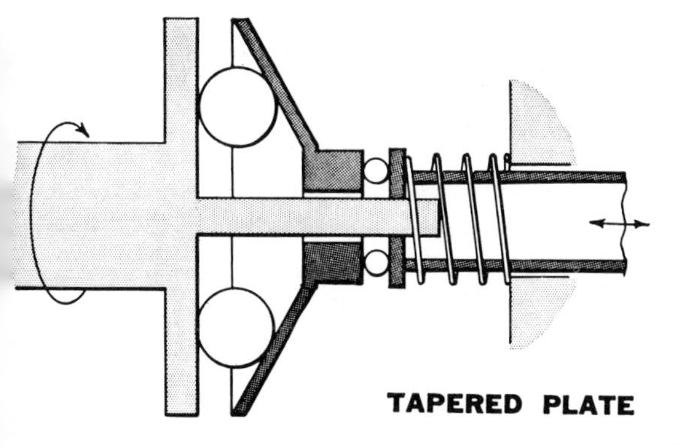

The speed sensors

By definition a speed governor is the speed-sensing part of a speed governing system. Its output — force, pressure, or voltage—is subsequently modified or amplified to do the actual controlling. The accompanying sketches of speed governors mainly are of sensors, although some include relays, servos, and other auxiliary actuating elements.

The governor title—*hydraulic, electrohydraulic,* etc—does not always indicate the speed-sensing method.

Centrifugal sensors are the most common—they are simple and sensitive and have high output force. There is more published information on centrifugal governors than on all other types combined.

In operation, centrifugal flyweights develop a force proportional to the square of the speed, modified by linkages as required. In small engines the flyweight movement can actuate the fuel throttle directly. Larger engines require amplifiers or relays, which gives rise to innumerable combinations of pilot pistons, linear actuators, dash-

Pneumatic governors

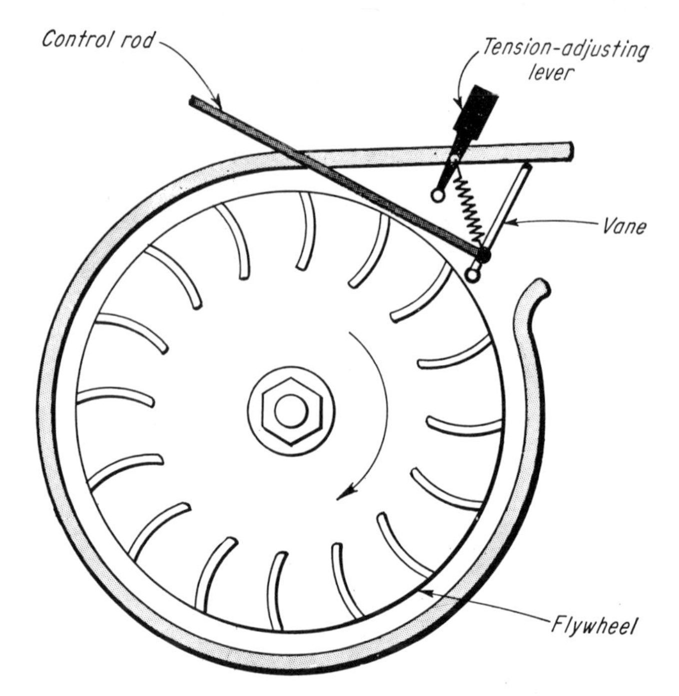

FAN-FLOW VELOCITY

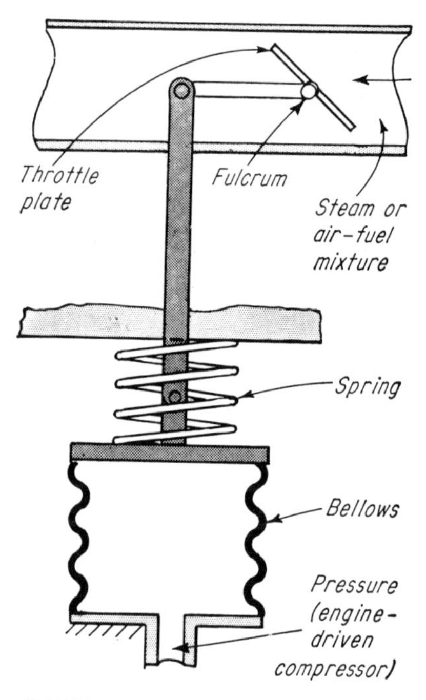

COMPRESSOR PRESSURE
(direct)

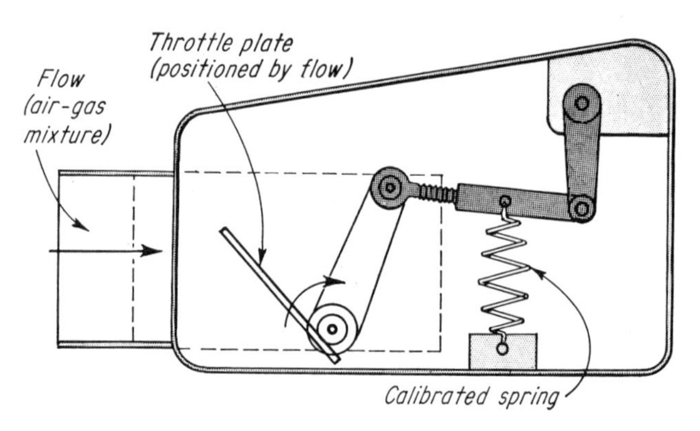

CARBURETOR-FLOW VELOCITY
(linkage)

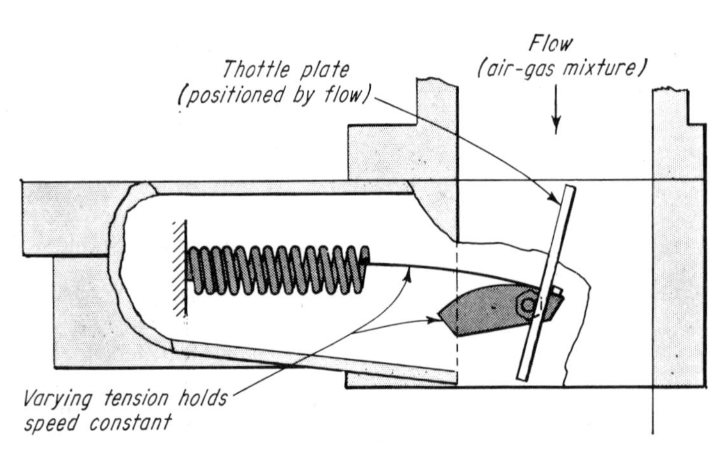

CARBURETOR-FLOW VELOCITY
(cam)

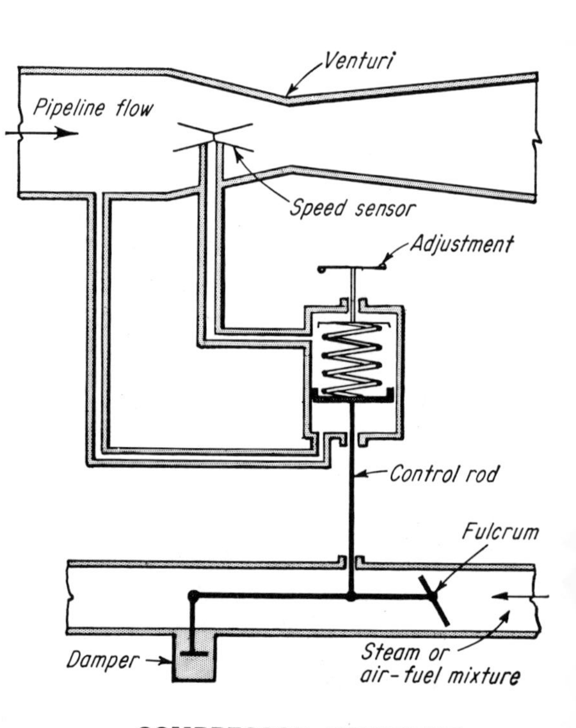

COMPRESSOR PRESSURE
(differential)

pots, compensators, and gear boxes.

Centrifugal flyweight governors have inherent speed droop, stabilizing the movement of the throttle valve. And they are simple and inexpensive. Their disadvantages are insufficient power for larger prime movers; no adjustment for droop, because speed droop is a fixed function of the regulating spring; limited variable-speed range; and friction in the linkage.

Stalled work capacity at the control arm of a centrifugal governor is the differential between the energy of the weights spinning in closed position and the energy of the weights spinning in open position. The initial setting or compression of the spring determines the speed at which the weights begin to move, and the spring rate determines the speed at which the weights are in full open position.

Where considerable power, sensitivity, and accuracy are required, a hydraulic servo system usually is added. It has a flyball head, oil pump, hydraulic relay system with a control valve and one or more amplifying pistons, and a compensating mechanism to stabilize speed changes. These adjustments are provided: speed range, speed droop, steady-state speed regulation, antihunting, and stability.

Centrifugal-hydraulic governors can use the oil to advantage in other ways, including an accumulator to reduce pump size, an oil bath for flyweights to damp their motion, and a dashpot timer to reduce hunting by setting the proper distance for the throttle to move for a given change in load. The disadvantages of oil are possible leaks, contamination, and sluggish flow at low temperatures.

One variant of the ballhead is the tapered-plate or thrust centrifugal governor. As the balls move radially outward one plate is moved axially to actuate the control. Usually there are grooves in the plates to maintain the balls in correct angular relation and dynamic balance. The grooves can be straight or spiral. Spiral grooves cause the balls to advance or retard in angular position as they move outward.

Another variant of the centrifugal governor is the inertial or acceleration governor. The flyweights sense the rate of change in speed in addition to sensing steady-state speed. The sketch of a steam engine governor on the previous page shows one way this is done. Reciprocating in a motion determined by the eccentricity of the flyweight arm, the control rod regulates a steam valve. The eccentricity changes with speed and rate of change of speed, shortening the valve response time.

Step-up gearing of differentials makes it possible to control a slow-speed shaft of less than 100 rpm by a high-speed and therefore more effective centrifugal governor. An interesting example is the differential drive on the previous page. The governor senses the difference in speed between the controlled shaft and the higher-speed shaft.

A brake-type centrifugal governor is used on some small electric motors to control speed. The brake shoe is the flyweight and is restrained by spring force as shown in the sketch. When speed is increased the brake shoe touches the housing and slows the motor. Applications include spring-driven motion picture cameras.

In a centrifugal air-bleed governor (sketch) the poppet valves are the flyweights. When the controlled speed is reached, the valves open and release air to reduce the pressure to an actuating piston, a diaphragm, or a servo valve. A centrifugal vacuum governor is similar except that the valve is held open by spring tension

Hydraulic governors

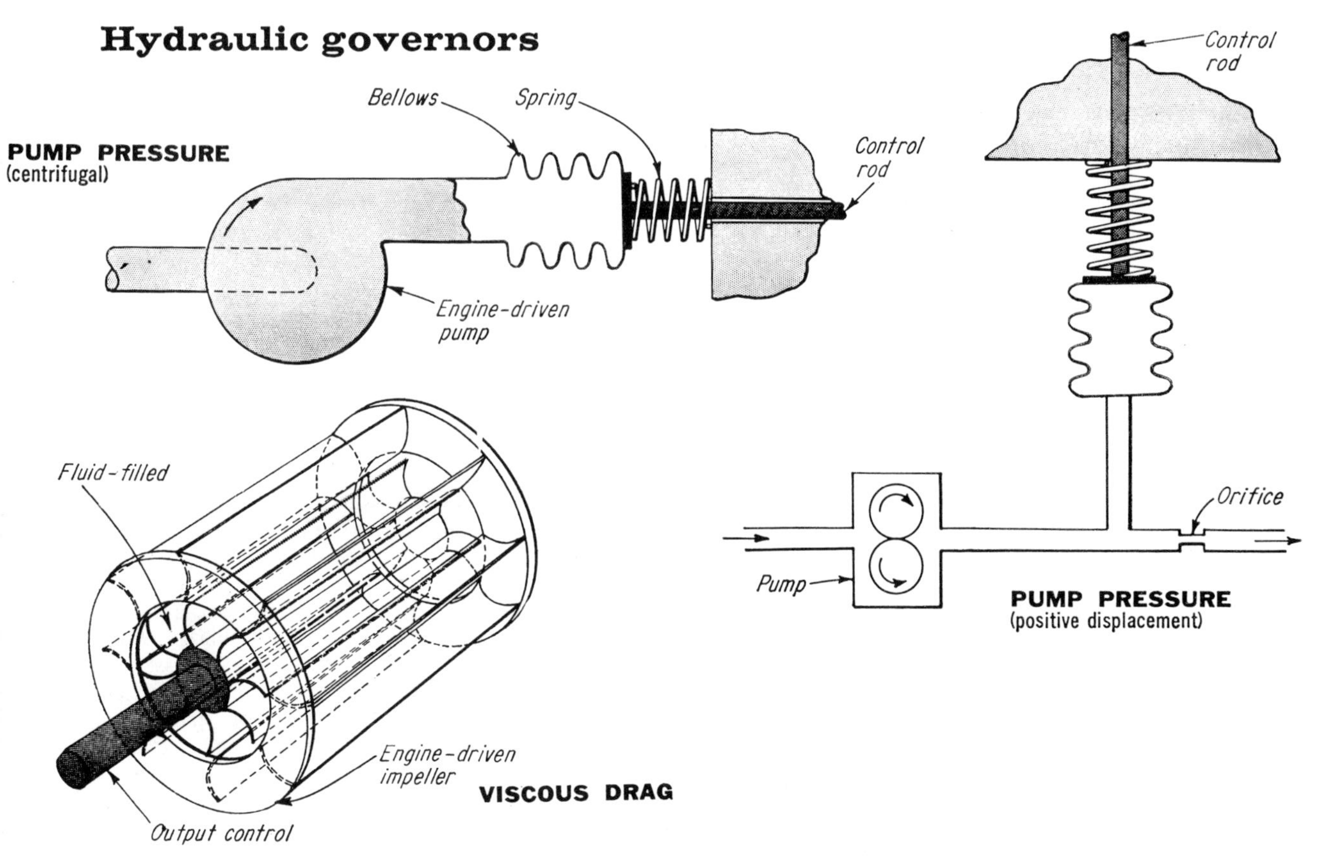

and the centrifugal force action tends to close it.

Switch contacts are the flyweights in another form of centrifugal-speed sensor. The restraining force is a leaf spring containing the contact point, and motor speed is controlled by switching the current on and off or inserting resistance in the armature circuit as the speed varies around the set point. A special form (shown in sketch on p. 282) has the centrifugal sensor separate from the contacts. At the set speed, the weights cause the carrier to snap over, moving an actuator axially to operate the contacts.

Centrifugal-switch governors are used chiefly for the starting winding cutout on single-phase electric motors. Safety protection and speed control are other possible applications.

Pneumatic sensing devices are the most inexpensive, and also the most inaccurate, of all speed-measuring and governing methods, yet they are entirely adequate for many applications. The pressure or velocity of cooling or combustion air is used to measure and govern the speed of the engine.

Best known is the vane type of air-velocity governor used on small gasoline engines **(p. 284).** The sensing vane is placed in the air stream of the engine cooling fan and is connected to the throttle so that an increase in speed, increasing air flow, will close the throttle valve.

Air-velocity governors also control the speed of larger gasoline engines. A vane or other obstruction is placed in the path of the gas-air flow between the carburetor and intake manifold, and as velocity increases the vane is moved to close the opening. In the type shown in the sketch, the force of the air-fuel mixture on the off-center throttle plate is balanced by spring tension holding the throttle in open position. Increased engine speed, and the resulting increased air-fuel velocity, tends to close the throttle. Some designs have a variable-force spring; others have an auxiliary manifold-vacuum sensor to close the governor throttle at idle speeds. One variant (sketch) achieves variable-spring force through a combination of cam and springs.

Pressure-sensing air governors utilize the static pressure of the air acting against a spring-loaded bellows, diaphragm, or valve. An ordinary pressure switch and a snap-operation valve are two of the relay techniques used. A venturi is another way: the differential pressure is sensed.

The pressure source is usually a compressor driven by the prime mover, and the pressure is approximately proportional to the square of compressor speed.

Electric governors

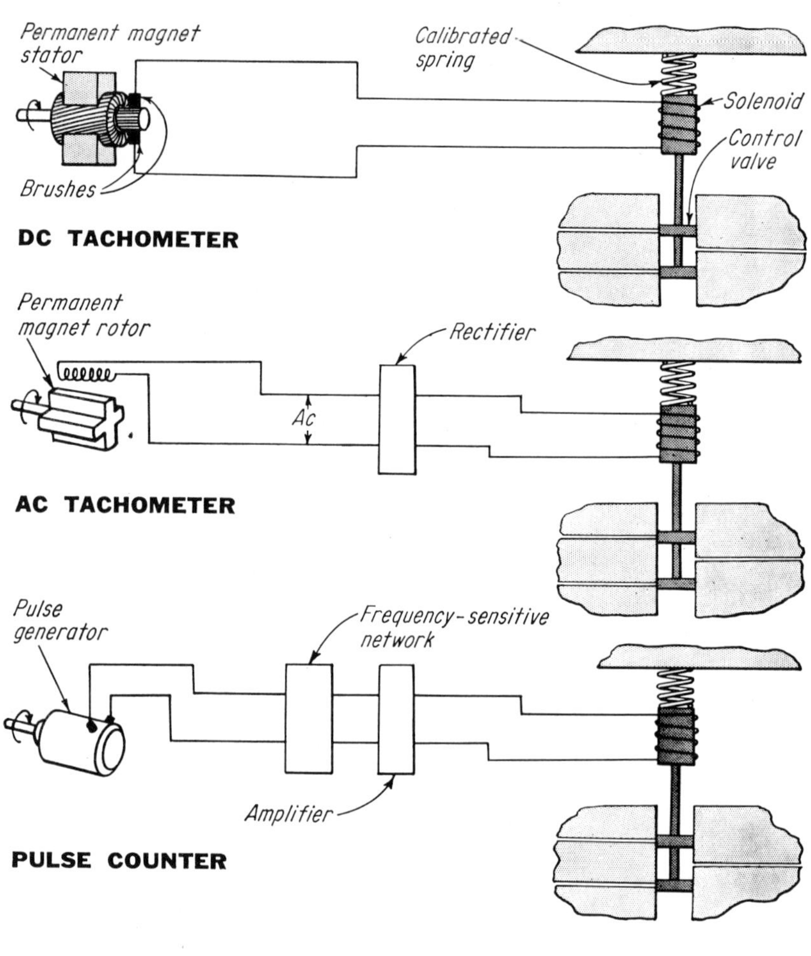

DC TACHOMETER

AC TACHOMETER

PULSE COUNTER

Typical applications of pressure-sensing governors are in process gas pumping, compressed air systems, and exhaust or inlet pressure control of steam turbines.

Pneumatic vacuum governors are common in reciprocating engines. They depend on a reduction in intake-manifold pressure as engine speed increases and an increase as engine load decreases. This vacuum is balanced against a spring-loaded diaphragm or other control element. It should be noted that if the engine has a vacuum-operated spark advance it may be necessary to correct for the effect the governor has on the vacuum sensed by the spark control.

Applied to diesel engines, a vacuum governor has a venturi in the air-intake line; the pressure differential or partial vacuum operates the control.

Hydraulic sensors

These measure discharge pressure of an engine-driven pump. Pressure is proportional to the square of the speed of the pump in most designs, although there are special impellers with linear pressure-speed characteristics.

Straight vanes are better than curved vanes because the pressure is less affected by the volume flow. Low pressures are preferred over high because fluid friction is less.

Typical applications include farm tractors using diesel or gas engines, larger diesel engines, and small steam turbines. Hydraulic-governor sensing using other than pressure has had limited application and success.

One type of hydraulic-pressure sensor is a dead-ended bellows fed by a centrifugal pump. Another is a fixed orifice fed by a constant-displacement pump; the pressure drop

ELECTRIC GOVERNOR for diesel generator senses current and frequency from three phases of alternator windings. Output is dc current to a proportional solenoid which actuates the fuel metering valve. *Made by Consolidated Controls Corp.*

across the orifice is proportional to the square of the speed (sketch).

Viscosity actuates one unusual speed-sensing device (sketch). The outer impeller is rotated by the engine; the inner impeller, connected to the throttle, is restrained by spring force.

Electric tachometers generate voltage proportional to speed; the output actuates a solenoid to move the control valve. Ac or dc designs are available. The ac version has no brushes on the rotating generator, but rectifiers must be added to the circuit. Dc tachometer generators have brushes, but no rectifiers are needed (see sketches).

In some instances an ac generator and a frequency-sensing circuit measure the speed; the output is amplified to actuate the control. In one version the sensor compares generator frequency with a reference frequency; the difference or error is amplified for throttle control.

Vibrating-contact governors are used on small dc motors to provide two regulated speeds in the 4000 to 15,000-rpm range. Each speed can be set by a separate adjusting screw. Two pairs of vibrating reed contacts control the adjusted speed.

Actually, any speed-measuring device can be used as the sensor of an electrical speed governor. This includes tachometers, speedometers, revolution counters, gear-tooth counters, pulse counters, vibrating reeds, and even stroboscopes.

Relays and servos

Amplifiers and follow-up controls, called relays or servos, add muscle where needed in many speed-governor systems. The speed sensor is relieved of heavy work, and droop can be greatly reduced.

The centrifugal-vacuum governor is an example of a sensor combined with a relay. A centrifugal valve is the pilot, admitting air to a working diaphragm with a partial vacuum on the opposite side.

Centrifugal - pneumatic governors, by the action of a centrifugal valve, release air from a balance chamber of a servo valve, moving the throttle toward the closed position.

Centrifugal - hydraulic governors have a centrifugal ballhead; a hydraulic servo piston admits or releases fluid from the throttle-actuating cylinder.

Other sensor-relay combinations include the hydraulic-relay governor, which is a hydraulic speed sensor and a hydraulic servomechanism control; electric-relay governors, with electric speed sensor and reversible electric-motor servo control; and electric-hydraulic governor, with electric speed-sensing and hydraulic servo control.

Hydraulic servo pistons are much faster than electric servomotors and are most frequently selected. Electric-hydraulic and centrifugal-hydraulic governors are very common in steam turbine and gas turbine power applications.

Example 1: gas engine for pumper

Requirements: Adjustable speed to control pipeline liquid flow. Speed changes must be slow to avoid engine stalling. No deadband is allowed because it would make control unstable, affecting pipeline and adjacent pumpers. Speed accuracy is not important. A flywheel is needed to provide kinetic energy during lag time or misfiring.

Choice of governor: Centrifugal-hydraulic, hydraulic-hydraulic, and electric-hydraulic are desirable, in that order.

The electric-hydraulic system is the least desirable because electric power is not always available or dependable at the pumping station, and a special engine-driven generator may be necessary. The hydraulic-hydraulic system is less dependent on electricity, but speed sensing is less accurate and flow variations will result. Also, force levels of most hydraulic sensors are somewhat less than for electric tachometers, and much less than for centrifugal ballheads.

The centrifugal-hydraulic system is best. It will cost less, partly because the sensing ballhead has enough force and accuracy to actuate a relatively simple hydraulic servo. The ballhead alone, however, is not sufficient—the pumper-speed control demands more actuator power than that.

Delayed response is achieved with dashpots, intermittent pumping of working oil, or flow-control valves. These are governor design problems, however.

Fuel control vs governor movement is a problem because the engine butterfly will have its own characteristics and the governor linkages must be designed accordingly. You can work this out with the governor designer. (See recommended performance.)

Torsional vibrations in the engine and gear trains must be analyzed to avoid later problems.

Special auxiliaries will be needed: solenoid shut-down for remote stopping, remote speed setting with a pneumatic or electric signal in conjunction with the pipeline control, and booster servo motors for setting fuel positions during the starting cycle.

MECHANICAL DEVICES FOR

Speed governors, designed to maintain speeds of machines within reasonably constant limits irrespective of loads, may depend for their action upon centrifugal force or cam linkages. Other types may utilize pressure differentials and fluid velocities as their actuating media. Examples of these governing devices are illustrated.

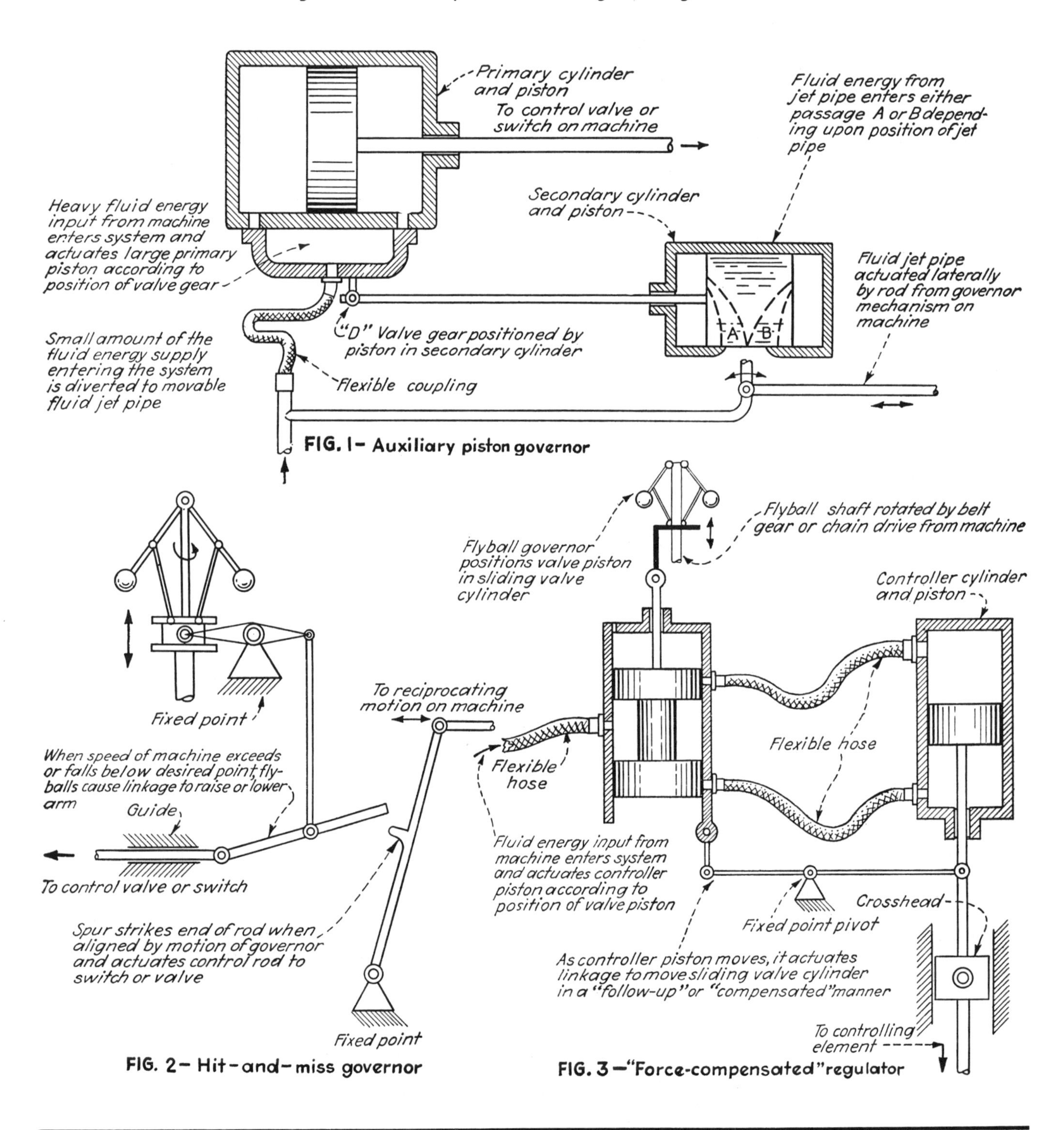

FIG. 1— Auxiliary piston governor

FIG. 2— Hit-and-miss governor

FIG. 3—"Force-compensated" regulator

AUTOMATICALLY GOVERNING SPEED

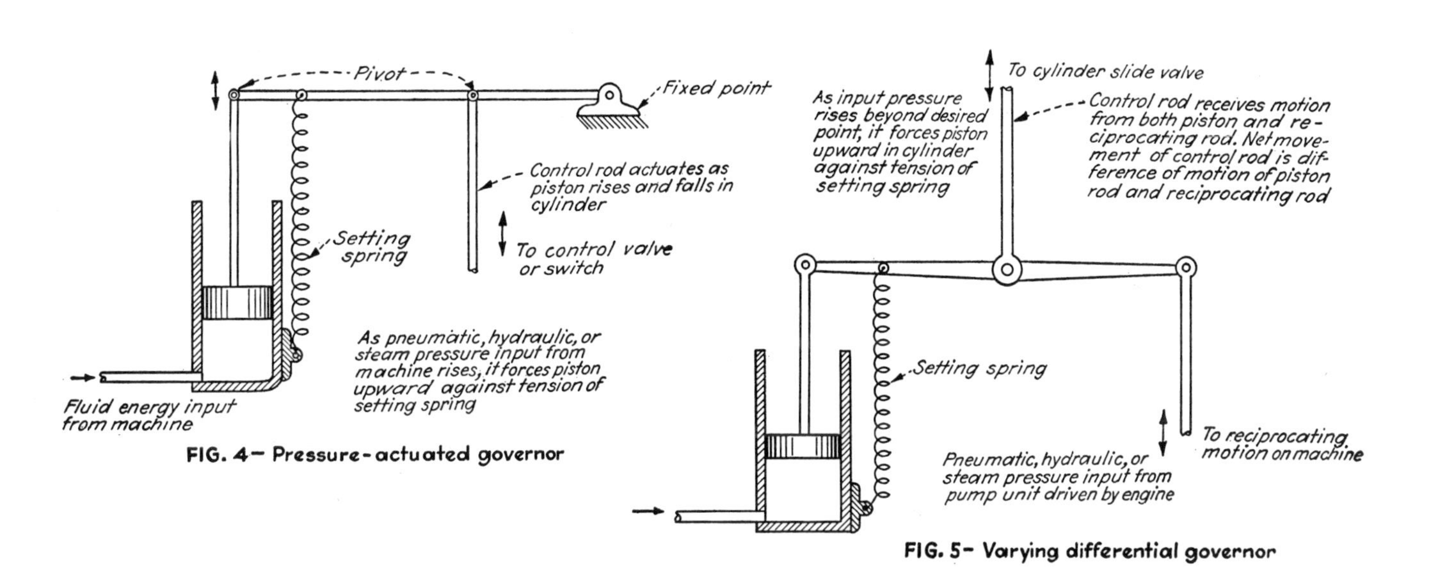

FIG. 4— Pressure-actuated governor

FIG. 5— Varying differential governor

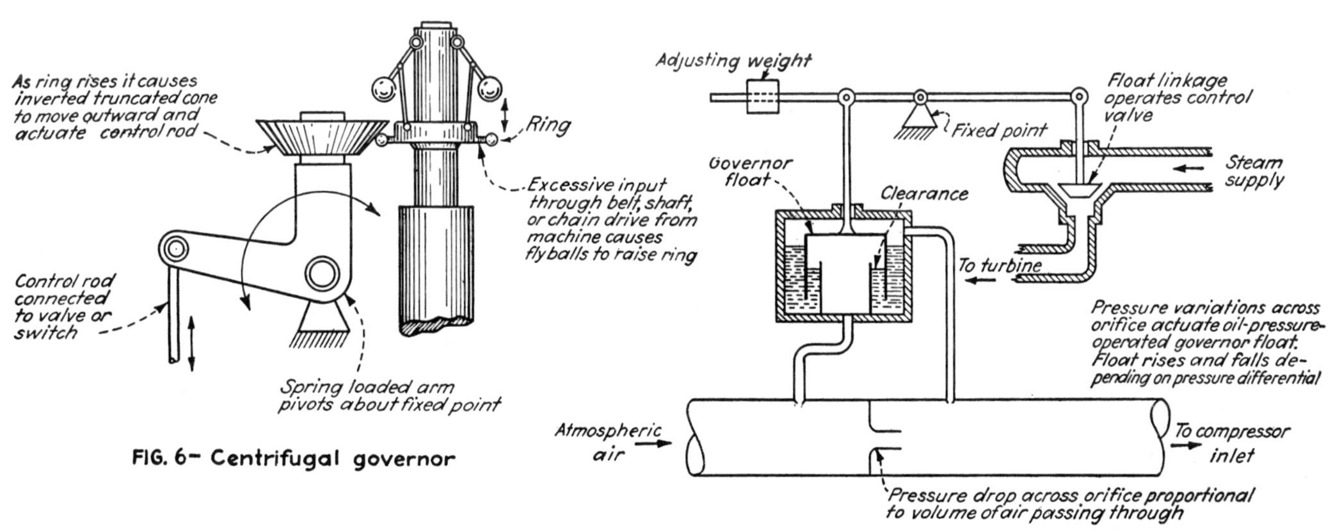

FIG. 6— Centrifugal governor

FIG. 7— Constant volume governor

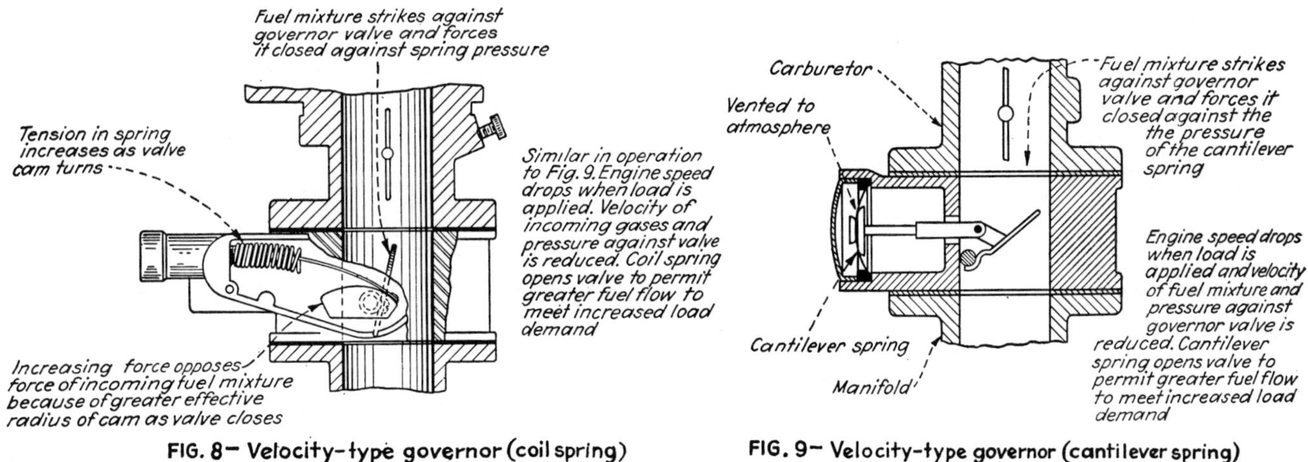

FIG. 8— Velocity-type governor (coil spring)

FIG. 9— Velocity-type governor (cantilever spring)

How to Obtain Constant Speed

ACCURATE METHODS for controlling constant speed devices operating at extremely low linear speeds (less than 1 in. per sec) are the basis of the development of a slow speed photographic recording machine. Attainment of constant linear velocity has presented a problem in the recording art since Edison invented the phonograph. The rigorous analysis of constant speed regulation—presented on the following pages—was made necessary by the requirements of geophysical exploration, in which time of arrival of seismic waves, excited by explosives, must be accurately recorded. The accuracy of the recorded information depends on the linear velocity of the medium, which is a function of the speed regulation of the drive mechanism.

At angular velocities below 10 rpm (1 in. per sec linear

METHOD 1—Flywheel with capstan string—drive from synchronous motor.

SPEED REGULATION: Excellent—most constant speed device in recording industry. Accuracy, 0.02 percent.
SYNCHRONISM: Not exact, since there is some slippage in string belt.
REFERENCE STANDARDS:
Primary—tuning fork; accuracy, 1 part in 500,000.
Secondary—synchronous motor; accuracy, 1 part in 50,000.
EFFECTS CONTRIBUTING TO ERROR:
Primary—flywheel bearing rumble; flywheel oscillation.
Secondary—synchronous motor bearings; standard frequency source; synchronous motor angular error; bearing eccentricities.
SIZE AND WEIGHT: Flywheel effect demands prohibitive mass.
NECESSARY SERVICING IN FIELD: Replacing string belt; electronic components; bearings; lubrication.
MANUFACTURING DIFFICULTIES: Few.
REMARKS: Used in recording to generate constant frequency tone of 0.05 percent accuracy.

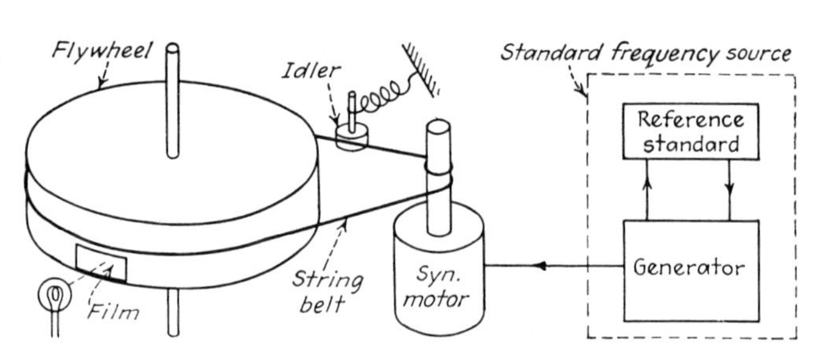

METHOD 2—Magnetic low pass torsional filter with worm gear reduction from synchronous motor.

SPEED REGULATION: Excellent—accuracy 0.01 percent.
SYNCHRONISM: Exact with secondary standard.
REFERENCE STANDARDS:
Primary—tuning fork; 1 part in 500,000.
Secondary—synchronous motor; accuracy, 1 part in 50,000.
EFFECTS CONTRIBUTING TO ERROR:
Primary—flywheel bearing rumble; flywheel oscillation; low pass filter characteristic.
Secondary—standard frequency source; reduction gear eccentricity; bearing eccentricities; magnetic filter eccentricities; synchronous motor angular error.
SIZE AND WEIGHT: Appreciable flywheel weight and size—about 10 in. dia and 35 lb for 1 cycle low frequency cutoff.
NECESSARY SERVICING IN FIELD: Electronic components; lubrication.
MANUFACTURING DIFFICULTIES: Precision cone type gear; magnetic filter symmetry; precision bearings.
REMARKS: Used in optical dial division to divide circle accurate to 0.01 percent. Eddy current effect damping can be used in combination with the hystereisis effect compliance of the magnetic low pass filter to provide superior filter characteristic. Loaded cone type worm gear is expensive.

Method 3—Flywheel—rim drive by roller in magnetic circuit from gear reduction and synchronous motor.

SPEED REGULATION: Fair—accuracy 0.2 percent.
SYNCHRONISM: Not synchronous because of slippage.
REFERENCE STANDARDS: Timing track necessary.
EFFECTS CONTRIBUTING TO ERROR:
Primary—wear of roller; flexure of shaft bearings.
Secondary—gear eccentricities.
SIZE AND WEIGHT: Appreciable size and weight.
NECESSARY SERVICING IN FIELD: Lubrication of drive components.
MANUFACTURING DIFFICULTIES: Precision gears; precision bearings.
REMARKS: Simple and effective low pass filter action of rim drive irons out gear ripple and eccentricities.

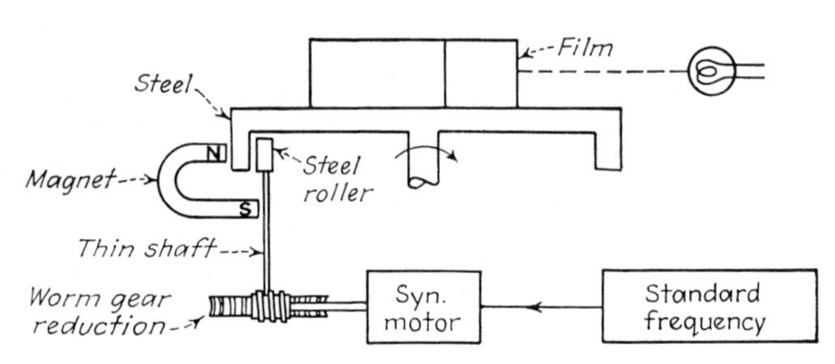

Motion Below 10 Rpm

WILLIAM HOTINE
Potter Instrument Company

speeds), a desired percentage accuracy of speed regulation is much more difficult to maintain than at comparatively high speeds. One reason for the difficulty can be attributed to the decreasing flywheel effect.

Another limitation in obtaining speed control at low velocities is created by the accuracy of frequency or time standards. The most accurate practical reference standard is the constant frequency signal broadcast by the National Bureau of Standards. This signal has an accuracy in the order of one part in 50 million. All of the techniques listed are based, directly or indirectly, on this standard.

In the following list of constant speed control techniques, it should be noted that the most inaccurate method is the one utilizing a mechanical drive.

METHOD 4—Synchronous friction drive. Servomechanism adjusts variable ratio friction drive to be in synchronism with reference gears.

Speed Regulation: Good—accuracy 0.1 percent.
Synchronism: In synchronism with secondary standard within 1 part in 1,000.
Reference Standards:
  Primary—synchronous motor.
  Secondary—reference gears.
Effects Contributing to Error:
  Primary—bearings; gear eccentricities.
  Secondary—friction drive eccentricity; synchronous motor.
Size and Weight: Fairly bulky.
Necessary Servicing in Field: Lubrication of drive components.
Manufacturing Difficulties: Precision machining required; precision gears; precision bearings.
Remarks: Quick starting. High torque.

Cone-type friction drive
Cam on pinion gear
Cam follower pin
Reference gear
Reference pinion, can rotate on shaft
Bellows coupling
Syn. motor
Standard frequency
Worm gear reduction

METHOD 5—Electronically controlled d-c motor. Optical pickup of mechanically generated rotating standard frequency record compared with reference standard frequency regulates d-c motor current.

Speed Regulation: Excellent—accuracy 0.01 percent.
Synchronism: In synchronism with standard to mechanical accuracy of optical tone generator.
Reference Standards:
  Primary—tuning fork; accuracy 1 part in 500,000.
  Secondary—tone generator; mechanical accuracy 1 part in 10,000.
Effects Contributing to Error:
  Primary—tone generator segment spacing accuracy and uniformity; temperature effects.
  Secondary—d-c motor bearings; d-c motor commutation; resolution of optical system; time constants of masses and regulating circuit.
Size and Weight: Light and compact.
Necessary Servicing in Field: Motor Brushes; electronic components.
Manufacturing Difficulties: Manufacture of mechanical tone generator to required tolerance; low inertia d-c motor.

Optical system
Photo cell
Lamp
Amp.
Standard frequency source
Phase det.
Reference standard
Film
Tone generator-grating lines on film
d-c motor
Regulated d-c output

METHOD 6—Magnetic recording controlled motor. Standard frequency is magnetically recorded on ferro-magnetic material. Picked up 360 deg out of phase and applied to phase comparator, d-c output of which regulates motor speed.

Speed Regulation: Excellent—accuracy, 0.01 percent.
Synchronism: Synchronous.
Reference Standards:
  Primary—tuning fork; accuracy 1 part in 500,000.
  Secondary—tone frequency recorded.
Effects Contributing to Error:
  Primary—definition of magnetic aperture and of ferro-magnetic material; main bearing eccentricity.
  Secondary—temperature.
Size and Weight: Fairly bulky.
Necessary Servicing in Field: Mainly electronic.
Manufacturing Difficulties: High precision machining.
Remarks: Very elaborate electronic equipment—high power consumption.

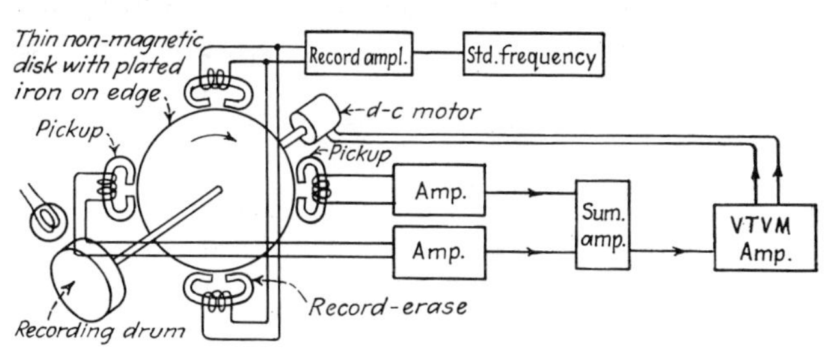

(continued on next page)

How to Obtain Constant Speed Motion Below 10 Rpm (continued)

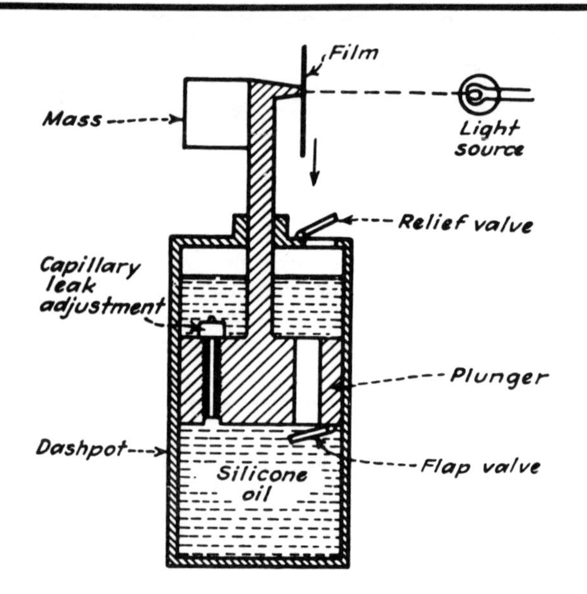

METHOD 7—Oil dashpot—falling mass controlled by oil viscosity.

Speed Regulation: Poor—accuracy 2 percent.

Synchronism: Not synchronous.

Reference Standards: Timing track necessary.

Effects Contributing to Error:
Primary—Variation in oil viscosity; temperature; friction, vibration; turbulence.
Secondary—wear.

Size and Weight: Small and compact.

Necessary Servicing in Field: Little.

Manufacturing Difficulties: High precision fits and fine finishes on some components.

Remarks: Viscosimeter art applies. Not used at present in high precision timing.

METHOD 8—Magnetic dashpot—falling mass controlled by hysteresis and eddy current effects.

Speed Regulation: Fair—accuracy 0.2 percent.

Synchronism: Not synchronous.

Reference Standards: Timing track necessary.

Effects Contributing to Error:
Primary—control voltage; vibration; friction.
Secondary—flexure; windage.

Size and Weight: Small and compact.

Necessary Servicing in Field: Little.

Manufacturing Difficulties: High precision machining; uniformity of cylinder material important.

Remarks: Considerably superior to oil dashpot. Inverse feedback used to improve linearity.

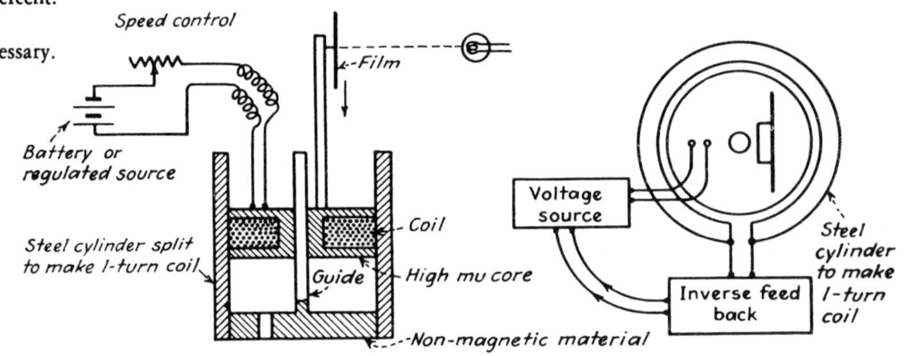

METHOD 9—Lathe—moving element driven by screw from synchronous motor.

Speed Regulation: Poor—accuracy 2 percent.

Synchronism: Not exact, affected by mechanical non-linearity.

Reference Standards:
Primary—tuning fork; accuracy 1 part in 500,000.
Secondary—lead screw; accuracy 1 part in 50,000.

Effects Contributing to Error:
Primary—backlash and wear in guides; nuts; thrust bearing; play in bearing; non-linearity of screw; wear on screw; variable friction; temperature effects.
Secondary—synchronous motor bearings; standard frequency source; synchronous motor error.

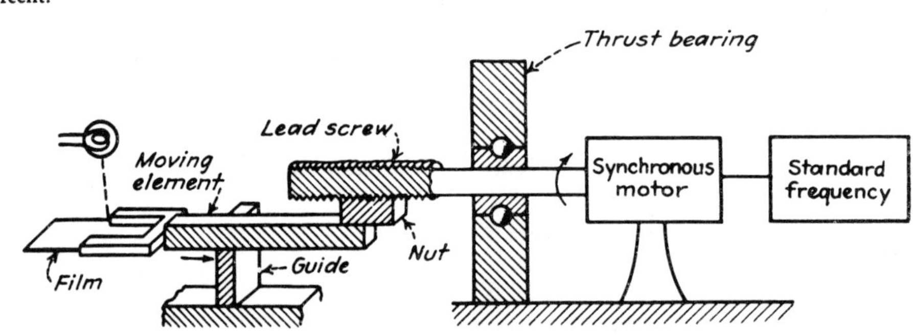

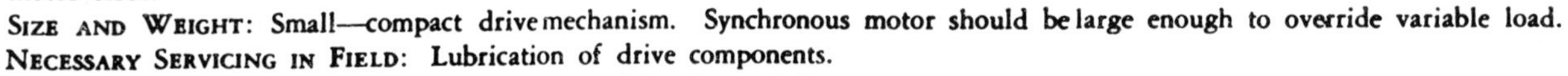
Size and Weight: Small—compact drive mechanism. Synchronous motor should be large enough to override variable load.

Necessary Servicing in Field: Lubrication of drive components.

Manufacturing Difficulties: Very high precision fits and fine finishes on drive components; precision bearings, temperature compensation construction.

Remarks: Best lead screw manufacturing practice introduces errors up to 0.01 percent.

METHOD 10—Galvanometer—mirror driven by moving coil.

SPEED REGULATION: Poor—accuracy 0.5 percent.
SYNCHRONISM: In synchronism with secondary standard, within 1 part in 5,000.
REFERENCE STANDARDS: Timing track necessary.
EFFECTS CONTRIBUTING TO ERROR:
Primary—non-linear magneto-motive torque; non-linear shaper amplifier; variable friction in bearing; play in bearings; flexure; windage, optical amplification.
Secondary—Standard frequency source.
SIZE AND WEIGHT: Small and compact; may require much electronic equipment and service.
NECESSARY SERVICING IN FIELD: Electronic components.
MANUFACTURING DIFFICULTIES: High precision required; delicate assembly.
REMARKS: Small mechanical errors in moving system multiplied by optical system. Optical difficulties. Electronic difficulties.

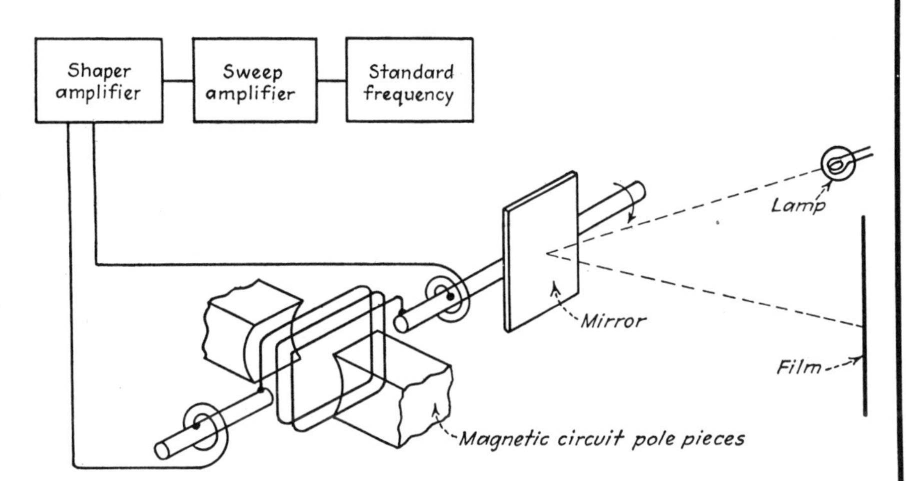

METHOD 11—Electrostatic voltmeter—mirror driven by moving vane.

SPEED REGULATION: Poor—accuracy 0.5 percent.

SYNCHRONISM: In synchronism with secondary standard within 1 part in 100.

REFERENCE STANDARDS: Timing track necessary.

EFFECTS CONTRIBUTING TO ERROR:
Primary—Vibration; flexure; windage.
Secondary—standard frequency source; linearity of variable voltage.

SIZE AND WEIGHT: Fairly bulky.

NECESSARY SERVICING IN FIELD: Electronic only.

MANUFACTURING DIFFICULTIES: High precision and delicate workmanship required.

REMARKS: Would need to be carefully handled. Small errors in moving system multiplied by optical system. Optical difficulties.

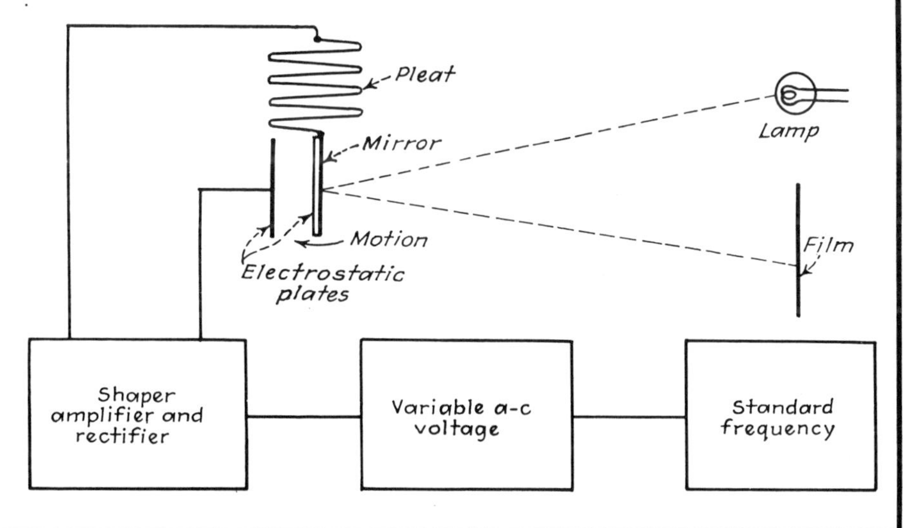

METHOD 12—Cathode ray recorder, flying spot type—raster (scan pattern) embodying desired line pattern intensity modulated by control grid of cathode ray tube.

SPEED REGULATION: Good—accuracy 0.1 percent.

SYNCHRONISM: Exact with secondary standard.

REFERENCE STANDARDS:
Primary—tuning fork.
Secondary—sweep circuit.

EFFECTS CONTRIBUTING TO ERROR:
Primary—non-linearity of sweep; stability of sweep components; thermal stability of cathode ray tube.
Secondary—halation on cathode ray tube face; spot size; optical system.

SIZE AND WEIGHT: Bulky and heavy.

NECESSARY SERVICING IN FIELD: Electronic only.

MANUFACTURING DIFFICULTIES: Linear sweep; cathode ray tube spot size; halation.

REMARKS: Unaffected by vibration; non-linearity may cancel out if same sweep is used for playback.

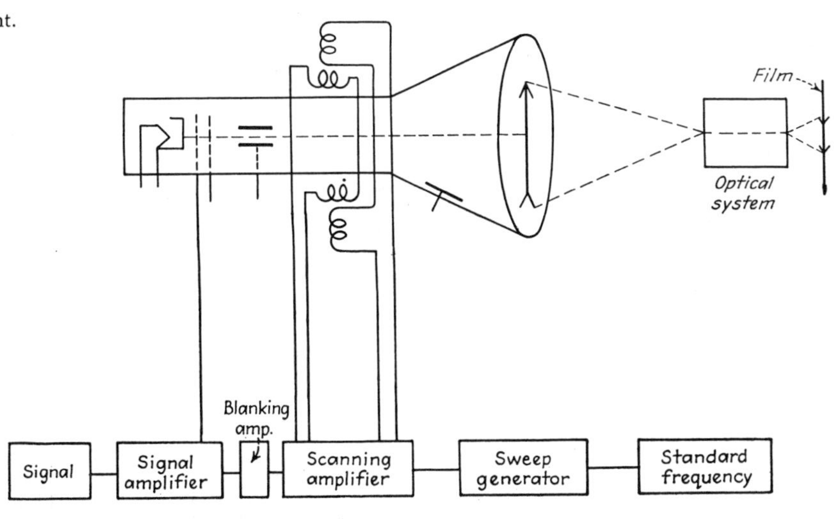

Governing Modifications for Engines

EDGAR J. KATES

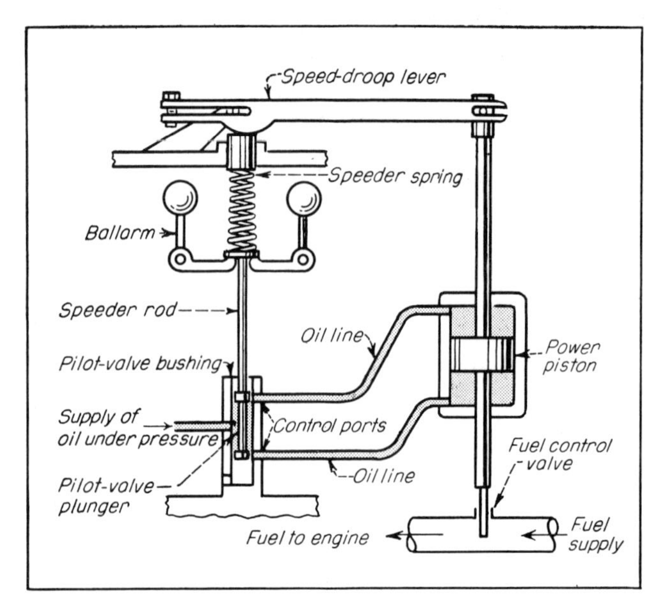

Hydraulic governor with speed droop lever to improve stability. Power piston multiplies limited output of sensing element so that a small fly ball can control rather large prime movers.

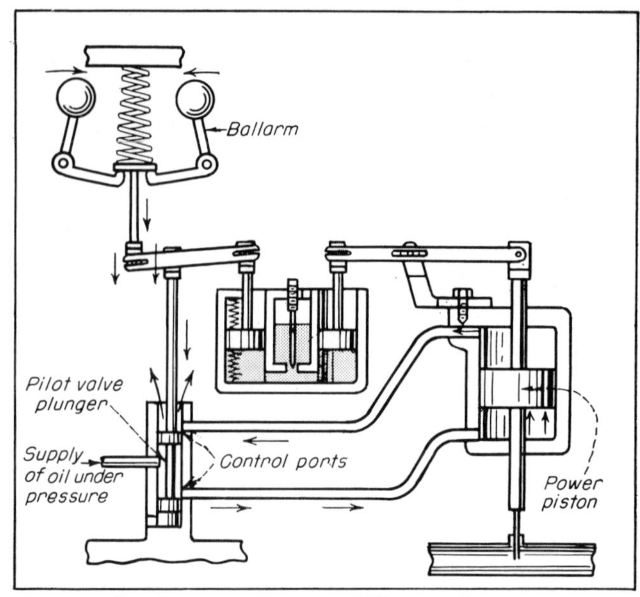

Synchronous governor. When load increases, speed decreases: fly balls move in, speeder rod moves downward opening pilot valve so it delivers oil under power piston. Then droop is applied and removed.

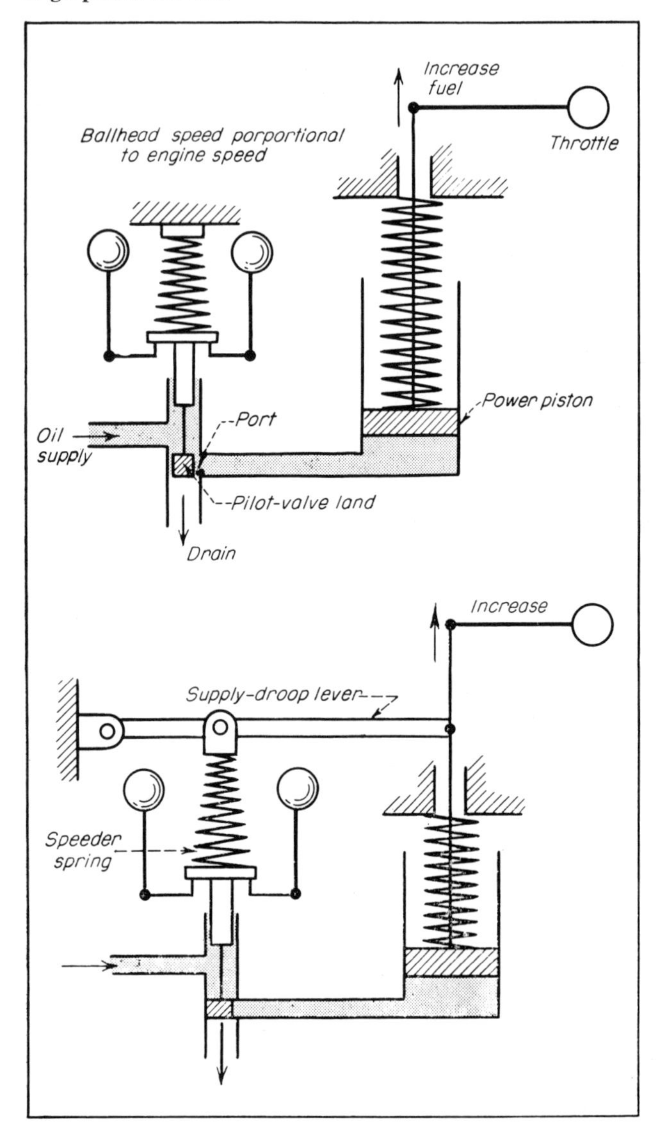

Comparison of hydraulic governors with (below) and without (above) permanent speed droop.

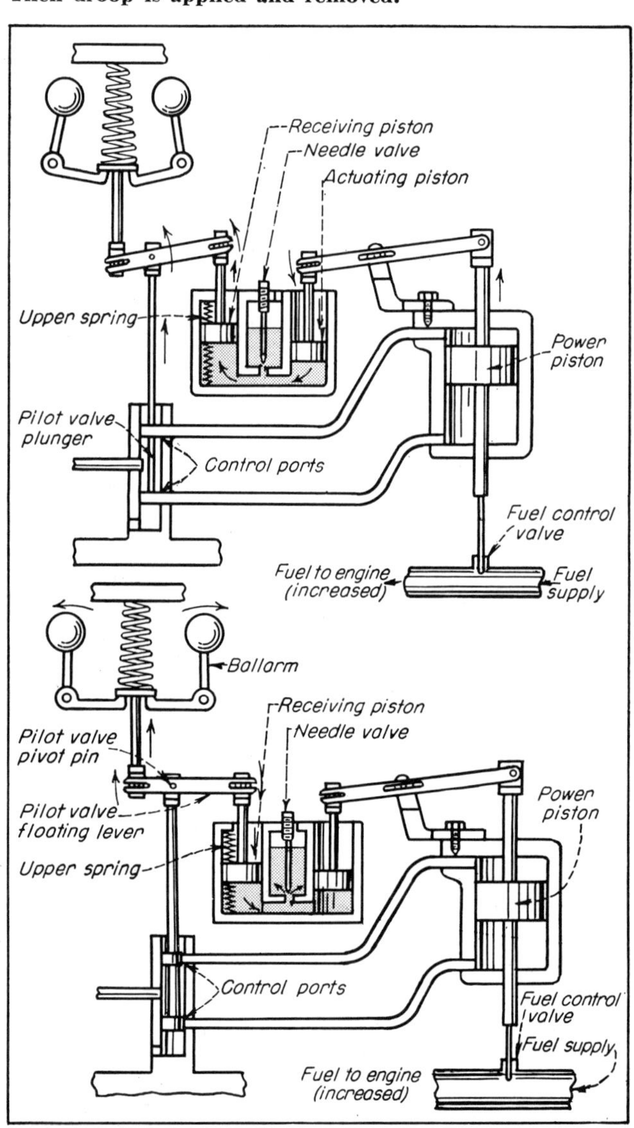

Applying droop (above): power piston moves up. Removing droop (below): fly balls straighten.

Ball-Type Transmission Is Self-Governing

The Gerritsen transmission, developed in England at the Tiltman Langley Laboratories Ltd., Redhill Aerodrome, Surrey, governs its own output speed within limits of plus/minus one percent. The usual difficulties of speed governing—lack of sensitivity, lag and hunting—associated with separate governor units are completely obviated because regulation is effected directly by the driving members through their own centrifugal force. The latter are precision bearing-steel balls that roll on four hardened-steel, cone-shaped rings, and these members can be used for different ratio arrangements.

The transmission can be used in three different ways: as a fixed "gear", as an externally controlled variable speed unit or as a self-governing drive that produces a constant output speed from varying input speeds.

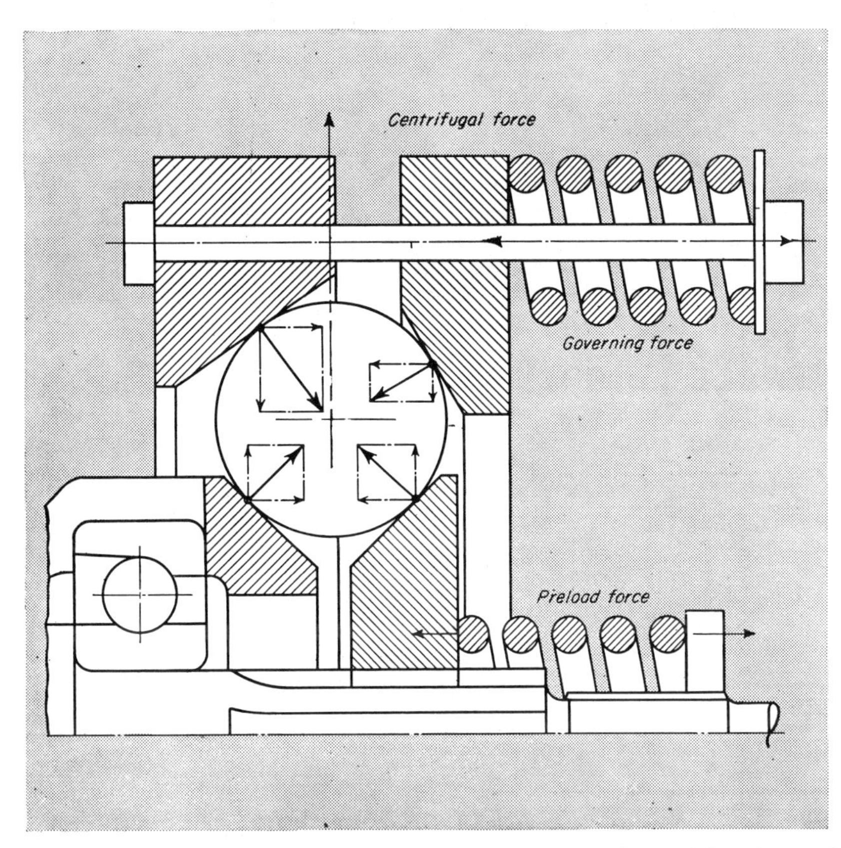

SELF-GOVERNING ACTION of the transmission is obtained by virtue of the centrifugal forces of the balls as they rotate. When the balls move outward radially, the input-output ratio changes. By properly arranging the rings and springs, the gear ratio can be controlled by the movement of the balls to maintain a constant value of output speed.

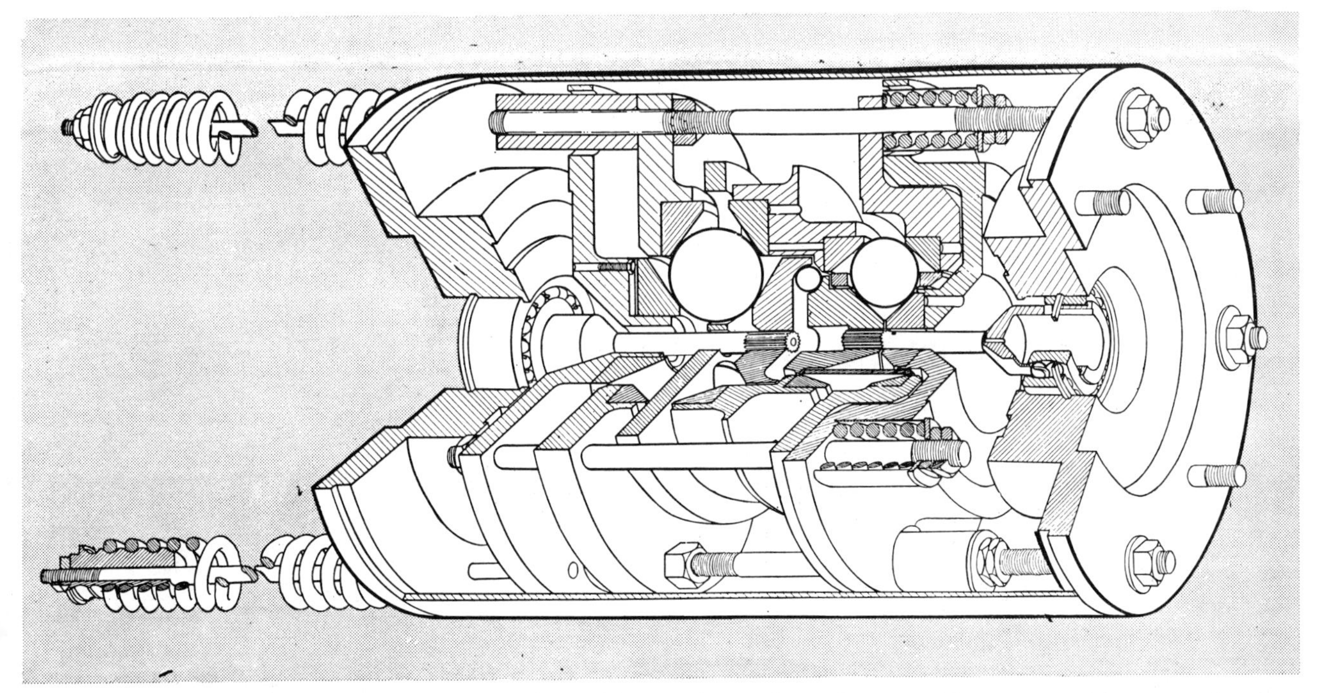

TWO-STAGE TRANSMISSION is typical of a self-governing unit used to provide constant-cycle power in alternator drives. Constant output speed of 8,000 rpm is maintained even though input speeds vary from 9,000 to 15,000 rpm. Axial movement of the coned surfaces that are in contact with the balls is controlled by the external governing springs. The two stage design has the advantage of increasing the input speed range using only one main thrust race between the input and output shafts, so it has only to cope with their relative speeds. Capacity is four horsepower.

Mechanical Systems for Controlling Tension and Speed

J. H. GEPFERT
Reeves Pulley Company

THE KEY TO THE SUCCESSFUL OPERATION of any continuous processing system that is linked together by the material being processed is positive speed synchronization of the individual driving mechanisms. Typical examples of such a system are steel strip lines, textile equipment, paper machines, rubber and plastic processers and printing presses. In each of these cases, the material will become wrinkled, marred, stretched or otherwise damaged if precise control is not maintained.

The automatic control for such a system contains three basic elements: The signal device or indicator, which senses the error to be corrected; the controller, which interprets the indicator signal and amplifies it, if necessary, to initiate control action; and the transmission, which operates from the controller to change the speed of the driving mechanism to correct the error.

Signal indicators for continuous systems fall into two general classifica-

TABLE I — PRIMARY INDICATORS

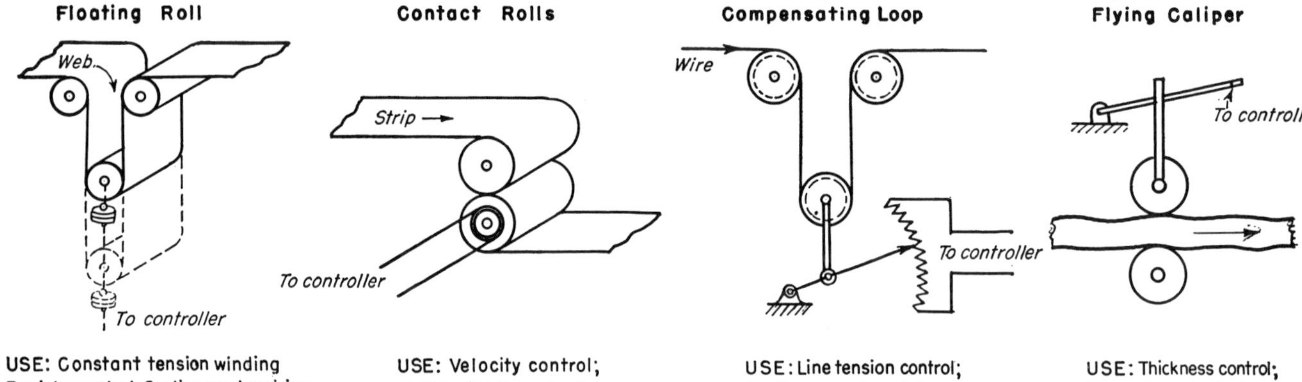

USE: Constant tension winding Registry control; Section synchronizing. | USE: Velocity control; Cutter feed control. | USE: Line tension control; Section synchronizing. | USE: Thickness control; Diameter control.

TABLE II — SECONDARY INDICATORS

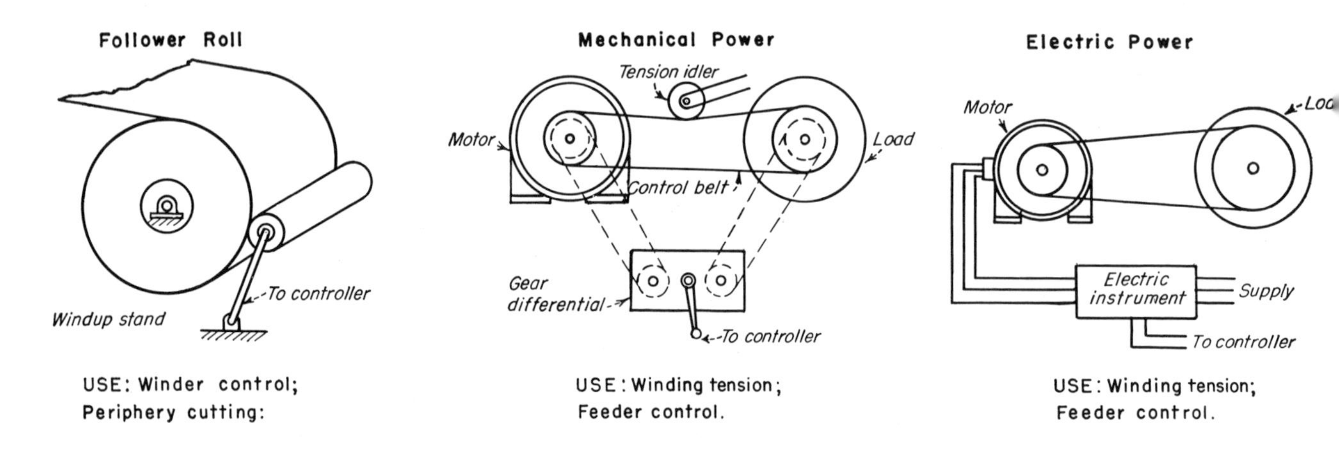

USE: Winder control; Periphery cutting: | USE: Winding tension; Feeder control. | USE: Winding tension; Feeder control.

TABLE III — CONTROLLERS AND ACTUATORS

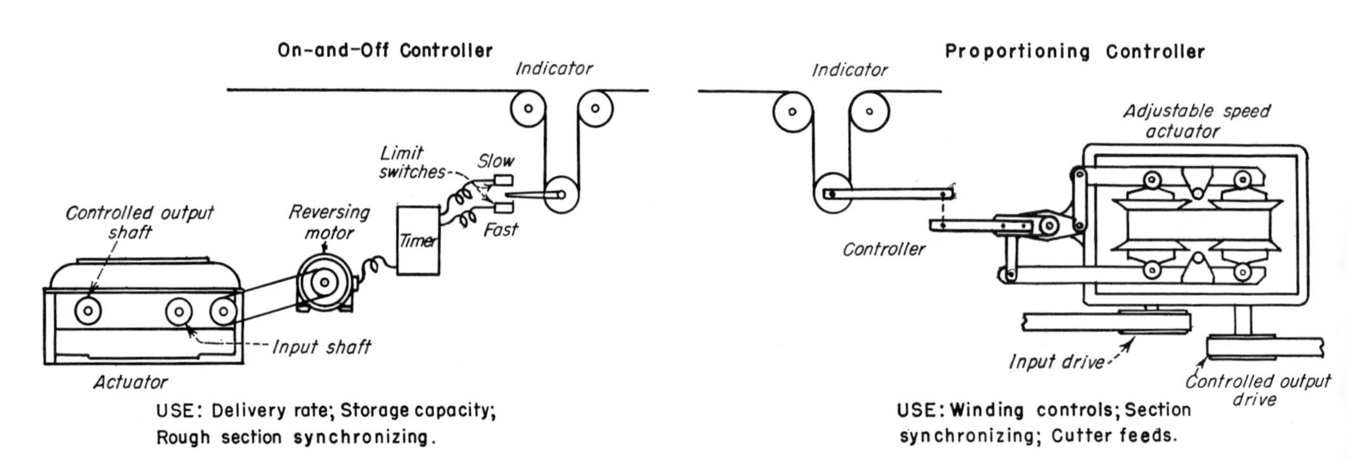

USE: Delivery rate; Storage capacity; Rough section synchronizing. | USE: Winding controls; Section synchronizing; Cutter feeds.

tions: Primary indicators that measure the change in speed or tension of the material by direct contact with the material; and secondary indicators, which measure a change in the material from some reaction in the system that is proportional to the change.

The primary type is inherently more accurate because of its direct contact with the material. These indicators take the form of contact rolls, floating or compensating rolls, resistance bridges and flying calipers, as illustrated in Table I. In each case, any change in the tension, velocity or pressure of the material is indicated directly and immediately by a displacement or change in position of the indicator element. The primary indicator, therefore, shows deviation from an established norm, regardless of the factors that have caused the change.

Secondary indicators, shown in Table II, are used in systems where the material cannot be in direct contact with the indicator or when the space limitations of a particular application make their use undesirable. This type indicator introduces into the control system a basic inaccuracy which is the result of measuring an error in the material from a reaction that is not exactly proportional to the error. The control follows the summation of the errors in the material and the indicator itself.

The controlling devices, which are operated by the indicators, determine the degree of speed change required to correct the error, the rate at which the correction must be made, and the stopping point of the control action after the error has been corrected. The manner in which the corrective action of the controller is stopped determines both the accuracy of the control system and the type of control equipment required.

Three general types of control action are illustrated in Table III. Their selection for any individual application is based on the degree of control action required, the amount of power available for initiating the control, that is: the torque amplification required, and the space limitations of the equipment.

The on-and-off control with timing action is the simplest of the three types. It functions on the basis that, when the indicator is displaced, the timer contact energizes the control in the proper direction for correcting the error. The control action continues until the timer stops the action. After a short interval, the timer again energizes the control system and, if the error still exists, control action is continued in the same direction. Thus, the control process is a step-by-step action to make the correction and to stop operation of the controller.

The proportionate type of controller corrects an error in the system, as shown by the indicator, by continuously adjusting the actuator to a speed that is in exact proportion to the displacement of the indicator. The diagram in Table III shows the proportionate controller in its simplest form as a direct link connection between the indicator and the actuating drive. However, the force amplification between the indicator and the drive is relatively low and hence limits this controller to applications where the indicator has sufficient operating force to adjust the speed of the vari-

Fluid Pressure

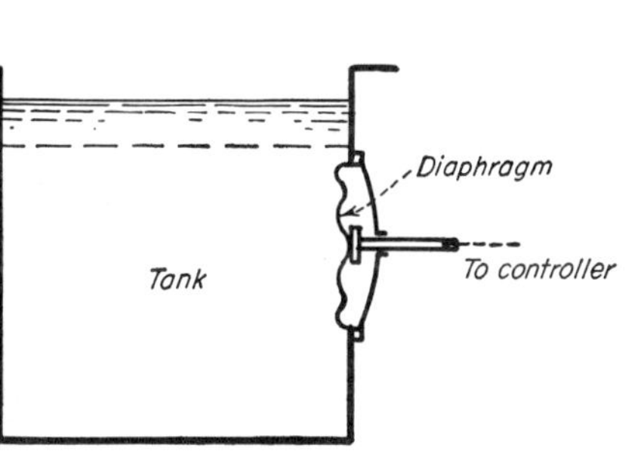

E: Fluid level control; Constant ssure control; Filtering rate control.

Liquid Level

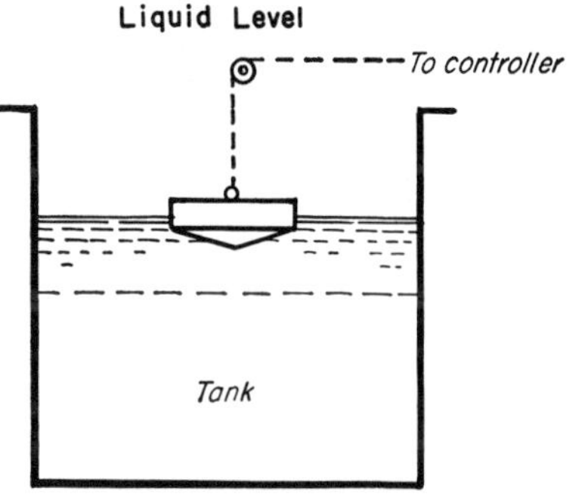

USE: Pumping rate control; System pressure control.

Temperature **Pressure**

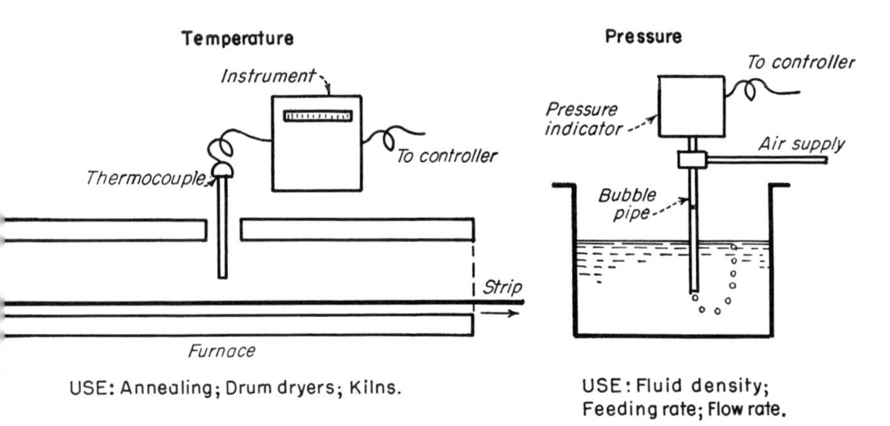

USE: Annealing; Drum dryers; Kilns.

USE: Fluid density; Feeding rate; Flow rate.

Proportioning-Throttling Controller

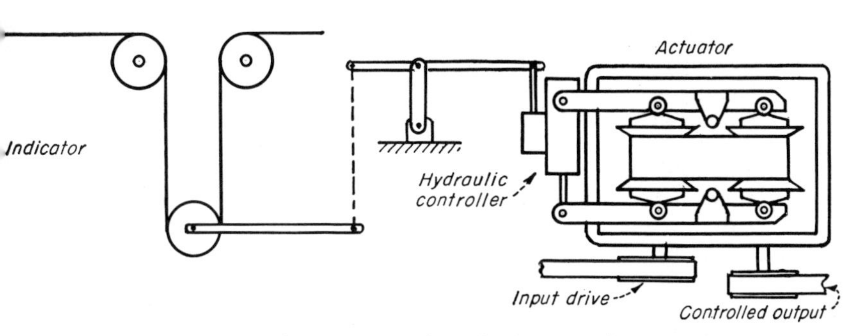

USE: Constant tension winding; Registry control; Exact section synchronizing.

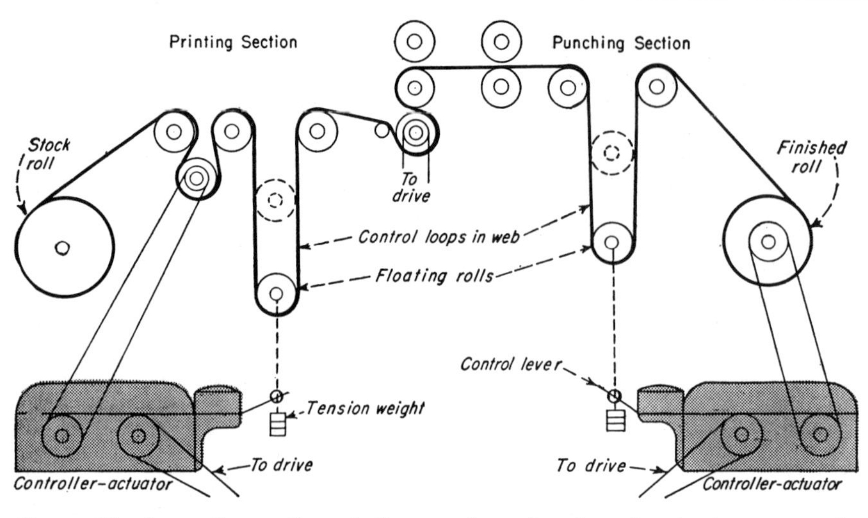

Fig. 1—Floating rolls are direct indicators of speed and tension in the paper web. Controller-actuators adjust feed and windup rolls to maintain registry during printing.

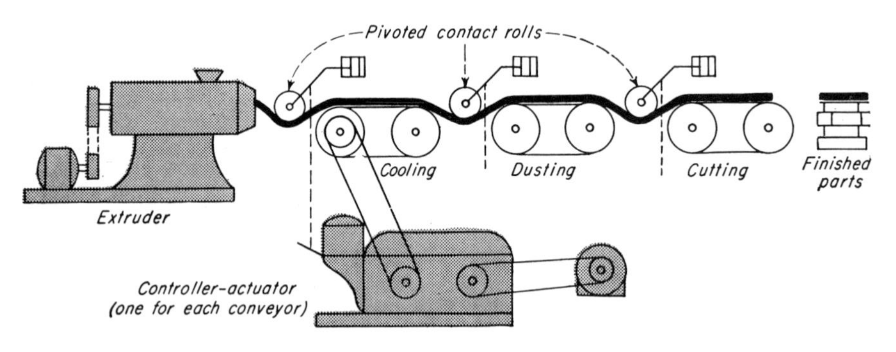

Fig. 2—Dimension control of extruded materials calls for primary indicators like the contact rolls shown. Their movements actuate conveyor control mechanisms.

able speed transmission directly.

The most accurate controller is the proportioning type with throttling action. Here, operation is in response to the rate of error indication. This type of controller, as shown in Table III, is connected to a throttling valve, which operates a hydraulic servo-mechanism for adjusting the variable speed transmission.

The throttling action of the valve provides a slow control action for small error correction or for continuous correction at a slow rate. For following large errors, as shown by the indicator, the valve opens to the full position and makes the correction as rapidly as the variable speed transmission will allow.

Many continuous processing systems can be automatically controlled by using a package unit consisting of a simple, mechanical, variable speed transmission and an accurate hydraulic controller.

This controller-transmission package can be used to change the speed relationship at the driving points in the continuous system from any indicator that signals for correction by a displacement. It has anti-hunting characteristics because of the throttling action on the control valve, and is self-neutralizing because the control valve is part of the transmission adjustment system.

An example of a continuous processing system that requires automatic control is the rotary type printing press. When making billing forms on such a press, the printing plates are rubber and the forms are printed on a continuous web of paper. The paper varies in texture, moisture content, flatness, elasticity and finish. In addition, the length of the paper changes as the ink is applied.

In a typical application of this type, the accuracy required for proper registry of the printing and hole punching must be held to a differential of 1/32 in. in 15 ft of web. For this degree of accuracy, a floating or compensating roll, as shown in Fig. 1, is used as the indicator, because it is the most accurate means of indicating changes in length of web by displacement. In this case, two floating rolls are used with two separate controllers and actuators, one to control the in-feed speed and tension of the paper stock, and the second to control the wind-up.

The in-feed is controlled by maintaining the turning speed of a set of feeding rolls that pull the paper off the stock roll. The second floating roll controls the speed of the wind-up mandrel. The web of paper is held to an exact value of tension between the feed rolls and the punching cylinder of the press by the in-feed control, and also between the punching cylinder and the wind-up roll. Hence, it is possible to not only control the tension in the web of different grades of paper, but also to adjust the relative length at these two points, thereby maintaining proper registry.

The secondary function of maintaining exact control of the tension in the paper as it is rewound after printing is to condition the paper and to obtain a uniformly wound roll so that the web is ready for subsequent operations.

Control of dimension or weight by tension and velocity regulation can be illustrated by applying the same general type of controller actuator to the take-off conveyors in an extruder line such as those used in rubber and plastics processing. The problem is twofold: First, to set the speed of the take-away conveyor at the extruder to match the variation in extrusion rate; and, second, to set the speeds of the subsequent conveyor sections to match the movement of the stock as it cools and tends to change dimension.

One way to handle this problem is to use as the indicators the pivoted idlers or contact rolls shown in Fig. 2. The rolls contact the extruded material between each of the conveyor sections and control the speed of the driving mechanism of the following section. The material forms a slight catenary between the stations and the change in the catenary length is used for indicating errors in driving speeds.

The plasticity of the material prevents the use of a complete control loop and hence the contact roll must operate with very little resistance or force through a small operating angle.

The problem of winding or coiling a strip of thin steel that has been plated or pre-coated for painting on a continuous basis is typical of processing systems in which primary indicators cannot be used. While it is important that no contact be made with the prepared surface of the steel, it is also desirable to rewind the strip after preparation in a coil that is sound and slip-free. An automatic, constant-tension winding control and a secondary type indicator are therefore used to initiate the control action.

The control system shown in Fig. 3 is used to wind coils from 16 in. core diameter to 48 in. maximum diameter. The power to wind the coil is used as the controlling medium because, by maintaining constant winding power as the coil builds up, a constant value of strip tension can be held within the limits required. Actually, this method is inaccurate to the extent that the losses in the driving equipment, which are a factor in the power being measured, are not constant and hence the strip tension changes slightly. This same factor enters into any control system that uses winding power as an index of control.

A torque measuring belt that operates a differential controller is used to measure the power of the winder. Then, in turn, the controller adjusts the variable speed transmission. The change in speed between the source of power and the transmission is measured by the three-shaft gear differential which is driven in tandem with the control belt. Any change in load across the control belt produces a change in speed between the driving and driven ends of the belt. The differential acts as the controller, since any change in speed between the two outside shafts of the differential results in a rotation or displacement of the center or control shaft. By connecting the control shaft of the differential directly to a screw-controlled variable speed transmission, a means is provided for adjusting the transmission to correct any change in speed and power as delivered by the belt.

This system is made completely automatic by establishing a neutralizing speed between the two input shafts of the differential (within the creep value of the belt). When there is no tension in the strip, for example when it is cut, the input speed to the actuator side of the differential is higher on the driven side than it is on the driving side of the differential. This unbalance reverses the rotation of the control shaft of the differential, which in turn resets the transmission to the high speed required for starting the next coil on the rewinding mandrel.

In operation, any element in the system that tends to change strip tension causes a change in winding power, which, in turn, is immediately compensated for by the rotation or tendency to rotate of the controlling shaft in the differential. Hence, the winding mandrel speed is continuously and automatically corrected to maintain constant tension in the strip.

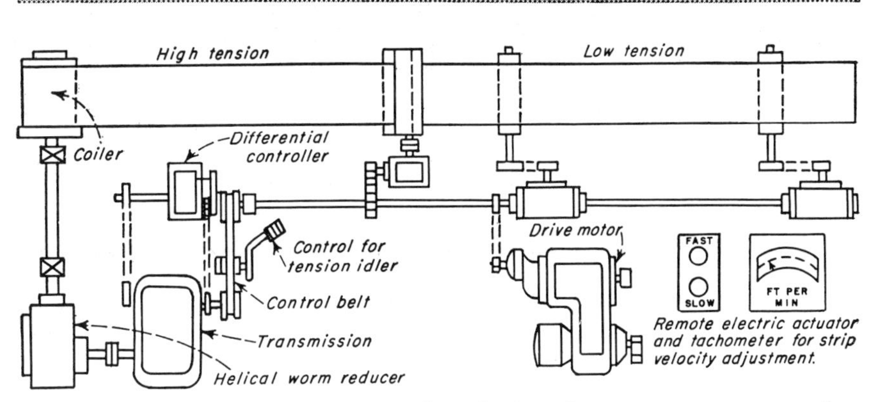

Fig. 3—Differential controller has third shaft that signals remote actuator when tension in sheet material changes. Coiler power is used as secondary control index.

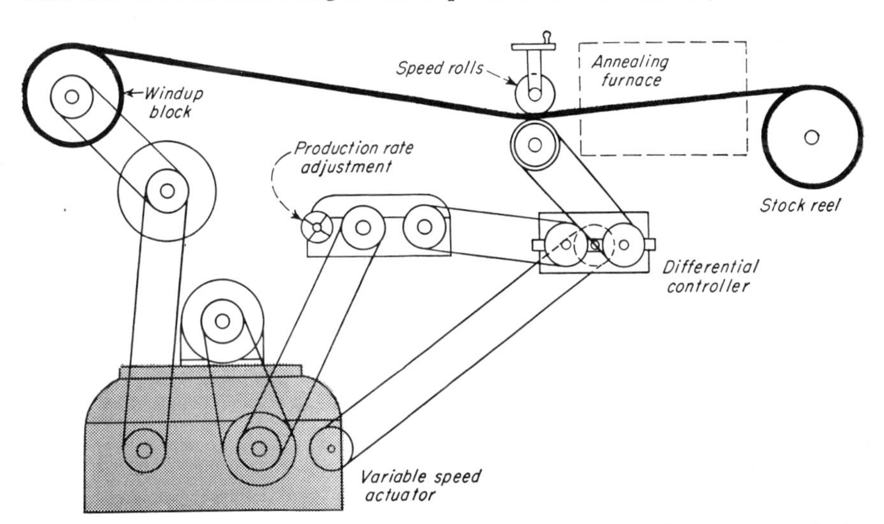

Fig. 4—Movement of wire through annealing furnace is regulated at constant velocity by continuously retarding the speed of the windup reels to allow for wire build-up.

When the correct speed relationships are established in the controller, the system operates automatically for all conditions of operation. In addition, tension in the strip can be adjusted to any value by moving the tension idler on the control belt to increase or decrease the load capacity of the belt to match a desired strip tension.

There are many continuous processing systems that require constant velocity of the material during processing, yet do not require accurate control of the tension in the material. An example of this type is the annealing of wire that is pulled off stock reels through an annealing furnace and then rewound on a wind-up block.

The problem is to pass the wire through the furnace at a constant rate, so that the annealing time is maintained at a fixed value. Since the wire is pulled through the furnace by the wind-up blocks shown in Fig. 4, its rate of movement through the furnace would increase as the wire builds up on the reels unless a control is used to slow down the reels.

A solution to this problem is to use a constant velocity type control with the wire as a direct indication to an indicator that takes the speed of initiate a control action for adjusting the speed of the wind-up reel. In this case, the wire can be contacted directly and a primary type indicator in the form of a contact roll can be used to register any change in speed. The contact roll drives one input shaft of the differential controller. The second input shaft is connected to the driving shaft of the variable speed transmission to provide a reference speed. The third or control shaft will then rotate when any difference in speed exists between the two input shafts. Thus, if the control shaft is connected to a screw-regulated actuator, an adjustment is obtained for slowing down the wind-up blocks as the coils build up and the wire progresses through the furnace at a constant speed.

DRIVES FOR CONTROLLING TENSION

Mechanical, electrical and hydraulic methods for obtaining controlled tension on winding reels and similar drives, or for driving independent units of a machine in synchronism

Mechanical Drives

Band Brake—Used on coil winders, insulation winders and similar applications wherein maintaining the tension within close limits is not required.

Simple and economical but tension will vary considerably. Friction drag at start may be several times that during running by virtue of the difference between coefficient of friction at starting and the coefficient of sliding friction, which latter will also be affected by moisture, foreign matter, and wear of surfaces.

Capacity—limited by the heat radiating capacity of the brake at the maximum permissible running temperature.

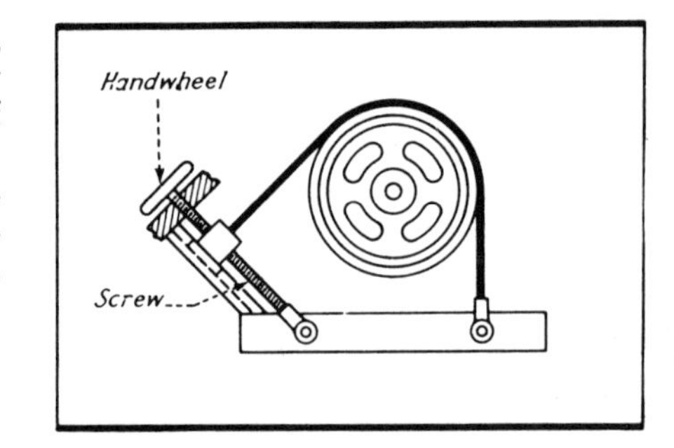

Differential Drives may be of various forms—epicylic spur gears, bevel gear differentials or worm gear differentials.

The braking device on the ring gear or spider may be a band brake, a fan, an impeller, an electric generator or an electric drag such as a copper disk rotating in a powerful magnetic field. A brake will give a drag or tension reasonably constant over a wide speed range. The other braking devices mentioned will exert a torque that will vary widely with speed but will be definite for any given speed of the ring gear or spider.

A definite advantage of any differential drive is that maximum driving torque can never exceed the torque developed by the braking device.

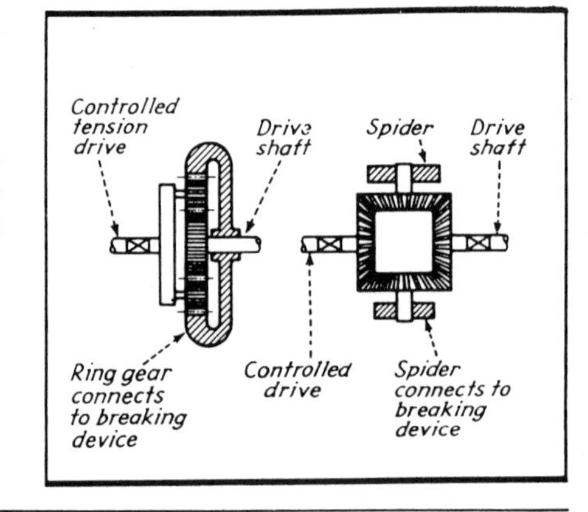

Differential gearing can be used to control a variable-speed transmission. With the ring gear and sun gear driven in opposite directions from the respective shafts to be held in synchronism, the gear train can be designed so that the spider on which the planetary gears are mounted will not rotate when the shafts are running at the desired relative speeds. If one or the other of the shafts speeds ahead, the spider rotates correspondingly. The spider rotation changes the ratio of the variable-speed transmission unit.

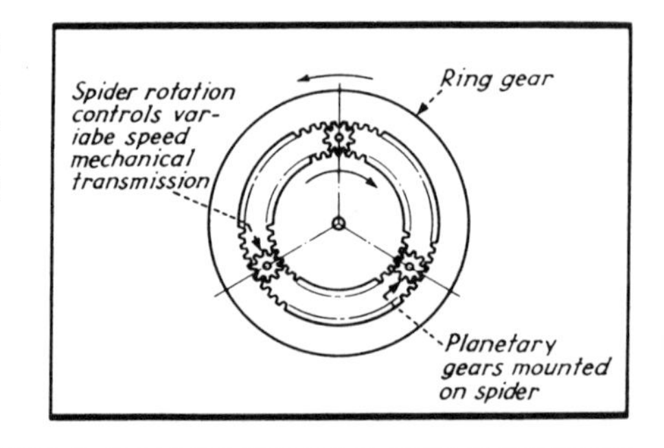

Electrical Drives

Shunt field rheostat in a d.c. motor drive can be used for synchronizing drives. Applied to a machine handling paper, cloth or similar material passing around a take-up roll, movement of the take-up roll moves a control arm that is connected to the rheostat. This type of drive is not suitable to wide changes of speed, above approximately 2½ to 1 ratio.

For wide ranges of speed, the rheostat is put in the shunt field of a d.c. generator which is driven by another motor. The voltage developed by the generator is controlled from zero to full voltage. The generator furnishes the current to the driving motor armature and the fields of the driving motor are separately excited. Thus the motor speed is controlled from zero to maximum.

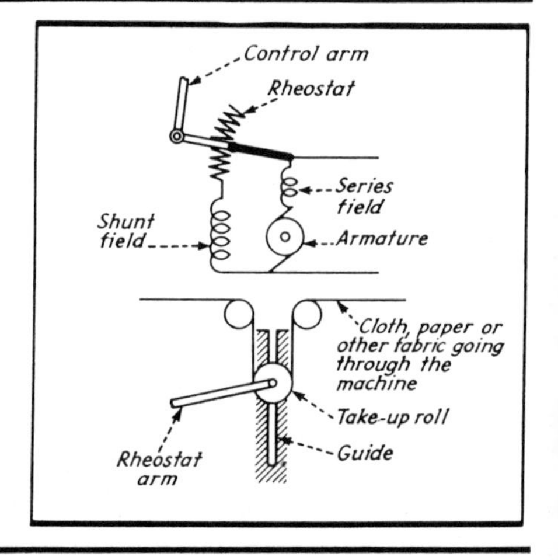

Selsyn motors can be used for direct drive to independent units in exact synchronism, provided inertias are not too great. But regardless of loads and speeds Selsyn motors can be used as the controlling units. As an example, variable-speed mechanical transmission units with built-in Selsyn motors are obtainable for furnishing constant-tension drives or synchronous driving of independent units.

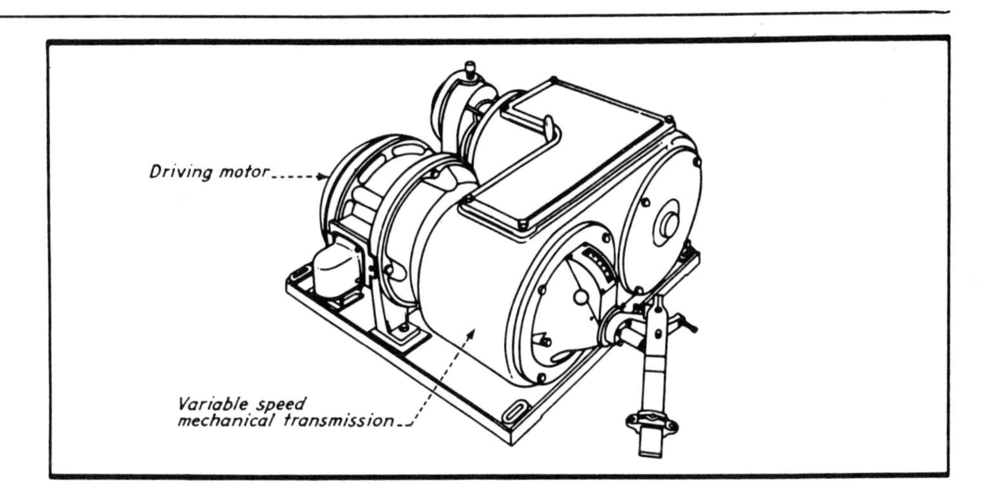

Hydraulic Drives

Hydraulic Control — Tension between successive pairs of rolls, or synchronism between successive units of a machine can be controlled automatically by hydraulic drives. Driving the variable delivery pump off of one of the pairs of rolls automatically maintains an approximately constant relative speed between the two units, at all speeds and loads. The variations caused by oil leakage and similar factors are compensated automatically by the idler roll and linkage which adjusts the pilot valve that controls the displacement of the variable delivery pump.

The counterweight on the idler roll is set for the desired tension in the felt, paper or other material. Increased tension resulting from the second pair of rolls going too fast, causes idler roll to be depressed, the control linkage thereby moving pilot valve to cause a decreased pump delivery which slows the speed of the second pair of rolls. The reverse operations take place when the tension in the paper decreases, allowing the idler roll to move upwards.

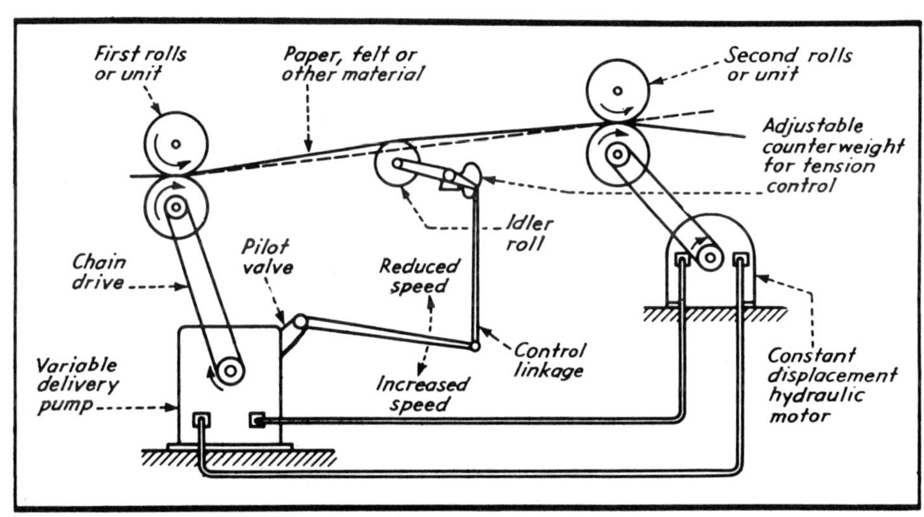

Courtesy The Oilgear Company

If the material passing through the machine is too weak to operate a mechanical linkage, the desired control can be obtained by photo-electric cells. The hydraulic operation is exactly the same as that described above.

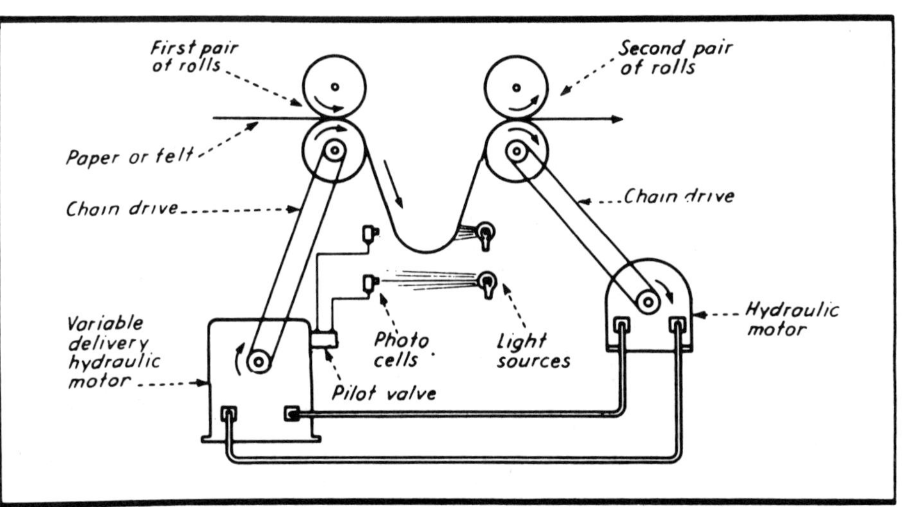

Courtesy The Oilgear Company

DRIVES FOR CONTROLLING TENSION

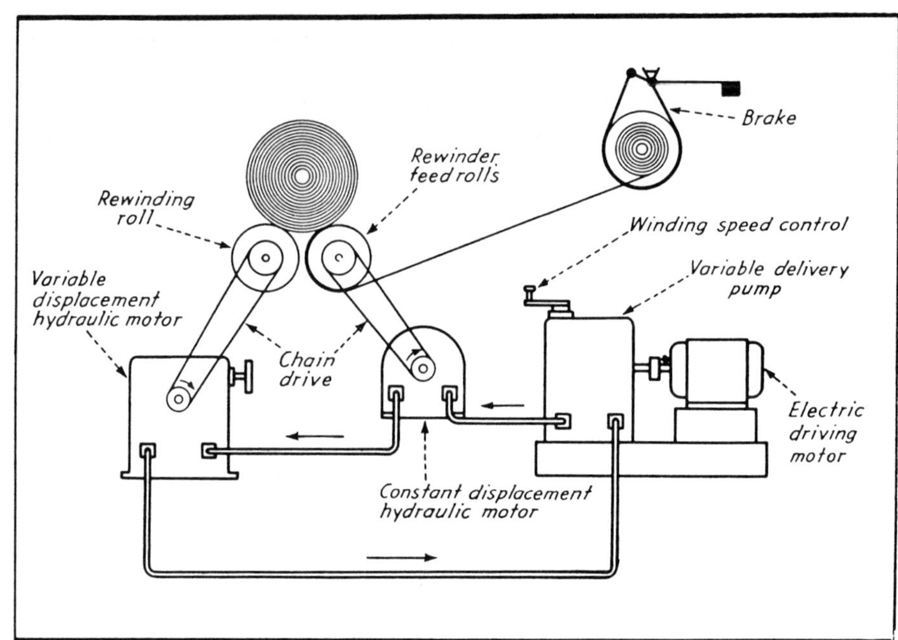

Courtesy The Oilgear Company

As mentioned, a band brake used to obtain a friction drag will give variable tension. In this hydraulic drive the winding tension is determined by the difference in torque exerted on the rewinder feed roll and the winding roll. The brake plays no part in establishing the tension.

The constant displacement hydraulic motor and variable displacement hydraulic motor are connected in series with the variable delivery pump. Thus the relative speeds of the two hydraulic motors will always remain substantially the same, the displacement of the variable speed motor being adjusted to an amount slightly greater than the displacement of the constant speed motor, thus tending to give the winding roll a speed slightly greater than the feed roll speed. This determines the tension, because the winding roll cannot go faster than the feed roll, both being in contact with the paper roll being wound. The pressure in the hydraulic line between the constant and variable displacement pump will increase correspondingly to the winding tension. For any setting of the winding speed controller on the variable delivery hydraulic pump the motor speeds are practically constant, hence the surface speed of winding will remain substantially constant, regardless of the diameter of the roll being wound.

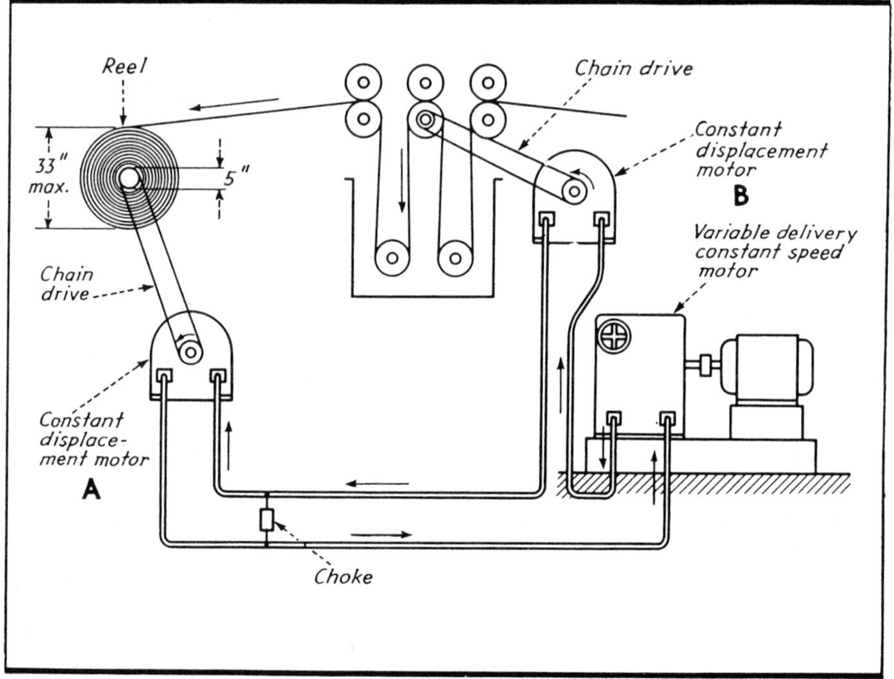

Courtesy The Oilgear Company

Hydraulic drive for fairly constant tension. The variable delivery constant speed pumping unit supplies the oil to two constant displacement motors, one driving the apparatus which carries the fabric through the bath at a constant speed and the other driving the winder. The two motors are in series. Motor *A* driving the winding reel, diameter of which ranges from about 5 in. when the reel is empty to about 33 in. when the reel is full, is geared to the reel so that even when the reel is empty the surface speed of the paper travel will tend to be somewhat faster than the mean rate of paper travel established by motor *B* driving the apparatus. When the reel is empty its speed nearly corresponds to that established by motor *B* driving the apparatus and only a small amount of oil will be by-passed through the choke interposed between the pressure and return line.

When roll is full the r.p.m. of the reel and its driving motor is only about one-seventh of the r.p.m. when the reel is empty. A much greater quantity of oil is forced through the choke when the reel is full because of the increased pressure in the line between the two motors. The pressure in this line increases as the reel diameter increases because the torque resistance encountered by the reel motor will be directly in proportion to the reel diameter, tension being constant. The greater the diameter of the fabric on the reel, the greater will be the torque exerted by the tension in the fabric. The installation is designed so that the torque developed by the motor driving the reel will be inversely proportional to the r.p.m. of the reel. Hence the tension on the fabric will remain at a fairly constant value regardless of the diameter of the reel. This drive is limited to about 3 hp. and is relatively inefficient.

Chapter 11 CLUTCH, COUPLING, AND BRAKING DEVICES

Springs, shuttle-gear, gliding-ball in

Novel one-way drives

These three devices change oscillating motion into one-way rotation for feeding operations and counters

R H MEIER, Engineer, Ottawa, Canada

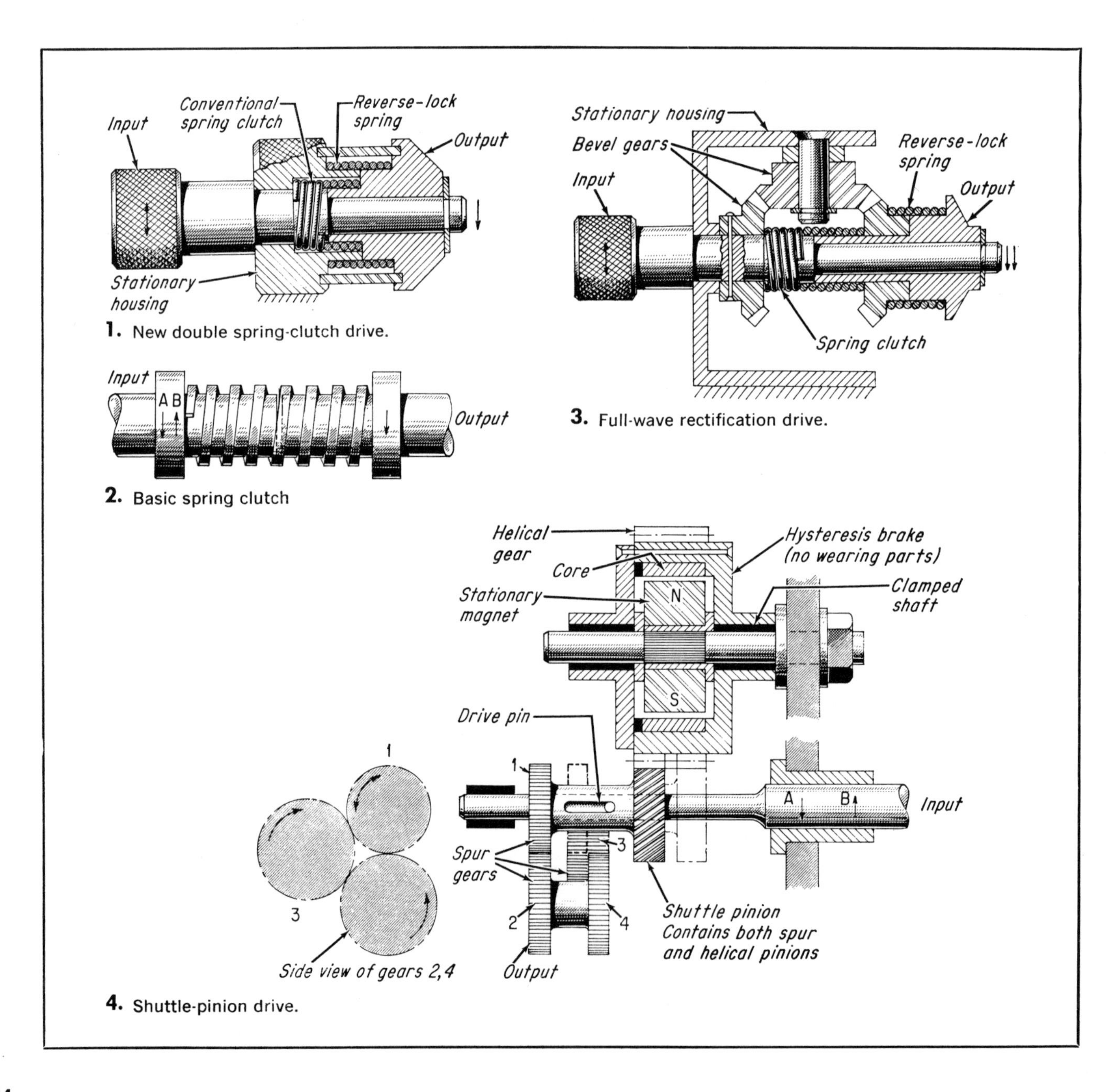

1. New double spring-clutch drive.

2. Basic spring clutch

3. Full-wave rectification drive.

4. Shuttle-pinion drive.

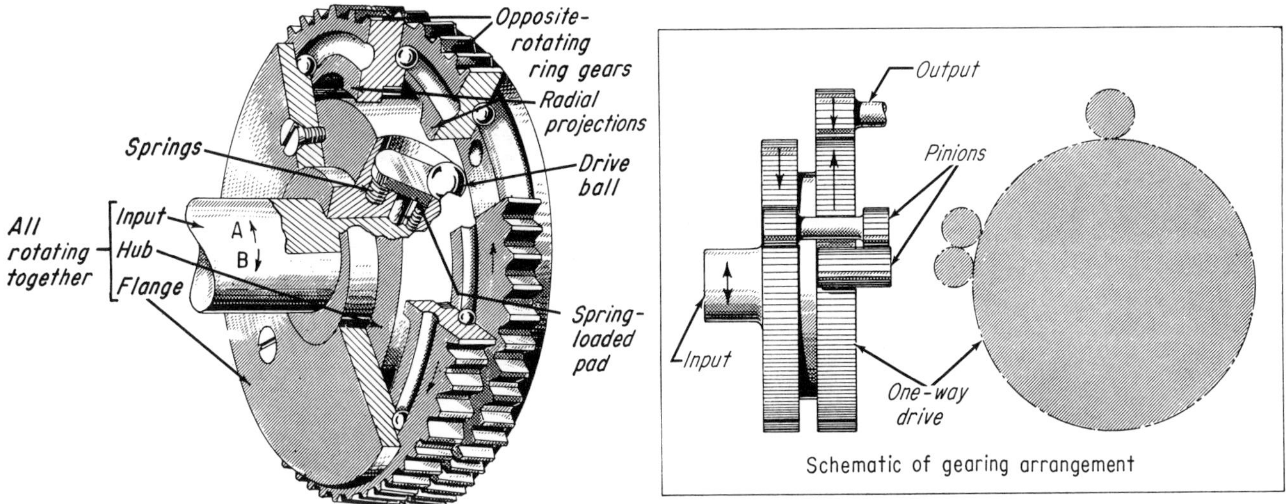

5. Reciprocating-ball drive.

WHILE engaged in the design of a money order imprinter for the Canada Post Office, I devised a one-way drive, Fig 1, which appears to have novel features.

The problem was to convert the oscillating motion of the input crank (20 deg in this case) into a one-way motion to advance the inking ribbon. Basically I used one of the simplest of known devices for obtaining a one-way drive—a spring clutch which is a helical spring joining two co-linear butting shafts (Fig 2). The spring is usually made of square or rectangular cross-section wire.

Such a clutch transmits torque in one direction only—it overrides when reversed. The helical spring, which bridges both shafts, need not be fastened at either end; a slight interference fit will do. Rotating the input shaft in the direction tending to wind the spring (direction *A* in Fig 2) will cause the spring to grip both shafts and thus transmit motion from the input to the output shaft. Reversing the input unwinds the spring, and it overrides the output shaft with a drag—but this drag, slight as it was, caused a problem in operation.

Double-clutch drive

The trouble with the simple approach of Fig 2 was that there was absolutely no friction in the tape drive which would have allowed the spring clutch to slip on the shafts on the return stroke. Thus the output moved in sympathy with the input, and the desired one-way drive was not obtained.

At first, we attempted to add friction artificially to the output, but this resulted in an awkward design. Finally the problem was elegantly solved, Fig 1, by using a second helical spring, slightly larger than the first and serving exactly the same purpose, viz to transmit motion in one direction only. This spring, however, joined the output shaft and a stationary cylinder. In this way, with the two springs of the same hand, the undesirable return motion of the ribbon drive was immediately arrested, and a positive one-way drive simply obtained.

This compact drive can be considered to be a *half wave rectifier* in that it transmits motion in one direction only, and suppresses motion in the reverse direction.

Full-wave rectifier

The principles described above will also produce a *full wave rectifier* by introducing some reversing gears, Fig 3. In this application the input drive in one direction is directly transmitted to the output as before, but on the reverse stroke the input is passed through reversing gears so that the output appears in the opposite sense, in other words, the original sense of the output is maintained. Thus the output moves forward twice for each back-and-forth movement of the input.

Shuttle-gear drive

A few years ago, I developed a one-way drive which harnesses the axial thrust of a pair of helical gears to shift a pinion, Fig 4. Although at first glance it may look somewhat complicated, the drive is inexpensive to make and has been operating successfully with little wear.

When the input rotates in direction *A*, it drives the output through spur gears *1* and *2*. The shuttle pinion is also driving the helical gear whose rotation is resisted by the magnetic flux built up between the stationary permanent magnet and the rotating core. This magnet-core arrangement is actually a hysteresis brake and its constant resisting torque produces an axial thrust in mesh of the helical pinion acting to the left. Reversing the input reverses the direction of thrust which shifts the shuttle pinion to the right. The drive then operates through gears *1*, *3* and *4* which nullifies the reversion to produce the same direction output.

Reciprocating-ball drive

When the input rotates in direction *A*, Fig 5, the drive ball trails to the right and its upper half engages one of the radial projections in the right ring gear to drive it in the direction as the input. The slot for the ball is milled at 45 deg to the shaft axes and extends to the flanges on each side.

When the input is reversed, the ball extends to the flanges on each side. trails to the left and deflects to permit the ball to ride over to the left ring gear, and engage its radial projection to drive the gear in the direction of the input.

Each gear, however, is constantly in mesh with a pinion, which in turn is in mesh with the other. Thus, no matter which direction the input is turned, the ball positions itself under one or another ring gear, and the gears will maintain their respective sense of rotation (the rotation shown in Fig 5). Hence, an output gear in mesh with one of the ring gears will rotate in one direction only.

1-WAY OUTPUT FROM SPEED REDUCERS

When input reverses, these 5 slow-down mechanisms continue supplying a non-reversing rotation.

LOUIS SLEGEL
Head, Dept of Mechanical Engineering
Oregon State College
Corvallis, Ore

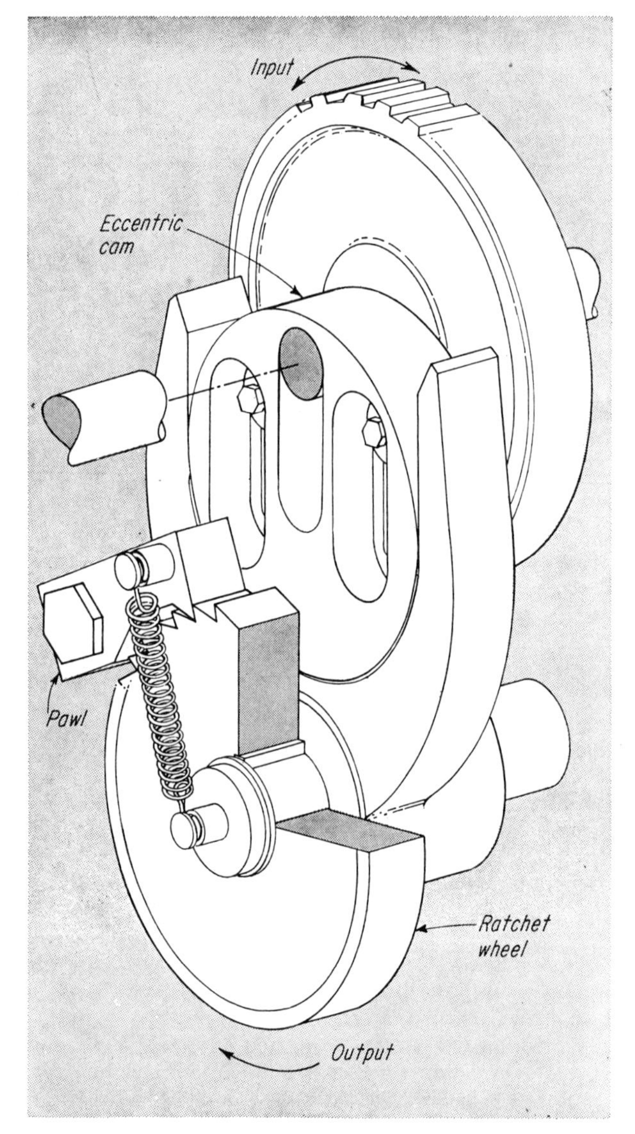

1 **ECCENTRIC CAM** adjusts over a range of high-reduction ratios, but unbalance limits it to low speeds. When direction of input changes, there is no lag in output rotation. Output shaft moves in steps because of ratchet drive through pawl which is attached to U-follower.

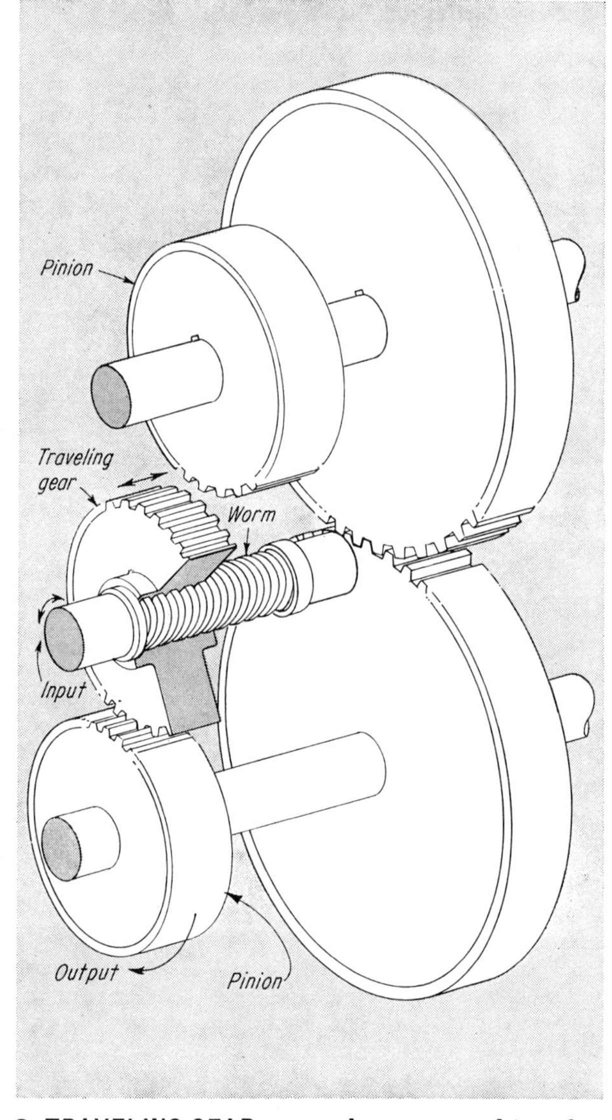

2 **TRAVELING GEAR** moves along worm and transfers drive to other pinion when input rotation changes direction. To ease engagement, gear teeth are tapered at ends. Output rotation is smooth, but there is a lag after direction changes as gear shifts. Gear cannot be wider than axial offset between pinions, or there will be interference.

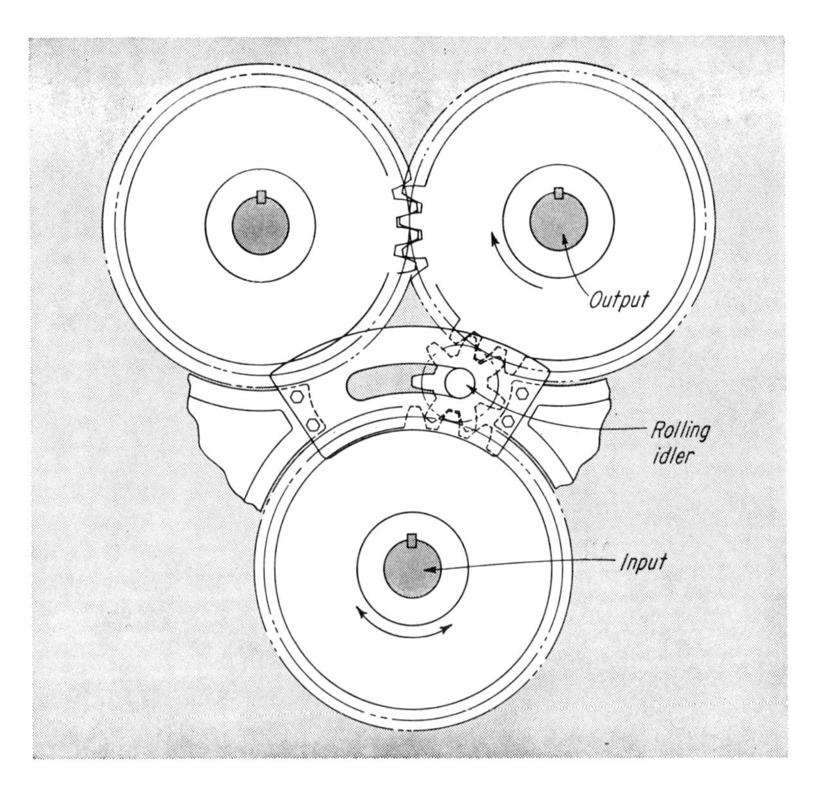

3 ROLLING IDLER also gives smooth output and slight lag after input direction changes. Small drag on idler is necessary, so that it will transfer into engagement with other gear and not sit spinning in between.

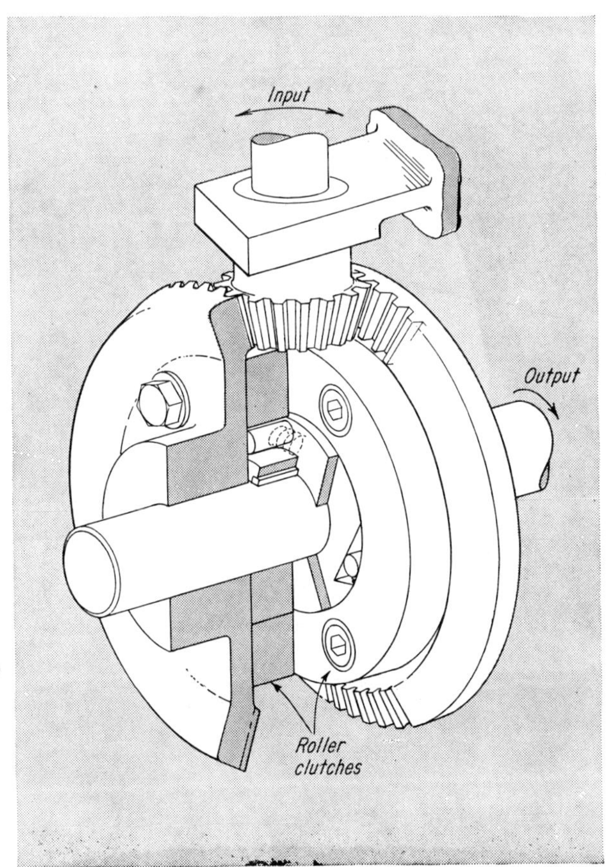

4 TWO BEVEL GEARS drive through roller clutches. One clutch catches in one direction; the other catches in the opposite direction. There is negligible interruption of smooth output rotation when input direction changes.

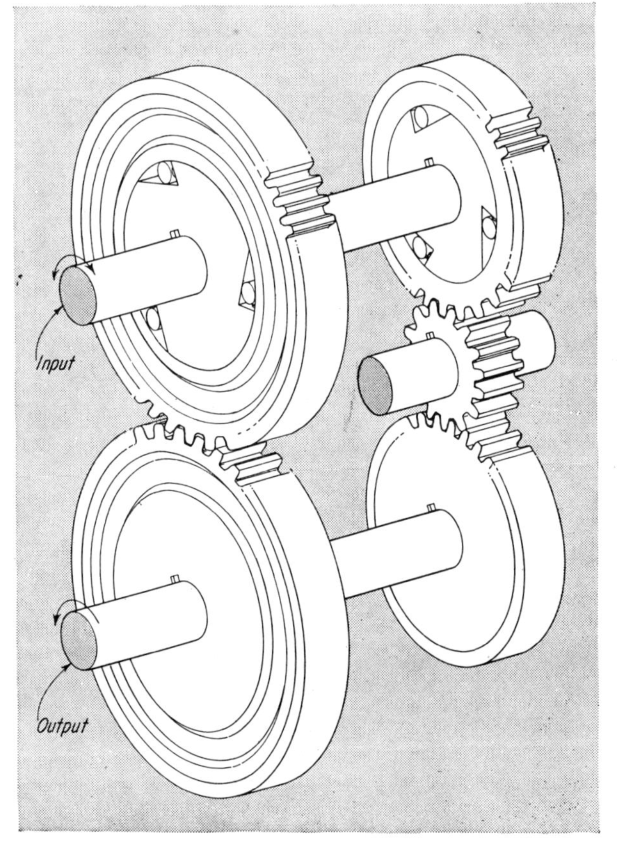

5 ROLLER CLUTCHES are on input gears in this drive, again giving smooth output speed and little output lag as input direction changes.

Two European Designs for Overrunning Clutches

Combination Bearing-clutch Transmits Torque, Carries Load

PARIS—Sprags combined with cylindrical rollers in a bearing assembly may be a simple, low-cost way to meet the torque and bearing requirements of most machine applications. Designed and built by Est. Nicot here, this new unit gives one-direction-only torque transmission in an overrunning clutch. In addition, it also serves as a roller bearing.

Torque rating of the clutch depends on the number of sprags. A minimum of three, equally spaced around the circumference of the races, is generally necessary to get acceptable distribution of tangential forces on the races. Torque ratings of standard models range up to 145 ft-lb; versions up to 1450 ft-lb are possible.

Two standard models are available: the smaller, with torque-handling capacity up to 43.5 ft-lb is about $30; the larger, rated at 145 ft-lb is about $37. ■

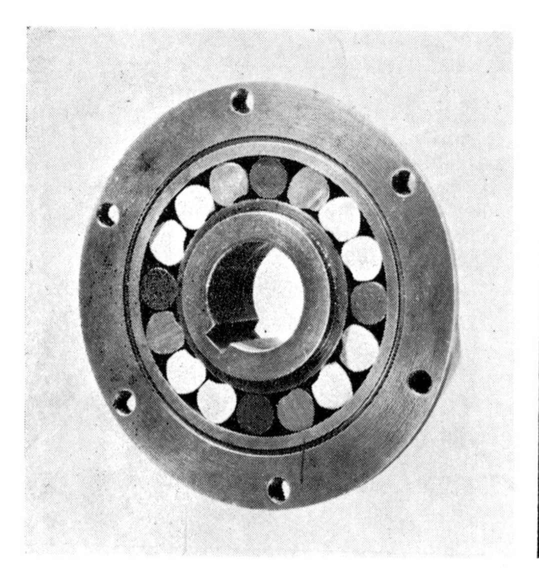

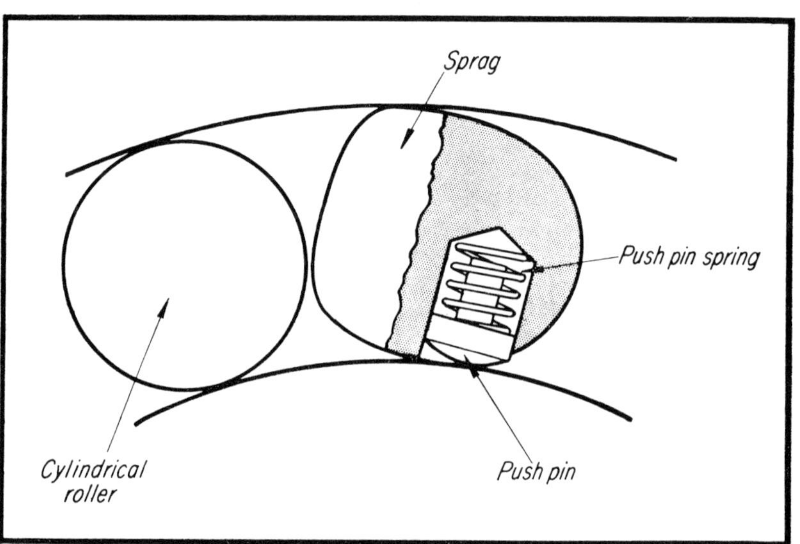

RACES ARE CONCENTRIC; locking ramp is provided by the sprag profile, which is composed of two nonconcentric curves of different radius. Spring-loaded pin holds the sprag in the locked position until torque is applied in the running direction. Stock roller bearing can not be converted, as the hard-steel races of bearing are too brittle to handle the locking impact of the sprags. Sprags and rollers can be mixed to give any desired torque value.

Roller-type clutch, designed in England . . .

. . . is adaptable to either electrical or mechanical activation, and will control ½ hp at 1500 rpm with only 7 watts of power in the solenoid. The rollers are positioned by a cage (integral with the toothed control wheel—see diagram) between the ID of the driving housing and the cammed hub (integral with the output gear).

When the pawl is disengaged, the drag of the housing on the friction spring rotates the cage and wedges the rollers into engagement. This permits the housing to drive the gear, through the cam.

When the pawl engages the control wheel while the housing is rotating, the friction spring slips inside the housing and the rollers are kicked back, out of engagement. Power is therefore interrupted.

According to the designer, Tiltman Langley Ltd, Redhill Aerodrome, Surrey, the unit will operate over the full temperature range of −40 to 200 F.

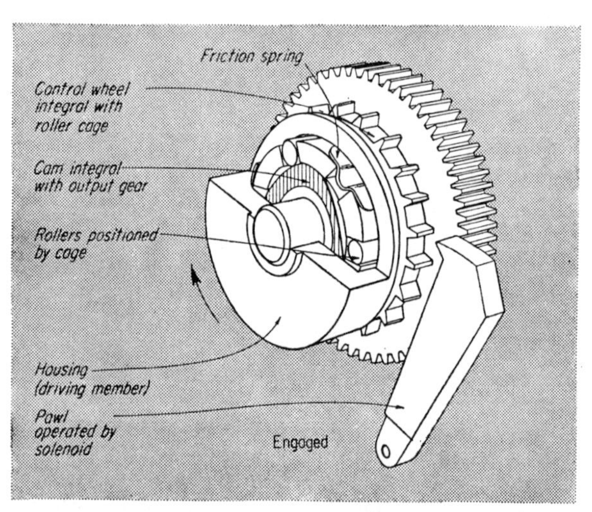

Positive drive is provided by this new British roller clutch design. Company says it's unusually small and compact.

7 Low-cost Designs for . . .

OVERRUNNING CLUTCHES

All are simple devices that can be constructed inexpensively in the laboratory workshop.

JAMES F MACHEN, *asst. professor, mechanical engineering, University of Toledo, Ohio*

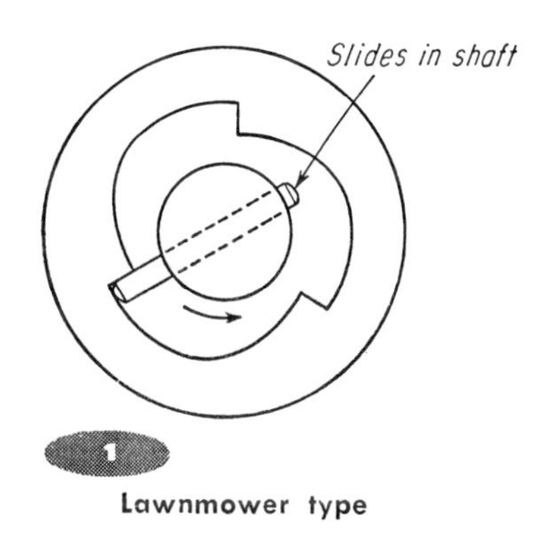

1 Lawnmower type

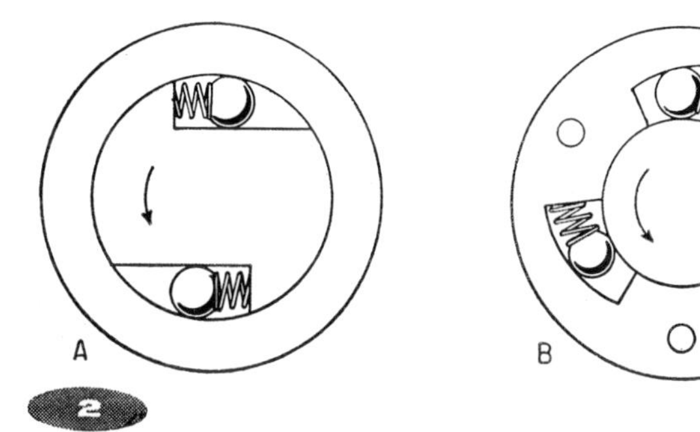

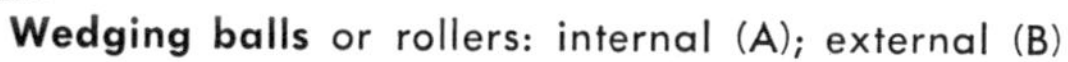

2 **Wedging balls** or rollers: internal (A); external (B)

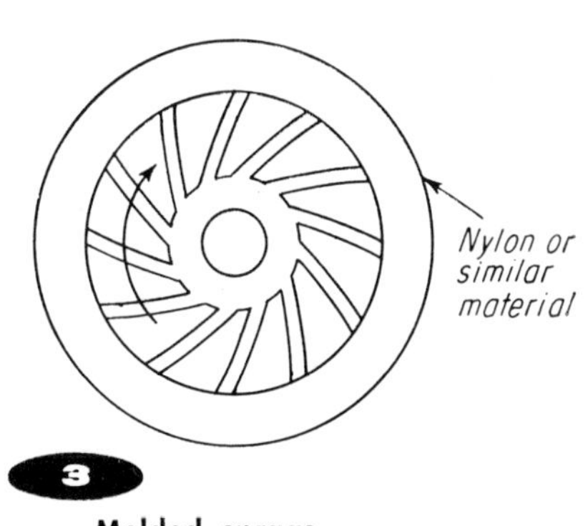

Molded sprags (for light duty)

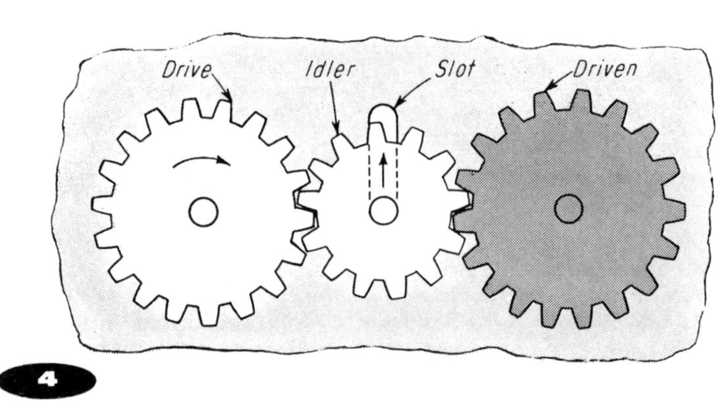

4 **Disengaging idler** rises in slot when drive direction is reversed

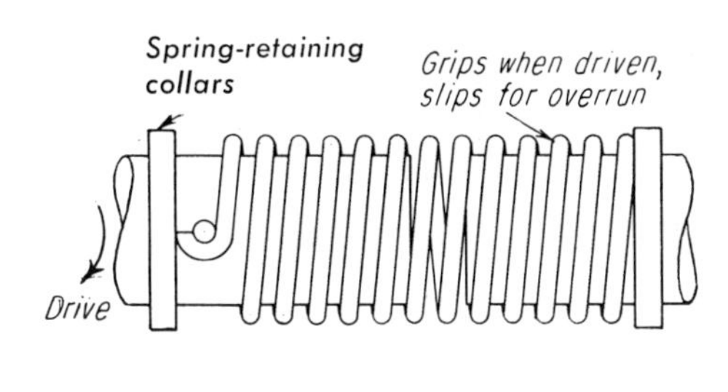

Slip-spring coupling

6 **Internal ratchet** and spring-loaded pawls

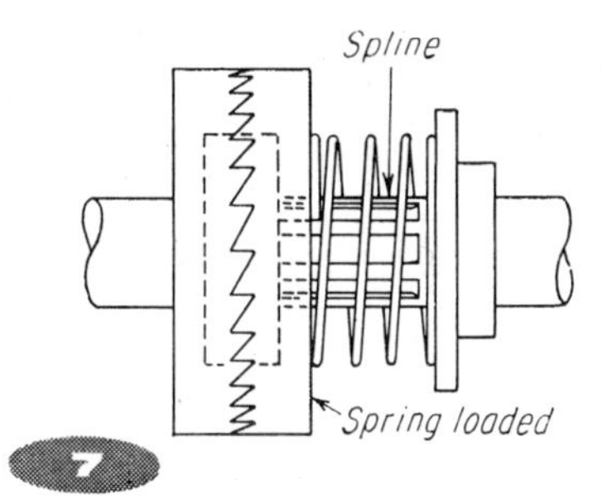

7 **One-way** dog clutch

Construction Details of

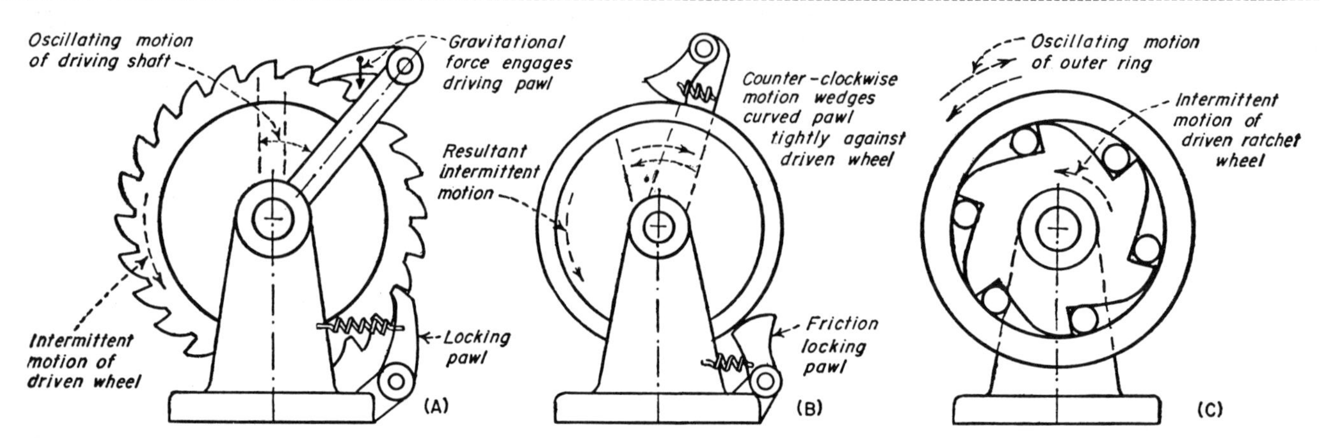

1 Elementary over-riding clutches: (A) Ratchet and Pawl mechanism is used to convert reciprocating or oscillating movement to intermittent rotary motion. This motion is positive but limited to a multiple of the tooth pitch. (B) Friction-type is quieter but requires a spring device to keep eccentric pawl in constant engagement. (C) Balls or rollers replace the pawls in this device. Motion of the outer race wedges rollers against the inclined surfaces of the ratchet wheel.

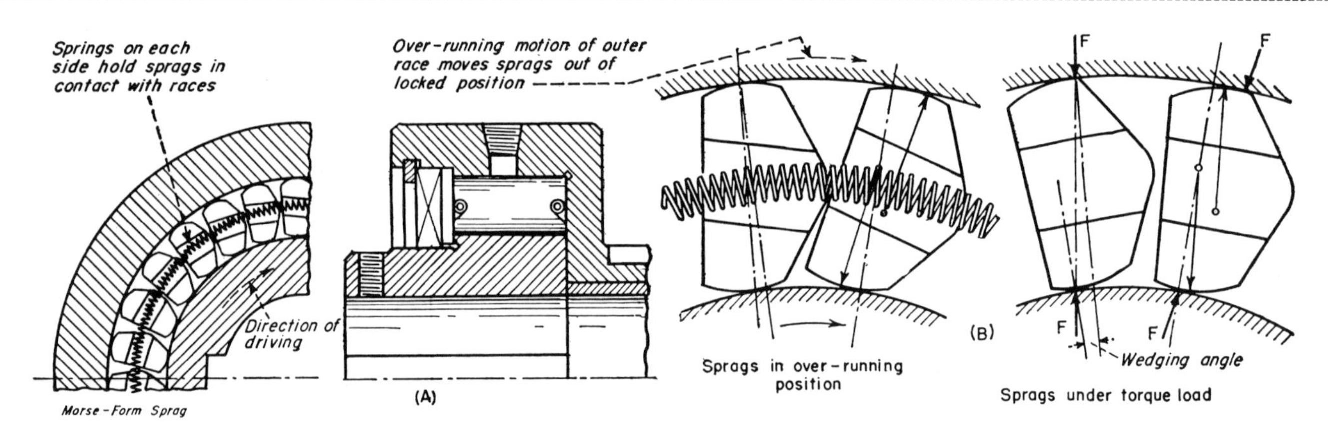

4 With cylindrical inner and outer races, sprags are used to transmit torque. Energizing springs serves as a cage to hold the sprags. (A) Compared to rollers, shape of sprag permits a greater number within a limited space; thus higher torque loads are possible. Not requiring special cam surfaces, this type can be installed inside gear or wheel hubs. (B) Rolling action wedges sprags tightly between driving and driven members. Relatively large wedging angle insures positive engagement.

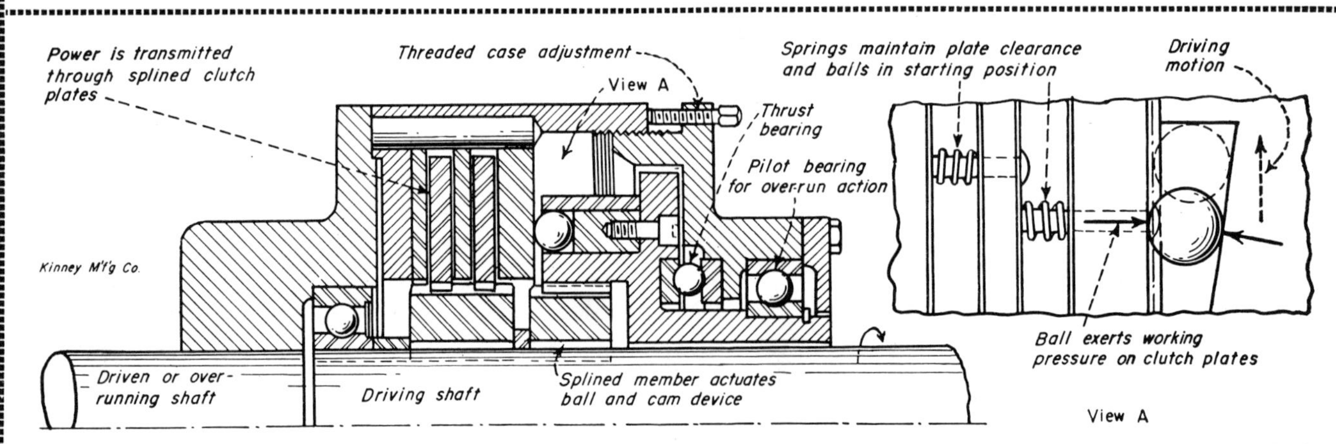

6 Multi-disk clutch is driven by means of several sintered-bronze friction surfaces. Pressure is exerted by a cam actuating device which forces a series of balls against a disk plate. Since a small part of the transmitted torque is carried by the actuating member, capacity is not limited by the localized deformation of the contacting balls. Slip of the friction surfaces determine the capacity and prevent rapid, shock loads. Slight pressure of disk springs insure uniform engagement.

Over-Riding Clutches

A. DeFEO
Design Engineer
Wright Aeronautical Corporation

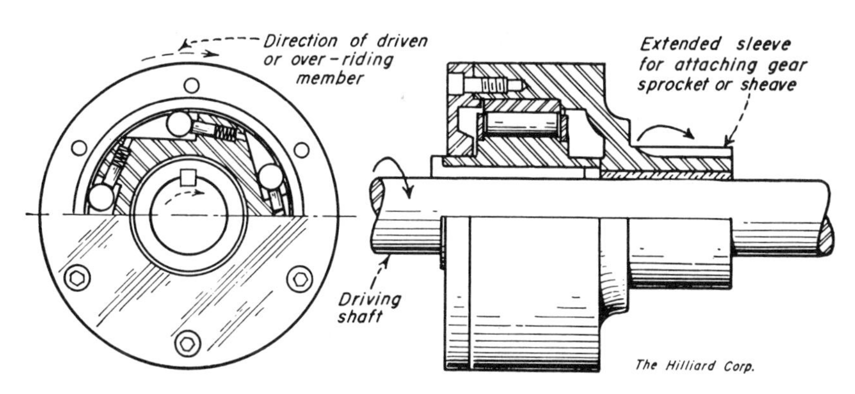

2 Commercial over-riding clutch has springs which hold rollers in continuous contact between cam surfaces and outer race; thus there is no backlash or lost motion. This simple design is positive and quiet. For operation in the opposite direction, the roller mechanism can easily be reversed in the housing.

3 Centrifugal force can be used to hold rollers in contact with cam and outer race. Force is exerted on lugs of the cage which controls the position of the rollers.

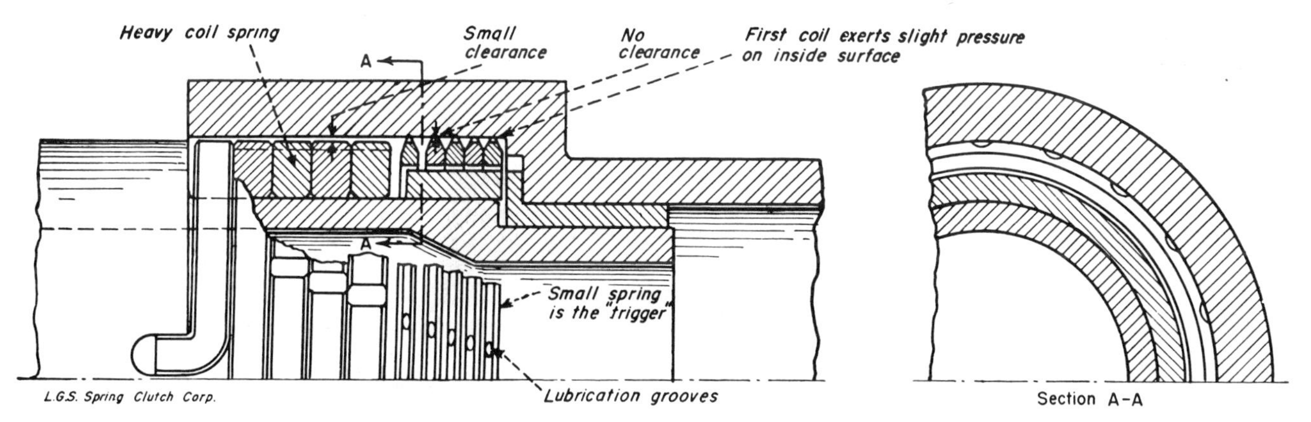

5 Engaging device consists of a helical spring which is made up of two sections: a light trigger spring and a heavy coil spring. It is attached to and driven by the inner shaft. Relative motion of outer member rubbing on trigger causes this spring to wind-up. This action expands the spring diameter which takes up the small clearance and exerts pressure against the inside surface until the entire spring is tightly engaged. Helix angle of spring can be changed to reverse the over-riding direction.

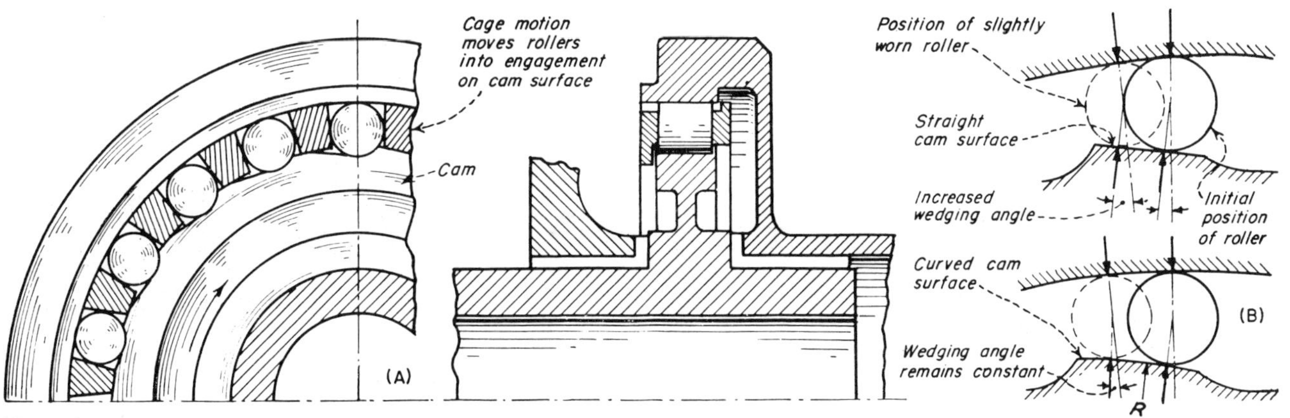

7 Free-wheeling clutch widely used in power transmission has a series of straight-sided cam surfaces. An engaging angle of about 3 deg is used; smaller angles tend to become locked and are difficult to disengage while larger ones are not as effective. (A) Inertia of floating cage wedges rollers between cam and outer race. (B) Continual operation causes wear of surfaces; 0.001 in. wear alters angle to 8.5 deg. on straight-sided cams. Curved cam surfaces maintain constant angle.

10 Ways to Apply OVERRUNNING

These clutches allow freewheeling, indexing and backstopping applicable to many design

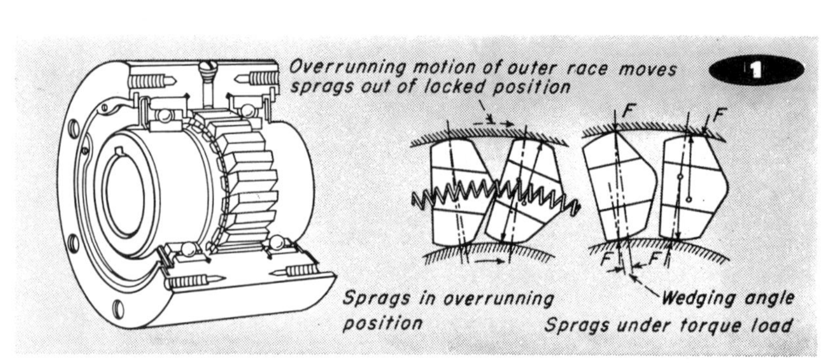

Precision Sprags . . .

act as wedges and are made of hardened alloy steel. In the Formsprag clutch, torque is transmitted from one race to another by wedging action of sprags between the races in one direction; in other direction the clutch freewheels.

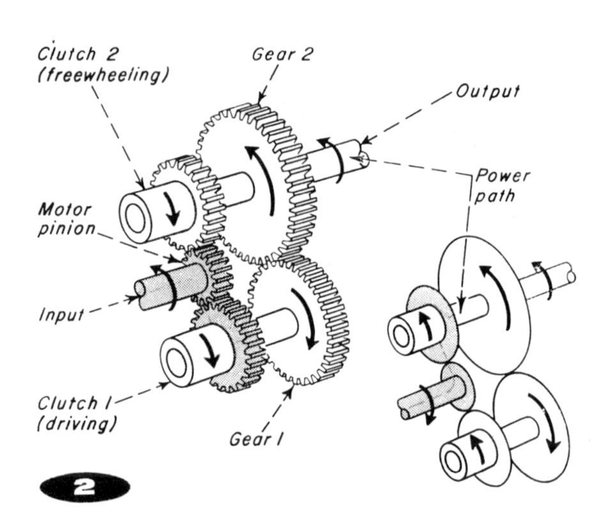

2-Speed Drive—I . . .

requires input rotation to be reversible. Counterclockwise input as shown in the diagram drives gear 1 through clutch 1; output is counterclockwise; clutch 2 over-runs. Clockwise input (schematic) drives gear 2 through clutch 2; output is still counterclockwise; clutch 1 over-runs.

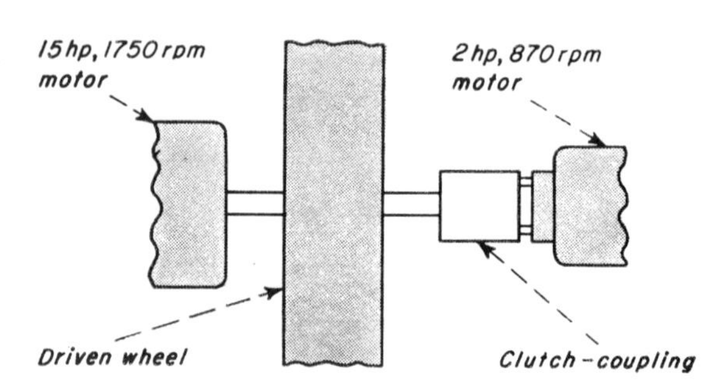

2-Speed Drive—II . . .

for grinding wheel can be simple, in-line design if over-running clutch couples two motors. Outer race of clutch is driven by gearmotor; inner race is keyed to grinding-wheel shaft. When gearmotor drives, clutch is engaged; when larger motor drives, inner race over-runs.

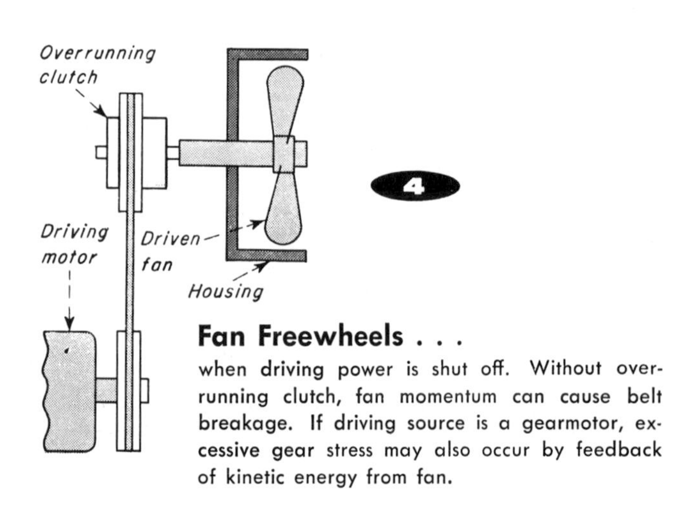

Fan Freewheels . . .

when driving power is shut off. Without over-running clutch, fan momentum can cause belt breakage. If driving source is a gearmotor, excessive gear stress may also occur by feedback of kinetic energy from fan.

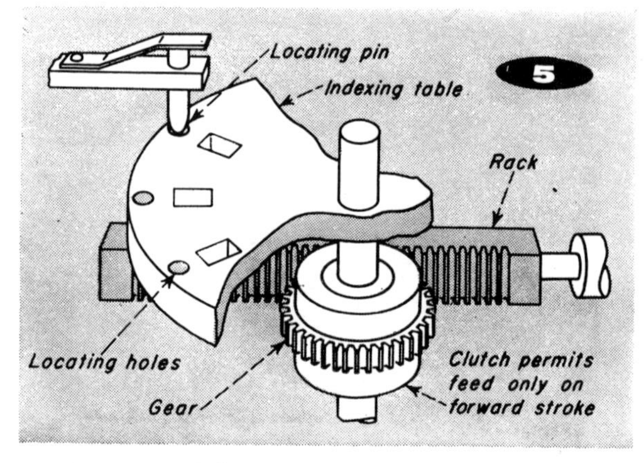

Indexing Table . . .

is keyed to clutch shaft. Table is rotated by forward stroke of rack, power being transmitted through clutch by its outer-ring gear only during this forward stroke. Indexing is slightly short of position required. Exact position is then located by spring-loaded pin, which draws table forward to final positioning. Pin now holds table until next power stroke of hydraulic cylinder

CLUTCHES

problems. Here are some clutch setups.

W. EDGAR MULHOLLAND, *Executive Sales Engineer*
JOHN L. KING, JR., *Project Engineer*
Formsprag Company, Warren, Mich.

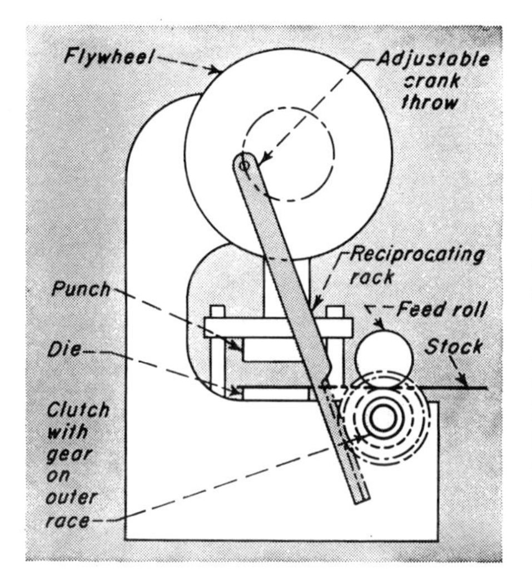

Punch Press Feed . .

is so arranged that strip is stationary on downstroke of punch (clutch freewheels); feed occurs during upstroke when clutch transmits torque. Feed mechanism can easily be adjusted to vary feed amount.

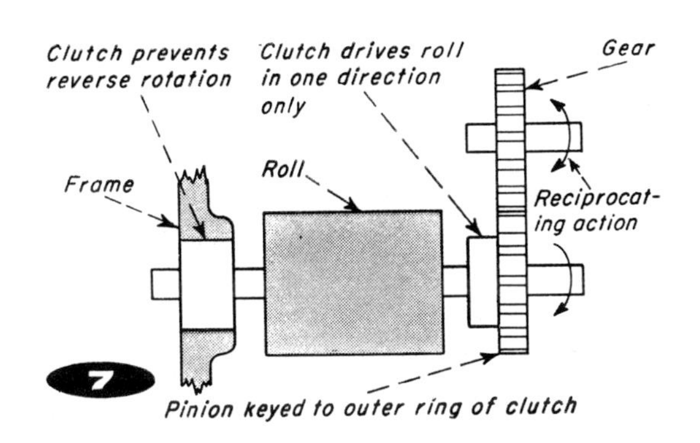

Indexing and Backstopping . . .

is done with two clutches so arranged that one drives while the other freewheels. Application here is for capsuling machine; gelatin is fed by the roll and stopped intermittently so blade can precisely shear material to form capsules.

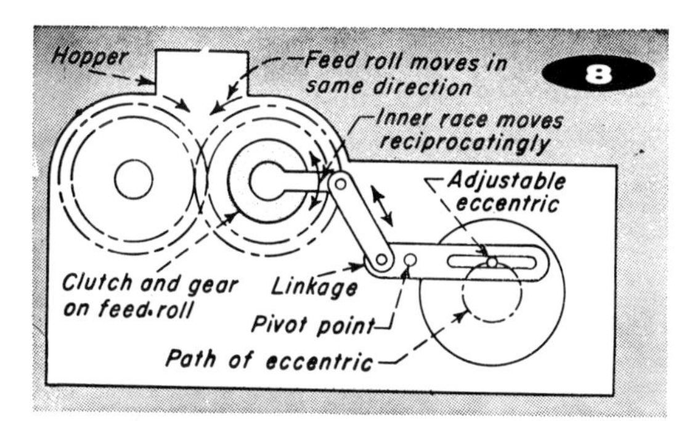

Intermittent Motion . . .

of candy machine is adjustable; function of clutch is to ratchet the feed rolls around. This keeps the material in the hopper agitated.

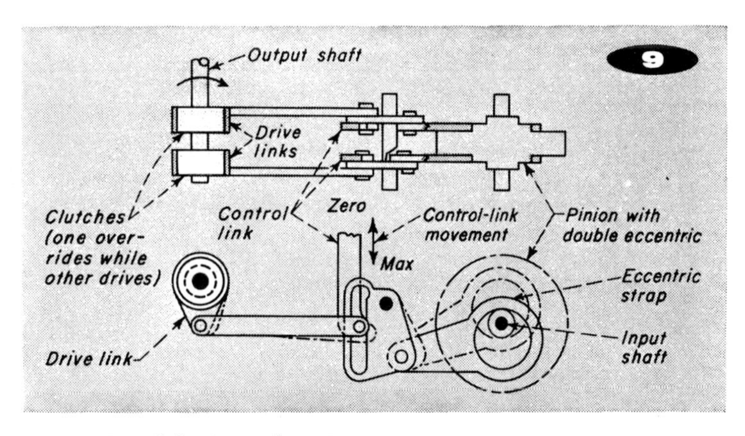

Double-impulse Drive . . .

employs double eccentrics and drive clutches. Each clutch is indexed 180° out of phase with the other. One revolution of eccentric produces two drive strokes. Stroke length, and thus the output rotation, can be adjusted from zero to max by the control link.

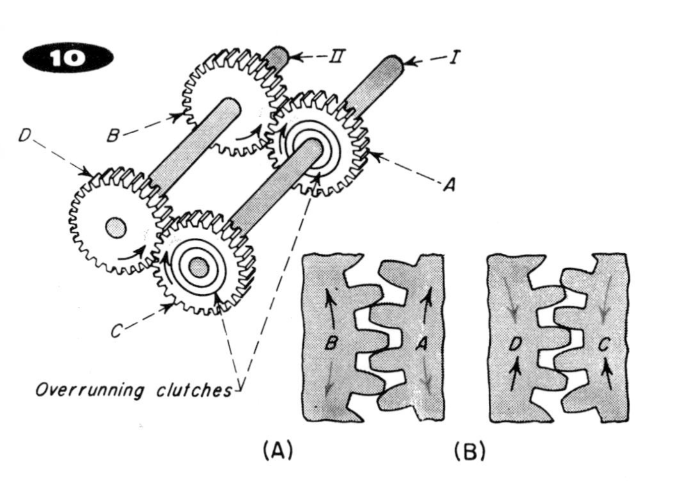

Anti-backlash Device . . .

uses over-running clutches to insure that no backlash is left in the unit. Gear *A* drives *B* and shaft II with the gear mesh and backlash as shown in (A). The over-running clutch in gear *C* permits gear *D* (driven by shaft II) to drive gear *C* and results in the mesh and backlash shown in (B). The over-running clutches never actually over-run. They provide flexible connections (something like split and sprung gears) between shaft I and gears *A*, *C* to allow absorption of all backlash.

Applications of sprag-type clutches

W. T. CHERRY, Manager, Application Engineering, Formsprag Company

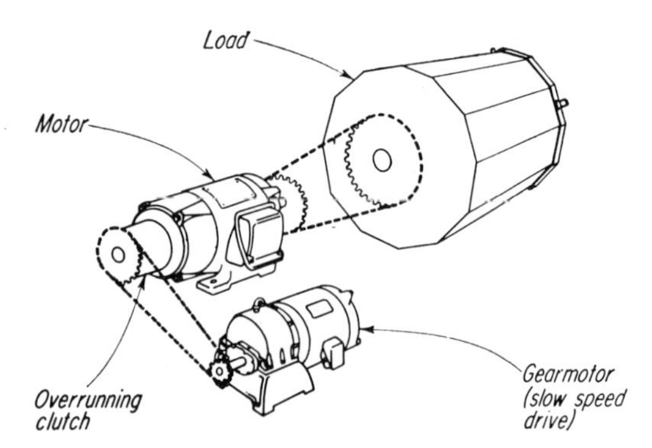

Fig. 1—Overrunning permits torque transmission in one direction; and free wheels or over-runs in the opposite direction. For example, gear motor drives the load by transmitting torque through the overrunning clutch and the high speed shaft. Energizing the high speed motor, causes the inner member to rotate at the rpm of the high speed motor, the gear motor continues to drive the inner member, but the clutch is freewheeling.

Sprag clutches of the overrunning type transmit torque in one direction and reduce speed, rest, hold, or free wheel in the reverse direction. Applications include overrunning, backstopping and indexing. Selection—similar to other mechanical devices—requires a review of torque to be transmitted, overrunning speed, type of lubrication, mounting characteristics, environmental conditions and shock conditions that may be met.

Fig. 2—Backstopping permits rotation in one direction only—clutch serves as a counter rotation holding device. An example is a clutch mounted on the headshaft of a conveyor, with the outer race restrained by means of torque-arming to the stationary frame of the conveyor. If for any reason power to the conveyor is interrupted, the backstopping clutch will prevent the buckets from running backwards and dumping the load.

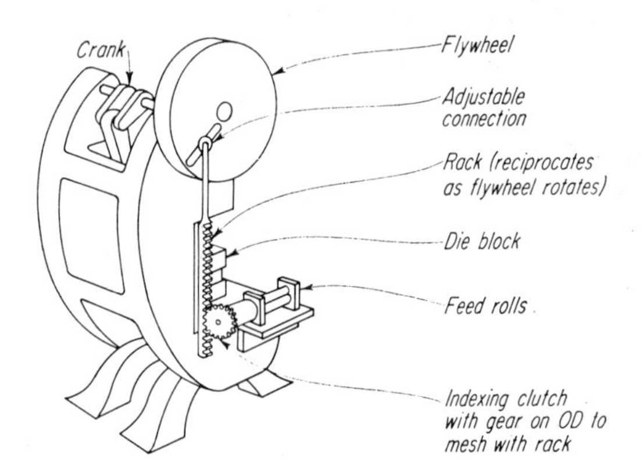

Fig. 3—Indexing is the transmission of intermittent rotary motion in one direction; an example, on the roll feeds of a punch press. On each stroke of the press crankshaft, a feed stroke on the roll feed is accomplished by means of the rack and pinion system. Rack and pinion system feeds the material into the dies of punch press.

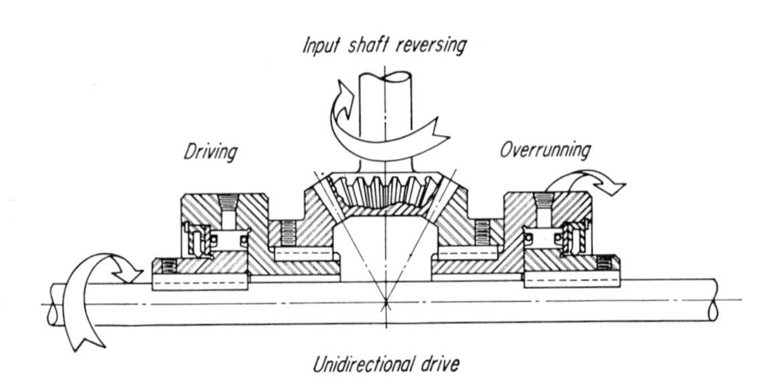

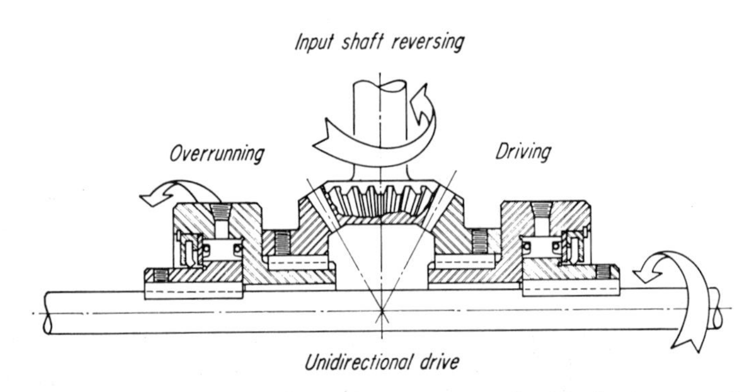

Fig. 4—Unidirectional drives with reverse mechanism by incorporating two overrunning clutches into the gears, sheaves or sprockets. Here, a 1:1 ratio right-angle drive is depicted having a reversing input shaft. The output shaft rotates clockwise regardless of input shaft direction. By changing gear sizes, combinations of continuous or intermittent unidirectional output relative to input is possible.

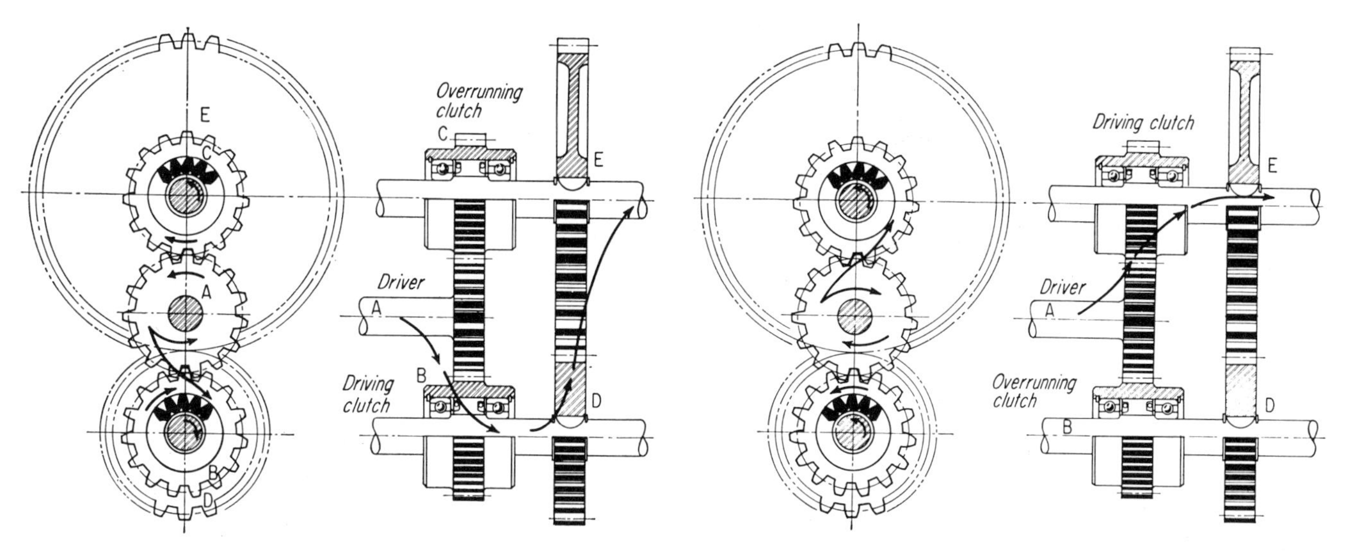

Fig. 5—Two speed unidirectional output is possible by using spur gears and reversing the direction of the input shaft. Rotation of shaft A power gears B, D, and E to output. Counterclockwise rotation engages lower clutch, freewheeling upper clutch as gear C is traveling at a faster rate than the shaft caused by reduction between gears B and E. Clockwise rotation of A engages upper clutch, while lower clutch freewheels because of the speed increase between gears D and E.

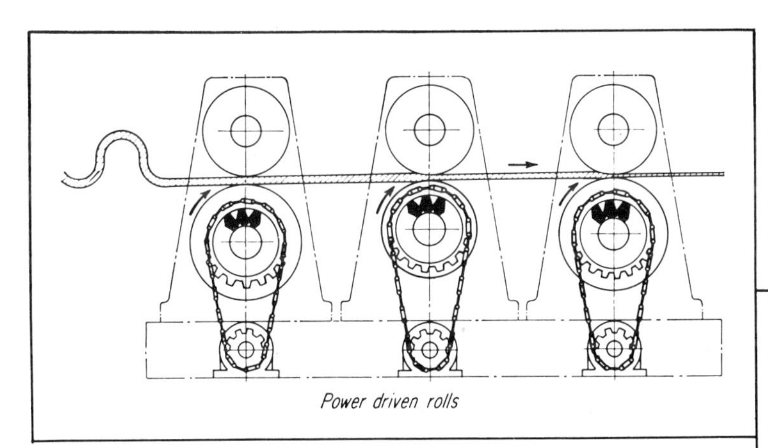

Fig. 6 (Left)—Speed-differential or compensation is required where a different speed range for a function is desired while retaining the same basic speed for all other functions. A series of individually power driven rolls may have different surface speeds because of drive or diameter variations of the rolls. An overrunning clutch permits the rolls of slower peripheral speed to overspeed and adjust to the material speed.

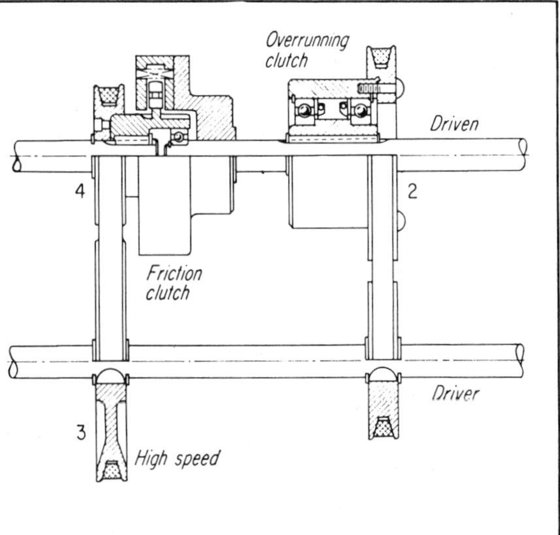

Fig. 7—Speed differential application permits operation of engine accessories within a narrow speed range while the engine operates over a wide range. Pulley No. 2 contains the overrunning clutch. When the friction or electric clutch is disengaged, driver pulley drives pulley No. 2 through the overrunning clutch, rotating the driven shaft. Engagement of the friction or electric clutch causes high speed driven shaft rotation causing an overrun condition in the clutch at pulley No. 2.

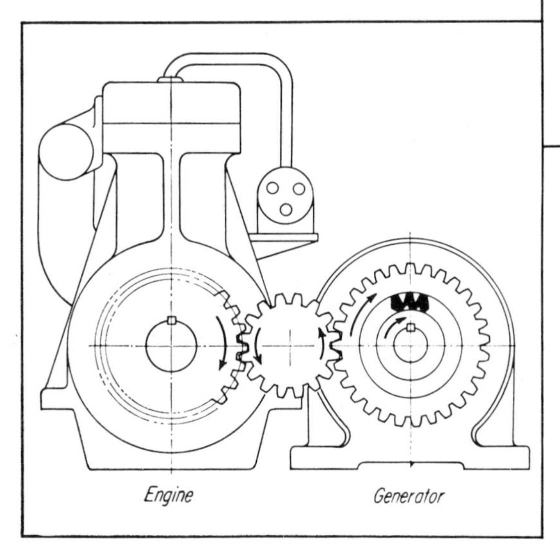

Fig. 8—High inertia dissipation is desirable to avoid driving back through a power system, or in machines having high resistances, to prevent power train damage. If the engine is shut down and the generator was under a "no-load" condition, there would be a tendency to twist off the generator shaft. Overrunning clutch allows generator deceleration at a slower rate than the engine deceleration.

Basic Types of Mechanical

Sketches include both friction and positive types. Figs. 1-7 are classified as externally controlled; Figs 8-12 are internally controlled. The latter are further divided into

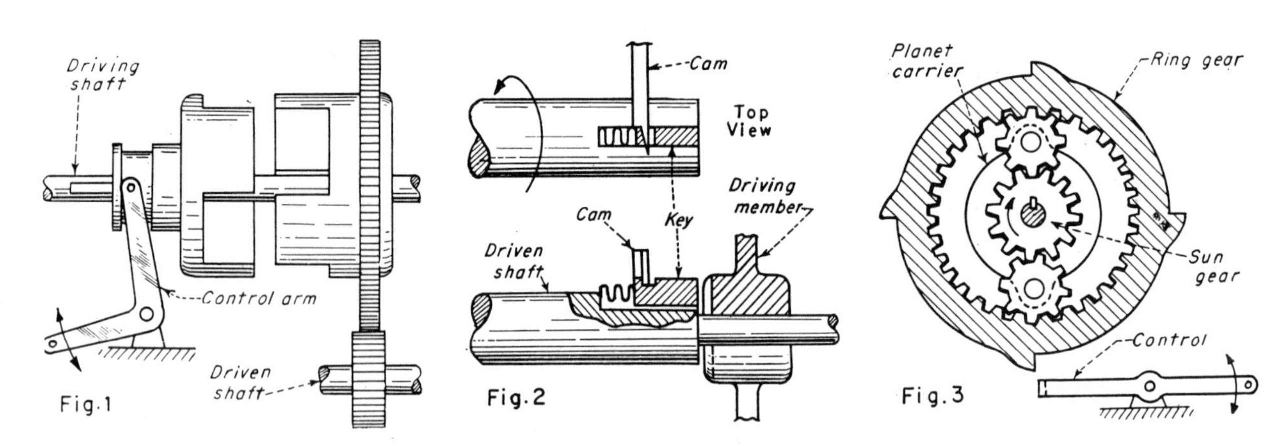

1. JAW CLUTCH. Left sliding half is keyed to the driving shaft while right half rotates freely. Control arm activates the sliding half to engage or disengage the drive. This clutch, though strong and simple, suffers from disadvantages of high shock during engagement, high inertia of the sliding half, and considerable axial motion required for engagement.

2. SLIDING KEY CLUTCH. Driven shaft with a keyway carries freely-rotating member which has radial slots along its hub; sliding key is spring loaded but is normally restrained from engaging slots by the control cam. To engage the clutch, control cam is raised and key enters one of the slots. To disengage, cam is lowered into the path of the key; rotation of driven shaft forces key out of slot until step on control cam prevents further motion.

3. PLANETARY TRANSMISSION CLUTCH. In disengaged position shown, driving sun gear will merely cause the free-wheeling ring gear to idle counter-clockwise, while the driven member, the planet carrier, remains motionless. If motion of the ring gear is blocked by the control arm, a positive clockwise drive is established to the driven planet carrier.

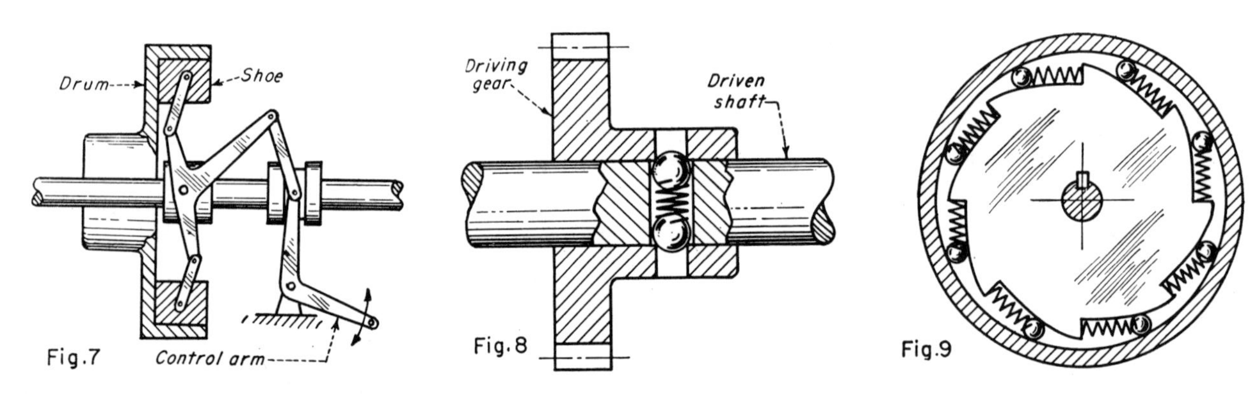

7. EXPANDING SHOE CLUTCH. In sketch above, engagement is obtained by motion of control arm which operates linkages to force friction shoes radially outward into contact with inside surface of drum.

8. SPRING AND BALL RADIAL DETENT CLUTCH. This design will positively hold the driving gear and driven shaft in a given timing relationship until the torque becomes excessive. At this point the balls will be forced inward against their spring pressure and out of engagement with the holes in the hub, thus permitting the driving gear to continue rotating while the driven shaft is stationary.

9. CAM AND ROLLER CLUTCH. This over-running clutch is suited for higher speed free-wheeling than the pawl and ratchet types. The inner driving member has camming surfaces at its outer rim and carries light springs that force rollers to wedge between these surfaces and the inner cylindrical face of the driven member. During driving, self-energizing friction rather than the springs forces the roller to tightly wedge between the members and give essentially positive drive in a clockwise direction. The springs insure fast clutching action. If the driven member should attempt to run ahead of the driver, friction will force the rollers out of a tight wedging position and break the connection.

Clutches

MARVIN TAYLOR
Mechanical Research & Development Department,
Monroe Calculating Machine Company

overload relief, over-riding, and centrifugal types. A second article in a subsequent issue will describe mechanical clutches for precise service in calculating machines.

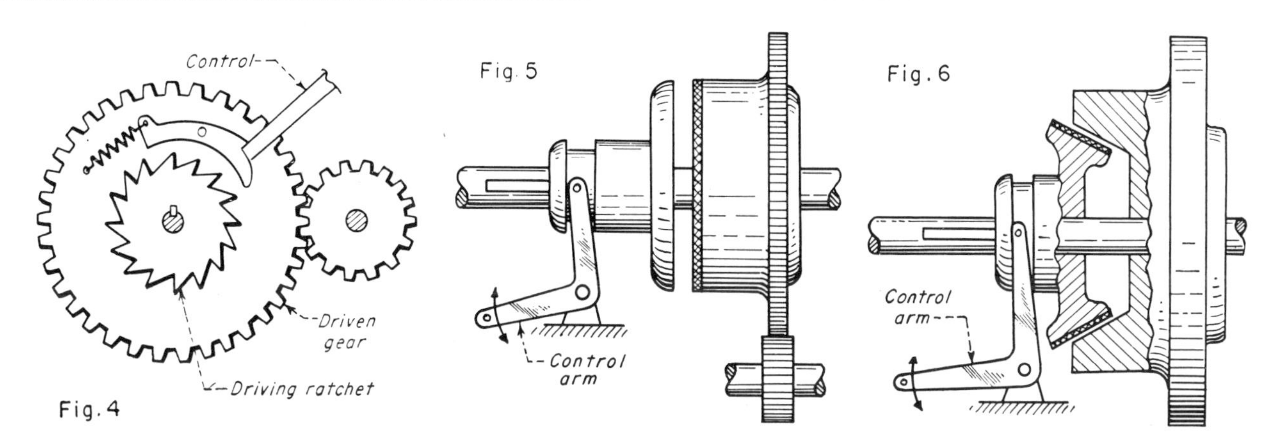

4. PAWL AND RATCHET CLUTCH. (External Control). Ratchet is keyed to the driving shaft; pawl is carried by driven gear which rotates freely on the driving shaft. Raising the control member permits the spring to pull the pawl into engagement with the ratchet and drive the gear. Engagement continues until control member is lowered into the path of a camming surface on the pawl. The motion of the driven gear will then force the pawl out of engagement and bring the driven assembly to a solid stop against the control member. This clutch can be converted into an internally controlled type of unit by removing the external control arm shown at the upper right.

5. PLATE CLUTCH. Available in many variations, with single and multiple plates, this unit transmits power through friction force developed between the faces of the left sliding half which is fitted with a feather key and the right half which is free to rotate on the shaft. Torque capacity depends upon the axial force exerted by the control member when it activates the sliding half.

6. CONE CLUTCH. This type also requires axial movement for engagement, but the axial force required is less than that required with plate clutches. Friction material is usually applied to only one of the mating surfaces.

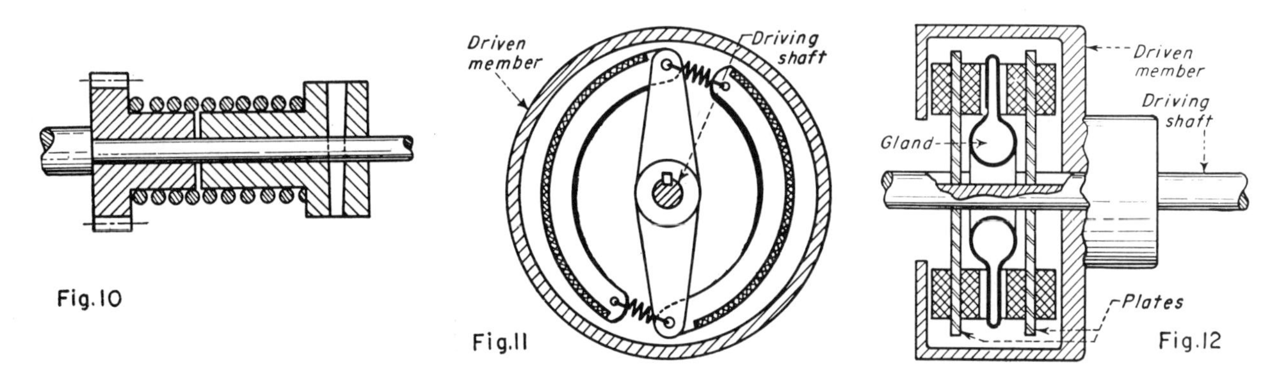

10. WRAPPED SPRING CLUTCH. Makes a simple and inexpensive uni-directional clutch consisting of two rotating members connected by a coil spring which fits snugly over both hubs. In the driving direction the spring tightens about the hubs producing a self energizing friction grip; in the opposite direction it unwinds and will slip.

11. EXPANDING SHOE CENTRIFUGAL CLUTCH. Similar in action to the unit shown in Fig. 7 with the exception that no external control is used. Two friction shoes, attached to the driving member, are held inward by springs until they reach the "clutch-in" speed, at which centrifugal force energizes the shoes outward into contact with the drum. As the driver rotates faster the pressure between the shoes and the drum increases thereby providing greater torque capacity.

12. MERCURY GLAND CLUTCH. Contains two friction plates and a mercury filled rubber gland, all keyed to the driving shaft. At rest, mercury fills a ring shaped cavity near the shaft; when revolved at sufficient speed, the mercury is forced outward by centrifugal force spreading the rubber gland axially and forcing the friction plates into driving contact with the faces of the driven housing.

Small Mechanical Clutches for

Clutches used in calculating machines must have: (1) Quick response—lightweight moving parts; (2) Flexibility—permit multiple members to

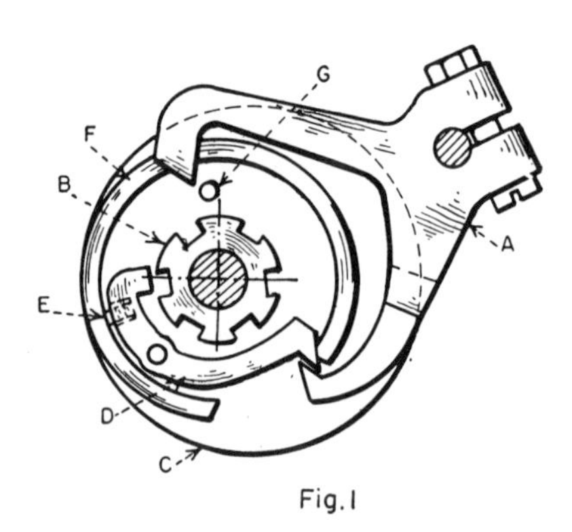

Fig. 1

PAWL AND RATCHET SINGLE CYCLE CLUTCH (Fig. 1). Known as Dennis Clutch, parts *B*, *C* and *D*, are primary components, *B*, being the driving ratchet, *C*, the driven cam plate and, *D*, the connecting pawl carryied by the cam plate. Normally the pawl is held disengaged by the lower portion of clutch arm *A*. When activated, arm *A* rocks counter-clockwise until it is out of the path of rim *F* on cam plate *C* and permits pawl *D* under the effect of spring *E* to engage with ratchet *B*. Cam plate *C* then turns clockwise until, near the end of one cycle, pin *G* on the plate strikes the upper part of arm *A* camming it clockwise back to its normal position. The lower part of *A* then performs two functions: (1) cams pawl *D* out of engagement with the driving ratchet *B* and (2) blocks further motion of rim *F* and the cam plate.

PAWL AND RATCHET SINGLE CYCLE DUAL CONTROL CLUTCH—(Fig. 2). Principal parts are: driving ratchet, *B*, directly connected to the motor and rotating freely on rod *A*; driven crank, *C*, directly connected to the main shaft of the machine and also free on A; and spring loaded ratchet pawl, *D*, which is carried by crank, *C*, and is normally held disengaged by latch *E*. To activate the clutch, arm *F* is raised, permitting latch *E* to trip and pawl *D* to engage with ratchet *B*. The left arm of clutch latch *G*, which is in the path of the lug on pawl *D*, is normally permitted to move out of interference by the rotation of the camming edge of crank *C*. For certain operations block *H* is temporarily lowered, preventing motion of latch *G*, resulting in disengagement of the clutch after part of the cycle until subsequent raising of block *H* permits motion of latch *G* and resumption of the cycle.

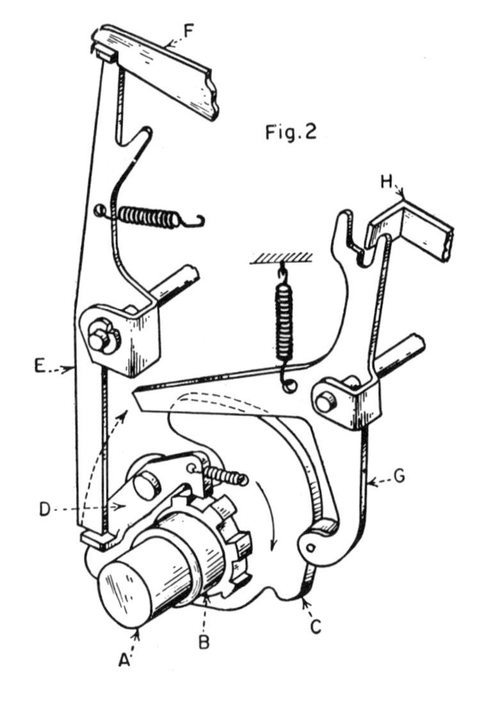

PLANETARY TRANSMISSION CLUTCH (Fig. 3). A positive clutch with external control, two gear trains to provide bi-directional drive to a calculator for cycling the machine and shifting the carriage. Gear *A* is the driver, gear *L* the driven member is directly connected to planet carrier *F*. The planet consists of integral gears *B* and *C*; *B* meshing with sun gear *A* and free-wheeling ring gear *G*, and *C* meshing with free-wheeling gear *D*. Gears *D* and *G* carry projecting lugs, *E* and *H* respectively, which can contact formings on arms *J* and *K* of the control yoke. When the machine is at rest, the yoke is centrally positioned so that the arms *J* and *K* are out of the path of the projecting lugs permitting both *D* and *G* to free-wheel. To engage the drive, the yoke rocks clockwise as shown, until the forming on arm *K* engages lug *H* blocking further motion of ring gear *G*. A solid gear train is thereby established driving *F* and *L* in the same direction as the drive *A* and at the same time altering the speed of *D* as it continues counter-clockwise. A reversing signal rotates the yoke counter-clockwise until arm *J* encounters lug *E* blocking further motion of *D*. This actuates the other gear train of the same ratio.

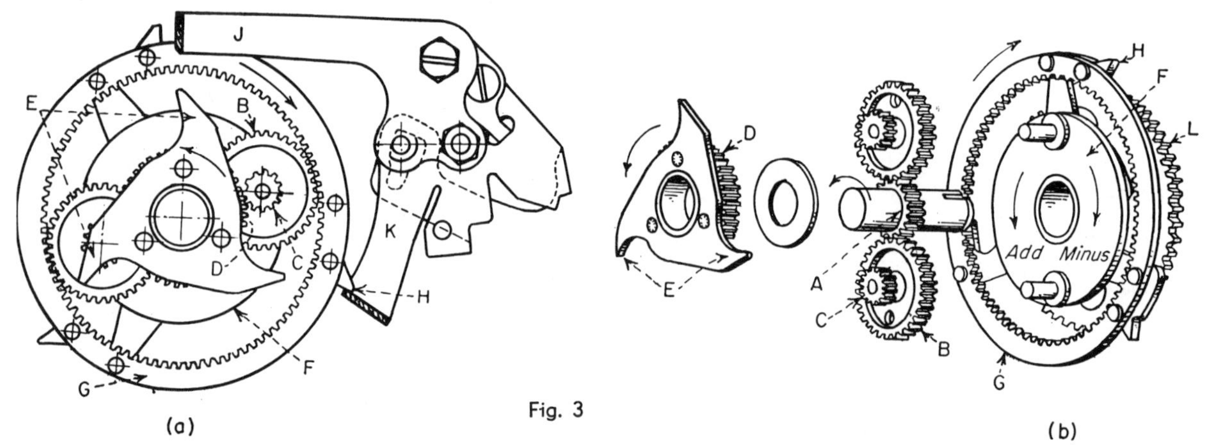

Fig. 3

Precise Service

MARVIN TAYLOR
Monroe Calculating Machine Company

control operation; (3) Compactness—for equivalent capacity positive clutches are smaller than friction; (4) Dependability; and (5) Durability.

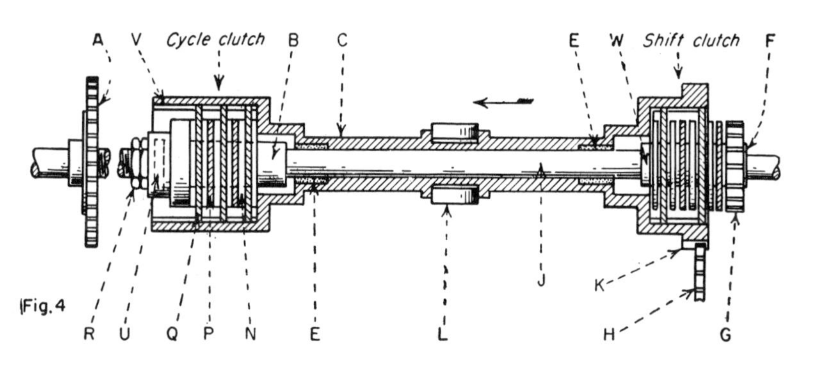

MULTIPLE DISK FRICTION CLUTCH (Fig. 4). Two multiple disk friction clutches are combined in a single two-position unit which is shown shifted to the left. A stepped cylindrical housing *C* enclosing both clutches is carried by self-lubricated bearing *E* on shaft *J* and is driven by the transmission gear *H* meshing with the housing gear teeth *K*. At either end, the housing carries multiple metal disks *Q* that engage keyways *V* and can make frictional contact with formica disks *N* which, in turn, can contact a set of metal disks *P* which have slotted openings for coupling with flats on sleeves *B* and *W*. In the position shown, pressure is exerted through rollers *L* forcing the housing to the left making the left clutch compact against adjusting nuts *R*, thereby driving gear *A* via sleeve *B* which is connected to jack shaft *J* by pin *U*. When the carriage is to be shifted, rollers *L* force the housing to the right, first relieving the pressure between the adjoining disks on the left clutch then passing through a neutral position in which both clutches are disengaged and finally making the right clutch compact against thrust bearing *F*, thereby driving gear *G* through sleeve *W* which rotates freely on the jack shaft.

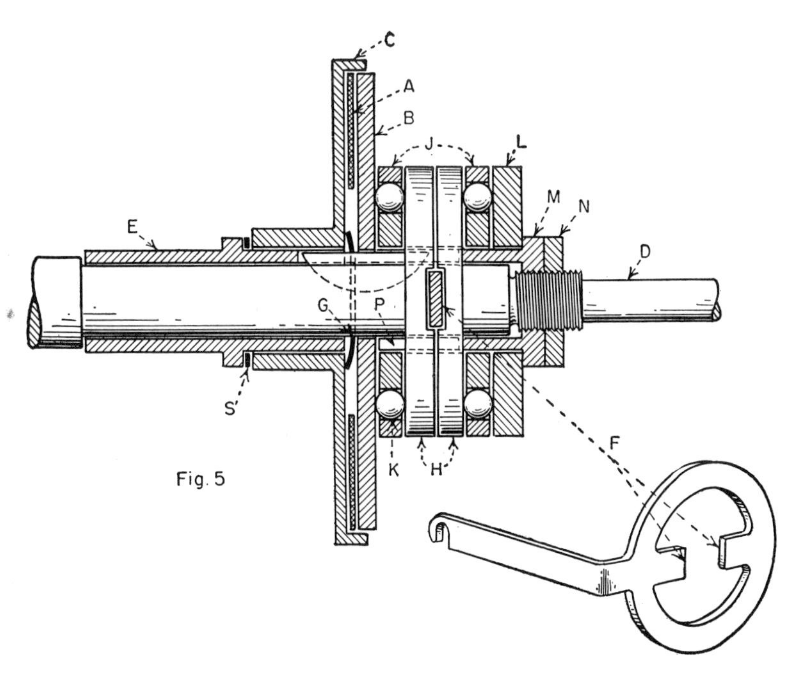

SINGLE PLATE FRICTION CLUTCH (Fig. 5). The basic clutch elements, formica disk *A*, steel plate *B* and drum *C*, are normally kept separated by spring washer *G*. To engage the drive, the left end of a control arm is raised, causing ears *F*, which sit in slots in plates *H*, to rock clockwise spreading the plates axially along sleeve *P*. Sleeves *E* and *P* and plate *B* are keyed to the drive shaft; all other members can rotate freely. The axial motion loads the assembly to the right through the thrust ball bearings *K* against plate *L* and adjusting nut *M*, and to the left through friction surfaces on *A*, *B* and *C* to thrust washer *S*, sleeve *E* and against a shoulder on shaft *D*, thus enabling plate *A* to drive the drum *C*.

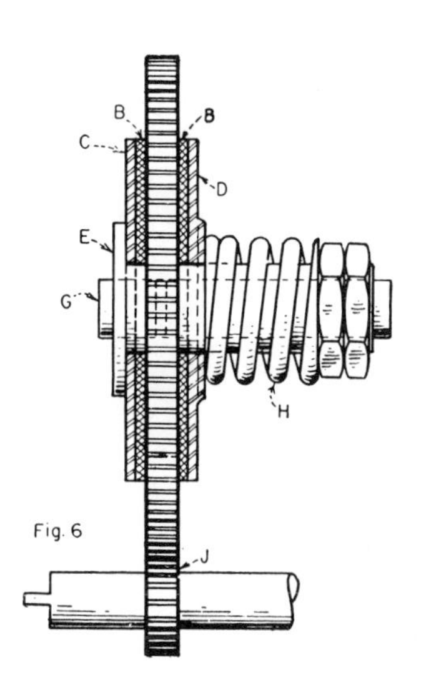

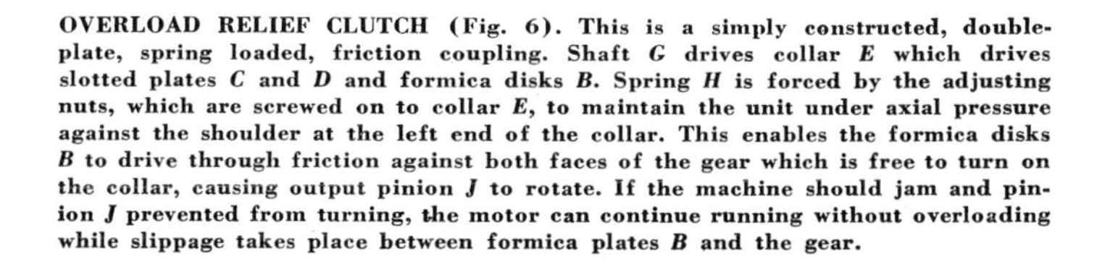
OVERLOAD RELIEF CLUTCH (Fig. 6). This is a simply constructed, double-plate, spring loaded, friction coupling. Shaft *G* drives collar *E* which drives slotted plates *C* and *D* and formica disks *B*. Spring *H* is forced by the adjusting nuts, which are screwed on to collar *E*, to maintain the unit under axial pressure against the shoulder at the left end of the collar. This enables the formica disks *B* to drive through friction against both faces of the gear which is free to turn on the collar, causing output pinion *J* to rotate. If the machine should jam and pinion *J* prevented from turning, the motor can continue running without overloading while slippage takes place between formica plates *B* and the gear.

FRICTION CLUTCHES

Types of automotive clutches showing how cones, single and multiple disks and centrifugally driven shoes are used as mechanism elements

HERBERT CHASE

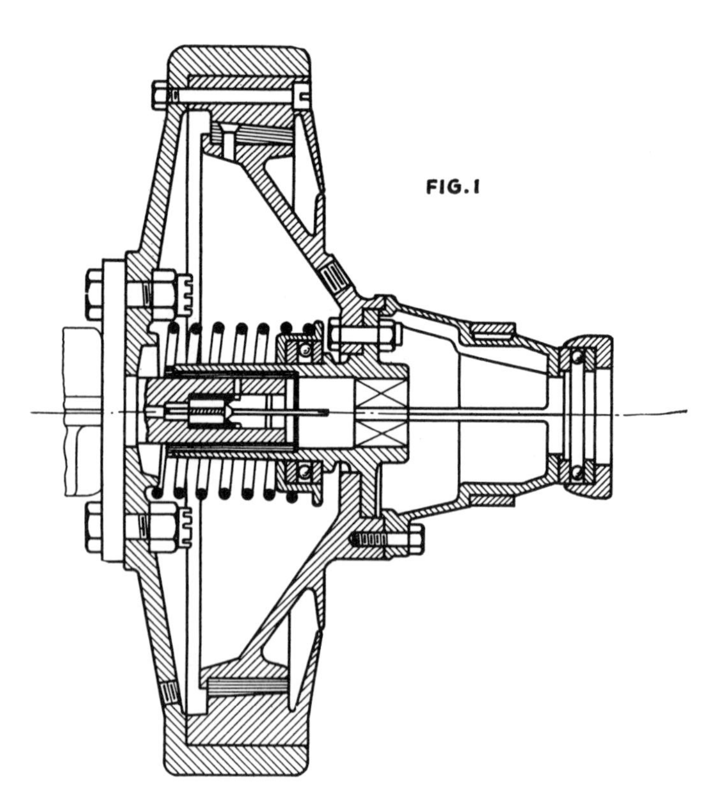

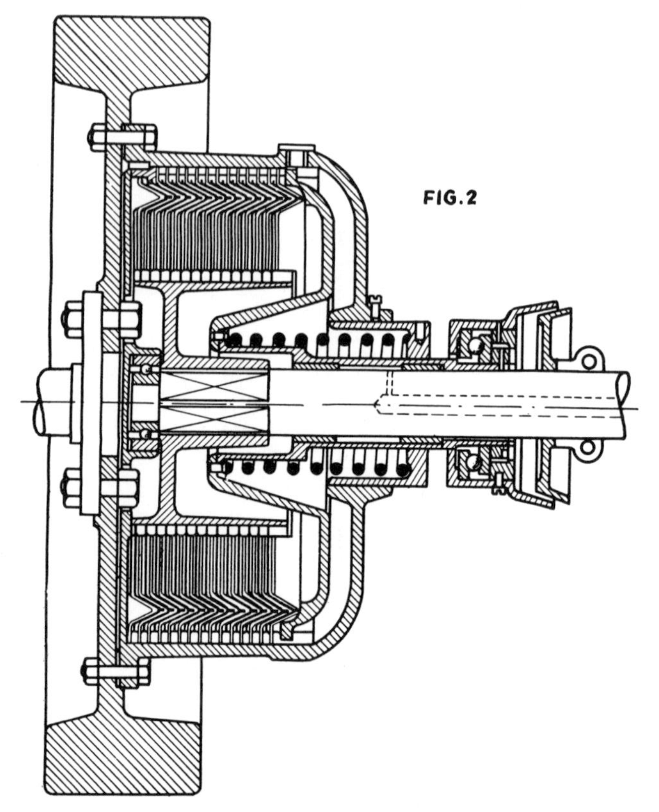

Fig. 1—Reversed cone clutch which moved toward engine to disengage the friction surface, on Thames car, employed a single central spring. Ball thrust bearings were used to take throw-out pressure. Leather friction material at that time had low coefficient of friction and was adversely affected by any considerable temperature rise, hence a relatively-large surface and heavy pressure was used to transmit relatively-small power outputs.

Fig. 2—Hele-Shaw clutch of multiple-disk type in which the disks were formed with a V-shape groove near their outer diameter. With this design sufficient torque capacity was obtained by nesting adjacent disks thus producing a wedging action with metal-to-metal contact at the V-grooves. Disks required accurate forming and were subject to distortion if overheated. Dragging was also difficult to prevent. A single axially-disposed spring provided engagement pressure.

Fig. 3—Hudson multiple-disk clutch fitted into a recess cast in the flywheel and was closed by a plate cover. Pressure was maintained by a single helical spring housed in a recess in end of crankshaft. Driving torque was applied to the disks by pins bolted to the flywheel and throw-out pressure was taken by a ball type thrust bearing.

Fig. 4—Development of friction facings capable of withstanding rather high temperatures without serious injury or change in friction coefficient made the single-plate clutch feasible. In the clutch shown, pressure of the engaging spring was applied through a set of levers, which were pivoted on a cover plate bolted to the flywheel, to a plate which forced the friction facings on the driven disk against the flywheel. The disk was splined to the clutch shaft.

Fig. 5—This single-plate clutch used in the Willys Six had direct-acting springs spaced around the pressure plate with short disengaging levers hooked into the plate in such a way that the latter was positively withdrawn when the clutch was disengaged. The driven plate was a light stamping to which the friction facings were riveted; its hub was splined to the clutch shaft.

Fig. 6—Known as the "Talbot Traffic" type, this clutch is of the centrifugal form with two shoes pivoted to the outer race of a central "reverse" free-wheel device, this race being the driving member bolted to flywheel. The driven member, splined to the transmission shaft, is the inner member of the free-wheeling element. Shoes are described as having a "trailing non-servo action" tending to prevent grab. Pivoted links with compression springs form the connecting members between the toe of one shoe and the heel of the other, the center of the links being pivoted to an annulus member capable of slight turning motion about the outer race of the free wheel. Under centrifugal force, shoes turn about their pivots until spring pressure is overcome, allowing shoes to contact and drive the drum. Only when the inner race tends to move clockwise in relation to the outer race is the drive transmitted.

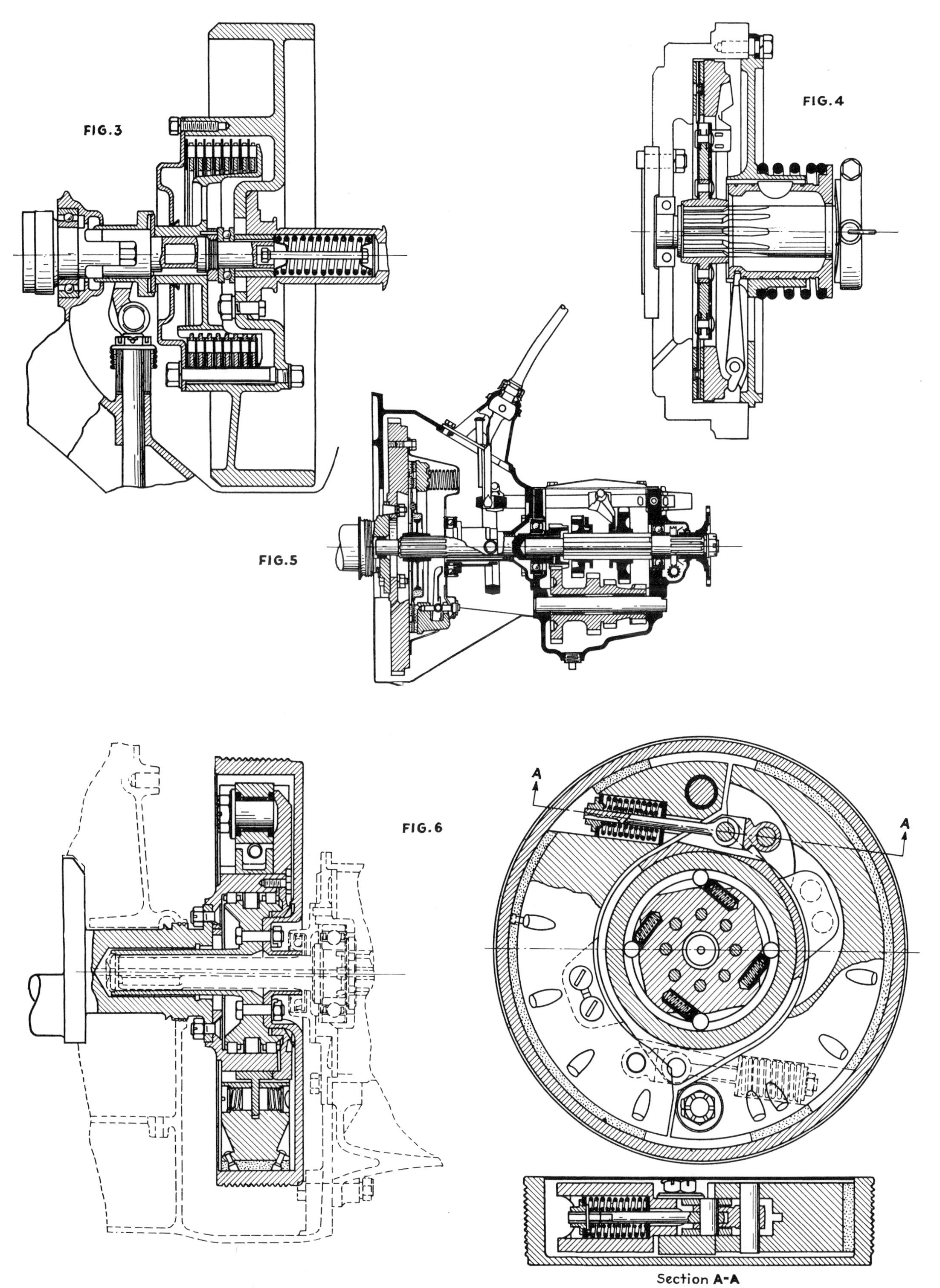
FIG.3
FIG.4
FIG.5
FIG.6
A
A
Section A-A

Mechanisms for STATION CLUTCHES

HERMANN HILL

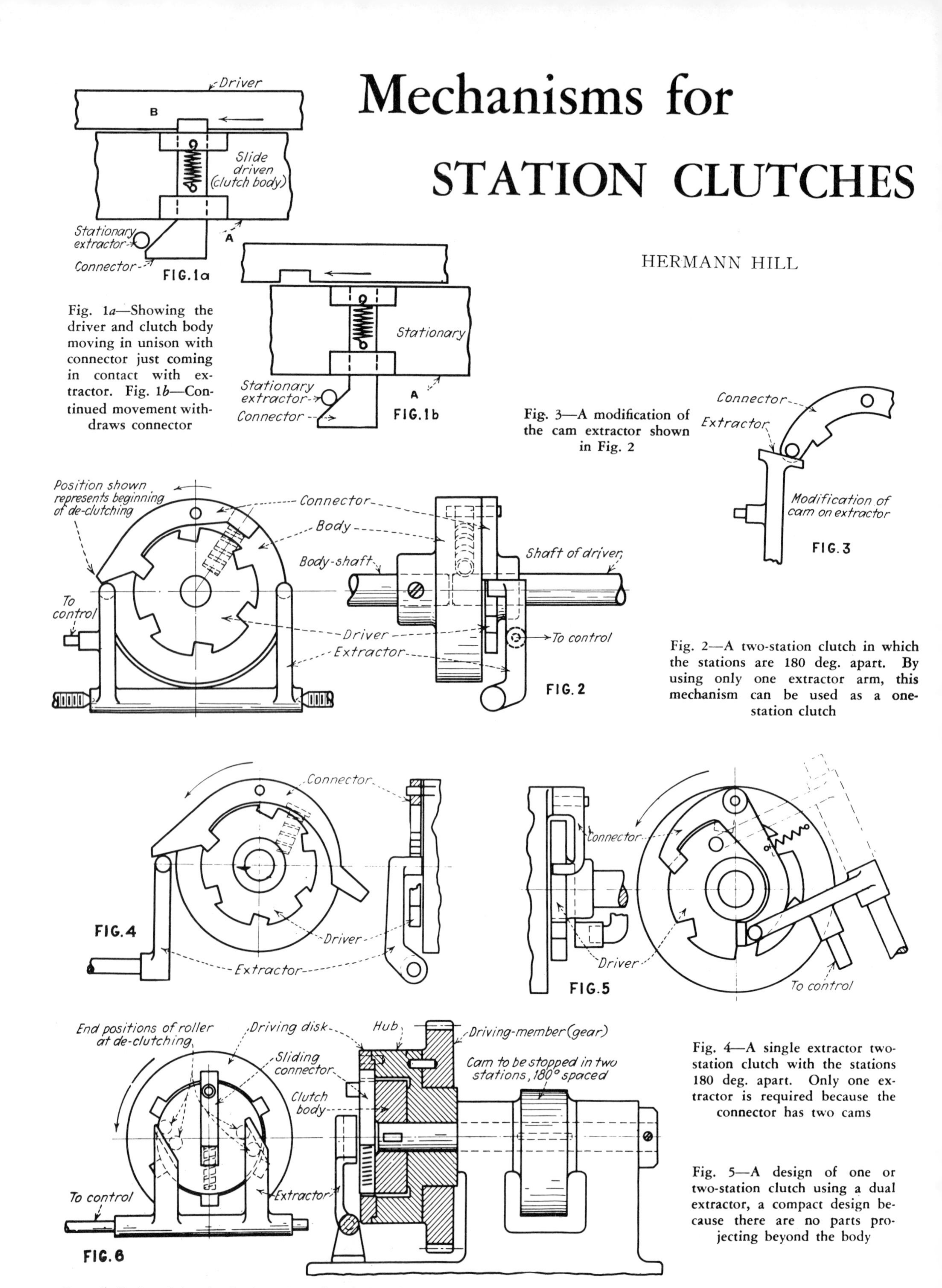

Fig. 1*a*—Showing the driver and clutch body moving in unison with connector just coming in contact with extractor. Fig. 1*b*—Continued movement withdraws connector

Fig. 3—A modification of the cam extractor shown in Fig. 2

Fig. 2—A two-station clutch in which the stations are 180 deg. apart. By using only one extractor arm, this mechanism can be used as a one-station clutch

Fig. 4—A single extractor two-station clutch with the stations 180 deg. apart. Only one extractor is required because the connector has two cams

Fig. 5—A design of one or two-station clutch using a dual extractor, a compact design because there are no parts projecting beyond the body

Fig. 6—End and longitudinal section of a design of a station clutch using internal driving recesses

Innumerable variations of these station clutches may be designed for starting and stopping machines at selected points in their cycle of operation

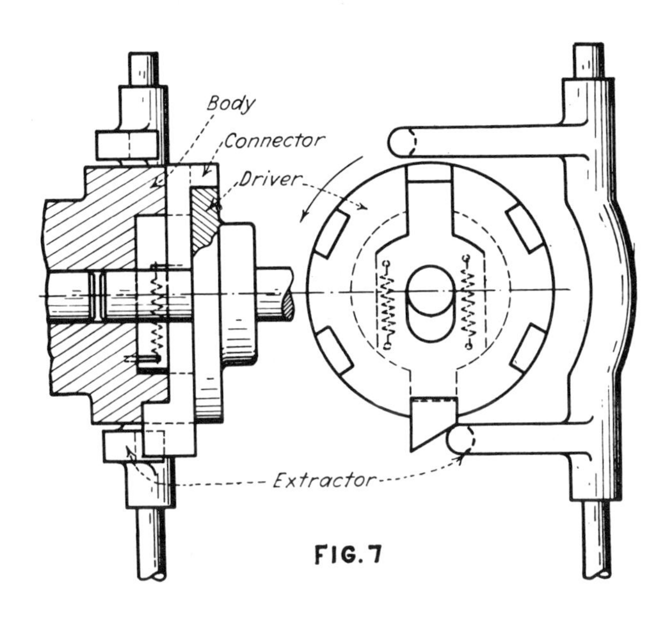

Fig. 7—A one or two-station clutch, depending on the use of a single or a dual extractor, the stations being spaced 180 deg. apart

Fig. 8—Another design of one or two-station clutch, using a single or dual extractor with stations spaced 180 deg. apart

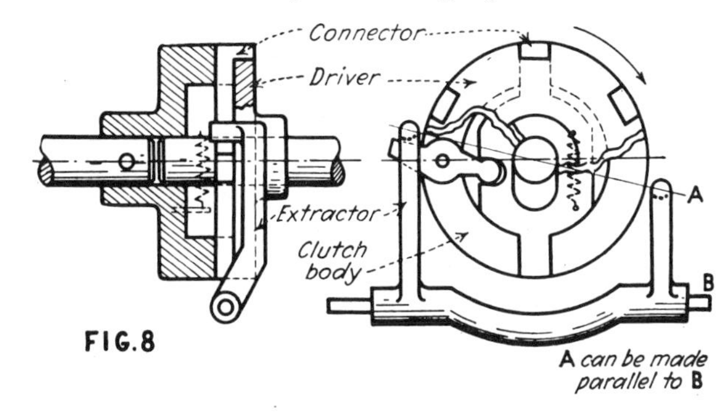

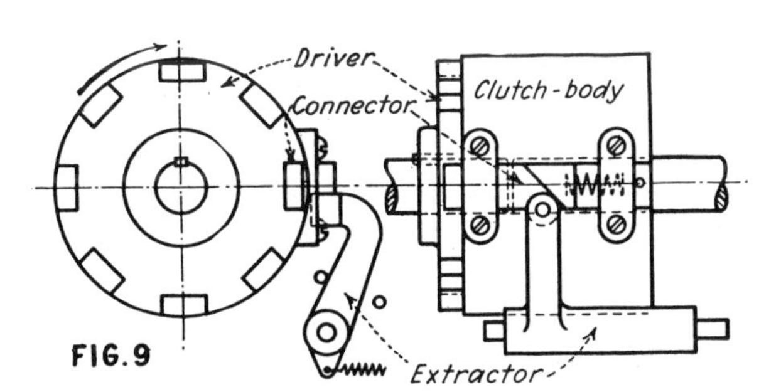

Fig. 9—A design of one-station clutch of the axail conector type

Fig. 11—A typical design of multi-station clutch of the non-selective type for instantaneous stopping at any position

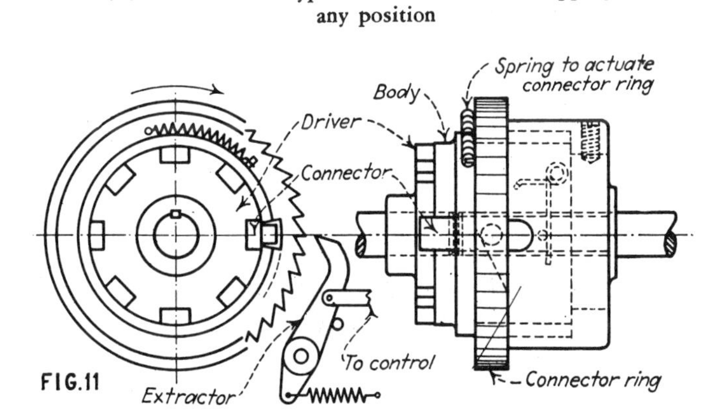

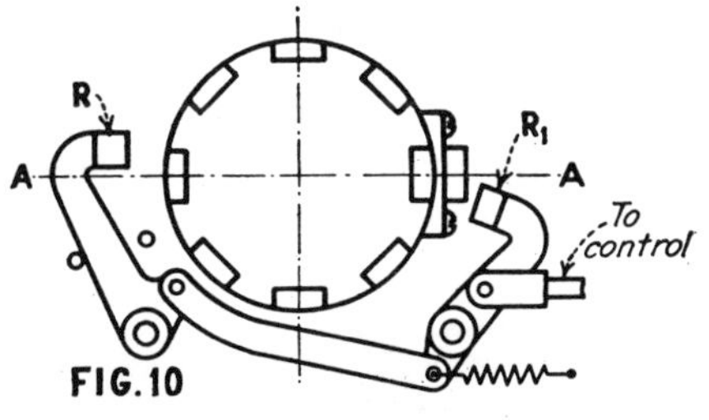

Fig. 10—In this design of two-station clutch the roller R and R_1 of the extractor may also be arranged on the center line A-A

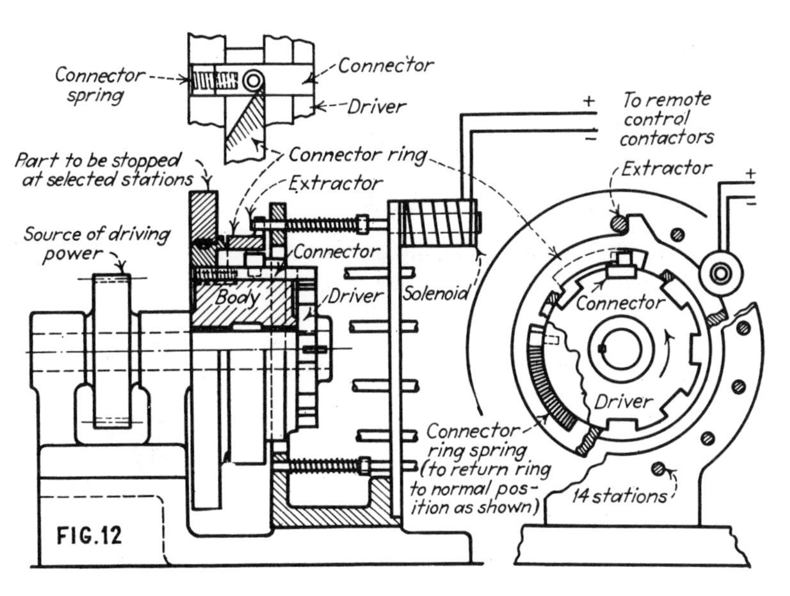

Fig. 12—A design of multi-station clutch with remote control. The extractor pins are actuated by solenoids which either hold the extractor pin in position against spring pressure, or release the pin

Parallel-link coupling

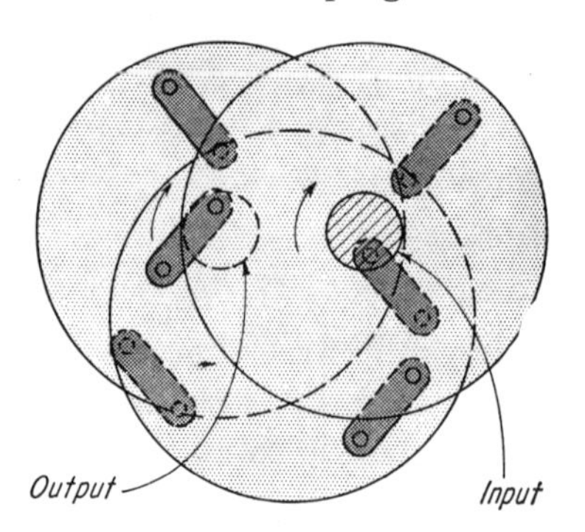

This arrangement of six links and three disks can synchronize the motion between adjacent, parallel shafts.

Bent-pin coupling

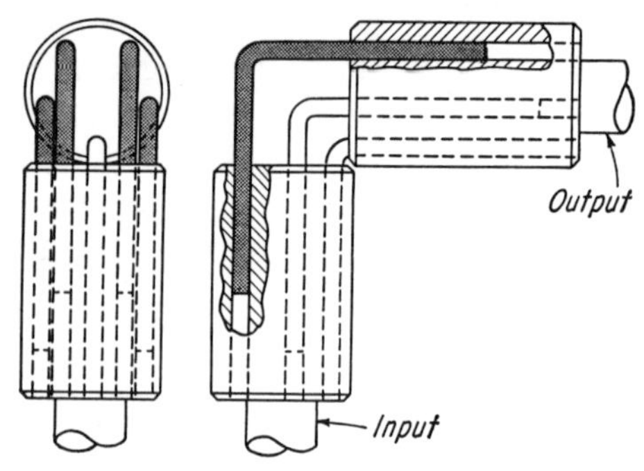

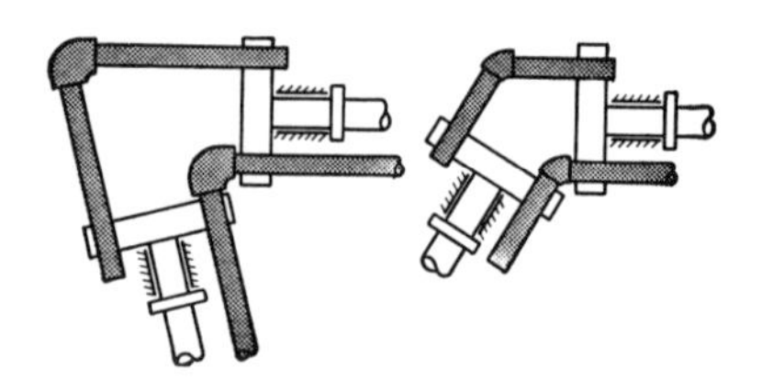

As the input rotates, the five bent pins will move in and out of the drilled holes to impart a constant velocity rotation to the right angle output shaft. The device can transmit constant velocity at angles other than 90 degrees, as shown.

Six-disk coupling

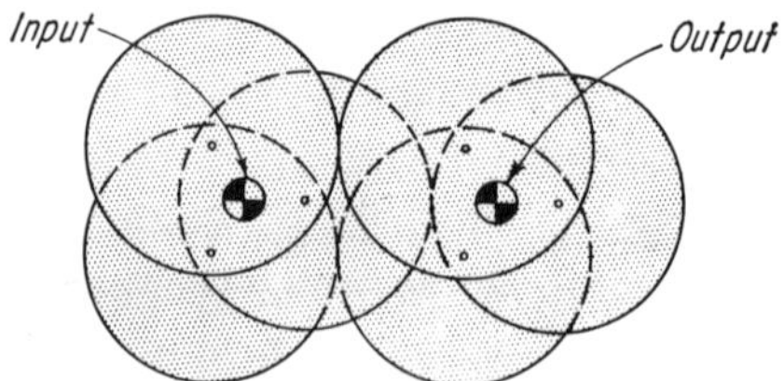

Here's another arrangement to synchronize shafts, but without need for links.

PREBEN W. JENSEN

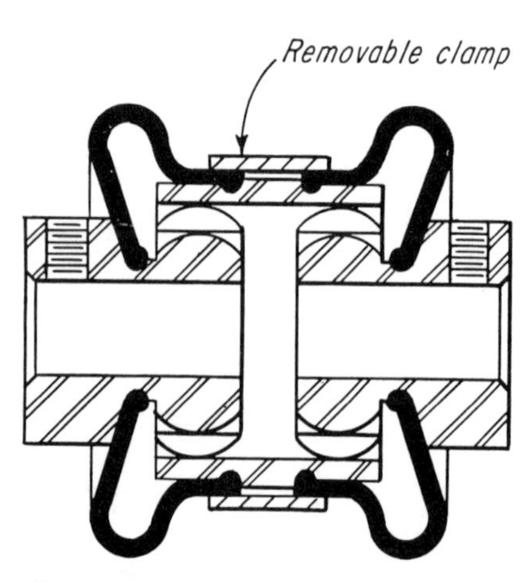

Heavy-duty flexible coupling . . . has neoprene seals that retain lubricant and exclude dust. Can handle angular offset to 15 deg, parallel offset to $\frac{3}{16}$ in. Torque rating is 5.2 hp at 1000 rpm. Bore sizes to ¾ in. dia. Fourdee Inc

Spring-wound coupling . . . also functions as a universal joint. This novel device by Lovejoy Flexible Coupling Co., Chicago, employs three to four layers of springs to combine high flexibility with high strength. Springs are of square wire, each layer triple-wound (with three interwoven strands). Layers are also wound in alternate directions—outer layer clockwise, second layer counterclockwise, etc.—and nested, one inside the other (for nested-spring design, see p. 61). Thus, under misalignment, one layer tightens, the other loosens. This permits angular misalignment to 20 deg, parallel misalignment to 5% of bore size, and axial displacement equal to shaft diameter. All-steel construction, requires no lubrication, and provides uniform velocity even during misalignment. Springs custom-designed to meet torsional-resiliency requirements. Horsepower to 250 hp at 1750 rpm, speed range to 6000 rpm, and bore from ½ to 2 in. Can carry 200% temporary overload and is suitable for low-speed high-torque applications.

Triple-strand spring

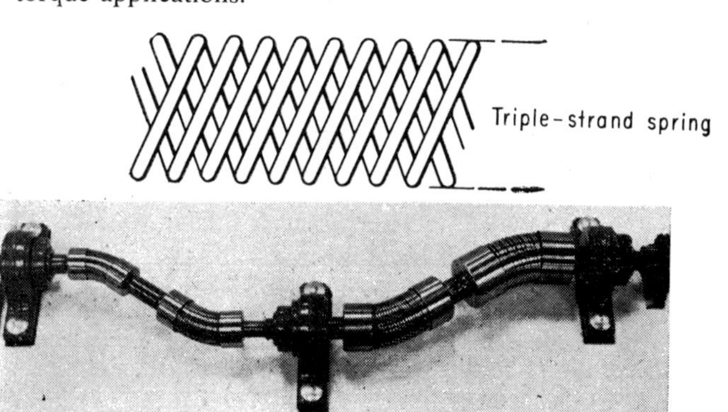

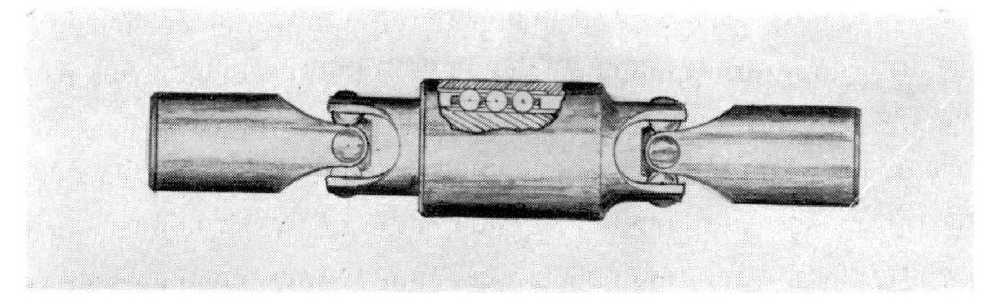

Telescoping universal joint . . .
for precise transmission of data consists of two preloaded ball splines sealed with lubricant in a slip joint. Backlash is said to be zero for entire assembly, which is made of stainless steels. Following are standard assemblies and all have $\frac{1}{4}$-in. lateral travel: $\frac{3}{16}$ in. body with $\frac{3}{32}$ or $\frac{1}{8}$ in. bores; $\frac{9}{32}$ in. body with $\frac{3}{16}$ in. bore; $\frac{3}{8}$ in. body with $\frac{1}{4}$ in. bore. Torque ratings are 16, 64 and 256 in.-oz, respectively.

Falcon Machine & Tool Co

The coupling is a centrifugal ball drive for asynchronous motors. It consists simply of a vaned rotor mounted on the motor shaft and enclosed in a housing which is coupled to the driven machine. The space between the rotor and housing is filled with small steel balls (see sketch).

The motor starts under light load, but then the rotating vanes set the balls in motion and centrifugal force drives them against the inner face of the housing, setting the housing—and the output shaft to which it is mounted—into rotation. In that way speed builds up until full power is attained.

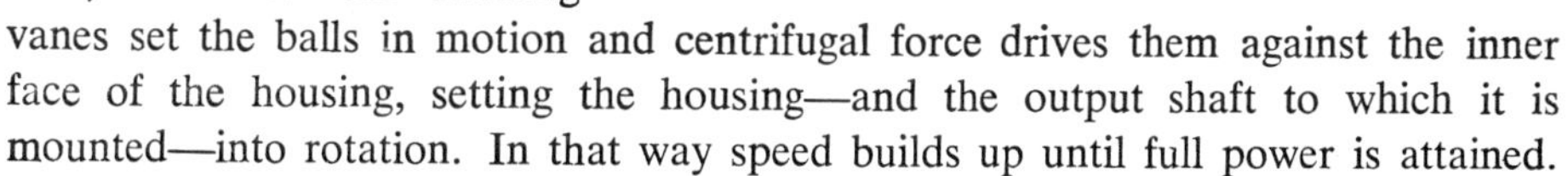

Ste Industrielle de Transmissions Colombes-Texrope, 116-120, rue Danton, Levallois-Perret (Seine), France, which is making the unit, says the advantage of the system is that the slippage between balls and housing absorbs sudden shock overloads, prevents damage to the motor, and makes for smoother starting. The couplings can be designed for motors ranging from ½ to 2000 hp, and design can be modified to fit various machine applications. One version has pulleys for V-belt and flat-belt drives.

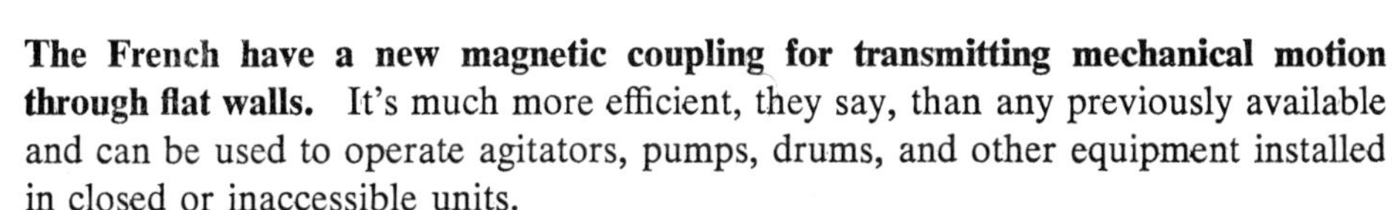

The French have a new magnetic coupling for transmitting mechanical motion through flat walls. It's much more efficient, they say, than any previously available and can be used to operate agitators, pumps, drums, and other equipment installed in closed or inaccessible units.

The design is simple and straightforward. There are two identical assemblies, one on each side of the wall (see diagram). Each consists of a horseshoe-shaped permanent magnet attached to a rectangular soft-iron plate, the long axis of which is at right angles to the axis of the magnet poles. The two magnet assemblies are coupled so that the magnetic flux crossing the poles of each magnet completes its circuit, after crossing the wall, through the soft-iron plate of the other.

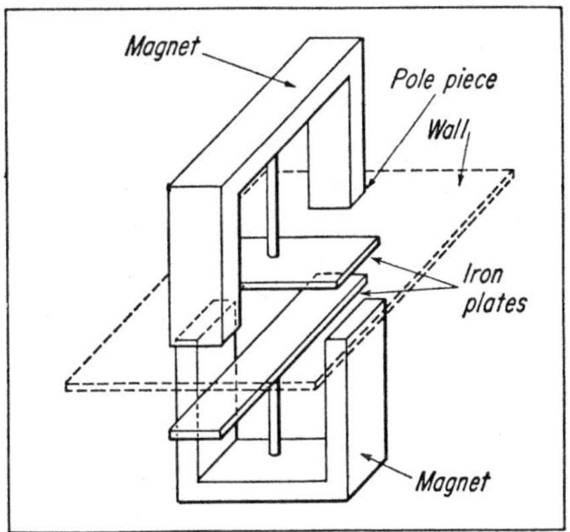

French Atomic Energy Commission engineer Guy Cherel, who designed the device, says that by using small, widely separated pole pieces rather than a single magnet it is possible to achieve a coupling that is three to five times stronger than can be attained with a single magnet.

10 UNIVERSAL

FEDERICO STRASSER

HOOKE'S JOINTS

The commonest form of a universal coupling is Hooke's joint. This can transmit torque efficiently up to a maximum shaft-alignment angle of about 36°. At slow speeds, on hand-operated mechanisms, the permissible angle can reach 45° Simplest arrangement for a Hooke's joint is two forked shaft-ends coupled by a cross-shaped piece. There are many variations and a few of them are included here.

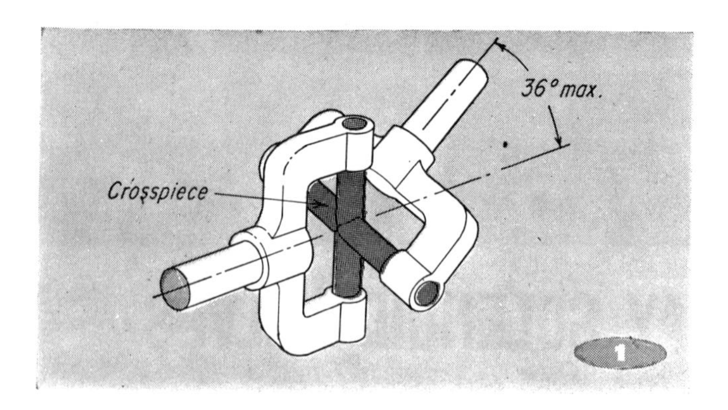

Basic design . . .

of Hooke's joint can transmit heavy loads. Anti-friction bearings are a refinement often used.

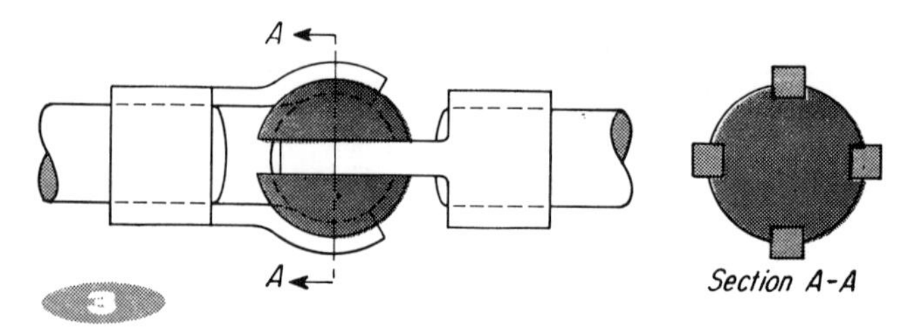

Grooved sphere . . .

is modification of pinned sphere. Tongues on fastening sleeves are bent over sphere on assembly. Greater sliding contact of tongues in grooves makes ample lubrication essential at high torques and alignment angles.

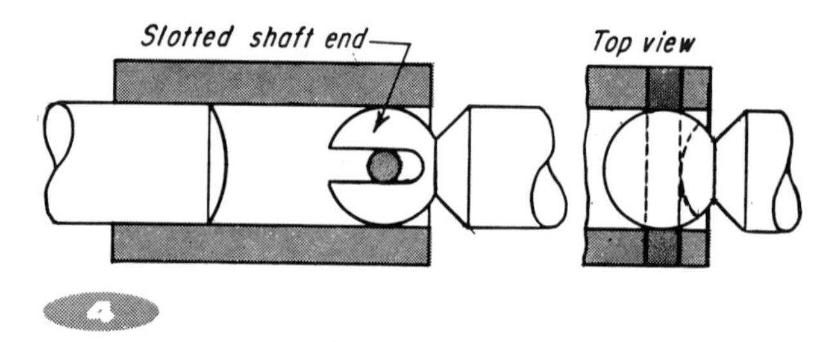

Pinned sleeve . . .

fastened to one shaft engages forked, spherical end on other shaft to provide joint that also allows for axial shaft-movement. In this example, however, angle between shafts can only be small. Also, joint is only suitable for low torques.

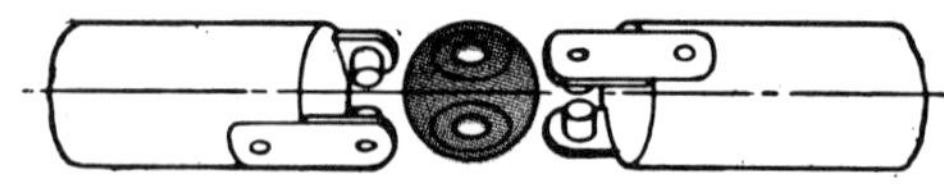

Pinned sphere . . .

replaces crosspiece in this design. Result is a more compact joint.

CONSTANT-VELOCITY COUPLINGS

The disadvantage of a single Hooke's joint is that velocity of driven shaft varies. Its maximum velocity can be found by multiplying driving-shaft speed by secant of the shaft angle; for minimum speed multiply by the cosine. An example of speed variation: Driving shaft rotates at 100 rpm; angle between shafts is 20°. Min output is 100 x 0.9397, which equals 93.97 rpm; max output is 1.0642 x 100, or 106.4 rpm. Thus the difference is 12.43 rpm. When output speed is high, output torque is low and vice versa. This is an objectionable feature in some mechanisms. However, two universal joints connected by an intermediate shaft solve the problem.

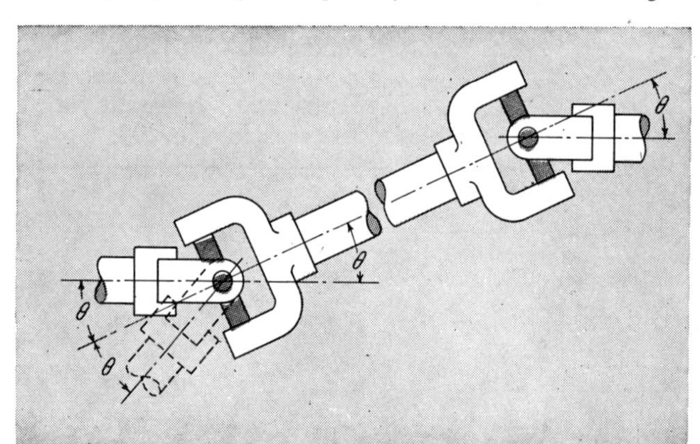

Constant-velocity . . .

joint made by coupling two Hooke's joints must have input and output angles equal for correct action. Also, the forks must be assembled so that they will always be in the same plane. Shaft-alignment angle may be double that for a single joint.

SHAFT-COUPLINGS

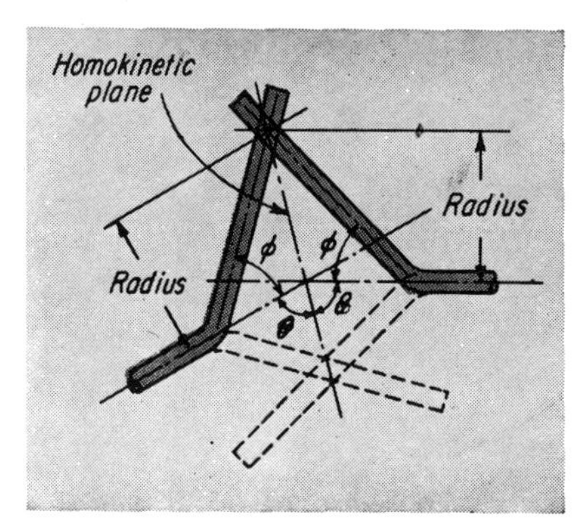

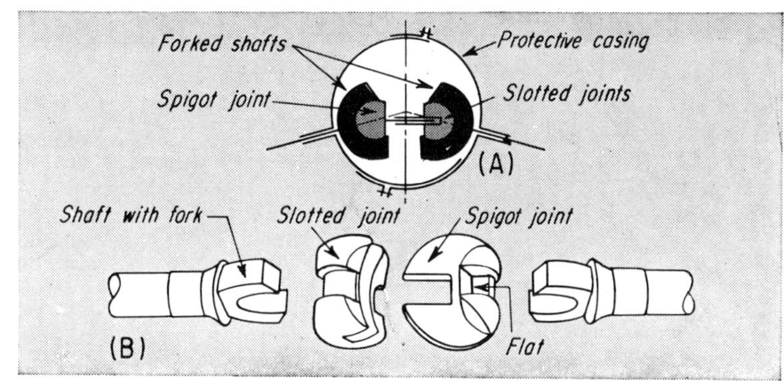

Single constant-velocity . . .

coupling is based on principle (6) that contact point of the two members must always lie on the homokinetic plane. Their rotation speed will then always be equal because radius to contact point of each member will always be equal. Such simple couplings are ideal for toy, instrument and other light-duty mechanisms. For heavy duty, such as front-wheel drive of military vehicles, a more complex coupling, shown diagramatically (7A) has two joints close coupled with a sliding member between them. Exploded view (7B) shows these members. There are other designs for heay-duty universal couplings; one, known as the Rzeppa, consists of a cage that keeps six balls in the homokinetic plane at all times. Another constant-velocity joint, the Bendix-Weiss, incorporates balls also.

MISCELLANEOUS COUPLINGS

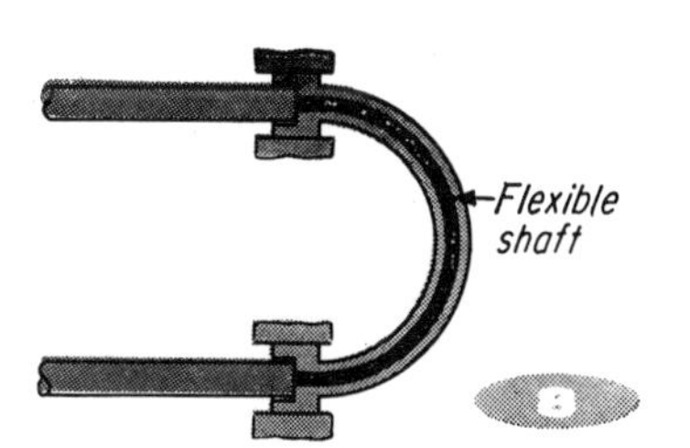

Flexible shaft . . .

allows any shaft angle. Such shafts, if long, should be supported to prevent backlash and coiling.

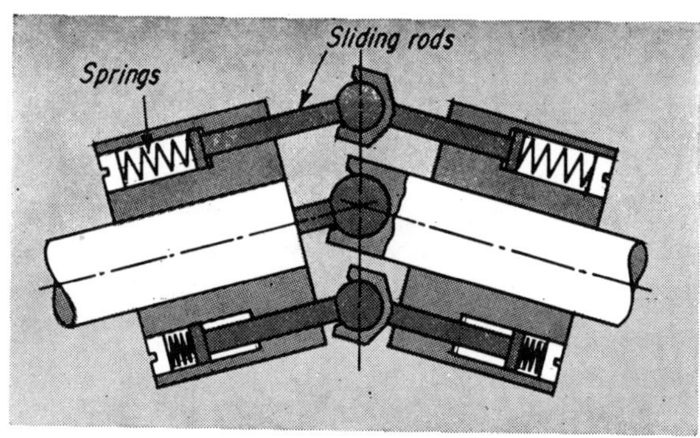

Pump-type coupling . . .

is so called because reciprocating action of sliding rods can be used to drive pistons in cylinders.

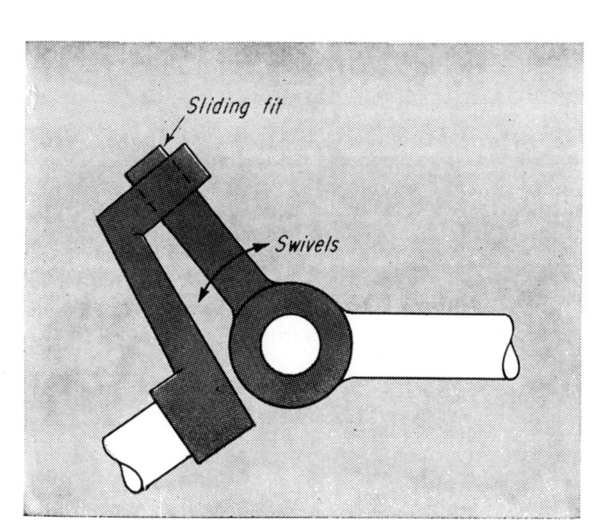

Light-duty coupling . . .

is ideal for many economical mechanisms. Sliding swivel-rod must be kept well lubricated.

6 FLEXIBLE SHAFT-COUPLINGS

Practical information on wobbler, flexible dog, laminated spring, bevel drive, spring pin and multi-spring couplings.

R. WARING-BROWN
Consulting Engineer
Anglesey, Gt Britain

It is frequently necessary, particularly with automation equipment, to couple shafts flexibly. This often occurs where dimensions of length and diameter are restricted and the use of standard coupling units is precluded and therefore couplings have to be specially designed.

Modern research indicates that every shaft coupling should provide some degree of flexibility, and while this may introduce some complexity with modern ideas and using modern materials it is far from impossible.

Our present purpose is to consider shaft couplings which are limited in external diameter and which must provide flexibility to cope with axial, radial and angular misalignments to obtain high efficiency and obviate noise and vibration. Further, due regard must be given to maintenance and adjustment.

Satisfactory selection of the shaft coupling depends upon the particular conditions under which it has to operate and the type of machinery involved, and is ultimately founded on experience. Below are given a few examples of current practice and modern ideas, where coupling diameters are comparatively small compared with shaft size.

WOBBLER COUPLINGS

Although a crude design, these couplings have qualities which make them unrivalled in the presence of considerable grit, water, steam and heat. In Fig. 1 is illustrated an assembly as adopted in steel rolling mills. It acts as a connection between a driving pinion and its corresponding roll. It will be seen that a "breaking spindle" is used which is designed to avoid other and more serious damage by fracture damage to fracture under the severe overload to which such equipment is often subjected.

The ends of the roller and the pinion shaft are of special section, in some instances resembling a four-leaved clover, variants being of cruciform construction. The ends of the breaking spindle are similarly loosely fitting sleeves. In the final assembly the two sleeves are held apart by pieces of wood placed in the grooves of the breaking spindle and bound together by wire, the great advantage of this primitive locking device being that it can be quickly dismounted and replaced at negligible cost.

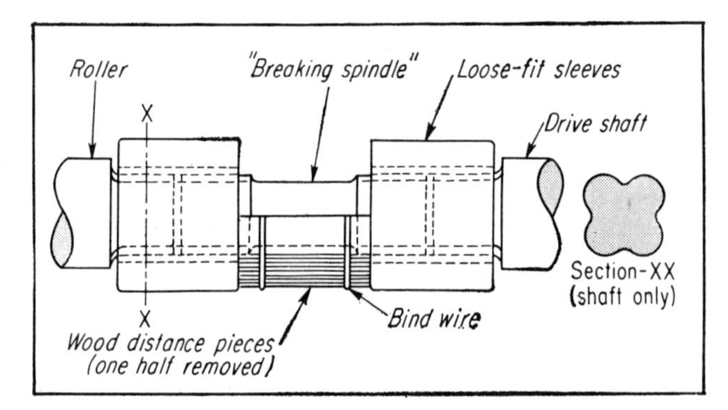

1. . WOBBLER flexible coupling.

The work such couplings are called on to perform is extremely arduous, but limited flexibility in any direction is attained at the expense of a fair amount of noise. However, this is not objectionable in the class of work involved, and excessive noise and vibration will not develop when the coupling is correctly designed to transmit the optimum torque. By sliding one of the sleeves toward the other, the overall length of the breaking spindle may be kept down to a dimension sufficient to permit its easy removal for severing the connection between the driving pinion and the roll.

A refined construction founded on the basic principle could be used in small mechanisms.

FLEXIBLE DOG COUPLINGS

Efforts to devise a simple and effective flexible coupling have resulted in the scheme illustrated in Fig. 2. This compact arrangement, while allowing limited flexibility in the various directions, satisfies the paramount consideration of small diameter. The component has a minimum of parts and has good anti-vibration qualities. The two shafts to be coupled carry keys to drive sleeves, which are held in position endwise by grub screws, the inner ends of the sleeves being provided with slots at right angles to each other. There is an intermediate shaft to the ends of which are fitted two special springs held in posi-

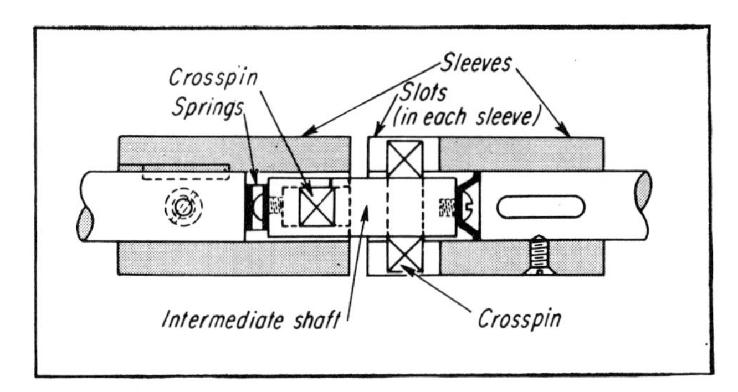

2. . **FLEXIBLE** dog coupling.

tion by screws, these springs being arranged to maintain contact between the driving and driven shafts at all times, thereby eliminating end movement, shock and vibration. At a little distance from the ends of the shaft two holes are drilled at 180°, and cross pins, formed with dog ends, are fitted to give a good sliding fit in the slots of the coupling sleeves. The intermediate shaft should be given a good clearance on the inside of the sleeves, and a gap of about $\frac{1}{4}$ in. should be allowed between the inner ends of the sleeves so that reasonable flexture and end motion may be accommodated. The coupling sleeves should be capable of being slid back on their keys for dismounting, while the springs to distribute the end pressure should have small arms, or alternatively a Belleville washer may be utilized.

This coupling has proved very satisfactory in practice over long periods of service.

LAMINATED SPRING COUPLING

It is sometimes advisable to install a shaft coupling in a limited space that will give improved flexibility, even though it is more expensive to produce. In Fig. 3 is illustrated a type that comes into this category, and it will be seen that the two shafts to be connected have their inner ends slotted out into holes drilled through the shafts, which latter take comparatively large diameter pins. These pins are also slotted to take a series of laminated spring plates which are fastened by rivets.

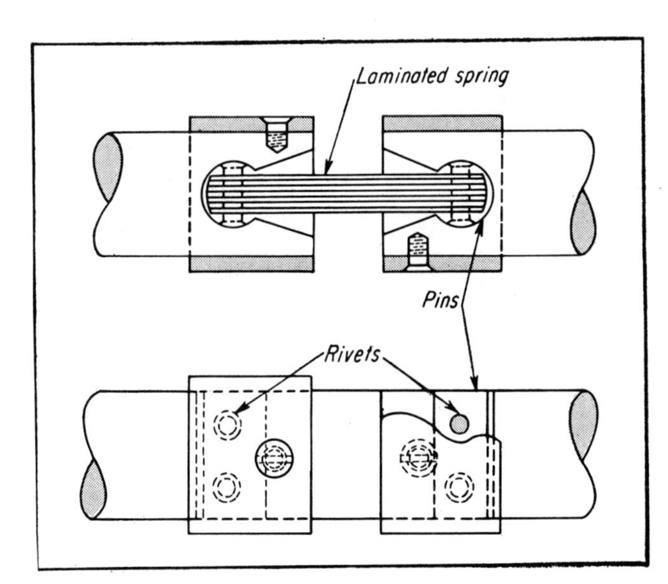

3. . **LAMINATED** spring flexible coupling.

The pins are hardened and ground and are slid into place in the shafts. Sleeves protect the pins, preserving a smooth finish without projections on the finished assembly. The coupling was primarily designed for textile machinery to eliminate keys and to transmit the drive but not vibration. It is easily disassembled.

LAMINATED SPRING BEVEL DRIVE COUPLING

In high-class work improved forms of the laminated spring coupling are being very successfully applied. In Fig. 4 is illustrated such an assembly utilized to drive an engine governor. A bevel pinion is splined to a shaft which is slotted to take a laminated spring, similarly connected at its other end to the driven shaft. The spring is firmly fixed to the bevel shaft and driven shaft by pins forced in under pressure. Slight axial movement is allowed, and a positive abutment drive is also provided. It is often necessary to detach a driven unit from a prime mover; this design facilitates this.

Another important feature is avoiding the transmission of cyclic variations, and here an improvement consists in making the slots in the shaft ends just wide enough to give sufficient clearance to the laminated spring to allow of some deflexion during operation, so that when the spring attains a resonant condition its increased deflection will bring the spring plates up to the edges of the slots, thereby shortening their effective length and immediately detuning the system.

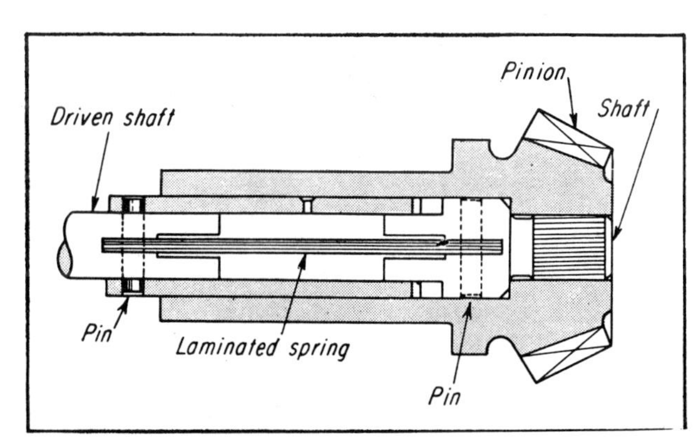

4. . **LAMINATED** spring bevel drive coupling.

In such a scheme misalignment is not allowed, but flexibility in transmission effectively eliminates vibration and noise. A positive abutment drive is also provided to prevent overstressing the spring plates during overload. To prevent cyclic variation it would be possible to incorporate some form of rubber coupling, but as these generally have a fixed natural frequency of oscillation, they would not prove satisfactory in the coupling required for the particular purpose of the bevel drive described.

The coupling arrangement is simple and compact while being easily dismountable, but it must be precision built. In certain cases it is sometimes advisable to arrange for the abutment drive to engage after only a slight deflection of the spring.

TYPICAL METHODS OF COUPLING

Methods of coupling rotating shafts vary from simple bolted flange constructions to complex spring and synthetic rubber mechanisms. Some types incorporating chain belts, splines, bands, and rollers are described and illustrated below.

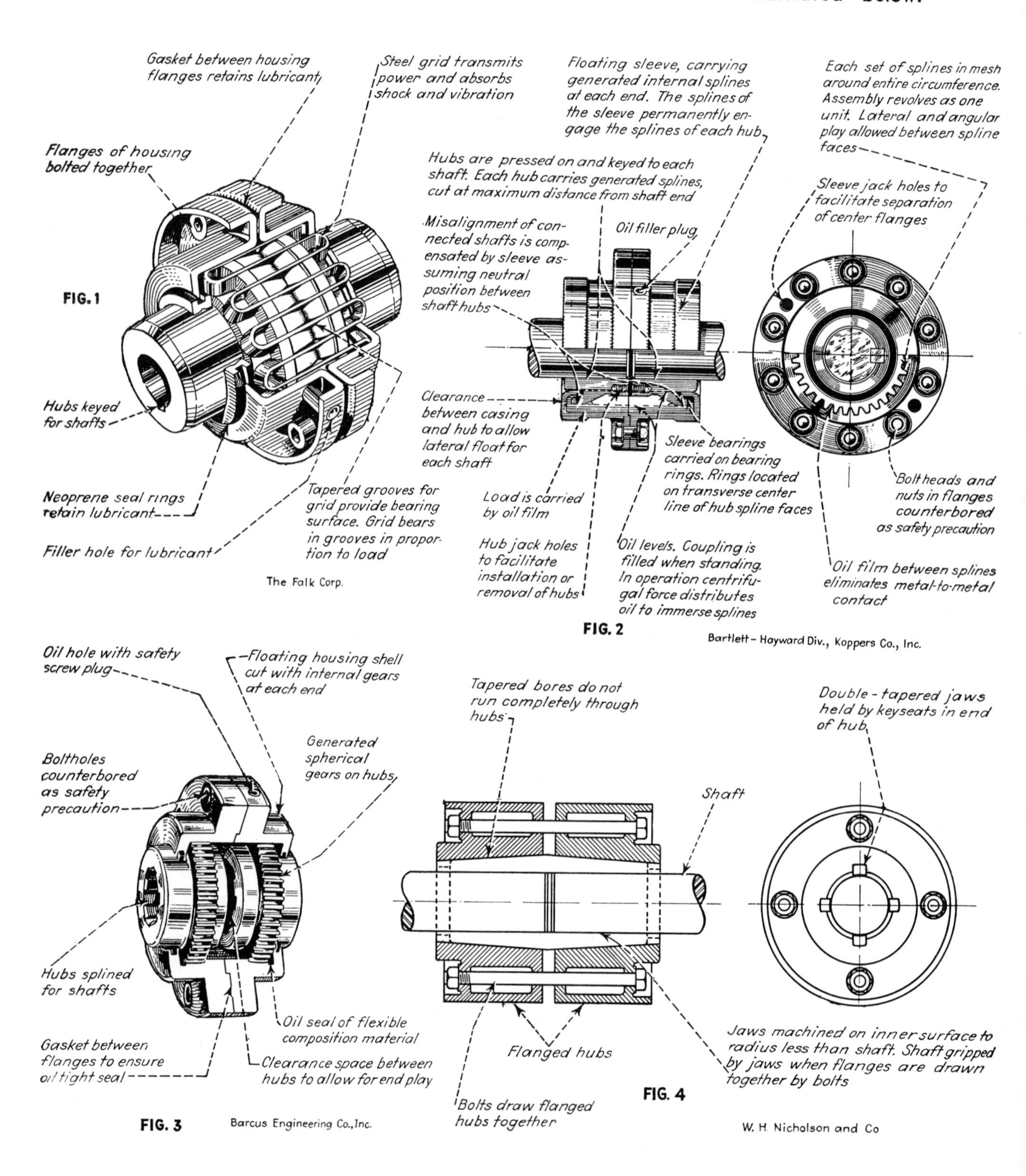

ROTATING SHAFTS—I

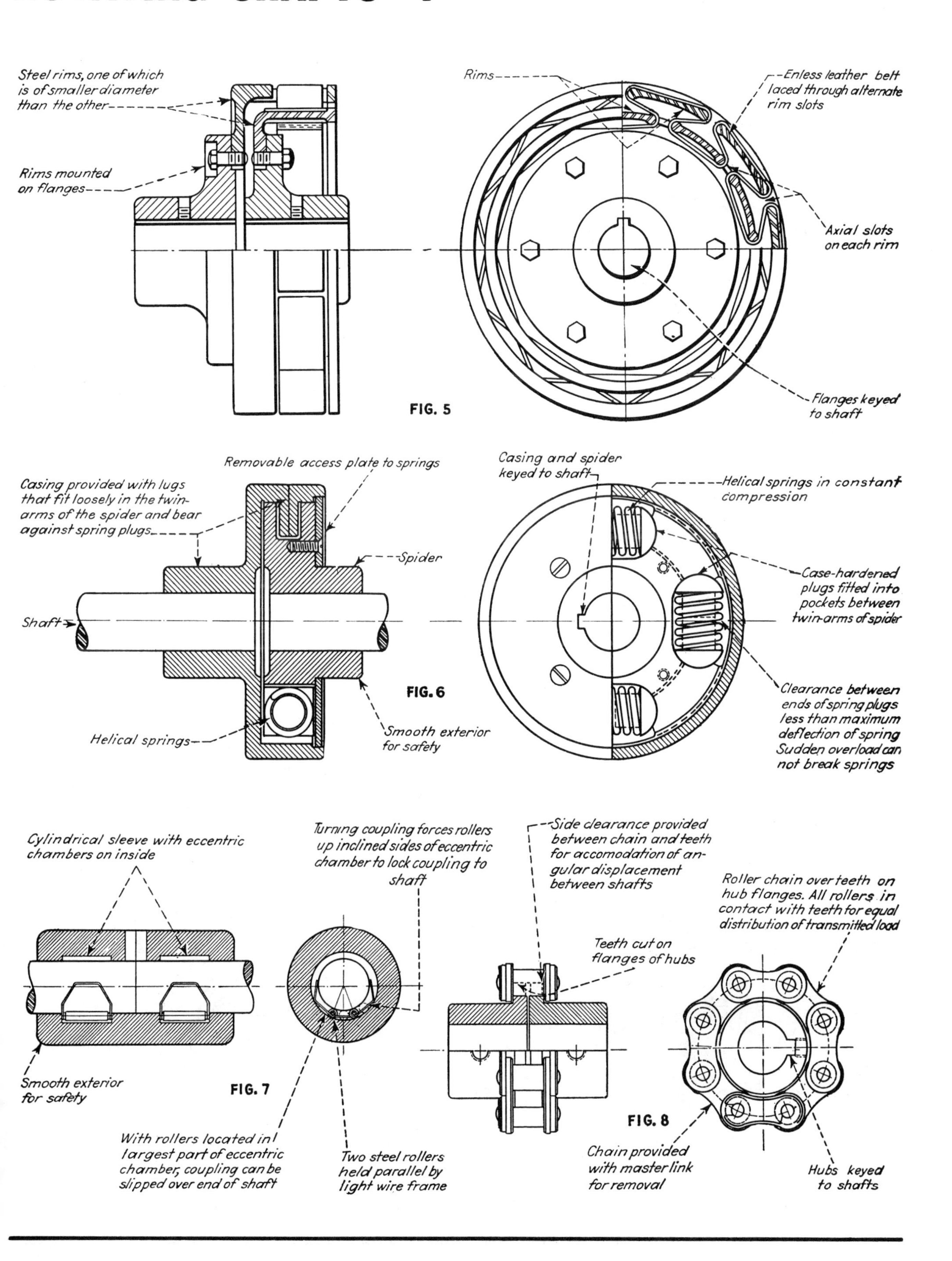

TYPICAL METHODS OF COUPLING

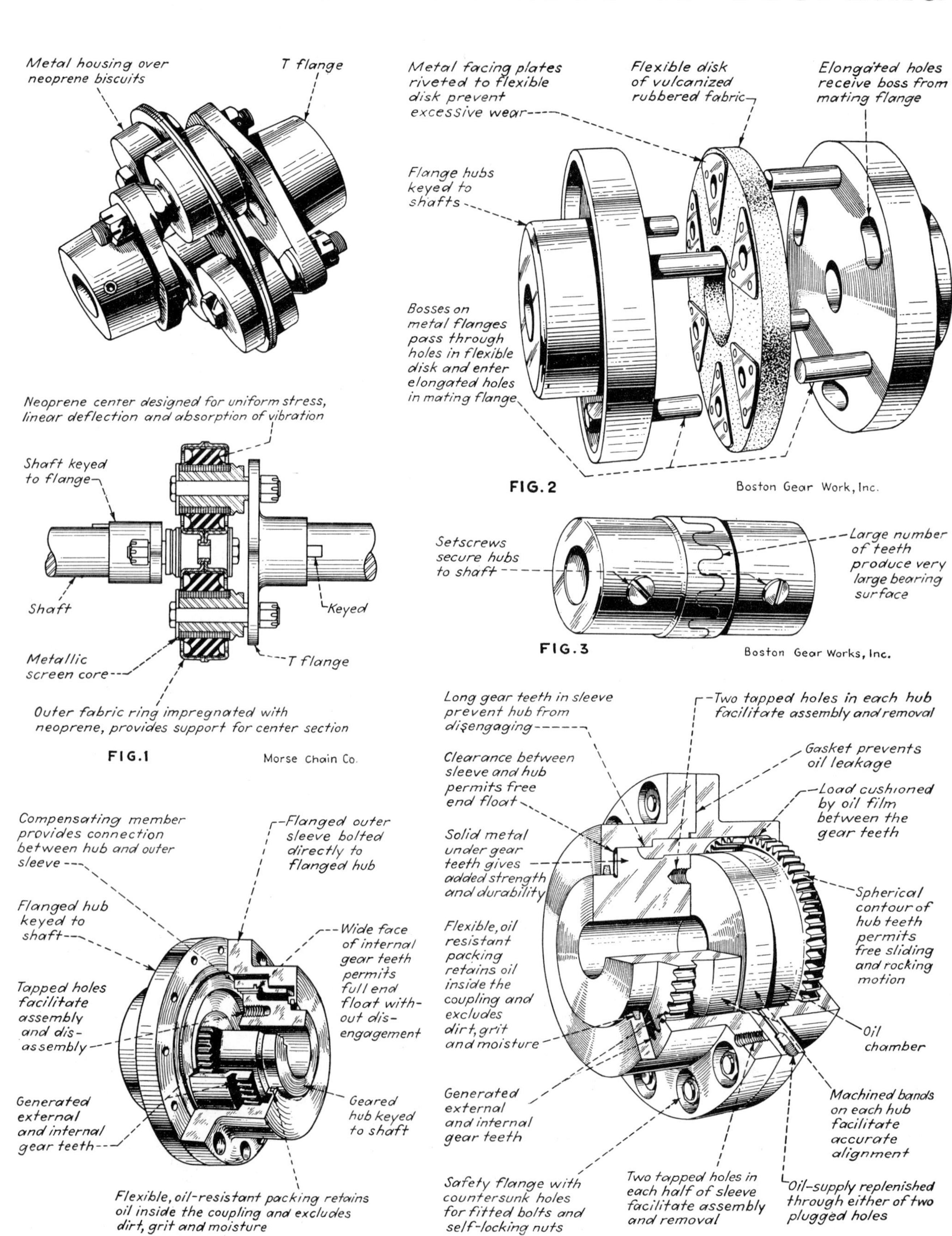

FIG. 1 Morse Chain Co.

FIG. 2 Boston Gear Work, Inc.

FIG. 3 Boston Gear Works, Inc.

FIG. 4 Farrel-Birmingham Co., Inc.

FIG. 5 Farrel-Birmingham Co., Inc.

ROTATING SHAFTS—II

Shaft couplings that utilize internal and external gears, balls, pins, and non-metallic parts to transmit torque are shown herewith.

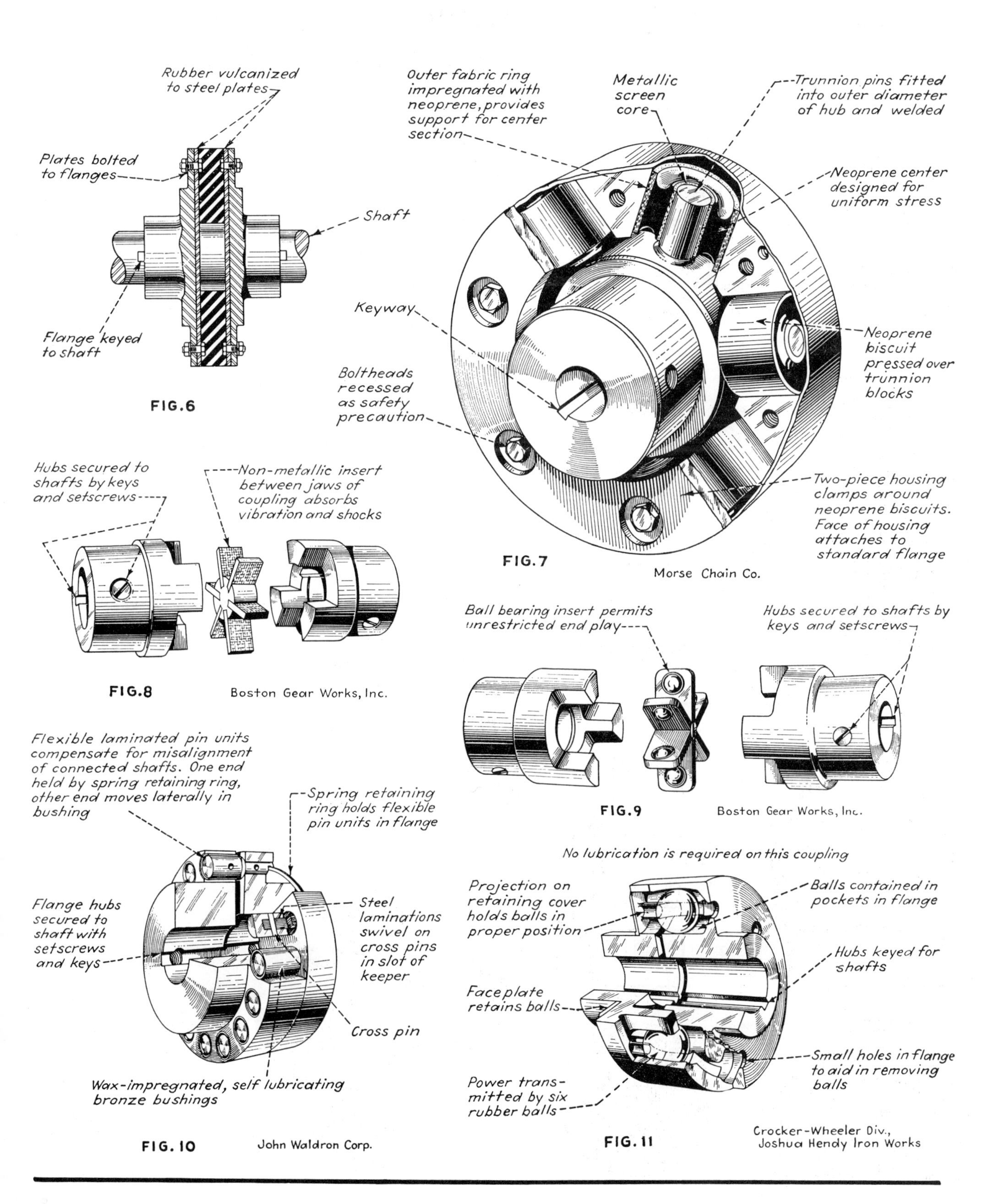

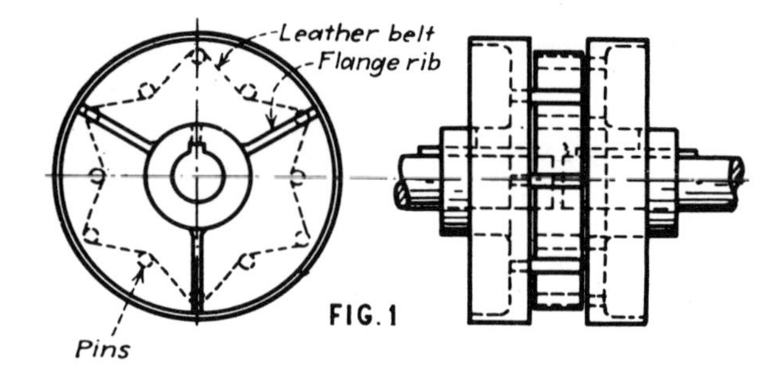

MORE NOVEL WAYS OF COUPLING SHAFTS

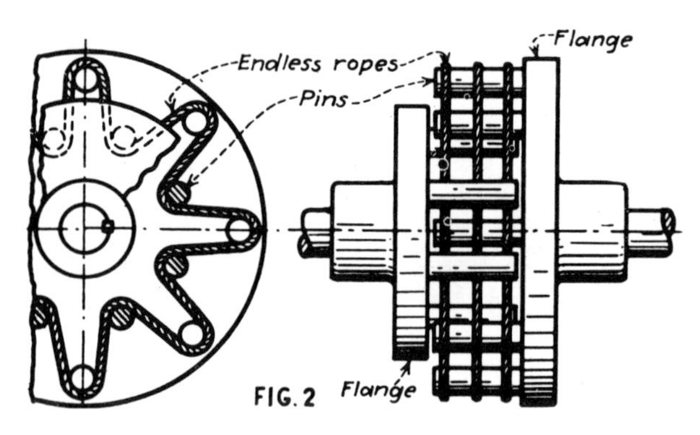

Fig. 1—This Webster Manufacturing Company coupling uses a single endless leather belt instead of a series of bands. The belt is looped over alternate pins in both flanges. Has good shock-resisting properties because of belt stretch and the tendency of the pins to settle back into the loops of the belt.

Fig. 2—This coupling made by the Weller Manufacturing Company is similar to the design in Fig. 1,. but instead of a leather belt uses hemp rope, made endless by splicing. The action under load is the same as in the endless belt type.

Fig. 3—This Bruce-Macbeth design uses leather links in-stead of endless wire cables. The load is transmitted from one flange to the other by direct pull of the links, which at the same time allows for the proper flexibility. Intended for permanent installations requiring a minimum of supervision.

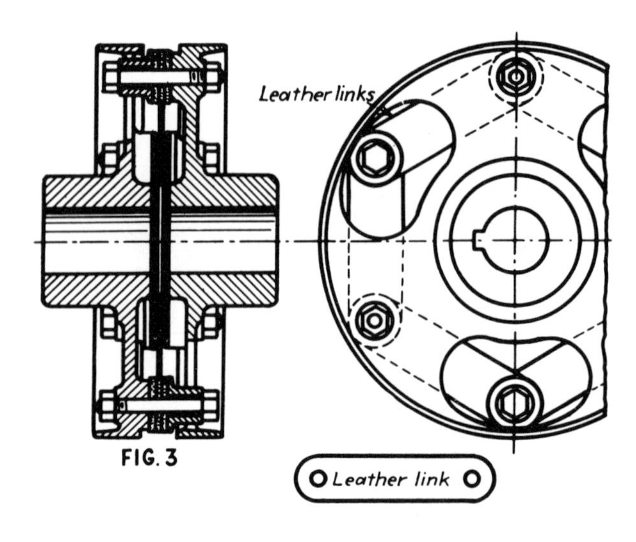

Fig. 4—The Oldham form of coupling made by W. A. Jones Foundry and Machine Company is of the two-jaw type with a metal disk. Is used for transmitting heavy loads at low speed.

Fig. 5—The Charles Bond Company star coupling is sim-ilar to the cross type. The star-shaped floating member is made of laminated leather. Has three jaws in each flange. Torque capacity is thus increased over the two-jaw or cross type. Coupling takes care of limited end play.

Fig. 6—Combination rubber and canvas disk is bolted to two metal spiders. Extensively used for low torques where compensation for only slight angular misalignment is required. Is quiet in operation and needs no lubrication or other attention. Offset misalignment shortens disk life.

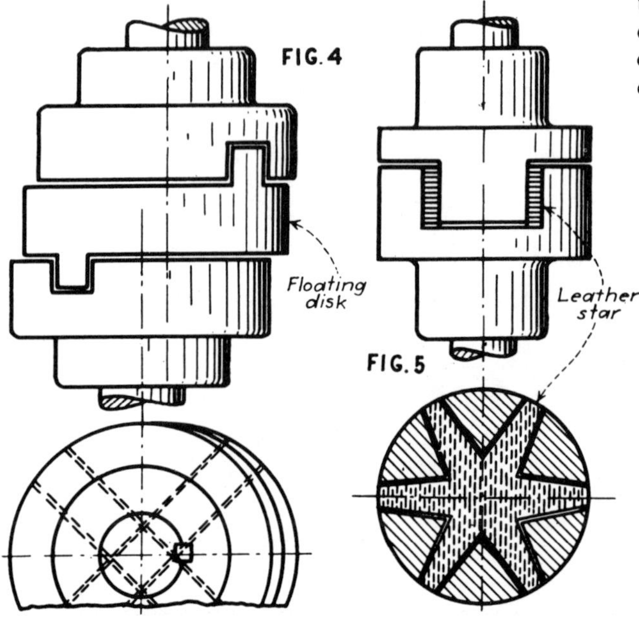

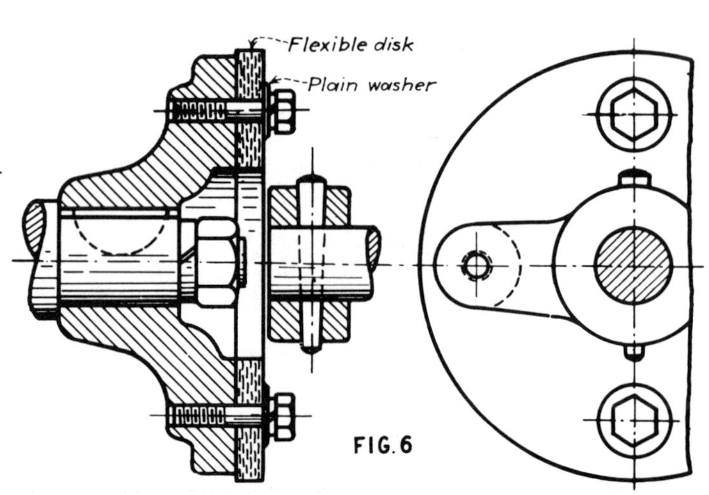

Fig. 7—A metal block as a floating center is used in this American Flexible Coupling Company design. Quiet operation is secured by facing the block with removable fiber strips and packing the center with grease. The coupling sets up no end thrusts, is easy to assemble and does not depend on flexible material for the driving action. Can be built in small sizes by using hardwood block without facings, for the floating member.

Fig. 8—This Westinghouse Nuttall Company coupling is an all-metal type having excellent torsional flexibility. The eight compression springs compensate for angular and offset misalignment. Allows for some free endwise float of the shafts. Will transmit high torques in either direction. No lubrication is needed.

Fig. 9—Similar to Fig. 6, but will withstand offset misalignment by addition of the extra disk. In this instance the center spider is free to float. By use of two rubber-canvas disks, as shown, coupling will withstand a considerable angular misalignment.

Fig. 10—In this Smith & Serrell coupling a flexible cross made of laminated steel strips floats between two spiders. The laminated spokes, retained by four segmental shoes, engage lugs integral with the flanges. Coupling is intended for the transmission of light loads only.

Fig. 11—This coupling made by Brown Engineering Company is useful for improvising connections between apparatus in laboratories and similar temporary installations. Compensates for misalignment in all directions. Will absorb varying degrees of torsional shocks by changing the size of the springs. Springs are retained by threaded pins engaging the coils. Overload protection is possible by the slippage or breakage of replacable springs.

Fig. 12—In another design by Brown Engineering Company, a series of laminated spokes transmit power between the two flanges without setting up end thrusts. This type allows free end play. Among other advantages are absorption of torsional shocks, has no exposed moving parts, and is well balanced at all speeds. Wearing parts are replacable and working parts are protected from dust.

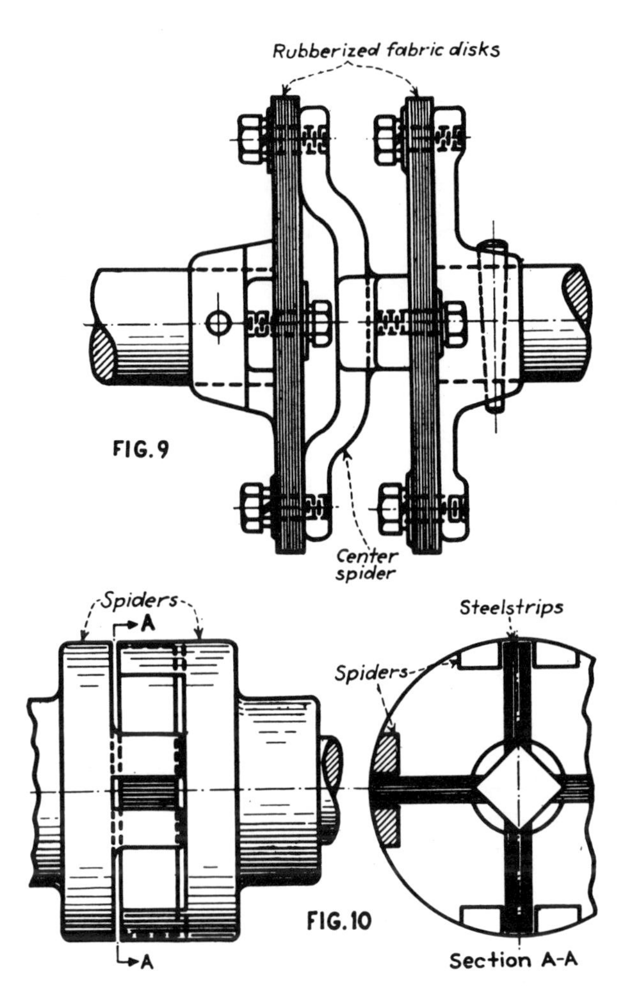

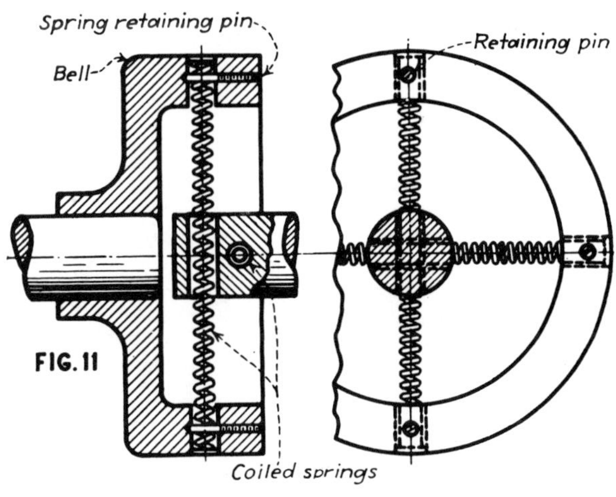

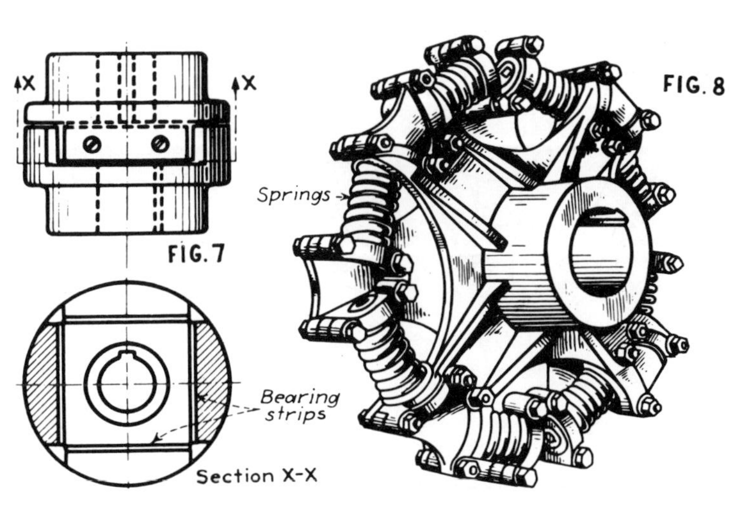

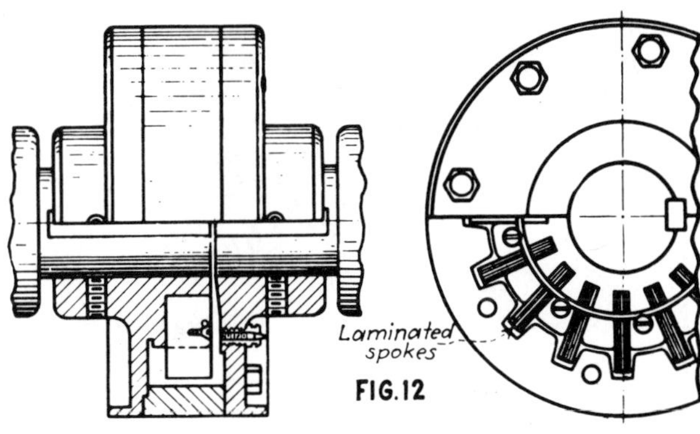

Linkages for Operating Band Clutches and Brakes

For the general run of conditions, one or another of these four designs of band clutches and four designs of band brakes will be found applicable. If necessary, of course, they can be modified for particular purposes

By A. C. RASMUSSEN
Mechanical Engineer

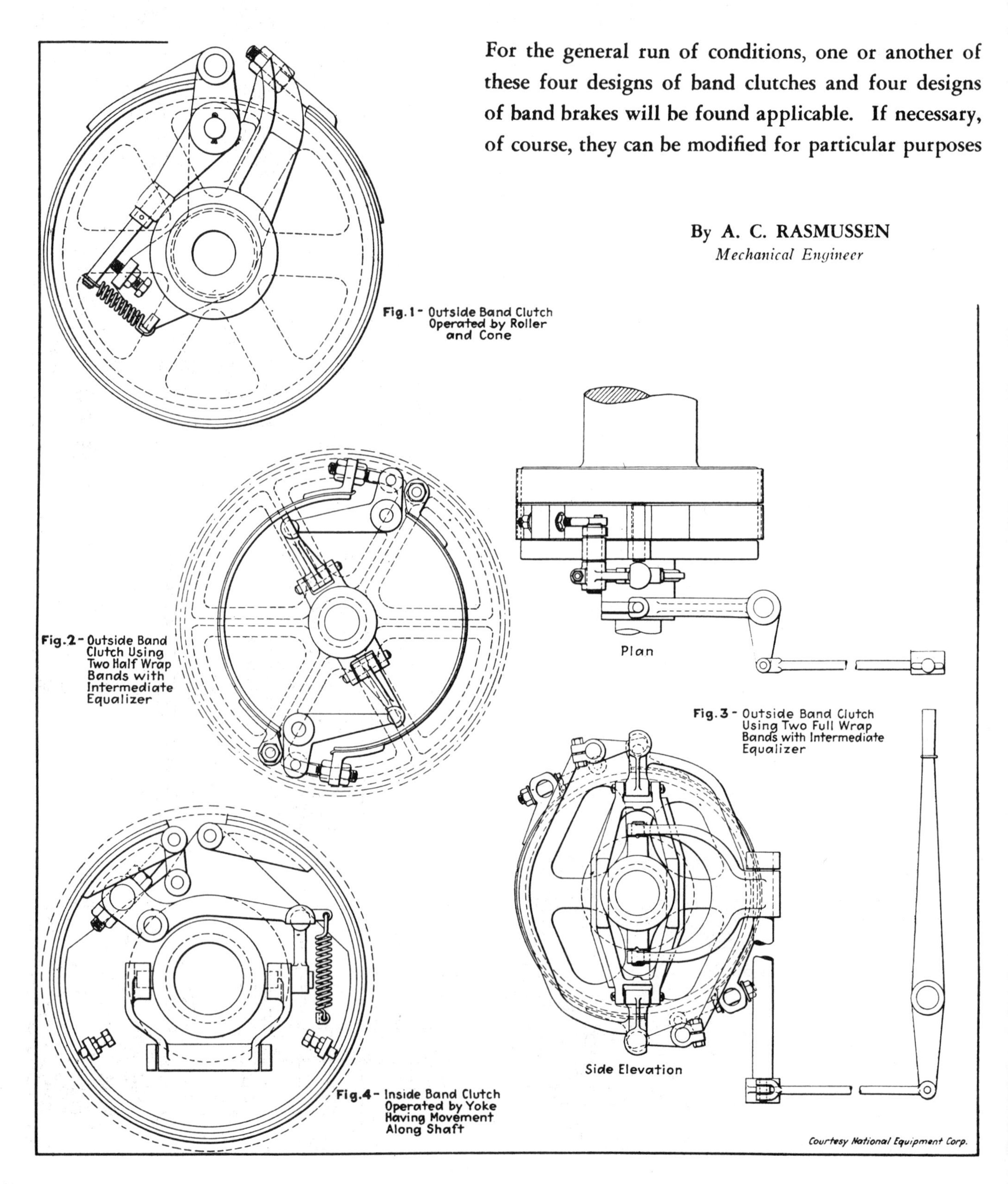

Fig. 1 - Outside Band Clutch Operated by Roller and Cone

Fig. 2 - Outside Band Clutch Using Two Half Wrap Bands with Intermediate Equalizer

Fig. 3 - Outside Band Clutch Using Two Full Wrap Bands with Intermediate Equalizer

Fig. 4 - Inside Band Clutch Operated by Yoke Having Movement Along Shaft

Linkages for Band Clutches and Brakes (continued)

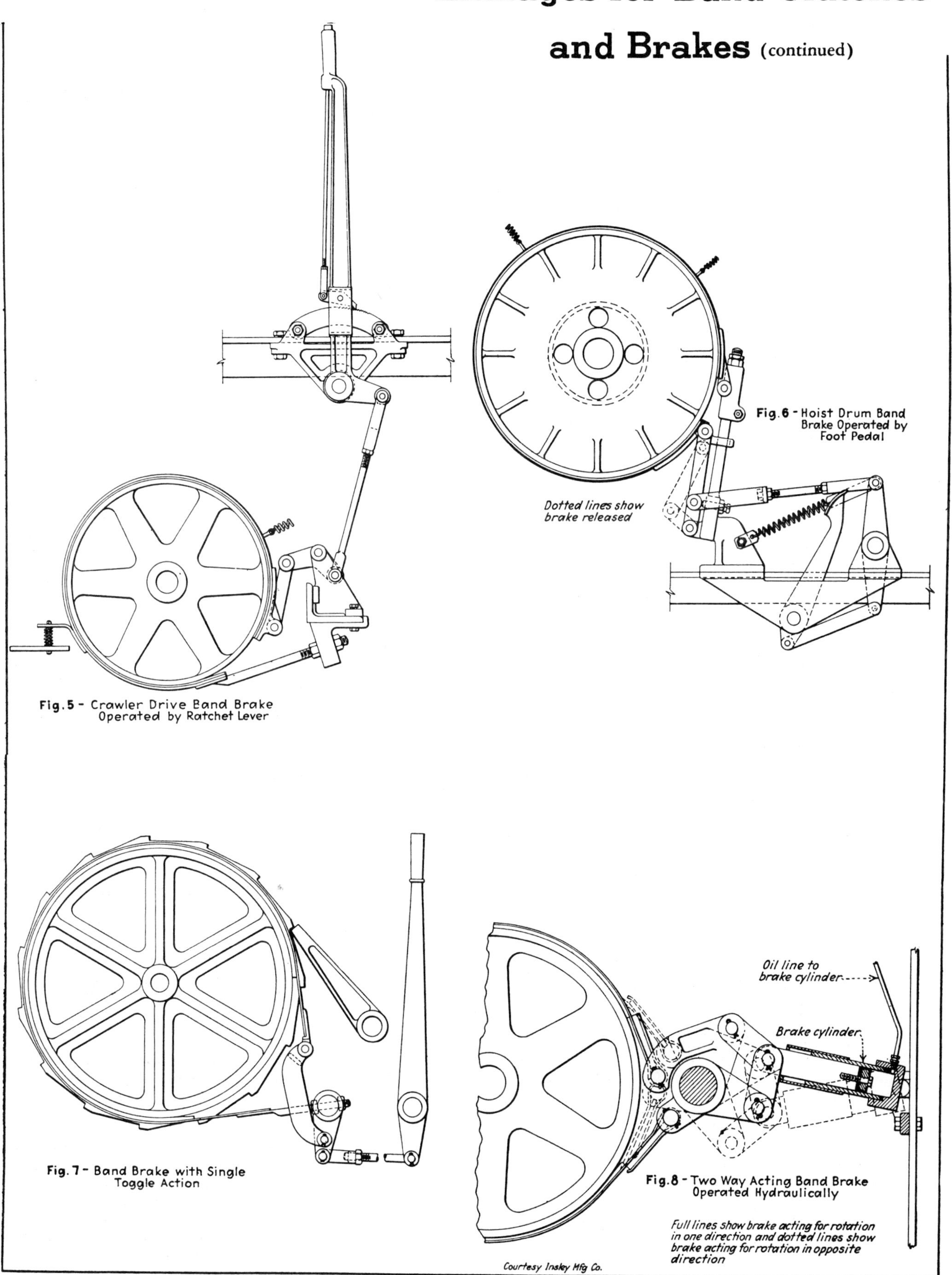

Fig. 5 - Crawler Drive Band Brake Operated by Ratchet Lever

Fig. 6 - Hoist Drum Band Brake Operated by Foot Pedal

Fig. 7 - Band Brake with Single Toggle Action

Fig. 8 - Two Way Acting Band Brake Operated Hydraulically

SPECIAL LINK COUPLING MECHANISMS

H. G. CONWAY
Cheltenham, England

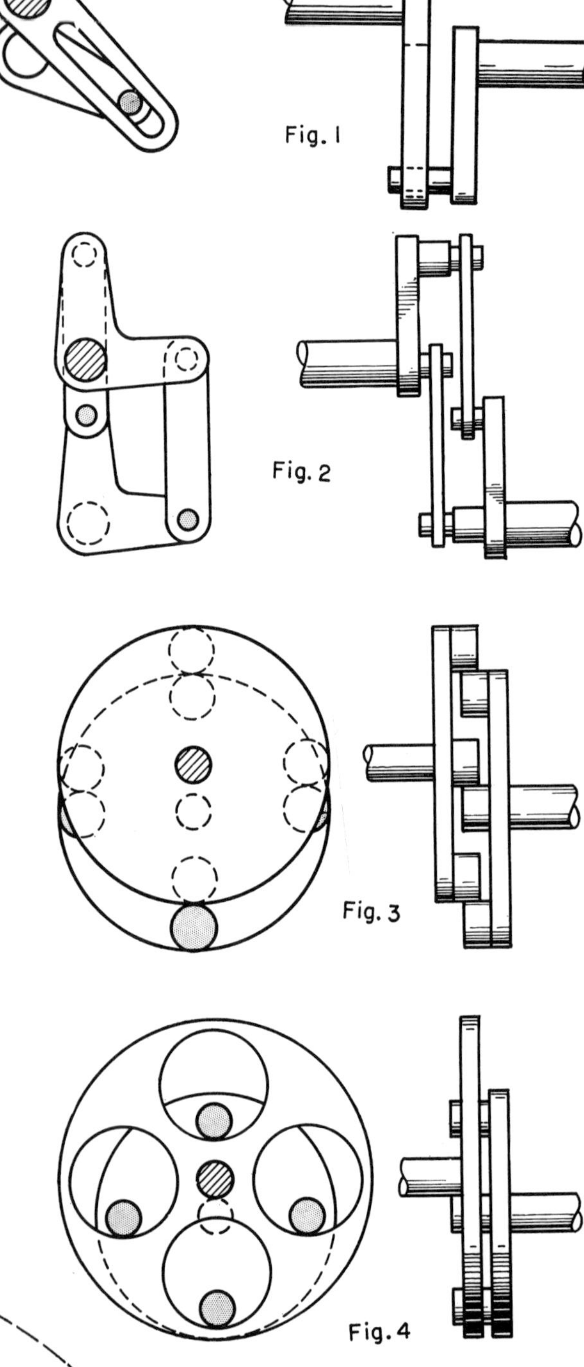

FIG. 1—If the velocity does not have to be constant a pin and slot coupling can be used. Velocity transmission is irregular as the effective radius of operation is continually changing, the shafts must remain parallel unless a ball joint is used between the slot and pin. Axial freedom is possible but any change in the shaft offset will further affect the fluctuation of velocity transmission.

FIG. 2—The parallel-crank mechanism is sometimes used to drive the overhead camshaft on engines. Each shaft has at least two cranks connected by links and with full symmetry for constant velocity action and to avoid dead points. By using ball joints at the ends of the links, displacement between the crank assemblies is possible.

FIG. 3—A mechanism kinematically equivalent to Fig. 2, can be made by substituting two circular and contacting pins for each link. Each shaft has a disk carrying three or more projecting pins, the sum of the radii of the pins being equal to the eccentricity of offset of the shafts. The lines of center between each pair of pins remain parallel as the coupling rotates. Pins do not need to be of equal diameter. Transmission is at constant velocity and axial freedom is possible.

FIG. 4—Similar to the mechanism in Fig. 3, but with one set of pins being holes. The difference of radii is equal to the eccentricity or offset. Velocity transmission is constant; axial freedom is possible, but as in Fig. 3, the shaft axes must remain fixed. This type of mechanism is sometimes used in epicyclic reduction gear boxes.

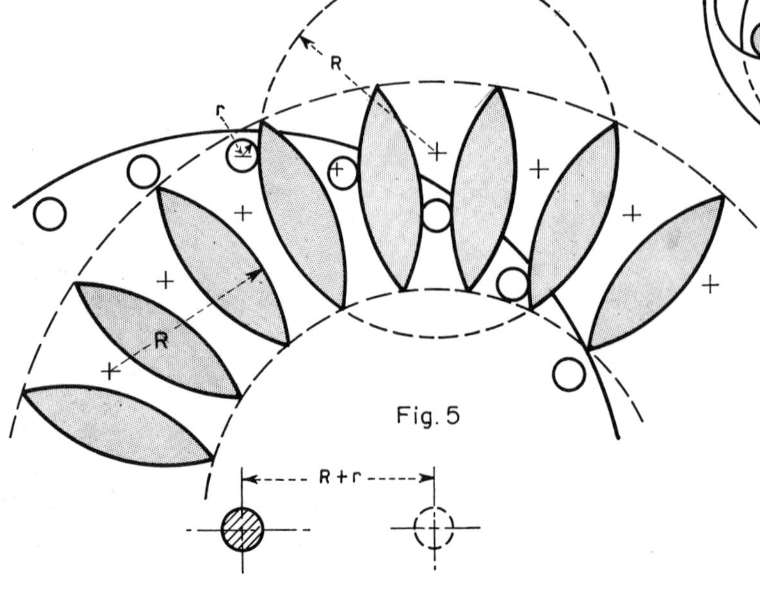

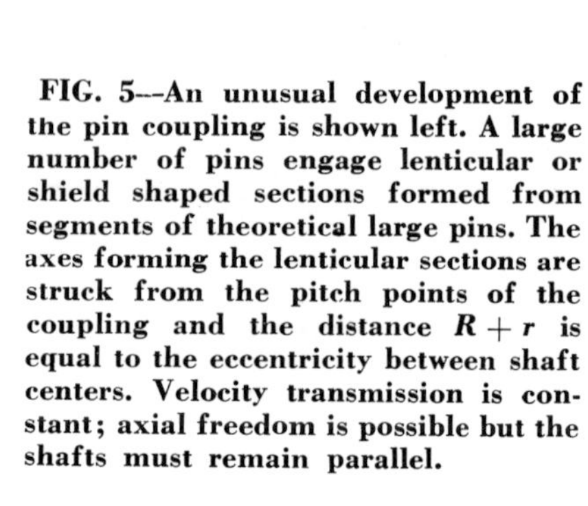

FIG. 5—An unusual development of the pin coupling is shown left. A large number of pins engage lenticular or shield shaped sections formed from segments of theoretical large pins. The axes forming the lenticular sections are struck from the pitch points of the coupling and the distance $R + r$ is equal to the eccentricity between shaft centers. Velocity transmission is constant; axial freedom is possible but the shafts must remain parallel.

Chapter 12 CLAMPING AND FASTENING DEVICES

Quick-Acting Clamps for Machines and Fixtures

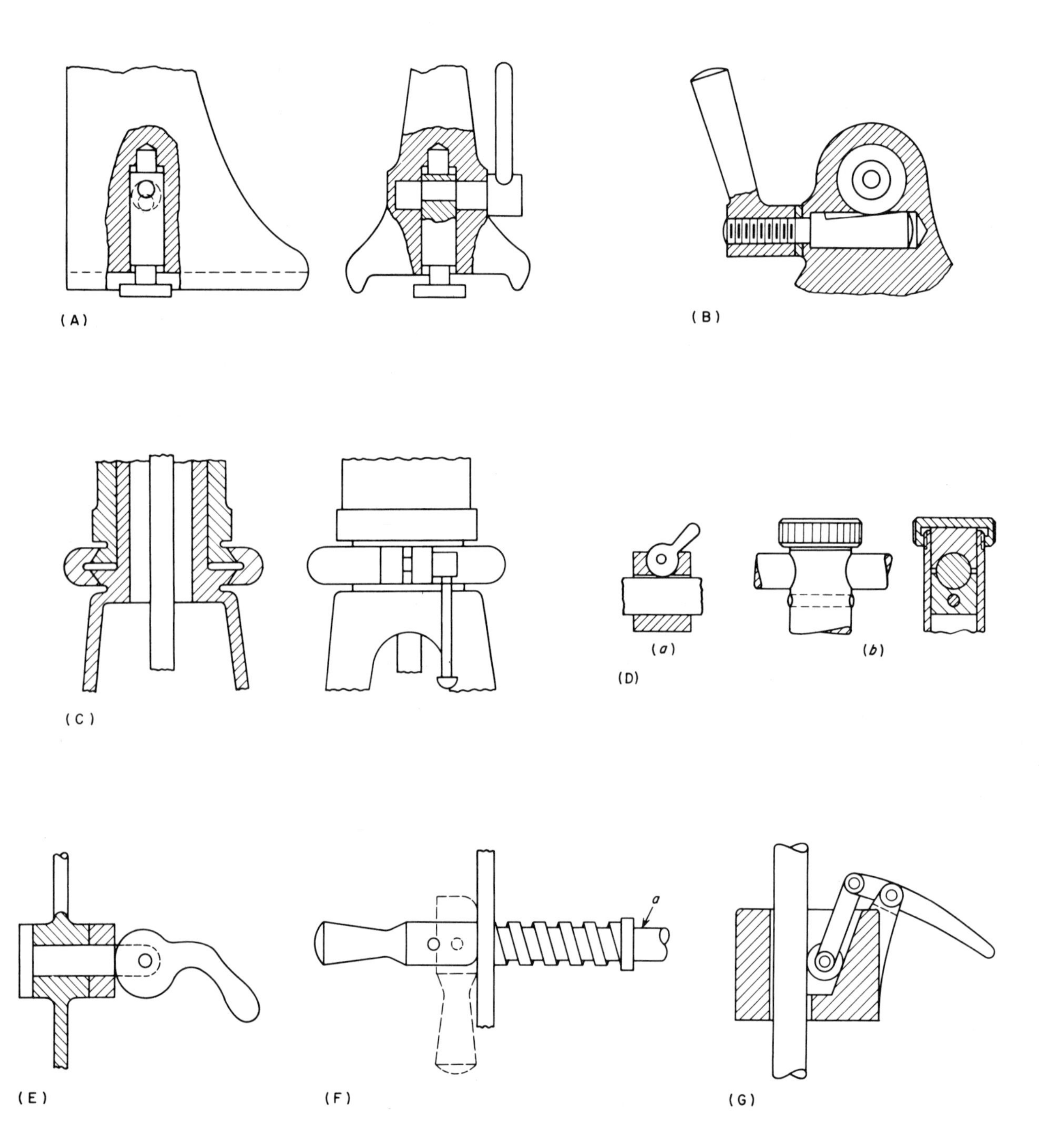

(A) An eccentric clamp. **(B)** Spindle clamping bolt. **(C)** Method of clamping hollow column to structure. Permits quick rotary adjustment of column. **(D)** **(a)** Cam catch for clamping rod or rope. **(b)** Method of fastening small cylindrical member to structure using thumb nut and clamp jaws. Permits quick longitudinal adjustment of shaft in structure. **(E)** Cam catch to lock a wheel or spindle. **(F)** Spring handle. Movement of handle in vertical or horizontal position provides movement at **A.** **(G)** Roller and inclined slot for locking a rod or rope. **(H)** Method of clamping light member to structure. Serrated edge on structure provides for quickly accommodating members of different thicknesses. **(I)** Spring taper holder with sliding ring. **(J)** Special clamp for holding member **a.** **(K)** Cone, nut, and levers grip member **a.** Grip may have two or more jaws. With only two jaws, device serves as a small vise. **(L)** Two different types of cam clamps. **(M)** Cam cover catch. Movement of handle downward locks cover tightly. **(N)** Clamping sliding member to slotted structure using wedge bolt. Provides means of quick adjustment of member on structure.

FROM *Handbook of Fastening and Joining of Metal Parts*

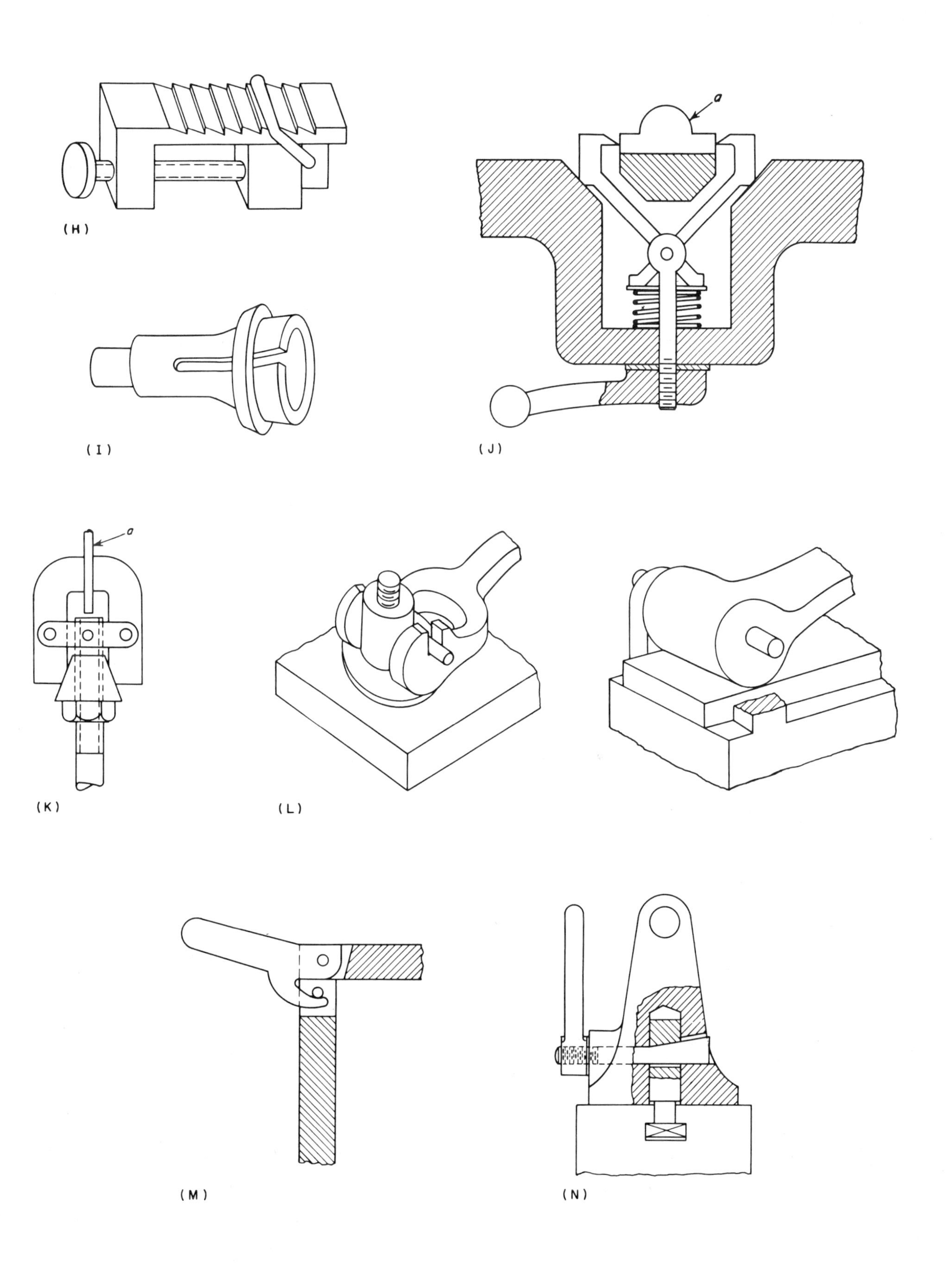

More Clamping Devices

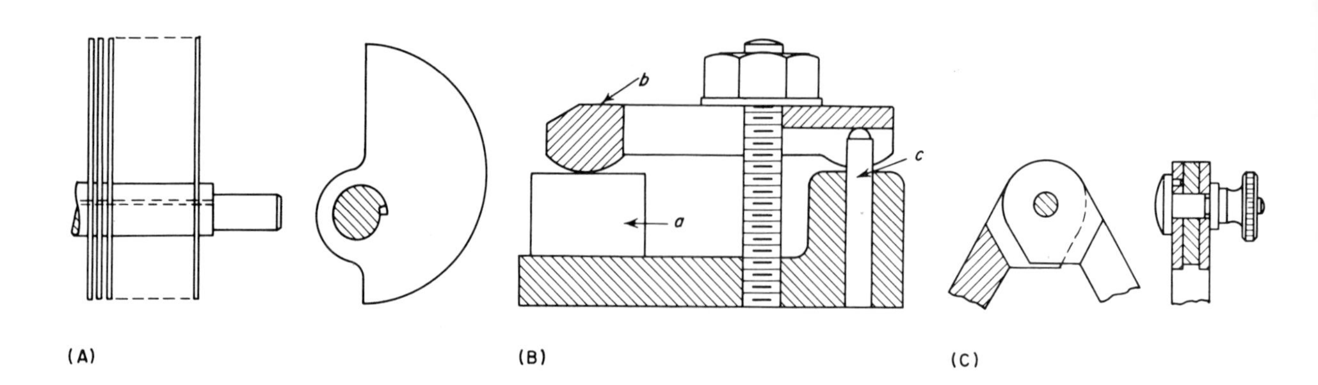

(A) (B) (C)

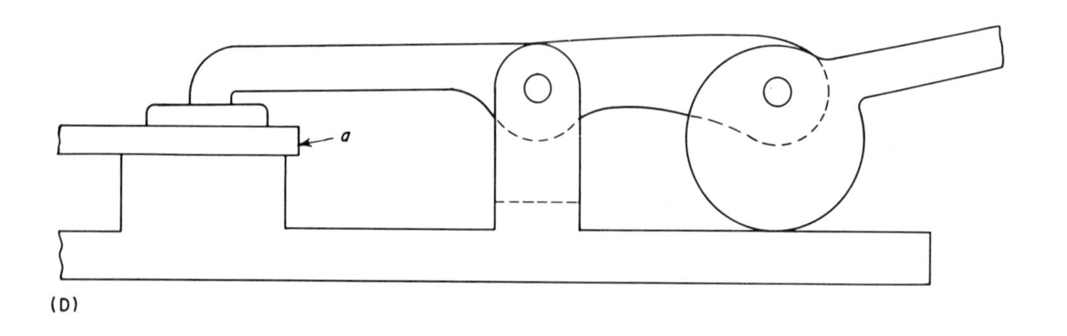

(D)

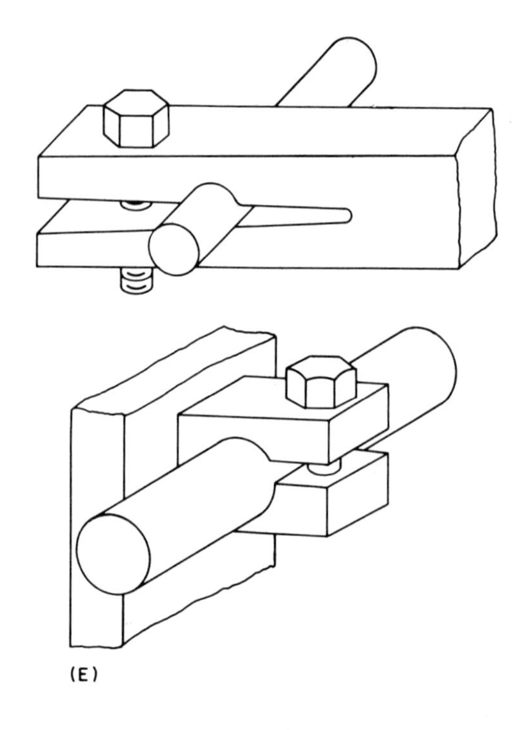

(E)

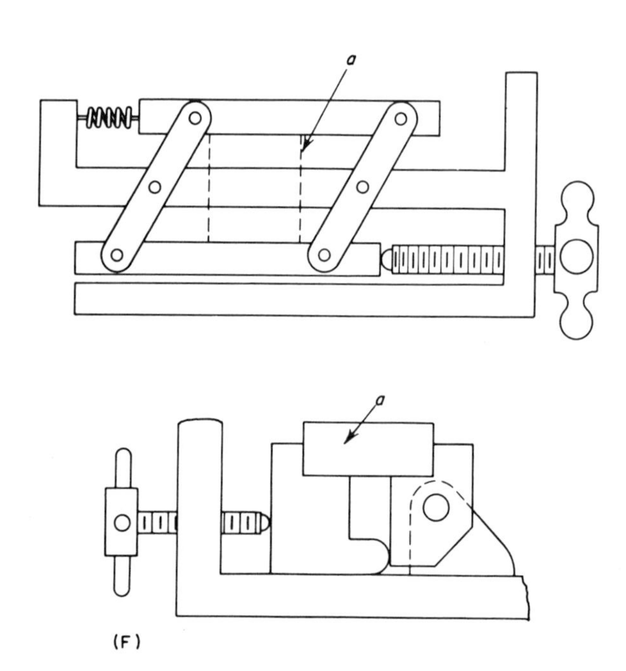

(F)

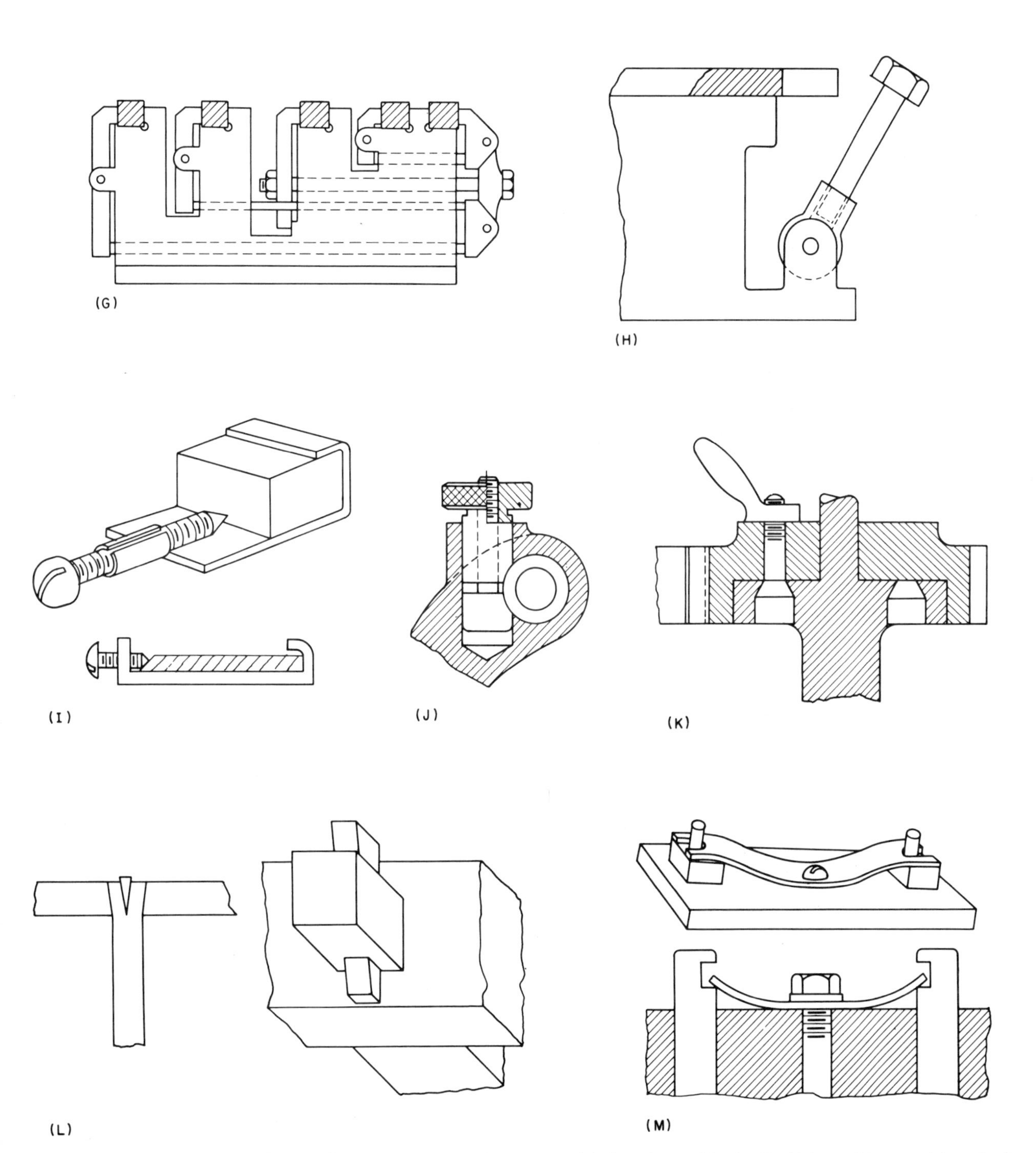

(A) Method of fastening condenser plates to structure using circular wedge. Rotation of plates in clockwise direction locks plates on structure. **(B)** Method of clamping member **a** using special clamp. Detail **b** pivots on pin **c**. **(C)** Method of clamping two movable parts so that they may be held in any angular position by use of clamping screw. **(D)** Cam clamp for clamping member **a**. **(E)** Methods of clamping cylindrical member. **(F)** Method of clamping member **a** using special clamps. **(G)** Special clamping device which permits parallel clamping of five parts by the tightening of one bolt. **(H)** Method of securing structure by use of bolt and movable detail which provides a quick method of fastening cover. **(I)** Method of quickly securing, adjusting, or releasing center member. **(J)** Method of securing bushing in structure by the use of clamp screw and thumb nut. **(K)** Method of securing attachment to structure by use of bolt and hand lever used as a nut. **(L)** Method of fastening member to structure using wedge. **(M)** Methods of fastening two members to structure using spring and one screw. Members may be removed without loosening screw.

FRICTION CLAMPING DEVICES

BERNARD J. WOLFE

ALL TYPES of mechanisms used for gaining mechanical advantage have probably been used in the design of friction clamps. This type of clamp can hold moderately large loads by friction grip on smooth surfaces even of comparatively small area and, in some designs, tightened or released with little effort and movement of the control. In the clamps illustrated here the mechanical advantage is gained by the use of the common devices: lever, toggle, screw, wedge, and combinations of these means.

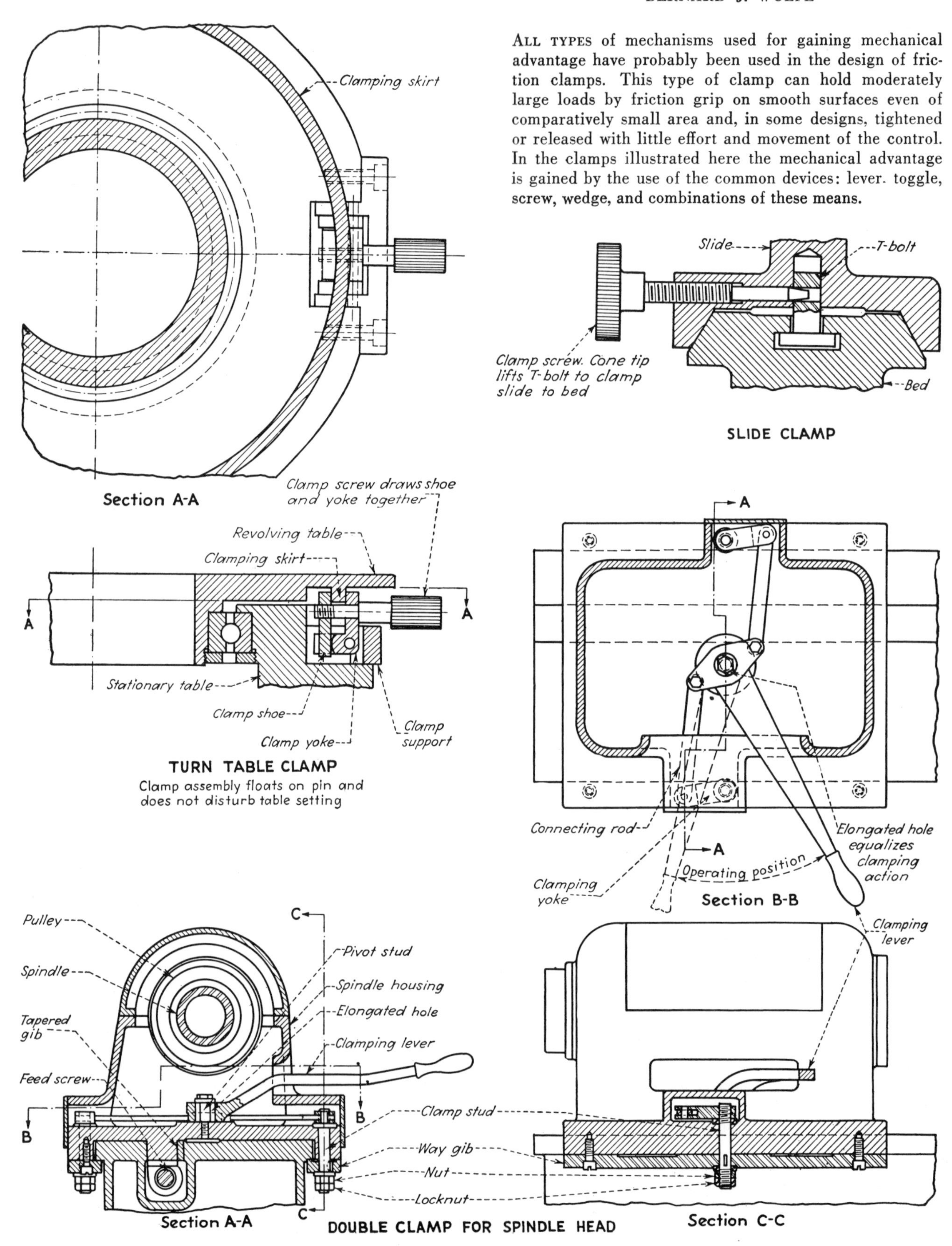

AND PRINCIPLES OF DESIGN

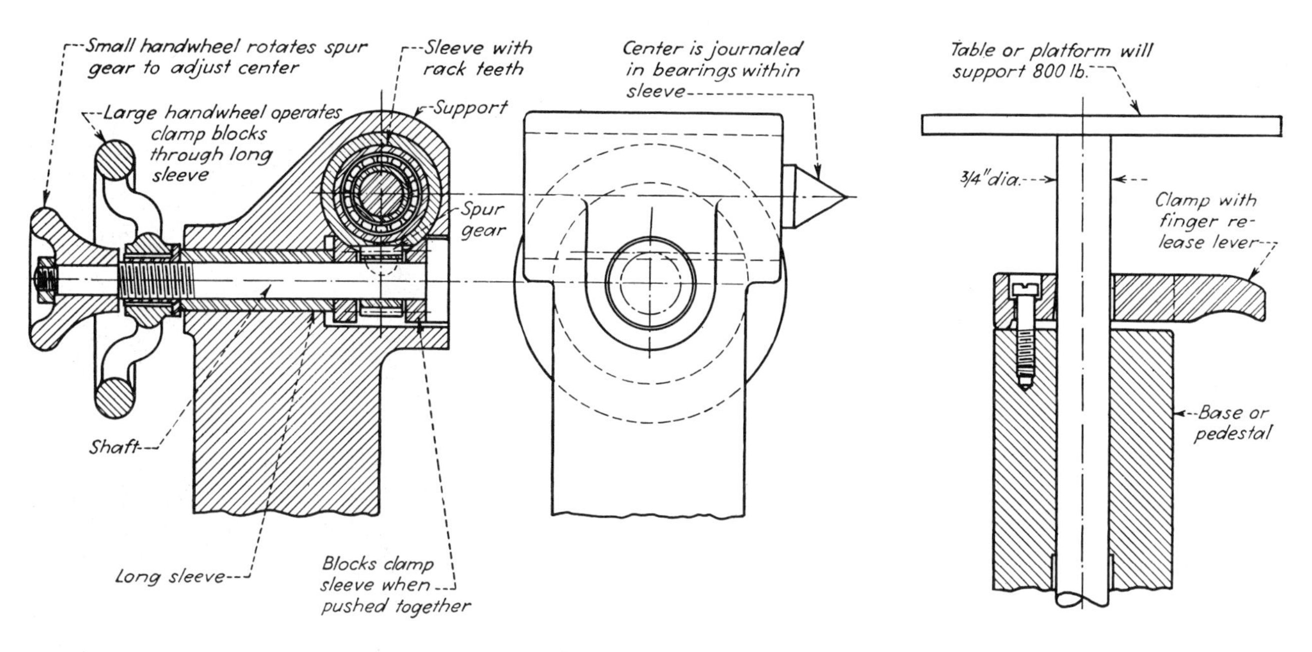

CENTER SUPPORT CLAMP

PEDESTAL CLAMP

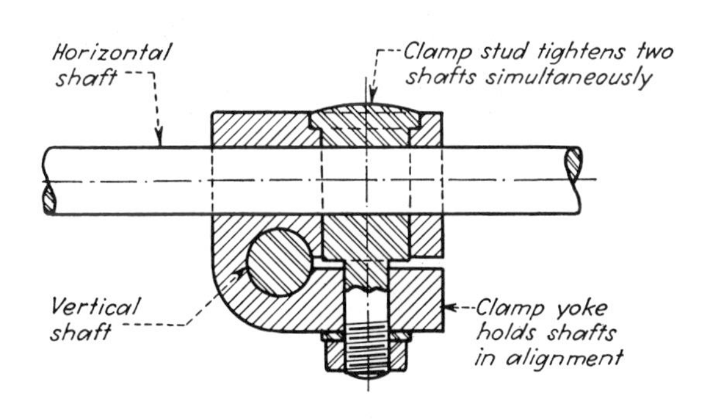

RIGHT ANGLE CLAMP

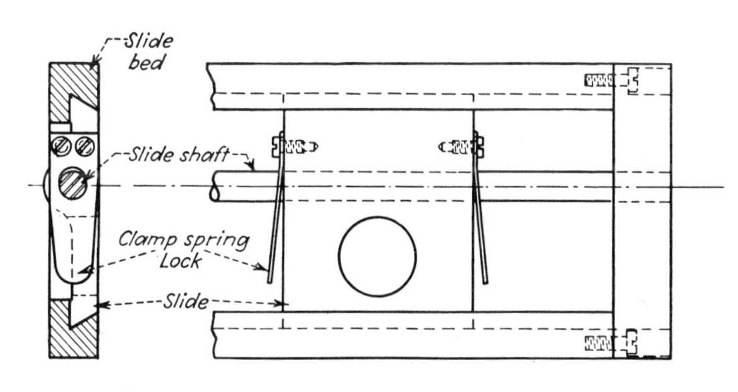

SLIDE CLAMP

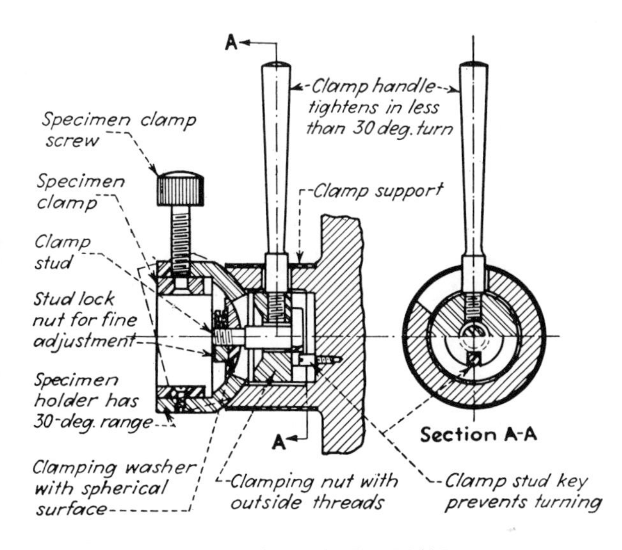

SPECIMEN HOLDER CLAMP

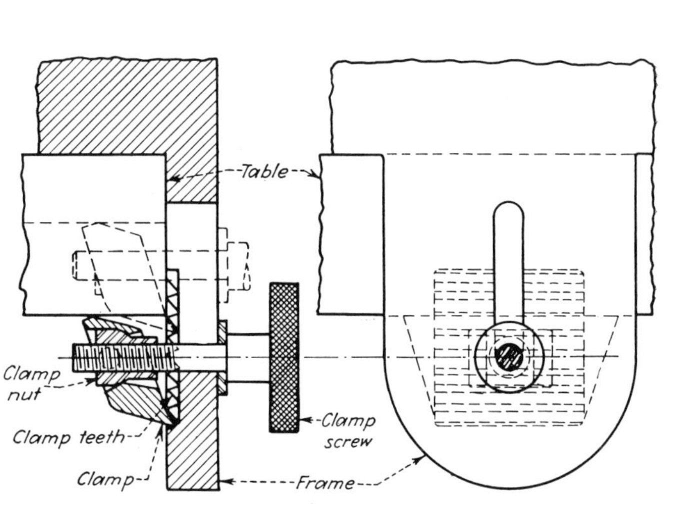

TABLE CLAMP

20 SIMPLE LOCKING

Springs, latches, eccentric cams, and other arrangements that hold two parts together

SPRINGS

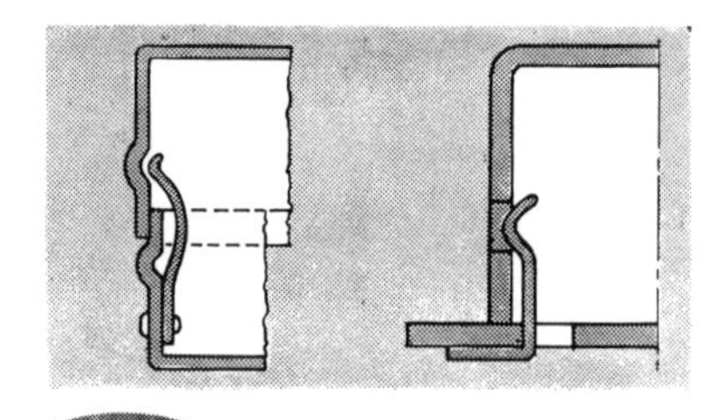

1

Spring catch acts as pawl-type catch for sliding members

2

Snap-on lid embodies pawl principle in parts themselves

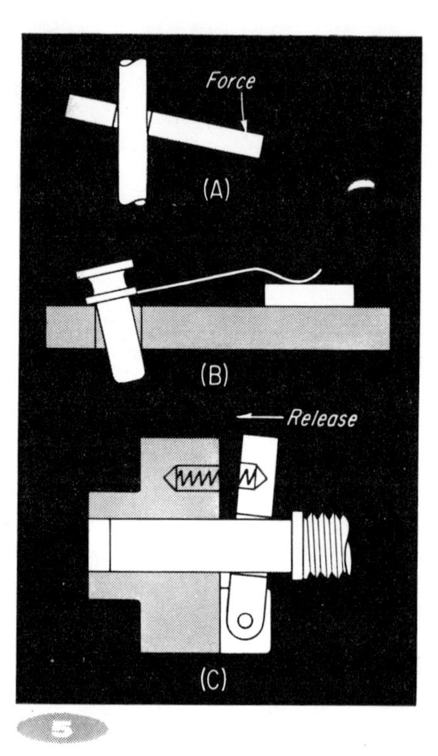

5

Tilt-lock principle (A) has many applications—examples are: holding specimen on microscope table (B); lock for vacuum-cleaner tube (C)

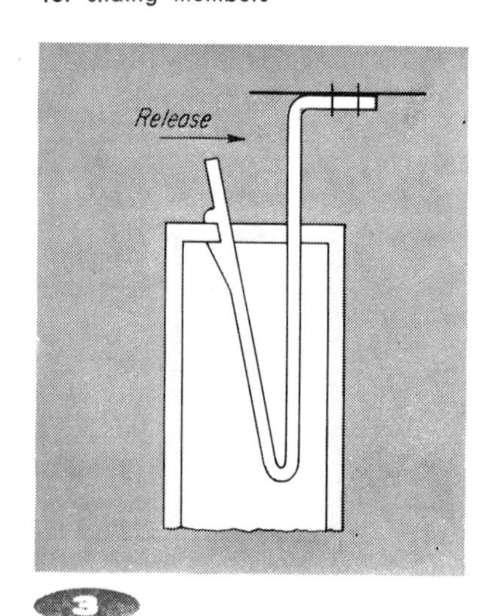

3

Spring hook locates itself, then locks to hold part

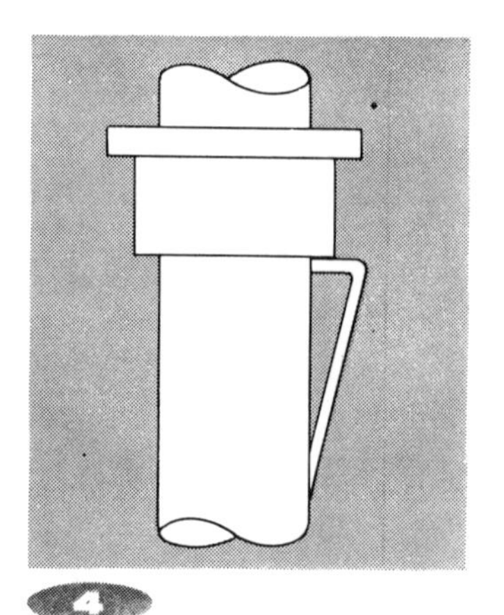

4

Umbrella-type catch is analagous to spring hook

LATCHES

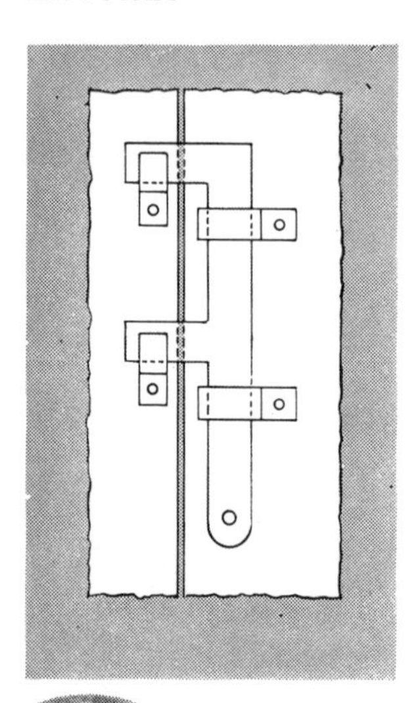

6

Double latch for high and narrow doors

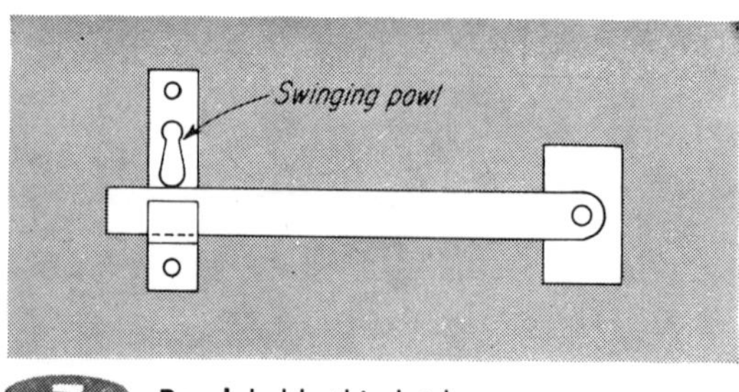

7 **Pawl** holds this latch

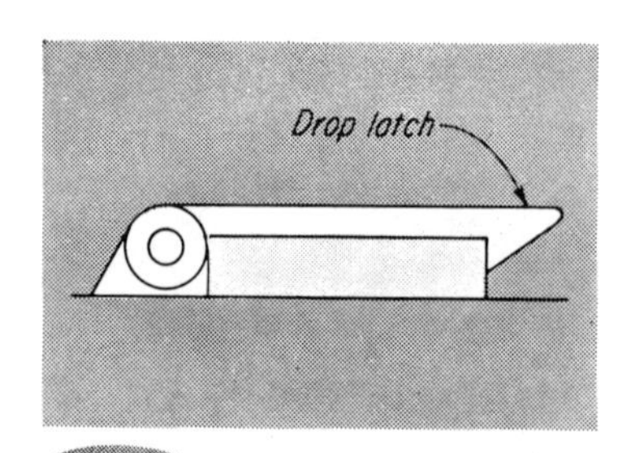

8

Drop-latch holds bar in position

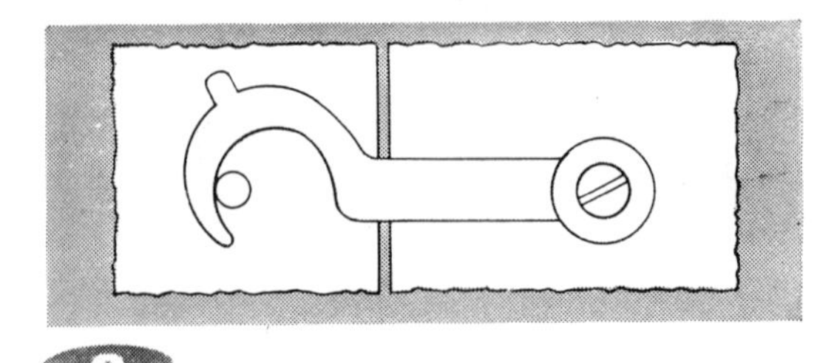

9

Hook-latch is inexpensive way to lock small lids

10

Hasp and staple (with or without pin)

DEVICES

FEDERICO STRASSER

ECCENTRICS

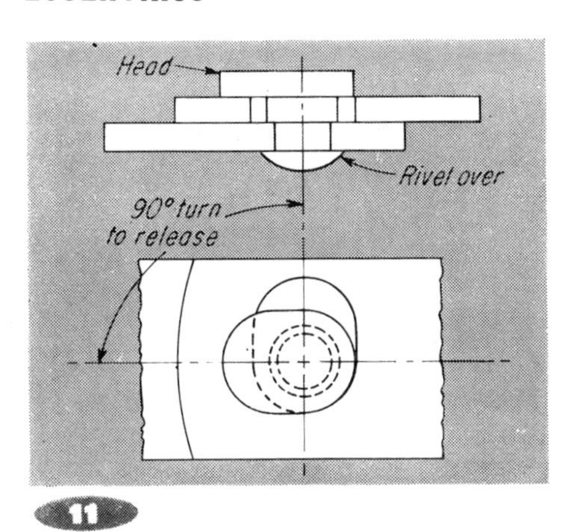

11

Off-center head holds parts together in position shown, releases when one part is given a turn of 90 degrees

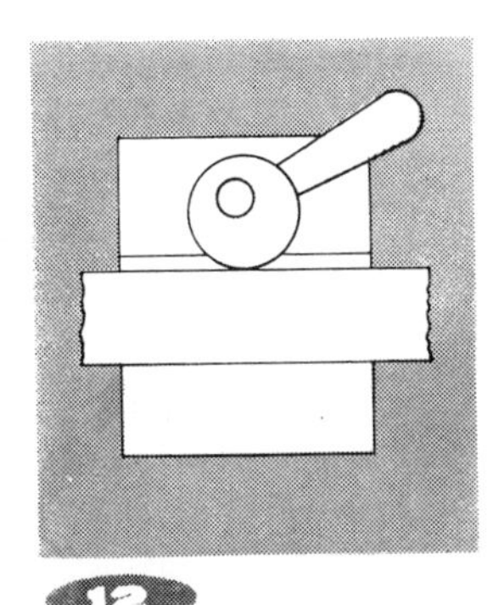

12

External cam exerts nigh friction-forces that provide clamping action

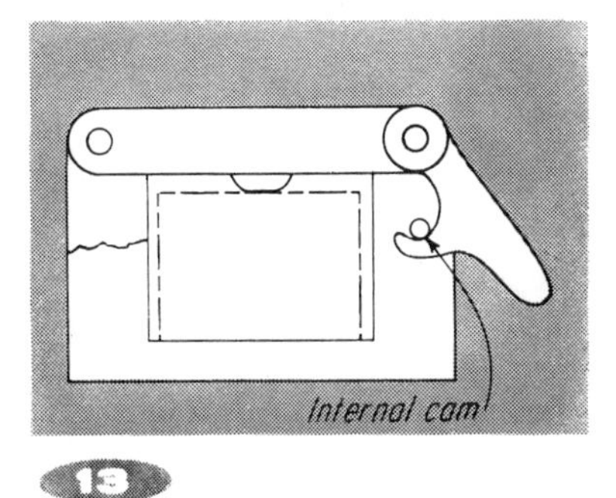

13

Internal cam here serves as lock for hinged cover

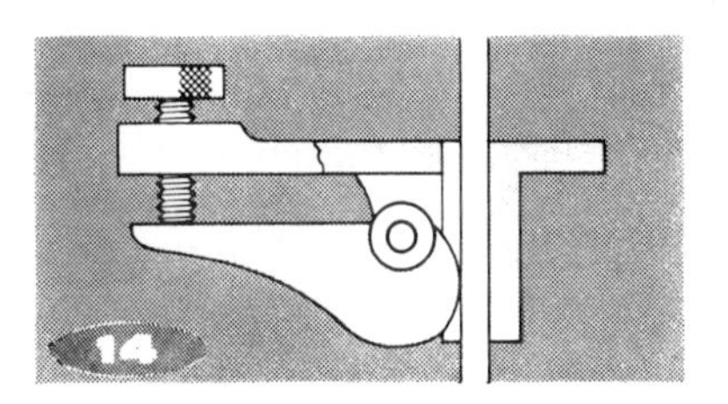

14

Screw-actuated cam exerts considerable clamping force

MISCELLANEOUS

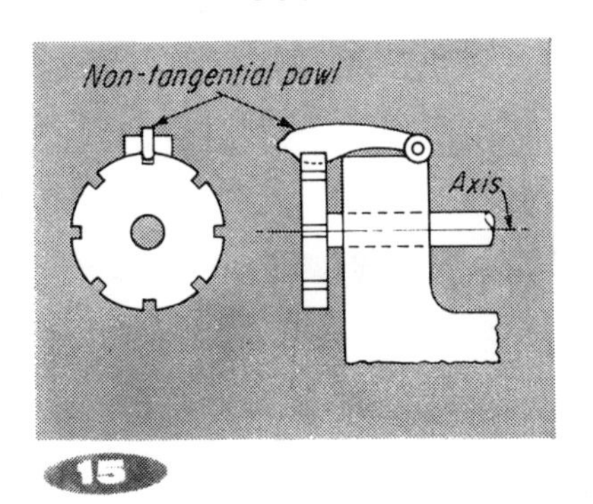

15

Pawl parallel to axis locks rotary device

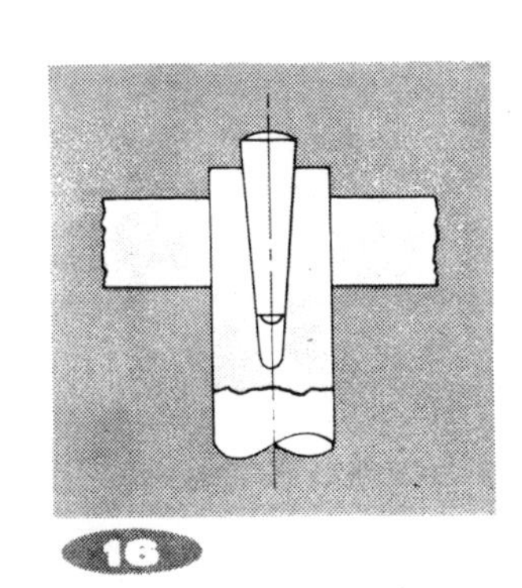

16

Taper pin expands rod, creates large holding forces

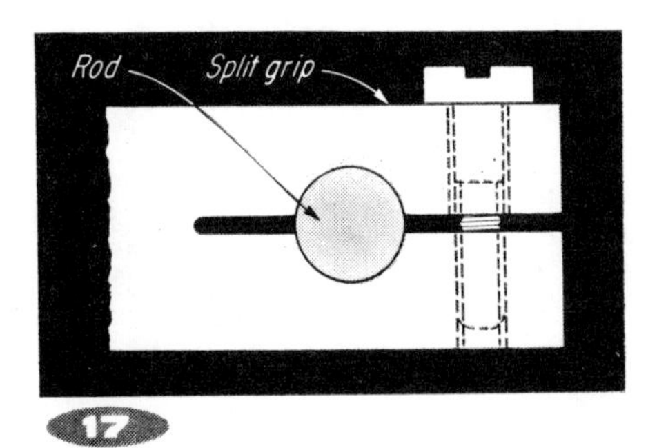

17

Split grip tightened by screw provides powerful rod-clamp

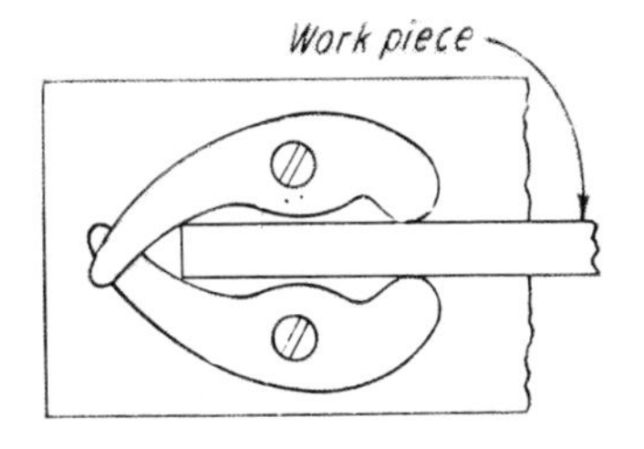

18

Carpenter clamps are self-locking, held tight by the workpiece itself

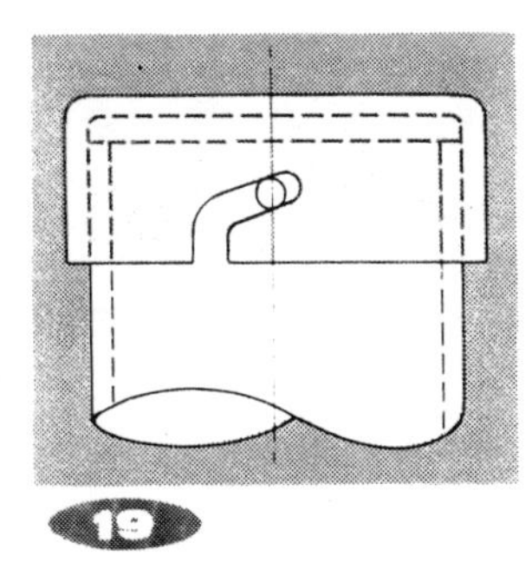

19

Bayonet-lock wedge action can't be shaken loose

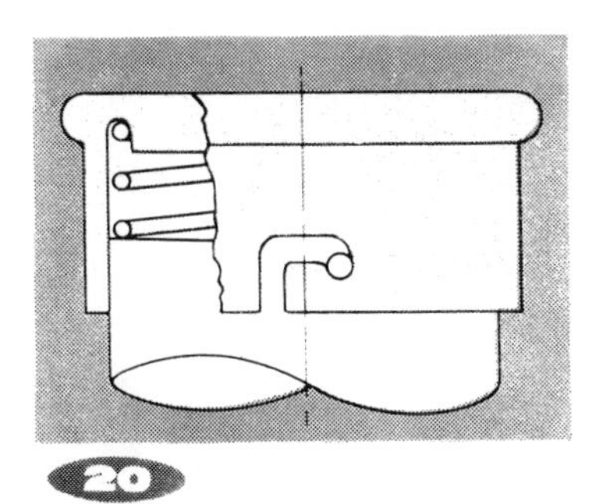

20

Spring-loaded bayonet lock provides positive grip

Linkages—the Why and How

HO CHOW, *Vice-president,*
Westchester Technical Corp

The ideal linkage joint should be a ball-and-socket type in order to be self-aligning. It should need support on only one side to have interrelated advantages of fewer parts, lighter weight, easier assembly and lower cost.

Arms and links are nonwearing members of a linkage; they are shaped to carry external loads. Pivots and joints are wearing members. A joint is a bearing unit connecting an arm and a link. A pivot is a stationary bearing unit attached to an arm. An arm is connected to a pivot and a joint; a link connects two joints.

In a 4-bar linkage the moving paths of the two joints at ends of the link are determined by movements of the arms rotating about the pivots. Distance between the joints is fixed by the link length. In transmitting force from driving to driven arm, a joint is a bearing member that transfers the force between arm and link. Relative movement between these elements could be in a single or variable plane.

Joint design leads to the selection of a cylinder or ball-type. Cylinder-type is a bearing with a straight centerline about which the arm and link must rotate—causing both the joint and pivot of the arm to be parallel. If not, bearing surfaces will bind—causing friction, undesirable strain and fast bearing wear.

Ball joint has a common center between arm and link at a point in the center of the ball. Link can move in any direction with respect to the ball. Advantage: high permissible misalignment of pivots.

It is almost impossible in production to make every part to the same dimension. Tolerance must be allowed or cost of the part will be too high. Cylinder-type joints require high precision to maintain the necessary linkage parallelism. Therefore, they should be avoided for economical and efficient design.

Ball-type joints permit generous tolerances because parallelism is not needed for proper operation. Arms can be in any relative position and will function without excessive wear caused by binding. Also, no bending stress is imposed on the link, permitting a lighter weight link.

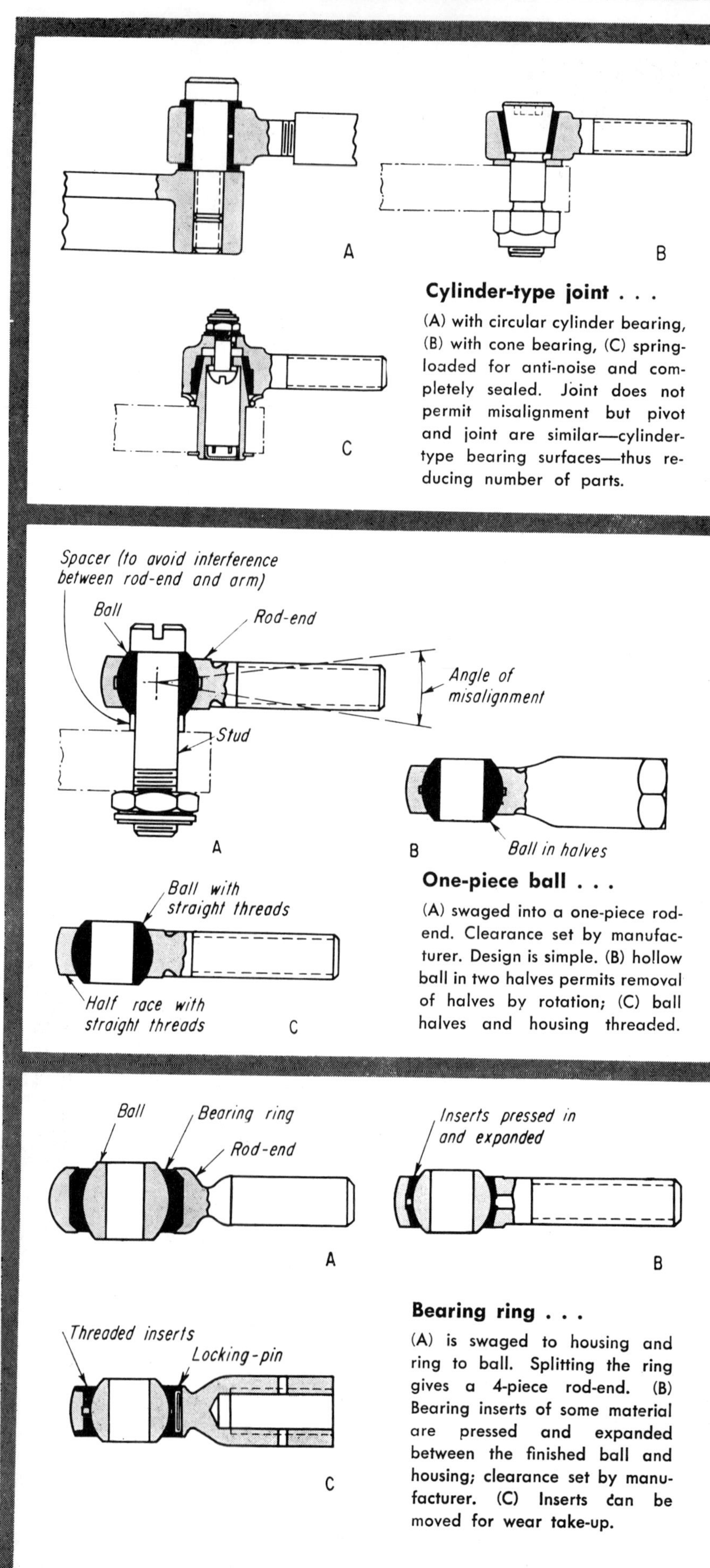

Cylinder-type joint . . . (A) with circular cylinder bearing, (B) with cone bearing, (C) spring-loaded for anti-noise and completely sealed. Joint does not permit misalignment but pivot and joint are similar—cylinder-type bearing surfaces—thus reducing number of parts.

One-piece ball . . . (A) swaged into a one-piece rod-end. Clearance set by manufacturer. Design is simple. (B) hollow ball in two halves permits removal of halves by rotation; (C) ball halves and housing threaded.

Bearing ring . . . (A) is swaged to housing and ring to ball. Splitting the ring gives a 4-piece rod-end. (B) Bearing inserts of some material are pressed and expanded between the finished ball and housing; clearance set by manufacturer. (C) Inserts can be moved for wear take-up.

AVAILABLE LINKAGE JOINTS

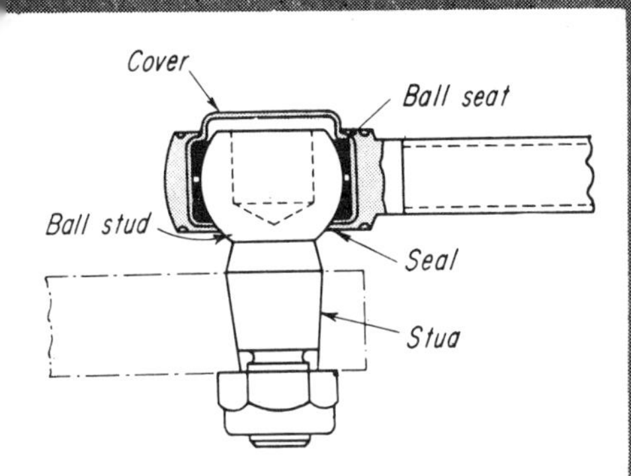

Ball integral with stud . . .
and a 2-piece ball seat staked into the rod-end. Cover and seal are optional features. Taper on stud gives largest dia where stresses caused by loading, are highest.

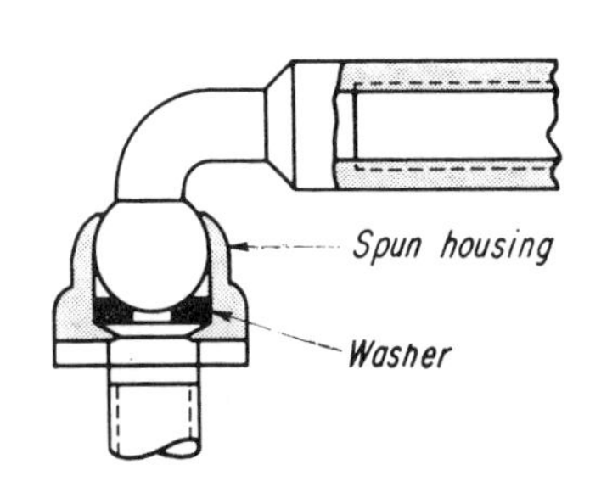

Ball integral with rod end . . .
retained by the spun housing and supported by the resonant washer. Washer cushions movement and maintains proper tension. No wear take-up adjustment possible.

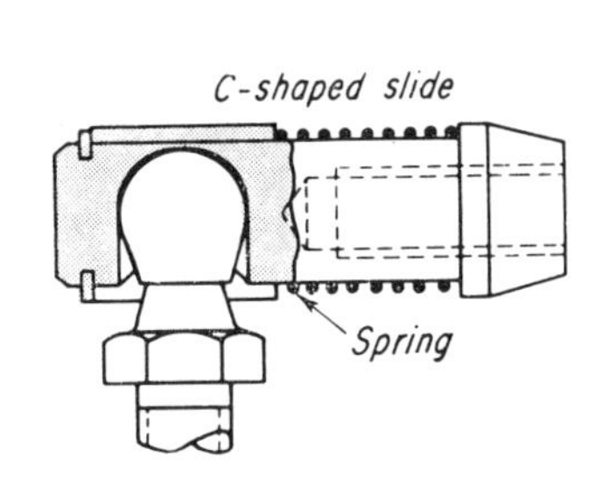

Specially shaped ball stud . . .
withstands higher transmission force by increasing stud-neck dia. Slide pushed against spring permits rapid disassembly. No provision for wear take-up or anti-rattling.

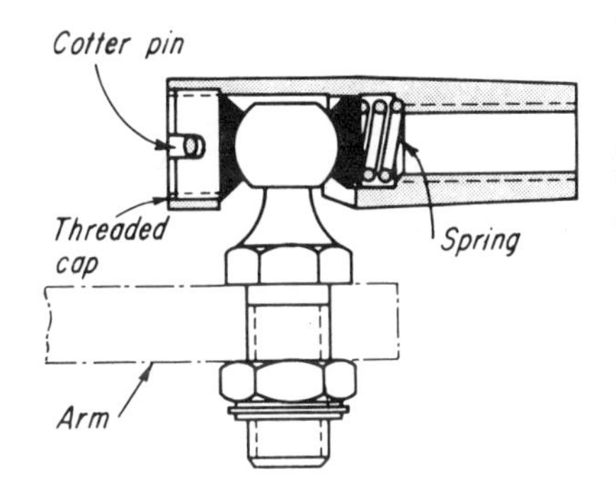

Threaded cap bearing . . .
is slotted and adjustable, to maintain a predetermined bearing force by the compression spring on the ball stud through the bearing seat. Wear take-up and anti-rattling are provided by the threaded cap.

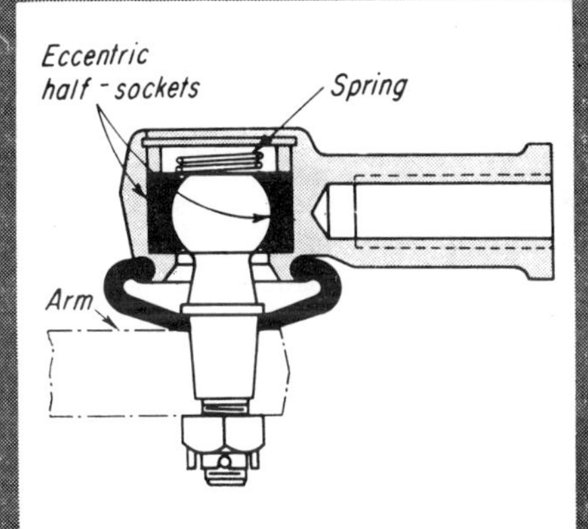

Steering linkage-joint . . .
for vehicles. Torsion spring forces two eccentric half sockets to tighten around the ball as wear occurs. Unit is sealed and enclosed and can be packed with lubricant or grease for maintenance-free application.

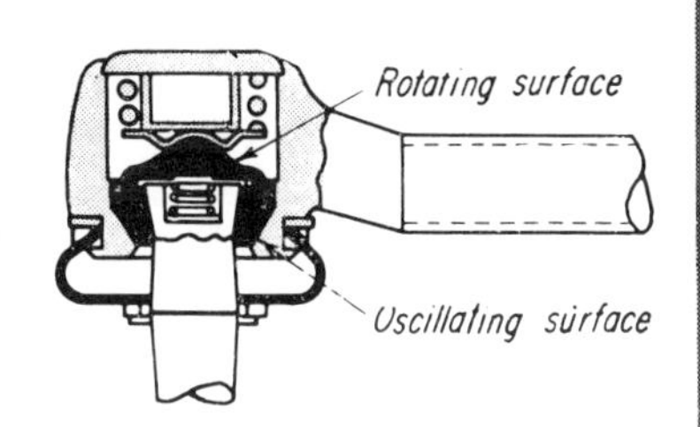

Vehicle joint . . .
for steering linkage. Note that the rotative bearing surface is separated from the oscillating bearing surface. Stud and ring ball can be one piece. Joint can be packed with lubricant or grease and then sealed.

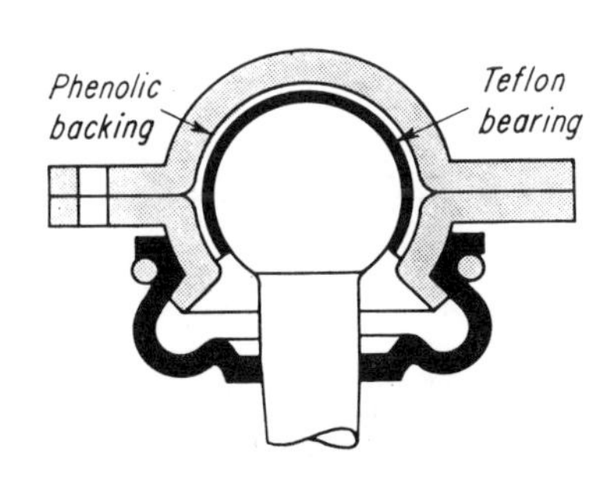

Front-wheel suspension . . .
for automobile. Kingpin not required. Thin layer of Teflon on a phenolic back reduces bearing friction. Ball is completely enclosed and sealed; can be grease-packed.

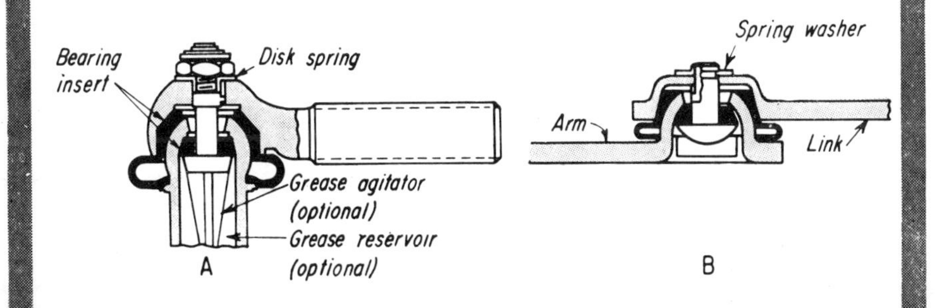

Coupling pin joins . . .
stud and housing (A). Clearance is set by nut on pin. Head rides on bearing ring which permits pin to float free in the stud. Grease agitator and fitting are optional. (B), Sheet metal arrangement using spring washer to hold coupling pin. Sealing is optional feature.

New Developments in Linkage Joints

A ball joint with a molded plastic socket that adjusts for wear has been designed for such heavy-duty applications as motor vehicle front wheel suspensions by Engineering Productions Ltd of Clevedon, Somerset, England. It's a patented design which seems relatively simple and easy to make.

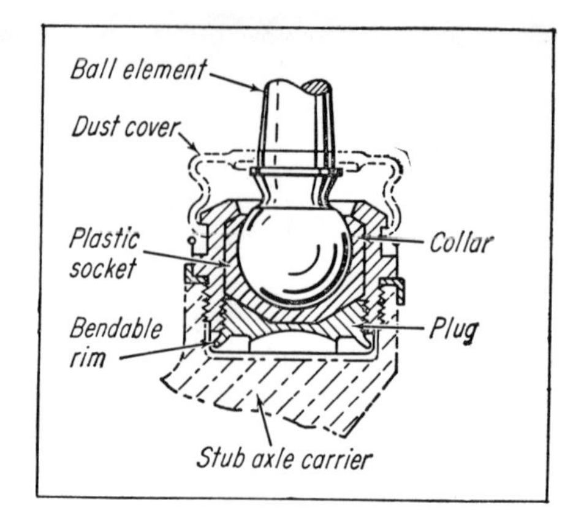

Major components of the joint are a steel ball; the socket itself, which may be any one of several plastics (nylon, polypropylene, fluorocarbon); and a housing which in its simplest version consists of a collar narrowed at one end by an overhanging shoulder, and closed at the other by a flat disk or a plug. The socket is molded closely around the ball element before it is inserted in the outer housing, and the combined element is then press-fitted to the housing to provide a preloaded joint.

In the front wheel suspension assembly shown, the plug screws into the collar and can be locked into position by bending over the rim so it fits into slots provided in the collar. In this version, preloading can be varied by adjusting the plug, and EP says the joint will remain free from hammer and backlash even during long working periods.

A ball joint that can be coupled directly to a plain link rod—without threading and without locknuts—has been designed by Coventry Movement Co (England). The joint carries a built-in, collet-type locking device that not only grips the link rod but also permits lengthwise adjustment.

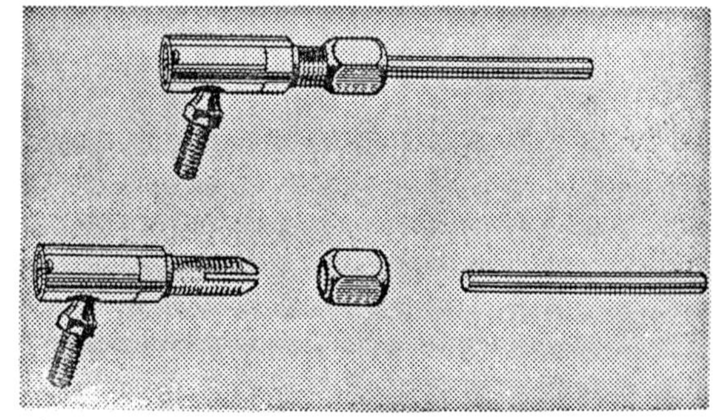

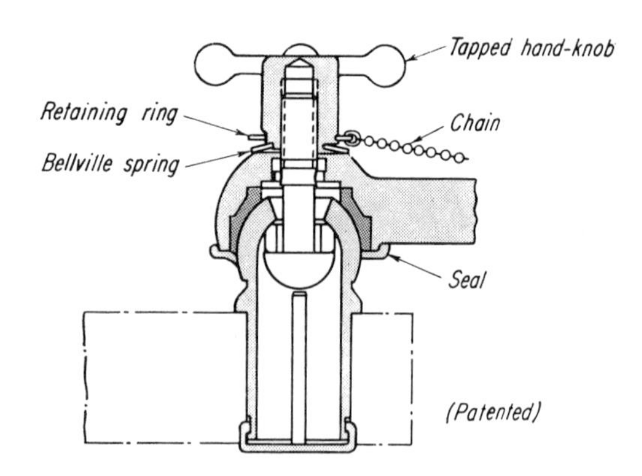

Trailer coupling . . .

using ideal joint arrangement. Vertical rod prevents coupling pin from dropping into stud upon accidental uncoupling. Coupling pin extends into hand-knob.

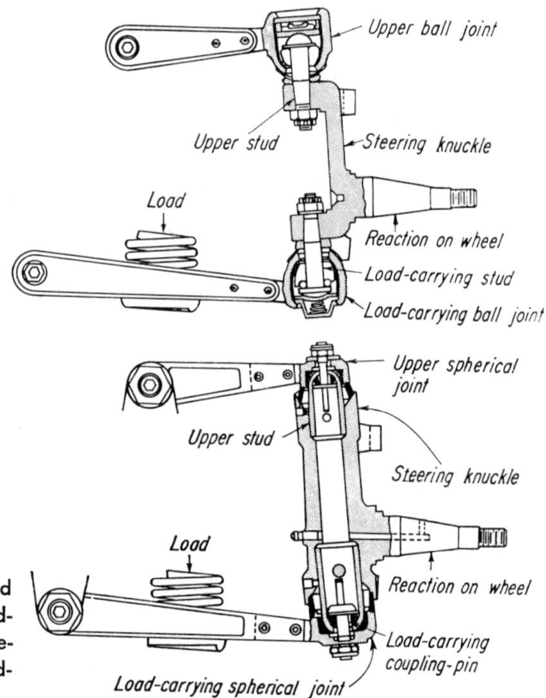

Automotive suspension system . . .

(Upper) replacing king-pin construction and threaded bushing. But studs are under tension, shear and bending. (Lower) improved design, using ideal joint arrangement, places only hollow stud under combined loading; coupling pin is subjected to tension only.

Fasteners that disconnect quickly

Ideal for linkages, these quick-disconnect designs can simplify installation and maintenance because no tools are needed.

FRANK W. WOOD JR, Senior Engineer, General Precision Inc – Link Div, Binghamton, NY

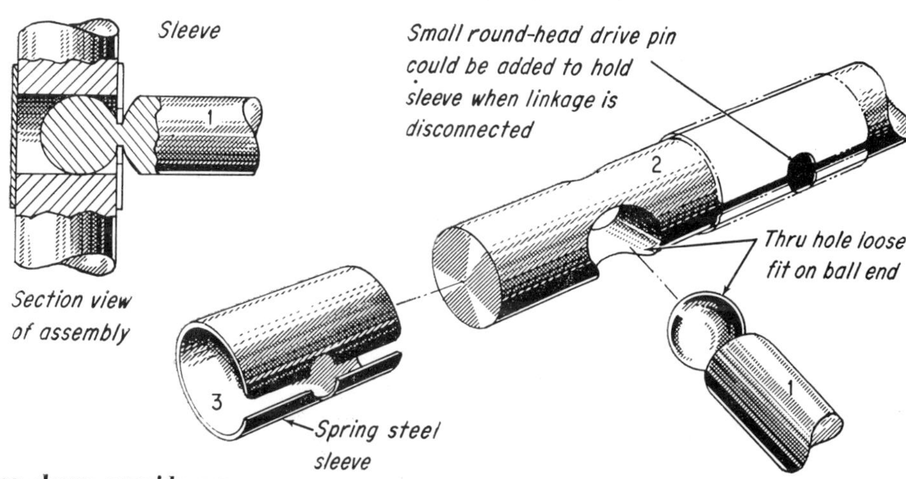

BALL JOINT and spring sleeve provide snug universal motion when hole in center of sleeve snaps over base diameter of ball. Ball diameter must always be less than that of the mating arm.

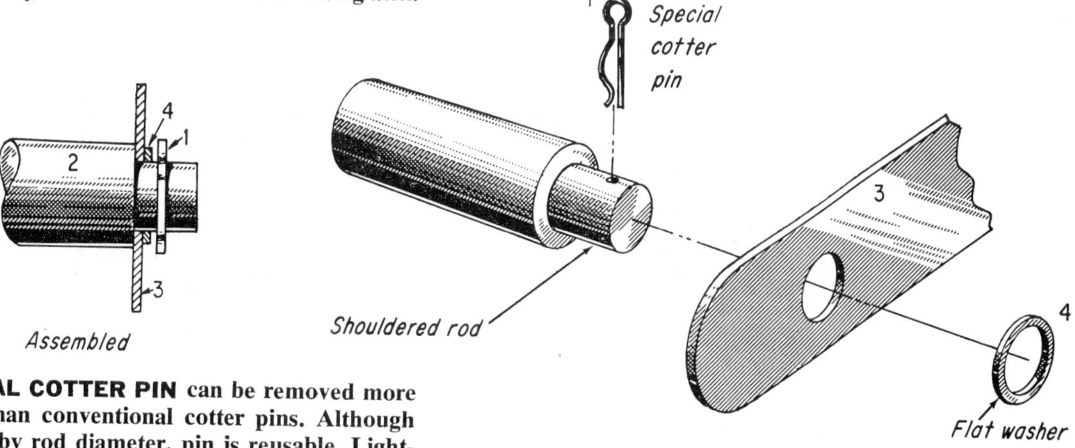

SPECIAL COTTER PIN can be removed more easily than conventional cotter pins. Although limited by rod diameter, pin is reusable. Light-duty applications only are recommended.

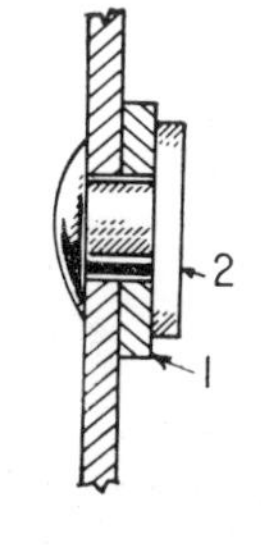

Assembly

ELONGATED FASTENER HEAD and slot can be disconnected only when head and slot are aligned. Phase the linkage to avoid alignment of slot and head; otherwise it may work free.

More quick-disconnect

These methods of fastening linkage arms allow them to be disassembled without tools. Snap slides, springs, pins, etc, are featured.

FRANK W. WOOD JR, Senior Engineer, General Precision Inc – Link Div, Binghamton, NY

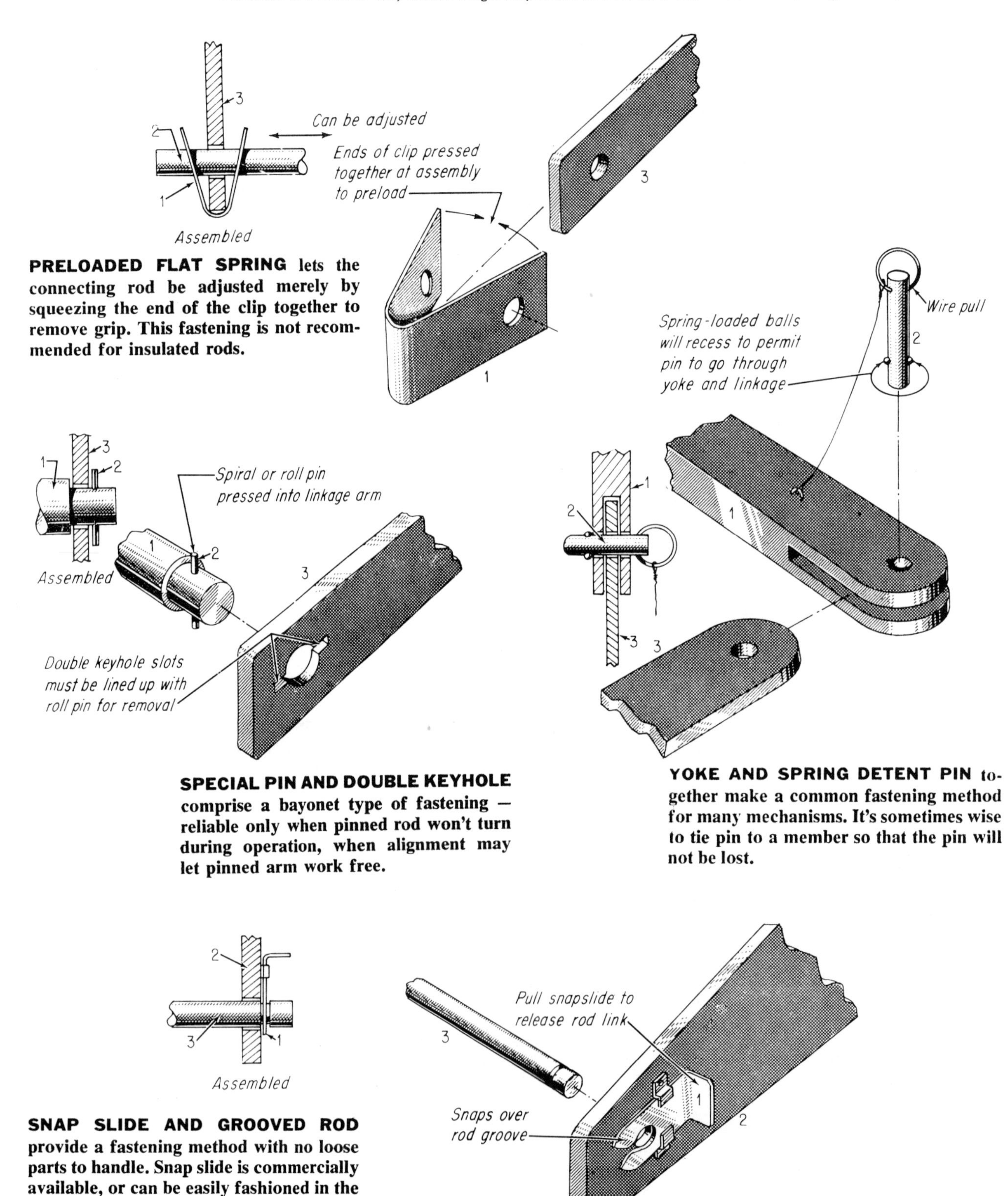

PRELOADED FLAT SPRING lets the connecting rod be adjusted merely by squeezing the end of the clip together to remove grip. This fastening is not recommended for insulated rods.

SPECIAL PIN AND DOUBLE KEYHOLE comprise a bayonet type of fastening – reliable only when pinned rod won't turn during operation, when alignment may let pinned arm work free.

YOKE AND SPRING DETENT PIN together make a common fastening method for many mechanisms. It's sometimes wise to tie pin to a member so that the pin will not be lost.

SNAP SLIDE AND GROOVED ROD provide a fastening method with no loose parts to handle. Snap slide is commercially available, or can be easily fashioned in the model room.

linkages

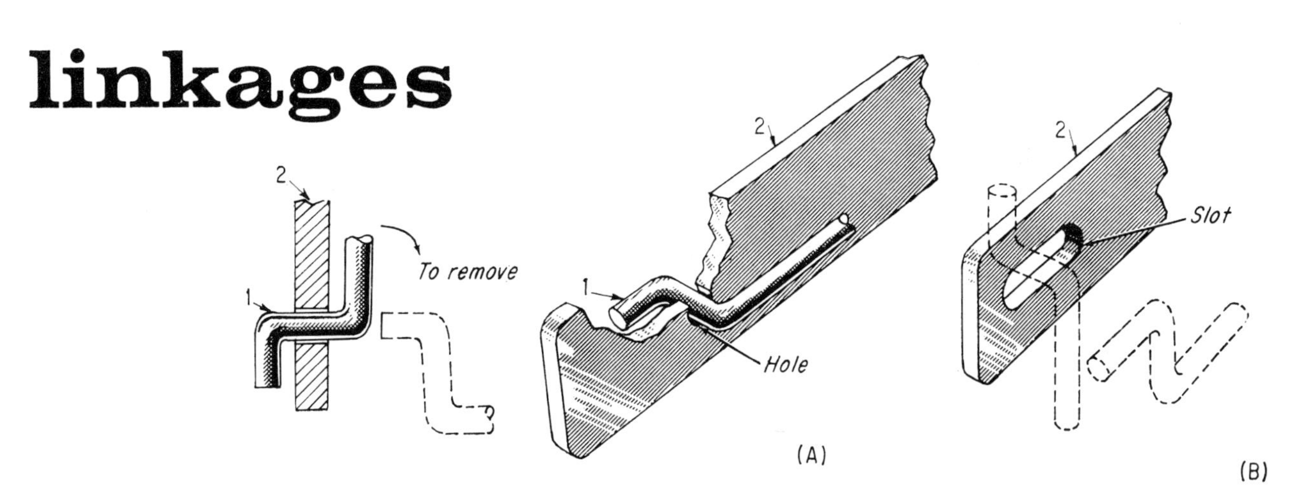

FORMED ROD in hole or slot allows disassembly only when the rod is free to be manipulated out of the arm. Slot design has play, which may or may not be advantageous.

RETAINER CLIP and formed rod are ideal when production quantities are high enough to warrant tooling costs necessary for clip. But the clip is relatively easy to make.

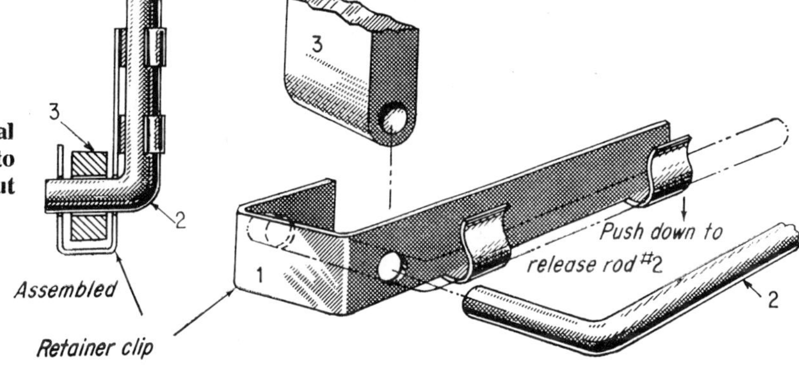

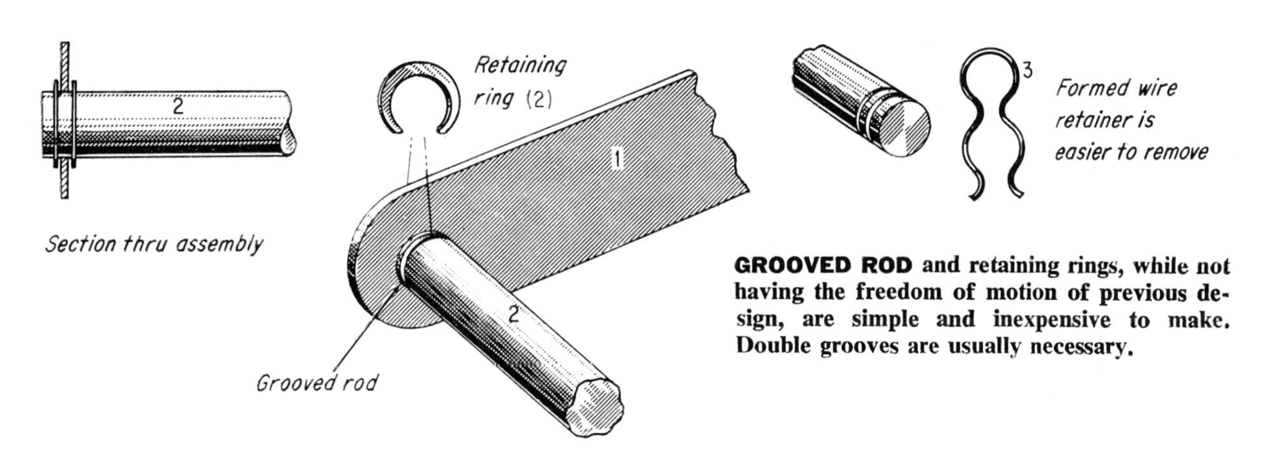

GROOVED ROD and retaining rings, while not having the freedom of motion of previous design, are simple and inexpensive to make. Double grooves are usually necessary.

NYLON ROD COUPLER allows typical ball-and-socket freedom combined with the self-lubricating properties of nylon. If load becomes excessive, the nylon will yield, preventing damage to linkage.

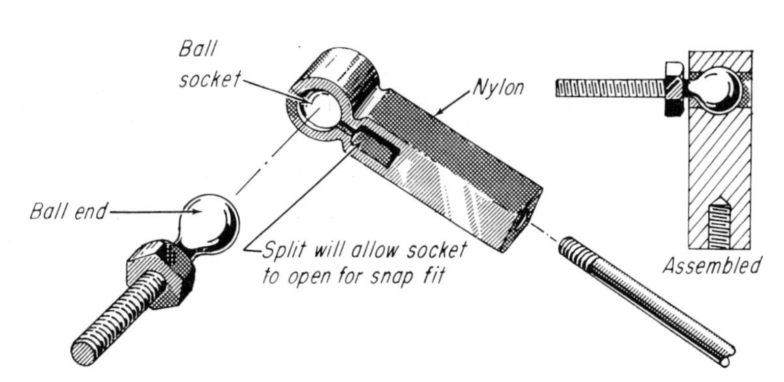

ADJUSTMENT LOCKS

(FOR ANGULAR AND AXIAL ADJUSTMENT)

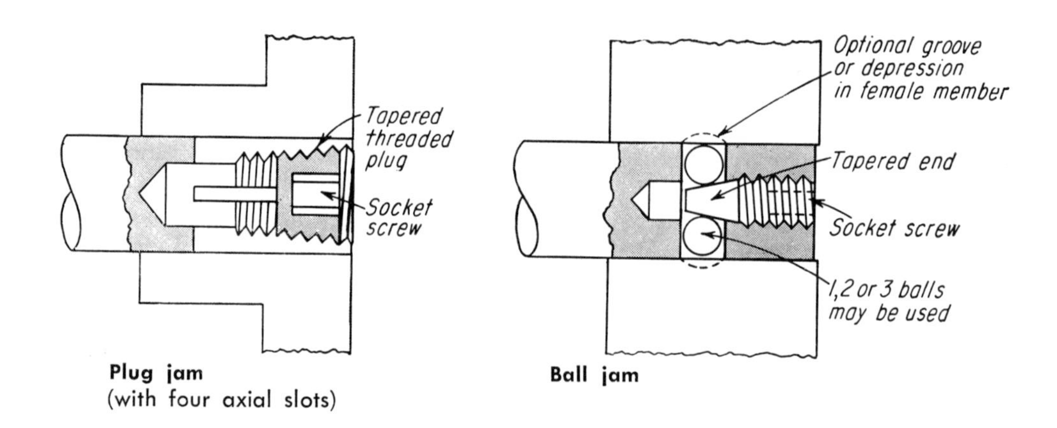

Plug jam
(with four axial slots)

Ball jam

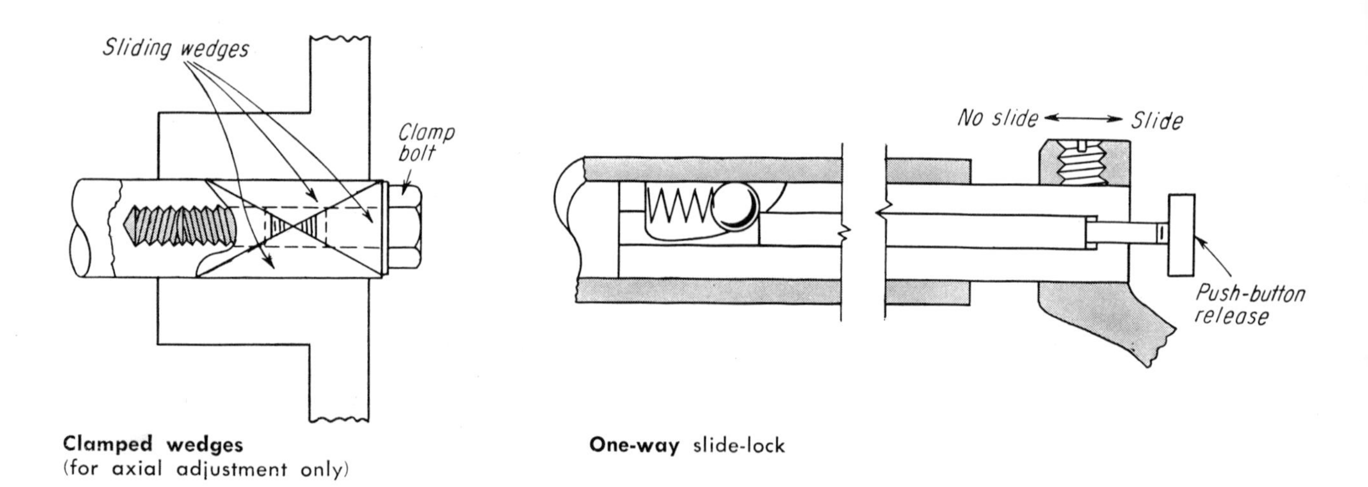

Clamped wedges
(for axial adjustment only)

One-way slide-lock

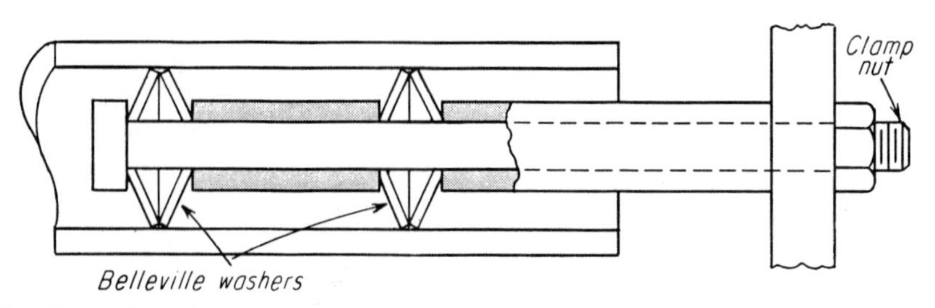

Belleville-washer clamp

INDEX